18th SIGBioMed Workshop on Biomedical Natural Language Processing (BioNLP 2019)

Florence, Italy
1 August 2019

ISBN: 978-1-5108-9103-6

BioNLP 2019

**SIGBioMed Workshop on
Biomedical Natural Language Processing**

Proceedings of the 18th BioNLP Workshop and Shared Task

August 1, 2019
Florence, Italy

Sesame Street at BioNLP 2019

Dina Demner-Fushman, Kevin Bretonnel Cohen, Sophia Ananiadou, and Junichi Tsujii

Recent years have seen an explosion of workshops, community challenges, corpora and publicly available tools in the biomedical and clinical language processing domain. That trend continues in 2019. In a significant advance, this year the original BioNLP-ST challenge matured into an open platform capable of providing technical support and sustaining any group that is interested in organizing a biomedical language processing challenge [1], while the BioNLP Special Interest Group continues supporting Shared Tasks in emerging areas of research through the annual meeting. This year, BioNLP-ST presents research directions explored by 72 teams for inference and entailment in the medical domain, and their contribution to domain-specific information retrieval and question answering systems [2].

The BioNLP meeting has now been ongoing for 18 years. BioNLP continues to stay the flagship and the generalist meeting in biomedical language processing, accepting noteworthy work independently of the tasks and sublanguages studied. BioNLP also continues promoting research in languages other than English, this year presenting work in Romanian, Portuguese, Spanish, and Chinese [3, 4, 5, 6], primarily covering development of resources for these languages.

The quality of submissions continues to impress the program committee and the organizers. BioNLP 2019 received 72 submissions to the workshop, and 21 for the Shared Task. Of the work submitted to the workshop, 14 papers were accepted for oral presentation and 24 as poster presentations. This year, various deep learning architectures are explored in all papers, with continuing focus on interesting new models and in-depth exploration of the state-of-the-art publicly available tools. Most of the work uses BERT [7] or BERT models trained on PubMed, with one paper exploring BERT and ELMo on ten biomedical benchmarking datasets [8] and many others using and exploring embeddings and neural networks for chemical recognition [9], concept extraction and coding [10], relation extraction [11, 12, 13], and phenotyping [14].

As for the past several years, the themes in this year's papers and posters continue to focus equally on clinical text and biological language processing. They also reveal sustained interest in social media and consumer language processing [15].

As it has been for the past 18 years, the workshop is truly a community-wide effort of the authors producing high quality work that is already contributing to acceleration of foundational biomedical research [16, 17, 18, 19] and clinical practice [20, 21, 22, 23] through improvements in information retrieval and extraction, question answering, diagnosis and clinical decision support [24]. We are equally happy to see sustained contributions from those who started forming the field of BioNLP research, and first-time contributions that show the increasing interest in the domain. We are particularly indebted to our reviewers who reviewed a higher than usual workload in a very short time. Their judgments resulted in a program that will undoubtedly advance both the BioNLP research and the practical areas that it serves. Due to space and time constraints, we could only accept the papers that were recommended for acceptance by at least two reviewers. We hope that the authors of the papers that could not be accepted received good feedback that will help them improve their work.

References

[1] *BioNLP-OST* https://2019.bionlp-ost.org. Last accessed 10 Jun 2019

[2] Ben Abacha A, Shivade C, Demner-Fushman D. *Overview of the MEDIQA 2019 Shared Task on Textual Inference, Question Entailment and Question Answering.*

[3] Mitrofan M, et al. *MoNERo: a Biomedical Gold Standard Corpus for the Romanian Language.*

[4] Lopes F, et al. *Contributions to Clinical Named Entity Recognition in Portuguese.*

[5] Campillos-Llanos L. *First Steps towards Building a Medical Lexicon for Spanish with Linguistic and Semantic Information.*

[6] Tian Y, et al. *ChiMed: A Chinese Medical Corpus for Question Answering.*

[7] Devlin J, et al. *BERT: Pre-training of Deep Bidirectional Transformers for Language Understanding.* NAACL 2019 Proc.

[8] Peng Y, et al. *Transfer Learning in Biomedical Natural Language Processing: An Evaluation of BERT and ELMo on Ten Benchmarking Datasets.*

[9] Zhai Z, et al. *Improving Chemical Named Entity Recognition in Patents with Contextualized Word Embeddings.*

[10] Wiegreffe S, et al. *Clinical Concept Extraction for Document-Level Coding.*

[11] Koroleva A, Paroubek P. *Extracting relations between outcomes and significance levels in Randomized Controlled Trials (RCTs) publications.*

[12] Chauhan G, et al. *REflex: Flexible Framework for Relation Extraction in Multiple Domains.*

[13] Khachatrian H, et al. *BioRelEx 1.0: Biological Relation Extraction Benchmark.*

[14] Liu D, et al. *Two-stage Federated Phenotyping and Patient Representation Learning.*

[15] Alhuzali H, Ananiadou S. *Improving classification of Adverse Drug Reactions through Using Sentiment Analysis and Transfer Learning.*

[16] Mezaoui H, et al. *Enhancing PIO Element Detection in Medical Text Using Contextualized Embedding.*

[17] Neumann M, et al. *ScispaCy: Fast and Robust Models for Biomedical Natural Language Processing.*

[18] Kotitsas S, et al. *Embedding Biomedical Ontologies by Jointly Encoding Network Structure and Textual Node Descriptors. BioNLP 2019 Proc.*

[19] Koptient A, et al. *Simplification-induced transformations: typology and some characteristics.*

[20] Ormerod M, et al. *Analysing Representations of Memory Impairment in a Clinical Notes Classification Model.*

[21] Yuwono SK., et al. *Learning from the Experience of Doctors: Automated Diagnosis of Appendicitis Based on Clinical Notes.*

[22] Newman-Griffis D, et al. *Classifying the reported ability in clinical mobility descriptions.*

[23] Soni S, Roberts K. *A Paraphrase Generation System for EHR Question Answering.*

[24] Apostolova E, et al. *Combining Structured and Free-text Electronic Medical Record Data for Real-time Clinical Decision Support.*

Organizers:

Sophia Ananiadou, National Centre for Text Mining and University of Manchester, UK
Kevin Bretonnel Cohen, University of Colorado School of Medicine, USA
Dina Demner-Fushman, US National Library of Medicine
Jun-ichi Tsujii, National Institute of Advanced Industrial Science and Technology, Japan

Program Committee:

Sophia Ananiadou, National Centre for Text Mining and University of Manchester, UK
Emilia Apostolova, Language.ai, USA
Eiji Aramaki, University of Tokyo, Japan
Asma Ben Abacha, US National Library of Medicine
Cosmin (Adi) Bejan, Vanderbilt University, Nashville, TN
Olivier Bodenreider, US National Library of Medicine
Leonardo Campillos Llanos, Universidad Autónoma de Madrid, Spain
Fenia Christopoulou, National Centre for Text Mining and University of Manchester, UK
Aaron Cohen, Oregon Health & Science University, USA
Kevin Bretonnel Cohen, University of Colorado School of Medicine, USA
Brian Connolly, Kroger Digital, USA
Viviana Cotik, University of Buenos Aires, Argentina
Dina Demner-Fushman, US National Library of Medicine
Travis Goodwin, The University of Texas at Dallas, USA
Natalia Grabar, CNRS, France
Cyril Grouin, LIMSI - CNRS, France
Tudor Groza, The Garvan Institute of Medical Research, Australia
Sadid Hasan, Philips Research, Cambridge, MA
Antonio Jimeno Yepes, IBM, Melbourne Area, Australia
Meizhi Ju, National Centre for Text Mining and University of Manchester, UK
William Kearns, University of Washington, USA
Halil Kilicoglu, US National Library of Medicine
Ari Klein, University of Pennsylvania, USA
Andre Lamurias, University of Lisbon, Portugal
Alberto Lavelli, FBK-ICT, Italy
Robert Leaman, US National Library of Medicine
Ulf Leser, Humboldt-Universität zu Berlin, Germany
Gal Levy-Fix, Columbia University, NY
Maolin Li, National Centre for Text Mining and University of Manchester, UK
Ramon Maldonado, The University of Texas at Dallas, USA
Timothy Miller, Children's Hospital Boston, USA
Yassine Mrabet, US National Library of Medicine
Aurelie Neveol, LIMSI - CNRS, France
Mariana Neves, German Federal Institute for Risk Assessment, Germany
Denis Newman-Griffis, Clinical Center, National Institutes of Health, USA
Nhung Nguyen, The University of Manchester, UK
Karen O'Connor, University of Pennsylvania, USA
Yifan Peng, US National Library of Medicine
Laura Plaza, UNED, Madrid, Spain
Sampo Pyysalo, University of Cambridge, UK
Francisco J. Ribadas-Pena, University of Vigo, Spain
Fabio Rinaldi, University of Zurich, Switzerland

Kirk Roberts, The University of Texas Health Science Center at Houston, USA
Roland Roller, DFKI GmbH, Berlin, Germany
Sumegh Roychowdhury, Indian Institute of Technology Kharagpur
Chaitanya Shivade, IBM Research, Almaden, USA
Noha Seddik Tawfik, Arab Academy for Science and Technology, Egypt
Thy Thy Tran, National Centre for Text Mining and University of Manchester, UK
Sumithra Velupillai, King's College London, UK
Davy Weissenbacher, University of Pennsylvania, USA
W John Wilbur, US National Library of Medicine
Amir Yazdavar, Wright State University, USA
Chrysoula Zerva, National Centre for Text Mining and University of Manchester, UK
Pierre Zweigenbaum, LIMSI - CNRS, France

Additional Reviewers:

Hadi Amiri, Harvard Medical School, USA
Siamak Barzegar, Barcelona Supercomputing Center, Spain
Qingyu Chen, US National Library of Medicine
Zfania Tom Korach, Harvard Medical School, USA
Majid Latifi, Trinity College Dublin, Ireland
Danielle Mowery, VA Salt Lake City Health Care System, USA
Claire Nedellec, INRA, France
Alastair Rae, US National Library of Medicine
Max Savery, US National Library of Medicine
Diana Sousa, University of Lisbon, Portugal
Shankai Yan, US National Library of Medicine
Ayah Zirikly, Clinical Center, National Institutes of Health, USA
Seyedjamal Zolhavarieh, The University of Auckland, NZ

Table of Contents

ix

Conference Program

Thursday August 1, 2019

8:30–8:45   **Opening remarks**

8:45–10:30   **Session 1: Clinical and Translational NLP**

8:45–9:00   *Classifying the reported ability in clinical mobility descriptions*
Denis Newman-Griffis, Ayah Zirikly, Guy Divita and Bart Desmet

9:00–9:15   *Learning from the Experience of Doctors: Automated Diagnosis of Appendicitis Based on Clinical Notes*
Steven Kester Yuwono, Hwee Tou Ng and Kee Yuan Ngiam

9:15–9:30   *A Paraphrase Generation System for EHR Question Answering*
Sarvesh Soni and Kirk Roberts

9:30–9:45   *REflex: Flexible Framework for Relation Extraction in Multiple Domains*
Geeticka Chauhan, Matthew B.A. McDermott and Peter Szolovits

9:45–10:00   *Analysing Representations of Memory Impairment in a Clinical Notes Classification Model*
Mark Ormerod, Jesús Martínez-del-Rincón, Neil Robertson, Bernadette McGuinness and Barry Devereux

10:00–10:15   *Transfer Learning in Biomedical Natural Language Processing: An Evaluation of BERT and ELMo on Ten Benchmarking Datasets*
Yifan Peng, Shankai Yan and Zhiyong Lu

10:15–10:30   *Combining Structured and Free-text Electronic Medical Record Data for Real-time Clinical Decision Support*
Emilia Apostolova, Tony Wang, Tim Tschampel, Ioannis Koutroulis and Tom Velez

10:30–11:00   ***Coffee Break***

11:00–12:00   Poster Session

Transfer Learning for Causal Sentence Detection
Manolis Kyriakakis, Ion Androutsopoulos, Artur Saudabayev and Joan Ginés i Ametllé

12:00–12:30	**Session 2: Ontology and Typology**

12:00–12:15   *Embedding Biomedical Ontologies by Jointly Encoding Network Structure and Textual Node Descriptors*
Sotiris Kotitsas, Dimitris Pappas, Ion Androutsopoulos, Ryan McDonald and Marianna Apidianaki

12:15–12:30   *Simplification-induced transformations: typology and some characteristics*
Anaïs Koptient, Rémi Cardon and Natalia Grabar

12:30–14:00   ***Lunch break***

14:00–15:30   **Session 3: Literature mining approaches and models**

14:00–14:15   *ScispaCy: Fast and Robust Models for Biomedical Natural Language Processing*
Mark Neumann, Daniel King, Iz Beltagy and Waleed Ammar

14:15–14:30   *Improving Chemical Named Entity Recognition in Patents with Contextualized Word Embeddings*
Zenan Zhai, Dat Quoc Nguyen, Saber Akhondi, Camilo Thorne, Christian Druckenbrodt, Trevor Cohn, Michelle Gregory and Karin Verspoor

14:30–14:45   *Improving classification of Adverse Drug Reactions through Using Sentiment Analysis and Transfer Learning*
Hassan Alhuzali and Sophia Ananiadou

14:45–15:00   *Exploring Diachronic Changes of Biomedical Knowledge using Distributed Concept Representations*
Gaurav Vashisth, Jan-Niklas Voigt-Antons, Michael Mikhailov and Roland Roller

15:00–15:15   *Extracting relations between outcomes and significance levels in Randomized Controlled Trials (RCTs) publications*
Anna Koroleva and Patrick Paroubek

15:30–16:00   ***Coffee Break***

16:00–17:00   Session 4: Shared Task

16:00–16:15   *Overview of the MEDIQA 2019 Shared Task on Textual Inference, Question Entailment and Question Answering*
Asma Ben Abacha, Chaitanya Shivade and Dina Demner-Fushman

16:15–16:30   *PANLP at MEDIQA 2019: Pre-trained Language Models, Transfer Learning and Knowledge Distillation*
Wei Zhu, Xiaofeng Zhou, Keqiang Wang, Xun Luo, Xiepeng Li, Yuan Ni and Guotong Xie

16:30–16:45   *Pentagon at MEDIQA 2019: Multi-task Learning for Filtering and Re-ranking Answers using Language Inference and Question Entailment*
Hemant Pugaliya, Karan Saxena, Shefali Garg, Sheetal Shalini, Prashant Gupta, Eric Nyberg and Teruko Mitamura

16:45–17:00   *DoubleTransfer at MEDIQA 2019: Multi-Source Transfer Learning for Natural Language Understanding in the Medical Domain*
Yichong Xu, Xiaodong Liu, Chunyuan Li, Hoifung Poon and Jianfeng Gao

17:00–18:00   Shared Task Poster Session

Surf at MEDIQA 2019: Improving Performance of Natural Language Inference in the Clinical Domain by Adopting Pre-trained Language Model
Jiin Nam, Seunghyun Yoon and Kyomin Jung

WTMED at MEDIQA 2019: A Hybrid Approach to Biomedical Natural Language Inference
Zhaofeng Wu, Yan Song, Sicong Huang, Yuanhe Tian and Fei Xia

KU_ai at MEDIQA 2019: Domain-specific Pre-training and Transfer Learning for Medical NLI
Cemil Cengiz, Ulaş Sert and Deniz Yuret

DUT-NLP at MEDIQA 2019: An Adversarial Multi-Task Network to Jointly Model Recognizing Question Entailment and Question Answering
Huiwei Zhou, Xuefei Li, Weihong Yao, Chengkun Lang and Shixian Ning

DUT-BIM at MEDIQA 2019: Utilizing Transformer Network and Medical Domain-Specific Contextualized Representations for Question Answering
Huiwei Zhou, Bizun Lei, Zhe Liu and Zhuang Liu

Dr.Quad at MEDIQA 2019: Towards Textual Inference and Question Entailment using contextualized representations
Vinayshekhar Bannihatti Kumar, Ashwin Srinivasan, Aditi Chaudhary, James Route, Teruko Mitamura and Eric Nyberg

Classifying the reported ability in clinical mobility descriptions

Denis Newman-Griffis[1,2]*, **Ayah Zirikly**[1]*, **Guy Divita**[1]*, **Bart Desmet**[1]

[1]Rehabilitation Medicine Dept., Clinical Center, National Institutes of Health, Bethesda, MD
[2]Dept. of Computer Science and Engineering, The Ohio State University, Columbus, OH
{denis.griffis, ayah.zirikly, guy.divita, bart.desmet}@nih.gov

Abstract

Assessing how individuals perform different activities is key information for modeling health states of individuals and populations. Descriptions of activity performance in clinical free text are complex, including syntactic negation and similarities to textual entailment tasks. We explore a variety of methods for the novel task of classifying four types of assertions about activity performance: *Able*, *Unable*, *Unclear*, and *None* (no information). We find that ensembling an SVM trained with lexical features and a CNN achieves 77.9% macro F1 score on our task, and yields nearly 80% recall on the rare *Unclear* and *Unable* samples. Finally, we highlight several challenges in classifying performance assertions, including capturing information about sources of assistance, incorporating syntactic structure and negation scope, and handling new modalities at test time. Our findings establish a strong baseline for this novel task, and identify intriguing areas for further research.

1 Introduction

Information on how individuals perform activities and participate in social roles informs conceptualizations of quality of life, disability, and social well-being. Importantly, activity performance and role participation are highly dependent on the environment in which they occur; for example, one individual may be able to walk around an office without issue, but experience severe difficulty walking along mountain paths. Thus, determining what level of performance an individual can achieve for activities in different environments is critical for identifying ability to meet work requirements, and designing public policy to support the participation of all people.

However, the interaction between individuals and environments makes modeling performance information a complex task. Assessments of activity performance within clinical healthcare settings are typically recorded in free text (Bogardus et al., 2004; Nicosia et al., 2019), and exhibit high flexibility in structure. Syntactic negation can be present, but is not necessarily indicative of inability to perform an action; for example, `Patient can walk with rolling walker` and `Patient cannot walk without rolling walker` are both likely to be used to assert the ability of the patient to walk with the use of an assistive device. Information about performance may also be given without a clear assertion, as in `the cane makes it difficult to walk`. Thus, extraction of performance information must not only distinguish between positive and negative assertions, but also those which cannot be clearly evaluated.

To the best of our knowledge, this is the first work to explore assertions of activity performance in health data. We explore a variety of methods for classifying assertion types, including rule-based approaches, statistical methods using common text features, and convolutional neural networks. We find that machine learning approaches set a strong baseline for discriminating between four assertion types, including rare negative assertions. While this work focuses on a relatively constrained and homogeneous corpus, error analysis suggests several broader directions for future research on classifying performance assertions.

2 Related Work

Though this is the first work focusing on the polarity of activity performance, three areas of prior work are particularly relevant to this research.

The first is concerned with applying NLP techniques and linguistic annotation to information about whole-person function, particularly activity

---

*These authors contributed equally to this work.

Proceedings of the BioNLP 2019 workshop, pages 1–10
Florence, Italy, August 1, 2019. ©2019 Association for Computational Linguistics

performance. Harris et al. (2003) experimented with term extraction for the purpose of terminology discovery to support information retrieval relating to functioning, disability and health, using linguistic, n-gram and hybrid techniques. Bales et al. (2005) and Kukafka et al. (2006) modified and applied the MedLEE NLP Extraction tool to code Rehabilitation Discharge Summaries using ICF (World Health Organization, 2001) encodings. Kuang et al. (2015) studied UMLS term coverage of functional status terms found in VA clinical notes and in social media sources, reporting that there is a need to extend existing terminologies to cover this area. Finally, Thieu et al. (2017) reported on an effort to build an annotated corpus of Physical Therapy (PT) notes from the Clinical Center of the National Institutes of Health (NIH) with functional status information. This corpus was also used for an investigation into using named entity recognition (NER) techniques to extract information about patient mobility (Newman-Griffis and Zirikly, 2018).

The second area is research on negation. Negation detection is a well-researched area (Morante and Sporleder, 2012), and both negation and uncertainty have historically been studied in the clinical NLP context (Mowery et al., 2012; Peng et al., 2018). Previous work studied the use of incorporating dependency parsers to help in identifying the scope (Sohn et al., 2012; Mehrabi et al., 2015). Recent work in this area involves the use of neural network models, where Long Short-Term Memory (LSTM), or variations of it, yielded competitive results on negation (cues and scope) detection (Taylor and Harabagiu, 2018).

One highly-related work to ours is Wu et al. (2014), which investigates detection of binary semantic negation status (i.e., the presence or absence of a finding, as opposed to syntactic negation) for clinical findings in EHR text. However, as Action Polarity is defined in terms of the interaction between an individual and a specific environment, it adds a layer of complexity to non-interactive physiological observations. Gkotsis et al. (2016) investigate using parsing-based scoping limitations for negation detection in complex clinical statements, though their focus is specifically on mentions of suicide.

Finally, classifying the assertion status of activity performance descriptions bears similarities to the problem of recognizing textual entailment (RTE) (Dagan et al., 2006; Marelli et al., 2014). RTE asks whether a given premise entails a specific hypothesis, and has historically been pursued in the general domain, though, recent efforts have developed datasets in biomedical literature (Ben Abacha et al., 2015; Ben Abacha and Demner-Fushman, 2016) and in clinical text (Romanov and Shivade, 2018). Our task, by asking whether a given description entails ability to perform an action in the an environment, is more constrained than RTE, but poses a related research challenge.

3 Data

We use an extended version of the dataset initially described by Thieu et al. (2017), consisting of 400 English-language Physical Therapy initial assessment and reassessment notes from the Rehabilitation Medicine Department of the NIH Clinical Center. These text documents have been annotated to identify descriptions and assessments of mobility status, typically including one or more specific Actions; for example, `Pt walked 300' with rolling walker` (Action underlined).

Each Action annotation was assigned one of four Polarity values, indicating what (if any) information the containing mobility description provides about the subject's ability to perform the given Action in the context of any described environmental factors.[1] The Polarity labels are defined in the following paragraphs.

Able The subject is able to complete the activity in the environment described. For example, `She states she can walk 20 minutes before tiring`; in the case of `now requires assistance of one person with transfers`, it is unknown whether the patient can perform the action independently, but they are able to do so with the assistance described.

Unable The subject is not able to complete the activity in the environment described; for example, `He is unable to walk`. More specific information may also be included, as in `Pt is now unable to walk more than 50 feet`.

Unclear Some information is provided about the subject's ability to perform the action, but not

---

[1] It is important to note that the Polarity label is dependent on the environmental factors described. For example, an individual may be able to walk a certain distance using an assistive device such as a rolling walker, but unable to walk that same distance independently.

Label	Train	Test	Total
Able	1,536	446	1,982
Unable	54	23	77
Unclear	158	48	206
None	1,784	478	2,262
Total	3,532	995	4,527

Table 1: Number of samples with each Polarity label in train and test data.

enough to make a definitive positive or negative judgment. For example, in `The cane makes it difficult to walk`, it is undetermined whether the subject can or cannot walk. This label also includes some cases of negated environmental factors; for example, `unable to propel wheelchair independently`.

None No direct information about ability to perform the action is provided. Common examples of this label refer to a scale that is either unavailable or distant in the document, as in `Ambulation: 1`. Other cases refer to a specific aspect of performing an action, without evaluation, as in `tendency during gait to quickly extend the leg from swing to stance`.

We randomly split the 400 documents into 320 training records and 80 testing records, stratified by distribution of Polarity labels. Table 1 provides frequencies of each label in these splits.

4 Methods

We investigate a variety of methods to classify the Polarity values of Action annotations. Rule-based methods have been used to great effect in clinical information extraction (Kang et al., 2013; Chapman et al., 2007), and form an important baseline for our task. We also make use of several common machine learning methods, such as support vector machines and k-nearest neighbors, along with more recent neural models such as convolutional neural networks (CNN). Finally, we experiment with ensembled combinations of our best-performing models. These approaches are described in the following subsections.

4.1 Rule-based

A UIMA (Ferrucci and Lally, 2004) based pipeline was constructed to identify action polarity from components of v3NLP-Framework (Divita et al., 2016). Leveraging the relationship of our task to detecting contextual attributes such as negation, the conTEXT (Chapman et al., 2007) algorithm embedded in the v3NLP-Framework was augmented with a few additional entries including "able" and "independent" as asserted evidence and "unable" as negative evidence.

The conTEXT algorithm relies on a lexicon of evidence and accompanying clues to indicate when evidence found to the right or left of a relevant entity within a bounded window should be applied. We used the sentence containing an Action mention as the bounds of its context window. An *Action Polarity* UIMA annotator was built to assign Polarity, given an Action annotation. This annotator is downstream from the conTEXT annotator that assigned negation, assertion, conditional, hypothetical, historical, and subject attributes to named entities. Within conTEXT-processed entities, we assigned *Unable* polarities to actions that had previously been attributed with negative and assigned *Able* polarities that had previously been assigned only asserted attributes. Actions that were tagged as conditional or hypothetical were not assigned a Polarity.

The v3NLP-Framework pipeline includes document decomposition annotators to identify sections, section names, sentences, slots and values, questions and their answers, and to a lesser extent checkboxes (Divita et al., 2014). Action mentions in clinical text occur within the boundaries of each of these elements. ConTEXT addresses action mentions within prose, but is not relevant for action mentions found in the semi-structured constructs. The Action Polarity annotator was thus augmented with additional rules to aid in polarity assignment based on where the mention was found. The most relevant rules are as follows:

- Action mentions that are in the slot part of a slot:value construct get their polarity assignment from positive or negative evidence in the value part of the construct. Table 2 provides guidelines to assigning polarity from slot:value and question and answer constructs.

- Action mentions that are within Goals or Education sections do not get a polarity. The section name is known for each named entity. For the time being, section names with "plan," "goals," "education," "intervention" and "recommendations" qualify. These are

3

Slot criteria	Value criteria	Assigned Polarity	Example
Asserted Action	Asserted Evidence	Able	`Transfers:  Independent`
Asserted Action	Negated Evidence	Unable	`Transfers:  Unable`
Negated Action	Negated Evidence	Able	`Difficulty Walking:  No`
Negated Action	Asserted Evidence	Unable	`Unable to Walk:  yes`
Asserted Action	Numbers	Unclear	`Transfers:  4`
Asserted Action	No context evidence	Unclear	`Sit to stand:  minimal assist`
Asserted Action	No value	None	`Stand to sit:`
Multiple Actions	Doesn't matter	None	`Difficulty with chores, shopping,` `driving:  Yes`

Table 2: Table of slot:value rules for Action Polarity

considered to be hypothetical constructs. The exception to this is if a goal is noted to have been met, it gets an *Able* Polarity.

- Action mentions within only the value part of the slot:value construct were handled the same way as Action mentions within prose.

4.2 Machine learning models

We evaluated the following common machine learning-based classification methods for our Polarity labeling task:[2]

- Random forest (RF), using 100 estimators;

- Naïve Bayes (NB), using Gaussian estimators;

- k-nearest neighbors (kNN), using k=5 with Euclidean distance;

- Support vector machine (SVM), with linear kernel;

- Deep neural network (DNN), using a 100-dimensional hidden layer followed by a 10-dimensional hidden layer.[3]

For a given Action mention a contained in a Mobility description m, we explored using both bag of binary unigram features[4] and word embedding features as model input. For both kinds of features, we experimented with using the context words in $m - a$ (i.e., all words in m except for the Action mention itself) only, and including the text of the Action mention a. Word embedding features were calculated by averaging the embeddings of all words used (either context alone or averaging context words and Action mention

Features	NB	RF	kNN	SVM	DNN
Unigrams	41.3	**77.3**	**67.0**	78.6	79.8
+Action	42.1	73.7	56.8	80.9	78.0
+Embeddings	41.6	64.3	66.3	78.8	**80.9**
+Both	**43.0**	65.1	65.2	**81.7**	79.6

Table 3: Macro F1 over Polarity classes in 5-fold cross validation feature selection experiments. All experiments start with binary unigram features using context words alone, and add Action words, embedding features from context words, or both (i.e., unigrams and embedding features from context and Action words combined). The best performing model configurations are marked in bold.

words together); we used pretrained FastText (Bojanowski et al., 2017) embeddings from Wikipedia and newswire, including subword information.[5] Where both unigram and embedding features are used, they are concatenated as a single feature vector.

4.2.1 Feature selection

In order to identify the best combination of features for the task, we performed five-fold cross validation experiments on the training data. As shown in Table 3, we found that three model configurations achieved statistically equivalent macro F1 in cross validation ($p \geq 0.001$ with bootstrap permutation test, $R = 10000$ (Berg-Kirkpatrick et al., 2012)).[6] These are RF with unigram features (78.5% F1), the 2-layer DNN with unigram and embedding features from context only (80.9%), and SVM with all features, i.e. unigrams and embeddings with both the mobility description and Action mention texts (81.7%).[7]

Given the class imbalance in our dataset,

---

[2]We used the implementations of each method in Scikit-Learn (Pedregosa et al., 2011).

[3]We experimented with $d \in 10, 100$, and number of layers $\in 1, 2, 3$.

[4]Binary unigram features consistently matched or outperformed unigram counts in our experiments.

[5]`https://fasttext.cc/docs/en/english-vectors.html`

[6]We use significance threshold $p = 0.001$ throughout this paper, as a conservative Bonferroni correction for multiple testing. To have sufficient resolution to those low threshold, we use 10,000 replicates in bootstrapping.

[7]Complete results tables will be made available online.

Model	Able	Unable	Unclear	None	Macro
NB (All)	68.2	15.1	25.6	62.9	43.0
RF (Uni)	84.5	68.1	69.9	86.7	77.3
KNN (Uni)	73.5	53.3	62.6	78.5	67.0
SVM (All)	**86.3**	76.2	**76.4**	**87.8**	**81.7**
DNN (Uni+Emb)	85.0	**76.8**	74.3	87.5	80.9

Table 4: Five-fold cross validation results (F1) by class with best configurations of learned baselines. *All* indicates using unigrams, embeddings, and Action mention features; *Uni* indicates using unigram features from context words only, and *Uni+Emb* indicates both unigram and embedding features from context words. The best result in each column is marked in bold.

Embeddings	Able	Unable	Unclear	None	Macro
prev_all	82.3	48.7	31.8	86.9	62.4
next_all	79.3	32.3	53.5	82.7	64.9
full_all	**87.6**	**63.4**	65.0	**89.4**	**76.4**
full_char	66.0	45.7	**72.9**	78.7	65.8
full_word	86.1	42.4	60.3	88.0	69.2

Table 5: CNN performance using different inputs.

we also analyzed per-class performance of each model. Interestingly, as Table 4 illustrates, we found that all models except Naïve Bayes were surprisingly robust to this imbalance, with both SVM and DNN achieving over 76% F1 on the smallest class (*Unable*). Across all four classes, the SVM and the 2-layer DNN yield statistically equivalent performance ($p \geq 0.001$); we therefore use absolute macro F1 to choose SVM as the best baseline model for comparing across approaches.

4.2.2 CNN model

We adopt the Convolutional Neural Network (CNN) architecture introduced in Kim (2014). In our architecture, shown in Figure 1, we combine word embeddings with character embeddings, to reduce the impact of out-of-vocabulary rate as opposed to using words alone. Additionally, character-level CNNs have been shown to improve the results of text classification (Zhang et al., 2015), but the improvement is more evident with larger data sizes.

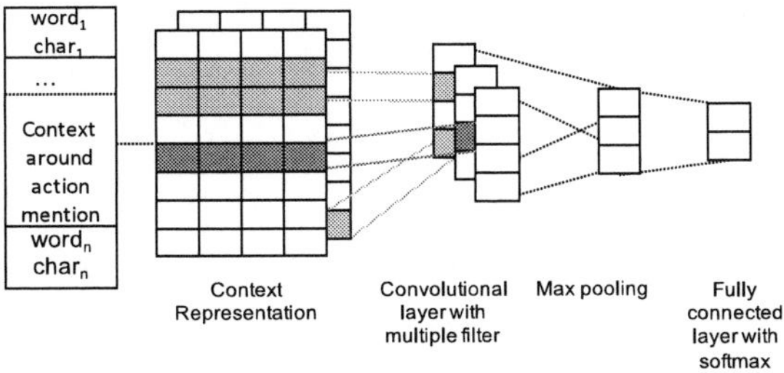

Figure 1: CNN architecture for Polarity classification.

Although our task is close to negation detection, it differs in that we do not need to detect the span of the Action: we take as inputs the Action mention and its parent mobility mention (a self-contained text span that can be considered a sentence). Unlike sequence tagging problems, where Long-Short Term Memory (LSTM) architectures would be a good fit (Fancellu et al., 2016), we treat the problem as a text classification task.

We experiment with character and word embeddings of the following inputs:

- previous context (*prev*): the set of words preceding and including the action mention.

- *next* context: the set of words following and including the action mention.

- *full* context: the union of *prev* and *next*.

We also compare the impact of using character (*full_char*) or word (*full_word*) embeddings only as opposed to combining both (**_all*), as shown in Table 5. We note that relying on part of the context significantly drops the *Unable* performance. However, as expected, *prev* outperforms *next*, given that the words preceding the Action mention carry most of the ability-related information. For the rare *Unable* class, character embeddings outperform word embeddings, with F1 72.9% on the test set; the highest across all systems.

Hyperparameters were optimized on a dev set (we used a 90/10 train/dev split), yielding a learning rate of 0.0001, dropout of 0.5, embeddings size 100, and Adam optimization (Kingma and Ba, 2014) with L2 regularization.

4.3 Ensemble models

Ensembling methods have been shown to improve performance in a variety of classification tasks (Buda et al., 2018), including in class-imbalanced tasks (Ju et al., 2018). In order to combine the strengths of each modeling approach, we therefore experimented with ensembling all three systems, using two ensembling strategies:

Majority voting Predictions from the single best configurations of the SVM and CNN models[8] were combined to make a single decision. When

---

[8] Adding rule-based predictions degraded performance in this case.

System	Able			Unable			Unclear			None			Macro F1
	Pr	Rec	F1	Pr	Rec	F1	Pr	Rec	F1	Pr	Rec	F1	
Rule-based	58.3	71.3	64.2	20.3	52.2	29.3	8.8	12.5	10.3	80.2	54.2	64.7	42.1
SVM	83.4	86.8	85.1	62.1	**78.3**	**69.2**	63.0	70.8	66.7	90.0	84.3	87.0	77.0
CNN	86.0	89.2	**87.6**	**72.2**	56.5	63.4	**81.2**	54.2	65.0	89.0	**89.7**	89.4	76.4
All (DNN chooser)	**87.5**	86.3	86.9	56.7	73.9	64.2	66.7	70.8	**68.7**	90.3	89.5	**89.9**	77.4
SVM+CNN (Voting)	82.3	**90.8**	86.4	62.1	**78.3**	**69.2**	62.1	**75.0**	67.9	**94.5**	82.2	87.9	**77.9**

Table 6: Precision (Pr), Recall (Rec), and F1 for each model evaluated on the test set. Top rows are individual models, bottom rows are ensembled results. The best result in each column is marked in bold.

the systems agreed, that label was chosen as output; in the case of disagreement, we chose the predicted label that was *less* frequent in training data, in order to prefer the strengths of individual models on rare classes.

DNN chooser Predictions from all three systems (rule-based and the best pretrained SVM and CNN models)[9] were passed as inputs to a DNN with a single 10-unit hidden layer.[10] In order to compensate for the class imbalance in our dataset, which would lead to preferring the CNN due to its higher precision, we identified all training samples that the three models disagreed on and grouped them by label, and identified the smallest of these disagreement sets. We then sampled no more than twice this number of points from each disagreement set, yielding a training sample of 182 points.

Using this downsampled training set, we trained the DNN to predict which, if any, of the systems chose the correct answer. As multiple systems may have made the correct prediction, this is a multi-label classification task. At test time, the system with highest probability output from the DNN was chosen as the reference decision for the final classification.

We also experimented with three approaches to predict the final class directly: using a DNN with the predictions of each system as input, using an SVM with predictions as input, and adding rule-based and CNN predictions as additional features to the SVM with lexical features. All variants underperformed the chooser in cross validation experiments on training data, thus we omit them from our results.

5   Results

The test results of the systems we compared are given in Table 6. The ensembled systems achieve the best overall performance, with 77.4% macro F1 with the DNN chooser and 77.9% with majority voting. Due in large part to the class imbalance in the dataset, the SVM, CNN, and ensemble methods do not yield statistically significantly different results in most cases ($p > 0.001$), although the voting ensemble does produce significantly higher precision on *None* samples than other methods ($p \ll 0.001$).

While performance is considerably better on the more frequent *Able* and *None* classes, the learned systems achieve good results on *Unclear* and the very rare *Unable*. Figure 2 shows the confusion matrices for all systems. The most common confusions are with *Able* and *None*, with only a small number of false positives for *Unable* and *Unclear* and no confusion between the two in the machine learning approaches.

Comparing between individual systems, the CNN is best at making the important distinction between *Able* and *Unable*. It consistently achieves high precision across all classes, but suffers large drops in recall for the rare labels. The SVM model reverses this tradeoff, yielding high recall for *Unable* and *Unclear*, but much lower precision. The ensembled methods are able to strike a good middle ground, keeping the high recall of the SVM without sacrificing too much of the CNN's precision.

6   Discussion

As is evident from the results, correctly classifying the minority classes *Unable* and *Unclear* is not trivial. This is not only caused by the lack of data for training those classes, but in the case of *Unclear*, also by its semantic ambiguity – even for humans.

An important area of confusion is when actions are hypothetical, as is the case for plans, recommendations or feelings towards an action (e.g. `eager to walk`), which should all be tagged as *None*. Semantic problems can also arise around

---

[9]For the chooser, adding rule-based predictions consistently improved results over just SVM and CNN.

[10]Experiments with a 64-unit hidden layer, to cover all possible label combinations, yielded the same results in cross validation.

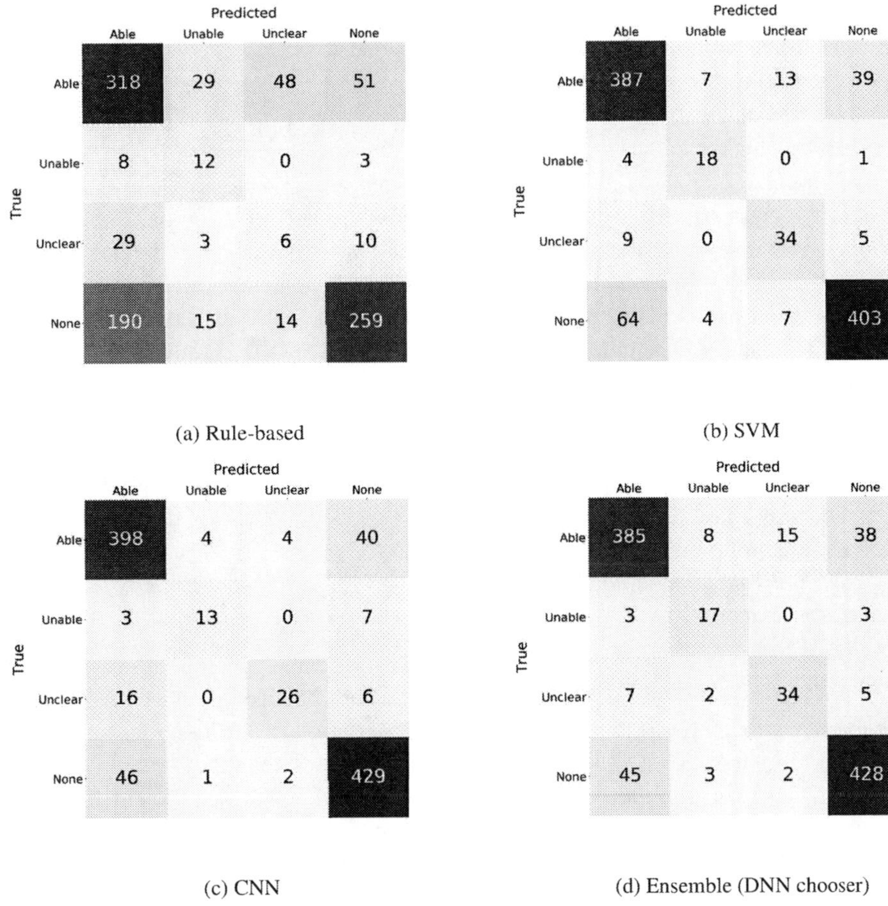

(a) Rule-based   (b) SVM

(c) CNN   (d) Ensemble (DNN chooser)

Figure 2: Confusion matrices for results on the test set.

the use of an assistive device. In the following synthetic example, the annotated polarity is *Able*:

```
she is unable to ambulate more
than a few feet without support.
```

Without the mention of assistance, it would have been *Unable*. In future work, assistance mentions will be modeled explicitly to better capture this.

Overall, we obtain models that perform well across the board, where each approach has different strengths as illustrated in Figure 2. Out of the 955 test instances, the rule-based approach classifies 37 correctly that no other system got right. Likewise, SVM and CNN have 27 and 25 unique true positives, respectively. 46 instances get misclassified by all classifiers. The ensemble is able to pick up on 31 of the unique true positives from the machine learning systems, but consistently ignores valid suggestions from the rule-based approach. This suggests that different ensembling parameters should be considered to take better advantage of the rule-based system's strengths.

Below, we discuss system-specific observations in more detail.

6.1 Rule-based

The following failures were observed in the training and testing output:

Scoping negation The scope for assigning negation attribution was set to be within sentential boundaries. Ideally, the scope should be tighter at the major phrase level. However, v3NLP-Framework does not currently employ a dependency graph parser. Breaking on phrasal boundaries was not successful, primarily due to the inability to distinguish between list markers such as commas, coordinating conjunctions (and/or), and true scope limiting phrasal boundaries. Several false negatives were due to the incorrect *Unable* assignment because of negation scoping.

Identifying variants of slots and values accurately Negation and assertion assignment are dependent upon whether the action is within prose, a slot or a value. A number of errors were due to multiple slot:value constructs within the same line making it difficult identifying the values, and/or nested constructs (i.e., the value of a slot:value construct was also a slot:value construct).

Nested sections A number of missed *None* errors were the result of mis-identifying what section the annotation was within, and picking up an inner section name. Several other issues arose from the use of spaces as delimiters between slots and values, as well as slots and values embedded within bulleted lists.

Pertinent negatives (Divita et al., 2014) A statement where the action mention had clear negative evidence really meant the patient could perform an action. For example, `no trouble walking`. An easy amelioration would be to gather constructs like "no trouble" and add them to the assertion evidence lexicon.

6.2 Machine learning

The machine learning systems are prone to failures in sentences that have multiple Action mentions, if their Polarity differs. This is because the systems do not take into account sentence structure. Similarly, sentence length seems to have a negative effect on performance, as it dilutes the information salient to the focus mention. In future work, we would limit the context information to exclude other mentions' contexts, add parse tree information relevant to the focus mention, or improve the neural network architecture to better model the sequential nature of the data.

The models would also benefit from better capturing semantic similarity. An example would be `Pt. is fearful to start walking again` (class: *None*), where the modality expressed by *fearful* might not have been learned from the training data. Additionally, lemmatization, stemming and character embeddings can blunt the impact of such unseen tokens, but using embeddings from large corpora would be more robust.

Finally, one potential limitation in our machine learning results is our use of pretrained embeddings from web text. As Newman-Griffis and Zirikly (2018) show, when only a small amount of text from the target domain is available, out-of-domain embeddings can roughly match performance with in-domain embedding features; however, developing or tuning more targeted word embeddings for use in this dataset is a useful area of future work.

6.3 Generalizability

It is important to note that the dataset used in this study was derived from one specialty – Physical Therapy – within a single institution – the NIH Clinical Center. Thus, the texts analyzed are likely to be more homogeneous than would be a broader dataset. Evaluating generalization of our findings to free text from other healthcare subdomains and other institutions, and describing ways in which performance assertions vary between these sources, is a valuable area of future work.

7 Conclusion

We have presented an evaluation of several approaches for the task of classifying whether a given description of an individual performing an activity indicates that they are able to perform it, unable, unclear, or insufficient information to determine. We found that machine learning approaches with lexical features perform surprisingly well on the task, including detecting the rarer labels of *Unable* and *Unclear*, and that an ensembled approach sets a strong baseline of 77.9% macro F1 for our dataset. In-depth analysis of system errors suggested several intriguing problems for future work. For instance, we intend to investigate hybrid models and test how information related to report formatting, section structure, slot info and assistive devices could improve the performance. To clarify the confusion of a patient's ability, we need models that can differentiate between factual and hypothetical statements (e.g. `Pt can run` vs. `Pt dislikes running`). Additionally, we would like to incorporate contextual representations such as ELMo (Peters et al., 2018) and BERT (Devlin et al., 2018) into our models.

To our knowledge, this is the first work expanding on the problem of clinical negation detection to complex interactions between individuals and their environments. This work joins a growing body of research on application of NLP techniques to information about activity performance and role participation, and identifies several research challenges in adapting NLP methods to this new domain.

Acknowledgments

The authors would like to thank Pei-Shu Ho, Jonathan Camacho Maldonado, and Maryanne Sacco for discussions about error analysis, and our anonymous reviewers for their helpful comments. This research was supported in part by the Intramural Research Program of the National Institutes of Health, Clinical Research Center and through an Inter-Agency Agreement with the US Social Security Administration.

References

Michael Bales, Rita Kukafka, Ann Burkhardt, and Carol Friedman. 2005. Extending a medical language processing system to the functional status domain. In *AMIA Annual Symposium Proceedings*, volume 2005, page 888. American Medical Informatics Association.

Asma Ben Abacha and Dina Demner-Fushman. 2016. Recognizing question entailment for medical question answering. *AMIA Annual Symposium proceedings. AMIA Symposium*, 2016:310–318.

Asma Ben Abacha, Duy Dinh, and Yassine Mrabet. 2015. Semantic analysis and automatic corpus construction for entailment recognition in medical texts. In *Artificial Intelligence in Medicine*, pages 238–242, Cham. Springer International Publishing.

Taylor Berg-Kirkpatrick, David Burkett, and Dan Klein. 2012. An empirical investigation of statistical significance in NLP. In *Proceedings of the 2012 Joint Conference on Empirical Methods in Natural Language Processing and Computational Natural Language Learning*, pages 995–1005. Association for Computational Linguistics.

Sidney T. Bogardus, Virginia Towle, Christianna S. Williams, Mayur M. Desai, and Sharon Inouye. 2004. What does the medical record reveal about functional status? *Journal of General Internal Medicine*, 16(11):728–736.

Piotr Bojanowski, Edouard Grave, Armand Joulin, and Tomas Mikolov. 2017. Enriching word vectors with subword information. *Transactions of the ACL*, 5:135–146.

Mateusz Buda, Atsuto Maki, and Maciej A Mazurowski. 2018. A systematic study of the class imbalance problem in convolutional neural networks. *Neural Networks*, 106:249–259.

Wendy W Chapman, David Chu, and John N Dowling. 2007. ConText: An algorithm for identifying contextual features from clinical text. In *Proceedings of the workshop on BioNLP 2007: biological, translational, and clinical language processing*, pages 81–88. Association for Computational Linguistics.

Ido Dagan, Oren Glickman, and Bernardo Magnini. 2006. The PASCAL Recognising Textual Entailment Challenge. *Lecture Notes in Computer Science (including subseries Lecture Notes in Artificial Intelligence and Lecture Notes in Bioinformatics)*, 3944 LNAI:177–190.

Jacob Devlin, Ming-Wei Chang, Kenton Lee, and Kristina Toutanova. 2018. Bert: Pre-training of deep bidirectional transformers for language understanding. *arXiv preprint arXiv:1810.04805*.

Guy Divita, Marjorie E Carter, Le-Thuy Tran, Doug Redd, Qing T Zeng, Scott Duvall, Matthew H Samore, and Adi V Gundlapalli. 2016. v3NLP framework: tools to build applications for extracting concepts from clinical text. *eGEMs*, 4(3).

Guy Divita, Shuying Shen, Marjorie Carter, Andrew Redd, Tyler Forbush, Miland N Palmer, Matthew H Samore, and Adi V Gundlapalli. 2014. Recognizing questions and answers in EMR templates using natural language processing. In *ICIMTH*, pages 149–152.

Federico Fancellu, Adam Lopez, and Bonnie Webber. 2016. Neural networks for negation scope detection. In *Proceedings of the 54th Annual Meeting of the Association for Computational Linguistics (Volume 1: Long Papers)*, volume 1, pages 495–504.

David Ferrucci and Adam Lally. 2004. UIMA: an architectural approach to unstructured information processing in the corporate research environment. *Natural Language Engineering*, 10(3-4):327–348.

George Gkotsis, Sumithra Velupillai, Anika Oellrich, Harry Dean, Maria Liakata, and Rina Dutta. 2016. Don't let notes be misunderstood: A negation detection method for assessing risk of suicide in mental health records. In *Proceedings of the Third Workshop on Computational Lingusitics and Clinical Psychology*, pages 95–105.

Marcelline R Harris, Guergana K Savova, Thomas M Johnson, and Christopher G Chute. 2003. A term extraction tool for expanding content in the domain of functioning, disability, and health: proof of concept. *Journal of biomedical informatics*, 36(4-5):250–259.

Cheng Ju, Aurélien Bibaut, and Mark van der Laan. 2018. The relative performance of ensemble methods with deep convolutional neural networks for image classification. *Journal of Applied Statistics*, 45(15):2800–2818.

Ning Kang, Bharat Singh, Zubair Afzal, Erik M van Mulligen, and Jan A Kors. 2013. Using rule-based natural language processing to improve disease normalization in biomedical text. *Journal of the American Medical Informatics Association : JAMIA*, 20(5):876–81.

Yoon Kim. 2014. Convolutional neural networks for sentence classification. *arXiv preprint arXiv:1408.5882*.

Diederik P Kingma and Jimmy Ba. 2014. Adam: A method for stochastic optimization. *arXiv preprint arXiv:1412.6980*.

Jinqiu Kuang, April F Mohanty, VH Rashmi, Charlene R Weir, Bruce E Bray, and Qing Zeng-Treitler. 2015. Representation of functional status concepts from clinical documents and social media sources by standard terminologies. In *AMIA Annual Symposium Proceedings*, volume 2015, page 795. American Medical Informatics Association.

Rita Kukafka, Michael E Bales, Ann Burkhardt, and Carol Friedman. 2006. Human and automated coding of rehabilitation discharge summaries according to the International Classification of Functioning, Disability, and Health. *Journal of the American Medical Informatics Association*, 13(5):508–515.

Marco Marelli, Luisa Bentivogli, Marco Baroni, Raffaella Bernardi, Stefano Menini, and Roberto Zamparelli. 2014. SemEval-2014 Task 1: Evaluation of compositional distributional semantic models on full sentences through semantic relatedness and textual entailment. In *Proceedings of the 8th International Workshop on Semantic Evaluation (SemEval 2014)*, pages 1–8, Dublin, Ireland. Association for Computational Linguistics.

Saeed Mehrabi, Anand Krishnan, Sunghwan Sohn, Alexandra M Roch, Heidi Schmidt, Joe Kesterson, Chris Beesley, Paul Dexter, C Max Schmidt, Hongfang Liu, et al. 2015. DEEPEN: A negation detection system for clinical text incorporating dependency relation into NegEx. *Journal of biomedical informatics*, 54:213–219.

Roser Morante and Caroline Sporleder. 2012. Modality and negation: An introduction to the special issue. *Comput. Linguist.*, 38(2):223–260.

Danielle L. Mowery, Sumithra Velupillai, and Wendy W. Chapman. 2012. Medical diagnosis lost in translation: Analysis of uncertainty and negation expressions in english and swedish clinical texts. In *Proceedings of the 2012 Workshop on Biomedical Natural Language Processing*, BioNLP '12, pages 56–64, Stroudsburg, PA, USA. Association for Computational Linguistics.

Denis Newman-Griffis and Ayah Zirikly. 2018. Embedding transfer for low-resource medical named entity recognition: A case study on patient mobility. In *Proceedings of the BioNLP 2018 workshop*, pages 1–11, Melbourne, Australia. Association for Computational Linguistics.

Francesca M Nicosia, Malena J Spar, Michael A Steinman, Sei J Lee, and Rebecca T Brown. 2019. Making function part of the conversation: Clinician perspectives on measuring functional status in primary care. *Journal of the American Geriatrics Society*, 67(3):493–502.

F Pedregosa, G Varoquaux, A Gramfort, V Michel, B Thirion, O Grisel, M Blondel, P Prettenhofer, R Weiss, V Dubourg, J Vanderplas, A Passos, D Cournapeau, M Brucher, M Perrot, and E Duchesnay. 2011. Scikit-learn: Machine learning in Python. *Journal of Machine Learning Research*, 12:2825–2830.

Yifan Peng, Xiaosong Wang, Le Lu, Mohammadhadi Bagheri, Ronald Summers, and Zhiyong Lu. 2018. NegBio: a high-performance tool for negation and uncertainty detection in radiology reports. *AMIA Joint Summits on Translational Science proceedings. AMIA Joint Summits on Translational Science*, 2017:188–196.

Matthew E Peters, Mark Neumann, Mohit Iyyer, Matt Gardner, Christopher Clark, Kenton Lee, and Luke Zettlemoyer. 2018. Deep contextualized word representations. *arXiv preprint arXiv:1802.05365*.

Alexey Romanov and Chaitanya Shivade. 2018. Lessons from natural language inference in the clinical domain. In *Proceedings of the 2018 Conference on Empirical Methods in Natural Language Processing*, pages 1586–1596, Brussels, Belgium. Association for Computational Linguistics.

Sunghwan Sohn, Stephen Wu, and Christopher G Chute. 2012. Dependency parser-based negation detection in clinical narratives. *AMIA Summits on Translational Science Proceedings*, 2012:1.

Stuart J Taylor and Sanda M Harabagiu. 2018. The role of a deep-learning method for negation detection in patient cohort identification from electroencephalography reports. In *AMIA Annual Symposium Proceedings*, volume 2018, page 1018. American Medical Informatics Association.

Thanh Thieu, Jonathan Camacho, Pei-Shu Ho, Julia Porcino, Min Ding, Lisa Nelson, Elizabeth Rasch, Chunxiao Zhou, Leighton Chan, Diane Brandt, Denis Newman-Griffis, Ao Yuan, and Albert M Lai. 2017. Inductive identification of functional status information and establishing a gold standard corpus: A case study on the mobility domain. In *2017 IEEE International Conference on Bioinformatics and Biomedicine (BIBM)*, pages 2300–2302. IEEE.

World Health Organization. 2001. *International Classification of Functioning, Disability, and Health: ICF*. World Health Organization, Geneva.

Stephen Wu, Timothy Miller, James Masanz, Matt Coarr, Scott Halgrim, David Carrell, and Cheryl Clark. 2014. Negation's not solved: Generalizability versus optimizability in clinical natural language processing. *PLoS ONE*, 9(11).

Xiang Zhang, Junbo Zhao, and Yann LeCun. 2015. Character-level convolutional networks for text classification. In *Advances in neural information processing systems*, pages 649–657.

Learning from the Experience of Doctors:
Automated Diagnosis of Appendicitis Based on Clinical Notes

Steven Kester Yuwono   **Hwee Tou Ng**
Department of Computer Science
National University of Singapore
sky@u.nus.edu
nght@comp.nus.edu.sg

Kee Yuan Ngiam
Department of Surgery
National University Hospital
kee_yuan_ngiam@nuhs.edu.sg

Abstract

The objective of this work is to develop an automated diagnosis system that is able to predict the probability of appendicitis given a free-text emergency department (ED) note and additional structured information (e.g., lab test results). Our clinical corpus consists of about 180,000 ED notes based on ten years of patient visits to the Accident and Emergency (A&E) Department of the National University Hospital (NUH), Singapore. We propose a novel neural network approach that learns to diagnose acute appendicitis based on doctors' free-text ED notes without any feature engineering. On a test set of 2,000 ED notes with equal number of appendicitis (positive) and non-appendicitis (negative) diagnosis and in which all the negative ED notes only consist of abdominal-related diagnosis, our model is able to achieve a promising $F_{0.5}$-score of 0.895 while ED doctors achieve $F_{0.5}$-score of 0.900. Visualization shows that our model is able to learn important features, signs, and symptoms of patients from unstructured free-text ED notes, which will help doctors to make better diagnosis.

1 Introduction

Medical diagnosis is an important task which requires high accuracy and efficiency, especially for patients admitted to the accident and emergency (A&E) department of a hospital. These patients have a wide range of medical conditions. However, it is highly improbable for a medical doctor to gain expertise in all medical fields. Therefore, it is very challenging for the attending doctors to perform quick and accurate diagnosis in order to prevent further complications.

Most of the relevant and useful information (e.g., signs and symptoms) is in the form of free text notes entered by medical doctors. The text does not consist of well-formed and well-structured sentences, but rather sentence fragments containing medical abbreviations and frequent misspelling (due to the time constraints imposed on doctors).

The task addressed in this paper is to diagnose *acute appendicitis*, a binary classification task. Appendicitis was chosen because of the fact that the lifetime risk of having appendicitis is high (8.6% for males and 6.7% for females (Addiss et al., 1990)). Furthermore, there would be high clinical impact if our system is successful. Besides reducing the number of misdiagnoses, our system is expected to help reduce cost by minimizing the number of patients requiring Computed Tomography (CT) scans. CT scans are performed by doctors when they are unsure whether a patient suffers from appendicitis. Although CT scans were found to be 98% accurate in diagnosing acute appendicitis (Rao et al., 1998), they are harmful to our body – one CT scan emits approximately 400[1] times the radiation of a regular chest X-ray. Moreover, there is an exponential increase (from 2.9% to 82.4% in 22 years) in CT scan utilization without any improvement in outcomes (Repplinger et al., 2016; Markar et al., 2014).

We propose a neural network model, which is a combination of a convolutional neural network (CNN) (LeCun et al., 1989), a recurrent neural network (RNN) (Elman, 1990), and a residual network (He et al., 2016) inspired by their recent successes in multiple tasks. RNN has proven to be successful in natural language processing (NLP) tasks such as machine translation (Bahdanau et al., 2015), automated essay scoring (Taghipour and Ng, 2016), and question answering (Kundu and

---

[1] https://www.fda.gov/
radiation-emittingproducts/
radiationemittingproductsandprocedures/
medicalimaging/medicalx-rays/ucm115329.
htm (Accessed on 7 June, 2019)

Proceedings of the BioNLP 2019 workshop, pages 11–19
Florence, Italy, August 1, 2019. ©2019 Association for Computational Linguistics

Ng, 2018). CNN has also been successfully used in NLP (Collobert et al., 2011; Chollampatt and Ng, 2018). The main strength of neural networks is that we can train the model without any feature engineering. Therefore, the model is scalable and generalizable to learn other diseases.

2 Automated Diagnosis

In this section, we define the diagnosis task and the evaluation metric used for measuring the performance of the automated diagnosis system.

2.1 Task Description

We formulate the task as a binary classification problem. Given a free-text ED note, and optional real-valued features (from the structured fields), the model is required to classify the ED note as positive appendicitis (represented by a 1), or negative appendicitis (represented by a 0). This is accomplished by producing a probability score, and comparing the score against a threshold, such that the class is positive if the probability score exceeds the threshold.

The corpus of hospital ED notes used in this paper is obtained from the National University Hospital (NUH), Singapore, spanning a period of ten years. However, the diagnosis stored in each ED note is not the true diagnosis. The ground truth is stored in the discharge summary (DS) of a patient after the patient is discharged from the hospital. Our corpus consists of about 180,000 ED notes and DS pairs. Each ED note contains 440 words on average.

The ED notes are written in sentence fragments and point forms, and very often contain abbreviations, symbols, and misspelled words. This adds to the difficulty in diagnosing appendicitis. Moreover, the free-text ED notes contain patients' personal health information (PHI) such as name, identification number, and contact number. The ED notes need to be anonymized (by removing the PHI) before they are used for research purposes. We have developed a simple and efficient algorithm to anonymize the ED notes (Yuwono et al., 2016) and it is used in this work.

2.2 Evaluation Metric

The standard evaluation metrics of binary classification are recall, precision, specificity, F_1-score, and $F_{0.5}$-score. The last two are shown in Equation 1.

$$F_1 = 2 \times \frac{precision \times recall}{precision + recall}$$

$$F_{0.5} = (1 + 0.5^2) \times \frac{precision \times recall}{(0.5^2 \times precision) + recall}$$

$$(1)$$

Let TP, FP, FN, and TN denote true positive, false positive, false negative, and true negative respectively. The positive class refers to class 1 (appendicitis), while the negative class refers to class 0 (not appendicitis). As clinicians favor precision and specificity over recall, we have adopted $F_{0.5}$-score as our main evaluation metric. We aim to have FP as low as possible to prevent patients from being operated on when they do not have appendicitis. Clinicians view FN as more tolerable (as compared to FP), because doctors are still required to investigate the condition of patients not diagnosed as appendicitis until they recover.

3 Neural Networks

3.1 Model Architecture

We have created a novel neural network architecture named convolutional residual recurrent neural network (CR2). Our architecture is illustrated in Figure 1.

Lookup Table Layer: The first layer of our neural network projects each word into a d_{LT} dimensional space. Given a sequence of words $\mathbf{W}$ represented by their *one-hot* representations $(\mathbf{w}_1, \mathbf{w}_2, ..., \mathbf{w}_M)$, the output of the lookup table layer (LT) is given by Equation 2.

$$\begin{aligned} LT(\mathbf{W}) &= (\mathbf{Ew}_1, \mathbf{Ew}_2, ..., \mathbf{Ew}_M) \\ &= (\mathbf{x}_1, \mathbf{x}_2, ..., \mathbf{x}_M) \end{aligned} \quad (2)$$

where $\mathbf{E}$ is the word embedding matrix which is learnt during training and M is the number of words in an ED note.

Convolution Layer: After the dense representation of the input sequence is computed from the lookup table layer, it is fed as the input to a convolution layer to extract *local features*. Given a window of word representations of length l, (i.e., $\mathbf{x}_1, \mathbf{x}_2, ..., \mathbf{x}_l$), they are first concatenated to form vector $\bar{\mathbf{x}}$, and then an output convolution vector $\mathbf{c}$ of length d_c is computed as shown in Equation 3.

$$\mathbf{c} = \mathbf{W}_v \bar{\mathbf{x}} + \mathbf{b} \quad (3)$$

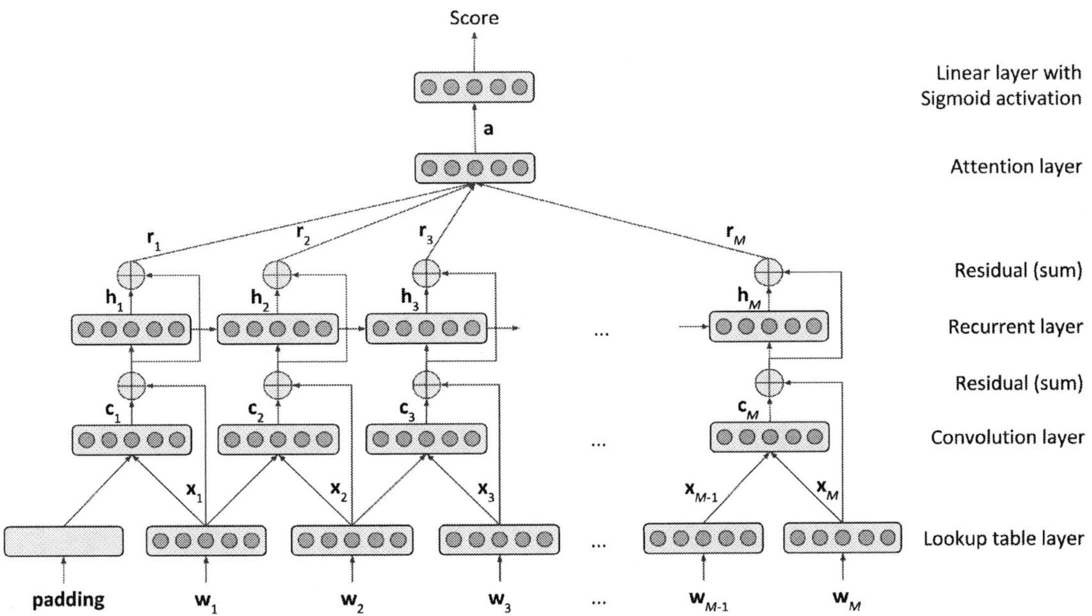

Figure 1: Our neural network architecture (CR2).

$\mathbf{W}_v$ and $\mathbf{b}$ are the trainable weight and bias parameters respectively, and they are shared across all windows in a sequence.

Residual Layer: We perform the sum operation on the sequence of the word embeddings ($\mathbf{X} = \mathbf{x}_1, \mathbf{x}_2, ..., \mathbf{x}_M$) and the output of the convolutional layer ($\mathbf{C} = \mathbf{c}_1, \mathbf{c}_2, ..., \mathbf{c}_M$) as shown in Equation 4.

$$\text{Sum}(\mathbf{X}, \mathbf{C}) = \mathbf{X} + \mathbf{C} \qquad (4)$$

To be able to perform the sum operation as shown above, the dimension of the word embeddings (d_{LT}) and the dimension of the output vectors of the convolution layer (or the number of filters) (d_c) have to be equal.

Recurrent Layer: After combining *local features* extracted by the convolution layer with the original dense word representations, the resulting vectors are fed as input to a recurrent layer. The recurrent layer processes the input to generate a representation of a given ED note. There are three well-known RNN units: basic recurrent units (Elman, 1990), gated recurrent units (GRU) (Cho et al., 2014), and long short-term memory units (LSTM) (Hochreiter and Schmidhuber, 1997). Based on our experimental results, LSTM outperforms the other two units and hence we only use LSTM as our RNN unit.

LSTM is able to learn to preserve or forget in-

formation. To control the flow of information, LSTM uses three gates to forget or pass information to the next time step. The formal definition of LSTM is described in Equation 5.

$$
\begin{aligned}
\mathbf{i}_t &= \sigma(\mathbf{W}_i\mathbf{x}_t + \mathbf{U}_i\mathbf{h}_{t-1} + \mathbf{b}_i) \\
\mathbf{f}_t &= \sigma(\mathbf{W}_f\mathbf{x}_t + \mathbf{U}_f\mathbf{h}_{t-1} + \mathbf{b}_f) \\
\tilde{\mathbf{c}}_t &= \tanh(\mathbf{W}_c\mathbf{x}_t + \mathbf{U}_c\mathbf{h}_{t-1} + \mathbf{b}_c) \\
\mathbf{c}_t &= \mathbf{i}_t \circ \tilde{\mathbf{c}}_t + \mathbf{f}_t \circ \mathbf{c}_{t-1} \\
\mathbf{o}_t &= \sigma(\mathbf{W}_o\mathbf{x}_t + \mathbf{U}_o\mathbf{h}_{t-1} + \mathbf{b}_o) \\
\mathbf{h}_t &= \mathbf{o}_t \circ \tanh(\mathbf{c}_t)
\end{aligned}
\qquad (5)
$$

$\mathbf{x}_t$ is the input vector at time t. LSTM produces one vector $\mathbf{h}_t$ at each time step t ($\mathbf{h}_0$ is the zero vector). $\mathbf{W}_i, \mathbf{W}_f, \mathbf{W}_c, \mathbf{W}_o, \mathbf{U}_i, \mathbf{U}_f, \mathbf{U}_c, \mathbf{U}_o$ are weight matrices and $\mathbf{b}_i, \mathbf{b}_f, \mathbf{b}_c, \mathbf{b}_o$ are the bias vectors. The circle symbol $\circ$ denotes element-wise multiplication and σ denotes the sigmoid function. The output of the recurrent layer is $\mathbf{H} = (\mathbf{h}_1, \mathbf{h}_2, ..., \mathbf{h}_M)$. Following (Taghipour and Ng, 2016), we use every output of the intermediate states of the RNN and perform summing (residual) and then pooling in the next layer to have a better representation of the entire ED note.

Residual Layer: We perform the sum operation on the sequence of the output vectors from the recurrent layer ($\mathbf{H} = \mathbf{h}_1, \mathbf{h}_2, ..., \mathbf{h}_M$) and the output vectors of the previous residual layer

(Sum($\mathbf{X}, \mathbf{C}$)) as shown in Equation 6.

$$\mathbf{R} = \text{Sum}(\mathbf{H}, \mathbf{X} + \mathbf{C}) = \mathbf{H} + \mathbf{X} + \mathbf{C} \qquad (6)$$

To be able to perform the sum operation as shown above, the dimension of the word embedding vectors (d_{LT}), output vectors of the convolution layer (d_c), and output vectors of the hidden RNN layer (d_r) have to be equal.

Attention layer: Visualizing the learned model is of high importance in the medical domain. By using an attention mechanism, we can show the degree of importance of words and phrases. Attention mechanism has been successful in many recent studies (Bahdanau et al., 2015; Hermann et al., 2015; Rush et al., 2015). The outputs of the previous residual layer $\mathbf{R} = (\mathbf{r}_1, \mathbf{r}_2, ..., \mathbf{r}_M)$ are used as inputs of the attention layer. In other words, this layer receives M vectors of size d_r, where d_r is the output dimension of the recurrent layer. $\mathbf{R}$ is a rich representation of the words in the ED note using a combination of word embeddings, CNN outputs, and RNN outputs. Each vector $\mathbf{r}_t$ is multiplied by a learnable real-valued weight s'_t between 0 and 1 before adding the elements of all M vectors into a single vector $\mathbf{a}$ as a form of *weighted average*. The functions of the attention layer are defined in Equation 7.

$$\begin{aligned}
s_t &= \mathbf{v} \cdot \tanh(\mathbf{W}_r \mathbf{r}_t) \\
s'_t &= \text{softmax}(\mathbf{s})_t \\
\mathbf{a} &= \sum_{t=1}^{M} s'_t \mathbf{r}_t
\end{aligned} \qquad (7)$$

$\mathbf{W}_r$ is a trainable matrix of size $d_r \times d_r$ and $\mathbf{v}$ is a trainable vector of size d_r. To learn more complex functions, $\mathbf{W}_r$ is introduced to increase the number of parameters and $\tanh$ is introduced to add non-linearity. $\mathbf{W}_r$ and $\mathbf{v}$ are *shared* across all time steps t. To make sure that the weights for all time steps sum to 1, the softmax function is performed on all the weights $\mathbf{s} = (s_1, s_2, ..., s_M)$. The attention layer is able to learn to assign varying weights to different time steps t depending on the input $\mathbf{r}_t$. The main advantage of having an attention layer is that we can retrieve the weight s'_t for each time step, and hence we are able to visualize and measure the importance of each word in the ED note.

Linear Layer with Sigmoid Activation: If there are no additional real-valued features, the input of this layer is the vector $\mathbf{a}$. Otherwise, it will be [$\mathbf{a}, \mathbf{l}$], the concatenation of $\mathbf{a}$ and $\mathbf{l}$, where $\mathbf{l}$ contains the additional real-valued features which will be described in the next subsection. The linear layer maps the input vector into a single scalar value. This mapping is a simple linear transformation, therefore the computed scalar value is unbounded. Since we are expected to predict either class 0 or 1, we will use a sigmoid function to ensure the scalar value is in the range $(0, 1)$. The mapping of the linear layer after applying the sigmoid function is shown in Equation 8.

$$s(\mathbf{x}) = \sigma(\mathbf{w} \cdot \mathbf{x} + b) \qquad (8)$$

where $\mathbf{x}$ is the input vector $\mathbf{a}$ or [$\mathbf{a}, \mathbf{l}$], $\mathbf{w}$ is the weight vector, and b is the bias value.

3.2 Additional Real-valued Features

Before using additional real-valued features such as lab results in the neural network, the values need to be normalized. We have adopted `normal sigmoid` to normalize the real-valued features which is shown in Equation 9. $\bar{x}$ and σ represent the mean and standard deviation for a particular feature (e.g., white blood cell count).

$$\begin{aligned}
\text{normal}(x) &= \frac{(x - \bar{x})}{\sigma} \\
\text{normal_sigmoid}(x) &= \frac{1}{1 + e^{-\text{normal}(x)}}
\end{aligned} \qquad (9)$$

There are also entries where ED notes are not accompanied by any lab results. To deal with missing values, we calculate the mean ($\bar{x}$) of all existing entries in the training set of that particular feature (e.g., white blood cell count) and then use the average value to fill in the gap.

In order to include the L real-valued normalized features $\mathbf{l} = (l_1, l_2, ..., l_L)$ in the model, we concatenate L real numbers (after normalizing them) to the output of the attention layer, before going into the next layer. The input of the final layer will be [$\mathbf{a}, \mathbf{l}$], a vector of size $d_r + L$. Figure 2 illustrates the process above.

3.3 Training

We use the RMSProp optimization algorithm (Dauphin et al., 2015) to minimize a loss function over the training data. Given N training ED notes and their corresponding true class s_i^* (either 0 or 1), the model computes the predicted score s_i in the range of $(0, 1)$ for all training ED notes and then updates the network weights such that the loss

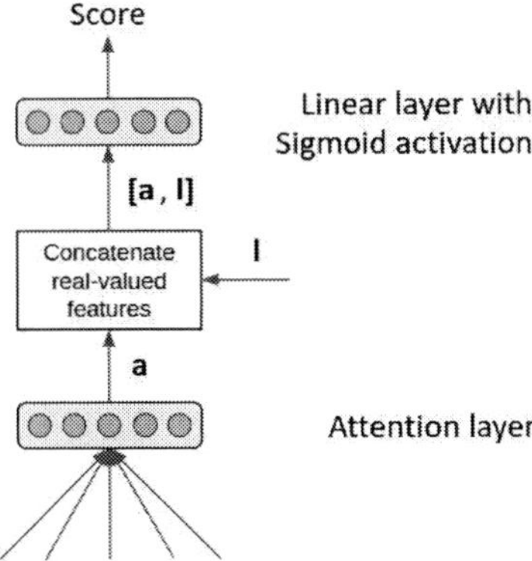

Figure 2: Concatenation of real-valued features before the final layer.

function is minimized. The loss function we have adopted in our system is the binary cross-entropy loss function as shown in Equation 10.

$$L(\mathbf{s}, \mathbf{s}^*) = -\sum_{i=1}^{N} s_i^* \log(s_i) + (1 - s_i^*) \log(1 - s_i)$$

$$(10)$$

In our data set, the distribution of the classes is highly imbalanced – the proportion of ED notes in class 0 can be as high as 98.4%, with the remaining 1.6% ED notes in class 1. To tackle this problem, we have adopted a *weighted* binary cross-entropy loss function, where each class is weighted *inversely proportional* to the class frequency in the training data to allocate more weight to the less frequent class, similar to the technique used by (Chollampatt et al., 2016) for rescaling.

To prevent overfitting, we have adopted dropout (Srivastava et al., 2014) regularization. We also clip the gradient if the gradient norm is larger than a certain threshold. We train the neural network for a specified number of epochs and evaluate the model on a validation set in every epoch. The epoch with the highest $F_{0.5}$-score on the validation set is then selected as the final model.

3.4 Threshold Adjustment

The output or score of the neural network is a real number between 0 and 1. However, we need to transform the score to either 1 (positive) or 0 (negative) to solve our binary classification problem. Therefore, there is a need to set a threshold as the decision boundary. The default threshold used to split the two classes is 0.5. For example, if the prediction score is greater than 0.5, then the predicted

class is positive (appendicitis); otherwise negative (not appendicitis).

The aforementioned threshold can be used to tune the model to have lower FP but higher FN, and vice versa. In this paper, we would like to achieve the lowest possible FP, trading for a higher FN. To achieve this, we use the validation set to search for a threshold with the best $F_{0.5}$-score. First, we use the model in the current epoch to predict the score of each instance in the validation set. Second, we sort the validation instances in ascending order of the predicted scores. Third, we perform a linear search to find the cut-off threshold to achieve the best $F_{0.5}$-score on the validation set. This is repeated in every epoch, resulting in a unique threshold for each epoch. The epoch with the best $F_{0.5}$-score (using its own unique threshold) on the validation set is used as the final model to evaluate the test set, using the same threshold used in the validation set.

4 Experiments

4.1 Setup

Our network has several hyper-parameters which need to be set. We use the RMSProp optimizer with decay rate of 0.9 and learning rate of 0.001. Mini-batch[2] size is 32 and the model is trained for 25 epochs. The vocabulary is created using all words in the training set. Out-of-vocabulary words are replaced by a special <unknown> token. Words that contain any digits are replaced by a special <num> token. The network is regularized by using dropout (Srivastava et al., 2014) with probability 0.5. During training, if the norm of the gradient exceeds 10, it will be clipped to a maximum value of 10. Word embedding dimension (d_{LT}), output dimension of the hidden layer for the RNN (d_r), and the number of filters for the CNN (d_c) are set to 300. The convolution window size (l) is set to 3. We initialize the lookup table layer with our custom pre-trained word embeddings which are trained using our entire corpus of 180,000 ED notes excluding the notes used as validation and test set. We use the word2vec skip-gram model (Mikolov et al., 2013) to train our word embeddings. Although the lookup table

---

[2]To create mini-batches for training, all the ED notes in a mini-batch are padded using a dummy token to have the same length. To remove the effect of padding tokens during training, they are masked to prevent the network from miscalculating the gradients.

layer is initialized with pre-trained word embeddings, the lookup table layer is trainable and not fixed. We utilize 4 additional features from the structured patient data, namely age, gender, and two lab test results (white blood cell count and neutrophils), and incorporate them into the network as described in Section 3.2.

4.2 Dataset

We have about 180,000 ED notes and DS pairs in total. The class distribution of the ED notes in the entire corpus is shown in Table 1 (second and third column). The first class listed in the first column is the class predicted by ED doctors in the ED notes, while the second class listed in the first column is the true diagnosis class obtained from the DS.

4.2.1 Dataset 1: Natural Distribution (Original Dataset)

Using the corpus shown in Table 1, we randomly sample 10% for training, 10% for validation, and 10% for test. The number of ED notes is 18,111, 18,108, and 18,107 respectively following its natural class distribution (about 1.6% positive ED notes). To speed up training, we only use ED notes with 750 words or less in the training set, resulting in 16,854 instead of 18,111 ED notes for training. We do not impose any length limit for both the validation and test set.

Class	Number of ED notes	Percentage
++ (TP)	2,194	1.2 %
+− (FP)	1,071	0.6%
−+ (FN)	796	0.4 %
−− (TN)	177,210	97.8 %
Total	181,271	100 %

Table 1: Class distribution of ED notes.

4.2.2 Dataset 2: Equal Class Distribution with Random Negative ED Notes

In our second dataset, we obtain a subset of the 181,271 ED notes (from Table 1) to create a dataset with 50% positive and 50% negative ED notes. There are 2,980, 1,000, and 2,000 ED notes for training, validation, and test respectively with equal distribution of positive and negative classes in each set. The negative ED notes consist of randomly sampled ED notes of all diagnosis classes that are not appendicitis.

4.2.3 Dataset 3: Equal Class Distribution with Abdominal-related Negative ED Notes

Our third dataset is very similar to our second dataset (in Section 4.2.2) with the same class distribution. The only difference is that the negative ED notes in this dataset only consist of abdominal-related diagnosis instead of any random diagnosis that is not appendicitis. The number of ED notes in the training, validation, and test set are the same as those in dataset 2. The 1,000 positive ED notes in this test set are identical to the 1,000 positive ED notes in the test set in dataset 2. Dataset 3 is more challenging than dataset 2 because the signs and symptoms of appendicitis are very similar to those of other abdominal conditions. The class distribution of all three test sets is shown in Table 2.

4.3 Results and Discussions

The experimental results of the best model (CR2, described in Sections 3 and 4.1) on the three datasets are summarized in Table 3.

We train the neural network model (end to end) on a single GPU (Nvidia TITAN X Pascal), and the training time is 3.2 hours for dataset 1, and 35 minutes for each of the datasets 2 and 3. After the model is trained, it is able to perform acute appendicitis diagnosis rapidly, at 400 ED notes per second. The *best* single CR2 model is chosen based on the highest $F_{0.5}$-score on the validation set over 50 runs with different seeds. The *average* score for the CR2 model in each column is calculated over 50 runs with different seeds. The $\pm$ sign represents the standard deviation over the 50 runs.

We have two baseline methods, namely a maxent (maximum entropy, also known as logistic regression) classifier and an Alvarado rule-based scoring system. This is inspired by prior work (Deleger et al., 2013) which performs appendicitis risk stratification using an Alvarado rule-based scoring system with features obtained from free text. Before using the aforementioned two methods, the texts are first tokenized, and negation are detected through Negex (Chapman et al., 2001), a simple regular expression rule-based algorithm which has been modified to suit our needs. For maxent, a list of words is built from the training ED notes and we obtain the bag-of-words representation for each ED note, add the lab results and other structured fields, and then use them as features to train a maxent classifier.

Class	Dataset 1		Dataset 2		Dataset 3	
	# ED notes	%	# ED notes	%	# ED notes	%
++ (TP)	216	1.2 %	734	36.7 %	734	36.7 %
+− (FP)	104	0.6 %	6	0.3 %	36	1.8 %
−+ (FN)	78	0.4 %	266	13.3 %	266	13.3 %
−− (TN)	17,709	97.8 %	994	49.7 %	964	48.2 %
Total	18,107	100 %	2,000	100 %	2,000	100 %

Table 2: Class distribution of ED notes in test sets.

model	TP	FP	FN	TN	Rec	Prec	Spec	F1	F0.5	Acc
					Dataset 1					
ED	216	104	78	17,709	0.735	0.675	0.994	0.704	0.686	0.990
ME	138	126	156	17,687	0.469	0.523	0.993	0.495	<u>0.511</u>	0.984
Alv	124	90	170	17,723	0.422	0.579	0.995	0.488	0.539	0.986
Best	141	90	153	17,723	0.480	0.610	0.995	0.537	**0.579***	0.987
Avg	154.8	109.2	139.2	17,703.8	0.527	0.588	0.994	0.553	0.573	0.986
	±16.9	±18.8	±16.9	±18.8	±0.058	±0.021	±0.0011	±0.030	±0.016	±0.00046
					Dataset 2					
ED	734	6	266	994	0.734	0.992	0.994	0.844	0.927	0.864
ME	952	62	48	938	0.952	0.939	0.938	0.945	<u>0.941</u>	0.945
Alv	617	12	383	988	0.617	0.981	0.988	0.758	0.877	0.803
Best	912	27	88	973	0.912	0.971	0.973	0.941	**0.959***	0.943
Avg	912.1	28.6	87.9	971.4	0.912	0.970	0.971	0.940	0.958	0.942
	±17.1	±6.1	±17.1	±6.1	±0.017	±0.0058	±0.0061	±0.0076	±0.0037	±0.0069
					Dataset 3					
ED	734	36	266	964	0.734	0.953	0.964	0.829	0.900	0.849
ME	880	125	120	875	0.880	0.876	0.875	0.878	<u>0.876</u>	0.878
Alv	617	72	383	928	0.617	0.896	0.928	0.731	0.821	0.773
Best	831	79	169	921	0.831	0.913	0.921	0.870	**0.895***	0.876
Avg	832.1	84.2	167.9	915.8	0.832	0.908	0.916	0.868	0.892	0.874
	+28.8	±12.2	±28.8	±12.2	±0.029	±0.0096	±0.0122	±0.0125	±0.0045	±0.0095

Table 3: Summary of the best model against ED doctors and the baselines on three datasets. The baseline for the statistical significance tests is underlined and statistically significant improvements ($p < 0.05$) are marked with '*'. ME stands for Maxent, Alv stands for Alvarado, Best stands for Best CR2, and Avg stands for Avg CR2.

In acute appendicitis diagnosis, there is an existing well-known scoring system, namely Alvarado score (Alvarado, 1986). It is also known as MANTRELS score, which is a mnemonic to remember the score factors (signs, symptoms, and lab readings) – **M**igration of pain to the right lower quadrant, **A**norexia, **N**ausea or vomiting, **T**enderness in the right lower quadrant, **R**ebound pain, **E**levated temperature (fever), **L**eukocytosis (high white blood cell count), and **S**hift of neutrophils to the left. The score for each factor is 1(M), 1(A), 1(N), 2(T), 1(R), 1(E), 2(L), and 1(S) respectively. The score for each factor present in a patient will be added together to obtain the final score. A higher score indicates that a patient is more likely to have appendicitis. The aforementioned 8 factors are detected through a regular expression (with negation) on the ED notes that have been preprocessed with Negex. Different threshold values (scores strictly greater than the threshold will be classified as positive, and negative otherwise) are explored and the threshold with the best $F_{0.5}$-score is chosen. The thresholds for Alvarado scoring in datasets 1, 2, and 3 are 6, 5, and 5 respectively.

Our neural network model (CR2) outperforms the two baselines in $F_{0.5}$-score on all three datasets. We also perform a statistical significance test ($p < 0.05$) to determine whether the obtained improvement is statistically significant. We found that our neural network improvements against maxent on all datasets are statistically significant. This shows that our neural network model is superior to the maxent classifier and Alvarado scoring system.

Based on the first row in Table 3, we can see that ED doctors' performance is better compared to our model. This is mainly caused by class imbalance (1.6% positive and 98.4% negative). Learning and predicting on a dataset with extremely skewed class distribution is challenging. However, as we can see from the results, the performance of our best model is close to that of ED doctors, with 14 fewer FP instances and 75 more FN instances out of 18,107 ED notes in the test set.

Based on the results of dataset 2 and 3, our model achieved lower FP+FN (in other words, higher accuracy) when compared to ED doctors. With equal distribution of positive and negative ED notes, our model performs better than ED doctors with much lower FN in exchange for slightly higher FP. Our model's $F_{0.5}$-score exceeds that of ED doctor on dataset 2 and is very close to that of ED doctor on dataset 3. Our model also consistently achieves better sensitivity (recall) than the ED doctor.

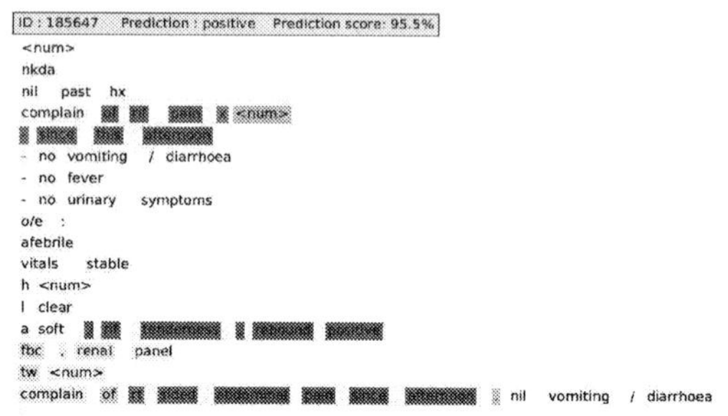

Figure 3: Visualization of how our model interprets a positive ED note.

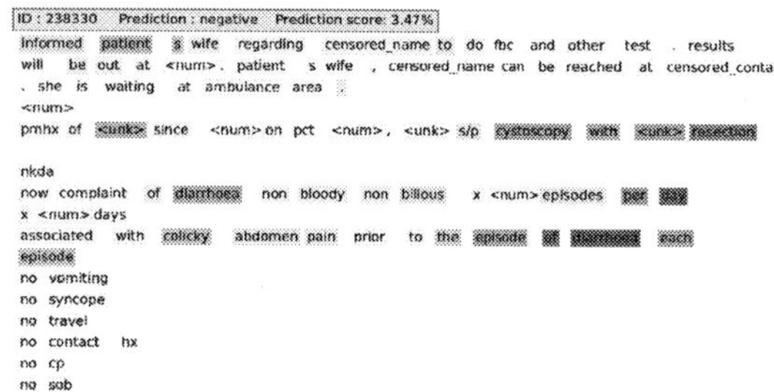

Figure 4: Visualization of how our model interprets a negative ED note.

To visualize the model and gain insights into how the model assigns importance to words and phrases, we retrieve the weights of the attention layer. The weights can be used to show the degree of importance of words and phrases in an ED note. From our observation, the model is able to pick up meaningful signs and symptoms of appendicitis most of the time. Figure 3 shows the visualization of our model, with appendicitis features highlighted, such as rif pain, and tenderness with rebound. In Figure 3, darker shade of red color indicates a higher weight assigned to a word. These signs and symptoms have been validated and used in practice as features of the Alvarado scoring scheme (Alvarado, 1986). On the other hand, the model is also able to pick up the features of non-appendicitis. In Figure 4, the model is able to pick up diarrhea and a few other features suggesting non-appendicitis.

We will explore other neural network architectures and more (deeper) layers in the future. We will also design our experiments to be able to fully utilize the entire 180,000 ED notes to train and validate our model.

5 Conclusion

In this paper, we tackle the task of automated diagnosis using free-text ED notes. We present a machine learning model which is able to learn from free text and optional additional features without any feature engineering. We show that the performance of our novel neural network architecture is promising and close to the performance of ED doctors. Analysis of the visualization shows that the attention layer is able to meaningfully learn the importance of words and phrases in ED notes and to change its emphasis depending on the context of the words. This is helpful in highlighting certain key description (i.e., signs and symptoms) that might have been missed otherwise by medical doctors in a real-life setting.

Acknowledgments

This research is supported by Singapore Ministry of Education Academic Research Fund Tier 1 grant T1-251RES1513.

References

David G Addiss, Nathan Shaffer, Barbara S Fowler, and Robert V Tauxe. 1990. The epidemiology of appendicitis and appendectomy in the United States. *Am. J. Epidemiol.*, 132(5):910–925.

Alfredo Alvarado. 1986. A practical score for the early diagnosis of acute appendicitis. *Ann. Emerg. Med.*, 15(5):557 – 564.

Dzmitry Bahdanau, Kyunghyun Cho, and Yoshua Bengio. 2015. Neural machine translation by jointly learning to align and translate. In *Proceedings of the 3rd International Conference on Learning Representations*.

Wendy W Chapman, Will Bridewell, Paul Hanbury, Gregory F Cooper, and Bruce G Buchanan. 2001. A simple algorithm for identifying negated findings and diseases in discharge summaries. *J. Biomed. Inform.*, 34(5):301–310.

Kyunghyun Cho, Bart van Merrienboer, Caglar Gulcehre, Dzmitry Bahdanau, Fethi Bougares, Holger Schwenk, and Yoshua Bengio. 2014. Learning phrase representations using RNN encoder-decoder for statistical machine translation. In *Proceedings of the 2014 Conference on Empirical Methods in Natural Language Processing*, pages 1724–1734.

Shamil Chollampatt and Hwee Tou Ng. 2018. A multilayer convolutional encoder-decoder neural network for grammatical error correction. In *Proceedings of the Thirty-Second AAAI Conference on Artificial Intelligence*, pages 5755–5762.

Shamil Chollampatt, Kaveh Taghipour, and Hwee Tou Ng. 2016. Neural network translation models for grammatical error correction. In *Proceedings of the Twenty-Fifth International Joint Conference on Artificial Intelligence*, pages 2768–2774.

Ronan Collobert, Jason Weston, Léon Bottou, Michael Karlen, Koray Kavukcuoglu, and Pavel Kuksa. 2011. Natural language processing (almost) from scratch. *J. Mach. Learn. Res.*, 12:2493–2537.

Yann Dauphin, Harm de Vries, and Yoshua Bengio. 2015. Equilibrated adaptive learning rates for non-convex optimization. In *Advances in Neural Information Processing Systems 28*, pages 1504–1512.

Louise Deleger, Holly Brodzinski, Haijun Zhai, Qi Li, Todd Lingren, Eric S Kirkendall, Evaline Alessandrini, and Imre Solti. 2013. Developing and evaluating an automated appendicitis risk stratification algorithm for pediatric patients in the emergency department. *J. Am. Med. Inform. Assoc.*, 20(e2):e212–e220.

Jeffrey L Elman. 1990. Finding structure in time. *Cognitive Science*, 14(2):179–211.

Kaiming He, Xiangyu Zhang, Shaoqing Ren, and Jian Sun. 2016. Deep residual learning for image recognition. In *Proceedings of the IEEE Conference on Computer Vision and Pattern Recognition*, pages 770–778.

Karl Moritz Hermann, Tomas Kocisky, Edward Grefenstette, Lasse Espeholt, Will Kay, Mustafa Suleyman, and Phil Blunsom. 2015. Teaching machines to read and comprehend. In *Advances in Neural Information Processing Systems 28*, pages 1693–1701.

Sepp Hochreiter and Jürgen Schmidhuber. 1997. Long short-term memory. *Neural Comput.*, 9(8):1735–1780.

Souvik Kundu and Hwee Tou Ng. 2018. A nil-aware answer extraction framework for question answering. In *Proceedings of the 2018 Conference on Empirical Methods in Natural Language Processing*, pages 4243–4252.

Yann LeCun, Bernhard Boser, John S. Denker, Donnie Henderson, Richard E. Howard, Wayne Hubbard, and Lawrence D. Jackel. 1989. Backpropagation applied to handwritten zip code recognition. *Neural Comput.*, 1(4):541–551.

Sheraz R Markar, Diluka Pinto, Marta Penna, Alan Karthikesalingam, Bulathsinghalage Kalana Sandun Bulathsinghala, Kumaralingam Kumaran, Majid Hashemi, and Ranil Fernando. 2014. A comparative international study on the management of acute appendicitis between a developed country and a middle income country. *Int. J. Surg.*, 12(4):357–360.

Tomas Mikolov, Ilya Sutskever, Kai Chen, Greg S Corrado, and Jeff Dean. 2013. Distributed representations of words and phrases and their compositionality. In *Advances in Neural Information Processing Systems 26*, pages 3111–3119.

Patrick M Rao, James T Rhea, Robert A Novelline, Amy A Mostafavi, and Charles J McCabe. 1998. Effect of computed tomography of the appendix on treatment of patients and use of hospital resources. *N. Engl. J. Med.*, 338(3):141–146.

Michael D Repplinger, Andrew C Weber, Perry J Pickhardt, Victoria P Rajamanickam, James E Svenson, William J Ehlenbach, Ryan P Westergaard, Scott B Reeder, and Elizabeth A Jacobs. 2016. Trends in the use of medical imaging to diagnose appendicitis at an academic medical center. *J. Am. Coll. Radiol.*, 13(9):1050–1056.

Alexander M. Rush, Sumit Chopra, and Jason Weston. 2015. A neural attention model for abstractive sentence summarization. In *Proceedings of the 2015 Conference on Empirical Methods in Natural Language Processing*, pages 379–389.

Nitish Srivastava, Geoffrey E. Hinton, Alex Krizhevsky, Ilya Sutskever, and Ruslan Salakhutdinov. 2014. Dropout: A simple way to prevent neural networks from overfitting. *J. Mach. Learn. Res.*, 15:1929–1958.

Kaveh Taghipour and Hwee Tou Ng. 2016. A neural approach to automated essay scoring. In *Proceedings of the 2016 Conference on Empirical Methods in Natural Language Processing*, pages 1882–1891.

Steven Kester Yuwono, Hwee Tou Ng, and Kee Yuan Ngiam. 2016. Automated anonymization as spelling variant detection. In *Proceedings of the COLING 2016 Workshop on Clinical Natural Language Processing*, pages 99–103.

A Paraphrase Generation System for EHR Question Answering

Sarvesh Soni, Kirk Roberts
School of Biomedical Informatics
University of Texas Health Science Center at Houston
Houston TX, USA
{sarvesh.soni, kirk.roberts}@uth.tmc.edu

Abstract

This paper proposes a dataset and method for automatically generating paraphrases for clinical questions relating to patient-specific information in electronic health records (EHRs). Crowdsourcing is used to collect 10,578 unique questions across 946 semantically distinct paraphrase clusters. This corpus is then used with a deep learning-based question paraphrasing method utilizing variational autoencoder and LSTM encoder/decoder. The ultimate use of such a method is to improve the performance of automatic question answering methods for EHRs.

1 Introduction

The useful information present in electronic health records (EHRs) is hard to access due to many of its usability issues (Zhang and Walji, 2014). Question answering (QA) systems have the potential to reduce the time it takes for users to access information present in the EHRs. However, the effectiveness of such QA systems largely depends on the variety of questions they are capable of handling. Automated paraphrasing techniques are known to improve the performance of QA models in general domain by generating different variations of a question (Duboue and Chu-Carroll, 2006; Fader et al., 2013; Berant and Liang, 2014; Bordes et al., 2014a,b; Dong et al., 2015; Narayan et al., 2016; Chen et al., 2016; Dong et al., 2017; Abujabal et al., 2018b). Thus, automatic generation of high quality paraphrases for patient-specific EHR questions has the potential to improve performance of the clinical QA systems.

Paraphrasing is a technique of rewording a given phrase such that its lexical and syntactic structure is different but its semantic information is retained (Bhagat and Hovy, 2013). For instance, the following two questions can be considered as paraphrases of each other.

- What *medications* am I currently taking?

- What are my current *medications*?

Such EHR-related questions are usually targeted toward specific clinical information (Roberts and Demner-Fushman, 2016). For example, the aforementioned questions are intended to get information regarding *medications*. In such a scenario, paraphrases can be considered as different ways of accessing the same medical data. As such, automatic clinical paraphrase generation can help in increasing the breadth of questions for training a clinical QA system.

While automated paraphrase generation is well-studied in the general domain (Madnani and Dorr, 2010; Androutsopoulos and Malakasiotis, 2010), very few studies have focused on clinical paraphrasing (Hasan et al., 2016; Adduru et al., 2018; Neuraz et al., 2018). On the other hand, clinical text simplification, which aims at generating easier to read paraphrases, has received relatively more attention (Zeng-Treitler et al., 2007; Elhadad and Sutaria, 2007; Deléger and Zweigenbaum, 2008; Kandula et al., 2010; Pivovarov and Elhadad, 2015; Qenam et al., 2017; Adduru et al., 2018; Bercken et al., 2019). However, these works in the clinical domain are not representative of QA needs as the usefulness of paraphrases is largely application-specific (Bhagat and Hovy, 2013). Also, existing datasets for clinical paraphrasing consist of either short phrases (Hasan et al., 2016) or webpage title texts (Adduru et al., 2018), both of which are not suitable to build a paraphrase generator for QA. One can resort to using external tools such as Google Translate for generating question paraphrases (Neuraz et al., 2018), but these general-purpose tools are not tailored to the medical domain (Liu and Cai, 2015).

In this paper, we propose a clinical paraphrasing corpus CLINIQPARA with questions which

20

Proceedings of the BioNLP 2019 workshop, pages 20–29
Florence, Italy, August 1, 2019. ©2019 Association for Computational Linguistics

can be answered using EHR data[1]. We further propose a deep learning-based automated clinical paraphrasing system utilizing a variational autoencoder (VAE) and a long short-term memory recurrent neural network (LSTM) (Gupta et al., 2018). To our knowledge, this is the first work aimed at automatically generating paraphrases without using any external resource for questions specifically focused on retrieving patient-specific information from EHRs. Our main contributions are summarized as follows:

- Crowdsourcing a large paraphrasing corpus of questions which are answerable using the data from EHR.

- Application of VAE in context to clinical paraphrasing task.

The rest of the paper is structured as follows. Section 2 explores related work in the domain of clinical paraphrasing. Then, Sections 3 and 4 discuss our dataset generation and model implementation details respectively. Next, Section 5 evaluates the results of our clinical paraphrasing system. Finally, Section 6 discusses our findings, and Section 7 provides a concluding summary.

2   Background

We begin this section by detailing work related to clinical text simplification and paraphrasing in Sections 2.1 and 2.2 respectively. Then, we highlight some of the current work in general-domain paraphrasing for QA as part of Section 2.3.

2.1   Clinical Text Simplification

As stated earlier, many studies have focused on clinical text simplification. Text simplification differs from paraphrasing as the former is a unidirectional task whereas the latter can be considered as bi-directional textual entailment (Androutsopoulos and Malakasiotis, 2010), but the methods nonetheless provide useful context for our work. Elhadad and Sutaria (2007) and Deléger and Zweigenbaum (2008) relied on parallel or comparable corpora to construct paraphrase pairs of specialized and lay medical texts. Zeng-Treitler et al. (2007) and Kandula et al. (2010) either replaced the difficult clinical phrases in text with simpler synonyms or included uncomplicated explanations for them. Qenam et al. (2017) concentrated on just substituting the difficult terms with

more comprehensible ones. Much of the simplification work in the clinical domain has been targeted toward lexical methods to convert or append the complex phrases present in the original sentence with their simpler alternatives (Pivovarov and Elhadad, 2015). Such simplification approaches usually make use of external vocabularies to map the difficult clinical terms. While these techniques reduce the complexity of a sentence at the lexical level, they generally leave the syntactic structure of a sentence unchanged. For instance,

- Patient suffered from *myocardial infarction.*

- Patient suffered from *heart attack.*

These variations correspond to a specific category of paraphrases named synonym substitution (Bhagat and Hovy, 2013) and amount to a smaller subset of possible paraphrases.

Alternatively, Adduru et al. (2018) and Bercken et al. (2019) constructed clinical simplification datasets from various web-based sources such as WebMD, MedicineNet, Wikipedia, and SimpleWikipedia utilizing sentence alignment techniques. While this approach is capable of generating more variations of a given sentence, it is still a simplification task and hence not suitable to be incorporated in a QA system (Bhagat and Hovy, 2013).

2.2   Clinical Text Paraphrasing

Comparatively, less focus has been drawn toward clinical paraphrase generation. Hasan et al. (2016) built their dataset by combining an existing general domain paraphrasing corpus PPDB 2.0 (Pavlick et al., 2015) with the UMLS (Unified Medical Language System) metathesaurus. Specifically, they utilized fully specified names of medical concepts present in UMLS. Though their corpus contains medical terms, it comprises of comparatively shorter length phrases rather than complete sentences.

Adduru et al. (2018) also created a paraphrasing corpus utilizing the titles of web articles from Mayo Clinic along with Wikipedia. While this dataset consists of complete clinical sentences, they are atypical of the patient-specific EHR questions.

Neuraz et al. (2018) used the Google Translate API to generate paraphrases for question templates in French. They utilized these generated template paraphrases to augment the size of their

---

Scenario 18: You've been having some low back pain recently, and want to make an appointment with your doctor's office through the doctor's website, but the system isn't clear. Write a short (up to 15 word), grammatical, one-sentence question asking how you make an appointment. No need to state it is confusing, simply ask a question.   **Question**: How do I make an appointment?
Scenario 41: Your elderly mother has been taking Metformin (a diabetes drug). She is forgetful and requires someone to organize her pills for each day. However, the person that normally organizes her pills hasn't done it for this week, and you need to know what the instructions are for your mother's Metformin prescription. Write a short (up to 15 word), grammatical, one-sentence question asking her doctor for this dosage information. You question must contain the word 'Metformin'.   **Question**: What are my mother's Metformin dosage instructions?
Scenario 43: You recently had an automobile accident, and you've started taking physical therapy to help recover. Your first appointment went well, but you forgot to write down when your next appointment was scheduled for. Write a short (up to 15 word), grammatical, one-sentence question asking your doctor for this information. Your question must contain 'physical therapy'.   **Question**: When is my next physical therapy appointment?

Table 1: Three scenarios used to build the CLINIQPARA corpus, along with a canonical question (not provided to annotators).

development dataset for natural language understanding task without evaluating the quality of the paraphrases. Such general-purpose machine translation systems lack the ability to capture the domain-specific nuances of biomedicine (Liu and Cai, 2015). This suggests the need for a question paraphrasing dataset targeted toward clinical domain.

As discussed earlier, existing clinical paraphrasing datasets are not suitable for building a paraphrase generation system for clinical questions. To the best of our knowledge, the proposed paraphrasing corpus is the first which aims at clinical questions.

2.3 Paraphrasing for Question Answering

There are several question paraphrasing corpora available for the general domain such as WikiAnswers (Fader et al., 2013), PPDB (Ganitkevitch et al., 2013), PPDB 2.0 (Pavlick et al., 2015), GraphQuestions (Su et al., 2016), and ComQA (Abujabal et al., 2018a). However, there is a scarcity of such datasets for clinical questions.

The proposed corpus consists of questions which can be answered using EHR data. Such a corpus would have utility beyond QA systems as well, like in question similarity (Luo et al., 2015; Nakov et al., 2017), and in particular could serve as a standard paraphrase corpus for the medical domain.

3 Dataset Construction

In order to quickly and efficiently collect hundreds of paraphrases, we utilized the crowdsourcing platform Amazon Mechanical Turk (AMT). Instead of prompting AMT workers with a question and directly asking for paraphrases–which could prime the workers and bias them toward very similar paraphrases–we presented them with a short, 3-6 sentence imaginary scenario that placed them in a situation where a specific piece of information was required (such as their current medications). The workers were then asked to provide questions directed to their doctor to answer that information need. After the crowd-sourced questions were collected, they were manually organized into distinct paraphrase clusters. This was necessary because some questions address the information need but are not logically equivalent paraphrases. These steps are discussed in more detail below.

3.1 Scenario Creation

To ensure a wide variety of EHR questions, we first came up with 11 top-level topic categories people might ask about: medications, other treatments, labs, immunizations, imaging, other exams, problem list, past medical history, family history, appointments, and documents. For each of these categories, 2-8 scenarios were created to capture relevant questions about the topic. In total, 50 scenarios were created. Table 1 shows three of these scenarios along with the canonical question expected by the scenario.

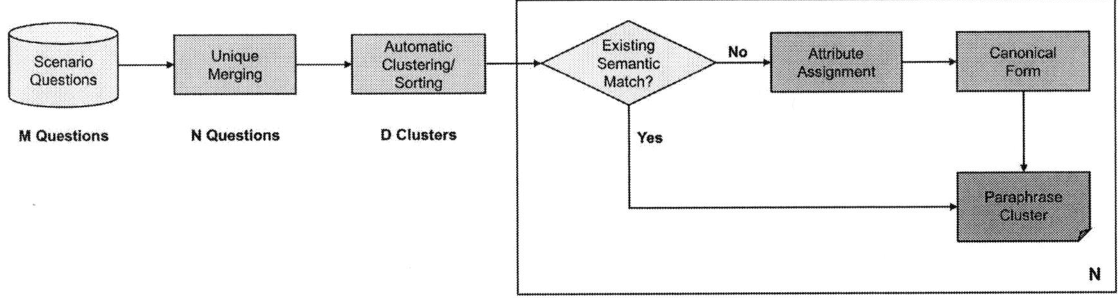

Figure 1: Paraphrase cluster creation process.

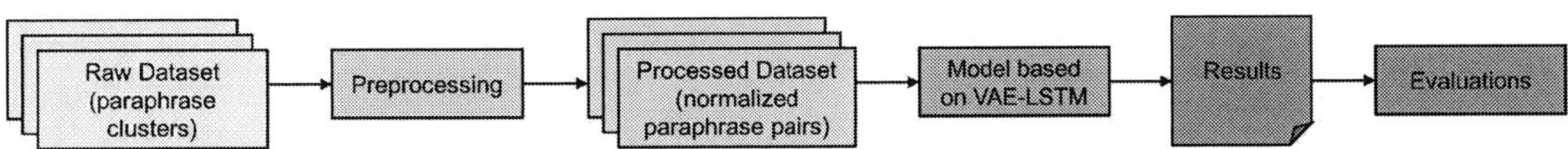

Figure 2: Framework of our paraphrasing system.

3.2 Crowdsourcing

The 50 scenarios were uploaded to AMT in three batches, one scenario per Human Intelligence Task (HIT). Workers were required to provide three questions per HIT, since first question might be obvious and not result in a particularly diverse set of paraphrases. Each HIT was assigned to 100 workers and the annotators were paid $0.08/HIT. Workers were required to be proficient in English, but otherwise no requirements were imposed and no demographic data was collected.

The initial validation process was minimal. HITs were rejected if the workers did not provide 3 questions, or if none of the questions were valid. 93% of submitted HITs were approved. Of the rejections, 73% were due to not providing 3 questions. Many of the rejections due to invalid questions were for questions that were completely unrelated to the scenario.

3.3 Paraphrase Cluster Creation

After collecting a set of questions for each scenario using crowdsourcing, the next step was to manually organize the questions into paraphrase clusters (Figure 1). We consider a paraphrase cluster to be composed only of exact paraphrases. That is, questions are paraphrases only if they should have the same semantic representation.

The first two steps in paraphrase construction were designed to ease the manual burden of paraphrase cluster assignment. First, questions were merged into case-independent unique sets. Second, questions were clustered using Dirichlet Process Mixture Model clustering with unigram and bigram features. This allowed us to sort the questions so that very similar questions, which are likely to be paraphrases, are annotated in succession. The remainder of this process required manual annotation for each question (with some computer assistance).

Each paraphrase cluster is represented by a canonical form. For each unique question, given the correct list of paraphrase clusters, the annotator selected a cluster that is the semantic match, or created a new cluster if none existed. Each new paraphrase cluster was assigned several values, notably including whether it was grammatical. Invalid questions (non-responsive, spurious responses that are common with crowdsourcing) were placed in either the INVALID-related cluster (invalid questions which were related to the scenario), or the INVALID-unrelated cluster. Finally, a canonical form was assigned to valid clusters.

The entire process in Figure 1 was repeated for each scenario. Since there were 100 workers per HIT, and 3 questions per worker, up to 300 questions needed to be clustered per scenario (with 50 scenarios, there were 15,000 questions). There were much fewer than 300 unique questions per scenario, and the process took between 30-40 minutes for most scenarios.

After ignoring casing and whitespace, there were an average of 240 unique questions per scenario. Three annotators manually clustered the questions (three scenarios were clustered as a group, with the remaining scenarios being clustering individually). Ignoring invalid questions (9%), and ungrammatical questions (6%), there were a

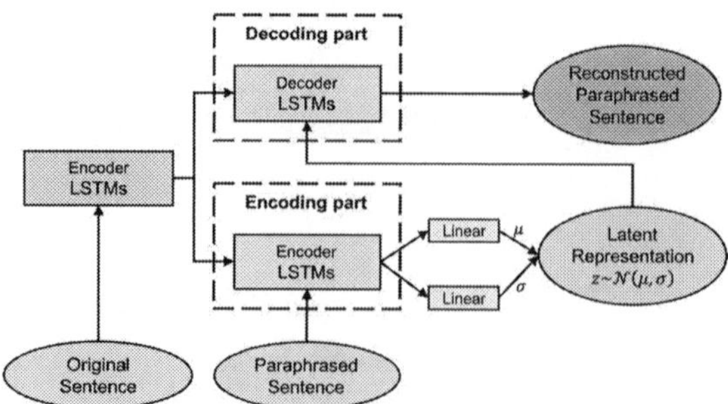

Figure 3: Architecture of the paraphrasing model based on VAE-LSTM.

median of 2.8 and mean of 5.6 paraphrase clusters, with a minimum of 5 questions, per scenario. Table 2 shows the paraphrase clusters for one of the scenarios.

4 Paraphrase Generator

An overall framework of our paraphrasing system is presented in Figure 2.

4.1 Preprocessing

First, we normalize the medical concepts and mask the person references and digits present in the question. This is carried out to make sure the questions from different scenarios are consistent. Consider the following questions and their masked versions:

- What types of *cancer* does my *father* have?
  $\rightarrow$ What types of *concept* does my *human* have?
- Was it in *2003* that I had my *appendectomy*?
  $\rightarrow$ Was it in *digits* that I had my *concept*?

After this step, we further deduplicate the questions and remove clusters with only 1 question (as a minimum of two questions are required for evaluating paraphrasing).

We then construct paraphrase pairs using the created clusters of paraphrases. Specifically, we generate all combinations of questions which are present in the same cluster. This results in over 258,000 unique question-paraphrase pairs for 10,578 questions distributed across 946 semantically distinct paraphrase clusters.

4.2 Model

We use a deep learning model based on VAE-LSTM (Gupta et al., 2018), the architecture of which is presented in Figure 3. One of the main characteristics of VAE that makes it a good choice

for paraphrasing task is that its latent representation is continuous. In other words, the encoder outputs a distribution rather than discrete values. This enables the decoder to produce naturalistic outputs even in the cases where latent code does not correspond to any of the already viewed inputs.

The model consists of two parts, namely, encoding and decoding. On the encoding side, the original sentence is first passed to an encoder LSTM which constructs a vector representation x for the sentence. Then, another encoder LSTM takes as input x along with the paraphrased sentence whose vector representation y is generated as the output. Finally, a feedforward neural network generates the VAE encoder's mean (μ) and standard deviation (σ) parameters using y.

Both original and paraphrased sentences are fed into their respective encoder LSTMs using word embeddings. We train the word embeddings on our paraphrasing corpus using word2vec (Mikolov et al., 2013) and keep them fixed while training the paraphrasing system.

In the decoding phase, we first generate a vector representation x by passing the original sentence to an encoder LSTM. Ultimately, a decoder LSTM reconstructs the paraphrased sentence using x and a latent code z which is sampled from $\mathcal{N}(\mu, \sigma)$. While x is fed to the decoder LSTM only at an initial stage, z is taken as input at each of its stages.

During training, we aim to maximize the objective function shown below in Equation 1, thereby learning the VAE parameters.

$$\mathcal{O}(\theta, \phi; x, y) = \mathbb{E}_{q_\phi(z|x,y)}[\log(p_\theta(y|z, x))] - \mathrm{KL}(q_\phi(z|x, y) \| p(z)) \quad (1)$$

where $q_\phi(z|x, y)$ is a posterior distribution (encoder model) on z that the VAE aims at keeping closer to its prior $p(z)$ (commonly a standard normal distribution). KL represents the Kullback-Leibler divergence which intuitively gives a similarity measure between the two distributions. At the decoder side, $p_\theta(y|z, x)$ is a distribution on y, given the latent code z and vector x, whose expectation $\mathbb{E}$ is taken with respect to $q_\phi(z|x, y)$. The objective function gives a lower bound on the true likelihood of the data. We follow the training mechanism proposed by Bowman et al. (2016).

During testing, the encoder part is ignored and paraphrases are generated for a given question using z sampled from a standard normal distribution.

Scenario:
You just realized you should have a doctor's appointment coming up soon, but cannot find it on your calendar. Write a short (up to 15 word), grammatical, one-sentence question asking your doctor about your next appointment.

Cluster 1 (229 questions, 164 unique): When is my next appointment?
When is my next appointment? (frequency = 32)
What time is my next appointment? (6)
When is my next scheduled appointment? (5)
Can you tell me when my next appointment is? (4)
When is my next appointment scheduled? (4)
When is my next appointment scheduled for? (4)
What is the date and time of my next appointment? (3)
(... 157 more ...)

Cluster 2 (38 questions, 33 unique): Do I have an appointment soon?
Do I have an appointment coming up? (3)
Do I have a doctor's appointment coming up soon? (2)
Do I have an appointment soon? (2)
Do I have an upcoming appointment scheduled? (2)
(... 29 more ...)

Cluster 3 (3 questions): Do I have an appointment this week?
Am I scheduled to come in to your office this week for an appointment?
Do I have an appointment this week?
Is my appointment scheduled for this week?

Cluster 4 (2 questions): Can I make an appointment?
Can I make an appointment?
Will you be able to make an appointment any soon?

Cluster 5 (1 question): How long until my next appointment?
How long until my next doctor's appointment?

Cluster 6 (1 question): Is my appointment this week or next?
Is my appointment scheduled for this week or next week?

Cluster 7 (1 question): Is my appointment next week?
Was my appointment scheduled for next week?

Cluster 8 (1 question): Is my appointment on Tuesday?
Is my scheduled appointment for Tuesday?

Cluster 9 (1 question): Is my appointment this month?
Do you have a record of my having made an appointment for later this month?

Cluster INVALID-related (34 questions)
Can you give me an appointment card?
How long will this appointment last?
What happens if I miss the appointment?
What will you be discussing in regards to my next check up?
Will I be meeting with you or with your assistant?
(... 29 more ...)

Cluster INVALID-unrelated (17 questions)
According to my lab results, what vitamins or supplements should I be taking?
Do you have the results of my mri?
How is my BMI?
What does this medicine do?
What symptoms should I watch for?
(... 12 more ...)

Table 2: Paraphrase Clusters for Scenario 3. Only a sample of questions are shown.

The presence of input question at the decoder side enables the model to generate its paraphrases.

We utilize the same model parameters as Gupta et al. (2018). Namely, the dimension of the word embedding is 300; the dimension of the encoder and decoder is 600; the latent space dimension is 1100; the encoder has 1 layer; the decoder has 2 layers; the learning rate is 5×10^{-5}; the dropout rate is 30%; the batch size is 32. We use PyTorch for implementing the model and run all our experiments on an NVidia Tesla V100 GPU (32G).

4.3 Evaluation

The paraphrased questions generated by the model are re-incorporated with the concept, person names, and digits which were extracted during the preprocessing step. The paraphrases are evaluated using standard paraphrase evaluation metrics such as BLEU (Papineni et al., 2002), METEOR (Lavie and Agarwal, 2007), and TER (Snover et al., 2006), which are shown to work well for the paraphrase identification task (Madnani et al., 2012). BLEU score assesses the lexical similarity of generated paraphrases with the reference ones using exact matching while METEOR additionally takes into account the word stems and synonyms. TER score measures the edit distance (number of edits required to convert one sentence into another) between generated and reference paraphrases. So, higher BLEU and METEOR scores are better whereas a lower TER score is preferable. Since we have multiple paraphrases for each question in our corpus, we calculate these metrics for the generated paraphrases against all the available ground truth paraphrases.

To evaluate the performance measures on all the parts of CLINIQPARA dataset, we perform 10-fold cross validation. Specifically, we split our dataset by scenarios (into 10 groups each containing 5 scenarios) and sequentially test the performance of model on each group of 5 scenarios after training it on the other 45. We report the individual and average scores from all these runs in our results.

We further evaluate the performance of our model on the Quora dataset[2], which contains over 400k pairs of questions of which around 150k pairs are paraphrases. We train on 90% of these paraphrase pairs and test on the remaining 10%.

We also perform human evaluation of the gen-

[2]https://data.quora.com/First-Quora-Dataset-Release-Question-Pairs

Dataset	Metric		
	BLEU	**METEOR**	**TER**
Quora	16.70	20.60	77.4
CLINIQPARA	13.25	21.47	91.93

Table 3: Performance of our paraphrasing system using automated evaluation metrics.

Fold (Scenarios)	Metric		
	BLEU	**METEOR**	**TER**
1-5	**19.25**	23.56	92.58
6-10	12.27	19.25	94.01
11-15	18.79	21.93	**78.17**
16-20	9.72	19.30	91.46
21-25	9.20	20.97	103.25
26-30	16.45	**23.66**	84.98
31-35	6.07	19.84	111.62
36-40	11.24	20.40	95.05
41-45	14.08	22.33	85.18
46-50	15.48	23.44	82.97
Average	**13.25**	**21.47**	**91.93**

Table 4: Results on CLINIQPARA using automated evaluation metrics for 10-fold cross validation. Each fold contains 5 scenarios over which the model is tested after being trained on the other 45 scenarios.

erated paraphrases for quantifying the aspects not covered solely by the automated evaluation metrics. For the CLINIQPARA dataset, we randomly select a set of 300 questions from all the scenarios. For each of these questions, we further choose a ground truth paraphrase as well as a system generated paraphrase in a random fashion. This result in a total of 600 pairs of question paraphrases, 300 from the gold dataset and 300 generated by the paraphrasing system. The constructed set is separately evaluated by two annotators who are asked to rate the paraphrases based on two parameters: *fluency* of the questions as natural language and their *relevance* to the original question. Both of these scores range from 1 (worse) to 5 (best). For each paraphrase, the final score is calculated by averaging the scores provided by the two annotators. The fact that a paraphrase is ground truth or generated by the model is hidden from the annotators to avoid bias. For the Quora dataset, we directly report the human evaluation results from Gupta et al. (2018).

Dataset	Type	Relevance	Fluency
Quora	Ground Truth	4.82	4.94
	VAE-LSTM	3.57	4.08
CLINIQPARA	Ground Truth	4.69	4.70
	VAE-LSTM	1.88	3.65

Table 5: Results of human evaluation. Range of scores is between 1 (worst) and 5 (best).

Input Question
Do you know when my next appointment is going to be?
Generated Paraphrases
1. Can you please confirm the date and time of my appointment?
2. On what day and what time do I have my appointment?
3. Do you have the date and time for my appointment?
4. Can you tell me when I am scheduled for my appointment.

Table 6: Example paraphrases generated by the model for an input question from Scenario 3 (Good).

5 Results

The results on CLINIQPARA (our dataset) and Quora dataset using automated evaluation metrics are shown in Table 3. More granular cross validation results on CLINIQPARA are presented in Table 4. Moreover, the results of our human evaluation process are shown in Table 5. Some of the system-generated paraphrases are included in Tables 6 and 7. Table 6 shows the examples from a fold which performed well during the cross validation step whereas Table 7 includes examples from a low-performing fold.

6 Discussion

The quality of generated paraphrases is promising, but further investigation is required to determine if performance is sufficient for use in training a downstream QA system. We note that the METEOR score on CLINIQPARA was comparable to that of the results on the Quora dataset. This shows the potential of our paraphrasing system in generating paraphrases similar to the ground truth paraphrases. Our system performed well on the Quora dataset in terms of BLEU score, which can be attributed to the larger size of the Quora dataset in terms of unique questions (150k in Quora vs. 10.5k in CLINIQPARA).

On analyzing the results of the qualitative evaluation, we observe that the majority of the errors are related to change in the person reference or asking about frequency-related information. For instance, the original question *"When shall I come for my next physical therapy?"* asking about the

Input Question
Is my latest CAT scan impression complete?
Generated Paraphrases
1. Was my CAT scan impression successful or not?
2. Was my CAT scan impression a success?
3. Was my diagnosis CAT scan impression?
4. does my father 's file show how many times he has CAT scan impression?

Table 7: Example paraphrases generated by the model for an input question from Scenario 32 (Moderate).

user's next appointment for a therapy is modified to a question *"May I have the number of times my father has physical therapy?"* asking about the number of times the user's father has undergone the therapy. A similar trend can be seen in the second example where the original question *"Can you please give me the dosage details on the metformin mom takes?"* is concerned about getting the dosage information for the user's mother whereas the system generated question *"Could you tell me the amount of time my father has metformin?"* is related to the frequency of metformin intake of the user's father. Further qualitative evaluation can help pointing out more specific problems with the model.

Our future work includes experimenting with more advanced embedding techniques (Peters et al., 2018; Devlin et al., 2018). We also plan to handle some of the aforementioned errors by incorporating additional constraints such as restricting the question paraphrase pairs in our corpus to contain only semantically similar masked references.

7 Conclusion

Automatic paraphrase generation of clinical questions can improve the performance of the QA systems. Little work has been focused on clinical paraphrasing, let alone concentrating on clinical questions. We have proposed a new clinical paraphrasing corpus CLINIQPARA, containing questions which can be answered using EHRs. Our model based on VAE-LSTM has the potential to generate quality clinical paraphrases.

Acknowledgments This work was supported by the U.S. National Library of Medicine, National Institutes of Health (NIH), under award R00LM012104; the Cancer Prevention and Research Institute of Texas (CPRIT), under award RP170668; as well as the Bridges Family Doctoral Fellowship Award.

References

Abdalghani Abujabal, Rishiraj Saha Roy, Mohamed Yahya, and Gerhard Weikum. 2018a. ComQA: A Community-sourced Dataset for Complex Factoid Question Answering with Paraphrase Clusters. *arXiv preprint arXiv:1809.09528.* Version 2.

Abdalghani Abujabal, Rishiraj Saha Roy, Mohamed Yahya, and Gerhard Weikum. 2018b. Never-Ending Learning for Open-Domain Question Answering over Knowledge Bases. In *Proceedings of the 2018 World Wide Web Conference*, pages 1053–1062.

Viraj Adduru, Sadid A. Hasan, Joey Liu, Yuan Ling, Vivek Datla, Kathy Lee, Ashequl Qadir, and Oladimeji Farri. 2018. Towards dataset creation and establishing baselines for sentence-level neural clinical paraphrase generation and simplification. In *CEUR Workshop Proceedings*, pages 45–52.

Ion Androutsopoulos and Prodromos Malakasiotis. 2010. A survey of paraphrasing and textual entailment methods. *Journal of Artificial Intelligence Research*, 38:135–187.

Jonathan Berant and Percy Liang. 2014. Semantic Parsing via Paraphrasing. In *Proceedings of the 52nd Annual Meeting of the Association for Computational Linguistics*, pages 1415–1425.

Laurens van den Bercken, Robert-Jan Sips, and Christoph Lofi. 2019. Evaluating Neural Text Simplification in the Medical Domain. In *The World Wide Web Conference*, pages 3286–3292.

Rahul Bhagat and Eduard Hovy. 2013. What Is a Paraphrase? *Computational Linguistics*, 39(3):463–472.

Antoine Bordes, Sumit Chopra, and Jason Weston. 2014a. Question Answering with Subgraph Embeddings. In *Proceedings of the 2014 Conference on Empirical Methods in Natural Language Processing (EMNLP)*, pages 615–620.

Antoine Bordes, Jason Weston, and Nicolas Usunier. 2014b. Open Question Answering with Weakly Supervised Embedding Models. In *Proceedings of the Joint European Conference on Machine Learning and Knowledge Discovery in Databases*, pages 165–180.

Samuel R. Bowman, Luke Vilnis, Oriol Vinyals, Andrew Dai, Rafal Jozefowicz, and Samy Bengio. 2016. Generating Sentences from a Continuous Space. In *Proceedings of The 20th SIGNLL Conference on Computational Natural Language Learning*, pages 10–21.

Bo Chen, Le Sun, Xianpei Han, and Bo An. 2016. Sentence Rewriting for Semantic Parsing. In *Proceedings of the 54th Annual Meeting of the Association for Computational Linguistics*, pages 766–777.

Louise Deléger and Pierre Zweigenbaum. 2008. Paraphrase acquisition from comparable medical corpora of specialized and lay texts. In *AMIA Annual Symposium Proceedings*, page 146.

Jacob Devlin, Ming-Wei Chang, Kenton Lee, and Kristina Toutanova. 2018. Bert: Pre-training of deep bidirectional transformers for language understanding. *arXiv preprint arXiv:1810.04805.* Volume 2.

Li Dong, Jonathan Mallinson, Siva Reddy, and Mirella Lapata. 2017. Learning to Paraphrase for Question Answering. In *Proceedings of the 2017 Conference on Empirical Methods in Natural Language Processing*, pages 875–886.

Li Dong, Furu Wei, Ming Zhou, and Ke Xu. 2015. Question answering over freebase with multi-column convolutional neural networks. In *Proceedings of the 53rd Annual Meeting of the Association for Computational Linguistics and the 7th International Joint Conference on Natural Language Processing*, pages 260–269.

Pablo Duboue and Jennifer Chu-Carroll. 2006. Answering the question you wish they had asked: The impact of paraphrasing for question answering. In *Proceedings of the Human Language Technology Conference of the North American Chapter of the ACL.*

Noemie Elhadad and Komal Sutaria. 2007. Mining a Lexicon of Technical Terms and Lay Equivalents. In *Proceedings of BioNLP*, pages 49–56.

Anthony Fader, Luke Zettlemoyer, and Oren Etzioni. 2013. Paraphrase-Driven Learning for Open Question Answering. In *Proceedings of the 51st Annual Meeting of the Association for Computational Linguistics*, pages 1608–1618.

Juri Ganitkevitch, Benjamin Van Durme, and Chris Callison-Burch. 2013. PPDB: The Paraphrase Database. In *Proceedings of the 2013 Conference of the North American Chapter of the Association for Computational Linguistics: Human Language Technologies*, pages 758–764.

Ankush Gupta, Arvind Agarwal, Prawaan Singh, and Piyush Rai. 2018. A Deep Generative Framework for Paraphrase Generation. In *Thirty-Second AAAI Conference on Artificial Intelligence*, pages 5149–5156.

Sadid A Hasan, Bo Liu, Joey Liu, Ashequl Qadir, Kathy Lee, Vivek Datla, Aaditya Prakash, and Oladimeji Farri. 2016. Neural Clinical Paraphrase Generation with Attention. In *Proceedings of the Clinical NLP Workshop*, pages 42–53.

Sasikiran Kandula, Dorothy Curtis, and Qing Zeng-Treitler. 2010. A semantic and syntactic text simplification tool for health content. In *Proceedings of the AMIA Annual Symposium*, pages 366–370.

Alon Lavie and Abhaya Agarwal. 2007. METEOR: An automatic metric for MT evaluation with high levels of correlation with human judgments. In *Proceedings of the Second Workshop on Statistical Machine Translation*, pages 228–231.

Weisong Liu and Shu Cai. 2015. Translating Electronic Health Record Notes from English to Spanish: A Preliminary Study. In *Proceedings of BioNLP*, pages 134–140.

Jake Luo, Guo-Qiang Zhang, Susan Wentz, Licong Cui, and Rong Xu. 2015. SimQ: Real-Time Retrieval of Similar Consumer Health Questions. *J Med Internet Res*, 17(2):e43.

Nitin Madnani and Bonnie J. Dorr. 2010. Generating Phrasal and Sentential Paraphrases: A Survey of Data-Driven Methods. *Computational Linguistics*, 36(3):341–387.

Nitin Madnani, Joel Tetreault, and Martin Chodorow. 2012. Re-examining Machine Translation Metrics for Paraphrase Identification. In *Proceedings of the 2012 Conference of the North American Chapter of the Association for Computational Linguistics: Human Language Technologies*, pages 182–190.

Tomas Mikolov, Ilya Sutskever, Kai Chen, Greg S Corrado, and Jeff Dean. 2013. Distributed representations of words and phrases and their compositionality. In *Advances in Neural Information Processing Systems 26*, pages 3111–3119.

Preslav Nakov, Doris Hoogeveen, Lluís Màrquez, Alessandro Moschitti, Hamdy Mubarak, Timothy Baldwin, and Karin Verspoor. 2017. SemEval-2017 Task 3: Community Question Answering. In *Proceedings of the 11th International Workshop on Semantic Evaluation (SemEval-2017)*, pages 27–48.

Shashi Narayan, Siva Reddy, and Shay B Cohen. 2016. Paraphrase Generation from Latent-Variable PCFGs for Semantic Parsing. In *Proceedings of the 9th International Natural Language Generation Conference*, pages 153–162.

Antoine Neuraz, Leonardo Campillos Llanos, Anita Burgun, and Sophie Rosset. 2018. Natural language understanding for task oriented dialog in the biomedical domain in a low resources context. *arXiv preprint arXiv:1811.09417*. Version 2.

Kishore Papineni, Salim Roukos, Todd Ward, and Wei-Jing Zhu. 2002. BLEU: a method for automatic evaluation of machine translation. In *Proceedings of the 40th Annual Meeting of the Association for Computational Linguistics*, pages 311–318.

Ellie Pavlick, Pushpendre Rastogi, Juri Ganitkevitch, Benjamin Van Durme, and Chris Callison-Burch. 2015. PPDB 2.0: Better paraphrase ranking, fine-grained entailment relations, word embeddings, and style classification. In *Proceedings of the 53rd Annual Meeting of the Association for Computational Linguistics and the 7th International Joint Conference on Natural Language Processing*, pages 425–430.

Matthew Peters, Mark Neumann, Mohit Iyyer, Matt Gardner, Christopher Clark, Kenton Lee, and Luke Zettlemoyer. 2018. Deep Contextualized Word Representations. In *Proceedings of the 2018 Conference of the North American Chapter of the Association for Computational Linguistics: Human Language Technologies*, pages 2227–2237.

Rimma Pivovarov and Nomie Elhadad. 2015. Automated methods for the summarization of electronic health records. *Journal of the American Medical Informatics Association*, 22(5):938–947.

Basel Qenam, Tae Youn Kim, Mark J Carroll, and Michael Hogarth. 2017. Text Simplification Using Consumer Health Vocabulary to Generate Patient-Centered Radiology Reporting: Translation and Evaluation. *J Med Internet Res*, 19(12):e417.

Kirk Roberts and Dina Demner-Fushman. 2016. Annotating logical forms for EHR questions. In *Proceedings of the Language Resources & Evaluation Conference*, pages 3772–3778.

Matthew Snover, Bonnie Dorr, Richard Schwartz, Linnea Micciulla, and John Makhoul. 2006. A Study of Translation Edit Rate with Targeted Human Annotation. In *Proceedings of 7th Conference of the Association for Machine Translation in the Americas*, pages 223–231.

Yu Su, Huan Sun, Brian Sadler, Mudhakar Srivatsa, Izzeddin Gur, Zenghui Yan, and Xifeng Yan. 2016. On Generating Characteristic-rich Question Sets for QA Evaluation. In *Proceedings of the 2016 Conference on Empirical Methods in Natural Language Processing*, pages 562–572.

Qing Zeng-Treitler, Sergey Goryachev, Hyeoneui Kim, Alla Keselman, and Douglas Rosendale. 2007. Making texts in electronic health records comprehensible to consumers: a prototype translator. In *Proceedings of the AMIA Annual Symposium*, pages 846–850.

Jiajie Zhang and Muhammad Walji. 2014. *Better EHR, Usability, Workflow and Cognitive Support in Electronic Health Records*. University of Texas School of Biomedical Informatics at Houston.

REflex: Flexible Framework for Relation Extraction in Multiple Domains

Geeticka Chauhan
MIT CSAIL
geeticka@mit.edu

Matthew B. A. McDermott
MIT CSAIL
mmd@mit.edu

Peter Szolovits
MIT CSAIL
psz@mit.edu

Abstract

Systematic comparison of methods for relation extraction (RE) is difficult because many experiments in the field are not described precisely enough to be completely reproducible and many papers fail to report ablation studies that would highlight the relative contributions of their various combined techniques. In this work, we build a unifying framework for RE, applying this on three highly used datasets (from the general, biomedical and clinical domains) with the ability to be extendable to new datasets. By performing a systematic exploration of modeling, pre-processing and training methodologies, we find that choices of pre-processing are a large contributor performance and that omission of such information can further hinder fair comparison. Other insights from our exploration allow us to provide recommendations for future research in this area.

1 Introduction

Relation Extraction (RE) has gained a lot of interest from the community with the introduction of the Semeval tasks from 2007 by (Girju et al., 2007) and 2010 by (Hendrickx et al., 2009). The task is a subset of information extraction (IE) with the goal of finding semantic relationships between concepts in a given sentence, and is an important component of Natural Language Understanding (NLU). Applications include automatic knowledge base creation, question answering, as well as analysis of unstructured text data. Since the introduction of RE tasks in the general and medical domains, many researchers have explored the performance of different neural network architectures on the datasets (Socher et al., 2012; Zeng et al., 2014; Liu et al., 2016b; Sahu et al., 2016).

However, progress in RE is hampered by reproducibility issues as well as the difficulty in assessing which techniques in the literature will generalize to novel tasks, datasets and contexts. To assess the extent of these problems, we performed a manual review of 53 relevant neural RE papers[1] citing the three datasets (Hendrickx et al., 2009; Segura-Bedmar et al., 2013; Uzuner et al., 2011). The procedure for finding these papers is highlighted in (Chauhan, 2019).

Reproducibility Reproducibility is important for validating previous work and building upon it (Fokkens et al., 2013). Lack of reproducibility can be attributed to many factors such as difficulty in availability of source code (Ince et al., 2012) and omission of sources of variability such as hyperparameter details (Claesen and De Moor, 2015). We found that only 16 out of the 53 relevant papers had released their source code. 14 out of 53 papers were evaluated on multiple datasets, but the source code was publicly available for only five of those. Despite this, much of this code was lacking in modularity to be easily extendable to new datasets. In many cases, the process of reproducing the paper results was often unclear and lack of documentation made this more difficult. Even though most papers mentioned some hyperparameter details, important details were missing such as number of epochs, batch size, random initialization seed, if any, and details about early stop if that technique was applied.

Ablation Studies Lack of generalizability is caused by a dearth of appropriate empirical evaluation to identify the source of modeling gains. Ablation studies are important for identifying sources of improvements in results. Among the 53 papers that we looked at, 20 of the 24 papers in the general domain performed ablation studies. However, only 10 out of 29 papers in the medical domain performed one. Among these ablation studies,

---

[1] The 53 papers were filtered from a list of 728 papers skimmed for relevance. Appendix A contains paper details.

Proceedings of the BioNLP 2019 workshop, pages 30–47
Florence, Italy, August 1, 2019. ©2019 Association for Computational Linguistics

key details related to pre-processing were missing, which we found critical in our experiments.

In the absence of such information about causes of large variability of results, *fair comparison* of models becomes difficult. In this paper, we present an open-source unifying framework enabling the comparison of various training methodologies, pre-processing, modeling techniques, and evaluation metrics. The code is available at `https://github.com/geetickachauhan/relation-extraction`.

The experimental goals of this framework are identification of sources of variability in results for the three datasets and provide the field with a strong baseline model to compare against for future improvements. The design goals of this framework are identification of best practices for relation extraction and to be a guide for approaching new datasets.

By performing systematic comparison on three datasets, we find that 1) pre-processing choices can cause the largest variations in performance, 2) reporting scores on one test set split is problematic due to split bias. We perform other analyses in section 5 and also include recommendations for future research in this field in section 7.

Upon testing various combinations of our approaches, we achieve results near state of the art ranges for the three datasets: 85.89% macro F1 for Semeval 2010 task 8 dataset (Hendrickx et al., 2009) i.e. `semeval`, 71.97% macro F1 for DDI Extraction 2013 (Segura-Bedmar et al., 2013) i.e. `ddi` and 71.01% micro F1 for i2b2/VA 2010 relation classification dataset (Uzuner et al., 2011) i.e. `i2b2`. We refer to `ddi` and `i2b2` as medical datasets, as they belong to the biomedical and clinical domains, respectively.

Dataset	Rel	Eval	Agreement	Det
semeval	18	Macro	0.6-0.95	No
ddi	5	Macro	>0.8; 0.55-0.72	Yes
i2b2	8	Micro	-	Yes

Table 1: Dataset information, with columns Rel = number of relations, Eval = evaluation metric (all F1 scores), Agreement = Inter-annotator agreement, Det = whether detection task from section 3.4 was evaluated on. Rel column only includes relations used in official evaluation metric. `ddi` was built from two separately annotated sources and therefore contains two inter-annotator agreements.

2 Datasets

We summarize important information about these datasets in table 1. We introduce *detection* and *classification* tasks in section 3.4, but also indicate the tasks evaluated for each dataset in table 1.

Semeval 2010 `semeval` consists of 8000 training sentences and 2,717 test sentences for the multi-way classification of semantic relations between pairs of nominals. Not included in the official evaluation is an *Other* class which is considered noisy, with annotators choosing this class if no fit was found in the other classes. It is important to note that this is a synthetically generated dataset, and *detection* scores were not calculated due to the noisy nature of the *Other* class.

DDI Extraction `ddi` consists of 1,017 texts with 18,491 pharmacological substances and 5,021 drug-drug interactions from Pubmed articles in the pharmacological literature. *None* class indicating no interaction between the drug pairs is included in the evaluation metric calculation.

i2b2/VA 2010 relations `i2b2` consists of discharge summaries from Partners Healthcare and the MIMIC II Database (Saeed et al., 2011). They released 394 training reports, 477 test reports and 877 unannotated reports. After the challenge, only a part of the data was publicly released for research. *None* relation was present in the data and not considered in the official evaluation.

3 Methodology

Our framework breaks up processing into different stages, allowing for future modular addition of components. First, a `formatter` converts the raw dataset into a common comma separated value (CSV) input format accepted by the `pre-processor`, and this information is then fed to the `model`, which performs the training, after which `evaluation` is performed on the test set. With our framework, we test the following variations in the main components:

3.1 Pre-Processing

We test various pre-processing methods after performing simple tokenization and lower-casing of the words: entity blinding used by Liu et al. (2016b), stop-word and punctuation removal, and digit normalization commonly applied for `ddi` in (Zhao et al., 2016), and named entity recognition related replacement (we call this NER blinding). We used the spaCy framework[2] for tokenization and to identify punctuation and digits.

Entity blinding and NER blinding are similar concept blinding techniques where the first is performed based on gold standard annotations, while the second is performed by running NER on the original sentence. We replace the words in the sentence matching the entity or named entity span with the target label and use those for training and testing.

Entity labels for `semeval` were not annotated with type information, whereas `ddi` identified drugs and `i2b2` identified medical problems, tests and treatments. Therefore, entity labels for `semeval` were *ENTITY*, for `ddi` were *DRUG* and for `i2b2` were *PROBLEM*, *TREATMENT* and *TEST*. In this paper, we use *fine-grained concept type* to refer to the presence of more than one concept type, as in the the the case of `i2b2`.

NER labels for `semeval` consisted of those provided by the large english model by spaCy and provided standard types such as *PERSON* and *ORGANIZATION*, whereas those for the medical datasets was provided by the scispacy medium size model[3] and did not provide types. In this case, blinding consisted of replacing the words in the sentence by *Entity*.

We chose the spaCy model for NER to complement the extendable design goals of `REflex`. Other options such as cTAKES (Savova et al., 2010) for clinical data and MetaMAP[4] for biomedical data are highly specific to the dataset type and require running additional scripts outside of the `REflex` pipeline.

3.2 Modeling

We employ a baseline model based upon (Zeng et al., 2014), (Santos et al., 2015) and (Jin et al., 2018), which is a convolutional neural network (CNN) with position embeddings and a ranking loss (referred to as `CRCNN` in this paper). We initialize the model with pre-trained word embeddings: the senna embeddings by Collobert et al. (2011) for the general domain dataset and the `PubMed-PMC-wikipedia` embeddings released by Pyssalo et al. (2013) for the medical domain. We test several perturbations on top of `CRCNN` model, such as piecewise max-pooling, as

suggested by Zeng et al. (2015) and the more recent ELMo embeddings by Peters et al. (2018). To compare different featurizations of contextualized embeddings, we also employ the embeddings generated by the BERT model (rather than the standard fine-tuning approach). For ELMo, we use the Original (5.5B) model weights in `semeval` and PubMed contributed model weights in the medical datasets released by (Peters et al., 2018). For BERT, we use the BERT-large uncased model (without whole word masking) in `semeval` released by (Devlin et al., 2018), BioBERT by (Lee et al., 2019) in `ddi` and Clinical BERT by (Alsentzer et al., 2019) in `i2b2`.

The fine-tuning approach, which tends to be computationally expensive, has been thoroughly explored for multiple tasks, including medical relation extraction by Lee et al. (2019), but the approach of featurizing them with an existing model has not been explored in the literature as much. We tested different ways of featurizing the BERT contextualized embeddings for researchers who want to utilize a less computationally intensive technique, while still aiming for performance gains for their task.

Because ELMo provides token level embeddings, we chose to concatenate them with the word and position embeddings from `CRCNN` before the convolution phase. However, BERT provides word-piece level as well as sentence level embeddings. The first was concatenated similar to ELMo (which we call BERT-tokens), while the second was concatenated with the fixed size sentence representation outputted after convolution of word and position embeddings (BERT-CLS).

3.3 Training

We explore two ways of doing hyperparameter tuning: manual tuning and random search (Bergstra and Bengio, 2012).

Evaluating on three datasets meant that we needed to identify a default list of hyperparameters by tuning on one of the datasets before we could identify the hyperparameter list for the other two. We chose `semeval` for initial tuning due to its larger literature and because the `CRCNN` model was originally evaluated on this dataset. We started with reference hyperparameters listed in Zeng et al. (2014) and Santos et al. (2015) and identified default hyperparameters after tuning on a dev set randomly sampled from the training data

---

[2]https://github.com/explosion/spaCy
[3]https://allenai.github.io/scispacy/
[4]https://metamap.nlm.nih.gov

of the `semeval` dataset. These default hyperparameters[5] were used as starting points for manual tuning on the medical datasets as well as random search for all datasets.

We perform manual tuning on a subset of the hyperparameters, mentioned in table 2. In order to avoid overfitting in cross validation pointed out by Cawley and Talbot (2010), we perform a nested cross validation procedure, keeping a dev fold for hyperparameter tuning and a held out fold for score reporting.

On these dev folds, we perform paired t-tests for each of the perturbations to the parameters listed in table 2. Our first pass involves changing one hyperparameter per experiment and noting the ones that cause a statistically significant improvement, which helps us identify a narrower list of hyperparameters to tune on. We further refine the hyperparameter values in our second pass by testing on values similar to those that were leading to statistically significant improvements in the first pass. For example, if we noticed that lower epoch values were helpful in the first pass, we tested them in combination with the other optimal hyperparameter values (from first pass) in the second pass.

For each of the datasets, we tuned based on their official challenge evaluation metrics listed in section 2. `ddi` and `i2b2` had 5-fold nested cross validation performed on them, whereas `semeval` had 10-fold cross validation performed.

Random search was performed based on the official evaluation metrics for each dataset, on a fixed dev set randomly sampled from the training data. Final distributions are listed in table 3.

3.4 Evaluation

The official challenge problems for all datasets compared models based on multi-class classification, but for the medical datasets, we were also interested in looking at the changes in model performance if we treated the task as a binary classification problem. This was based on the rationale that in the drug literature, for example, pharmacologists would not want to sacrifice the ability to identify a potentially life threatening drug interaction pair, even if the type of the drug pair is not known. Therefore, we report results for both multi-class and binary classification scenarios. For clarity, we refer to them in the rest of the paper as *classification* and *detection* respectively.

Detection results were obtained using our evaluation scripts by treating existing relations as one class, ignoring the types outputted by the model. The other class in this task was the *None* or *Other* class, representing non-existing relations. Note that we did not re-train our model for this.

In addition to evaluating on two tasks for the medical and one task for the general dataset, we comment on the implications of different evaluation metrics in section 5.5.

4 Results

For experiments on the medical datasets i.e. `i2b2` and `ddi`, we used hyperparameters found from manual search individually performed on them. `semeval` had the default hyperparameters used for its experiments. These sets of hyperparameters were used in all experiments other than those reported in table 6, where we compare hyperparameter tuning methodologies.

Once we had a fixed set of hyperparameters for each dataset, we tested the perturbations for preprocessing as well as modeling in tables 4 and 5. Perturbations on the hyperparameter search are listed in table 6 and compare performance with different hyperparameter values found using different tuning strategies.

We generate the standard *classification* and the additional *detection* scores by the procedure described in section 3.4, and report these results under the *Class* and *Detect* columns.

We also report additional experiments in tables 7 and 8 based on the improvements found in tables 4 and 5. For all results tables, we report official test set results at the top, with accompanying cross validated results (averaged over all folds with their standard deviation) in smaller font below them.[6]

5 Discussion

Recently, CNNs have achieved strong performance for text classification and are typically more efficient than recurrent architectures (Bai et al., 2018; Kalchbrenner et al., 2014; Wang et al., 2015; Zhang et al., 2015b). The speed of our baseline CRCNN model allows us to explore multiple alternatives for every stage of our pipeline. We discuss these results pertaining to the *classification* task for all datasets and the *detection* task for

---

[5]listed in source code

[6]Results tables for metrics other than the official ones were omitted in the interest of space, but their analysis exists in section 5.5.

Hyperparameter	Values
epoch	{50,100,150,200}
lr decay	[1e-3, 1e-4, 1e-5]
sgd momentum	{T, F}
early stop	{T, F}
pos embed	{10, 50, 80, 100}
filter dimension	{50, 150}
filter size	2-3-4, 3-4-5
batch size	{70, 30}

Table 2: Hyperparameters explored for the first pass of manual search. lr decay means learning rate decay at [60, 120] epochs, pos embed refers to the position embedding size.

Hyperparameter	Distributions
epoch	uniform(70, 300)
lr	{constant, decay}
lr init	uniform(1e-5, 0.001)
filter size	2-3, 2-3-4, 2-3-4-5 3-4-5, 3-4-5-6
early stop	{T, F}
batch size	uniform(30, 70)

Table 3: Hyperparameter distributions for random search. Those written in {} are picked with equal probabilities. The learning rate (lr) was uniformly initialized, and decayed from 0.001 to the intialized value at half of the number of epochs. If early stop was true, patience was set to a fifth of the number of epochs. We ran 100-120 experiments for each dataset to search for optimal hyperparameters.

the medical datasets.

5.1 Pre-processing

Often, papers fail to mention the importance of pre-processing in performance improvements. Experiments in table 4 reveal that they can cause larger variations in performance than modeling.

We applied pre-processing changes with the CRCNN model with default hyperparameters for semeval and manual hyperparameters for the medical datasets. All comparisons are performed against the original pre-processing technique, which involved using the original dataset sentences in training and test.

Punctuation and digits hold more importance for the ddi dataset, which is a biomedical dataset, compared to the other two datasets. We looked at examples where this technique led to an incorrect prediction, but original pre-processing led to a correct one to investigate the source of performance further. The examples indicate that removal

of punctuation is driving worse performance compared to the normalization of digits. A detailed analysis for these is present in (Chauhan, 2019).

Stop word removal is a common technique in Natural Language Processing (NLP) to simplify the sentence by cutting out commonly used words such as *the* and *is* in order to simplify the sentence. We found that stop words seem to be important for relation extraction for all three datasets that we looked at, to a smaller degree for i2b2 compared to the other two datasets. Looking at examples misclassified by this technique revealed important stop words for different relations, which indicates that the removal of stop words is not beneficial in the relation extraction setting. Example types are shown in (Chauhan, 2019).

The availability of fine-grained concept types is likely to boost performance in relation extraction settings. The i2b2 dataset provided fine-grained concept types in the form of medical problem, test and treatments. Entity blinding causes almost 9% improvement in *classification* performance and 1% improvement in *detection* performance. In contrast, ddi only provided gold standard annotations for drug types in the sentence, and while this does not cause statistically significant improvements for cross validation, it does improve test set classification performance by about 1.5% and detection performance by 1%. For these medical datasets, NER blinding consisted of replacing the detected named entities by *Entity* because named entity types were not available. Due to the coarse-grained nature of the entities, it hurts *classification* performance significantly, and *detection* performance a little.

While entity blinding hurts performance for semeval, possibly due to the coarse-grained nature of the replacement, NER blinding does not hurt performance. Looking at misclassified examples for entity blinding and NER blinding techniques supports this hypothesis (Chauhan, 2019).

To recall, entity blinding involved replacement of entity words by *Entity*, while NER blinding involved replacing named entities in the sentence with labels such as *ORGANIZATION* and *PERSON*. In settings where fine-grained entity blinding may not be helping, they may be helpful as added features into the model, as shown by (Socher et al., 2012).

For the medical datasets, while *classification* performance varies highly with different pre-

Dataset	semeval	ddi		i2b2	
Preprocess		Class	Detect	Class	Detect
Original	**81.55**	65.53	81.74	59.75	83.17
	80.85 (1.31)	82.23 (0.32)	88.40 (0.48)	70.10 (0.85)	86.45 (0.58)
Entity Blinding	72.73	**67.02**	**82.37**	**68.76**	**84.37**
	71.31 (1.14)	83.56 (2.05)•	89.45 (1.05)•	76.59 (1.07)	88.41 (0.37)
Punct and Digit	81.23	63.41	80.49	58.85	81.96
	80.95 (1.21)•	80.44 (1.77)	87.52 (0.98)	69.37 (1.43)•	85.82 (0.43)
Punct, Digit and Stop	72.92	55.87	76.57	56.19	80.47
	71.61 (1.25)	78.52 (1.99)	85.65 (1.21)	68.14 (2.05)•	84.84 (0.77)
NER Blinding	81.63	57.22	79.03	50.41	81.61
	80.85 (1.07)•	78.06 (1.45)	86.79 (0.65)	66.26 (2.44)	86.72 (0.57)•

Table 4: Pre-processing techniques with CRCNN model. Row labels Original = simple tokenization and lower casing of words, Punct = punctuation removal, Digit = digit removal and Stop = stop word removal. Test set results at the top with cross validated results (average with standard deviation) below. All cross validated results are statistically significant compared to Original pre-processing ($p < 0.05$) using a paired t-test except those marked with a •

Dataset	semeval	ddi		i2b2	
Modeling		Class	Detect	Class	Detect
CRCNN	81.55	65.53	81.74	59.75	83.17
	80.85 (1.31)	82.23 (0.32)	88.40 (0.48)	70.10 (0.85)	86.45 (0.58)
Piecewise pool	81.59	63.01	80.62	60.85	83.69
	80.55 (0.99)•	81.99 (0.38)•	88.47 (0.48)•	73.79 (0.97)	89.29 (0.61)
BERT-tokens	85.67	**71.97**	**86.53**	63.11	**84.91**
	85.63 (0.83)	85.35 (0.53)	90.70 (0.46)	72.06 (1.36)	87.57 (0.75)
BERT-CLS	82.42	61.3	79.63	56.79	81.91
	80.83 (1.18)•	82.71 (0.68)•	88.35 (0.77)•	67.37 (1.08)	85.43 (0.36)
ELMo	**85.89**	66.63	83.05	**63.18**	84.54
	84.79 (1.08)	84.53 (0.96)	90.11 (0.56)	72.53 (0.80)	87.81 (0.34)

Table 5: Modeling techniques with original pre-processing. Test set results at the top with cross validated results (average with standard deviation) below. All cross validated results are statistically significant compared to CRCNN model ($p < 0.05$) using a paired t-test except those marked with a •. In terms of statistical significance, comparing contextualized embeddings with each other reveals that BERT-tokens is equivalent to ELMo for i2b2, but for semeval BERT-tokens is better than ELMo and for ddi BERT-tokens is better than ELMo only for detection.

processing techniques, *detection* is relatively unaffected. In a setting where one cares more about detection of relationships rather than multi-class classification, one would be able to get away with using non-complicated pre-processing techniques to maintain reasonable performance.

5.2 Split Bias

All three datasets evaluate models based on one score on the test set, which is common practice for NLP challenges. Reporting one score as opposed to a distribution of scores has been shown to be problematic by Reimers and Gurevych (2017) for sequence tagging. Recently, Crane (2018) discuss similar problems for question-answering. We show that even if you keep the same random ini-

tialization seed (all our experiments have a fixed random initialization seed), split bias can be another source of variation in scores.

In our experiments, significance testing of some cross validated results reveals no significance even when the test set result improves in performance. This is particularly concerning for ddi where entity blinding (called drug blinding in the literature) is used as a standard pre-processing technique without ablation studies demonstrating its effectiveness. Our results suggest the contrary: entity blinding seems to help test set performance for ddi in table 4, but shows no statistical significance. Table 8 further demonstrates that using this in conjunction with other techniques results in test score variations despite being statistically insignif-

Dataset / Hyperparam Tuning	semeval	ddi		i2b2	
		Class	Detect	Class	Detect
Default	81.55	62.55	80.29	55.15	81.98
	80.85 (1.31)	81.62 (1.35)	87.76 (1.03)	67.28 (1.83)	86.57 (0.58)
Manual Search	-	**65.53**	**81.74**	**59.75**	**83.17**
		82.23 (0.32)•	88.40 (0.48)•	70.10 (0.85)	86.45 (0.58)•
Random Search	**82.2**	62.29	79.04	55.0	80.77
	81.10 (1.26)•	75.43 (1.48)	83.54 (0.60)	60.66 (1.43)	82.73 (0.49)

Table 6: Hyperparameter tuning methods with original pre-processing and fixed CRCNN model. Test set results at the top with cross validated results (average with standard deviation) below. All cross validated results are statistically significant compared to Default with $p < 0.05$ except those marked with a •. Note that hyperparameter tuning can involve much higher performance variation depending on the distribution of the data. Therefore, even though there is no statistical significance in the manual search case for the held out fold in the ddi dataset, there was statistical significance for the dev fold which drove those set of hyperparameters. For both ddi and i2b2 datasets, manual search is better than random search with $p < 0.05$.

Technique / Task	Classification	Detection
E + ent	70.46	86.17
	77.70(1.26)	89.36 (0.50)
B + ent	70.56	85.66
	76.72 (1.04)	88.63 (0.33)
E + piece + ent	70.62	86.14
	79.41 (0.53)	90.37 (0.44)
B + piece + ent	**71.01**	**86.26**
	79.51 (1.09)	90.34 (0.53)
piece + ent	69.73	85.44
	78.12 (1.10)	89.74 (0.44)
E + piece	63.19	84.92
	74.76 (0.68)	89.90 (0.37)
B + piece	63.23	85.45
	74.67 (0.89)	89.61 (0.68)

Table 7: Additional experiments for i2b2. E = ELMo, B = BERT-tokens, ent = entity blinding, piece = piece-wise pooling. All results are statistically significant compared to BERT-tokens and ELMo models respectively from table 5 and piece + ent row is statistically significant compared to piecewise pool model as well as entity blinding model. These are all statistically significantly better than the CRCNN model from table 5

Technique / Task	Classification	Detection
E + ent	68.69	83.72
	86.25 (1.54)	91.35 (0.90)
B + ent	**70.66**	**85.35**
	85.79 (1.54)	91.26 (0.63)

Table 8: Additional experiments for ddi. E = ELMo, B = BERT-tokens, ent = entity blinding. Results are not statistically significant compared to BERT-tokens and ELMo models respectively from table 5 and not from each other either.

icant.

No statistical significance is seen even when the test set result worsens in performance for BERT-CLS in table 5 where it hurts test set performance on ddi but is not statistically significant when cross validation is performed.

5.3 Modeling

In table 5, we tested the generalizability of the commonly used piecewise pooling technique proposed in (Zeng et al., 2015), a variant of which was applied in the model by Luo et al. for i2b2. We also tested the improvements offered by different featurizations of contextualized embeddings, which has not been explored much for relation extraction.

Modeling changes were applied with the original pre-processing technique for the CRCNN model with default hyperparameters for semeval and manual hyperparameters for the medical datasets. All comparisons are performed with the baseline performance of the CRCNN model.

While piecewise pooling helps i2b2 by 1%, it hurts test set performance on ddi and doesn't affect performance on semeval. While it may be intuitive to split pooling by entity location, this technique is not generalizable to other datasets.

We also found that while contextualized embeddings generally boost performance, they should be concatenated with the word embeddings before the convolution stage to cause a significant boost in performance. We found ELMo and BERT-tokens to boost performance significantly for all datasets, but that BERT-CLS hurt performance for the medical datasets. While BERT-CLS boosted test set performance for `semeval`, this was not found to be a statistically significant difference for cross validation. Note that we featurized ELMo similarly to BERT-tokens and the details are present in section 3.2.

This indicates that the technique of featurizing the contextualized embeddings is important for a CNN architecture. Concatenating the contextualized embeddings with the word embeddings keeps a tighter coupling, which is helpful for relation extraction where the word level associations are essential in predicting the relation type.

5.4 Hyperparameter Tuning

Bergstra and Bengio (2012) show the superiority of random search over grid search in terms of faster convergence, but leave to future work automating the procedure of manual tuning, i.e. sequential optimization. Bayesian optimization strategies could help with this (Snoek et al., 2012) but often require expert knowledge for correct application. We tested how manual tuning, requiring less expert knowledge than Bayesian optimization, would compare to the random search strategy in table 6. For both `i2b2` and `ddi` corpora, manual search outperformed random search.

5.5 Evaluation Metrics

Picking the right evaluation metric for a dataset is critical, and it is important to choose a metric that has the biggest delta between different model performances for example types we care about. Tables for different metric results for all datasets are provided in Appendix B.

When using micro and macro statistics (precision, recall and F1), class imbalance dictates the one to pick. Macro statistics are highly affected by imbalance, whereas micro statistics are able to recover well. Despite suffering due to class imbalance, though, macro statistics may be more appropriate than micro as they provide stronger discriminative capabilities by providing equal importance to classes of smaller sizes. However, micro statistics are as discriminative as macro statistics in settings when the classes are relatively balanced. We are going to talk about the *classification* tasks in the next two paragraphs.

Compared to `semeval`, `ddi` and `i2b2` suffer from stark class imbalances. `semeval` has a number of examples in non-*Other* classes ranging from 200 or 300 to 1000. *Other* class has about 3000 examples which are not included in the official metric calculations. `ddi` has one class with 228 examples, while the others have about 1000 examples. The *None* class has 21,948 examples which is included for the official score calculations. `i2b2` has five classes in the 100-500 range, while the others contain about 2000 examples. *None* is the largest class with 19,934 examples.

Using micro statistics is reasonable for `i2b2` because the highly imbalanced class is not included in the calculations. Therefore, this metric is able to be as discriminative as macro statistics. For example, test set micro F1 between baseline and entity blinding techniques is 59.75 and 68.76, while that for macro F1 is 36.44 and 43.76. In contrast, using micro statistics is a bad idea for `ddi` because the performance on the *None* class would drive most of the predictive results of the model. For example, micro-F1 between baseline and NER blinding is 88.69 and 86.18, whereas macro-F1 is 65.53 and 57.22. `semeval` does not have a stark contrast between micro and macro scores due to *Other* class not being included in the calculation. Using either metric to evaluate models is reasonable for this dataset.

The detection task does not suffer from such variations due to the lower class imbalance. For example, `ddi` dataset micro-F1 between baseline and NER blinding model is 90.01 and 88.74, while macro-F1 is 81.74 and 79.03. This further suggests that modeling differences and pre-processing differences cause more variation in performance in settings when the class imbalance is higher.

6 Comparison with SOTA

The best *classification* test set results found are listed in table 9. Note that we do not compare the *extraction* task for datasets other than `ddi` because the official challenges only compared *classification* results. Even though the official challenge did not rank models based on the *detection* task, recent papers in the `ddi` literature mention these results.

Wang et al. (2016) report a result of 88% on `semeval` and do not provide any public source code for replication purposes. Despite being below the state of the art range, `REflex` provides the best performing publicly available model for

Dataset	Result	Technique
semeval	85.89	E
ddi	71.97, 86.53	B
i2b2	71.01	B + piece + ent

Table 9: Best test set *classification* results for all datasets, except `ddi` where *detection* results are mentioned after the classification results. piece = Piecewise pooling, ent = entity blinding, E = ELMo, B = BERT-tokens. Result corresponds to F1 scores, macro for `semeval` and `ddi`, but micro for `i2b2`.

this dataset. Zheng et al. (2017) report the best result on `ddi` (77.3%) but perform negative instance filtering, which is a highly specific pre-processing technique that does not fit with the flexible nature of `REflex`. This technique cuts specific examples from the dataset, but the paper is unclear about whether train as well as test data are shortened. If the test data is being shortened, the performance comparison becomes unfair due to evaluation on different test samples. Unfortunately, source code was not publicly available to answer these questions.

Note that Zhao et al. (2016) show that negative instance filtering causes a 4.1% improvement in test set performance. If `REflex` were to use this pre-processing technique, it would reach close to the state-of-the-art (SOTA) number on the *classification* task. On the other hand, results from the *detection* results *outperform* this model by 2.53%.

Sahu et al. (2016) (code unavailable) report a state of the art result of 71.16% on `i2b2`, which the results in table 9 are able to match. Note that (Rink et al., 2011) report a result of 73.7% with a support vector machine, but they used a larger version of the dataset. Comparison against different subsets of the dataset would not be fair.

Comparison against these numbers demonstrates that `REflex` is the only open-source framework, providing performance near SOTA ranges for the three datasets. Therefore, `REflex` can be used as a strong baseline model in future relation extraction studies.

7   Conclusion

Our findings reveal variations offered by pre-processing and training methodologies, which often go unreported. They indicate that comparing models without having these techniques standardized can make it difficult to assess the true source of performance gains. Our key findings are:

1.   Pre-processing can have a strong effect on performance, sometimes more than modeling techniques, as is the case of `i2b2`. Concept types seem to offer useful information, perhaps revealing more general semantic information in the sentence that can help with predictions. Fine-grained Gold standard annotated concept types are most beneficial, but those from automatically extracted packages may also be useful as long as they consist of multiple types. Punctuation and digits may hold more importance in biomedical settings, but stop words hold significance in all settings.

2. Reporting on one test set score can be problematic due to split bias, and a cross validation approach with significance tests may help ease some of this bias. Drug blinding for `ddi` is commonly used in the literature but does not seem to offer any statistically significant improvements. Therefore, it is unnecessary to use in this domain.

3.   Contextualized embeddings are generally helpful but the featurizing technique is important: for CNN models, concatenating them with the word embeddings before convolution is most beneficial.

4.   Picking the right hyperparameters for a dataset is important to performance. We suggest an initial manual hyperparameter search based on cross validation significance tests because that may be sufficient in most cases. If one is not pressed for time, random search is a reasonable automated option for hyperparameter tuning, but requires more experience for picking the right search space and the right distributions for the hyperparameters.

5. Picking the right evaluation metrics for a new dataset should be driven by class imbalance issues for the classes chosen to be evaluated on.

Acknowledgments

This work was funded in part by a collaborative agreement between MIT and Wistron Corp, the National Institutes of Health (National Institutes of Mental Health grant P50-MH106933), and a Mitacs Globalink Research Award. Finally, the authors would like to thank Di Jin and Elena Sergeeva from the MIT-CSAIL Clinical Decision Making Group for providing helpful feedback.

References

Heike Adel, Benjamin Roth, and Hinrich Schütze. 2016. Comparing convolutional neural networks to traditional models for slot filling. In *Proceedings of the 2016 Conference of the North American Chapter of the Association for Computational Linguistics: Human Language Technologies*, pages 828–838. Association for Computational Linguistics.

Emily Alsentzer, John R Murphy, Willie Boag, Wei-Hung Weng, Di Jin, Tristan Naumann, and Matthew McDermott. 2019. Publicly available clinical bert embeddings. *arXiv preprint arXiv:1904.03323*.

Shaojie Bai, J Zico Kolter, and Vladlen Koltun. 2018. An empirical evaluation of generic convolutional and recurrent networks for sequence modeling. *arXiv preprint arXiv:1803.01271*.

James Bergstra and Yoshua Bengio. 2012. Random search for hyper-parameter optimization. *Journal of Machine Learning Research*, 13(Feb):281–305.

Rui Cai, Xiaodong Zhang, and Houfeng Wang. 2016. Bidirectional recurrent convolutional neural network for relation classification. In *Proceedings of the 54th Annual Meeting of the Association for Computational Linguistics (Volume 1: Long Papers)*, volume 1, pages 756–765.

Gavin C Cawley and Nicola LC Talbot. 2010. On overfitting in model selection and subsequent selection bias in performance evaluation. *Journal of Machine Learning Research*, 11(Jul):2079–2107.

Geeticka Chauhan. 2019. *REflex: Flexible Framework for Relation Extraction in Multiple Domains*. Master's thesis, Massachusetts Institute of Technology.

Veera Raghavendra Chikka and Kamalakar Karlapalem. 2018. A hybrid deep learning approach for medical relation extraction. *CoRR*.

Marc Claesen and Bart De Moor. 2015. Hyperparameter search in machine learning. *arXiv preprint arXiv:1502.02127*.

Ronan Collobert, Jason Weston, Léon Bottou, Michael Karlen, Koray Kavukcuoglu, and Pavel Kuksa. 2011. Natural language processing (almost) from scratch. *Journal of Machine Learning Research*, 12(Aug):2493–2537.

Matt Crane. 2018. Questionable answers in question answering research: Reproducibility and variability of published results. *Transactions of the Association of Computational Linguistics*, 6:241–252.

Jacob Devlin, Ming-Wei Chang, Kenton Lee, and Kristina Toutanova. 2018. Bert: Pre-training of deep bidirectional transformers for language understanding. *arXiv preprint arXiv:1810.04805*.

Javid Ebrahimi and Dejing Dou. 2015. Chain based rnn for relation classification. In *Proceedings of the 2015 Conference of the North American Chapter of the Association for Computational Linguistics: Human Language Technologies*, pages 1244–1249.

Antske Fokkens, Marieke Van Erp, Marten Postma, Ted Pedersen, Piek Vossen, and Nuno Freire. 2013. Offspring from reproduction problems: What replication failure teaches us. In *Proceedings of the 51st Annual Meeting of the Association for Computational Linguistics (Volume 1: Long Papers)*, volume 1, pages 1691–1701.

Roxana Girju, Preslav Nakov, Vivi Nastase, Stan Szpakowicz, Peter Turney, and Deniz Yuret. 2007. Semeval-2007 task 04: Classification of semantic relations between nominals. In *Proceedings of the 4th International Workshop on Semantic Evaluations*, pages 13–18. Association for Computational Linguistics.

Kazuma Hashimoto, Makoto Miwa, Yoshimasa Tsuruoka, and Takashi Chikayama. 2013. Simple customization of recursive neural networks for semantic relation classification. In *Proceedings of the 2013 Conference on Empirical Methods in Natural Language Processing*, pages 1372–1376.

Bin He, Yi Guan, and Rui Dai. 2018a. Classifying medical relations in clinical text via convolutional neural networks. *Artificial intelligence in medicine*.

Bin He, Yi Guan, and Rui Dai. 2018b. Convolutional gated recurrent units for medical relation classification. *CoRR*, abs/1807.11082.

Iris Hendrickx, Su Nam Kim, Zornitsa Kozareva, Preslav Nakov, Diarmuid Ó Séaghdha, Sebastian Padó, Marco Pennacchiotti, Lorenza Romano, and Stan Szpakowicz. 2009. Semeval-2010 task 8: Multi-way classification of semantic relations between pairs of nominals. In *Proceedings of the Workshop on Semantic Evaluations: Recent Achievements and Future Directions*, pages 94–99. Association for Computational Linguistics.

Degen Huang, Zhenchao Jiang, Li Zou, and Lishuang Li. 2017. Drug-drug interaction extraction from biomedical literature using support vector machine and long short term memory networks. *Information Sciences*, 415.

Darrel C Ince, Leslie Hatton, and John Graham-Cumming. 2012. The case for open computer programs. *Nature*, 482(7386):485.

Di Jin, Franck Dernoncourt, Elena Sergeeva, Matthew McDermott, and Geeticka Chauhan. 2018. MIT-MEDG at SemEval-2018 Task 7: Semantic Relation Classification via Convolution Neural Network. In *Proceedings of The 12th International Workshop on Semantic Evaluation*, pages 798–804. Association for Computational Linguistics.

Wen Juan Hou and Bamfa Ceesay. 2018. Extraction of drug-drug interaction using neural embedding. *Journal of Bioinformatics and Computational Biology*, 16.

Nal Kalchbrenner, Edward Grefenstette, and Phil Blunsom. 2014. A convolutional neural network for modelling sentences. In *Proceedings of the 52nd Annual Meeting of the Association for Computational Linguistics (Volume 1: Long Papers)*, pages 655–665. Association for Computational Linguistics.

Ramakanth Kavuluru, Anthony Rios, and Tung Tran. 2017. Extracting drug-drug interactions with word and character-level recurrent neural networks. *2017 IEEE International Conference on Healthcare Informatics (ICHI)*, pages 5–12.

Jinhyuk Lee, Wonjin Yoon, Sungdong Kim, Donghyeon Kim, Sunkyu Kim, Chan Ho So, and Jaewoo Kang. 2019. Biobert: pre-trained biomedical language representation model for biomedical text mining. *arXiv preprint arXiv:1901.08746*.

Omer Levy, Minjoon Seo, Eunsol Choi, and Luke Zettlemoyer. 2017. Zero-shot relation extraction via reading comprehension. In *Proceedings of the 21st Conference on Computational Natural Language Learning (CoNLL 2017)*, pages 333–342. Association for Computational Linguistics.

Fei Li, Meishan Zhang, Guohong Fu, and Donghong Ji. 2017. A neural joint model for entity and relation extraction from biomedical text. *BMC bioinformatics*, 18(1):198.

Jiwei Li and Dan Jurafsky. 2015. Do multi-sense embeddings improve natural language understanding? *arXiv preprint arXiv:1506.01070*.

Q. Li, Z. Yang, L. Luo, L. Wang, Y. Zhang, H. Lin, J. Wang, L. Yang, K. Xu, and Y. Zhang. 2018a. A multi-task learning based approach to biomedical entity relation extraction. In *2018 IEEE International Conference on Bioinformatics and Biomedicine (BIBM)*, pages 680–682.

Yifu Li, Ran Jin, and Yuan Luo. 2018b. Classifying relations in clinical narratives using segment graph convolutional and recurrent neural networks (seg-gcrns). *Journal of the American Medical Informatics Association*, 26(3):262–268.

Sangrak Lim and Jaewoo Kang. 2018a. Chemicalgene relation extraction using recursive neural network. In *Database*.

Sangrak Lim and Jaewoo Kang. 2018b. Drug drug interaction extraction from the literature using a recursive neural network. In *PloS one*.

Sangrak Lim, Kyubum Lee, and Jaewoo Kang. 2018. Drug drug interaction extraction from the literature using a recursive neural network. *Plos one*, 13:1–17.

Shengyu Liu, Kai Chen, Qingcai Chen, and Buzhou Tang. 2016a. Dependency-based convolutional neural network for drug-drug interaction extraction. *2016 IEEE International Conference on Bioinformatics and Biomedicine (BIBM)*, pages 1074–1080.

Shengyu Liu, Buzhou Tang, Qingcai Chen, and Xiaolong Wang. 2016b. Drug-drug interaction extraction via convolutional neural networks. *Computational and mathematical methods in medicine*, 2016.

Yuan Luo. 2017. Recurrent neural networks for classifying relations in clinical notes. *Journal of Biomedical Informatics*, 72.

Yuan Luo, Yu Cheng, Özlem Uzuner, Peter Szolovits, and Justin Starren. 2017. Segment convolutional neural networks (seg-cnns) for classifying relations in clinical notes. *Journal of the American Medical Informatics Association*, 25(1):93–98.

Xinbo Lv, Yi Guan, Jinfeng Yang, and Jiawei Wu. 2016. Clinical relation extraction with deep learning. In *International Journal of Hybrid Information Technology*.

Makoto Miwa and Mohit Bansal. 2016. End-to-end relation extraction using lstms on sequences and tree structures. In *Proceedings of the 54th Annual Meeting of the Association for Computational Linguistics (Volume 1: Long Papers)*, pages 1105–1116. Association for Computational Linguistics.

Dat Quoc Nguyen and Karin Verspoor. 2018. Convolutional neural networks for chemical-disease relation extraction are improved with character-based word embeddings. *arXiv preprint arXiv:1805.10586*.

Thien Huu Nguyen and Ralph Grishman. 2015. Combining neural networks and log-linear models to improve relation extraction. *CoRR*, abs/1511.05926.

Matthew Peters, Mark Neumann, Mohit Iyyer, Matt Gardner, Christopher Clark, Kenton Lee, and Luke Zettlemoyer. 2018. Deep contextualized word representations. In *Proceedings of the 2018 Conference of the North American Chapter of the Association for Computational Linguistics: Human Language Technologies, Volume 1 (Long Papers)*, pages 2227–2237. Association for Computational Linguistics.

Sampo Pyssalo, Filip Ginter, Hans Moen, Tapio Salakoski, and Sophia Ananiadou. 2013. Distributional semantics resources for biomedical text processing. In *Proceedings of the 5th International Symposium on Languages in Biology and Medicine, Tokyo, Japan*, pages 39–43.

Pengda Qin, Weiran Xu, and Jun Guo. 2016. An empirical convolutional neural network approach for semantic relation classification. *Neurocomput.*, 190:1–9.

Chanqin Quan, Lei Hua, Xiao Sun, and Wenjun Bai. 2016. Multichannel convolutional neural network for biological relation extraction. In *BioMed research international*.

Nils Reimers and Iryna Gurevych. 2017. Reporting score distributions makes a difference: Performance study of lstm-networks for sequence tagging. In *Proceedings of the 2017 Conference on Empirical Methods in Natural Language Processing*, pages 338–348, Copenhagen, Denmark. Association for Computational Linguistics.

Bryan Rink, Sanda Harabagiu, and Kirk Roberts. 2011. Automatic extraction of relations between medical concepts in clinical texts. *Journal of the American Medical Informatics Association*, 18(5):594–600.

Jonathan Rotsztejn, Nora Hollenstein, and Ce Zhang. 2018. Eth-ds3lab at semeval-2018 task 7: Effectively combining recurrent and convolutional neural networks for relation classification and extraction. In *Proceedings of The 12th International Workshop on Semantic Evaluation*, pages 689–696. Association for Computational Linguistics.

Mohammed Saeed, Mauricio Villarroel, Andrew T Reisner, Gari Clifford, Li-Wei Lehman, George Moody, Thomas Heldt, Tin H Kyaw, Benjamin Moody, and Roger G Mark. 2011. Multiparameter intelligent monitoring in intensive care ii (mimic-ii): a public-access intensive care unit database. *Critical care medicine*, 39(5):952.

Sunil Sahu, Ashish Anand, Krishnadev Oruganty, and Mahanandeeshwar Gattu. 2016. Relation extraction from clinical texts using domain invariant convolutional neural network. In *Proceedings of the 15th Workshop on Biomedical Natural Language Processing*, pages 206–215. Association for Computational Linguistics.

Sunil Kumar Sahu and Ashish Anand. 2018. Drug-drug interaction extraction from biomedical texts using long short-term memory network. *Journal of Biomedical Informatics*, 86:15 – 24.

Cicero Nogueira dos Santos, Bing Xiang, and Bowen Zhou. 2015. Classifying relations by ranking with convolutional neural networks. *arXiv preprint arXiv:1504.06580*.

Guergana K Savova, James J Masanz, Philip V Ogren, Jiaping Zheng, Sunghwan Sohn, Karin C Kipper-Schuler, and Christopher G Chute. 2010. Mayo clinical text analysis and knowledge extraction system (ctakes): architecture, component evaluation and applications. *Journal of the American Medical Informatics Association*, 17(5):507–513.

Isabel Segura-Bedmar, Paloma Martínez, and María Herrero Zazo. 2013. Semeval-2013 task 9: Extraction of drug-drug interactions from biomedical texts (ddiextraction 2013). In *Second Joint Conference on Lexical and Computational Semantics (* SEM), Volume 2: Proceedings of the Seventh International Workshop on Semantic Evaluation (SemEval 2013)*, volume 2, pages 341–350.

Jasper Snoek, Hugo Larochelle, and Ryan P Adams. 2012. Practical bayesian optimization of machine learning algorithms. In *Advances in neural information processing systems*, pages 2951–2959.

Richard Socher, Brody Huval, Christopher D Manning, and Andrew Y Ng. 2012. Semantic compositionality through recursive matrix-vector spaces. In *Proceedings of the 2012 joint conference on empirical methods in natural language processing and computational natural language learning*, pages 1201–1211. Association for Computational Linguistics.

Simon Suster, Madhumita Sushil, and Walter Daelemans. 2018. Revisiting neural relation classification in clinical notes with external information. In *Proceedings of the Ninth International Workshop on Health Text Mining and Information Analysis*, pages 22–28. Association for Computational Linguistics.

Özlem Uzuner, Brett R South, Shuying Shen, and Scott L DuVall. 2011. 2010 i2b2/va challenge on concepts, assertions, and relations in clinical text. *Journal of the American Medical Informatics Association*, 18(5):552–556.

Ngoc Thang Vu, Heike Adel, Pankaj Gupta, and Hinrich Schütze. 2016. Combining recurrent and convolutional neural networks for relation classification. In *Proceedings of the 2016 Conference of the North American Chapter of the Association for Computational Linguistics: Human Language Technologies*, pages 534–539. Association for Computational Linguistics.

Linlin Wang, Zhu Cao, Gerard De Melo, and Zhiyuan Liu. 2016. Relation classification via multi-level attention CNNs. In *Proceedings of the 54th annual meeting of the Association for Computational Linguistics (volume 1: long papers)*, volume 1, pages 1298–1307.

Peng Wang, Jiaming Xu, Bo Xu, Chenglin Liu, Heng Zhang, Fangyuan Wang, and Hongwei Hao. 2015. Semantic clustering and convolutional neural network for short text categorization. In *Proceedings of the 53rd Annual Meeting of the Association for Computational Linguistics and the 7th International Joint Conference on Natural Language Processing (Volume 2: Short Papers)*, volume 2, pages 352–357.

Wei Wang, Xi Yang, Canqun Yang, Xiao-Wei Guo, Xiang Zhang, and Chengkun Wu. 2017. Dependency-based long short term memory network for drug-drug interaction extraction. *BMC Bioinformatics*, 18.

Kun Xu, Yansong Feng, Songfang Huang, and Dongyan Zhao. 2015a. Semantic relation classification via convolutional neural networks with simple negative sampling. *arXiv preprint arXiv:1506.07650*.

Yan Xu, Ran Jia, Lili Mou, Ge Li, Yunchuan Chen, Yangyang Lu, and Zhi Jin. 2016. Improved relation classification by deep recurrent neural networks with data augmentation. In *Proceedings of COLING 2016, the 26th International Conference on Computational Linguistics: Technical Papers*, pages 1461–1470. The COLING 2016 Organizing Committee.

Yan Xu, Lili Mou, Ge Li, Yunchuan Chen, Hao Peng, and Zhi Jin. 2015b. Classifying relations via long short term memory networks along shortest dependency paths. In *Proceedings of the 2015 conference on empirical methods in natural language processing*, pages 1785–1794.

Wenpeng Yin, Katharina Kann, Mo Yu, and Hinrich Schütze. 2017. Comparative study of cnn and rnn for natural language processing. *CoRR*, abs/1702.01923.

Mo Yu, Matthew Gormley, and Mark Dredze. 2014. Factor-based compositional embedding models. In *NIPS Workshop on Learning Semantics*, pages 95–101.

Daojian Zeng, Kang Liu, Yubo Chen, and Jun Zhao. 2015. Distant supervision for relation extraction via piecewise convolutional neural networks. In *Proceedings of the 2015 Conference on Empirical Methods in Natural Language Processing*, pages 1753–1762.

Daojian Zeng, Kang Liu, Siwei Lai, Guangyou Zhou, and Jun Zhao. 2014. Relation classification via convolutional deep neural network. In *Proceedings of COLING 2014, the 25th International Conference on Computational Linguistics: Technical Papers*, pages 2335–2344.

Dongxu Zhang and Dong Wang. 2015. Relation classification via recurrent neural network. *arXiv preprint arXiv:1508.01006*.

Shu Zhang, Dequan Zheng, Xinchen Hu, and Ming Yang. 2015a. Bidirectional long short-term memory networks for relation classification. In *Proceedings of the 29th Pacific Asia Conference on Language, Information and Computation*, pages 73–78.

Xiang Zhang, Junbo Zhao, and Yann LeCun. 2015b. Character-level convolutional networks for text classification. In *Advances in neural information processing systems*, pages 649–657.

Yuhao Zhang, Peng Qi, and Christopher D. Manning. 2018. Graph convolution over pruned dependency trees improves relation extraction. In *Proceedings of the 2018 Conference on Empirical Methods in Natural Language Processing*, pages 2205–2215. Association for Computational Linguistics.

Zhehuan Zhao, Zhihao Yang, Ling Luo, Hongfei Lin, and Jian Wang. 2016. Drug drug interaction extraction from biomedical literature using syntax convolutional neural network. *Bioinformatics*, 32(22):3444–3453.

Wei Zheng, Hongfei Lin, Ling Luo, Zhehuan Zhao, Zhengguang Li, Yijia Zhang, Zhihao Yang, and Jian Wang. 2017. An attention-based effective neural model for drug-drug interactions extraction. In *BMC Bioinformatics*.

A Quantitative Literature Review

paper	cite	code	ablation	hyperparam	cross val	word-embed	datasets
Socher et al. (2012)	890	y	•	y	•	y	2
Zeng et al. (2014)	477	•	y	y	y	y	1
Santos et al. (2015)	220	•	y	y	y	y	1
Nguyen and Verspoor (2018)	146	•	y	y	y	•	2
Miwa and Bansal (2016)	175	•	y	y	y	•	3
Li and Jurafsky (2015)	107	y	y	y	•	y	6
Xu et al. (2015a)	108	•	y	y	•	y	1
Wang et al. (2016)	102	•	y	•	•	y	1
Hashimoto et al. (2013)	64	•	y	y	•	y	1
Zhang and Wang (2015)	68	•	y	•	y	y	2
Vu et al. (2016)	57	•	y	y	•	y	1
Yin et al. (2017)	116	•	n	y	•	•	7
Yu et al. (2014)	45	y	y	y	y	y	1
Xu et al. (2016)	54	y	y	y	•	•	1
Zhang et al. (2015a)	51	•	•	•	•	y	1
Nguyen and Grishman (2015)	42	•	y	y	•	y	2
Qin et al. (2016)	39	•	•	y	y	y	1
Cai et al. (2016)	44	•	y	y	•	y	1
Sahu et al. (2016)	32	•	y	y	y	y	1
Adel et al. (2016)	29	y	y	•	•	y	1
Zeng et al. (2015)	190	•	y	y	•	y	1
Xu et al. (2015b)	171	•	y	y	•	y	1
Zhang et al. (2018)	3	•	y	y	•	y	2
Levy et al. (2017)	20	y	y	y	•	y	1
Liu et al. (2016b)	48	•	•	y	•	y	1
Zhao et al. (2016)	41	y	y	y	•	y	1
Ebrahimi and Dou (2015)	30	•	•	•	•	•	2
Li et al. (2017)	27	y	y	y	y	y	2
Quan et al. (2016)	23	y	•	y	y	y	2

Paper	cite	code	ablation	hyperparam	cross val	word-embed	datasets
Sahu and Anand (2018)	13	y	y	y	•	y	1
Liu et al. (2016a)	9	•	•	y	•	y	1
Lim and Kang (2018b)	4	•	•	•	•	•	1
Zheng et al. (2017)	12	•	y	y	y	y	1
Wang et al. (2017)	5	n	y	y	•	y	1
Lim et al. (2018)	1	y	y	y	y	y	2
Kavuluru et al. (2017)	8	•	•	y	•	•	1
Huang et al. (2017)	4	•	•	y	•	y	1
Juan Hou and Ceesay (2018)	1	•	•	•	•	y	1
Lim and Kang (2018a)	4	y	•	y	•	y	1
Rotsztejn et al. (2018)	2	•	•	y	y	y	1
Jin et al. (2018)	0	•	y	y	y	y	1
Sahu et al. (2016)	31	•	y	y	y	y	1
Luo (2017)	21	•	•	y	•	y	1
Lv et al. (2016)	15	•	•	•	•	•	1
Jin et al. (2018)	14	•	y	y	•	y	1
Chikka and Karlapalem (2018)	1	y	•	y	•	•	1
Li et al. (2018b)	0	y	•	y	y	y	1
Li et al. (2018a)	0	•	•	•	•	•	5
Suster et al. (2018)	0	y	•	y	•	y	1
Luo et al. (2017)	16	y	•	y	•	y	1
He et al. (2018a)	2	•	•	y	•	y	1
He et al. (2018b)	0	•	•	y	y	y	2
Nguyen and Verspoor (2018)	1	•	y	y	•	y	1

Table 10: Following are the columns in this table: **cite** = number of papers that cited the paper; **code** = whether code was publicly available (y for yes and • for no); **ablation** = whether an ablation study was performed; **hyperparam** = whether hyperparameter details were mentioned; **cross val** = whether cross validation details were mentioned; **word-embed** = whether information about word embeddings used was mentioned; **datasets** = number of datasets evaluated on

B   Evaluation Metric Results on Test Data

Each row represents a pre-processing, modeling technique or combination based on the additional experiments run on each dataset. Only test set results (as opposed to cross validation) are reported for ease of analysis. In all the tables, Baseline refers to the CRCNN model with original pre-processing and default hyperparameters for `semeval` and manual hyperparameters for the medical datasets (`ddi` and `i2b2`). The following short forms are used as row labels:

B = BERT-tokens
E = ELMo
Ent Blind = Entity Blinding
Piece Pool = Piecewise Pooling

Technique \ Metric	acc	micro-P	micro-R	micro-F1	macro-P	macro-R	macro-F1
Baseline	77.11	79.95	85.11	82.45	79.25	84.06	81.55
Entity Blinding	67.94	70.72	77.15	73.8	69.77	76.31	72.73
Punct and Digit	76.48	79.19	85.42	82.19	78.33	84.51	81.23
Punct, Digit and Stop	68.28	73.0	74.78	73.88	72.84	73.48	72.92
NER Blinding	77.25	79.3	86.03	82.53	78.49	85.13	81.63
Piecewise pool	77.0	79.54	85.55	82.44	78.86	84.71	81.59
ELMo	77.77	81.87	84.62	83.22	81.24	83.71	82.42
BERT-CLS	77.77	81.87	84.62	83.22	81.24	83.71	82.42
BERT-tokens	81.3	86.63	86.74	86.69	86.08	85.61	85.67

Table 11: Different Evaluation Metric results on test set of `semeval` dataset. Only test set results are reported for ease of analysis. Metric short forms used are **acc** = accuracy; **P** = precision; **R** = recall.

Metric / Technique	acc		micro-P		micro-R		micro-F1		macro-P		macro-R		macro-F1	
	Class	Detect	Class	Detect	Class	Detect	Class	Detect	Class	Detect	Class	Detect	Class	Detect
Baseline	88.69	90.01	88.69	90.01	88.69	90.01	88.69	90.01	72.32	82.06	63.48	81.43	65.53	81.74
Entity Blinding	89.22	90.44	89.22	90.44	89.22	90.44	89.22	90.44	71.26	82.99	64.63	81.79	67.02	82.37
Punct and Digit	88.31	89.61	88.31	89.61	88.31	89.61	88.31	89.61	69.49	81.7	60.81	79.43	63.41	80.49
Punct, Digit and Stop	86.58	87.86	86.58	87.86	86.58	87.86	86.58	87.86	67.4	78.59	52.72	74.98	55.87	76.57
NER Blinding	86.18	88.74	86.18	88.74	86.18	88.74	86.18	88.74	59.13	79.9	55.93	78.24	57.22	79.03
Piecewise pool	88.14	89.54	88.14	89.54	88.14	89.54	88.14	89.54	70.49	81.39	60.38	79.91	63.01	80.62
E	89.76	90.97	89.76	90.97	89.76	90.97	89.76	90.97	73.41	84.36	63.65	81.9	66.63	83.05
BERT-CLS	87.84	89.05	87.84	89.05	87.84	89.05	87.84	89.05	68.2	80.51	59.31	78.84	61.3	79.63
B	91.31	92.72	91.31	92.72	91.31	92.72	91.31	92.72	77.66	87.34	69.27	85.78	71.97	86.53
E + Entity Blinding	89.97	91.18	89.97	91.18	89.97	91.18	89.97	91.18	72.44	84.42	66.41	83.06	68.69	83.72
B + Entity Blinding	90.93	92.15	90.93	92.15	90.93	92.15	90.93	92.15	76.79	86.57	63.39	84.26	70.66	85.35

Table 12: Different Evaluation Metric results on test set of ddi dataset. Only test set results are reported for ease of analysis. Metric short forms used are **acc** = accuracy; **P** = precision; **R** = recall.

Metric / Technique	acc		micro-P		micro-R		micro-F1		macro-P		macro-R		macro-F1	
	Class	Detect	Class	Detect	Class	Detect	Class	Detect	Class	Detect	Class	Detect	Class	Detect
Baseline	78.68	83.17	61.39	83.17	58.19	83.17	59.75	83.17	49.24	81.16	34.2	80.29	36.44	80.69
Entity Blinding	81.92	84.37	68.88	84.37	68.65	84.37	68.76	84.37	53.33	82.32	40.72	82.27	43.76	82.29
Punct and Digit	77.25	81.96	58.09	81.96	59.64	81.96	58.85	81.96	49.28	79.53	33.56	79.92	34.93	79.71
Punct, Digit and Stop	76.05	80.47	57.15	80.47	55.27	80.47	56.19	80.47	43.26	77.96	31.16	77.47	32.99	77.7
NER Blinding	75.12	81.61	52.58	81.61	48.42	81.61	50.41	81.61	39.44	79.45	26.3	78.17	29.15	78.73
Piecewise pool	78.63	83.69	59.41	83.69	62.37	83.69	60.85	83.69	46.16	81.41	35.77	82.17	36.44	81.76
E	80.4	84.54	64.56	84.54	61.86	84.54	63.18	84.54	59.28	82.69	36.17	81.97	38.1	82.31
BERT-CLS	76.94	81.91	57.66	81.91	55.95	81.91	56.79	81.91	49.88	76.61	32.4	79.15	34.05	79.37
B	80.79	84.91	64.92	84.91	61.4	84.91	63.11	84.91	58.05	83.08	36.8	82.1	39.31	82.55
E + Entity Blinding	83.62	86.17	72.43	86.17	68.6	86.17	70.46	86.17	60.79	84.65	40.11	83.67	42.99	84.13
E + Piece Pool + Ent Blind	83.46	86.14	71.11	86.14	70.14	86.14	70.62	86.14	54.87	84.37	42.41	84.13	44.43	84.25
Ent Blind + Piece Pool	82.72	85.44	69.49	85.44	69.98	85.44	69.73	85.44	48.82	83.49	41.97	83.61	42.89	83.55
E + Piece Pool	80.1	84.92	61.98	84.92	64.45	84.92	63.19	84.92	49.68	82.79	36.91	83.43	37.52	83.09
B + Ent Blind	83.27	85.66	71.52	85.66	69.63	85.66	70.56	85.66	55.62	83.9	38.82	83.44	41.83	83.66
B + Ent Blind + Piece pool	83.57	86.26	70.9	86.26	71.13	86.26	71.01	86.26	55.6	84.43	42.58	84.49	44.4	84.46
B + Piece pool	80.59	85.45	63.08	85.45	63.39	85.45	63.23	85.45	56.01	83.51	36.84	83.59	38.84	83.55

Table 13: Different Evaluation Metric results on test set of i2b2 dataset. Only test set results are reported for ease of analysis. Metric short forms used are **acc** = accuracy; **P** = precision; **R** = recall.

Analysing Representations of Memory Impairment in a Clinical Notes Classification Model

Mark Ormerod, Jesús Martínez del Rincón, Neil Robertson

Bernadette McGuinness, Barry Devereux

Queen's University Belfast

{mormerod01, j.martinez-del-rincon, n.robertson, b.mcguinness, b.devereux}@qub.ac.uk

Abstract

Despite recent advances in the application of deep neural networks to various kinds of medical data, extracting information from unstructured textual sources remains a challenging task. The challenges of training and interpreting document classification models are amplified when dealing with small and highly technical datasets, as are common in the clinical domain. Using a dataset of de-identified clinical letters gathered at a memory clinic, we construct several recurrent neural network models for letter classification, and evaluate them on their ability to build meaningful representations of the documents and predict patients' diagnoses. Additionally, we probe sentence embedding models in order to build a human-interpretable representation of the neural network's features, using a simple and intuitive technique based on perturbative approaches to sentence importance. In addition to showing which sentences in a document are most informative about the patient's condition, this method reveals the types of sentences that lead the model to make incorrect diagnoses. Furthermore, we identify clusters of sentences in the embedding space that correlate strongly with importance scores for each clinical diagnosis class.

1 Introduction

While the majority of clinical data is made up of structured information (Jee and Kim, 2013), which can often be readily integrated into data models for research, there is a significant amount of semi-structured and unstructured data which is increasingly being targeted by machine learning practitioners for analysis. As a general rule, this unstructured data is more difficult to analyse due to an absence of a standardised data model (Ann Alexander and Wang, 2018). Unstructured clinical data includes a variety of media, such as video, audio, image and text-based data, with the majority of such data being made up of text

and images. Recently, there has been a series of breakthroughs in the application of machine learning techniques for medical imaging data in order to achieve expert-level performance on diagnosis tasks (Rajpurkar et al., 2017). However, machine learning models using semi-structured and unstructured textual data from the clinical domain have received less attention and to date have not seen the same degree of successful application. Examples of unstructured medical data featuring "free text" include discharge summaries, nursing reports and progress notes. Historically, one of the challenges of applying natural language processing (NLP) methods to clinical data has been the often limited amount of data available, which has traditionally necessitated a reliance on manual feature engineering and relatively shallow textual features (Shickel et al., 2018).

Taking a novel dataset of labelled clinical letters compiled at a memory clinic as the target data domain, we build state-of-the-art deep learning models for the task of clinical text classification, and evaluate them on their ability to predict a clinician's diagnosis of the patient. However, deep learning models generally require very large training datasets. Our approach to the problem therefore incorporates transfer learning, and we make use of embedding data from pre-trained models trained on large corpora. In order to investigate the relative usefulness of word-level and sentence-level information, we train and evaluate several models, including a ULMFiT model (Howard and Ruder, 2018) and two long short-term memory (LSTM) (Hochreiter and Schmidhuber, 1997) models: one trained on word embedding representations of the documents and one trained on sentence embedding representations (Basile et al., 2012).

An infamous problem of deep neural networks is that they are "black boxes", with the details of how they represent and process information

Proceedings of the BioNLP 2019 workshop, pages 48–57

Florence, Italy, August 1, 2019. ©2019 Association for Computational Linguistics

being uninterpretable to humans. To shed light on how a recurrent neural network models clinical documents in order to correctly predict a patient's diagnosis, we investigate two complementary approaches to model interpretation. Firstly, we develop a simple measure of sentence importance and demonstrate its effectiveness in interpreting a complex LSTM model's decision making process. Secondly, we discover clusters in the high-dimensional space of the sentence embedding model and test their correlation with feature importance scores for a given diagnosis class. This analysis yields insights into a model's representation of the clinical notes, allowing us to automatically extract clusters of sentences that are most relevant to the model's predictions.

2 Related Work

Document classification is a well-researched task in NLP that has been tackled using a wide variety of machine learning models, such as support-vector machines (SVMs) (Manevitz and Yousef, 2001), convolutional neural networks (CNNs) (Conneau et al., 2016) and recurrent neural networks (RNNs) (Yogatama et al., 2017). In the clinical domain, document classification models have been used in diverse tasks such as predicting cancer stage information in clinical records (Yim et al., 2017), extracting patient smoker-status from health records (Wang et al., 2019) and classifying radiology reports by their ICD-9CM code (Garla and Brandt, 2013). The problem of categorising clinical free text documents is closely related to several subtasks in the area of Electronic Health Record (EHR) analysis, including information extraction and representation learning. Information extraction is an umbrella term that covers diverse subtasks such as expanding abbreviations using contextual information, and the automatic annotation of temporal events (e.g. mapping from inputs such as "The patient was given stress dose steroids prior to his surgery" to output "[stress dose steroids] BEFORE [his surgery]" (Sun et al., 2013). Other NLP problems in this field that are relevant to free text analysis are outcome prediction and de-identification.

There are many ways to construct a representation of the input data that can be provided to a document classification model. A popular alternative to older approaches to text representations, such as bag-of-words (BoW), is to embed the input tokens in a high-dimensional vector space, resulting in each word being mapped to a list of real-valued numbers (a "word embedding"). One simple method of extracting word embeddings involves concatenating the hidden layer activations observed in a trained language model after processing all words up to the target word. As language models automatically learn rich semantic and syntactic features of words, these embeddings can provide valuable input features for downstream information extraction tasks. While the dimensions in the embedding space can correspond to interpretable features, this is not generally the case. However, a major motivation for using word embeddings is the ability to re use pre trained embeddings, essentially resulting in a form of transfer learning (Pan et al., 2010). In this study we use 300-dimensional fastText word embeddings (Bojanowski et al., 2017) which were pretrained on the Common Crawl dataset using the skipgram schema (Mikolov et al., 2013), which involves predicting a target word based on nearby words.

Similar to word embeddings, sentence embeddings are high-dimensional vectors that can represent features of a sequence of words. Our use of sentence embeddings is motivated by the fact that, for small amounts of data, it may be more difficult for a recurrent neural network to capture diagnosis-relevant dependencies over many word vectors than it is to classify a document made up of a smaller number of semantically richer sentence vectors. In this study we use 4096-dimensional InferSent embeddings (Conneau et al., 2017) that were extracted from a model pre-trained on the Common Crawl dataset.

After training recurrent models using these state-of-art NLP techniques to predict the diagnosis class associated with each document, we explore ways of visualizing and understanding how the models incorporate these vectors in order to make accurate predictions.

Class	# doc.	# sent.	# sent. (masked)
D	32	1420	1225
M	30	1140	985
N	44	1767	1547

Table 1: Number of documents and sentences in the clinical notes dataset. *D*: Dementia, *M*: MCI, *N*: Non-impaired.

Model	Accuracy	Precision	Recall	F1 Score
Random	0.333	0.333	0.333	0.333
Max. class	0.415	0.138	0.333	0.196
BoW+Random Forest	0.425	0.417	0.413	0.414
LSTM (*fastText word emb.*)	0.543	0.636	0.502	0.502
LSTM (*InferSent sentence emb.*)	**0.690**	**0.702**	**0.669**	**0.674**
ULMFiT	0.571	0.437	0.500	0.440

Table 2: Results (average over 5 folds) for the diagnosis classification task for the masked dataset. Precision, recall and F1 score are macro-averaged across the classes.

3 Data

We collected a corpus of consultation reports compiled by clinicians at a memory clinic to use as the data domain for the document classification task. Each report is anonymised and describes the clinican's review of a patient who suffers from memory or cognitive issues. Each report is labelled by one of three classes, corresponding to the diagnoses of *dementia*, *mild cognitive impairment (MCI)* and *non-impaired*. The documents can be considered semi-structured, as they are made up of free-text details that follow a loose narrative trajectory. The notes typically begin with a description of the patient's history and symptoms, and ultimately conclude with recommendations on how to proceed which may include scheduling a follow-up appointment, arranging further tests, or organising a treatment course based on the available evidence.

From this corpus, we build a version of the notes in which explicit diagnostic information is masked out. For example, the sentence *"We would recommend commencing on a Rivastigmine patch 4.6 mg for 24 hours and then to be increased to 9.5 mg for 24 hours once daily if tolerated."* would not be included in the masked diagnosis dataset, as the drug Rivastigmine is used to treat mild to moderate Alzheimer's disease and Parkinson's, and so its mention here trivially identifies the diagnosis. In this work, we are interested in the ability to make predictions from more subtle diagnostic signals, requiring our model to build semantic representations of cognitive impairment that go beyond counting the occurrence of single words. Table 1 presents summary metrics of the datasets.

Deep learning models are generally trained and tested on very large datasets, in contrast to the small corpus of demential letters that we have gathered, and in contrast to clinical note databases generally. This motivates our use of transfer learning.

Tackling the problems of training and interpreting models trained on datasets of this scale is directly relevant to the real world challenges of using natural language processing to support clinical decisions, such as identifying patients who may be applicable to participate in a clinical trial (Sarmiento and Dernoncourt, 2016). Annotating gold-standard training examples for such problems is resource intensive (Savkov et al., 2016). We would therefore like to build robust and general models given a small amount of samples. Recent work on training large language models on massive amounts of data thus has much potential for zero-shot classification of natural language documents (Yogatama et al., 2017).

4 Models and Evaluation

We investigate the relative performance of LSTM models trained with a sequence of word embeddings, LSTM models trained with a sequence of sentence embeddings, and a state-of-the-art document classification model, ULMFiT. One motivation for choosing these experimental models is to investigate which models can capture long-term dependencies across a clinical document, given a relatively small amount of samples ($n=106$). In addition to these three models, we also test a random forest baseline model, a model that randomly selects the class and a model that chooses the most common class (which is *non-impaired*). The random forest model is trained to classify a document based on its bag-of-words representation. All models are cross-validated using 5 folds of

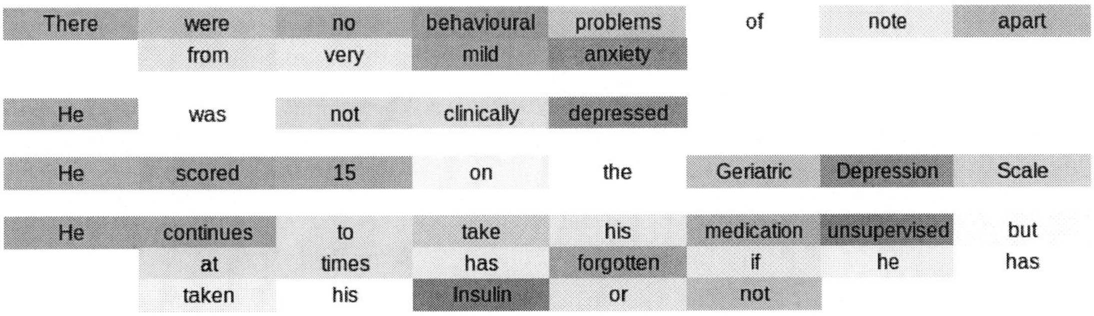

Figure 1: Visualisation of sentence importance with respect to the successful classification of *non-impaired* for a subset of a document. Sentences that were found to be important for the classification of *non-impaired* are coloured green while a sentence that increases the chance of a misclassification (i.e. an incorrect *MCI* diagnosis) is coloured red. The saturation of the colours corresponds to how much a given word contributes to a sentence's InferSent embedding

the dataset, ensuring that the class distribution is equal across all folds. The ULMFiT model is pre-trained on the Wikitext-103 dataset (Merity et al., 2017) and fine-tuned using default hyperparameters (*fine-tuning epochs=25, fine-tuning batch size=8, fine-tuning learning rate=0.004, training epochs=50, training batch size=32, training learning rate=0.01*) which have been shown to be robust across various tasks (Howard and Ruder, 2018). The LSTM model's hyperparameters were chosen by a grid-search. Both the sentence embedding LSTM and the word embedding LSTM were made up of one hidden layer with 256 hidden units.

The classification results for the models for the masked dataset are presented in Table 2. Each of our three models perform significantly better than chance and better than the random forest baseline model, with the LSTM model trained with sentence-embedding sequential input achieving the best performance. For this amount of training data, we would expect models that are trained on shorter sequences of more semantically enriched pre-trained vectors (i.e. sentence embeddings) to perform better than much longer sequences of vectors with less dimensions (i.e. word embeddings). This is because much of the work of combining word-level tokens into a contextual representation that is relevant to a statistical model of human language has already been done when training with pretrained representations extracted at the sentence-level. Somewhat surprisingly, the model trained on sentence embeddings outperformed the fine-tuned ULMFiT. Future work may shed light on how the amount of training samples

can affect the choice of whether to use fine-tuning or pre-trained embedding representations as model input.

5 Model Interpretability: Calculating Sentence Importance Scores

After demonstrating the effectiveness of using pre-trained sentence embeddings to classify the clinical documents, we investigated model interpretability by calculating a measure of the importance of each sentence in the sequence of sentences to the model's prediction for a document. We propose a measure of feature importance based on perturbative approaches to variable importance (Breiman, 2001), which estimate the importance of variables by iteratively randomly perturbing each variable and observing the change in loss. This technique is similar to measuring information gain (Quinlan, 1986), but rather than selecting important components of fixed input, we rate the importance of a sentence vector in the sequence of sentence vectors presented to our sequential LSTM classifier. For example, in order to generate the importance score for the first sentence in a document made up of *m* sentence embeddings, we construct an augmented version of the document containing all but the first sentence, and examine the resulting change in the prediction for that document. More formally, for sentence *n*, we generate the following version of the document *d* (with ground truth label *c*) with sentence *n* removed:

$$d_n = [s_0, s_1, \ldots, s_{n-1}, s_{n+1}, \ldots, s_{m-1}]$$

Next, the augmented document d_n is fed into the trained LSTM (using the best-in-fold model

51

Ratio	Sentence
-3.469	"He and his wife both report agitation disinhibition and irritability"
0.078	"He would say that he feels depressed at times"
0.149	"She was tremulous which <NAME> felt was most likely due to anxiety"
	. . .
2.108	"He had an equivalent score of 19 / 30 on the MMSE"
8.105	"He had an equivalent score of 29 / 30 on the MMSE"
12.887	"He had an equivalent score of 22 / 30 on the MMSE"

Table 3: Sentences sorted by feature importance for a correct diagnosis of *non-impaired*. Sentences with low scores do not support a prediction of *non-impaired* within the context of the corresponding clinical letter.

from Section 3, which achieved an accuracy of 73%) and we measure the network's output logit for the correct class. The importance score is calculated as the ratio of the model's output for the correct class excluding the sentence to the model's output for the correct class including a given sentence.

$$ratio_n = \frac{logit(c \mid d_n)}{logit(c \mid d)}$$

The most important sentences minimise this ratio. When the ratio is over 1, the inclusion of the sentence in the document leads to a smaller probability of selecting the correct class, and so sentences that maximise the ratio are the most misleading sentences with respect to the correct classification. Examples of highly important and highly misleading sentences across the corpus for a diagnosis of *non-impaired* are presented in Table 3. The average sentence importance trajectory over each class was also investigated and is presented in in Figure 2.

Figure 1 presents a section of a clinical letter for a patient with a diagnosis of *non-impaired*, with sentences coloured green or red depending on whether they increase or decrease the chance of correctly classifying the document. Within each sentence, the contribution of a word to the InferSent sentence embedding is visualised by colour saturation. We can see that the importance measure provides intuitive insights into how the recurrent neural network models the document. For example, the final sentence in Figure 1 decreases the chance of classifying the document as *non-impaired* because it states that the patient sometimes forgets to take their medicine – in isolation this sentence could naively be considered to imply a diagnosis of memory impairment, but as the model processes the full document it is able to

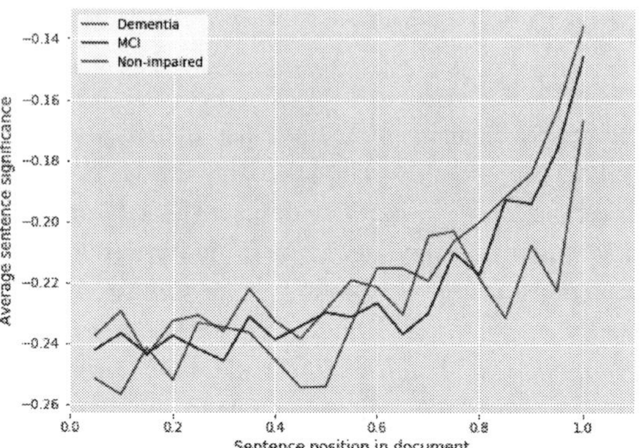

Figure 2: Average sentence importance over each class, as a function of sentences' position in the texts. Sentence importance ratios are normalised within each document and split by in-document position into 20 bins. For each class, we plot the negative of the average for each bin.

accumulate evidence and predict the correct diagnosis. By examining the contribution of each word to the InferSent vectors, we can see that negating words such as "not" are handled appropriately within the sentence embedding (e.g. "not clinically depressed" increases the probability of a correct *non-impaired* classification). Our model interpretation technique therefore demonstrates how the LSTM sentence embedding model improves on the simple bag-of-words baseline, where the word "depressed" would be incorrectly taken as negative evidence for a non-impaired diagnosis.

6 Cluster Analysis

In order to investigate the relationship between sentence importance and the sentence embedding space, we performed a cluster analysis. The 4096-dimensional sentence embeddings were projected

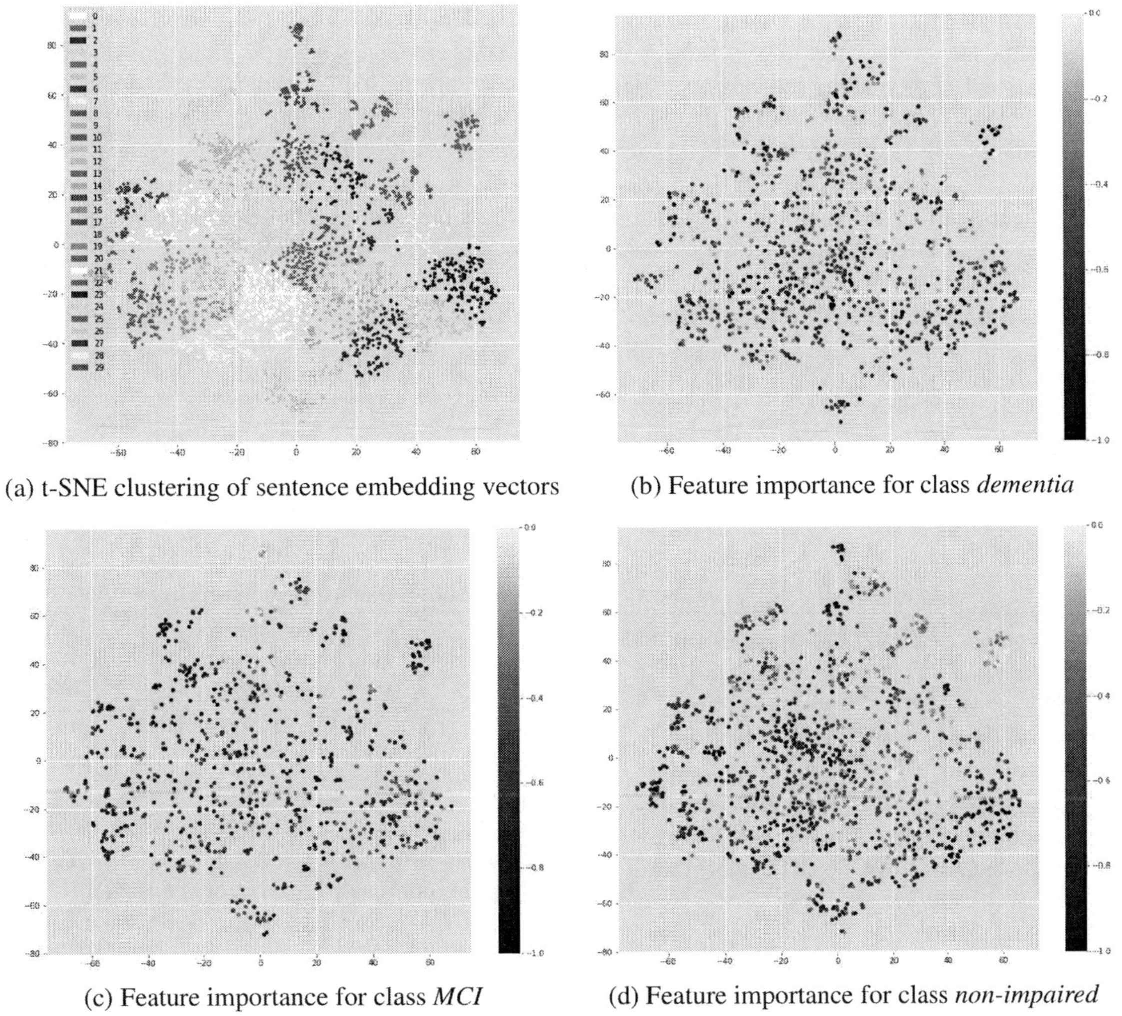

(a) t-SNE clustering of sentence embedding vectors   (b) Feature importance for class dementia

(c) Feature importance for class MCI   (d) Feature importance for class non-impaired

Figure 3: 2-dimensional projection of sentence embedding vectors. **(a)**: 30 clusters were identified and labelled using mean shift clustering. **(b) - (d)**: Heat maps of sentence vectors coloured by sentence importance for each class reveal clusters of sentences that are relevant to a given diagnosis. Colour scales indicate normalised values; brighter colours indicate more important sentences.

to two dimensions using t-SNE (van der Maaten and Hinton, 2008). We used the mean shift clustering technique (Yizong Cheng, 1995), an algorithm that does not require the number of clusters to be specified in advance, to discover clusters of similarly represented sentences in this space (Fig. 3(a)). Sentences that are important for the model's classification of a specific diagnosis are visualised by colouring the sentences using the corresponding importance score. This step was performed for each of the three classes (Fig. 3(b)-(d)).

Correlation tests were used to investigate the relationship between sentence clusters and their importance to a model's prediction for each class.

For each class c and for each cluster cl, we first gather the sentences that appear in documents of class c. Next, we assign each sentence a value of 1 or 0 depending on whether the sentence is in cluster cl. Using Spearman's Rho, we calculate the correlation between this value and the sentences' importance scores for the given class. In each trial, sentences that do not appear in documents of the target class are excluded. The results reported in Table 4 show the clusters that were found to be significantly correlated with at least one of the classes' importance scores. It was found that 15 out of the 30 automatically discovered sentence clusters can be considered significantly important in the model's decision making.

To assist in interpreting the information captured by each cluster, we depict the clusters using the most frequent bigrams across all of that cluster's sentences (Table 4). For example, one cluster (corresponding to cluster 20 in Figure 3(a)) contains sentences that mention the individual's family (significantly positively associated with a *non-impaired* diagnosis), while cluster 22 corresponds to sentences about the patient's blood pressure and heart rate (significantly negatively associated with a *non-impaired* diagnosis). Again, these results show the utility of combining sentence importance measures with sentence embeddings to reveal the clinically relevant detail in the documents.

7   Discussion

The results presented in Table 3 demonstrate the sentences that are most significant and most misleading for the LSTM InferSent model with respect to the diagnosis of non-impairment. We can see that the most significant sentences are those that refer to patients' mood and anxiety disorders. These types of sentences are over-represented in the *non-impaired* group. The types of sentences that are most misleading to the diagnosis of *non-impaired* are those of the format "*[pronoun] had an equivalent score of [score] / 30 on the MMSE*". An obvious question regarding this result is whether information about MMSE scores can be represented by the InferSent embeddings in such a way as to distinguish it from other sentences that differ only, but importantly, by a single integer value. We can see that the relationship between the significance of the sentence to the actual results in the sentence is non-linear. The 84 mentions of the Mini-Mental State Examination (MMSE) test are equally divided across the 3 classes; as there are more *non-impaired* documents in the dataset overall, the model benefits from learning not to predict this diagnosis when it encounters any sentence embedding in the MMSE cluster (cluster 17 in Figure 3(a); the corresponding points in Figure 3(d) indicate their decreased importance for this category). Further analysis may include using *diagnostic classifiers* (Hupkes et al., 2018) to test whether a model can accurately decide whether the first of two given sentence embeddings reports a larger score.

Figure 2 shows the average sentence significance across the documents for each of the three classes. For all classes, we can see that the importance of sentences tends to increase with their in-document position. This trend may correspond to the semi-structured nature of the documents, reflecting information becoming more relevant to a diagnosis towards the end of a document. Another possible explanation could be that the recurrent neural network is unable to capture long-distance dependencies given the small amount of samples in the dataset, resulting in a kind of recency bias in the model's processing (since the model only makes its prediction at the end of the sequence of sentences). Further work may involve systematically changing the position of each sentence within each document in order to investigate the effect that this has on the importance scores associated with each sentence.

Table 4 shows that no clusters were significantly correlated with the class *dementia*, with all reported clusters being significantly correlated with at least one of *MCI* or *non-impaired*. Excluding cluster 18, all of the clusters that are significant for both *MCI* and *non-impaired* form pairs of negative vs. positive correlations between these two classes, suggesting that the model learns primarily to discriminate between these classes. Examining the confusion matrix for the model, we found that the model has a true positive rate of 1.0 and 0.89 for *MCI* and *non-impaired*, and minimises the amount of false positives between these two classes. However, the model performs poorly when the actual document corresponds to a diagnosis of *dementia* (with a true positive rate of 0.29). This is consistent with the observation that none of the clusters significantly correlate with this class. While this insight could be gained from examining the confusion matrix alone, the advantage of employing the interpretation methods developed in this paper is that they allow us to gain an understanding of how the model's processing of sentences over time leads to these inequalities, suggesting avenues of attack for constructing more accurate representations of the documents going forward.

In future work, we plan to gather more clinical documents that describe patients with memory impairment and continue our analysis of language modelling and classification in this distribution. We hope to subsequently apply state of the art contextualised embeddings such as ELMO (Peters et al., 2018) and BERT (Devlin et al., 2018)

Cluster	Top bigrams in cluster	Rho$_D$	Rho$_M$	Rho$_N$
2	"behavioural problems", "neurological deficit", "extra pyramidal"	0.036	**−0.215*****	0.142***
3	"short term", "years ago", "poor short"	−0.022	0.032	**0.119*****
5	"family history", "disease dementia", "alzheimers disease"	0.039	**−0.147*****	0.146***
7	"activities daily", "daily living", "remains independent"	−0.013	−0.121*	**0.155*****
9	"medical history", "ischaemic heart", "heart disease"	−0.003	**0.161*****	−0.131***
10	"memory fluency", "verbal fluency", "points lost"	−0.033	**0.133****	−0.097*
12	"misplacing items", "cognitive checklist", "disorientation time"	0.000	**0.143*****	−0.098*
17	"30 mmse", "mmse equivalent", "29 30"	−0.040	**−0.193*****	0.112***
18	"cognitive testing", "100 ace", "addenbrooke cognitive"	−0.010	−0.171***	**−0.181*****
20	"unaccompanied morning", "four children", "two children"	0.022	−0.029	**0.165*****
22	"blood pressure', "bpm regular", "examination pulse"	−0.045	0.130**	**−0.124*****
23	"b12 folate", "screening bloods", "thyroid function"	−0.022	0.022	**−0.089***
24	"current medications", "mg daily", "40 mg"	0.021	**0.181*****	−0.106**
25	"geriatric depression", "depression scale", "scored 15"	0.010	**0.182*****	−0.229
27	"onset progression", "progression described", "physical examination"	−0.064	0.132**	**−0.113*****

Table 4: Automatically discovered sentence clusters that significantly correlate with sentence importance for at least one class. For each cluster and for each class, we use Spearman's Rho to test the correlation between a sentence's importance with respect to the class of interest, and whether or not the sentence is in the given cluster. The most frequent within-cluster bigrams were extracted after removing stop words from the sentences. * $p<0.05$, ** $p<0.01$, *** $p<0.001$, Bonferroni corrected. D: Dementia, M: MCI, N: Non-impaired.

to a larger corpus in order to further use feature extraction to build and understand meaningful semantic representations of cognitive impairment as described by clinicians. As part of this work, we aim to examine how models trained on the writing style of one clinician apply to those written by others, as the corpus used in this study was sourced from a small number of clinicians. We suspect that analysing a model's inter- and intra-clinician performance metrics will yield useful insights into how well the model has generalised, and how clinicians may differ in terms of the subtle but diagnosis-relevant information they include in the documents.

8 Conclusion

We showed the effectiveness of using pre-trained sentence embeddings and recurrent neural networks for a document classification task using a corpus of natural language clinical reports. The sentence-level LSTM model performed better than both an LSTM trained on word embeddings and a simple bag-of-words baseline. Following this result, we developed a simple and intuitive perturbative measure of sentence importance for the sentences in the corpus. After demonstrating how this measure can be used to interpret the success and failure cases of a trained model, we used cluster analysis to identify regions in the sentence embedding space that are significantly correlated with sentence importance for specific diagnosis classes.

By reviewing the most frequent bigrams in each cluster and examining the sign of Spearman's Rho for each corresponding correlated class, we can interpret how differential processing of sentence vectors within each cluster can lead to class imbalances in the model's predictions, demonstrating the power of our approach for model interpretability and evaluation.

Acknowledgements

We would like to thank the three anonymous reviewers, Stuart Millar, and Steven Derby for their feedback and suggestions. This work was partfunded by a Data Analytics Dementia Pathfinder Programme Grant from the Northern Ireland HSCB eHealth Directorate.

References

Cheryl Ann Alexander and Lidong Wang. 2018. Big Data and Data-Driven Healthcare Systems. *Journal of Business and Management Sciences*, 6(3):104–111.

Pierpaolo Basile, Annalina Caputo, and Giovanni Semeraro. 2012. A Study on Compositional Semantics of Words in Distributional Spaces. In *2012 IEEE Sixth International Conference on Semantic Computing*, pages 154–161. IEEE.

Piotr Bojanowski, Edouard Grave, Armand Joulin, and Tomas Mikolov. 2017. Enriching Word Vectors with Subword Information. *Transactions of the Association for Computational Linguistics*, 5:135–146.

Leo Breiman. 2001. Random Forests. *Machine Learning*, 45(1):5–32.

Alexis Conneau, Douwe Kiela, Holger Schwenk, Loic Barrault, and Antoine Bordes. 2017. Supervised Learning of Universal Sentence Representations from Natural Language Inference Data. *arXiv preprint arXiv:1705.02364*.

Alexis Conneau, Holger Schwenk, Loïc Barrault, and Yann Lecun. 2016. Very deep convolutional networks for text classification. *arXiv preprint arXiv:1606.01781*.

Jacob Devlin, Ming-Wei Chang, Kenton Lee, and Kristina Toutanova. 2018. Bert: Pre-training of deep bidirectional transformers for language understanding. *arXiv preprint arXiv:1810.04805*.

Vijay N Garla and Cynthia Brandt. 2013. Knowledge-based biomedical word sense disambiguation: an evaluation and application to clinical document classification. *Journal of the American Medical Informatics Association*, 20(5):882–886.

Sepp Hochreiter and Jürgen Schmidhuber. 1997. Long short-term memory. *Neural computation*, 9(8):1735–1780.

Jeremy Howard and Sebastian Ruder. 2018. Universal Language Model Fine-tuning for Text Classification. *Proceedings of the 56th Annual Meeting of the Association for Computational Linguistics (Volume 1: Long Papers)*, 1:328–339.

Dieuwke Hupkes, Sara Veldhoen, and Willem Zuidema. 2018. Visualisation and 'diagnostic classifiers' reveal how recurrent and recursive neural networks process hierarchical structure. *Journal of Artificial Intelligence Research*.

Kyoungyoung Jee and Gang-Hoon Kim. 2013. Potentiality of Big Data in the Medical Sector: Focus on How to Reshape the Healthcare System. *Healthcare Informatics Research*, 19(2):79.

Laurens van der Maaten and Geoffrey Hinton. 2008. Visualizing Data using t-SNE. *Journal of Machine Learning Research*, 9(Nov):2579–2605.

Larry M Manevitz and Malik Yousef. 2001. One-class SVMs for document classification. *Journal of machine Learning research*, 2(Dec):139–154.

Stephen Merity, Nitish Shirish Keskar, and Richard Socher. 2017. Regularizing and Optimizing LSTM Language Models. *arXiv preprint arXiv:1708.02182*.

Tomas Mikolov, Kai Chen, Greg Corrado, and Jeffrey Dean. 2013. Efficient Estimation of Word Representations in Vector Space. *preprint arXiv:1301.3781*.

Sinno Jialin Pan, Qiang Yang, and Others. 2010. A survey on transfer learning. *IEEE Transactions on knowledge and data engineering*, 22(10):1345–1359.

Matthew E Peters, Mark Neumann, Mohit Iyyer, Matt Gardner, Christopher Clark, Kenton Lee, and Luke Zettlemoyer. 2018. Deep contextualized word representations. *arXiv preprint arXiv:1802.05365*.

J. R. Quinlan. 1986. Induction of decision trees. *Machine Learning*, 1(1):81–106.

Pranav Rajpurkar, Jeremy Irvin, Kaylie Zhu, Brandon Yang, Hershel Mehta, Tony Duan, Daisy Ding, Aarti Bagul, Curtis Langlotz, Katie Shpanskaya, Matthew P. Lungren, and Andrew Y. Ng. 2017. CheXNet: Radiologist-Level Pneumonia Detection on Chest X-Rays with Deep Learning. *arXiv preprint arXiv:1711.05225*.

Raymond Francis Sarmiento and Franck Dernoncourt. 2016. Improving patient cohort identification using natural language processing. In *Secondary analysis of electronic health records*, pages 405–417. Springer.

Aleksandar Savkov, John Carroll, Rob Koeling, and Jackie Cassell. 2016. Annotating patient clinical records with syntactic chunks and named entities: the harvey corpus. *Language resources and evaluation*, 50(3):523–548.

Benjamin Shickel, Patrick James Tighe, Azra Bihorac, and Parisa Rashidi. 2018. Deep EHR: A Survey of Recent Advances in Deep Learning Techniques for Electronic Health Record (EHR) Analysis. *IEEE Journal of Biomedical and Health Informatics*, 22(5):1589–1604.

Weiyi Sun, Anna Rumshisky, and Ozlem Uzuner. 2013. Temporal reasoning over clinical text: the state of the art. *Journal of the American Medical Informatics Association : JAMIA*, 20(5):814–9.

Yanshan Wang, Sunghwan Sohn, Sijia Liu, Feichen Shen, Liwei Wang, Elizabeth J. Atkinson, Shreyasee Amin, and Hongfang Liu. 2019. A clinical text classification paradigm using weak supervision and deep representation. *BMC Medical Informatics and Decision Making*, 19(1):1.

Wen-Wai Yim, Sharon W Kwan, Guy Johnson, and Meliha Yetisgen. 2017. Classification of hepatocellular carcinoma stages from free-text clinical and radiology reports. *AMIA ... Annual Symposium proceedings. AMIA Symposium*, 2017:1858–1867.

Yizong Cheng. 1995. Mean shift, mode seeking, and clustering. *IEEE Transactions on Pattern Analysis and Machine Intelligence*, 17(8):790–799.

Dani Yogatama, Chris Dyer, Wang Ling, and Phil Blunsom. 2017. Generative and discriminative text classification with recurrent neural networks. *arXiv preprint arXiv:1703.01898*.

Transfer Learning in Biomedical Natural Language Processing: An Evaluation of BERT and ELMo on Ten Benchmarking Datasets

Yifan Peng   Shankai Yan   Zhiyong Lu
National Center for Biotechnology Information
National Library of Medicine, National Institutes of Health
Bethesda, MD, USA
{yifan.peng, shankai.yan, zhiyong.lu}@nih.gov

Abstract

Inspired by the success of the General Language Understanding Evaluation benchmark, we introduce the Biomedical Language Understanding Evaluation (BLUE) benchmark to facilitate research in the development of pre-training language representations in the biomedicine domain. The benchmark consists of five tasks with ten datasets that cover both biomedical and clinical texts with different dataset sizes and difficulties. We also evaluate several baselines based on BERT and ELMo and find that the BERT model pre-trained on PubMed abstracts and MIMIC-III clinical notes achieves the best results. We make the datasets, pre-trained models, and codes publicly available at https://github.com/ncbi-nlp/BLUE_Benchmark.

1   Introduction

With the growing amount of biomedical information available in textual form, there have been significant advances in the development of pre-training language representations that can be applied to a range of different tasks in the biomedical domain, such as pre-trained word embeddings, sentence embeddings, and contextual representations (Chiu et al., 2016; Chen et al., 2019; Peters et al., 2017; Lee et al., 2019; Smalheiser et al., 2019).

In the general domain, we have recently observed that the General Language Understanding Evaluation (GLUE) benchmark (Wang et al., 2018a) has been successfully promoting the development of language representations of general purpose (Peters et al., 2017; Radford et al., 2018; Devlin et al., 2019). To the best of our knowledge, however, there is no publicly available benchmarking in the biomedicine domain.

To facilitate research on language representations in the biomedicine domain, we present the

Biomedical Language Understanding Evaluation (BLUE) benchmark, which consists of five different biomedicine text-mining tasks with ten corpora. Here, we rely on preexisting datasets because they have been widely used by the BioNLP community as shared tasks (Huang and Lu, 2015). These tasks cover a diverse range of text genres (biomedical literature and clinical notes), dataset sizes, and degrees of difficulty and, more importantly, highlight common biomedicine text-mining challenges. We expect that the models that perform better on all or most tasks in BLUE will address other biomedicine tasks more robustly.

To better understand the challenge posed by BLUE, we conduct experiments with two baselines: One makes use of the BERT model (Devlin et al., 2019) and one makes use of ELMo (Peters et al., 2017). Both are state-of-the-art language representation models and demonstrate promising results in NLP tasks of general purpose. We find that the BERT model pre-trained on PubMed abstracts (Fiorini et al., 2018) and MIMIC-III clinical notes (Johnson et al., 2016) achieves the best results, and is significantly superior to other models in the clinical domain. This demonstrates the importance of pre-training among different text genres.

In summary, we offer: (i) five tasks with ten biomedical and clinical text-mining corpora with different sizes and levels of difficulty, (ii) codes for data construction and model evaluation for fair comparisons, (iii) pretrained BERT models on PubMed abstracts and MIMIC-III, and (iv) baseline results.

2   Related work

There is a long history of using shared language representations to capture text semantics in biomedical text and data mining research. Such re-

58

Proceedings of the BioNLP 2019 workshop, pages 58–65
Florence, Italy, August 1, 2019. ©2019 Association for Computational Linguistics

search utilizes a technique, termed transfer learning, whereby the language representations are pretrained on large corpora and fine-tuned in a variety of downstream tasks, such as named entity recognition and relation extraction.

One established trend is a form of word embeddings that represent the semantic, using high dimensional vectors (Chiu et al., 2016; Wang et al., 2018c; Zhang et al., 2019). Similar methods also have been derived to improve embeddings of word sequences by introducing sentence embeddings (Chen et al., 2019). They always, however, require complicated neural networks to be effectively used in downstream applications.

Another popular trend, especially in recent years, is the context-dependent representation. Different from word embeddings, it allows the meaning of a word to change according to the context in which it is used (Melamud et al., 2016; Peters et al., 2017; Devlin et al., 2019; Dai et al., 2019). In the scientific domain, Beltagy et al. released SciBERT which is trained on scientific text. In the biomedical domain, BioBERT (Lee et al., 2019) and BioELMo (Jin et al., 2019) were pretrained and applied to several specific tasks. In the clinical domain, Alsentzer et al. (2019) released a clinical BERT base model trained on the MIMIC-III database. Most of these works, however, were evaluated on either different datasets or the same dataset with slightly different sizes of examples. This makes it challenging to fairly compare various language models.

Based on these reasons, a standard benchmarking is urgently required. Parallel to our work, Lee et al. (2019) introduced three tasks: named entity recognition, relation extraction, and QA, while Jin et al. (2019) introduced NLI in addition to named entity recognition. To this end, we deem that BLUE is different in three ways. First, BLUE is selected to cover a diverse range of text genres, including both biomedical and clinical domains. Second, BLUE goes beyond sentence or sentence pairs by including document classification tasks. Third, BLUE provides a comprehensive suite of codes to reconstruct dataset from scratch without removing any instances.

3 Tasks

BLUE contains five tasks with ten corpora that cover a broad range of data quantities and difficulties (Table 1). Here, we rely on preexisting datasets because they have been widely used by the BioNLP community as shared tasks.

3.1 Sentence similarity

The sentence similarity task is to predict similarity scores based on sentence pairs. Following common practice, we evaluate similarity by using Pearson correlation coefficients.

BIOSSES is a corpus of sentence pairs selected from the Biomedical Summarization Track Training Dataset in the biomedical domain (Soğancıoğlu et al., 2017).[1] To develop BIOSSES, five curators judged their similarity, using scores that ranged from 0 (no relation) to 4 (equivalent). Here, we randomly select 80% for training and 20% for testing because there is no standard splits in the released data.

MedSTS is a corpus of sentence pairs selected from Mayo Clinic's clinical data warehouse (Wang et al., 2018b). To develop MedSTS, two medical experts graded the sentence's semantic similarity scores from 0 to 5 (low to high similarity). We use the standard training and testing sets in the shared task.

3.2 Named entity recognition

The aim of the named entity recognition task is to predict mention spans given in the text (Jurafsky and Martin, 2008). The results are evaluated through a comparison of the set of mention spans annotated within the document with the set of mention spans predicted by the model. We evaluate the results by using the strict version of precision, recall, and F1-score. For disjoint mentions, all spans also must be strictly correct. To construct the dataset, we used spaCy[2] to split the text into a sequence of tokens when the original datasets do not provide such information.

BC5CDR is a collection of 1,500 PubMed titles and abstracts selected from the CTD-Pfizer corpus and was used in the BioCreative V chemical-disease relation task (Li et al., 2016).[3] The diseases and chemicals mentioned in the articles were annotated independently by two human experts with medical training and curation experience. We use the standard training and test set in the

---

59

Corpus	Train	Dev	Test	Task	Metrics	Domain	Avg sent len
MedSTS, sentence pairs	675	75	318	Sentence similarity	Pearson	Clinical	25.8
BIOSSES, sentence pairs	64	16	20	Sentence similarity	Pearson	Biomedical	22.9
BC5CDR-disease, mentions	4182	4244	4424	NER	F1	Biomedical	22.3
BC5CDR-chemical, mentions	5203	5347	5385	NER	F1	Biomedical	22.3
ShARe/CLEFE, mentions	4628	1075	5195	NER	F1	Clinical	10.6
DDI, relations	2937	1004	979	Relation extraction	micro F1	Biomedical	41.7
ChemProt, relations	4154	2416	3458	Relation extraction	micro F1	Biomedical	34.3
i2b2 2010, relations	3110	11	6293	Relation extraction	F1	Clinical	24.8
HoC, documents	1108	157	315	Document classification	F1	Biomedical	25.3
MedNLI, pairs	11232	1395	1422	Inference	accuracy	Clinical	11.9

Table 1: BLUE tasks

BC5CDR shared task (Wei et al., 2016).

ShARe/CLEF eHealth Task 1 Corpus is a collection of 299 deidentified clinical free-text notes from the MIMIC II database (Suominen et al., 2013).[4] The disorders mentioned in the clinical notes were annotated by two professionally trained annotators, followed by an adjudication step, resulting in high inter-annotator agreement. We use the standard training and test set in the ShARe/CLEF eHealth Tasks 1.

3.3 Relation extraction

The aim of the relation extraction task is to predict relations and their types between the two entities mentioned in the sentences. The relations with types were compared to annotated data. We use the standard micro-average precision, recall, and F1-score metrics.

DDI extraction 2013 corpus is a collection of 792 texts selected from the DrugBank database and other 233 Medline abstracts (Herrero-Zazo et al., 2013).[5] The drug-drug interactions, including both pharmacokinetic and pharmacodynamic interactions, were annotated by two expert pharmacists with a substantial background in pharmacovigilance. In our benchmark, we use 624 train files and 191 test files to evaluate the performance and report the micro-average F1-score of the four DDI types.

ChemProt consists of 1,820 PubMed abstracts with chemical-protein interactions annotated by domain experts and was used in the BioCreative VI text mining chemical-protein interactions shared task (Krallinger et al., 2017).[6] We use the

standard training and test sets in the ChemProt shared task and evaluate the same five classes: CPR:3, CPR:4, CPR:5, CPR:6, and CPR:9.

i2b2 2010 shared task collection consists of 170 documents for training and 256 documents for testing, which is the subset of the original dataset (Uzuner et al., 2011).[7] The dataset was collected from three different hospitals and was annotated by medical practitioners for eight types of relations between problems and treatments.

3.4 Document multilabel classification

The multilabel classification task predicts multiple labels from the texts.

HoC (the Hallmarks of Cancers corpus) consists of 1,580 PubMed abstracts annotated with ten currently known hallmarks of cancer (Baker et al., 2016).[8] Annotation was performed at sentence level by an expert with 15+ years of experience in cancer research. We use 315 (~20%) abstracts for testing and the remaining abstracts for training. For the HoC task, we followed the common practice and reported the example-based F1-score on the abstract level (Zhang and Zhou, 2014; Du et al., 2019).

3.5 Inference task

The aim of the inference task is to predict whether the premise sentence entails or contradicts the hypothesis sentence. We use the standard overall accuracy to evaluate the performance.

MedNLI is a collection of sentence pairs selected from MIMIC-III (Romanov and Shivade, 2018).[9] Given a premise sentence and a hy-

---

[4] https://physionet.org/works/ ShAReCLEFeHealth2013/

[5] http://labda.inf.uc3m.es/ddicorpus

[6] https://biocreative. bioinformatics.udel.edu/news/corpora/ chemprot-corpus-biocreative-vi/

[7] https://www.i2b2.org/NLP/DataSets/

[8] https://www.cl.cam.ac.uk/~sb895/HoC. html

[9] https://physionet.org/physiotools/ mimic-code/mednli/

pothesis sentence, two board-certified radiologists graded whether the task predicted whether the premise entails the hypothesis (entailment), contradicts the hypothesis (contradiction), or neither (neutral). We use the same training, development, and test sets in Romanov and Shivade (Romanov and Shivade, 2018).

3.6 Total score

Following the practice in Wang et al. (2018a) and Lee et al. (2019), we use a macro-average of F1-scores and Pearson scores to determine a system's position.

4 Baselines

For baselines, we evaluate several pre-training models as described below. The original code for the baselines is available at `https://github.com/ncbi-nlp/NCBI_BERT`.

4.1 BERT

4.1.1 Pre-training BERT

BERT (Devlin et al., 2019) is a contextualized word representation model that is pre-trained based on a masked language model, using bidirectional Transformers (Vaswani et al., 2017).

In this paper, we pre-trained our own model BERT on PubMed abstracts and clinical notes (MIMIC-III). The statistics of the text corpora on which BERT was pre-trained are shown in Table 2.

Corpus	Words	Domain
PubMed abstract	> 4,000M	Biomedical
MIMIC-III	> 500M	Clinical

Table 2: Corpora

We initialized BERT with pre-trained BERT provided by (Devlin et al., 2019). We then continue to pre-train the model, using the listed corpora.

We released our BERT-Base and BERT-Large models, using the same vocabulary, sequence length, and other configurations provided by Devlin et al. (2019). Both models were trained with 5M steps on the PubMed corpus and 0.2M steps on the MIMIC-III corpus.

4.1.2 Fine-tuning with BERT

BERT is applied to various downstream text-mining tasks while requiring only minimal architecture modification.

For sentence similarity tasks, we packed the sentence pairs together into a single sequence, as suggested in Devlin et al. (2019).

For named entity recognition, we used the BIO tags for each token in the sentence. We considered the tasks similar to machine translation, as predicting the sequence of BIO tags from the input sentence.

We treated the relation extraction task as a sentence classification by replacing two named entity mentions of interest in the sentence with pre-defined tags (e.g., @GENE$, @DRUG$) (Lee et al., 2019). For example, we used "@CHEMICAL$ protected against the RTI-76-induced inhibition of @GENE$ binding." to replace the original sentence "Citalopram protected against the RTI-76-induced inhibition of SERT binding." in which "citalopram" and "SERT" has a chemical-gene relation.

For multi-label tasks, we fine-tuned the model to predict multi-labels for each sentence in the document. We then combine the labels in one document and compare them with the gold-standard.

Like BERT, we provided sources code for fine-tuning, prediction, and evaluation to make it straightforward to follow those examples to use our BERT pre-trained models for all tasks.

4.2 Fine-tuning with ELMo

We adopted the ELMo model pre-trained on PubMed abstracts (Peters et al., 2017) to accomplish the BLUE tasks.[10] The output of ELMo embeddings of each token is used as input for the fine-tuning model. We retrieved the output states of both layers in ELMo and concatenated them into one vector for each word. We used the maximum sequence length 128 for padding. The learning rate was set to 0.001 with an Adam optimizer. We iterated the training process for 20 epochs with batch size 64 and early stopped if the training loss did not decrease.

For sentence similarity tasks, we used bag of embeddings with the average strategy to transform the sequence of word embeddings into a sentence embedding. Afterward, we concatenated two sentence embeddings and fed them into an architecture with one dense layer to predict the similarity of two sentences.

---

[10]`https://allennlp.org/elmo`

Task	Metrics	SOTA*	ELMo	BioBERT	Our BERT			
					Base (P)	Base (P+M)	Large (P)	Large (P+M)
MedSTS	Pearson	83.6	68.6	84.5	84.5	**84.8**	84.6	83.2
BIOSSES	Pearson	84.8	60.2	82.7	89.3	**91.6**	86.3	75.1
BC5CDR-disease	F	84.1	83.9	85.9	**86.6**	85.4	82.9	83.8
BC5CDR-chemical	F	93.3	91.5	93.0	**93.5**	92.4	91.7	91.1
ShARe/CLEFE	F	70.0	75.6	72.8	75.4	**77.1**	72.7	74.4
DDI	F	72.9	78.9	78.8	78.1	79.4	**79.9**	76.3
ChemProt	F	64.1	66.6	71.3	72.5	69.2	**74.4**	65.1
i2b2	F	73.7	71.2	72.2	74.4	**76.4**	73.3	73.9
HoC	F	81.5	80.0	82.9	85.3	83.1	**87.3**	85.3
MedNLI	acc	73.5	71.4	80.5	82.2	**84.0**	81.5	83.8
Total			78.8	80.5	82.2	**82.3**	81.5	79.2

* SOTA, state-of-the-art as of April 2019, to the best of our knowledge: MedSTS, BIOSSES (Chen et al., 2019); BC5CDR-disease, BC5CDR-chem (Yoon et al., 2018); ShARe/CLEFE (Leaman et al., 2015); DDI (Zhang et al., 2018). Chem-Prot (Peng et al., 2018); i2b2 (Rink et al., 2011); HoC (Du et al., 2019); MedNLI (Romanov and Shivade, 2018). P: PubMed, P+M: PubMed + MIMIC-III

Table 3: Baseline performance on the BLUE task test sets.

For named entity recognition, we used a Bi-LSTM-CRF implementation as a sequence tagger (Huang et al., 2015; Si et al., 2019; Lample et al., 2016). Specifically, we concatenated the GloVe word embeddings (Pennington et al., 2014), character embeddings, and ELMo embeddings of each token and fed the combined vectors into the sequence tagger to predict the label for each token. The GloVe word embeddings[11] and character embeddings have 100 and 25 dimensions, respectively. The hidden sizes of the Bi-LSTM are also set to 100 and 25 for the word and character embeddings, respectively.

For relation extraction and multi-label tasks, we followed the steps in fine-tuning with BERT but used the averaged ELMo embeddings of all words in each sentence as the sentence embedding.

5 Benchmark results and discussion

We pre-trained four BERT models: BERT-Base (P), BERT-Large (P), BERT-Base (P+M), BERT-Large (P+M) on PubMed abstracts only, and the combination of PubMed abstracts and clinical notes, respectively. We present performance on the main benchmark tasks in Table 3. More detailed comparison is shown in the Appendix A.

Overall, our BERT-Base (P+M) that were pre-trained on both PubMed abstract and MIMIC-III achieved the best results across five tasks, even though it is only slightly better than the one pre-trained on PubMed abstracts only. Compared to the tasks in the clinical domain and biomedical domain, BERT-Base (P+M) is significantly superior to other models. This demonstrates the importance of pre-training among different text genres.

When comparing BERT pre-trained using the base settings against that using the large settings, it is a bit surprising that BERT-Base is better than BERT-Large except in relation extraction and document classification tasks. Further analysis shows that, on these tasks, the average length of sentences is longer than those of others (Table 1). In addition, BERT-Large pre-trained on PubMed and MIMIC is worse than other models overall. However, BERT-Large (P) performs the best in the multilabel task, even compared with the feature-based model utilizing enriched ontology (Yan and Wong, 2017). This is partially because the MIMIC-III data are relatively smaller than the PubMed abstracts and, thus, cannot pre-train the large model sufficiently.

In the sentence similarity tasks, BERT-Base (P+M) achieves the best results on both datasets. Because the BIOSSES dataset is very small (there

[11]https://nlp.stanford.edu/projects/glove/

are only 16 sentence pairs in the test set), all BERT models' performance was unstable. This problem has also been noted in the work of Devlin et al. (2019) when the model was evaluated on the GLUE benchmarking. Here, we obtained the best results by following the same strategy: selecting the best model on the development set after several runs. Other possible ways to overcome this issue include choosing the model with the best performance from multiple runs or averaging results from multiple fine-tuned models.

In the named entity recognition tasks, BERT-Base (P) achieved the best results on two biomedical datasets, whereas BERT-Base (P+M) achieved the best results on the clinical dataset. In all cases, we observed that the winning model obtained higher recall than did the others. Given that we use the pre-defined vocabulary in the original BERT and that this task relies heavily on the tokenization, it is possible that using BERT as pertaining to a custom sentence piece tokenizer may further improve the model's performance.

6 Conclusion

In this study, we introduce BLUE, a collection of resources for evaluating and analyzing biomedical natural language representation models. We find that the BERT models pre-trained on PubMed abstracts and clinical notes see better performance than do most state-of-the-art models. Detailed analysis shows that our benchmarking can be used to evaluate the capacity of the models to understand the biomedicine text and, moreover, to shed light on the future directions for developing biomedicine language representations.

Acknowledgments

This work was supported by the Intramural Research Programs of the NIH National Library of Medicine. This work was supported by the National Library of Medicine of the National Institutes of Health under award number K99LM013001-01. We are also grateful to shared task organizers and the authors of BERT and ELMo to make the data and codes publicly available.

References

Emily Alsentzer, John R. Murphy, Willie Boag, Wei-Hung Weng, Di Jin, Tristan Naumann, and Matthew B. A. McDermott. 2019. Publicly available clinical BERT embeddings. *arXiv:1904.03323*.

Simon Baker, Ilona Silins, Yufan Guo, Imran Ali, Johan Högberg, Ulla Stenius, and Anna Korhonen. 2016. Automatic semantic classification of scientific literature according to the hallmarks of cancer. *Bioinformatics (Oxford, England)*, 32:432–440.

Iz Beltagy, Arman Cohan, and Kyle Lo. Scibert: Pretrained contextualized embeddings for scientific text. *arXiv preprint arXiv:1903.10676*.

Qingyu Chen, Yifan Peng, and Zhiyong Lu. 2019. BioSentVec: creating sentence embeddings for biomedical texts. In *Proceedings of the 7th IEEE International Conference on Healthcare Informatics*.

Billy Chiu, Gamal Crichton, Anna Korhonen, and Sampo Pyysalo. 2016. How to train good word embeddings for biomedical NLP. In *Proceedings of BioNLP Workshop*, pages 166–174.

Zihang Dai, Zhilin Yang, Yiming Yang, William W Cohen, Jaime Carbonell, Quoc V Le, and Ruslan Salakhutdinov. 2019. Transformer-XL: Attentive language models beyond a fixed-length context. *arXiv preprint arXiv:1901.02860*.

Jacob Devlin, Ming-Wei Chang, Kenton Lee, and Kristina Toutanova. 2019. BERT: Pre-training of deep bidirectional transformers for language understanding. *arXiv preprint arXiv:1810.04805*.

Jingcheng Du, Qingyu Chen, Yifan Peng, Yang Xiang, Cui Tao, and Zhiyong Lu. 2019. ML-Net: multi-label classification of biomedical texts with deep neural networks. *Journal of the American Medical Informatics Association (JAMIA)*.

Nicolas Fiorini, Robert Leaman, David J Lipman, and Zhiyong Lu. 2018. How user intelligence is improving pubmed. *Nature biotechnology*, 36:937–945.

María Herrero-Zazo, Isabel Segura-Bedmar, Paloma Martínez, and Thierry Declerck. 2013. The DDI corpus: an annotated corpus with pharmacological substances and drug-drug interactions. *Journal of biomedical informatics*, 46:914–920.

Chung-Chi Huang and Zhiyong Lu. 2015. Community challenges in biomedical text mining over 10 years: success, failure and the future. *Briefings in Bioinformatics*, 17(1):132–144.

Zhiheng Huang, Wei Xu, and Kai Yu. 2015. Bidirectional LSTM-CRF models for sequence tagging. *arXiv preprint arXiv:1508.01991*.

Qiao Jin, Bhuwan Dhingra, William W. Cohen, and Xinghua Lu. 2019. Probing biomedical embeddings from language models. *arXiv:1904.02181*.

Alistair E. W. Johnson, Tom J. Pollard, Lu Shen, Li-Wei H. Lehman, Mengling Feng, Mohammad Ghassemi, Benjamin Moody, Peter Szolovits, Leo Anthony Celi, and Roger G. Mark. 2016. MIMIC-III, a freely accessible critical care database. *Scientific data*, 3:160035.

Daniel Jurafsky and James H. Martin. 2008. *Speech and language processing: an introduction to natural language processing, computational linguistics, and speech recognition*, 2 edition. Prentice Hall.

Martin Krallinger, Obdulia Rabal, Saber A. Akhondi, Martín Pérez Pérez, Jesús Santamaría, Gael Pérez Rodríguez, Georgios Tsatsaronis, Ander Intxaurrondo, José Antonio López1 Umesh Nandal, Erin Van Buel, Akileshwari Chandrasekhar, Marleen Rodenburg, Astrid Laegreid, Marius Doornenbal, Julen Oyarzabal, Analia Lourenço, and Alfonso Valencia. 2017. Overview of the BioCreative VI chemical-protein interaction track. In *Proceedings of BioCreative*, pages 141–146.

Guillaume Lample, Miguel Ballesteros, Sandeep Subramanian, Kazuya Kawakami, and Chris Dyer. 2016. Neural architectures for named entity recognition. In *Proceedings of NAACL-HLT*, pages 260–270.

Robert Leaman, Ritu Khare, and Zhiyong Lu. 2015. Challenges in clinical natural language processing for automated disorder normalization. *Journal of biomedical informatics*, 57:28–37.

Jinhyuk Lee, Wonjin Yoon, Sungdong Kim, Donghyeon Kim, Sunkyu Kim, Chan Ho So, and Jaewoo Kang. 2019. BioBERT: a pre-trained biomedical language representation model for biomedical text mining. *arXiv:1901.08746*.

Jiao Li, Yueping Sun, Robin J Johnson, Daniela Sciaky, Chih-Hsuan Wei, Robert Leaman, Allan Peter Davis, Carolyn J Mattingly, Thomas C Wiegers, and Zhiyong Lu. 2016. BioCreative V CDR task corpus: a resource for chemical disease relation extraction. *Database: the journal of biological databases and curation*, 2016.

Oren Melamud, David McClosky, Siddharth Patwardhan, and Mohit Bansal. 2016. The role of context types and dimensionality in learning word embeddings. In *Proceedings of NAACL-HLT*, pages 1030–1040.

Yifan Peng, Anthony Rios, Ramakanth Kavuluru, and Zhiyong Lu. 2018. Extracting chemical-protein relations with ensembles of SVM and deep learning models. *Database: the journal of biological databases and curation*, 2018.

Jeffrey Pennington, Richard Socher, and Christopher Manning. 2014. Glove: Global vectors for word representation. In *Proceedings of EMNLP*, pages 1532–1543.

Matthew E. Peters, Waleed Ammar, Chandra Bhagavatula, and Russell Power. 2017. Semi-supervised sequence tagging with bidirectional language models. In *Proceedings of ACL*, pages 1756–1765.

Alec Radford, Karthik Narasimhan, Tim Salimans, and Ilya Sutskever. 2018. Improving language understanding by generative pre-training.

Bryan Rink, Sanda Harabagiu, and Kirk Roberts. 2011. Automatic extraction of relations between medical concepts in clinical texts. *Journal of the American Medical Informatics Association*, 18:594–600.

Alexey Romanov and Chaitanya Shivade. 2018. Lessons from natural language inference in the clinical domain. In *Proceedings of EMNLP*, pages 1586–1596.

Yuqi Si, Jingqi Wang, Hua Xu, and Kirk Roberts. 2019. Enhancing clinical concept extraction with contextual embedding. *arXiv preprint arXiv:1902.08691*.

Neil R Smalheiser, Aaron M Cohen, and Gary Bonifield. 2019. Unsupervised low-dimensional vector representations for words, phrases and text that are transparent, scalable, and produce similarity metrics that are not redundant with neural embeddings. *Journal of biomedical informatics*, 90:103096.

Gizem Soğancıoğlu, Hakime Öztürk, and Arzucan Özgür. 2017. BIOSSES: a semantic sentence similarity estimation system for the biomedical domain. *Bioinformatics (Oxford, England)*, 33:i49–i58.

Hanna Suominen, Sanna Salanterä, Sumithra Velupillai, Wendy W Chapman, Guergana Savova, Noemie Elhadad, Sameer Pradhan, Brett R South, Danielle L Mowery, Gareth JF Jones, et al. 2013. Overview of the ShARe/CLEF eHealth evaluation lab 2013. In *International Conference of the Cross-Language Evaluation Forum for European Languages*, pages 212–231. Springer.

Özlem Uzuner, Brett R South, Shuying Shen, and Scott L DuVall. 2011. 2010 i2b2/va challenge on concepts, assertions, and relations in clinical text. *Journal of the American Medical Informatics Association (JAMIA)*, 18:552–556.

Ashish Vaswani, Noam Shazeer, Niki Parmar, Jakob Uszkoreit, Llion Jones, Aidan N Gomez, Łukasz Kaiser, and Illia Polosukhin. 2017. Attention is all you need. In *Advances in Neural Information Processing Systems*, pages 5998–6008.

Alex Wang, Amanpreet Singh, Julian Michael, Felix Hill, Omer Levy, and Samuel R. Bowman. 2018a. GLUE: A multi-task benchmark and analysis platform for natural language understanding. *arXiv preprint arXiv:1804.07461*.

Yanshan Wang, Naveed Afzal, Sunyang Fu, Liwei Wang, Feichen Shen, Majid Rastegar-Mojarad, and Hongfang Liu. 2018b. MedSTS: a resource for clinical semantic textual similarity. *Language Resources and Evaluation*, pages 1–16.

Yanshan Wang, Sijia Liu, Naveed Afzal, Majid Rastegar-Mojarad, Liwei Wang, Feichen Shen, Paul

Kingsbury, and Hongfang Liu. 2018c. A comparison of word embeddings for the biomedical natural language processing. *Journal of biomedical informatics*, 87:12–20.

Chih-Hsuan Wei, Yifan Peng, Robert Leaman, Allan Peter Davis, Carolyn J Mattingly, Jiao Li, Thomas C Wiegers, and Zhiyong Lu. 2016. Assessing the state of the art in biomedical relation extraction: overview of the BioCreative V chemical-disease relation (CDR) task. *Database: the journal of biological databases and curation*, 2016.

Shankai Yan and Ka-Chun Wong. 2017. Elucidating high-dimensional cancer hallmark annotation via enriched ontology. *Journal of biomedical informatics*, 73:84–94.

Wonjin Yoon, Chan Ho So, Jinhyuk Lee, and Jaewoo Kang. 2018. CollaboNet: collaboration of deep neural networks for biomedical named entity recognition. *arXiv preprint arXiv:1809.07950*.

Min-Ling Zhang and Zhi-Hua Zhou. 2014. A review on multi-label learning algorithms. *IEEE Transactions on Knowledge and Data Engineering*, 26(8):1819–1837.

Yijia Zhang, Qingyu Chen, Zhihao Yang, Hongfei Lin, and Zhiyong Lu. 2019. BioWordVec, improving biomedical word embeddings with subword information and mesh. *Scientific data*, 6:52.

Yijia Zhang, Wei Zheng, Hongfei Lin, Jian Wang, Zhihao Yang, and Michel Dumontier. 2018. Drug-drug interaction extraction via hierarchical RNNs on sequence and shortest dependency paths. *Bioinformatics (Oxford, England)*, 34:828–835.

A Appendices

TP: true positive, FP: false positive, FN: false negative, P: precision, R: recall, F1: F1-score

A.1 Named Entity Recognition

BC5CDR-disease	TP	FP	FN	P	R	F1
(Yoon et al., 2018)	-	-	-	85.6	82.6	84.1
ELMo	3740	749	684	83.3	84.5	83.9
BioBERT	3807	637	617	85.7	86.1	85.9
Our BERT						
Base (P)	3806	635	564	85.9	87.3	86.6
Base (P+M)	3788	655	636	85.3	85.6	85.4
Large (P)	3729	847	695	81.5	84.3	82.9
Large (P+M)	3765	799	659	82.5	85.1	83.8

BC5CDR-chemical	TP	FP	FN	P	R	F1
(Yoon et al., 2018)	-	-	-	94.3	92.4	93.3
ELMo	4864	386	521	92.6	90.3	91.5
BioBERT	5029	404	356	92.6	93.4	93.0
Our BERT						
Base (P)	5027	336	358	93.7	93.4	93.5
Base (P+M)	4914	341	471	93.5	91.3	92.4
Large (P)	4941	454	444	91.6	91.8	91.7
Large (P+M)	4905	484	480	91.0	91.1	91.1

ShARe/CLEFE	TP	FP	FN	P	R	F1
(Leaman et al., 2015)	-	-	-	79.7	71.3	75.3
ELMo	3928	1117	1423	77.9	73.4	75.6
BioBERT	3898	1024	1453	79.2	72.8	75.9
Our BERT						
Base (P)	4032	1010	1319	80.0	75.4	77.6
Base (P+M)	4126	948	1225	81.3	77.1	79.2
Large (P)	3890	1441	1461	73.0	72.7	72.8
Large (P+M)	3980	1456	1371	73.2	74.4	73.8

A.2 Relation extraction

DDI	TP	FP	FN	P	R	F1
(Zhang et al., 2018)	-	-	-	74.1	71.8	72.9
ELMo	-	-	-	79.0	78.9	78.9
BioBERT	786	229	193	77.4	80.3	78.8
Our BERT						
Base (P)	737	172	242	81.1	75.3	78.1
Base (P+M)	775	198	204	79.7	79.2	79.4
Large (P)	788	206	191	79.3	80.5	79.9
Large (P+M)	748	234	231	76.2	76.4	76.3

Chem-Prot	TP	FP	FN	P	R	F1
(Peng et al., 2018)	1983	746	1475	72.7	57.4	64.1
ELMo	-	-	-	66.7	66.6	66.6
BioBERT	2359	803	1099	74.6	68.2	71.3
Our BERT						
Base (P)	2443	834	1015	74.5	70.6	72.5
Base (P+M)	2354	996	1104	70.3	68.1	69.2
Large (P)	2610	948	848	73.4	75.5	74.4
Large (P+M)	2355	1423	1103	62.3	68.1	65.1

i2b2	TP	FP	FN	P	R	F1
(Rink et al., 2011)	-	-	-	72.0	75.3	73.7
ELMo	-	-	-	71.2	71.1	71.1
BioBERT	4391	1474	1902	74.9	69.8	72.2
Our BERT						
Base (P)	4592	1459	1701	75.9	73.0	74.4
Base (P+M)	4683	1291	1610	78.4	74.4	76.4
Large (P)	4684	1805	1609	72.2	74.4	73.3
Large (P+M)	4700	1719	1593	73.2	74.7	73.9

A.3 Document classification

HoC	P	R	F1
(Du et al., 2019)	81.3	81.7	81.5
ELMo	78.2	81.9	80.0
BioBERT	83.4	82.4	82.9
Our BERT			
Base (P)	86.2	84.4	85.3
Base (P+M)	84.0	82.3	83.1
Large (P)	91.0	83.9	87.3
Large (P+M)	88.8	82.1	85.3

Combining Structured and Free-text Electronic Medical Record Data for Real-time Clinical Decision Support

Emilia Apostolova[1], Tony Wang[2], Ioannis Koutroulis[3], Tim Tschampel[4], Tom Velez[4]

[1] Language.ai, Chicago, IL *emilia@language.ai*

[2] Imedacs, Ann Arbor, MI *xwang@imedacs.com*

[3] Children's National Health System, Washington, DC *ikoutrouli@childrensnational.org*

[4] Computer Technology Associates, Ridgecrest, CA *tim.tschampel@cta.com, tom.velez@cta.com*

Abstract

The goal of this work is to utilize Electronic Medical Record (EMR) data for real-time Clinical Decision Support (CDS). We present a deep learning approach to combining in real time available diagnosis codes (ICD codes) and free-text notes: *Patient Context Vectors*. Patient Context Vectors are created by averaging ICD code embeddings, and by predicting the same from free-text notes via a Convolutional Neural Network. The Patient Context Vectors were then simply appended to available structured data (vital signs and lab results) to build prediction models for a specific condition. Experiments on predicting ARDS, a rare and complex condition, demonstrate the utility of Patient Context Vectors as a means of summarizing the patient history and overall condition, and improve significantly the prediction model results.

1 Introduction

A key goal in critical care medicine is the early identification and timely treatment of rapidly progressive, life-threatening conditions, such as Sepsis, Septic Shock, and Acute Respiratory Distress Syndrome (ARDS). Such life-threatening conditions, are both rare, and at the same time, complex and heterogeneous, involving the interaction of multiple risk factors, comorbidities, and current symptoms. Hospital alert systems typically rely on screening of structured data such as vital signs and lab results, and, in the case of such rare conditions, are often associated with "alert fatigue" and require manually entered clinical judgement.

The information needed for a reliable risk evaluation of such rare and complex conditions is typically dispersed across the patient EMR, and available at different times throughout the patient stay. The patient demographics, past medical and visit history, chronic conditions, risk factors, current signs and symptoms can be found in the form of clinical notes (e.g. nursing notes, radiology reports, etc.), diagnosis and procedure codes, vital signs, lab orders and results. The challenge of real-time CDS systems is the variability and the availability of real-time EMR data, resulting from different charting behaviors, health care delivery models, hospital settings, etc.

The goal of this work is to utilize all available EMR patient information for real-time predictive modelling. While our experiments are focused on identifying ARDS cases, the described method is applicable to a variety of use cases needing information dispersed across the EMR patient record. The primary contribution of this work is the use of low-dimensional representation of the patient's history, current symptoms and conditions, which we refer to as *Patient Context Vector*. At prediction time, Patient Context Vectors are generated from the combination of available up-to-date ICD codes (if any) and available nursing notes. Patient Context Vectors (vectors of real numbers) are then simply added to the list of existing structured data variables (vital signs and lab results) and used to identify patients at risk of developing life-threatening conditions that require rapid intervention.

2 Method

In this work, we combine ICD codes, clinical notes, vital signs, lab results, and demographic information to build a real-time ARDS prediction model. Low-dimensional representation of ICD codes (ICD embeddings) is generated from a large corpus of patient ICD records. Patient visit EMR data is used to look up recorded up-to-date ICD codes, clinical notes, vital signs, and lab results. The visit ICD codes are converted to embeddings and averaged to produce Patient Context Vectors.

Pertinent patient information might not be necessarily "ICD-coded" during prediction time, but

Proceedings of the BioNLP 2019 workshop, pages 66–70
Florence, Italy, August 1, 2019. ©2019 Association for Computational Linguistics

can be available in the form of nursing notes. A deep learning model was trained to predict the patient's Patient Context Vector from nursing notes. The Patient Context Vectors obtained from available in the system ICD codes, and from free-text notes are then used in conjunction with vital signs, and lab results to predict the patient's outcome. Details for each step of the approach are provided in subsequent sections.

2.1 Dataset

We utilized the freely available database comprising deidentified health-related data associated with over 40,000 patients who stayed in critical care units of the Beth Israel Deaconess Medical Center between 2001 and 2012: the MIMIC3 Intensive Care Unit (ICU) database (Johnson et al., 2016). The dataset contains over 2 million free-text clinical notes and over 650,000 diagnosis codes for over 58,000 visits. Included ICUs are medical, surgical, trauma-surgical, coronary, and cardiac surgery recovery units. EMR data includes vital signs, laboratory results, diagnosis codes, free text nursing notes, radiology reports, medications, discharge summaries, treatments, etc.

2.2 ICD Embeddings and Patient Context Vectors

Clinicians viewing properly coded patient diagnosis codes (ICD9 and ICD10 codes[1]) are typically capable of deducing the overall condition, history, and risk factors associated with a patient. Intuitively, the totality of patient's diagnosis codes represent a meaningful medical summary of the patient. Diagnosis codes are used to describe both current diagnoses (e.g. *Community-acquired Pneumonia*), but also a variety of additional facts. For example, ICD codes can describe patient's history and chronic conditions (e.g. *Chronic kidney disease; Personal history of traumatic fracture;* etc.); information regarding past and current treatments and procedures (e.g. *Infection due to other bariatric procedure*). In some cases, ICD codes contain information such as the patient age group (e.g. *Sepsis of newborn; Elderly multigravida*); expected outcome (*Encounter for palliative care*); patient's social history (e.g. *Adult emotional/psychological abuse*); the reason for the visit, (e.g. *Railway accidents; Motor Vehicle accidents*, etc).

While there are a large number of ICD codes (around 15,000 ICD9 codes and around 68,000 ICD10 codes), they tend to be interdependent, and to co-occur. For example, *Pneumonia* ICD codes are often accompanied with ICD codes describing *Cough, Fever, Pleural effusion*, etc. Inspired by word embeddings (Mikolov et al., 2013), it has been suggested that this medical code co-occurence can be exploited to generate low-dimensional representations of ICD codes: ICD Embeddings (Choi et al., 2016b,a; Kartchner et al., 2017).

All available MIMIC3 patient data was used to generate the ICD embeddings following the approach of (Choi et al., 2016b). In our approach, we attempted to generate a low-dimensional representation of the patient history, symptoms, risk factors, diagnosis, etc, by averaging the patient ICD code embeddings (creating Patient Context Vectors). The optimum size of the vectors was determined to be 50.

2.3 Predicting Patient Context Vectors from Clinical Texts

While averaged ICD embeddings appear to be a useful summary of the overall patient history, condition, symptoms, and risk factors, ICD code data is not necessarily available for real-time CDS systems. Some ICD codes associated with patients' history and symptoms might be entered early on in the EMR system. However, diagnosis ICD codes are typically obtained after tests and lab results and might not be available during prediction time. Similarly, not all relevant patient history and symptoms are necessarily ICD-coded.

At the same time, nursing notes typically contain all currently available information, even if not present in the form of ICD codes. Nursing notes include information such as past medical history, reason for visit, current symptoms, summary of test outcomes, etc.

In order to capture information present in free-text notes, we also built a word-level CNN model that predicts the patient Patient Context Vector from the note text. The model was trained on available nursing and discharge notes and achieved a mean squared error of 0.179 on the validation set. The network was trained on 1,081,176 free-text notes, with pre-trained word-embeddings of size 100. The texts were truncated/padded to the 90th percentile length (785 tokens). The network consists of a Convolutional, Max Pooling layers, followed by 2 hidden layers of size 500. The last layer uses linear activation with loss func-

---

[1] The International Classification of Diseases, ©The World Health Organization.

tion of mean squared error to predict the Patient Context Vector[2].

2.4 Patient Context Vectors in Prediction Models

In order to test the utility of the Patient Context Vectors for predicting patient outcomes, we focused on building a real-time ARDS prediction model. ARDS is a rare and life-threatening condition that require an early intervention (Fan et al., 2017).

ARDS patients were limited to adult patients only (age 18 or older). The patients inclusion criteria consist of the presence of acute respiratory failure and continuous mechanical ventilation, excluding patients with acute exacerbation of asthma or chronic obstructive pulmonary disease (Bime et al., 2016) [3]. This resulted in 4,624 ARDS admissions from a total of 48,399 admissions.

An ARDS prediction model was built utilizing a combination of vital signs, lab results, ICD codes and free-text notes. Features considered in the baseline predictive model building include: 1) vital signs: heart rate, respiratory rate, body temperature, systolic blood pressure, diastolic blood pressure, mean arterial pressure, oxygen saturation, tidal volume, BMI; 2) laboratory tests: white blood cell count, bands, hemoglobin, hematocrit, lactate, creatinine, bicarbonate, pH, PT, INR, BUN, blood gas measurements (partial pressure of arterial oxygen, fraction of inspired oxygen, and partial pressure of arterial carbon dioxide); 4) motor, verbal, and eye sub-score of Glasgow Coma Scale ; and 5) demographics: gender and age.

In addition to the baseline features (available in structured format in MIMIC), we also included as features the patient's Patient Context Vectors computed from ICD codes and from notes. In real-time CDS systems, it is likely that not all ICD or nursing notes will be available at prediction times. To test this most realistic scenario, we also built a Patient Context Vector by averaging the first half of the patient's ICD codes, and the first half of the patient's nursing notes CNN model predictions.

A Gradient Boosting Machine (GBM) model (Friedman, 2001) and a Distributed Random Forrest Model (DRF) (Geurts et al., 2006) were used to predict ARDS patients from the total population of adult patients. In all cases default model parameters were used (h2o). All results were produced via 10-fold cross evaluation. Table 1 shows the result from the experiments.

---

[3]Inclusion ICD9 Codes: 51881, 51882, 51884, 51851, 51852, 51853, 5184, 5187, 78552, 99592, 9670, 9671, 9672; Exclusion ICD9 Codes: 49391, 49392, 49322, 4280

GBM				
Features	**AUC**	**P**	**R**	**F1**
Baseline	90.42	41.76	67.80	51.68
Baseline + ICD Patient Context Vector	93.30	53.02	68.44	59.75
Baseline + Notes Patient Context Vector	91.88	48.25	64.25	55.11
Baseline + first half of notes/ICD	**93.59**	56.35	66.52	**61.01**

DRF				
Features	**AUC**	**P**	**R**	**F1**
Baseline	89.14	38.58	66.43	48.81
Baseline + ICD Patient Context Vector	92.08	51.87	63.75	57.20
Baseline + Notes Patient Context Vector	91.18	47.89	62.11	54.08
Baseline + first half of notes/ICD	**92.61**	57.02	61.08	**58.98**

Table 1: 10-fold cross-validation GBM and DRF results of predicting ARDS patients. P=Precision, R=Recall, F1= F1-score for the positive (ARDS) class. The Baseline set of features consists of vital signs, lab results, Glasgow Coma Scale score, gender and age, in the form of structured data. "Baseline + ICD Patient Context Vector" includes all baseline features, plus the Patient Context Vector (of size 50). "Baseline + Notes" includes all baseline features, plus Patient Context Vectors predicted from all visit nursing notes. "Baseline + first half of notes/ICD" includes the average of the first half of entered visit ICD codes embeddings, and Patient Context Vectors predicted from the first half of the visit nursing notes.

Introducing information from both ICD codes and nursing notes data significantly increased the overall performance. Most importantly, the combination of the use of half of the visit notes (used to predict Patient Context Vectors) and the first half of the patient ICD codes produced the best results in both models (GBM and DRF), and proves the utility of the method for combining structured and free-text data for prediction models.

The benefit of averaged ICD-code embeddings, and using notes to predict the same embedding vectors is also illustrated by the model variable importances shown in Figures 1 - 4. As shown, the predictive value of certain embedding dimensions is on a par with important vital signs, such as Tidal Volume, Glasgow Coma Scale, and Mean Respiratory Rate. Intuitively, clinicians' experience utilizes all information present in nursing notes (also coded as ICD codes) to evaluate a patient's condition. Our approach demonstrates that it is possible to summarize that knowledge by combining nursing and ICD codes in the form of predicted and averaged ICD embeddings.

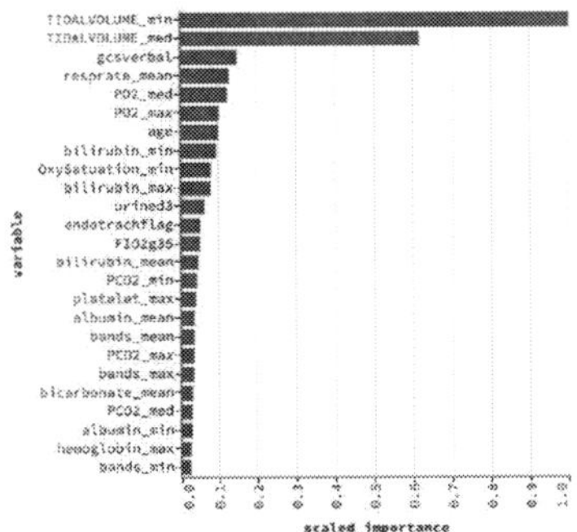

Figure 1: GBM scaled variable importance of Baseline model features.

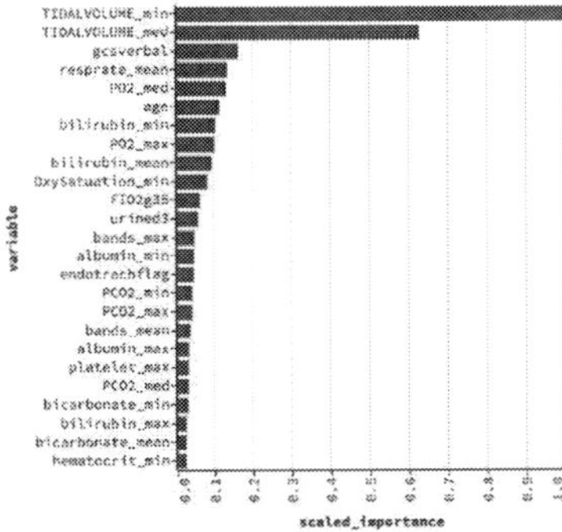

Figure 2: DRF scaled variable importance of Baseline model features.

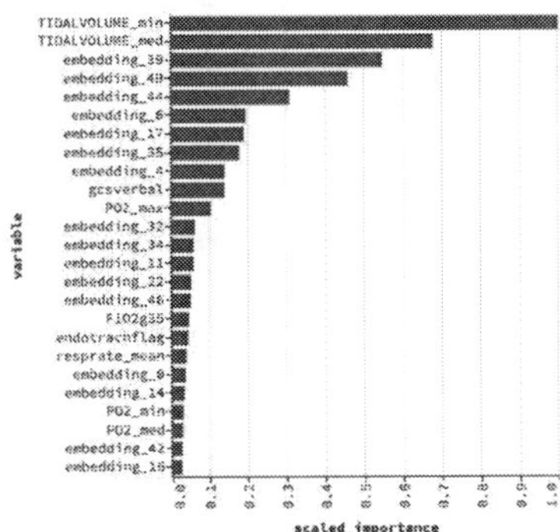

Figure 3: GBM scaled variable importance of Baseline model features plus Patient Context Vectors from first half of ICD codes/notes.

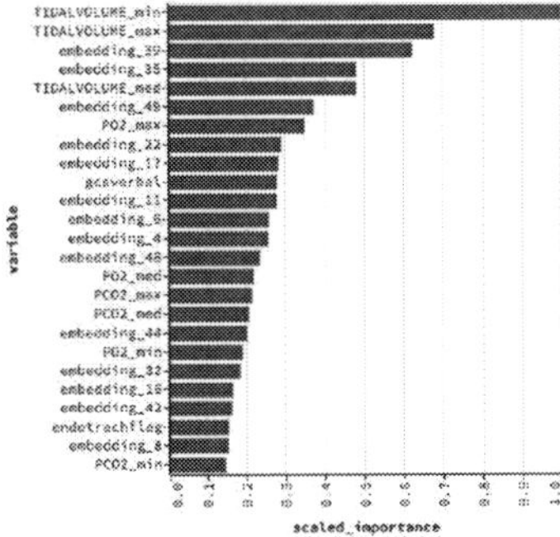

Figure 4: DRF scaled variable importance of Baseline model features plus Patient Context Vectors from first half of ICD codes/notes.

3 Related Work

A large volume of literature on combining structured and free-text EMR data pre-processes the free-text data by applying some information extraction (IE) technique (most frequently, Medical Concept detection). For example, DeLisle et al.(2010) and Zheng et al. (2014) apply free-text search on the notes to find a set of hand-crafted non-negated symptoms, later used as variables in their ML models. Ford et al. (2016) present a review of various approaches to IE from free-text notes for the purpose of detecting cases of a clinical condition, often in conjunction with structured data. The majority of approaches extract UMLS[4] or SNOMED-CT[5] concepts from free-text with their negation status with various off-the-shelf tools (Gundlapalli et al., 2008; Carroll et al., 2011; Karnik et al., 2012; Ananthakrishnan et al., 2013; Zheng et al., 2014).

More recently, deep learning has been used to combine free-text and structured EMR data. Relevant ICD embeddings work was mentioned in Section 2.2. Shickel et al. (2018) present a survey of various deep learning techniques. Most notably, Miotto et al. (2016) convert notes to concepts, which are then used in conjunction with structured data to build a *Deep Patient* representation in an unsupervised manner via denoising autoencoders.

4 Conclusion

Intuitively, the information available in notes and ICD codes, enhances the knowledge of the overall patient condition, which is indicative of the patient outcome. Results show that Patient Context Vectors can be easily combined with structured data in the form of vital signs an lab results and improve significantly the prediction model results. Results also indicate that Patient Context Vectors are suitable for real-time CDS as they perform equally well when only the first half of available ICD codes and notes is used.

Acknowledgements

Research reported in this publication was supported by a NIH SBIR award to CTA by NIH National Heart, Lung, and Blood Institute, of the National Institutes of Health under award number 1R43HL135909-01A1.

---

[4]Unified Medical Language System, ©The U.S. National Library of Medicine.

[5]Systematized Nomenclature of Medicine - Clinical Terms, ©IHTSDO.

References

h2o.ai. `https://www.h2o.ai/`. Accessed: 2019-01-30.

Ashwin N Ananthakrishnan, Tianxi Cai, Guergana Savova, Su-Chun Cheng, Pei Chen, Raul Guzman Perez, Vivian S Gainer, Shawn N Murphy, Peter Szolovits, Zongqi Xia, et al. 2013. Improving case definition of crohn's disease and ulcerative colitis in electronic medical records using natural language processing: a novel informatics approach. *Inflammatory bowel diseases*, 19(7):1411–1420.

Christian Bime, Chithra Poongkunran, Mark Borgstrom, Bhupinder Natt, Hem Desai, Sairam Parthasarathy, and Joe GN Garcia. 2016. Racial differences in mortality from severe acute respiratory failure in the united states, 2008–2012. *Annals of the American Thoracic Society*, 13(12):2184–2189.

Robert J Carroll, Anne E Eyler, and Joshua C Denny. 2011. Naïve electronic health record phenotype identification for rheumatoid arthritis. In *AMIA annual symposium proceedings*, volume 2011, page 189. American Medical Informatics Association.

Edward Choi, Mohammad Taha Bahadori, Elizabeth Searles, Catherine Coffey, Michael Thompson, James Bost, Javier Tejedor-Sojo, and Jimeng Sun. 2016a. Multi-layer representation learning for medical concepts. In *Proceedings of the 22nd ACM SIGKDD International Conference on Knowledge Discovery and Data Mining*, pages 1495–1504. ACM.

Youngduck Choi, Chill Yi-I Chiu, and David Sontag. 2016b. Learning low-dimensional representations of medical concepts. *AMIA Summits on Translational Science Proceedings*, 2016:41.

Sylvain DeLisle, Brett South, Jill A Anthony, Ericka Kalp, Adi Gundlapallli, Frank C Curriero, Greg E Glass, Matthew Samore, and Trish M Perl. 2010. Combining free text and structured electronic medical record entries to detect acute respiratory infections. *PloS one*, 5(10):e13377.

Eddy Fan, Lorenzo Del Sorbo, Ewan C Goligher, Carol L Hodgson, Laveena Munshi, Allan J Walkey, Neill KJ Adhikari, Marcelo BP Amato, Richard Branson, Roy G Brower, et al. 2017. An official american thoracic society/european society of intensive care medicine/society of critical care medicine clinical practice guideline: mechanical ventilation in adult patients with acute respiratory distress syndrome. *American journal of respiratory and critical care medicine*, 195(9):1253–1263.

Elizabeth Ford, John A Carroll, Helen E Smith, Donia Scott, and Jackie A Cassell. 2016. Extracting information from the text of electronic medical records to improve case detection: a systematic review. *Journal of the American Medical Informatics Association*, 23(5):1007–1015.

Jerome H Friedman. 2001. Greedy function approximation: a gradient boosting machine. *Annals of statistics*, pages 1189–1232.

Pierre Geurts, Damien Ernst, and Louis Wehenkel. 2006. Extremely randomized trees. *Machine learning*, 63(1):3–42.

Adi V Gundlapalli, Brett R South, Shobha Phansalkar, Anita Y Kinney, Shuying Shen, Sylvain Delisle, Trish Perl, and Matthew H Samore. 2008. Application of natural language processing to va electronic health records to identify phenotypic characteristics for clinical and research purposes. *Summit on translational bioinformatics*, 2008:36.

Alistair EW Johnson, Tom J Pollard, Lu Shen, H Lehman Li-wei, Mengling Feng, Mohammad Ghassemi, Benjamin Moody, Peter Szolovits, Leo Anthony Celi, and Roger G Mark. 2016. Mimic-iii, a freely accessible critical care database. *Scientific data*, 3:160035.

Shreyas Karnik, Sin Lam Tan, Bess Berg, Ingrid Glurich, Jinfeng Zhang, Humberto J Vidaillet, C David Page, and Rajesh Chowdhary. 2012. Predicting atrial fibrillation and flutter using electronic health records. In *Engineering in Medicine and Biology Society (EMBC), 2012 Annual International Conference of the IEEE*, pages 5562–5565. IEEE.

David Kartchner, Tanner Christensen, Jeffrey Humpherys, and Sean Wade. 2017. Code2vec: Embedding and clustering medical diagnosis data. In *Healthcare Informatics (ICHI), 2017 IEEE International Conference on*, pages 386–390. IEEE.

Tomas Mikolov, Kai Chen, Greg Corrado, and Jeffrey Dean. 2013. Efficient estimation of word representations in vector space. *arXiv preprint arXiv:1301.3781*.

Riccardo Miotto, Li Li, Brian A Kidd, and Joel T Dudley. 2016. Deep patient: an unsupervised representation to predict the future of patients from the electronic health records. *Scientific reports*, 6:26094.

Benjamin Shickel, Patrick James Tighe, Azra Bihorac, and Parisa Rashidi. 2018. Deep ehr: a survey of recent advances in deep learning techniques for electronic health record (ehr) analysis. *IEEE journal of biomedical and health informatics*, 22(5):1589–1604.

Hongzhang Zheng, Holly Gaff, Gary Smith, and Sylvain DeLisle. 2014. Epidemic surveillance using an electronic medical record: an empiric approach to performance improvement. *PloS one*, 9(7):e100845.

MoNERo: a Biomedical Gold Standard Corpus for the Romanian Language

Maria Mitrofan
RACAI, Bucharest, Romania
maria@racai.ro

Verginica Barbu Mititelu
RACAI, Bucharest, Romania
vergi@racai.ro

Grigorina Mitrofan
NIDMD "N.C. Paulescu"
Bucharest, Romania
grigorinam@gmail.com

Abstract

In an era when large amounts of data are generated daily in various fields, the biomedical field among others, linguistic resources can be exploited for various tasks of Natural Language Processing. Moreover, increasing number of biomedical documents are available in languages other than English. To be able to extract information from natural language free text resources, methods and tools are needed for a variety of languages. This paper presents the creation of the MoNERo corpus, a gold standard biomedical corpus for Romanian, annotated with both part of speech tags and named entities. MoNERo comprises 154,825 morphologically annotated tokens and 23,188 entity annotations belonging to four entity semantic groups corresponding to UMLS Semantic Groups.

1 Introduction

Natural Language Processing (NLP) is a research area that provides methods to convert (human-understandable) unstructured textual information into (machine-readable) structured data and uses it for different objectives. NLP techniques can be used to process and exploit the large amount of biomedical information which is continuously generated. Examples of such repositories are MEDLINE[1], which contains more than 25 million documents belonging to the biomedical domain, or PubMed Central[2], which is an archive of biomedical journal literature and contains more than 5 million full-text articles. These resources can be exploited and used together with different NLP systems previously adapted to the biomedical field to improve the quality of the health care

process, to further develop research in the field and benefit both physicians and patients. Information Extraction (IE) tools can be used to extract relevant information from biomedical textual resources (Goeuriot et al., 2017; Li et al., 2017). Reaching suitable results for this NLP subtask is not trivial and there is still room for improvement of results. Advances of these IE tools depend on the existence of annotated resources specific to the field of study (Wilbur et al., 2006; Thompson et al., 2009; Kilicoglu, 2017), annotated corpora being relevant in both phases: development of the models that will determine the behaviour of the system and system performance evaluation. Even though the availability of these resources has increased lately, the main part of the efforts have been directed to the development of annotated corpora for English in different subdomains. However, MoNERo is a resource created for the Romanian language that helps the development of named entity recognition and classification task especially for this language. Romanian benefits from the existence of other corpora created in our institute: the representative corpus of contemporary language (CoRoLa) (Barbu Mititelu et al., 2018), a balanced corpus (ROMBAC) (Ion et al., 2012), the corpus annotated with verbal multiword expressions (Barbu Mititelu et al., 2019). Just like all of these, MoNERo is annotated at the morphological level. However, it stands out given its annotation with four types of Named Entities (NEs) for the medical domain, which are relevant to the identification of: anatomy parts, diseases and disorders, chemicals and drugs, and medical procedures.

This paper has four main objectives: (i) to present the construction of a biomedical gold standard corpus annotated both with part-of-speech tags and named entities; (ii) to present general statistics over the corpus; (iii) to release the final

---

[1]https://www.nlm.nih.gov/bsd/medline.html

[2]https://www.nlm.nih.gov/bsd/medline.html

71

Proceedings of the BioNLP 2019 workshop, pages 71–79
Florence, Italy, August 1, 2019. ©2019 Association for Computational Linguistics

version of the corpus to the scientific community, (iv) to show the contribution in the development of NLP tools for Romanian language. All the results are discussed in parallel for the two types of annotations.

2 Related Work

This section reviews relevant corpora annotated with NEs specific to the biomedical domain.

1. For English we mention:

   - CLEF corpus (Roberts et al., 2009) – it contains 150 documents of clinical narratives, histopathology reports and imaging reports. It was subtracted from a corpus of 565,000 documents and manually annotated with six types of NEs (condition, intervention, investigation, result, drug or device, locus);
   - i2b2 corpus (Uzuner et al., 2010) – it contains 1243 discharge summaries automatically pre-annotated, out of which a subset of 251 was manually revised. This corpus contains seven types of NEs (medications, dosages, modes, frequencies, durations, reasons of administration, list/narrative);
   - NCBI corpus (Doğan et al., 2014) – a gold-standard corpus for disease mentions and concepts that contains 793 abstracts extracted from PubMed;
   - CHEMDNER corpus (Krallinger et al., 2015) – a corpus of 10,000 abstracts collected from PubMed annotated with two types of NEs: chemicals and drugs.

2. For French there is the Quaero corpus (Névéol et al., 2014) which contains 103,056 words collected from three types of documents: texts with information on drugs extracted from European Medicines Agency (EMEA), titles from research articles comprised in MEDLINE and patents. This corpus was annotated with ten types of NEs defined using UMLS: anatomy, chemical and drugs, devices, disorders, geographic areas, living beings, objects, phenomena, physiology, procedures.

3. For Spanish the following corpora exist:

   - IxaMedGS corpus (Oronoz et al., 2015) – it is composed of 142,154 discharge records out of which 75 were annotated with two types of NEs: diseases and drugs;
   - DrugSemantics corpus (Moreno et al., 2017) – it has 226,729 tokens annotated with ten types of NEs: chemical composition, disease, drug, excipient, food, medicament, pharmaceutical form, route, therapeutic action, and unit of measurement.

All these corpora are available and have had a significant role in information extraction research, especially in named entity recognition (NER) research and were developed for well-established purposes, having in mind the possibility of reusability.

3 Corpus Development Description

3.1 Selection of Corpus Documents

The gold standard morphologically and named entity annotated Romanian medical corpus (MoNERo) was extracted from the BioRo corpus (Mitrofan and Tufiş, 2018), a Romanian biomedical corpus. MoNERo contains texts extracted from three types of documents: scientific medical literature books, scientific medical journal articles and medical blog posts, but predominant are those coming from medical literature. The medical books were chosen as the main source because they contain descriptive materials, full of domain-specific terms. In addition, the texts are of good quality and the use of medical terms is correct. The medical journal[3] from which a part of the texts were extracted is a scientific journal that addresses the specialists, so the language used is specific to the medical domain. In the case of blog posts those collected were texts of popularization and awareness of various medical problems.

The texts were selected so that they belong to three medical subdomains: cardiology, diabetes and endocrinology (see table 3). The main motivation behind choosing these three medical domains is that our textual resources available were centered around the pathology of Diabetes. Since Diabetes is an endocrine disorder it is naturally included in the Endocrinology category. In the same

---

[3] https://rmj.com.ro/

time because of a very close relation between diabetes and cardiovascular diseases we also obtain a significant category from Cardiology field. Other categories such as neurology, nephrology would have had a very low contribution and we chose not to take them separately but in Diabetes field, because the terms were related to diabetes complications.

The selection was made based on the metadata scheme associated with each document present in the BioRo corpus. The order of the sentences was preserved.

All these texts are Intellectual Property Right (IPR) cleared, thus enabling us to make it available to the community (see section 6).

3.2 MoNERo Annotation Scheme

The annotation scheme of MoNERo has two different levels: (i) a morphologic level at which all part of speech tags were revised by an experienced linguist; and (ii) a named entity level at which NEs were identified and classified in the corresponding semantic group.

3.2.1 Part of Speech Annotation Scheme

The process of the annotation of the corpus with part of speech tags had two phases: automatic annotation (all the texts comprised in this corpus were previously processed when included in BioRo, the source from which MoNERo was extracted) and manual verification of the tags allocated by the tool used (see below section 3.3.1). Here we present the manual verification phase which was done by an expert linguist. The annotation scheme used for morphologic annotation was based on the MSD tagset developed in the Multext-East project (Dimitrova et al., 1998), which contains 715 tags for Romanian. This tagset is very complex and precise, containing fourteen classes of words (noun, verb, adjective, adverb, pronoun, determiner, article, adposition, conjunction, numeral, interjection, abbreviation, residual and particle), each class having a set of attributes such as: type, gender, number, case, definiteness, clitic, verb form, tense, person, degree, etc. (Tufiş et al., 1997).

3.2.2 Named Entities Annotation Scheme

In the case of named entities identification the annotation scheme was based on UMLS[4] (Unified Medical Language System) semantic groups. This

resource contains concepts from different terminologies specific to the biomedical domain. Moreover, UMLS is organized as a hierarchical semantic network that comprises semantic types and semantic relations. All the semantic types are grouped in 15 semantic groups (McCray et al., 2001). For this work the annotation scheme contains four semantic groups chosen from the UMLS scheme: anatomy, chemicals and drugs, disorders and procedures. The attributes of each entity type are described below:

1. **Anatomy (ANAT)**: body location or region, body part, organ, or organ component, body substance, body system, cell, fully formed anatomical structure, tissue;

2. **Chemicals and Drugs (CHEM)**: amino acid, peptide, protein, antibiotic, biologically active substance, chemical, clinical drug, hormone, organic chemical, pharmacologic substance, receptor, steroid, vitamin;

3. **Disorders (DISO)**: acquired abnormality, anatomical abnormality, cell or molecular dysfunction, congenital abnormality, disease or syndrome, experimental model of disease, finding, injury or poisoning, sign or symptom;

4. **Procedures (PROC)**: diagnostic procedure, health care activity, laboratory procedure, molecular biology research technique, therapeutic or preventive procedure.

Examples for each type can be seen in Table 1.

Named Entity	Example
Anatomy	*pancreas* ("pancreas") *nerv optic* ("optic nerve")
Chemicals and Drugs	*paracetamol* ("paracetamol") *acid folic* ("folic acid")
Disorders	*diabet* ("diabetes") *fibrilaţie* ("fibrillation")
Procedures	*EKG* ("EKG") *CT* ("CT")

Table 1: Examples of named entities extracted from MoNERo.

[4]https://semanticnetwork.nlm.nih.gov/

The main reason for choosing these four types of entities was a trade off between the minimum number of entities (due to an increased complexity of the annotation process) and the maximum relevance for our corpus. However we had some challenges. For example, Physiology was a category that could be included, but due to the fact that the medical texts available were mainly related to pathology, the contribution would have been limited (less than 5%).

Having a tokenized corpus with each token on a separate line, we chose IOB2 (Insid-Outside-Beginning) (Sang and Veenstra, 1999) as the annotation format for named entities. Lately, this format has become popular within the scientific community, being also supported by the CoNLL challenges [5]. The B-tag is used for the first token of every NE, I-tag indicates the token that is inside a named entity and O-tag is used for surrounding tokens that do not belong to a NE (*În/O schimb/O, HDL-colesterolul/ B-CHEM, apolipoproteinele/B-CHEM A/I-CHEM şi/O B/I-CHEM sunt/O superiori/O ca/O indicatori/O de/O risc/B-DISO cardiovascular/I-DISO ./O*) ("On the other hand, HDL-cholesterol and lipoproteins A and B are superior as cardiovascular risk indicators."). For ease of reading, in all the examples below we chose not to mention the O-tag, but only the B- and I-tags.

3.3 Annotation Guidelines

3.3.1 Part of Speech

In the initial phase the corpus was automatically preprocessed (sentence split, tokenized, lemmatized) and annotated with POS tags using the TTL annotator (Ion, 2007; Mitrofan and Tufiş, 2018), which was trained on news corpora of about 200,000 tokens with POS labeling checked by trained linguists (Tufiş, 2000). The accuracy for this task was 98.23%. When TTL was trained in order to perform domain adaptation for biomedical domain the accuracy was 97.83% (Mitrofan and Ion, 2017). Therefore, in order to annotate this corpus with POS tags the baseline model was chosen. The second phase of the annotation process, which makes the focus of this paper, was to manually check all the automatically assigned labels. A trained and experienced linguist revised all the tokens included in MoNERo. For this task the guidelines were:

1. correct the token if needed;

2. correct the lemma if needed;

3. correct the POS tag if needed;

4. compounds written as separate words should be split.

3.3.2 Named Entities

The guidelines for named entity annotation were:

1. a complex entity will not be decomposed into simpler entities belonging to different semantic groups; only one semantic group will be associated to the longest entity (*cancer de ficat* ("liver cancer") will be annotated only as a disorder, not as a disorder (*cancer/B-DISO de/I-DISO ficat/I-DISO* ("cancer"/B-DISO) "of"/I-DISO "liver"/I-DISO) and an anatomical part (*ficat/B-ANAT* ("liver"/B-ANAT)); so, there is no embedded annotation;

2. in cases when one head noun is shared by two or more biomedical named entities (coordinations or disjunctions) the annotation will be done as follows: in case of coordinations *ateroscleroza aortei şi a vaselor periferice* ("atherosclerosis of the aorta and peripheral vessels"), should be annotated as *ateroscleroza/B-DISO aortei/I-DISO şi vaselor/I-DISO periferice/I-DISO* or in case of disjunctions *celule beta pancreatice sau hepatice* ("pancreatic beta or hepatic cells") should be annotated as *celule/B-ANAT beta/I-ANAT pancreatice/I-ANAT sau hepatice/I-ANAT*);

3. discontinuous entities will be annotated as contiguous terms and classified in the same semantic group: in the examples *Anevrismele/B-DISO pot fi fusiforme/I-DISO (aspect cilindric al vasului/B-ANAT sangvin/I-ANAT) sau sacciforme/I-DISO* ("Aneurysms/B-DISO may be fusiforms/I-DISO (cylindrical appearance of the blood/B-ANAT vessel/I-ANAT) or sacciforms/I-DISO") the NEs *Anevrismele fusiforme* and *anevrismele sacciforme* are discontinuous;

4. in case of cascaded constructions when one entity is incorporated in another entity (eg. parenthetical constructions) the annotation will be done as: *Anevrismele/B-DISO pot fi fusiforme/I-DISO (aspect cilindric al vasului/B-ANAT sangvin/I-ANAT) sau*

sacciforme/I-DISO ("Aneurysms/B-DISO may be fusiforms/I-DISO (cylindrical appearance of the blood/B-ANAT vessel/I-ANAT) or sacciforms/I-DISO"). Within the discontinuous NE *Anevrismele sacciforme* there is another NE, *vasului sangvin.*

3.4 Annotation Development

3.4.1 Part of Speech Tags

Even though the accuracy of the automatic annotation with POS tags was very high (subsection 3.3.1), given the high number of POS tags in the Romanian MSD tagset, there was a lot of manual work to be done by the linguist. This task involved manual validation of tokenization, lemmatization, and also correcting the errors of part of speech and errors of morphological categories (see 3.5.1) for each token.

3.4.2 Named Entities

For the named entities annotation task two annotators were employed: one physician and one experienced annotator, both having Romanian as native language. The physician was chosen as annotator due to her capacity of understanding the medical field. Prior to the annotation process there was a training period for both annotators. In this phase they debated issues such as whether or not to annotate overlapping terms, when and if complex terms should be decomposed, how conjunctions should be treated.

Even though the initial guidelines gave them instructions on what should and should not be annotated, they collaborated and discussed throughout the annotation process. Even if the identification of a biomedical entity was a relatively easy task, fitting it into the correct semantic group sometimes required prior knowledge of the biomedical vocabulary. Therefore the experienced annotator has accessed various terminological resources in order to better understand the terms and to categorized them into the correct semantic group. In a post-annotation phase, the two annotators discussed the annotation differences in order to reach agreement.

3.5 Discussion Over the Annotation Process

3.5.1 Part of Speech

During the manual correction process of the part of speech tags the annotator encountered several types of errors generated by the tool used:

1. tokenization errors: wrong segmentation of time intervals (*2000-2001*) was annotated as a single token), typos that led to wrong tokenization of the word (*fi cat* instead of *ficat* ("liver"));

2. lemmatization errors: in case of the unknown words (*adenoamă* instead of *adenom* ("adenoma")) or in case of morphologically ambiguous forms: the form *copii* can be the plural indefinite of either the masculine noun *copil* ("child") or of the feminine noun *copie* ("copy"); given this homography, the lemmatizer mistakes one of the words with the other one;

3. tagging errors where classified in two categories:

   - errors of part of speech – wrong automatic identification of the part of speech (nouns as adjectives, adjectives as adverbs and vice versa, verbs as adjectives);
   - wrong identification of the morphological class – the part of speech is correctly identified but some of the specifications are wrongly identified: gender, number, case, etc.

Even though the overall error rate of the tool used was low (1.77% see section 3.3.1) and pre-annotation with POS tags of the corpus was useful, the task of correcting it was a difficult one due to the complexity of the tag set and the laborious manual work needed to determine if the token, lemma and POS tag are correct for each word in the corpus. Annotation time ranges between 17 tokens per minute (at the beginning of the task) and 33 tokens per minute (after the annotator became accustomed with the task and the types of errors). The use of only one annotator for correcting the POS tags is justified, on the one hand, by the low error rate and, on the other, by the expense of the task. However, we are aware of the limitation represented by the lack of inter-annotator agreement measurements (even on a sample) on the morphological annotation.

3.6 Named Entities

The task of annotating the corpus with named entities had an increased difficulty due to several factors such as:

- the need to understand specialized terminology. Several cases can be identified here:

    - completeness of NEs: given the lack of expertise in the biomedical domain, the expert annotator sometimes omitted components of the complex entities, thus attributing the NE a wrong class;

    - ambiguity: both annotators needed to agree upon the cases when to annotate conjunctions present in some entities: for example, although in the vast majority of cases, the conjunction *şi* ("and") is not part of an NE, there are a few cases when it is: one such example is the NE *ocluzia/B-DISO arterelor/I-DISO mici/I-DISO şi/I-DISO mijlocii/I-DISO* ("occlusion of small and medium sized arteries") in which the conjunction *şi* is part of the entity (see its annotation as *I-DISO*) and does not get unannotated as in an example such as *ateroscleroza/B-DISO aortei/I-DISO şi vaselor/I-DISO periferice/I-DISO* ("atherosclerosis of aorta and peripheral vessels");

    - abbreviations: this challenge was encountered especially by the experienced annotator. It is known that biomedical literature is very rich in abbreviations (Federiuk, 1999). Unless their meaning is clear to the annotator, a wrong type can be assigned to it. What is more, many abbreviations are difficult to correctly classify because of their multiple meanings. For example, depending on the context, *ACE* can be *angiotensin convertază* ("angiotensin-converting enzyme") and it belongs to "Chemicals and Drugs" semantic class or *electroforeză capilară de afinitate* ("affinity capillary electrophoresis") and in this case it is correctly labeled as "Procedure"; notice also that the abbreviation is borrowed from English, thus posing challenges to the annotator lacking medical background;

- four different entities types.

Annotating all relevant entities was itself a challenge. One reason for this is the lack of prior knowledge of biomedical terminologies by the experienced annotator, some of the terms encountered not being covered in the terminological resources used for this task or being present with other senses than the one needed;

- the use of IOB2 format, which is an elaborated type of annotation format.

Estimated annotation time for this task was about 15 tokens per minute (for the experienced annotator) and 30 tokens per minute (for the physician).

The consistency of the annotations was established computing the (Carletta, 1996) coefficient on a sample of 1,628 tokens annotated by the two annotators, especially for this, after they finished the annotation. For this set the Kappa coefficient was 92.8%, denoting high agreement between the two annotators and indicates that the annotation was reliable.

4   Corpus General Statistics

Table 2 presents general corpus statistics offering an overview of the MoNERo corpus. Currently it contains 154,825 tokens (including the punctuation) distributed in 4,989 sentences, all of them annotated with POS tags and NEs. It can be seen that the average sentence length, 31 tokens/sentence, is above 16.06 tokens/sentence, the average sentence length in a balanced Romanian corpus, containing legal, news, medical (i.e. pharmacological), fiction and biographical texts (Ion et al., 2012).

Sentences	4,987
Tokens	154,825
Tokens/Sentence	31.04
Punctuation	20,741
Punctuation/Sentence	4.15

Table 2: MoNERo statistics.

Table 3 presents the distribution of sentences across the domains addressed. As can be seen, the distribution of sentences is not balanced, this being the result of the fact that due to copyright restrictions, the same number of sentences could not be collected for each of the selected domains, especially in the case of endocrinology.

Table 4 presents the distribution of content words. As can be seen, nouns are the most frequent ones, followed by adjectives: medical literature (especially medical literature books) has a

Domain	#tokens	#sentences
Cardiology	63,043	2,028
Diabetes	69,085	2,136
Endocrinology	22,697	823
Total	**154,825**	**4,987**

Table 3: Distribution of corpus sentences corresponding to each medical field.

descriptive structure, there are cases when nouns are modified by two or more adjectives: *bronşită cronică obstructivă* ("chronic obstructive bronchitis"). We notice quite an important number of abbreviations: the scientific subcomponent of the Romanian reference corpus, CoRoLa (Barbu Mititelu et al., 2018), contains 1.16% abbreviations, whereas MoNERo contains 1.9%. In the medical domain, as opposed to other scientific domains, it is common practice to designate concepts by abbreviated forms.

Tag	Percentage
Noun	27.8%
Verb	10.4%
Adjective	11.5%
Adverbs	3.5 %
Abbreviations	1.9 %
Total	55.1 %

Table 4: Percentages of content words.

Table 5 presents the distribution of entity annotation over each of the four semantic groups. This table highlights the fact that the most frequent NE categories are CHEM and DISO, PROC and ANAT being less frequent.

NE type	No. of entities
ANAT	1,964
CHEM	4,156
DISO	6,611
PROC	1,402
Total	**14,133**

Table 5: NEs distribution.

5 Corpus format

The corpus is available in a tabular format that contains four columns, UTF-8 encoded, with LF character as line break. Each line contains annotations of a token in four fields separated by a tab character: word form or punctuation symbol (token), lemma of the word form, NER tag and POS tag. We show below the annotation of the sentence: *Abordul arterei iliace comune se face retroperitoneal, iar grefonul folosit este unul sintetic din Dracon sau PTFE.* ("The access to the common iliac artery is retroperitoneal, and the graft used is a synthetic one from Dracon or PTFE.")

```
Abordul abord B-PROC Ncmsry
arterei arter I-PROC Ncfsoy
iliace iliac I-PROC Afpfson
comune comun I-PROC Afpfson
se sine O Px3--a--------w
face face O Vmip3s
retroperitoneal retroperitoneal O
Rgp
, ,O COMMA
iar iar O Rc
grefonul grefon O Ncmsry
folosit folosit O Afpms-n
este fi O Vmip3s
unul unul O Pi3msr
sintetic sintetic O Afpms-n
din din O Spsa
Dacron dacron O Ncms-n
sau sau O Ccssp
PTFE PTFE O Yn
. . O PERIOD
```

6 Utility of the corpus

There are several reasons for which MoNERo has an important contribution in named entity recognition and information extraction:

- it is the first Romanian gold standard biomedical corpus annotated with both part of speech tags and named entities;

- it was annotated with four types of named entities, making it very useful for training and testing NER systems based on supervised learning;

- it is pre-processed: tokenized, lemmatized and annotated with part of speech tags;

- it has a tabular format that makes it easy to use and the annotations are compliant with IOB2 format standards;

- it is a resource in a language other than English, which can help to train and test NER systems to perform language and domain adaptation;

- it is freely available for download[6] and non-commercial use. The archive contains three files, one for each medical domain, and another file containing all the other ones.

To prove the maturity and utility of this resource we used it to train and test a NER system (Boroş et al., 2018) for biomedical named entity recognition task for Romanian language. The architecture used is based on Bidirectional Long-Short-Term Memory (BDLSTM) networks (Graves, 2012). The system is trained to produce fully connected subgraphs. The feature-set is composed of word embeddings and character-level embeddings. In order to train the system the corpus was split in three sets: training set 80%, development set 10% and test set 10%. The evaluation of the performance of the system was done computing the F1 score and a score of 81.4 was obtained [7]. This experiment represents a starting point for the development/adaptation of NER systems for biomedical domain in Romanian.

7   Conclusions

We presented the MoNERo corpus, a gold standard biomedical corpus for Romanian language enhanced with two types of annotations: morphological and named entities specific to the biomedical field. To our knowledge this is the first biomedical corpus of this type for the Romanian language. This resource has already proven its value and utility, having been used in the development of the NER systems for the Romanian language. The MoNERo corpus is freely available for download and non-commercial use, which makes it even more valuable for the community.

8   Acknowledgements

Part of the work presented here was supported by a grant of the Romanian Ministry of Research and Innovation, PCCDI - UEFISCDI, project number PN-III-P1-1.2-PCCDI-2017-0818/72, within PNCDI III. We would like to thank the three anonymous reviewers for the valuable suggestions and hard work.

---

[6]The corpus is accessible at `http://www.racai.ro/en/tools/text/`

[7]The tool trained can be accessed at the following address: `http://89.38.230.23/teprolin/index.php?path=teprolin/custom`

References

Verginica Barbu Mititelu, Mihaela Cristescu, and Mihaela Onofrei. 2019. The Romanian Corpus Annotated with Verbal Multiword Expressions. In *Proceedings of the Joint Workshop on Multiword Expressions and WordNet (MWE-WN 2019)*, Florence, Italy.

Verginica Barbu Mititelu, Dan Tufiş, and Elena Irimia. 2018. The Reference Corpus of the Contemporary Romanian Language (CoRoLa). In *Proceedings of the Eleventh International Conference on Language Resources and Evaluation (LREC 2018)*, pages 1178–1185, Miyazaki, Japan. European Language Resources Association (ELRA).

Tiberiu Boroş, Stefan Daniel Dumitrescu, and Ruxandra Burtica. 2018. NLP-cube: End-to-end raw text processing with neural networks. In *Proceedings of the CoNLL 2018 Shared Task: Multilingual Parsing from Raw Text to Universal Dependencies*, pages 171–179, Brussels, Belgium. Association for Computational Linguistics.

Jean Carletta. 1996. Assessing agreement on classification tasks: the kappa statistic. *Computational linguistics*, 22(2):249–254.

Ludmila Dimitrova, Nancy Ide, Vladimir Petkevic, Tomaz Erjavec, Heiki Jaan Kaalep, and Dan Tufiş. 1998. Multext-east: Parallel and comparable corpora and lexicons for six central and eastern european languages. In *Proceedings of the 36th Annual Meeting of the Association for Computational Linguistics and 17th International Conference on Computational Linguistics - Volume 1*, ACL '98/COLING '98, pages 315–319, Stroudsburg, PA, USA. Association for Computational Linguistics.

Rezarta Islamaj Doğan, Robert Leaman, and Zhiyong Lu. 2014. Ncbi disease corpus: a resource for disease name recognition and concept normalization. *Journal of biomedical informatics*, 47:1–10.

Carol S Federiuk. 1999. The effect of abbreviations on medline searching. *Academic emergency medicine*, 6(4):292–296.

Lorraine Goeuriot, Liadh Kelly, Hanna Suominen, Aurélie Névéol, Aude Robert, Evangelos Kanoulas, Rene Spijker, Joao Palotti, and Guido Zuccon. 2017. Clef 2017 ehealth evaluation lab overview. In *International Conference of the Cross-Language Evaluation Forum for European Languages*, pages 291–303. Springer.

Alex Graves. 2012. Supervised sequence labelling. In *Supervised sequence labelling with recurrent neural networks*, pages 5–13. Springer.

Radu Ion. 2007. *Word Sense Disambiguation Methods Applied to English and Romanian (in Romanian)*. Ph.D. thesis, Romanian Academy.

Radu Ion, Elena Irimia, Dan Stefanescu, and Dan Tufiş. 2012. Rombac: The romanian balanced annotated corpus. In *LREC*, pages 339–344. Citeseer.

Halil Kilicoglu. 2017. Biomedical text mining for research rigor and integrity: tasks, challenges, directions. *Briefings in bioinformatics*, 19(6):1400–1414.

Martin Krallinger, Obdulia Rabal, Florian Leitner, Miguel Vazquez, David Salgado, Zhiyong Lu, Robert Leaman, Yanan Lu, Donghong Ji, Daniel M Lowe, et al. 2015. The chemdner corpus of chemicals and drugs and its annotation principles. *Journal of cheminformatics*, 7(1):S2.

Fei Li, Meishan Zhang, Guohong Fu, and Donghong Ji. 2017. A neural joint model for entity and relation extraction from biomedical text. *BMC bioinformatics*, 18(1):198.

Alexa T McCray, Anita Burgun, and Olivier Bodenreider. 2001. Aggregating umls semantic types for reducing conceptual complexity. *Studies in health technology and informatics*, 84(0 1):216.

Maria Mitrofan and Radu Ion. 2017. Adapting the ttl romanian pos tagger to the biomedical domain. In *BiomedicalNLP@ RANLP*, pages 8–14.

Maria Mitrofan and Dan Tufiş. 2018. Bioro: The biomedical corpus for the romanian language. In *Proceedings of the Eleventh International Conference on Language Resources and Evaluation (LREC-2018)*.

Isabel Moreno, Ester Boldrini, Paloma Moreda, and M Teresa Romá-Ferri. 2017. Drugsemantics: a corpus for named entity recognition in spanish summaries of product characteristics. *Journal of biomedical informatics*, 72:8–22.

Aurélie Névéol, Cyril Grouin, Jeremy Leixa, Sophie Rosset, and Pierre Zweigenbaum. 2014. The quaero french medical corpus: A ressource for medical entity recognition and normalization. In *In Proc BioTextM, Reykjavik*. Citeseer.

Maite Oronoz, Koldo Gojenola, Alicia Pérez, Arantza Díaz de Ilarraza, and Arantza Casillas. 2015. On the creation of a clinical gold standard corpus in spanish: Mining adverse drug reactions. *Journal of biomedical informatics*, 56:318–332.

Angus Roberts, Robert Gaizauskas, Mark Hepple, George Demetriou, Yikun Guo, Ian Roberts, and Andrea Setzer. 2009. Building a semantically annotated corpus of clinical texts. *Journal of biomedical informatics*, 42(5):950–966.

Erik F Sang and Jorn Veenstra. 1999. Representing text chunks. In *Proceedings of the ninth conference on European chapter of the Association for Computational Linguistics*, pages 173–179. Association for Computational Linguistics.

Paul Thompson, Syed A Iqbal, John McNaught, and Sophia Ananiadou. 2009. Construction of an annotated corpus to support biomedical information extraction. *BMC bioinformatics*, 10(1):349.

Dan Tufiş. 2000. Using a large set of eagles-compliant morpho-syntactic descriptors as a tagset for probabilistic tagging. In *Proceedings of LREC*.

D Tufiş, AM Barbu, V Pătraşcu, G Rotariu, and C Popescu. 1997. Corpora and corpus-based morpho-lexical processing. *Recent Advances in Romanian Language Technology, Editura Academiei*, pages 35–56.

Özlem Uzuner, Imre Solti, Fei Xia, and Eithon Cadag. 2010. Community annotation experiment for ground truth generation for the i2b2 medication challenge. *Journal of the American Medical Informatics Association*, 17(5):519–523.

W John Wilbur, Andrey Rzhetsky, and Hagit Shatkay. 2006. New directions in biomedical text annotation: definitions, guidelines and corpus construction. *BMC bioinformatics*, 7(1):356.

Domain Adaptation of SRL Systems for Biological Processes

Dheeraj Rajagopal♠* **Nidhi Vyas**♠* **Aditya Siddhant**♠ **Anirudha Rayasam**♠
Niket Tandon♣ **Eduard Hovy**♠
♠Carnegie Mellon University
♣Allen Institute for Artificial Intelligence
{dheeraj,nkvyas,asiddhan,arayasam}@andrew.cmu.edu
nikett@allenai.org, hovy@cmu.edu

Abstract

Domain adaptation remains one of the most challenging aspects in the wide-spread use of Semantic Role Labeling (SRL) systems. Current state-of-the-art methods are typically trained on large-scale datasets, but their performances do not directly transfer to low-resource domain-specific settings. In this paper, we propose two approaches for domain adaptation in biological domain that involve pre-training LSTM-CRF based on existing large-scale datasets and adapting it for a low-resource corpus of biological processes. Our first approach defines a mapping between the source labels and the target labels, and the other approach modifies the final CRF layer in sequence-labeling neural network architecture. We perform our experiments on ProcessBank (Berant et al., 2014) dataset which contains less than 200 paragraphs on biological processes. We improve over the previous state-of-the-art system on this dataset by 21 F1 points. We also show that, by incorporating event-event relationship in ProcessBank, we are able to achieve an additional 2.6 F1 gain, giving us possible insights into how to improve SRL systems for biological process using richer annotations.

1 Introduction

Semantic Role Labeling (SRL) is shallow semantic representation of a sentence, that allows us to capture the roles of arguments that anchor around an event. Despite significant recent progress in Deep SRL systems (He et al., 2017; Tan et al., 2017), there has been limited work in adapting such systems to low resource domain-specific scenarios where the label space of both domains are completely different. Additionally, existing domain adaptation for SRL requires an overhead of annotating the new corpus using guidelines similar

to the source dataset, and every domain-specific corpora might not necessarily adhere to the same label structure and similar annotation guidelines.

We present two different domain adaptation strategies that rely on training the model on a large corpora (source dataset) and fine-tuning on a low-resource domain-specific corpus (target dataset), more specifically biological processes domain. The first approach uses mappings from the source label space to the target label space. For this, we present DeepSRL-CRF, which incorporates a CRF layer over the DeepSRL model (He et al., 2017) with an intermediate step of mapping labels from source to target domain. For the second approach, we use a CNN-LSTM-CRF model to pre-train the neural network weights on the source domain, and adapt the final CRF layer of the network based on the target label space. We then fine-tune the model on the target dataset.

For empirical evaluation, we explore the challenge of SRL in ProcessBank dataset, where the target domain (biological processes) is drastically different compared to the source domain (news). Both of our approaches are effective for adapting SRL systems for biological processes. Compared to the previous best system, we get an improvement of about 24 F1 points when we use label-mapping approach, and about 21 F1 point improvement when we adapt the final CRF layer. Our contributions can be summarized as follows:

1. Two different approaches for domain adaptation of SRL for biological processes, with our code and models publicly available [1]

2. An in-depth analysis for each of the domain adaptation strategies, both perform significantly better in low-resource SRL for biological processes

3. Analysis of the model performance when the

---

*Both authors equally contributed to the paper.

[1]https://github.com/dheerajrajagopal/SciQA

Proceedings of the BioNLP 2019 workshop, pages 80–87
Florence, Italy, August 1, 2019. ©2019 Association for Computational Linguistics

target corpus is annotated with event-event relationships to the SRL corpus

2 Models

To label the event-argument relationships, we propose two models inspired from the current state-of-art on the SRL and NER literature. Since our downstream task lends itself to a low-resource setting, we hypothesize that an LSTM-CRF architecture would be better suited for the role-labeling task.

DeepSRL-CRF : We introduce DeepSRL-CRF, that is inspired from DeepSRL (He et al., 2017). The DeepSRL-CRF model uses a stacked BiLSTM network structure as its representation layer with a CRF layer on top. The overall model uses stacked BiLSTMs using an interleaved structure, as proposed in Zhou and Xu (2015). As described in the original model, we use gated highway connections (Zhang et al., 2016; Srivastava et al., 2015) to prevent over-fitting.

CNN-LSTM-CRF : We adapt the state-of-art sequence-labeling model by Ma and Hovy (2016). This is an end-to-end model, which uses a BiLSTM, Convolutional Neural Network (CNN) and CRF to capture both word- and character-level representations. The model first uses a CNN to capture character-level representation. These embeddings are concatenated with the word-level embeddings and fed into a BiLSTM to capture the contextual information at word-level. Here, we adapt this model to additionally concatenate 100-dimensional predicate indicators for every word before feeding the result into a BiLSTM. The output vectors from the BiLSTM are fed into the CRF layer, which jointly decodes the best sequence. The model uses dropout layers for both CNN and BiLSTM to prevent overfitting.

3 Domain Adaptation

Label Mapping : In our first approach, we perform domain by mapping each label from the target label-space to the source label-space by aligning it to the closest label from the source dataset. Since we used the CoNLL-2005 and CoNLL-2012 datasets for pre-training, we used the PropBank labels to map each relation in ProcessBank according to the PropBank annotation guidelines. Although there is human intervention in the pipeline, it is time-efficient since this process has to be done only once for a target dataset. We asked three independent annotators to perform the mapping of these labels, and the majority voted mapping was used as the final mapping scheme. In case of no majority vote, the annotators discussed to reach a consensus. We had an inter-annotator agreement of 0.8. The entire process for ProcessBank dataset took approximately two hours. The mapping for individual relationships are given in Table 1. The network architecture did not change throughout the training process for both source and target domains. The final CRF layer of the neural network maintains the same dimensions as the source domain.

PropBank	ProcessBank
ARG0	Agent
ARGM-LOC	Location
ARG2	Theme
ARG3	Source
ARG4	Destination
ARG1	Result
ARGM-MNR	Other

Table 1: Label Mapping: PropBank to ProcessBank

Adapting the CRF Layer : In the second approach, we maintain the network weights for the BiLSTM layers constant from the pre-training and we learn the transition and emission probabilities from scratch in the target domain dataset. More specifically, we first train the entire model on CoNLL-2005 and CoNLL-2012 SRL data. Next, we replace the final CRF layer with the label-space dimensions in our target domain, and keep the remaining weights in the model as is. Finally, we start fine-tuning the entire model by training it on the target data. Contrary to the previous approach, this approach does not require any manual intervention.

Event Interactions : The ProcessBank dataset is also annotated with event-event interactions. In our model, we also study whether event-event structure is important in predicting the event-argument structure. We leverage this additional event-event interaction annotations, and add them to the input to predict the event-argument role-labels. From an annotation perspective, this experiment helps us analyze whether the event-event structure labels are the bottle-neck for better SRL performance - especially in domain specific settings.

4 Experiments

Experimental Setup : For evaluation, we use the CoNLL-2005 (Carreras and Màrquez, 2005) and CoNLL-2012 (Pradhan et al., 2013) datasets as our primary large-scale datasets with the standard splits. For the domain adaptation scenario, we use the ProcessBank dataset (Berant et al., 2014)[2]. We used 134 annotated paragraphs for training, 19 for development and 50 for testing. Each passage in the ProcessBank dataset describes a *process*, defined by a directed graph (T, A, E_{tt}, E_{ta}), where nodes T denote event triggers and A denote their corresponding arguments. E_{tt} represents labeled edges event-event relations and E_{ta} describe event-argument relations. The edges E_{ta} are annotated with semantic roles AGENT, THEME, SOURCE, DESTINATION, LOCATION, RESULT and OTHER. Each E_{tt} edge between event and another event is annotated with the relations CAUSE, ENABLE and PREVENT. Our experiments primarily focus on the prediction of the event-argument structures E_{ta} since the source datasets that we use for domain adaption do not contain any event-event relationship annotation.

Baselines : In our first set of baselines, we compare our models on the CoNLL-2005 and CoNLL-2012 tasks. We use the previous state-of-the-art SRL system from He et al. (2018) as our baseline.[3]. Since our model is based on LSTM-CRF hybrid architecture, we implement two other baselines for our approach. We use a standard BiLSTM-CRF model (Huang et al., 2015), and a model based on the structured attention proposed in Liu and Lapata (2017) which uses CRF style structure in the intermediate layer. For a fair comparison, we augmented this structured attention based network with a CRF layer on top. We use 300D GLoVe embeddings (Pennington et al., 2014) across all models. For domain adaptation, we use the original system from Berant et al. (2014) as the baseline. It uses the approach in Punyakanok et al. (2008), where for each trigger, a set of argument candidates are first determined, and then a binary classifier uses argument identification features to prune

---

[2]For dataset statistics, we refer readers to Berant et al. (2014), Table 1. We use the same training and test split provided in the original dataset. We further split the training set into training and development set.

[3]Due to resource limitations, we were unable to run the same model for 500 epochs, so we report results from their paper for CoNLL-2005 and CoNLL-2012 datasets

this set with high recall.

5 Results

Semantic Role Labeling : Table 2 shows the SRL results[4] for the CoNLL-2005 and CoNLL-2012 datasets across all baseline models. From the table, it is evident that our DeepSRL-CRF model with ELMo embeddings performs slightly lesser than the current state-of-the-art SRL model DeepSRL with ELMo. We were able to perform significantly better than the other baselines – BiLSTM-CRF and Strucutured Attention model. Our DeepSRL-CRF model without ELMo performed significantly lower and the improvement was notably higher with ELMo.

Domain Adaptation : For all our domain adaptation experiments, we found that the DeepSRL and DeepSRL-CRF models reach similar F1 scores without any pre-training. Table 3 shows the results for the set of models that were trained for domain adaptation using label mapping. After pre-training it on the CoNLL 2005 and CoNLL-2012 dataset for 50 epochs, we fine-tuned the weights on the ProcessBank dataset without making any changes to the network. The results signify that the models that were effective for a large dataset, might not achieve similar gains when restricted to specific low-resource domains. The DeepSRL-CRF model, after incorporating event-event relationships, outperforms the previous system from Berant et al. (2014) by about 24 F1 points.

In our second domain adaptation approach, we test the CNN-LSTM-CRF model by learning the final CRF layer with transition and emission probabilities for the target label space. The CNN-LSTM-CRF model, without any pre-training achieves 40.62 F1 which is similar to previous performance from Berant et al. (2014). However, after pre-training it on CoNLL 2005 and CoNLL-2012 dataset for 50 epochs, the models outperforms by about 21.7 F1 points. Adapting the CRF layer, with transition and emission probabilities for the target domain data in its label space, shows impressive gains in the low-resource setting, specially when there is a limitation for using any human-intervention in the domain adaptation process. Although empirically effective, we believe that there is immense scope to understanding the impact of better initialization from a theoretical perspective. We also observe that pretraining

---

[4]We use span-based precision, recall and F1 measure

Model	CoNLL-2005 (WSJ)			CoNLL-2012 (OntoNotes)		
	P	R	F1	P	R	F1
BiLSTM-CRF	80.9	79.4	80.3	80.0	77.8	78.9
Structured Attention	81.0	80.1	80.5	79.6	77.9	78.8
CNN-LSTM-CRF	82.1	82.7	82.4	81.7	83.0	82.3
DeepSRL	81.6	81.6	81.6	81.8	81.4	81.6
DeepSRL-ELMo	-	-	**87.4**	-	-	**85.5**
DeepSRLCRF	35.0	46.3	40.0	51.6	78.1	62.2
DeepSRLCRF-ELMo	84.7	83.6	84.1	84.4	85.8	85.1

Table 2: SRL results for CoNLL-2005 and CoNLL-2012 datasets. DeepSRL-ELMo resuls from He et al. (2018)

Model	Development			Test		
	P	R	F1	P	R	F1
Berant et al. (2014)	-	-	-	43.4	34.4	38.3
			CoNLL-2005			
DeepSRL	46.7	53.7	50.0	46.1	51.0	48.5
DeepSRL-ELMo	55.0	48.0	51.7	48.8	41.7	44.5
DeepSRLCRF	51.4	58.1	54.5	50.8	57.0	53.7
DeepSRLCRF-ELMo	53.5	66.2	59.2	49.1	63.2	55.3
+ Event relations	63.0	63.7	63.3	61.0	62.2	61.6
			CoNLL-2012			
DeepSRL	51.1	56.9	53.9	43.9	49.0	46.3
DeepSRL-ELMo	52.6	50.0	51.2	48.1	43.2	44.6
DeepSRLCRF	45.9	63.1	53.1	40.3	56.7	47.2
DeepSRLCRF-ELMo	44.6	**67.5**	53.7	36.9	62.1	46.3
+ Event relations	**65.0**	65.0	**65.0**	**62.1**	**63.0**	**62.6**

Table 3: SRL results for ProcessBank dataset - Domain adaptation using label mapping.

Model	Test		
	P	R	F1
Berant et al. (2014)	43.4	34.4	38.3
No pre-training	40.6	40.6	40.6
		CoNLL-2005	
BiLSTM-CRF	44.7	42.3	43.4
CNN-LSTM-CRF	56.8	55.5	56.1
+Event relations	55.3	53.4	54.4
		CoNLL-2012	
BiLSTM-CRF	42.8	41.0	42.3
CNN-LSTM-CRF	**59.7**	**60.2**	**60.0**
+ Event relations	58.8	57.7	58.3

Table 4: Results for ProcessBank - Domain adaption by replacing the CRF layer

on CoNLL-2012 dataset was more effective compared to pre-training on CoNLL-2005 dataset for this model. The former has about 35000 more training data instances than later.

Which domain adaptation technique works best? Our results show that the DeepSRL-CRF model based on label mapping approach perform the best overall (improvement of 24 F1 points) assuming we have event-event relationship anno-

tations. In a setting where there are multiple datasets of different domains, training different network for each of the datasets might be cumbersome. We believe that the domain adaptation based on label mapping would suit such situations better. However, in the cases where there is no explicit label mapping possible or no readily available event-event interaction annotations in target domains, resorting to replacing the CRF layer would be the most effective for domain adaption gains. Our CNN-LSTM-CRF model achieves an improvement of 21 F1 points by replacing the CRF layer without event-event annotations. One of the drawbacks of this system is that it cannot be trained end-to-end. Given that there is limited overhead in modifying the architecture, we believe this wouldn't be a bottleneck for NLP systems. If end-to-end training is a hard constraint, we resort to our DeepSRL-CRF model. In terms of generalization capability and performance, pre-training on the CoNLL-2012 dataset and fine-tuning on the ProcessBank dataset with explicit label mapping with additional event-event relations gives us the best results.[5]

6 Related Work

Domain adaptation leverages on large-scale datasets to help improve the performance on other smaller and similar tasks. From the SRL perspective, one of the earliest work from Daume III and Marcu (2006) showed simple but effective ways for 'transferring the learning' from a source to a target domain. Building on strong feature-rich models, Dahlmeier and Ng (2010) analyzed various features and techniques that are used for domain adaptation and conducted an extensive study for biological SRL task. Later, Lim et al. (2014) proposed a model that uses structured learning for domain adaptation. Although effective, these methods rely on hand-annotated features. Recently, there have been neural-network based attempts at Domain adaptation in SRL. Do et al. (2015) combined the knowledge from a neural language model and external linguistic resource for domain adaptation for biomedical corpora. Our work closely aligns to this work from a modeling stand-point. Our target domain is biological process descriptions from high-school biology without restrictions of PropBank style annotations.

Our work builds on multiple existing works, especially the dataset from Berant et al. (2014), using the thematic roles defined in Palmer et al. (2005). Our approach is inspired by the recent success in including structured representations in deep neural networks (He et al., 2017; Ma and Hovy, 2016) for structured prediction tasks. Our primary motivation is to improve the system performance for low-resource domain-specific event-argument labeling tasks, particularly biological processes. Argument labeling, specifically, SRL as been used for biomedical domain previously. E.g. Shah and Bork (2006) applied SRL in the LSAT system to identify sentences with gene transcripts, and Bethard et al. (2008) applied SRL to extract information about protein movement. However, developing annotated SRL data for each task can be expensive.

7 Conclusion

In this work, we present two new approaches to adapt deep learning models trained on large scale datasets, to smaller domain-specific biological process dataset. We present a LSTM-CRF based architectures which perform on-par with the state-of-the-art models for SRL but significantly better than them in low-resource domain-specific settings. We show significant improvement of approximately 24 F1 points over current best model for role-labeling on the ProcessBank - notably different in nature compared to CoNLL dataset.

References

Jonathan Berant, Vivek Srikumar, Pei-Chun Chen, Abby Vander Linden, Brittany Harding, Brad Huang, Peter Clark, and Christopher D Manning. 2014. Modeling biological processes for reading comprehension. In *EMNLP*, pages 1499–1510.

Steven Bethard, Zhiyong Lu, James H Martin, and Lawrence Hunter. 2008. Semantic role labeling for protein transport predicates. *BMC bioinformatics*, 9(1):277.

Xavier Carreras and Lluís Màrquez. 2005. Introduction to the conll-2005 shared task: Semantic role labeling. pages 152–164. Association for Computational Linguistics.

Daniel Dahlmeier and Hwee Tou Ng. 2010. Domain adaptation for semantic role labeling in the biomedical domain. *Bioinformatics*, 26(8):1098–1104.

Hal Daume III and Daniel Marcu. 2006. Domain adaptation for statistical classifiers. *Journal of artificial Intelligence research*, 26:101–126.

Quynh Ngoc Thi Do, Steven Bethard, and Marie-Francine Moens. 2015. Domain adaptation in semantic role labeling using a neural language model and linguistic resources. *IEEE/ACM Trans. Audio, Speech & Language Processing*, pages 1812–1823.

Xavier Glorot and Yoshua Bengio. 2010. Understanding the difficulty of training deep feedforward neural networks. In *AISTATS*, pages 249–256.

Kaiming He, Xiangyu Zhang, Shaoqing Ren, and Jian Sun. 2015. Delving deep into rectifiers: Surpassing human-level performance on imagenet classification. In *ICCV*, pages 1026–1034.

Luheng He, Kenton Lee, Omer Levy, and Luke Zettlemoyer. 2018. Jointly predicting predicates and arguments in neural semantic role labeling. *arXiv preprint arXiv:1805.04787*.

Luheng He, Kenton Lee, Mike Lewis, and Luke Zettlemoyer. 2017. Deep semantic role labeling: What works and whats next. In *ACL*, volume 1, pages 473–483.

Zhiheng Huang, Wei Xu, and Kai Yu. 2015. Bidirectional lstm-crf models for sequence tagging. *arXiv preprint arXiv:1508.01991*.

---

[5]Please refer to the supplemental material 9 for a detailed discussion on results

Soojong Lim, Changki Lee, Pum-Mo Ryu, Hyunki Kim, Sang Kyu Park, and Dongyul Ra. 2014. Domain-adaptation technique for semantic role labeling with structural learning. *ETRI Journal*, 36(3):429–438.

Yang Liu and Mirella Lapata. 2017. Learning structured text representations. *CoRR*, abs/1705.09207.

Xuezhe Ma and Eduard Hovy. 2016. End-to-end sequence labeling via bi-directional lstm-cnns-crf. In *ACL*, volume 1, pages 1064–1074.

Martha Palmer, Daniel Gildea, and Paul Kingsbury. 2005. The proposition bank: An annotated corpus of semantic roles. *Computational linguistics*, 31(1):71–106.

Jeffrey Pennington, Richard Socher, and Christopher Manning. 2014. Glove: Global vectors for word representation. In *Proceedings of the 2014 conference on empirical methods in natural language processing (EMNLP)*, pages 1532–1543.

Sameer Pradhan, Alessandro Moschitti, Nianwen Xue, Hwee Tou Ng, Anders Björkelund, Olga Uryupina, Yuchen Zhang, and Zhi Zhong. 2013. Towards robust linguistic analysis using ontonotes. In *CoNLL*, pages 143–152.

Vasin Punyakanok, Dan Roth, and Wen-tau Yih. 2008. The importance of syntactic parsing and inference in semantic role labeling. *Computational Linguistics*, 34(2):257–287.

Parantu K Shah and Peer Bork. 2006. Lsat: learning about alternative transcripts in medline. *Bioinformatics*, 22(7):857–865.

Rupesh K Srivastava, Klaus Greff, and Jürgen Schmidhuber. 2015. Training very deep networks. In *Advances in neural information processing systems*, pages 2377–2385.

Zhixing Tan, Mingxuan Wang, Jun Xie, Yidong Chen, and Xiaodong Shi. 2017. Deep semantic role labeling with self-attention. *arXiv preprint arXiv:1712.01586*.

Yu Zhang, Guoguo Chen, Dong Yu, Kaisheng Yaco, Sanjeev Khudanpur, and James Glass. 2016. Highway long short-term memory rnns for distant speech recognition. In *ICASSP*.

Jie Zhou and Wei Xu. 2015. End-to-end learning of semantic role labeling using recurrent neural networks. In *ACL*, volume 1, pages 1127–1137.

8 Appendices

8.1 Parameter Settings

CNN-LSTM-CRF : The words that are absent in GloVe embeddings are replaced with <UNK> and intialized randomly. The character-embeddings are intialized with uniform samples as proposed in He et al. (2015). Weight matrices are initialized using Glorot initialization (Glorot and Bengio, 2010). Bias vectors are initialized to zero except the bias vector of Bi-LSTM (b_f) which is initialized to 1. Parameter optimization is performed using Adam optimizer with batch size of 32 and learning rate of 0.0001. We use a non-variational dropout of 0.5 on CNN and BiLSTM layers. We use a hidden size of 512, and use 5 layers for the BiLSTM. For character embeddings, we use a hidden size of 30. The CNN's use 30 filters.

DeepSRL-CRF : We maintain most of the experimental settings similar to He et al. (2017). We convert all tokens to lower-case, initialize with the embeddings. We use the Adadelta with $\epsilon = 1e^{-6}$ and $\rho = 0.95$ with mini-batch size 64. The dropout probability was set to 0.1 and gradient clipping at 1. The models are trained for 50 epochs (compared to 500 epochs in the original DeepSRL model) and use the best model from 50 epochs for pretraining. We do not add any constraints for decoding and we use the viterbi decoding to get our output tags.

9 Supplemental Material

9.1 Additional Discussion

DeepSRL-CRF: The DeepSRL-CRF model achieves comparable but slightly lower performance compared to the current state-of-the-art in the CoNLL-2005 and CoNLL-2012 SRL datasets. We observed that these performances did not directly translate to the ProcessBank dataset. In the limited-resource domain of ProcessBank, the final CRF layer had a more pronounced performance improvements. Adding CRF layer to DeepSRL model improves performance by atleast 4 F1 points when pre-trained using CoNLL-2005 and 1 F1 point when pre-training using CoNLL-2012 dataset. Adding ELMo embeddings to the DeepSRL and DeepSRL-CRF models did not result in performance gains in ProcessBank except for one experimental setup (DeepSRL-CRF pre-trained on CoNLL-2005). Across both datasets, we acheived our best results when we incorporated event-event relations in the SRL annotation. Although a performance improvement is expected, the best results for domain adaptation was achieved after adding the event relations. The tags that gain most from the event relationships are *Agent*, *Destination*, *Source* and *Location*. The improvements primarily come from the gain in precision with a slight drop in recall. We believe that the reason for this improvement is the artifact of the dataset's event-event relationships tend to correlate often with these entities given the nature of these biological processes. Across CoNLL-2005 and CoNLL-2012, it did not make a considerable difference as to which dataset we used for pre-training. Although CoNLL-2012 has slighly better performance (shown in table 3, there could be additional hyper-parameter tuning that could lead to slightly different results between the two datasets.

CNN-LSTM-CRF: The CNN-LSTM-CRF model on ProcessBank achieves 40.62 F1 without any pre-training. This result is comparable to the baseline, showing the importance of initialization of weights while training a neural network based model. However, we achieve substantial improvement of about 21.7 F1 with pre-training on CoNLL data and later adapting only the final CRF layer for the target label space. In contrast to DeepSRL-CRF, we notice that performance difference between pre-training on CoNLL-2005 and CoNLL-2012 is considerable (4 F1 points). We have to note that CoNLL-2012 dataset has about 35000 more training data instances than CoNLL-2005. We hypothesize that these additional training instances might have contributed to the final F1 score while training using CoNLL-2012 dataset. We also observe that pre-training improves the performance of tags that have less number of instances in the target domain (ProcessBank). One of the unique cases is shown in table 7, where *Source* tag prediction shows huge improvements (57.0 F1) after the model was pre-trained using the CoNLL data. However, we do not see the same trend for the *Other* tag. Further, as per table 5 and 6, the model particularly confused the *Other* tag with the O tag of the BIO scheme. In the original ProcessBank dataset, the tags that do not belong to the original proposed categories, were classified as one single *Other* category and this category

had the least number of annotated examples. We believe that the combination of these factors made it challenging for the model to predict this particular category. According to table 5 and 6, the most frequent tags – *Theme* and *Agent* have high prediction accuracy. However, their spans are sometimes incorrectly identified. For instance the *Theme* tags are identified incorrectly as O or vice-versa. Overall B tags have higher precision than the I tags, and the model is able to better predict the start of a span than the end of a span.

From table 7, we also notice that annotating a dataset with event-event relationships does not consistently improve the performance which we observed in DeepSRL-CRF. These results also show that adding the CNN-layer of character embeddings to the BiLSTM-CRF model helps the model perform better across all the labels. emphasizing the relevance of these character embeddings.

%	Agt.	Dest	Loc	Oth.	Res.	Src.	The.	O
Agt.	**71.1**	1.0	0.0	0.0	0.0	1.0	6.2	20.6
Dest.	0.0	**53.9**	7.7	0.0	7.7	0.0	15.4	15.4
Loc	0.0	3.0	**45.5**	0.0	3.0	0.0	9.1	39.4
Oth.	0.0	25.0	25.0	0.0	0.0	0.0	0.0	**50.0**
Res.	0.0	0.0	2.2	0.0	31.1	0.0	24.4	**42.2**
Src.	0.0	15.8	0.0	0.0	0.0	**68.4**	15.8	0.0
The.	4.0	1.6	0.0	0.0	1.2	0.8	**85.9**	6.5

Table 5: Best performing CNN-LSTM-CRF model's breakdown of true (rows) and predicted (columns) *B* tags with BIO tagging scheme. (Agt.=Agent; Dest.=Destination; Loc.=Location; Oth.=Other; Res.=Result; Src.=Source; The.=Theme; O=*O* tag in BIO tagging)

%	Agt.	Dest	Loc	Oth.	Res.	Src.	The.	O
Agt.	**65.6**	1.1	0.0	0.0	0.0	0.0	7.1	26.2
Dest.	0.0	**43.0**	15.8	0.0	5.3	0.0	16.7	19.3
Loc	0.0	9.2	**48.7**	0.0	5.3	0.0	6.6	30.3
Oth.	0.0	16.7	16.7	0.0	0.0	0.0	0.0	**66.7**
Res.	0.0	0.0	0.8	0.0	**43.0**	0.0	20.3	35.9
Src.	0.0	6.5	0.0	0.0	0.0	**71.0**	22.6	0.0
The.	3.2	2.7	3.4	0.0	1.7	1.2	**73.2**	14.6

Table 6: Best performing CNN-LSTM-CRF model's breakdown of true (rows) and predicted (columns) *I* tags with BIO tagging scheme. (Agt.=Agent; Dest.=Destination; Loc.=Location; Oth.=Other; Res.=Result; Src.=Source; The.=Theme; O=*O*tag in BIO tagging)

		BiLSTM-CRF		CNN-LSTM-CRF		
	#Instances	PB only	+Pretrain. +Dom. adp.	PB only	Pretrain. +Dom. adp.	+Verb Relations
Agent	280	25.8	37.0	35.5	62.1	**63.3**
Destination	153	8.0	2.7	38.5	51.3	**53.1**
Location	109	4.8	1.8	26.1	**44.1**	38.8
Other	11	0.0	0.0	0.0	0.0	0.0
Result	173	2.8	12.0	11.1	**34.7**	25.0
Source	50	8.7	15.4	0.0	57.9	**59.1**
Theme	838	44.9	57.3	52.1	**67.2**	66.0

Table 7: F1 scores on different tags in ProcessBank with BiLSTM-CRF and CNN-LSTM-CRF model (PB=ProcessBank). Pre-training was done on CoNLL-2012 dataset

Deep Contextualized Biomedical Abbreviation Expansion

Qiao Jin
University of Pittsburgh
qiao.jin@pitt.edu

Jinling Liu
University of Pittsburgh
jil172@pitt.edu

Xinghua Lu
University of Pittsburgh
xinghua@pitt.edu

Abstract

Automatic identification and expansion of ambiguous abbreviations are essential for biomedical natural language processing applications, such as information retrieval and question answering systems. In this paper, we present DEep Contextualized Biomedical Abbreviation Expansion (**DECBAE**) model. DECBAE automatically collects substantial and relatively clean annotated contexts for 950 ambiguous abbreviations from PubMed abstracts using a simple heuristic. Then it utilizes BioELMo (Jin et al., 2019) to extract the contextualized features of words, and feed those features to abbreviation-specific bidirectional LSTMs, where the hidden states of the ambiguous abbreviations are used to assign the exact definitions. Our DECBAE model outperforms other baselines by large margins, achieving average accuracy of 0.961 and macro-F1 of 0.917 on the dataset. It also surpasses human performance for expanding a sample abbreviation, and remains robust in imbalanced, low-resources and clinical settings.

1 Introduction

Abbreviations are shortened forms of text-strings. They are prevalent in biomedical literature such as scientific articles, clinical notes and user queries in information retrieval systems. Abbreviations can be ambiguous (e.g.: ER can refer to estrogen receptor, endoplasmic reticulum, emergency room etc.), especially when they appear in short or professional texts where the definitions are not given. For instance, about 15% of PubMed queries include abbreviations (Islamaj Dogan et al., 2009), and about 14.8% of all tokens in a clinical note dataset are abbreviations (Xu et al., 2007). In both cases, the definitions of the abbreviations are rarely provided. Thus, automatic expansion of ambiguous abbreviations to their full forms is vital

in biomedical natural language processing (NLP) systems.

In this paper, we focus on the cases where definitions of ambiguous abbreviations are not directly available in the contexts, so reasoning over the contexts is required for disambiguation. Under the conditions where definitions are provided in the contexts, one can easily extract them using rule-based methods.

We present DEep Contextualized Biomedical Abbreviation Expansion (DECBAE) model. DECBAE uses a simple heuristic to automatically construct large supervised disambiguation datasets for 950 abbreviations from PubMed abstracts: In scientific writing, authors define abbreviations the first time they are used, and the same abbreviations in the following sentences have the same definitions as those of the first ones. We extract all the sentences containing the same abbreviations in each PubMed abstract, and use the definition given in the first sentence as the full form label of abbreviations in the following sentences. We group the definitions for each abbreviation and formulate abbreviation expansion as a classification task, where input is an ambiguous abbreviation with its context, and the output is one of its possible definitions.

Recent breakthroughs of language models (LM) pre-trained on large corpora like ELMo (Peters et al., 2018) and BERT (Devlin et al., 2018) clearly show that unsupervised LM pre-training can vastly improve performance of downstream models. To fully utilize the knowledge encoded in PubMed abstracts, DECBAE uses BioELMo (Jin et al., 2019), a domain adapation verison of ELMo, to embed the words. After the embedding layer, DECBAE applies abbreviation-specific bidirectional LSTM (biLSTM) classifiers to do the abbreviation expansion, where the biLSTM parameters are trained separately for each abbrevi-

Proceedings of the BioNLP 2019 workshop, pages 88–96
Florence, Italy, August 1, 2019. ©2019 Association for Computational Linguistics

ation. We train DECBAE from the automatically collected dataset of 950 ambiguous abbreviations.

At inference time, DECBAE feeds the BioELMo embeddings of the whole sentence and uses the corresponding abbreviation-specific biLSTM classifiers to perform disambiguation of abbreviations in the sentence. We show that DECBAE outperforms other baselines by large margins and even performs better than single human expert. Although training instances of DECBAE are collected from PubMed, it covers 85% of clinically related abbreviations mentioned in a previous work (Xu et al., 2012). Moreover, DECBAE remains robust in low-resource and imbalanced settings.

2 Related Work

Contextualized word embeddings: Recently, contextualized word representations pre-trained by large corpora like ELMo (Peters et al., 2018) and BERT (Devlin et al., 2018) significantly improve the performance of various NLP tasks. ELMo is a pre-trained biLSTM language model. ELMo word embeddings are calculated by a weighted sum of the hidden states of each biLSTM layer. The weights are task-specific learnable parameters while biLSTM layers are fixed. In-domain trained contextual embeddings further improve the performance on domain-specific tasks. In this paper, we use BioELMo, which is a biomedical version of ELMo trained on 10M PubMed abstracts (Jin et al., 2019). BioELMo outperforms general ELMo by large margins on several biomedical NLP tasks.

We don't use BERT for contextualized embeddings due to its fine-tuning nature: users just need to download 1 BioELMo and N abbreviation-specifc biLSTM weights to run DECBAE locally, which takes significantly less disk size than N fine-tuned BERTs for each abbreviation. N is the number of abbreviations.

Word sense disambiguation (WSD): The goal of WSD is to determine the correct sense of words in different contexts. Abbreviation expansion is a specific case of WSD where the ambiguous words are abbreviations. In this paper, we use abbreviation expansion and abbreviation disambiguation interchangeably. Several human-annotated datasets are available for supervised WSD (Navigli et al., 2013; Camacho-Collados et al., 2016; Raganato et al., 2017b). However, human anno-

tations could be expensive, especially in domain specific settings. To address this problem, some automatic dataset collection methods have been proposed (Yu et al., 2007; Ciosici et al., 2019), where abbreviations are automatically labeled if they are defined previously in the same documents. We use a similar approach in this work.

Peters et al. (2018) report that just matching the ELMo embedding of the target words with the nearest sense representations, calculated by averaging their ELMo embeddings, leads to comparable WSD performance with state-of-the-art models using hand crafted features (Iacobacci et al., 2016) or task-specific biLSTM trained with multiple tasks (Raganato et al., 2017a). Instead of searching the nearest contextualized embeddings neighbors of the abbreviation and definitions, we model abbreviation expansion as classification.

Biomedical abbreviation expansion: Various methods have been introduced for automatically expanding biomedical abbreviations. Yu et al. (2007) train naive Bayes and SVM classifiers with bag-of-word features on an automatically collected dataset from PubMed. Some works disambiguate abbreviations to their senses in controlled vocabularies like Medical Subject Headings[1] (MeSH) and Unified Medical Language System[2] (UMLS). Xu et al. (2015) use pooled neighbor word embeddings of the abbreviations as features to train SVM classifiers for clinical abbreviaiton disambiguation. Jimeno-Yepes et al. (2011) introduced MSH WSD dataset to test the performance of supervised biomedical WSD systems and several supervised models have been proposed on it (Antunes and Matos; Yepes, 2017). Recently Pesaranghader et al. (2019) presented deep-BioWSD which sets new state-of-the-art performance on it. DeepBioWSD uses a single biLSTM encoder for disambiguation of all abbreviations by calculating the pairwise similarity between context representations and sense representations.

To the best of our knowledge, DECBAE is the first model that uses deep contextualized word embeddings for biomedical abbreviation expansion.

3 Methods

Figure 1 shows the architecture of DECBAE. During training, we first construct abbreviation ex-

---

[1] https://www.nlm.nih.gov/mesh
[2] https://www.nlm.nih.gov/research/umls/

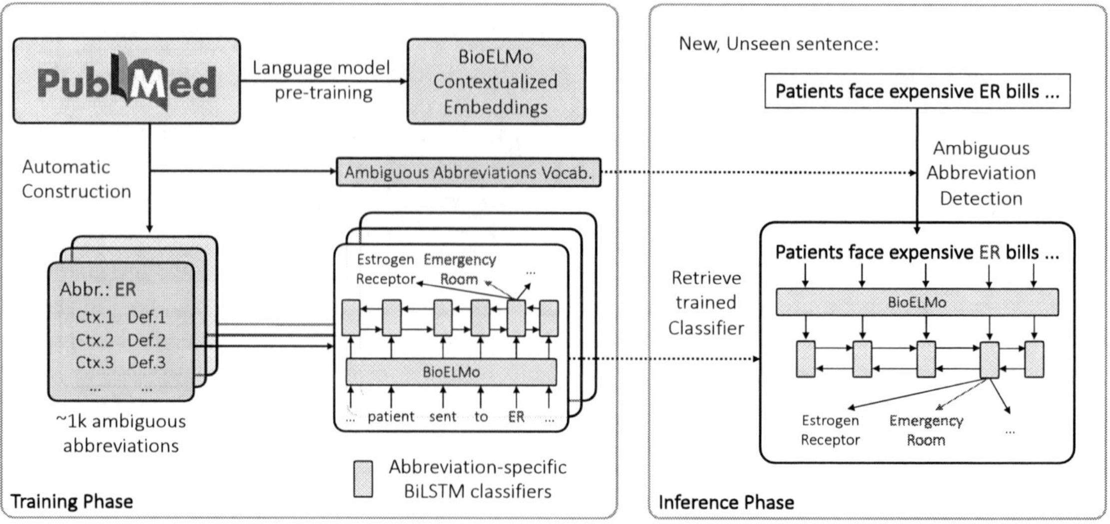

Figure 1: Architecture of DECBAE. Training and inference phases are illustrated in the left and right boxes, respectively. The PubMed corpus is used in training BioELMo (Jin et al., 2019) and collecting the disambiguation dataset. We train a separate biLSTM classifier for each abbreviation, and the specific pre-trained classifier is retrieved in inference phase.

pansion datasets from PubMed (§3.1). We use BioELMo (§3.2) to get the contextualized representations of words, and train a specific biLSTM classifier (§3.3) for each abbreviation. During inference (§3.5), we first detect whether there are ambiguous abbreviations in input sentences by the expert-curated ambiguous abbreviation vocabulary. If so, we use BioELMo and the corresponding abbreviation-specific biLSTM classifiers to do the disambiguation.

3.1 Dataset Collection

Figure 2 shows our approach of automatically collecting disambiguation dataset. For each abstract, we first detect and extract the pattern of "*Definition (Abbreviation)*", e.g.: "endoplasmic reticulum (ER)". Then we collect all the following sentences that contain the abbreviation, and label them with the definition.

This would generate a noisy label set due to the variations of writing the same definition (e.g.: emergency department and emergency departments). To group the same definitions together, we use MetaMap-derived MeSH terms (Demner-Fushman et al., 2017) as features of definitions and define the MeSH similarity between definition a and definition b as:

$$s = \frac{|\mathcal{M}_a \cap \mathcal{M}_b|}{\sqrt{|\mathcal{M}_a|\,|\mathcal{M}_b|}}$$

where $\mathcal{M}_a$ and $\mathcal{M}_b$ are the MeSH term sets of definition a and b, respectively. We group those definitions with high MeSH similarity and close edit distance by heuristic thresholds.

We collected 1970 abbreviations. However, due to the unsupervised nature of the collection process, some abbreviations are invalid or not ambiguous. For this, one biomedical expert[3] filtered the abbreviations we found, based on 1) **Validity**: abbreviations should be biomedically meaningful; 2) **Ambiguity**: abbreviations should have multiple possible definitions, and prevalence of the dominant one should be $< 99\%$. After the filtering, there are 950 valid ambiguous abbreviations. Their statistics are shown in Table 1. We split the instances of each abbreviation into training, development and test sets: If there is more than 10k instances, we randomly select 1k for both development and test sets. Otherwise, we randomly select 10% of all instances for both development and test sets.

3.2 BioELMo

BioELMo is a biomedical version of ELMo pretrained on 10 millions of PubMed abstracts (Jin et al., 2019). It serves as a contextualized feature extractor in DECBAE: given an input sentence of

---

[3]A post-doctoral fellow with a Ph.D. degree in biology.

The endoplasmic reticulum: structure, function and response to cellular signaling.

Abstract
The endoplasmic reticulum (ER) is a large, dynamic structure that serves many roles in the cell including calcium storage, protein synthesis and lipid metabolism. The diverse functions of the ER are performed by distinct domains; consisting of tubules, sheets and the nuclear envelope. Several proteins that contribute to the overall architecture and dynamics of the ER have been identified, but many questions remain as to how the ER changes shape in response to cellular cues, cell type, cell cycle state and during development of the organism. Here we discuss what is known about the dynamics of the ER, what questions remain, and how coordinated responses add to the layers of regulation in this dynamic organelle.

Context: ... functions of the ER are performed by ...   *Definition*: endoplasmic reticulum
Context: ... dynamics of the ER have been identified ...   *Definition*: endoplasmic reticulum
Context: ... as to how the ER changes shape ...   *Definition*: endoplasmic reticulum
Context: ... dynamics of the ER, what questions ...   *Definition*: endoplasmic reticulum

Figure 2: An example of automatically generated training instances for disambiguation from the abstract of Schwarz and Blower (2016). In this case, we extract "endoplasmic reticulum" as the definition for all ER mentions in the abstract, and store those instances to the dataset.

Statistic	Whole	Random	Imbalanced	Low-resources	Clinical	Human
# of all abbreviations	950	100	42	28	11	1
Average # of instances	8790.0	6564.3	19493.1	958.8	28642.8	8312.0
Average # of possible definitions	4.1	3.7	2.3	2.2	8.5	4.0
Average % of dominant definition	64.1	63.5	96.7	66.7	53.3	63.8

Table 1: Statistics of the automatically generated abbreviation disambiguation dataset and its subsets.

L tokens:

$$\text{input} = [t_1; t_2; ...; t_L]$$

We use BioELMo to embed it to

$$\mathbf{E} = [\mathbf{e_1}; \mathbf{e_2}; ...; \mathbf{e_L}] \in \mathbb{R}^{L \times D}$$

where $\mathbf{e} \in \mathbb{R}^D$ is the token embedding and D is the embedding dimension[4].

3.3 Abbreviation-specific biLSTM Classifiers

For each abbreviation, we train a specific biLSTM classifier, denoted as biLSTM_i for abbreviation i. We feed the BioELMo representations of sentences containing abbreviation i to biLSTM_i:

$$\text{biLSTM}_i(\mathbf{E}) = [\mathbf{h_1}; \mathbf{h_2}; ...; \mathbf{h_L}] \in \mathbb{R}^{L \times 2H}$$

where $\mathbf{h} \in \mathbb{R}^{2H}$ is the concatenation of forward and backward hidden states of the biLSTM. We take as input the concatenated hidden states of the abbreviation i (i.e. the ambiguous token) $\mathbf{h_a}$ and use several feed-forward neural network (FFN)

layers with softmax output unit to predict its definition:

$$p(\text{def}_k \,|\, \text{input}) \propto \exp(\mathbf{w_k^T} \, \text{FFN}_i(\mathbf{h_a}))$$

where $\mathbf{w_k}$ is the learnt weight vector corresponding to definition k, and def_k is the k-th definition of abbreviation i in our dataset. Similarly, we train FFN separately for different abbreviations.

3.4 Training

The weights of BioELMo are pre-trained and fixed, while the averaging weights and scaling factor of BioELMo embeddings are trained separately for each abbreviation along with the abbreviation-specific biLSTM classifiers. We use Adam (Kingma and Ba, 2014) to optimize the cross-entropy loss of the predicted label and ground-truth label.

3.5 Inference

At inference time, we denote the tokenized input sentence as $[t_1; t_2; ...; t_L]$ and our ambiguous abbreviation set as $\mathcal{A}$. If $\exists t_j \in \mathcal{A}$, we run DECBAE to expand the t_j: First, we use BioELMo to compute the representations of all the input tokens to

---

[4]Note that it's after scaling and averaging the 3 BioELMo layers using task-specific weights.

$\mathbf{E} = [\mathbf{e_1}; \mathbf{e_2}; ...; \mathbf{e_L}]$. The trained biLSTM for abbreviation t_j, denoted as biLSTM_{t_j}, is retrieved and used to calculate the hidden states given the BioELMo embeddings of the input sentence:

$$\text{biLSTM}_{t_j}(\mathbf{E}) = [\mathbf{h_1}; \mathbf{h_2}; ...; \mathbf{h_L}] \in \mathbb{R}^{L \times 2H}$$

Then $\mathbf{h_{t_j}}$, which is the concatenated hidden states of the ambiguous abbreviation t_j, is used for disambiguation through the trained abbreviation-specific FFN:

$$\text{Definition}(t_j) = \text{def}_{\operatorname{argmax}_k \mathbf{w_k^T} \text{FFN}_{t_j}(\mathbf{h_{t_j}})}$$

4 Experiments

4.1 Baseline Settings

A trivial baseline is to predict the majority of definition for all cases, which could still lead to high accuracy in severely imbalanced datasets. We denote this method as **Majority**. We also test other baseline settings of different feature learning schemes. They are all followed by several FFN layers and a softmax output unit.

Bag-of-words: Following most of the previous works, we use bag-of-words features to represent the context by $\mathbf{c} \in \mathbb{R}^{|\mathcal{V}|}$, where $|\mathcal{V}|$ is the vocabulary size.

BioELMo: We take the BioELMo embeddings of the ambiguous abbreviations as input features.

biLSTM: We use biomedical w2v (Moen and Ananiadou) as word embeddings and train task-specific biLSTMs and use the hidden states of the ambiguous abbreviations as input features.

We also measure the **human performance**: due to limitation of resources, we just study single-expert performance on one sampled abbreviation. For this, the expert is shown with the test sentences, and asked to classify the ambiguous abbreviation to its possible definitions. An ensemble of experts will obviously generate better results, so our single-human results just represent the lower bound of human performance.

4.2 Subset Settings

We report the model performance on different subsets of our dataset. Statistics of those datasets are shown in Table 1.

Random samples: It's computationally expensive[5] and unnecessary to test the models on all 950

abbreviations. Instead, we use randomly sampled 100 abbreviations to represent the whole set.

Imbalanced samples: We define abbreviations whose dominant definitions have over 95% frequency as imbalanced samples. Multi-label classification with imbalanced classes is considered as a hard machine learning task.

Low-resources samples: We define abbreviations that have less than 1k training instances as low-resources samples. It's motivated by the fact that most biomedical datasets are typically limited by scale, so models that can still perform well under low-resources settings have the potential to be applied in real world settings.

Clinical samples: Though our abbreviations are collected from PubMed abstracts, we have included 11 out of 13 of clinical ambiguous abbreviations mentioned in a previous work of clinical abbreviation disambiguation (Xu et al., 2012). We also test our models on the subset of these 11 clinically related abbreviations.

Testing sample for human expert: We test human performance on one abbreviation (DAT), due to limited resources. The statistics of DAT abbreviation expansion dataset are close to the averages of the whole dataset, as shown in Table 1. Possible definitions of DAT include: 1) Dopamine transporter (63.9%); 2) Direct antiglobulin test (5.8%); 3) Direct agglutination test (5.8%); 4) Dementia of the Alzheimer type (24.5%).

4.3 Evaluation Metrics

We model abbreviation expansion as a multi-label classification task, and use the following metrics to measure the performance of different models:

Accuracy: Accuracy is defined as the proportion of right predictions in all predictions. Most of the definition labels are imbalanced, so accuracy could be misleadingly high for a trivial majority solution in these cases, thus may not reflect the real capability of models.

Macro-F1: In multi-label classification, macro-F1 is calculated as an unweighted average of F1 score for each class. Class-wise F1 score is defined as follows:

$$F1 = 2 \cdot \frac{\text{precision} \cdot \text{recall}}{\text{precision} + \text{recall}}$$

where precision and recall are calculated for each class.

Kappa Statistic: Cohen's kappa was originally introduced as a metric to measure inter-rater

---

[5] The rate-determining step is BioELMo due to its large size and recurrent nature.

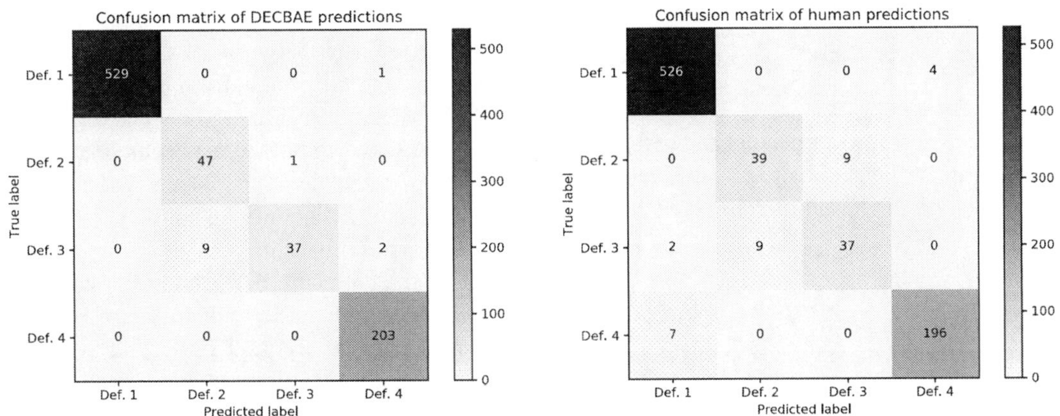

Figure 3: Confusion matrix for the predictions of DECBAE (left) and the human expert (right). Def. 1: dopamine transporter; Def. 2: direct antiglobulin test; Def. 3: direct agglutination test; Def. 4: dementia of the Alzheimer type.

Model	Random Set	Imbalanced Set	Low-resources Set	Clinical Set	Human Set
Majority					
Accuracy	$63.6 \pm 21.0^\dagger$	$96.7 \pm 1.0^\dagger$	$67.0 \pm 15.6^\dagger$	$53.3 \pm 25.7^\dagger$	63.9
Macro-F1	$28.3 \pm 14.9^\dagger$	$45.4 \pm 8.8^\dagger$	$37.2 \pm 8.8^\dagger$	$12.0 \pm 10.6^\dagger$	19.5
Kappa Statistic	$0.0 \pm 0.0^\dagger$	$0.0 \pm 0.0^\dagger$	$0.0 \pm 0.0^\dagger$	$0.0 \pm 0.0^\dagger$	0.0
BoW-FFN					
Accuracy	$84.4 \pm 11.2^\dagger$	$97.5 \pm 1.7^\dagger$	$89.6 \pm 7.5^\dagger$	$76.1 \pm 12.5^\dagger$	84.3
Macro-F1	$73.1 \pm 17.1^\dagger$	$71.5 \pm 19.9^\dagger$	$83.4 \pm 14.6^\dagger$	$57.9 \pm 14.2^\dagger$	71.9
Kappa Statistic	$63.8 \pm 25.3^\dagger$	$50.4 \pm 33.7^\dagger$	$71.1 \pm 24.8^\dagger$	$60.6 \pm 8.9^\dagger$	69.6
BioELMo					
Accuracy	$94.1 \pm 7.2^\dagger$	96.3 ± 15.3	98.1 ± 2.7	91.1 ± 8.4	97.1
Macro-F1	$86.0 \pm 17.4^\dagger$	$81.3 \pm 23.5^\dagger$	95.4 ± 9.3	75.5 ± 21.7	92.6
Kappa Statistic	$86.1 \pm 19.8^\dagger$	$73.2 \pm 34.2^\dagger$	$93.2 \pm 10.8^\dagger$	86.6 ± 9.3	94.6
biLSTM					
Accuracy	$88.0 \pm 16.8^\dagger$	$98.0 \pm 1.9^\dagger$	$92.7 \pm 10.5^\dagger$	$88.2 \pm 8.2^\dagger$	97.3
Macro-F1	$77.1 \pm 26.0^\dagger$	$70.2 \pm 27.0^\dagger$	$82.9 \pm 24.5^\dagger$	68.8 ± 26.1	93.2
Kappa Statistic	$69.3 \pm 37.2^\dagger$	$49.1 \pm 45.7^\dagger$	$70.4 \pm 41.5^\dagger$	70.5 ± 35.3	94.9
DECBAE					
Accuracy	$\mathbf{96.1 \pm 5.5}$	$\mathbf{98.9 \pm 1.4}$	$\mathbf{98.7 \pm 2.2}$	$\mathbf{95.1 \pm 3.3}$	**98.4**
Macro-F1	$\mathbf{91.7 \pm 13.2}$	$\mathbf{87.2 \pm 17.8}$	$\mathbf{98.3 \pm 3.5}$	$\mathbf{83.0 \pm 21.9}$	**93.9**
Kappa Statistic	$\mathbf{90.9 \pm 15.5}$	$\mathbf{79.6 \pm 30.2}$	$\mathbf{96.8 \pm 6.8}$	$\mathbf{91.7 \pm 5.5}$	**97.0**
Human Expert					
Accuracy	–	–	–	–	96.3
Macro-F1	–	–	–	–	89.0
Kappa Statistic	–	–	–	–	92.8

Table 2: Mean and standard deviation of model performance on different subsets. †Significantly lower than the corresponding metric of DECBAE. Significance is defined by $p < 0.05$ in paired t-test. All numbers are in percentages. High deviations are expected due to the variety of abbreviations in each subset.

agreement (Cohen, 1960). It can also be used to evaluate predictions of multi-label classification:

$$\kappa = \frac{p_o - p_e}{1 - p_e}$$

where p_o is the observed agreement and in the case of classification $p_o = $ accuracy, p_e is the expected agreement which can be achieved by pure chance:

$$p_e = \sum_c p_c \hat{p}_c$$

p_c and $\hat{p}_c$ refer to the proportion of class c in ground truth labels and predictions, respectively. Empirical results in Table 2 show that Kappa statistics are often lower than accuracy and macro-F1, and thus serving as a more distinctive metric for our task.

4.4 Results

In Table 2, we report means and standard deviations of each model's performance on different subsets evaluated by the three metrics. In all subsets, DECBAE performs significantly better than most other models by large margins. A general trend of DECBAE > BioELMo > biLSTM > BoW-FFN > Majority conserves across subsets.

In the **Random** subset which represents the whole dataset, all metrics of DECBAE exceed 0.90, setting very promising state-of-the-art performance despite the potential noise of the dataset.

In the **Imbalanced** subset where the most frequent definitions consist of over 95% of all the labels, a trivial Majority solution gets over 95% accuracy. However, for macro-F1 and kappa statistic, performance of the baselines drop dramatically while DECBAE can still generate decent results.

DECBAE and BioELMo alone remain robust in **Low-resources** setting. This is due to the transfer learning nature of BioELMo, which utilizes the knowledge encoded in the PubMed abstracts.

Our abbreviation expansion dataset covers roughly 85% of clinical abbreviations mentioned in Xu et al. (2012). On this **Clinical** subset, DECBAE gets pretty good results and vastly outperform other baselines despite its variety in possible definitions (8.5 possible definitions per abbreviation, as shown in Table 1).

On the testset for human performance (i.e.: abbreviation expansion for DAT), DECBAE and even some neural baselines outperform single human expert.

5 Analysis

In Fig. 3, we use confusion matrices to visualize the differences between DECBAE or the human expert and the ground truth labels, for disambiguation of abbreviation "DAT". The high agreement level between human expert predictions and the automatically assigned labels indicates that our pipeline of collecting the abbreviation disambiguation dataset is valid.

In general, both DECBAE and the human expert perform well in the task, with only few misclassifications. Specifically, DECBAE, and even other neural baselines like biLSTM and BioELMo, outperform the human expert in all metrics. Compared to DECBAE, the human expert is more likely to misclassify direct agglutination test with direct antiglobulin test (9 v.s. 1), and misclassify dementia of the Alzheimer type with dopamine transporter (7 v.s. 0). We show several instances of human and DECBAE's errors in Table 3.

One limitation of this work is that we just test DECBAE on our automatically collected dataset. Since the proposed model can also be used on other biomedical abbreviation expansion datasets as well, evaluating on other datasets like MSH WSD is a clear future work to do.

Another potential direction for improvement is to accelerate the inference speed. Currently DECBAE uses BioELMo for embedding and abbreviation-specific biLSTM for classification, resulting in two recurrent models in total. Our results show that just BioELMo with several FFN layers also generates decent results, so in some cases we might use only BioELMo as a compromise for faster inference.

6 Conclusion

We present DECBAE, a state-of-the-art biomedical abbreviation expansion model on the automatically collected dataset from PubMed. The results show that, with only minimum expert involvement, we can still perform well in such a domain-specific task by automatically collecting training data from a large corpus and utilize embeddings from pre-trained biomedical language models.

7 Acknowledgement

We are grateful for the annonymous reviewers of BioNLP 2019 who gave us very insightful comments and suggestions. J.L. is supported by NLM training grant 5T15LM007059-32.

Test sentence	Label	Human	DECBAE
The reduction of the number of different segments in **DAT** compared to controls and patients suffering from depression may be helpful for differential diagnosis.	Def. 4	Def. 1	**Def. 4**
Reliance on objective brain phenotype measures, for example, those afforded by brain imaging, might critically improve detection of **DAT** genotype-phenotype association.	Def. 1	**Def. 1**	Def. 4
DAT was more commonly positive among BO incompatible (21.5% in BO vs. 14.8% in AO , P=0.001) and black (18.8% in blacks vs. 10.8% in nonblacks , P=0.003) infants.	Def. 2	Def. 3	**Def. 2**
NPY-LI showed a significant reduction in **DAT** but not in FTD.	Def. 4	Def. 1	**Def. 4**
The study included 122 healthy subjects, aged 18-83 years, recruited in the multicentre 'ENC-**DAT**' study (promoted by the European Association of Nuclear Medicine).	Def. 1	Def. 4	**Def. 1**

Table 3: Some samples of errors made by the human expert and DECBAE. Def. 1: dopamine transporter; Def. 2: direct antiglobulin test; Def. 3: direct agglutination test; Def. 4: dementia of the Alzheimer type.

References

Rui Antunes and Sérgio Matos. Supervised learning and knowledge-based approaches applied to biomedical word sense disambiguation. *Journal of integrative bioinformatics*, 14(4).

José Camacho-Collados, Mohammad Taher Pilehvar, and Roberto Navigli. 2016. Nasari: Integrating explicit knowledge and corpus statistics for a multilingual representation of concepts and entities. *Artificial Intelligence*, 240:36–64.

Manuel R Ciosici, Tobias Sommer, and Ira Assent. 2019. Unsupervised abbreviation disambiguation. *arXiv preprint arXiv:1904.00929*.

Jacob Cohen. 1960. A coefficient of agreement for nominal scales. *Educational and psychological measurement*, 20(1):37–46.

Dina Demner-Fushman, Willie J Rogers, and Alan R Aronson. 2017. Metamap lite: an evaluation of a new java implementation of metamap. *Journal of the American Medical Informatics Association*, 24(4):841–844.

Jacob Devlin, Ming-Wei Chang, Kenton Lee, and Kristina Toutanova. 2018. Bert: Pre-training of deep bidirectional transformers for language understanding. *arXiv preprint arXiv:1810.04805*.

Ignacio Iacobacci, Mohammad Taher Pilehvar, and Roberto Navigli. 2016. Embeddings for word sense disambiguation: An evaluation study. In *Proceedings of the 54th Annual Meeting of the Association for Computational Linguistics (Volume 1: Long Papers)*, volume 1, pages 897–907.

Rezarta Islamaj Dogan, G Craig Murray, Aurélie Névéol, and Zhiyong Lu. 2009. Understanding pubmed® user search behavior through log analysis. *Database*, 2009.

Antonio J Jimeno-Yepes, Bridget T McInnes, and Alan R Aronson. 2011. Exploiting mesh indexing in medline to generate a data set for word sense disambiguation. *BMC bioinformatics*, 12(1):223.

Qiao Jin, Bhuwan Dhingra, William W Cohen, and Xinghua Lu. 2019. Probing biomedical embeddings from language models. *arXiv preprint arXiv:1904.02181*.

Diederik P Kingma and Jimmy Ba. 2014. Adam: A method for stochastic optimization. *arXiv preprint arXiv:1412.6980*.

SPFGH Moen and Tapio Salakoski2 Sophia Ananiadou. Distributional semantics resources for biomedical text processing.

Roberto Navigli, David Jurgens, and Daniele Vannella. 2013. Semeval-2013 task 12: Multilingual word sense disambiguation. In *Second Joint Conference on Lexical and Computational Semantics (* SEM), Volume 2: Proceedings of the Seventh International Workshop on Semantic Evaluation (SemEval 2013)*, volume 2, pages 222–231.

Ahmad Pesaranghader, Stan Matwin, Marina Sokolova, and Ali Pesaranghader. 2019. deepbiowsd: effective deep neural word sense disambiguation of biomedical text data. *Journal of the American Medical Informatics Association*, 26(5):438–446.

Matthew E Peters, Mark Neumann, Mohit Iyyer, Matt Gardner, Christopher Clark, Kenton Lee, and Luke Zettlemoyer. 2018. Deep contextualized word representations. *arXiv preprint arXiv:1802.05365*.

Alessandro Raganato, Claudio Delli Bovi, and Roberto Navigli. 2017a. Neural sequence learning models for word sense disambiguation. In *Proceedings of the 2017 Conference on Empirical Methods in Natural Language Processing*, pages 1156–1167.

Alessandro Raganato, Jose Camacho-Collados, and Roberto Navigli. 2017b. Word sense disambiguation: A unified evaluation framework and empirical comparison. In *Proceedings of the 15th Conference of the European Chapter of the Association for Computational Linguistics: Volume 1, Long Papers*, pages 99–110.

Dianne S Schwarz and Michael D Blower. 2016. The endoplasmic reticulum: structure, function and response to cellular signaling. *Cellular and Molecular Life Sciences*, 73(1):79–94.

Hua Xu, Peter D Stetson, and Carol Friedman. 2007. A study of abbreviations in clinical notes. In *AMIA annual symposium proceedings*, volume 2007, page 821. American Medical Informatics Association.

Hua Xu, Peter D Stetson, and Carol Friedman. 2012. Combining corpus-derived sense profiles with estimated frequency information to disambiguate clinical abbreviations. In *AMIA Annual Symposium Proceedings*, volume 2012, page 1004. American Medical Informatics Association.

Jun Xu, Yaoyun Zhang, Hua Xu, et al. 2015. Clinical abbreviation disambiguation using neural word embeddings. *Proceedings of BioNLP 15*, pages 171–176.

Antonio Jimeno Yepes. 2017. Word embeddings and recurrent neural networks based on long-short term memory nodes in supervised biomedical word sense disambiguation. *Journal of biomedical informatics*, 73:137–147.

Hong Yu, Won Kim, Vasileios Hatzivassiloglou, and W John Wilbur. 2007. Using medline as a knowledge source for disambiguating abbreviations and acronyms in full-text biomedical journal articles. *Journal of biomedical informatics*, 40(2):150–159.

RNN Embeddings for Identifying Difficult to Understand Medical Words

Hanna Pylieva[1], Artem Chernodub[1, 2], Natalia Grabar[3], and Thierry Hamon[4,5]

[1]Faculty of Applied Sciences, Ukrainian Catholic University, Lviv, Ukraine
{pylieva,chernodub}@ucu.edu.ua
[2]Grammarly, Kyiv, Ukraine
[3]CNRS, Univ. Lille, UMR 8163 - STL - Savoirs Textes Langage, F-59000, Lille, France
natalia.grabar@univ-lille.fr
[4]LIMSI, CNRS, Université Paris-Saclay, F-91405, Orsay, France
hamon@limsi.fr
[5]Université Paris 13, Sorbonne Paris Cité, F-93430, Villetaneuse, France

Abstract

Patients and their families often require a better understanding of medical information provided by doctors. We currently address this issue by improving the identification of difficult to understand medical words. We introduce novel embeddings received from RNN - FrnnMUTE (French RNN Medical Understandability Text Embeddings) which allow to reach up to 87.0 F1 score in identification of difficult words. We also note that adding pre-trained FastText word embeddings to the feature set substantially improves the performance of the model which classifies words according to their difficulty. We study the generalizability of different models through three cross-validation scenarios which allow testing classifiers in real-world conditions: understanding of medical words by new users, and classification of new unseen words by the automatic models. The RNN - FrnnMUTE embeddings and the categorization code are being made available for the research.

1 Introduction

Specialized areas, such as medical area, convey and use technical words, or terms, which are typically related to knowledge developed within these areas. In the medical area, this specific knowledge often corresponds to fundamental medical notions related to disorders, procedures, treatments, and human anatomy. For instance, technical terms like *blepharospasm* (abnormal contraction or twitch of the eyelid), *alexithymia* (inability to identify and describe emotions in the self), *appendicectomy* (surgical removal of the vermiform appendix from intestine), or *lombalgia* (low back pain) are frequently used by experts in the medical area texts. As in any specialized areas, two main kinds of users exist in the medical area: experts of the domain, i.e. medical doctors, both researchers or practitioners; consumers of the healthcare process, i.e. patients and their relatives. The latter usually do not have expert knowledge in the medical domain, while it is important that they understand the purpose and issues of their healthcare process.

The existing literature provides several studies dedicated to the understanding of medical notions and terms by non-expert users, and of how the level of health literacy of patients impacts on a successful healthcare process (McCray, 2005; Eysenbach, 2007), as indeed it is quite common that patients and their relatives must face technical health documents and information. These observations provide the main motivation for our work. Hence, we address the need of non-specialized users in the medical domain to understand medical and health information.

In this paper, we propose to apply deep learning techniques to improve identification of readability and understandability of medical words by non-expert users. In particular, we will tackle the word categorization task and compare the performance of classification model on different feature sets: standard linguistic and non-linguistic features described in section 4, features obtained using different deep learning approaches, and the combinations of all of them. We also investigate how different feature sets perform with three different cross-validation settings, described in section 5. The medical data used in this work are in French. Three human annotators participated in the creation of the reference data (specifying the understandability of words).

2 Related Work

Related work is globally positioned in the text simplification task which involves the detection of complex contents in documents and their adaptation for the target population. We are also interested in the first aspect with additional constraints:

Proceedings of the BioNLP 2019 workshop, pages 97–104
Florence, Italy, August 1, 2019. ©2019 Association for Computational Linguistics

detection and diagnosis of technical contents in texts from medical domain.

Traditional readability measures rely on two main factors: the familiarity of semantic units such as words or phrases, and the complexity of syntax. To make these measures straightforward for applications, some simplifying assumptions were used. As a result, final formulas mostly rely on the number of letters and/or of syllables a word contains and on linear regression models (Flesch, 1948; Gunning, 1973). While such readability measures are easy to compute, they are based on shallow characteristics of text, ignoring deeper levels of text processing which are important factors in readability, such as cohesion, syntactic ambiguity, rhetorical organization, and propositional density (Collins-Thompson, 2014). Moreover, traditional readability measures were demonstrated to be unreliable for non-traditional documents (Si and P. Callan, 2001). As a result of such limitations and due to the recent growth of computational and data resources, the focus of NLP researchers moved to *computational* readability measurements, which rely on the use of machine learning algorithms on richer linguistic features (Malmasi et al., 2016; Ronzano et al., 2016; Bingel et al., 2016).

Not so much effort has been devoted to the exploitation of NLP potential in the measurement of readability of medical texts. In the biomedical domain, as well as in general language, the readability assessment is currently approached as a classification task. The difference is that in the former a much smaller variety of features has been tested: a combination of classical readability formulas with medical terminologies (Kokkinakis and Toporowska Gronostaj, 2006); n-grams of characters (Poprat et al., 2006); stylistic (Grabar et al., 2007) or discursive (Goeuriot et al., 2008) features; morphological features (Chmielik and Grabar, 2011); combinations of different features from those listed above (Zeng-Treiler et al., 2007). Among the recent experiments dedicated to readability study in the medical domain are, for example, manual rating of medical words (Zheng et al., 2002), automatic rating of medical words on the basis of their presence in different vocabularies (Borst et al., 2008), exploitation of machine learning approach with various features (Grabar et al., 2014). The last experiment achieved up to 85.0 F-score on individual annotations.

Due to the recent significant advance in readability study in general language and relatively slow progress with the task in the medical area, there is a great potential of experimenting with the machine learning-based approaches on medical texts. This fact motivated us for choosing this kind of methodology.

3 Materials

For the experiments, we used the publicly available set of words with annotations[1]. The process of words collection and annotation is briefly described below.

3.1 Linguistic data description

The set of required biomedical terms was obtained from the French part of Snomed International[2] (Côté et al., 1993). Snomed Int contains 151,104 medical terms organized into eleven semantic axes such as disorders and abnormalities, procedures, chemical products, living organisms, anatomy, social status, etc. For the word understandability study, five axes related to the main medical notions were chosen: disorders, abnormalities, procedures, functions, and anatomy. These categories are assumed to be faced frequently by layman. In contrast, chemical products and living organisms are excluded because they mainly correspond to Latin borrowings and are typically non-understandable by laypeople.

The 104,649 selected terms were then processed. First, they were tokenized, POS-tagged and lemmatized using TreeTagger (Schmid, 1994). Then the lemmatization was checked with FLEMM (Namer, 2000). After that we received 29,641 unique words. For instance, the term *trisulfure d'hydrogène* provided three words (*trisulfure, de, hydrogène*). The final dataset contains compound words which contain several bases (*abdominoplastie* (abdominoplasty), *dermabrasion* (dermabrasion)), constructed words which contain one base and at least one affix (*lipoïde* (lipoid), *cardiaque* (cardiac)), simple words which contain one base, no affixes and possibly infections when the lemmatization fails (*acné* (acne), *fragment* (fragment)).

---

[1]`http://natalia.grabar.free.fr/resources.php#rated`
[2]`https://esante.gouv.fr/terminologie-snomed-35vf`

Annotators / Categories	Cat1	Cat2	Cat3	Total
O1 (%)	8,099 (28)	1,895 (6)	19,647 (66)	29,641
O2 (%)	8,625 (29)	1,062 (4)	19,954 (67)	29,641
O3 (%)	7,529 (25)	1,431 (5)	20,681 (70)	29,641

Table 1: Number (and percentage) of words assigned to reference categories by seven annotators (O1, O2, O3).

3.2 Annotation process

The set of 29,641 unique words was annotated by three French speakers, 25-40-year-old, without medical training, without specific medical problems, but with the linguistic background. The annotators were expected to represent the average knowledge of medical words among the population as a whole. They were presented with a list of terms and asked to assign each word to one of the three categories: (Cat1) *I can understand the word*; (Cat2) *I am not sure about the meaning of the word*; (Cat3) *I cannot understand the word*. The annotators were asked not to use dictionaries during the annotation process. The interannotator agreement shows substantial agreement: Fleiss' Kappa 0.735 and Cohen's Kappa 0.736. This is a very good result, especially when working with linguistic data for which the agreement is usually difficult to obtain. The annotation results are represented in Table 1.

4 Method

We aim to categorize medical words according to whether they can be understood or not by non-specialized people, using features obtained with NLP tools and with deep learning methods. The manual annotations of these words described in the previous section provide the reference data. The proposed method includes calculation of NLP features associated with the annotated words, training machine learning models for word classification, and evaluation of classification using cross-validation.

4.1 Feature sets

We distinguish and use two kinds of features: standard features provided by the NLP analysis of words, and features issued from existing or specifically trained word embeddings. These two types of features are first opposed and then combined.

4.1.1 Standard NLP features

The standard NLP features include 24 linguistic and extra-linguistic features related to general and specialized languages. The features are computed automatically and can be grouped into ten classes:

- *Syntactic categories.* Syntactic categories and lemmas are computed by TreeTagger (Schmid, 1994) and then enriched by FLEMM (Namer, 2000).

- *Presence of words in reference lexica.* Two reference lexica of the French language were exploited: TLFi[3] and *lexique.org*[4]. TLFi is a dictionary of the French language covering XIX and XX centuries. It contains almost 100,000 entries. *lexique.org* is a lexicon created for psycholinguistic experiments. It contains over 135,000 entries, among which inflectional forms of verbs, adjectives and nouns, and almost 35,000 lemmas.

- *Frequency of words through a non specialized search engine.* Each word were queried on Google to find out the frequency of the word on the web.

- *Frequency of words in the medical terminology.* The frequency of words in the medical terminology Snomed Int corresponds to the number of different terms containing a given word.

- *Number and types of semantic categories associated to words.* The information on the semantic categories of Snomed Int was exploited.

- *Length of words in number of their characters and syllables.* For each word, the number of its characters and syllables was computed.

- *Number of bases and affixes.* Each lemma was analyzed by the morphological analyzer Dérif (Namer and Zweigenbaum, 2004), adapted to the treatment of medical words. It performs the decomposition of lemmas into bases and affixes known in its database

---

[3] http://www.atilf.fr/
[4] http://www.lexique.org/

and it provides also semantic explanation of the analyzed lexemes. The morphological decomposition information (number of affixes and bases) was exploited. For instance, *hématomètre* (*haemometer*) is analyzed and decomposed into two basis (*hémato* meaning *blood* and *mètre* meaning *measure*, while *myélite* (*myelitis*) is decomposed into *myél* meaning *marrow* and *ite* meaning *inflammation*.

- *Initial and final substrings of the words.* Initial and final substrings of different length, from three to five characters, were computed.

- *Number and percentage of consonants, vowels and other characters.* The number and the percentage of consonants, vowels and other characters (i.e., hyphen, apostrophe, comas) was computed.

- *Classical readability scores.* Two classical readability measures were applied: Flesch (Flesch, 1948) and its variant Flesch-Kincaid (Kincaid et al., 1975). Such measures are typically used for evaluating the difficulty level of a text.

4.1.2 FastText word embeddings usage.

FastText word embeddings (Bojanowski et al., 2017) is a good candidate feature for the detection of word difficulty because they are able to use the morphological information of words and generalize over it. Since the word embeddings capture context and morphological information, we assume that using them as features will improve classification accuracy for our specific problem.

We note that FastText word embeddings trained on Wikipedia and Common Crawl[5] texts have an important part of words from our dataset. According to our analysis, the currently published FastText[6] model for French contains 44.26% (13,118 out of 29,641) medical words from our dataset and up to 56.00% (16,598 out of 29,641) lowercased medical words from our dataset.

4.1.3 French RNN Medical Understandability Text Embeddings (FrnnMUTE).

According to the general functionality of RNNs, the final hidden state aggregates the informa-

tion about all input sequence. This idea is frequently used to receive hidden representations of sequences. Sequence-to-sequence model is a well-known example of how this idea works in practice (Sutskever et al., 2014). Such models consist of two parts: an *encoder* is an RNN which encodes the input sequence into a representation in hidden space (which is also called *thought vector*), and a *decoder* which generates a new sequence out of the hidden representations.

We used this idea for representing words from our dataset. To receive words representations from an RNN, we first trained it to classify words based on labels by one annotator (we chose $O1$), then for each word we collect values of the last hidden state of the RNN and use this vector as features during the detection of words understandability for different users (or annotators). Train/test split was 70%/30% of randomly shuffled samples.

As a direct classifier, we trained a character-level RNN using PyTorch framework[7] and one GPU Tesla K80. We lowercased all words, lemmatized them and substituted all Unicode symbols with their ASCII analogs. We tested several RNN architectures and hyperparameter sets. The best performance was reached with a model consisting of two unidirectional long short-term memory (LSTM) units, each with 50 hidden units. The dropout of the model is 0.7. The input size is 57 as the number of unique characters in lowercased and converted to ASCII input words. The output size is 3 as the number of classes in our data. This model reached the best performance on the eighth epoch with $F1 = 78.94$ and $accuracy = 81.21\%$ on development set. Using this model we received 50-dimensional word representations which we called FrnnMUTE (French RNN Medical Understandability Text Embeddings).

5 Experiments and Results

We study the impact of adding words embeddings as features for identifying difficult for understanding words. First, we observe how FastText word embeddings influence the quality of classification in different cross-validation scenarios. Then, we study how FrnnMUTE used as features impact on classification quality in all the same cross-validation scenarios. The quality of the classifications is evaluated using four standard *macroaveraging* (Sebastiani, 2002) measures: accuracy A,

---

[5] http://commoncrawl.org/
[6] https://fasttext.cc

[7] https://pytorch.org/

precision P, recall R and F1-measure F.

5.1 Cross-validation scenarios

For a thorough study of generalization abilities of the classification models, we propose to consider three distinct cross-validation scenarios based on different combinations of users and vocabulary in train and test sets.

5.1.1 User-in vocabulary-out cross-validation

The cross-validation is performed on each dataset (i.e., each user annotation) separately. We aim to measure the ability of the classification model to generalize class recognition on the *known user* and to predict annotations for *unknown words*. From the practical perspective, *user-in* means learning the profile of a user. Hence, a model trained by such scenario represents the word understanding or knowledge of the annotator.

The experiments use (i) the standard features only, (ii) the FastText word embeddings only and (iii) their combination. The experiments with isolated FastText word embeddings as features resulted in poor F1 scores (Table 2), that can be explained by the fact that contextual information, which is dominant in these word embeddings, is not enough to define the word understandability. Adding the FastText word embeddings to the standard feature set resulted in up to 1.0 higher F1 score due to higher Precision (up to 1.8), meaning that contextual information slightly impacts on the understandability of a word by a given person.

5.1.2 User-out vocabulary-in cross-validation

We then learn from all the annotations of one user and then test the model on annotations of another user. Thereby, in such a setting, we measure the ability of the classifier to generalize on all known words, but for unknown users. This scenario is realistic to a real-world situation: the reference annotations can be obtained only from a couple of users, presumably representing the overall population, but not from all the possible users. In this scenario, the model learns the profile of a user and we want to identify whether a new user has the same profile as another user. Then it can be used for identification of not understandable words for the new users.

These experiments show a substantial improvement of combined features in comparison to the standard features (Table 3). When knowledge of words understandability of one user is used to predict it for another user, adding the FastText word embeddings provides up to 2.9 better F1 score. Used separately, standard features and embeddings show similar performance as in user-in vocabulary-out cross-validation (Table 2). We assume that there exists a robust nonlinear dependency between some subsets of standard features and subword-level components of FastText word embeddings. Testing this hypothesis is the topic of future work.

5.1.3 User-out vocabulary-out cross-validation

Finally, we consider (k-1) folds of data from one user for training and use k-th fold for testing from the remaining user. We aim to measure the ability of the method to generalize both on *unknown users* and *unknown vocabulary*. This experiment should be helpful in identifying the number of words needed for determining whether the profile of one user is the same as profile of other users in case the model achieves good performance.

In these experiments, FastText word embeddings provide approximately 0.5% higher F1 score in case of learning on users O1 and O3 (Table 4). When learning on user O2, embeddings decrease F by 0.5, which means that annotations and health literacy of user O2 are different from users O1 and O3. It seems that adding embeddings makes overfitting the machine learning model to the dataset. As a result, tests on other "kind of word understandability" and combined features are less successful compared to using standard features only for learning. This may also be due to the lack of systematicity in annotations of O2.

5.2 FrnnMUTE impact study

The FrnnMUTE embeddings were used separately and in combination with standard features and with FastText word embeddings for classifying medical words with the decision tree algorithm. To simplify the process of analyzing and comparing the results of this and the previous part, we aggregated the resulting F1 scores for combinations of a feature set and cross-validation scenario over all available users (Table 5). We observed that, in all cross-validation scenarios, our FrnnMUTE performs better when used separately by comparison with the FastText word embeddings used separately. FrnnMUTE provides the maximal F1 score (79.5) among user pairs versus the F1 score provided by the FastText word embeddings in user-in

Train user	Test user	Standard features				FastText embeddings				Standard features + FastText embeddings			
		A	P	R	F	A	P	R	F	A	P	R	F
O1	O1	**82.5**	77.2	**82.5**	79.8	72.5	67	72.5	69.3	82.4	**79**	82.4	**80.2**
O2	O2	**82**	78.9	**82**	80	73.5	69.9	73.5	71.3	81.9	**79.5**	81.9	**80.3**
O3	O3	85.5	81.2	85.5	83.2	74.9	70.4	74.9	72.3	**85.9**	**83**	**85.9**	**84.2**

Table 2: Experiments on user-in vocabulary-out cross-validation. The best score for a combination of quality measure and experiment is in bold.

Train user	Test user	Standard features				FastText embeddings				Standard features + FastText embeddings			
		A	P	R	F	A	P	R	F	A	P	R	F
O1	O2	81.7	78.6	81.7	80.1	74	70.3	74	71.2	**84.2**	**82**	**84.2**	**82.8**
O1	O3	85	81.2	85	83	75.4	70.7	75.4	72.6	**87.6**	**84.9**	**87.6**	**85.9**
O2	O1	82.2	77	82.2	79.1	72.8	67.3	72.8	69.6	**83.9**	**80.2**	**83.9**	**81.1**
O2	O3	85.4	81.1	85.4	83	75.3	71.1	75.3	73	**86.8**	**83.5**	**86.8**	**84.7**
O3	O1	82.8	77.4	82.8	79.7	72.7	67.1	72.7	69.4	**84.9**	**81.3**	**84.9**	**82.4**
O3	O2	82.2	79	82.2	80.2	74.1	70.4	74.1	71.6	**84.2**	**82.1**	**84.2**	**82.8**

Table 3: Experiments on user-out vocabulary-in cross-validation.

Train user	Test user	Standard features				FastText embeddings				Standard features + FastText embeddings			
		A	P	R	F	A	P	R	F	A	P	R	F
O1	O2	81.7	78.6	81.7	80.1	73.6	69.9	73.6	71.3	**81.8**	**79.8**	**81.8**	**80.6**
O1	O3	**85**	81.2	**85**	83	74.8	70.4	74.8	72.4	84.9	82.2	84.9	83.4
O2	O1	**82.2**	76.9	**82.2**	**79.1**	72.5	66.9	72.5	69.3	81.7	**77.5**	81.7	**79.1**
O2	O3	**85.3**	81	**85.3**	**83**	75.1	70.7	75.1	72.7	84.4	**81.3**	84.4	82.5
O3	O2	**82.7**	77.3	**82.7**	79.7	72.5	66.9	72.5	69.2	82.6	**78.9**	82.6	**80.2**
O3	O3	82.1	79	82.1	80.1	73.8	70.2	73.8	71.4	**82.2**	**80**	**82.2**	**80.7**

Table 4: Experiments on user-out vocabulary-out cross-validation.

	user-in vocabulary-out		user-out vocabulary-in		user-out vocabulary-out	
	$\mu \pm \sigma$	max	$\mu \pm \sigma$	max	$\mu \pm \sigma$	max
Standard features	77.7 ± 5.2	83.4	77.7 ± 4.9	84.4	77.6 ± 4.9	84.3
FT emb	67.9 ± 5.7	75.1	67.6 ± 5.3	75.3	67.3 ± 5.2	74.9
FrnnMUTE	75.1 ± 3.9	79.5	77.1 ± 3.9	82.4	74.5 ± 3.9	79.6
Standard features + FT emb	78.9 ± 5.1	85.2	79.5 ± 4.6	86.9	77.1 ± 4.6	84.6
Standard features + FrnnMUTE	80.0 ± 5.1	85.8	80.3 ± 4.3	87.0	78.6 ± 4.4	85.2
Standard features + FT emb + FrnnMUTE	79.9 ± 5.0	85.8	80.4 ± 4.3	87.4	78.1 ± 4.3	85.2

Table 5: Mean, standard deviation and maximum of F1 scores

vocabulary-out cross-validation (75.1). Similarly, the F1 score is higher on the user-out vocabulary-in experiment (82.4 versus 75.3), and in the user-out vocabulary-out experiment (79.6 versus 74.9). The FrnnMUTE results have the smallest dispersion (3.8-3.9) among all considered "solo" feature sets types (4.8-5.3) when aggregated by all available users. This means that FrnnMUTE are more robust in generalizing information from user to user and between different subsets of vocabulary. For the user-in vocabulary-out and the user-out vocabulary-out experiments the combination of standard features and FrnnMUTE in almost all cases shows the best performance among all feature sets. We can observe that the difference in F1 reaches 2.9 for some user pairs and that the maximum improvement achieved by combining standard features with FrnnMUTE over using standard features only hits 5.2 in F-measure. This testifies that FrnnMUTE helps standard linguistic and non-linguistic features to capture word understandability better than FastText embeddings. The fact that the combination of all three feature sets performs insignificantly better of even worse than standard features with only FrnnMUTE can be explained by the overfitting of the classification model in the first case because the resulting feature vector has the biggest dimensionality.

6 Conclusion

We tackle the prediction of understanding of French medical words by using FastText word embeddings as features. Yet, the embeddings solely as features are not enough for good word categorization. Whereas adding FastText word embeddings to standard features results in a substantial improvement of classification model performance when generalizing them to unknown users. We also proposed a novel type of embeddings trained on reference data from one annotator, and called them FrnnMUTE (French RNN Medical Understandability Text Embeddings). Compared with the case of using only standard features with and without FastText word embeddings, the combination of our FrnnMUTE with standard features substantially improves the performance of classification model. This indicates that FrnnMUTE capture better the specifics of medical words required for identifying their understandability by users, than FastText word embeddings. The FrnnMUTE embeddings and the categorization code

are being made publicly available for scientific non-commercial purposes[8].

We have several directions for future work. Currently we use the existing word embeddings pre-trained on Wikipedia and Web Crawl. We assume that training words embeddings on medical data may improve their impact on the results from categorization of medical terms. Another issue is that, after analysis of results of the application of FastText word embeddings in a categorization task, we assumed the existence of a robust nonlinear dependency between some subsets of standard features and subword-level components of FastText word embeddings. We plan to test this hypothesis in further research. Finally, while the annotations go forward, the annotators usually show *learning* progress in decoding the morphological structure of terms and their understanding. This progress is not taken into account in the current experiments, and is also the topic of our future research.

Acknowledgments

This work has been partly founded by the French ANR (grant number ANR-17-CE19-0016-01) as part of the project CLEAR (Communication, Literacy, Education, Accessibility, Readibility).

References

Joachim Bingel, Natalie Schluter, and Héctor Martínez Alonso. 2016. Coastalcph at semeval-2016 task 11: The importance of designing your neural networks right. In *Proceedings of the 10th International Workshop on Semantic Evaluation (SemEval-2016)*, pages 1028–1033. Association for Computational Linguistics.

Piotr Bojanowski, Edouard Grave, Armand Joulin, and Tomas Mikolov. 2017. Enriching word vectors with subword information. *Transactions of the Association for Computational Linguistics*, 5:135–146.

A Borst, A Gaudinat, C Boyer, and N Grabar. 2008. Lexically based distinction of readability levels of health documents. In *MIE 2008*. Poster.

J Chmielik and N Grabar. 2011. Détection de la spécialisation scientifique et technique des documents biomédicaux grâce aux informations morphologiques. *TAL*, 51(2):151–179.

Kevyn Collins-Thompson. 2014. Computational assessment of text readability: A survey of current and future research. *International Journal of Applied Linguistics*, 165(2):97–135.

---

[8]`https://github.com/hpylieva/FrnnMUTE`

Roger A. Côté, D. J. Rothwell, J. L. Palotay, R. S. Beckett, and Louise Brochu. 1993. *The Systematised Nomenclature of Human and Veterinary Medicine: SNOMED International*. College of American Pathologists, Northfield.

Gunther Eysenbach. 2007. Poverty, human development, and the role of ehealth. *J Med Internet Res*, 9(4):34–4.

R Flesch. 1948. A new readability yardstick. *Journ Appl Psychol*, 23:221–233.

L Goeuriot, N Grabar, and B Daille. 2008. Characterization of scientific and popular science discourse in french, japanese and russian. In *LREC*.

N Grabar, S Krivine, and MC Jaulent. 2007. Classification of health webpages as expert and non expert with a reduced set of cross-language features. In *Ann Symp Am Med Inform Assoc*, pages 284–288.

Natalia Grabar, Thierry Hamon, and Dany Amiot. 2014. Automatic diagnosis of understanding of medical words. In *EACL PITR Workshop*, pages 11–20.

Robert Gunning. 1973. *The technique of clear writing*. McGraw Hill, New York, NY.

JP Kincaid, RP Jr Fishburne, RL Rogers, and BS Chissom. 1975. Derivation of new readability formulas (automated readability index, fog count and flesch reading ease formula) for navy enlisted personnel. Technical report, Naval Technical Training, U. S. Naval Air Station, Memphis, TN.

D Kokkinakis and M Toporowska Gronostaj. 2006. Comparing lay and professional language in cardiovascular disorders corpora. In *WSEAS Transactions on BIOLOGY and BIOMEDICINE*, pages 429–437.

Shervin Malmasi, Mark Dras, and Marcos Zampieri. 2016. Ltg at semeval-2016 task 11: Complex word identification with classifier ensembles. In *Proceedings of the 10th International Workshop on Semantic Evaluation (SemEval-2016)*, pages 996–1000. Association for Computational Linguistics.

Alexa T. McCray. 2005. Promoting health literacy. *Journal of the American Medical Informatics Association*, 12(2):152–163.

F Namer. 2000. FLEMM : un analyseur flexionnel du français à base de règles. *Traitement automatique des langues (TAL)*, 41(2):523–547.

Fiammetta Namer and Pierre Zweigenbaum. 2004. Acquiring meaning for French medical terminology: contribution of morphosemantics. In *Ann Symp Am Med Inform Assoc*.

M Poprat, K Markó, and U Hahn. 2006. A language classifier that automatically divides medical documents for experts and health care consumers. In *Int Congress of the European Federation for Medical Informatics*, pages 503–508, Maastricht.

Francesco Ronzano, Ahmed Abura'ed, Luis Espinosa Anke, and Horacio Saggion. 2016. Taln at semeval-2016 task 11: Modelling complex words by contextual, lexical and semantic features. In *Proceedings of the 10th International Workshop on Semantic Evaluation (SemEval-2016)*, pages 1011–1016. Association for Computational Linguistics.

H Schmid. 1994. Probabilistic part-of-speech tagging using decision trees. In *Int Conf on New Methods in Language Processing*, pages 44–49.

Fabrizio Sebastiani. 2002. Machine learning in automated text categorization. *ACM Computing Surveys*, 34(1):1–47.

Luo Si and James P. Callan. 2001. A statistical model for scientific readability. In *Proceedings of the Tenth International Conference on Information and Knowledge Management (CIKM '01)*, pages 574–576.

Ilya Sutskever, Oriol Vinyals, and Quoc V. Le. 2014. Sequence to sequence learning with neural networks. In *Proceedings of the 27th International Conference on Neural Information Processing Systems - Volume 2*, NIPS'14, pages 3104–3112.

Q Zeng-Treiler, H Kim, S Goryachev, A Keselman, L Slaugther, and CA Smith. 2007. Text characteristics of clinical reports and their implications for the readability of personal health records. In *MEDINFO*, pages 1117–1121, Brisbane, Australia.

W Zheng, E Milios, and C Watters. 2002. Filtering for medical news items using a machine learning approach. In *Ann Symp Am Med Inform Assoc*, pages 949–53.

A distantly supervised dataset for automated data extraction from diagnostic studies

Christopher Norman,[1,2] Mariska Leeflang,[2] René Spijker,[3,4]
Evangelos Kanoulas,[5] and Aurélie Névéol[1]
[1] LIMSI, CNRS, Université Paris-Saclay
[2] KEBB, Amsterdam Public Health, Amsterdam UMC, University of Amsterdam
[3] Medical Library, Amsterdam Public Health, Amsterdam UMC, University of Amsterdam
[4] Cochrane Netherlands, Julius Center for Health Sciences and Primary Care, UMCU, Utrecht University
[5] Informatics Institute and Amsterdam Business School, University of Amsterdam
`norman@limsi.fr`, `m.m.leeflang@amc.uva.nl`,
`R.Spijker-2@umcutrecht.nl`, `E.Kanoulas@uva.nl`,
`neveol@limsi.fr`

Abstract

Systematic reviews are important in evidence based medicine, but are expensive to produce. Automating or semi-automating the data extraction of index test, target condition, and reference standard from articles has the potential to decrease the cost of conducting systematic reviews of diagnostic test accuracy, but relevant training data is not available. We create a distantly supervised dataset of approximately 90,000 sentences, and let two experts manually annotate a small subset of around 1,000 sentences for evaluation. We evaluate the performance of BioBERT and logistic regression for ranking the sentences, and compare the performance for distant and direct supervision. Our results suggest that distant supervision can work as well as, or better than direct supervision on this problem, and that distantly trained models can perform as well as, or better than human annotators.

1 Background

Evidence based medicine is founded on systematic reviews, which synthesize all published evidence addressing a given research question. By examining multiple studies, a systematic review can examine the variation between different studies, the discrepancies between them, as well as look at the quality of evidence across studies in a way that is difficult in a single trial. Since a systematic review needs to consider the entire body of published literature, producing a systematic review is expensive and labor-intensive process, often requiring months of manual work (O'Mara-Eves et al., 2015).

To ensure that the results of a systematic review are as comprehensive and unbiased as possible, their production follows a strict and systematic procedure. To catch and resolve disagreements, all steps of the process are performed in duplicate by at least two reviewers. There have recently been examples of systematic reviews using automation in a limited capacity (Bannach-Brown et al., 2019; Przybyła et al., 2018; Lerner et al., 2019), but the impact of automation on the reliability of systematic reviews is not yet fully understood. Automation is not part of accepted practice in current guidelines (De Vet et al., 2008).

After a set of potentially included studies have been identified, systematic reviewers complete a so-called *data extraction form* for each study. These forms comprise a semi-structured summary of the studies, identifying and extracting a consistent, pre-specified set of data items from abstracts or full-text articles in a coherent format (see the left part of Table 1 for sample exerpts). The coherent format allows the data from the studies to be synthesized qualitatively or quantitatively to address the research question of the review.

In this study we will focus on systematic reviews of diagnostic test accuracy (DTA), which examine the accuracy of tests and procedures for diagnosing medical conditions, and which have seen little attention in previous literature on automated data extraction. To compare and synthesize results across studies, reviewers extract diagnostic accuracy from each study, but also determine the *index test* (the specific diagnostic test or procedure that is being tested), what *target condition* the test seeks to diagnose, and the *reference standard* (the diagnostic test or procedure that is being used as the gold standard) (see Fig 1 for an example). These data must be determined for each study to know if the diagnostic accuracy in different studies can be compared.

Proceedings of the BioNLP 2019 workshop, pages 105–114
Florence, Italy, August 1, 2019. ©2019 Association for Computational Linguistics

Original		Cleaned	
Review: CD008892, study: Dutta 2006			
Index tests:	TUBEX Typhidot	Index test:	TUBEX
		Index test:	Typhidot
Target condition and reference standard(s):	Target condition Salmonella Typhi Reference standard: peripheral blood culture		
		Target condition:	Salmonella Typhi
		Target condition:	Typhoid fever
		Reference standard:	Peripheral blood culture
Note: These are the data items corresponding to the example text in Fig. 1			
Review: CD010502, study: Schwartz 1997b			
Index tests:	Throat swab: not reported Commercial name of the RADT: QuickVue In-Line Strep A (Quidel) Type of RADT: EIA	Index test:	QuickVue In-Line Strep A
		Index test:	EIA
		Index test:	ELISA Immunoassays
Target condition and reference standard(s):	See Schwartz 1997a	Target condition:	Group A streptococcus
		Target condition:	Group A streptococcal infection
		Reference standard:	Microbial culture
		Reference standard:	Bacterial culture
Note: Neither the target condition nor the reference standard were mentioned in the table for Schwartz 1997a, but assumed the same for all studies included in this systematic review (they were presumably considered obvious by the authors).			

Table 1: Examples of raw data from three data extractions forms in unstructured format (left) and a structured summary of the data intended for distant supervision by pattern matching (right).

Although typhoid fever is confirmed by culture of Salmonella enterica serotype Typhi, rapid and simple diagnostic serologic tests would be useful in developing countries. We examined the performance of Widal test in a community field site and compared it with Typhidot and Tubex tests for diagnosis of typhoid fever. [...] Sensitivity, specificity, positive predictive value (PPV), and negative predictive value (NPV) of the 3 serologic tests were calculated using culture-confirmed typhoid fever cases as "true positives" and paratyphoid fever and malaria cases as "true negatives". [...] The sensitivity, specificity, PPV, and NPV of Typhidot and Tubex were not better than Widal test. There is a need for more efficient rapid diagnostic test for typhoid fever especially during the acute stage of the disease. Until then, culture remains the method of choice.

Legend: Target condition Index Test Reference standard

Figure 1: Examples of data items highlighted in text, with supporting context underlined. Based on the manual annotation by one expert (ML) on a study by Dutta et al. (2006).

1.1 BERT

BERT (Bidirectional Encoder Representations from Transformers) is a deep learning model that is unsupervisedly pretrained on a large general language corpus, then supervisedly fine-tuned on natural language processing tasks (Devlin et al., 2018). Despite being a general approach, with almost no task-specific modifications, BERT achieves state-of-the-art performance across a number of natural language processing tasks, including text classification, question answering, inference, and named entity recognition.

Pretrained models like BERT can be used directly for screening automation or automated data extraction. However, by default BERT is trained on a general language corpus, which differs radically in word choice and grammar from the special language found in biomedicine and related fields (Sager et al., 1980). Pretraining on biomedical corpora, rather than general corpora, has been demonstrated to improve performance on several biomedical natural language processing tasks (Lee et al., 2019; Beltagy et al., 2019; Si et al., 2019).

1.2 Objectives

In this study we seek to:

1. Construct a dataset for training machine learning models to identify and extract data from full-text articles on diagnostic test accuracy. We focus on the target condition, index test, and reference standard.

2. Train models to identify specific data items in full-text articles on diagnostic test accuracy

One of the main aims of our study is to determine how such a dataset should be constructed to allow for training well performing models. In particular, do we need directly supervised data, or can we build reliable models with distantly supervised data? If we do need directly supervised data, how much is necessary?

2 Related Work

There have been attempts to extract several types of data relevant to systematic reviews, most notably extracting PICO[1] statements from article text (Wallace et al., 2016; Kiritchenko et al., 2010; Kim et al., 2011; Nye et al., 2018). Other data items include background and study design (Kim et al., 2011), as well as automatically performing risk of bias assessments (Marshall et al., 2014). There is also a recent TAC track for data extraction in systematic reviews of environmental agents.[2] Similarly, previous work by Kiritchenko et al. (2010) aimed to extract 21 different kinds of data from articles, including treatment name, sample size, as well as the primary and secondary outcome from article text. Furthermore, the key criterion for extraction in a systematic review is not the actual data, but the context it appears in. For instance, both intervention studies and a diagnostic studies have target conditions, but these refer to different things: the intervention study seek to *treat* the condition while the diagnostic study seeks to *diagnose* it. As a consequence, in an intervention study the inclusion criterion often mentions the disease, while in a diagnostic study inclusion criteria may mention symptoms rather than the actual disease. This means that a data extraction system trained on interventions may not work as well (or at all) for systematic reviews of diagnostic test accuracy, even though it may seem that the same data is extracted in both. Furthermore, unlike the data required in diagnostic reviews, many previously considered data items are mentioned once in articles, often using formulaic expressions (e.g. sex, blinding, randomization).

Conventional methods for automated data extraction split articles into sentences and classify these individually using conventional machine learning methods (e.g. SVM, Naive Bayes) (Jonnalagadda et al., 2015), or label spans in the text and classify these using sequence tagging (e.g. CRF, LSTM) (Nye et al., 2018).

Despite the body of previous work on automation, many data items relevant to systematic reviews have been overlooked. A 2015 systematic review of data extraction found 26 articles describing the attempted extraction of 52 different data items, but almost all focused on interventions (Jonnalagadda et al., 2015). No study considered any data item specific to diagnostic studies, except for general data items common to both interventions and diagnostic studies, such as age, sex, blinding, or the generation of random allocation sequences. The likely reason for this is that traditional data extraction systems require bespoke training data for each particular data item to extract, which is generally only available through expensive, manual annotation by experts.

A cheaper way to construct datasets for data extraction is to use distant supervision, where the dataset is annotated per article or per review, rather than per sentence or per text span. Supervised methods are then trained on fuzzy annotations derived heuristically for each sentence. For instance, Wallace et al. (2016) used supervised distant supervision to learn to identify PICO statements in full text, and Marshall et al. (2014) used supervised distant learning with SVMs to identify risk of bias assessments.

There is likely a trade-off between quality and data size. All else being equal, direct supervision is generally better than distant supervision (distantly supervised training data adds a source of noise not present for direct supervision). At the same time, it may not be feasible for experts to annotate large amounts of data. Crowd-sourcing is sometimes used as an alternative to a group of known experts, but if a high degree of expertise is necessary to annotate, crowd-sourcing may not give sufficient guarantees about the expertise of the annotators.

3 Material

We used data from a previous dataset, the LIMSI-Cochrane dataset (Norman et al., 2018),[3] to identify references included in previous systematic reviews of diagnostic test accuracy. The LIMSI-Cochrane dataset comprises 1,738 references to DTA studies from 63 DTA systematic reviews. The dataset includes the data extraction forms for each

---

[1] Population, intervention, control group, and outcome.
[2] https://tac.nist.gov/2018/SRIE/index.html
[3] DOI: 10.5281/zenodo.1303259

Target Condition

	pos	neg	total
Distant train	11,336	63,204	74,540
test	2,884	13,572	16,456
total	14,220	77,776	90,996
Annotated by ML	92	889	981
Annotated by RS	48	983	1,031

Index Test

	pos	neg	total
Distant train	14,280	63,343	77,623
test	2,675	13,992	16,667
total	16,955	77,335	94,290
Annotated by ML	93	888	981
Annotated by RS	87	944	1,031

Reference Standard

	pos	neg	total
Distant train	7,006	56,638	63,644
test	1,258	14,602	15,860
total	8,264	71,240	79,504
Annotated by ML	26	955	981
Annotated by RS	26	1,005	1,031

Table 2: The number of sentences in our dataset, broken into distantly annotated training and test sets, as well as a manually annotated subset. Distant annotations for each data type were not available for all studies, and the total number of labelled sentences are therefore different for each data type.

study completed by the systematic review authors.

The dataset itself does not contain abstracts or full-texts, but include identifiers in the form of PubMed IDs and DOIs which can be used to retrieve abstracts or full-texts.

We used the reference identifiers (PMID and/or DOI) taken from the LIMSI-Cochrane dataset to construct a collection of PDF articles. We used EndNote's 'find full text' feature, which retrieves PDF articles from a range of publishers.[4] The PDF articles were then converted into XML format using Grobid (Lopez, 2009).

We randomly split the dataset into dedicated training and evaluation sets, where we used 48 of the systematic reviews as the training set, and we kept the remaining 15 systematic reviews for evaluation. For each of the 15 systematic reviews in the evaluation set, we randomly selected one article to be annotated manually. The remaining articles in the evaluation set were not used for training, since training and testing on the same system-

atic review is known to overestimate classification performance (Cohen, 2008). The goal of this work is to learn the semantics of the context, rather than the semantics of particular terms, and these contexts should be consistent across reviews.

3.0.1 Distant annotation

The data forms from the systematic reviews were intended to be read by and be useful to the human systematic review authors. The contents are therefore usually semi-structured rather than structured, and will include different kinds of data depending on what is relevant to the systematic review (see Table 1).

We create a dataset of distant annotations from the LIMSI-Cochrane dataset by manually converting the semi-structured data into structured data items, and by ensuring that these items can be found in the corresponding article using pattern matching (see Table 1).

We split each of the XML documents into sentences using the nltk sentence splitter.[5] The sentences are then divided into positive and negative depending on whether the relevant data items occur as a partial match in the sentence. Partial matches were calculated using *tf·idf* cosine similarity between the data item and the sentence, where we took the 20 top ranking sentences for each pair of data item and article, with a similarity score of 0.1 or higher. We chose 20 as a target number of sentences since we felt this was a reasonable upper limit on the number of relevant sentences in a single article. We added an absolute threshold of 0.1 to keep the system from annotating obviously non-relevant sentences (scores close to zero) when no matches could be found in the article. For articles that have multiple data items we used the concatenation of all data items. For example, in Table 1, the data items for 'Schwartz 1997b' would be: target condition: 'Group A streptococcus; Group A streptococcal infection', index test: 'QuickVue In-Line Strep A; EIA; ELISA Immunoassays', and reference standard 'Microbial culture; Bacterial culture'.

We excluded all articles where the data items were not provided in the data form (because the reviewers did not extract this data), or where data forms were missing from the systematic review. Since we do not know which sentences were relevant or not in these articles we did not use these

---

[4]https://endnote.com/

[5]https://www.nltk.org/

articles as either positive or negative data. As a consequence the total amount of sentences differ for the target condition, index test and reference standard.

We repeated the matching precedure for the target condition, the index test and the reference standard, resulting in three distinct datasets.

3.0.2 Expert annotation

We randomly split the evaluation set into three sets of five systematic reviews. Two experts (ML and RS) on systematic reviews of diagnostic test accuracy manually annotated the 15 articles by highlighting all sentences in the text that 1) mentions the target condition, index test, and reference standard 2) makes it clear that these are the target condition, index test and reference standard, and 3) do not simply mention these same items in an unrelated context. The annotation instructions were written and adjusted twice to remove ambiguity, and the reasons for disagreement were discussed and resolved after two rounds of annotation. As a compromise between getting more data and being able to use the agreement between the experts as baseline for the performance, one expert annotated the first five studies, the second expert annotated the next five studies, and both annotated the last five studies.

4 Method

We construct three pipelines, one for each of the target condition, index test, and reference standard, and we train and evaluate these separately.

We varied our experiments in three dimensions: We tried A) two machine learning algorithms, B) two levels of preprocessing, and C) distantly supervised training data versus directly supervised training data. The directly and distantly supervised models were evaluated on the same data.

4.0.1 A1: BioBERT

We here used a pointwise learning-to-rank approach, where we trained a sentence ranking model by using BioBERT, a version of BERT pre-trained on PubMed and PMC (Lee et al., 2019), and fine-tuned the model by training it to regress probability scores. This model was thus trained to map sentences to relevance scores.

To train and evaluate, we used the default BERT setup for the GLUE datasets,[6] modified to output

---

[6]https://github.com/google-research/bert

a relevance score rather than a binary value. We used default parameters.

4.0.2 A2: Logistic Regression

We here used a pairwise learning-to-rank approach, where we trained a logistic regression model using stochastic gradient descent (sklearn). As features we used 1) lowercased, *tf·idf* weighted word n-grams, 2) lowercased, binary word n-grams, 3) lowercased, *tf·idf* weighted, stemmed word n-grams, 4) lowercased, stemmed, binary word n-grams, as well as *i*) lowercased, *tf·idf* weighted character n-grams, and *ii*) *non*-lowercased, *tf·idf* weighted character n-grams. We used word n-grams up to length 3, and character n-grams up to length 6. The first set of features is intended to capture contextual information ('for the diagnosis of ...'); the second set of features is intended to capture medical technical terms, which are often distinctive at the morpheme level (e.g. 'ischemia', 'anemia'). We deliberately did not use stop-words, since doing so would discard almost all the contextual information. This results in a sparse feature matrix consisting of approximately 1.8 million features for the distantly supervised experiments, and approximately 300,000 features for the directly supervised experiments.

We handled class imbalance by setting the weight for the positive class to 80. This was previously determined to be a reasonable weight in experiments on screening automation in diagnostic test accuracy systematic reviews, a problem with similar class imbalance.

4.0.3 B1: Raw Sentences

Here we used the sentences as they appear in the articles.

4.0.4 B2: Sentences with UMLS Concepts

In this setup we used the *Unified Medical Language System*, a large ontology of medical concepts maintained by the National Library of Medicine (Bodenreider, 2004; Lindberg et al., 1993). We used MetaMap[7] to locate concept mentions in the sentences, and to replace these with their corresponding UMLS semantic types. For instance the sentence *'Typhoid fever is a febrile and often serious systemic illness caused by Salmonella enterica serotype Typhi'* was transformed into *'DSYN is a FNDG and TMCO serious DSYN caused by BACT enterica BACT'*.

---

[7]https://metamap.nlm.nih.gov/

Target condition

	Auto	ML	RS
Auto	1.00	0.07	0.04
ML	0.90	1.00	0.38
RS	1.00	0.62	1.00

Index test

	Auto	ML	RS
Auto	1.00	0.09	0.07
ML	1.00	1.00	0.61
RS	0.93	0.70	1.00

Reference standard

	Auto	ML	RS
Auto	1.00	0.01	0.03
ML	1.00	1.00	0.86
RS	1.00	0.40	1.00

Table 3: Agreement in terms of recall where columns are considered ground truth, e.g. annotator RS chose 62% of ML's annotations for the target condition.

4.0.5 C1: Directly Supervised Training

We here trained and evaluated on the articles manually annotated by our two experts (ML and RS), using leave-one-out cross-validation. In other words, to evaluate on each of the ten articles annotated by each annotator we used the remaining 9 articles annotated by the same expert as training data. This was done separately for each expert, and the annotations from the other expert was not used.

4.0.6 C2: Distantly Supervised Training

We here trained on the distant annotations from the 48 systematic reviews in the training set, and evaluated on the 15 manually annotated articles in the evaluation set, where each annotator provided annotation data for 10 articles (with a 5 article overlap). The articles used for evaluation were the same as in C1.

4.1 Evaluation

Since our model output ranked sentences, rather than a binary classification, we evaluated all experiments in terms of average precision.

As a comparison, we also evaluated the average precision using the ranking given by the other annotator. In plain language, we tried to evaluate how useful it would have been for the expert to highlight sentences for each other. The expert annotations were binary (Yes/No), rather than a ranking score, so we calculated the average precision by interpolating ties in the ranking.

5 Results

Out of the 1,738 references in the LIMSI-Cochrane dataset, 1152 had either a PMID or DOI assigned. EndNote was able to retrieve PDF articles for 666 of these references. A total of 90,996 sentences were distantly labeled for target condition, 94,290 sentences were distantly labeled for index test, and 79,504 sentences were distantly labeled for reference standard. The first annotator (ML) annotated 981 sentences and the second annotator (RS) annotated 1,031 sentences (Table 2).

We present the results of our algorithm evaluated on the annotations by ML in Table 4, and evaluated on the annotations by RS in Table 5.

The ranking performance exhibited large variations. Neither BioBERT or logistic regression were consistently better than the other, neither distant supervision or direct supervision were consistently better than the other, and neither raw sentence nor sentences augmented with UMLS concepts were consistently better than the other. For the target condition, the best performance was achieved by logistic regression on raw sentences using either distant or direct supervision, with a maximum at 0.412 compared to human performance at 0.376 and 0.386 respectively. For the index test, the performance fell within the range 0.344–0.468 compared to human performance at 0.525 and 0.516 respectively. For the reference standard, BioBERT exhibited substantially inferior results on the reference standard compared to logistic regression, while logistic regression performance fell within the range 0.345–0.467, compared to human performance at 0.267 and 0.381 respectively.

The performance also varied between systematic reviews, with consistently close to perfect performance on a few reviews (CD007394 and CD0008782), and consistently very low performance on a few (CD009647 and CD010339). These also correspond to the articles with the highest and lowest inter-annotator agreement. The consensus of the two experts is that CD010339 is not a diagnostic test accuracy study.

6 Discussion

Raw sentences worked consistently better for logistic regression on the target condition (8/8), and worked better than UMLS concepts as a general trend (20/24). While general concepts could theoretically improve performance by help-

| | | BioBERT | | | | Logistic Regression | | | | As ranked by the other expert (RS) |
| | | Distant | | Supervised | | Distant | | Supervised | | |
	n pos	Raw	UMLS	Raw	UMLS	Raw	UMLS	Raw	UMLS	
Target condition										
CD007394	1	1.000	0.500	0.143	0.250	1.000	0.500	1.000	0.500	0.500
CD007427	14	0.228	0.267	0.500	0.588	0.423	0.573	0.462	0.509	—
CD008054	10	0.197	0.353	0.060	0.182	0.167	0.118	0.170	0.148	—
CD008782	2	1.000	1.000	0.283	0.567	0.500	0.417	0.500	0.583	0.700
CD008892	29	0.182	0.274	0.384	0.247	0.368	0.439	0.290	0.333	0.338
CD009372	29	0.110	0.117	0.461	0.543	0.328	0.250	0.378	0.276	—
CD010339	16	0.192	0.179	0.642	0.513	0.537	0.432	0.482	0.495	0.154
CD010653	2	0.053	0.035	0.023	0.015	0.107	0.112	0.062	0.086	—
CD011420	6	0.070	0.074	0.239	0.175	0.189	0.138	0.254	0.157	0.190
mean:		0.336	0.311	0.304	0.342	0.402	0.331	0.400	0.343	0.376
Index test										
CD007394	2	1.000	1.000	0.643	0.361	0.750	0.500	0.583	0.583	1.000
CD007427	17	0.354	0.225	0.580	0.568	0.551	0.526	0.534	0.484	—
CD008054	10	0.388	0.305	0.449	0.281	0.170	0.161	0.195	0.218	—
CD008782	2	0.833	1.000	0.079	0.523	0.750	0.750	0.750	0.750	0.700
CD008892	34	0.342	0.473	0.458	0.391	0.471	0.484	0.496	0.529	0.524
CD009372	8	0.269	0.351	0.194	0.225	0.261	0.270	0.303	0.390	—
CD010339	1	0.167	0.050	0.067	0.067	0.071	0.100	0.013	0.017	0.010
CD011420	19	0.251	0.342	0.284	0.218	0.288	0.266	0.280	0.256	0.391
mean:		0.450	0.468	0.344	0.329	0.414	0.382	0.394	0.403	0.525
Reference standard										
CD007394	0	n/a	n/a	n/a	n/a	n/a	n/a	n/a	n/a	n/a
CD007427	2	0.145	0.032	0.081	0.034	0.052	0.037	0.035	0.041	—
CD008054	6	0.215	0.108	0.239	0.076	0.635	0.619	0.525	0.515	—
CD008892	13	0.112	0.097	0.152	0.154	0.408	0.351	0.264	0.255	0.201
CD009372	3	0.052	0.095	0.253	0.414	0.681	0.692	0.679	0.729	—
CD010653	1	0.020	0.016	0.020	0.059	0.029	0.034	0.067	0.067	—
CD011420	1	0.034	0.100	1.000	0.014	1.000	1.000	0.500	0.500	0.333
mean:		0.097	0.075	0.291	0.125	0.467	0.455	0.345	0.351	0.267

Table 4: Average precision results for the 8 different machine learning models on the data annotated by the first annotator (ML), compared to the performance of an independent human expert (annotator RS). The 'Raw' columns denote results for models trained and evaluated on raw sentences. The 'UMLS' columns denote results for models trained and evaluated on sentences where the concept mentions have been replaced with their corresponding UMLS semantic types. The '*n* pos' column denotes the number of positive sentences labeled by ML for each article. Rows were omitted for which no sentences were labeled positive. In the baseline results, cells are marked '—' if the article was not annotated by the other expert (RS).

ing the models generalize, this may also remove important semantic information from the sentences, keeping the models from ranking accurately. We also note that BioBERT already encodes a language model (similar to word embeddings), and concepts may therefore be unhelpful for the model.

BioBERT performed consistently better than logistic regression on the index test when using distant supervision (4/4), but not when using direct supervision (0/4). Logistic regression performed consistently better than BioBERT on both the target condition and the reference standard (16/16). On the reference standard the difference in performance is substantial, with BioBERT scoring very poorly, and logistic regression performing much better than human performance. The reason for BioBERT's poor performance on the reference standard may be due to the relative sparsity of the annotations for this subtask (see Table 2).

Distant supervision was consistently on par with or better than direct supervision. The top performing models also outperformed the human annotators on the target condition and the reference standard, and came comparatively close on the index test (0.468 versus 0.525 and 0.444 versus 0.516).

6.1 Limitations

We only manually annotated a small sample of the dataset. The small size is further compounded by problems with converting PDF to text, which may also bias the training and evaluation in favor of articles where the conversion works better (mainly articles from big publishers).

The dataset was constructed from articles included in previous systematic reviews of diagnostic test accuracy. These include articles that con-

Target condition

	n pos	BioBERT				Logistic Regression				As ranked by the other expert (ML)
		Distant		Supervised		Distant		Supervised		
		Raw	UMLS	Raw	UMLS	Raw	UMLS	Raw	UMLS	
CD007394	2	0.750	0.500	0.667	0.040	0.833	0.500	1.000	0.833	0.667
CD008081	8	0.136	0.198	0.213	0.371	0.504	0.380	0.394	0.388	—
CD008760	5	0.200	0.144	0.283	0.163	0.252	0.300	0.481	0.300	—
CD008782	1	1.000	1.000	0.500	1.000	0.500	0.333	1.000	0.500	0.500
CD008892	15	0.170	0.270	0.088	0.342	0.440	0.505	0.667	0.542	0.564
CD009647	2	0.036	0.021	0.021	0.047	0.020	0.026	0.012	0.023	—
CD010339	2	0.061	0.040	0.066	0.062	0.044	0.029	0.063	0.023	0.019
CD010360	2	0.089	0.080	0.093	0.261	0.181	0.083	0.244	0.064	—
CD010705	7	0.189	0.269	0.127	0.341	0.382	0.359	0.254	0.402	—
CD010420	4	0.036	0.044	0.209	0.097	0.210	0.214	0.302	0.132	0.178
mean:		0.267	0.257	0.227	0.273	0.337	0.273	0.412	0.321	0.386

Index test

	n pos	Raw	UMLS	Raw	UMLS	Raw	UMLS	Raw	UMLS	
CD007394	2	1.000	1.000	0.417	0.393	0.750	0.500	0.700	0.750	1.000
CD008081	11	0.464	0.229	0.463	0.454	0.431	0.412	0.394	0.447	—
CD008760	9	0.357	0.411	0.512	0.475	0.457	0.470	0.481	0.476	—
CD008782	1	1.000	1.000	1.000	0.500	1.000	1.000	1.000	1.000	0.500
CD008892	27	0.499	0.539	0.717	0.758	0.740	0.666	0.667	0.474	0.692
CD009647	1	0.053	0.015	0.020	0.006	0.006	0.009	0.012	0.040	—
CD010339	6	0.085	0.054	0.040	0.047	0.053	0.041	0.063	0.047	0.058
CD010360	8	0.154	0.119	0.233	0.278	0.222	0.202	0.244	0.242	—
CD010705	14	0.599	0.533	0.292	0.270	0.352	0.327	0.254	0.327	—
CD010420	8	0.234	0.296	0.280	0.251	0.259	0.235	0.302	0.257	0.328
mean:		0.444	0.420	0.397	0.343	0.427	0.386	0.412	0.406	0.516

Reference standard

	n pos	Raw	UMLS	Raw	UMLS	Raw	UMLS	Raw	UMLS	
CD008081	3	0.254	0.132	0.134	0.177	0.867	0.698	1.000	1.000	—
CD008760	2	0.101	0.553	0.529	0.013	0.667	0.833	0.667	0.833	—
CD008892	11	0.110	0.212	0.283	0.108	0.356	0.286	0.334	0.225	0.417
CD010339	1	0.012	0.010	0.029	0.009	0.224	0.031	0.071	0.028	n/a
CD010360	1	0.200	0.037	0.111	0.038	0.810	0.023	0.167	0.143	—
CD010705	5	0.150	0.152	0.194	0.086	0.224	0.122	0.172	0.125	—
CD010420	3	0.167	0.347	0.358	0.019	0.810	0.806	0.692	0.694	0.345
mean:		0.142	0.206	0.234	0.064	0.428	0.400	0.443	0.435	0.381

Table 5: Average precision results for the 8 different machine learning models on the data annotated by the second annotator (RS), compared to the performance of an independent human expert (annotator ML). Abbreviations are the same as in Table 4. In the baseline results, cells are marked '—' if the article was not annotated by the other expert (ML).

tain diagnostic results, while not being diagnostic test accuracy studies. Arguably, these should be excluded from training or evaluation, and possibly even from the dataset.

7 Conclusions

Our results suggest that distant supervision is sufficient to train models to identify target condition, index test, and reference standard in diagnostic articles. Our results also suggest that such models can perform on par with human annotators.

We constructed a dataset of full-text articles of diagnostic test accuracy studies, with distant annotations for target condition, index test and reference standard, that can be used to train machine learning models. We also provide a subset of the data manually annotated by experts for evaluation. Our dataset cannot be publicly distributed due to copyright restrictions, but will be available upon request. We also plan to distribute the code for the distant annotations and data preprocessing, as well as the cleaned data extraction forms.

7.1 Future Work

The dataset is being updated, and we plan to increase the amount of manually annotated data to improve the statistical reliability of the experiments. We also plan to let all experts annotate the same articles to simplify the comparisons.

Acknowledgments

This project has received funding from the European Union's Horizon 2020 research and innovation programme under the Marie Sklodowska-Curie grant agreement No 676207.

References

Alexandra Bannach-Brown, Piotr Przybyła, James Thomas, Andrew SC Rice, Sophia Ananiadou, Jing Liao, and Malcolm Robert Macleod. 2019. Machine learning algorithms for systematic review: reducing workload in a preclinical review of animal studies and reducing human screening error. *Systematic reviews*, 8(1):23.

Iz Beltagy, Arman Cohan, and Kyle Lo. 2019. SciBERT: Pretrained contextualized embeddings for scientific text. *arXiv preprint arXiv:1903.10676*.

Olivier Bodenreider. 2004. The unified medical language system (UMLS): integrating biomedical terminology. *Nucleic acids research*, 32(suppl_1):D267–D270.

Aaron M Cohen. 2008. Optimizing feature representation for automated systematic review work prioritization. *AMIA Annual Symposium proceedings*, pages 121–5.

HCW De Vet, A Eisinga, II Riphagen, B Aertgeerts, D Pewsner, and R Mitchell. 2008. Chapter 7: searching for studies. *Cochrane handbook for systematic reviews of diagnostic test accuracy version 0.4 [updated September 2008]. The Cochrane Collaboration*.

Jacob Devlin, Ming-Wei Chang, Kenton Lee, and Kristina Toutanova. 2018. BERT: Pre-training of deep bidirectional transformers for language understanding. *arXiv preprint arXiv:1810.04805*.

Shanta Dutta, Dipika Sur, Byomkesh Manna, Bhaswati Sen, Alok Kumar Deb, Jacqueline L Deen, John Wain, Lorenz Von Seidlein, Leon Ochiai, John D Clemens, et al. 2006. Evaluation of new-generation serologic tests for the diagnosis of typhoid fever: data from a community-based surveillance in calcutta, india. *Diagnostic microbiology and infectious disease*, 56(4):359–365.

Siddhartha R Jonnalagadda, Pawan Goyal, and Mark D Huffman. 2015. Automating data extraction in systematic reviews: a systematic review. *Systematic reviews*, 4(1):78.

Su Nam Kim, David Martinez, Lawrence Cavedon, and Lars Yencken. 2011. Automatic classification of sentences to support evidence based medicine. In *BMC bioinformatics*, volume 12, page S5. BioMed Central.

Svetlana Kiritchenko, Berry de Bruijn, Simona Carini, Joel Martin, and Ida Sim. 2010. ExaCT: automatic extraction of clinical trial characteristics from journal publications. *BMC medical informatics and decision making*, 10:56.

Jinhyuk Lee, Wonjin Yoon, Sungdong Kim, Donghyeon Kim, Sunkyu Kim, Chan Ho So, and Jaewoo Kang. 2019. BioBERT: pre-trained biomedical language representation model for biomedical text mining. *arXiv preprint arXiv:1901.08746*.

Ivan Lerner, Perrine Créquit, Philippe Ravaud, and Ignacio Atal. 2019. Automatic screening using word embeddings achieved high sensitivity and workload reduction for updating living network meta-analyses. *Journal of clinical epidemiology*, 108:86–94.

Donald AB Lindberg, Betsy L Humphreys, and Alexa T McCray. 1993. The unified medical language system. *Yearbook of Medical Informatics*, 2(01):41–51.

Patrice Lopez. 2009. Grobid: Combining automatic bibliographic data recognition and term extraction for scholarship publications. In *International conference on theory and practice of digital libraries*, pages 473–474. Springer.

Iain J. Marshall, Joël Kuiper, and Byron C. Wallace. 2014. Automating risk of bias assessment for clinical trials. *Proceedings of the 5th ACM Conference on Bioinformatics, Computational Biology, and Health Informatics - BCB '14*, pages 88–95.

Christopher Norman, Mariska Leeflang, and Aurélie Névéol. 2018. Data extraction and synthesis in systematic reviews of diagnostic test accuracy: A corpus for automating and evaluating the process. In *AMIA Annual Symposium Proceedings*, volume 2018, page 817. American Medical Informatics Association.

Benjamin Nye, Junyi Jessy Li, Roma Patel, Yinfei Yang, Iain J Marshall, Ani Nenkova, and Byron C Wallace. 2018. A corpus with multi-level annotations of patients, interventions and outcomes to support language processing for medical literature. In *Proceedings of the conference. Association for Computational Linguistics. Meeting*, volume 2018, page 197. NIH Public Access.

Alison O'Mara-Eves, James Thomas, John McNaught, Makoto Miwa, and Sophia Ananiadou. 2015. Using text mining for study identification in systematic reviews: a systematic review of current approaches. *Systematic reviews*, 4(1):5.

Piotr Przybyła, Austin J Brockmeier, Georgios Kontonatsios, Marie-Annick Le Pogam, John McNaught, Erik von Elm, Kay Nolan, and Sophia Ananiadou. 2018. Prioritising references for systematic reviews with robotanalyst: A user study. *Research synthesis methods*, 9(3):470–488.

Juan C Sager, David Dungworth, Peter F McDonald, et al. 1980. *English special languages: principles and practice in science and technology*. John Benjamins Pub Co.

Yuqi Si, Jingqi Wang, Hua Xu, and Kirk Roberts. 2019. Enhancing clinical concept extraction with contextual embedding. *arXiv preprint arXiv:1902.08691*.

Byron C Wallace, Joël Kuiper, Aakash Sharma, Mingxi Brian Zhu, and Iain J Marshall. 2016. Extracting PICO sentences from clinical trial reports using supervised distant supervision. *Journal of Machine Learning Research*, 17(132):1–25.

Query selection methods for automated corpora construction with a use case in food-drug interactions

Georgeta Bordea[1], Tsanta Randriatsitohaina[2], Natalia Grabar[3], Fleur Mougin[1] and Thierry Hamon[2,4]

[1]Univ. Bordeaux, Inserm UMR 1219, Bordeaux Population Health, team ERIAS, Bordeaux, France
Email: name.surname@u-bordeaux.fr
[2]LIMSI, CNRS UPR 3251, Université Paris-Saclay, Orsay, France
[3]CNRS UMR 8163 - STL - Savoirs Textes Langage, Univ. Lille, Lille, France
[4]Université Paris 13, Sorbonne Paris Cité, Villfetaneuse, France

Abstract

In this paper, we address the problem of automatically constructing a relevant corpus of scientific articles about food-drug interactions. There is a growing number of scientific publications that describe food-drug interactions but currently building a high-coverage corpus that can be used for information extraction purposes is not trivial. We investigate several methods for automating the query selection process using an expert-curated corpus of food-drug interactions. Our experiments show that index terms features along with a decision tree classifier are the best approach for this task and that feature selection approaches and in particular gain ratio outperform frequency-based methods for query selection.

1 Introduction

Unexpected Food-Drug Interactions (FDIs) occasionally result in treatment failure, toxicity and an increased risk of side-effects. While drug-drug interactions can be investigated systematically, there is a much larger number of possible FDIs. Therefore, these interactions are generally discovered and reported only after a drug is administered on a wide scale during post-marketing surveillance. A notable example is the discovery that grapefruit contains bioactive furocoumarins and flavonoids that activate or deactivate many drugs in ways that can be life-threatening (Dahan and Altman, 2004). This effect was first noticed accidentally during a test for drug interactions with alcohol that used grapefruit juice to hide the taste of ethanol.

Currently, information about FDIs is available to medical practitioners from online databases such as DrugBank[1] and compendia such as the Stockley's Drug Interactions (Baxter and Preston, 2010), but these resources have to be regularly updated to keep up with a growing body of evidence from biomedical articles. Recent advances in information extraction are a promising direction to partially automate this work by extracting information about drug interactions. This approach has already shown promising results in the context of drug-drug interactions (Segura-Bedmar et al., 2013) but in the case of FDIs, similar progress is currently hindered by a lack of annotated corpora. The work presented in (Jovanovik et al., 2015) for inferring interactions between drugs and world cuisine is based on a largely manual effort of extracting food-drug interactions from descriptions provided in DrugBank.

Although a first corpus of MEDLINE abstracts about FDIs called POMELO was recently made available (Hamon et al., 2017), this corpus has a low coverage of relevant documents for FDIs. The authors made use of PubMed to retrieve all the articles indexed with the *Food-Drug Interactions* term from the MeSH thesaurus[2], but the challenge is that while articles annotated with *Drug Interactions* are abundant, there is a much smaller number of documents indexed with *Food-Drug Interactions*. A bibliographic analysis of the references cited in the Stockley's Drug Interactions in relation to foods shows that only 11% of these articles are indexed with the MeSH term *Food-Drug Interactions*, while almost 70% of the articles are available in MEDLINE (Bordea et al., 2018).

Constructing a high-coverage corpus of FDIs using MeSH terms and PubMed is not trivial because there is a large number of articles that describe food interactions that were published before the introduction of the *Food-Drug Interactions* MeSH term in the early nineties. At the same time, MeSH terms are assigned to scientific articles based on their main topics of interest, miss-

---

[1]https://www.drugbank.ca

[2]https://www.nlm.nih.gov/mesh/

Proceedings of the BioNLP 2019 workshop, pages 115–124
Florence, Italy, August 1, 2019. ©2019 Association for Computational Linguistics

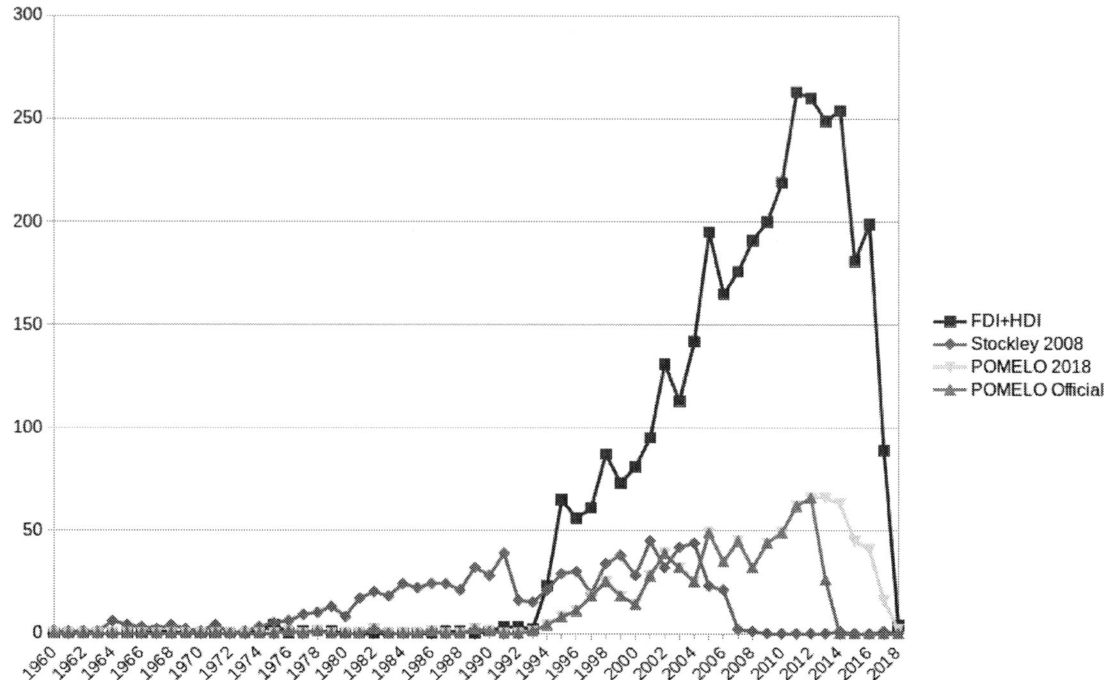

Figure 1: Timeline of MEDLINE articles cited in Stockley 2008 and retrieved using relevant MeSH terms

ing a considerable amount of articles that briefly mention interactions with food. Furthermore, the POMELO corpus has an even more narrow focus on articles related to adverse effects, therefore it covers only 3% of the references provided in the Stockley compendium.

Figure 1 shows a comparison of scientific articles cited in a reference compendium (*Stockley 2008*), with the articles annotated with the *Food-Drug Interactions* MeSH term and the *Herb-Drug Interactions* MeSH term (*FDI+HDI*). It is worth noticing the overall ascending trend of scientific articles that address FDIs, showing an increased interest in this type of interactions. This makes increasingly more costly the effort to manually summarise related information in specialised compendia. The figure also shows the timeline of the articles gathered in the official POMELO corpus (*POMELO Official*) and a more recent retrieval result of the POMELO query (*POMELO 2018*).

We address these limitations by considering several approaches for automatically selecting queries that can be used to retrieve domain-specific documents using an existing search engine. The approach takes as input a sample set of relevant documents that are cited in the Stockley compendium. In this way, the problem of FDI discovery from biomedical literature is limited to the task of interaction candidates search, that is the task of finding documents that describe FDIs from a large bibliographic database. We make use of a large corpus of relevant publications to investigate index terms used to annotate articles about FDIs and we propose an automated method for query selection that increases recall.

The main contributions of this work are:

- a discriminative model for automatically constructing high-coverage and domain-specific corpora for information extraction,

- an approach for automatically selecting queries using index terms as candidates,

- an automated method to evaluate queries based on a sample corpus.

The paper is structured as follows. We begin by discussing several design decisions for the subtask of classifying documents based on relevance, adopting a discriminative model for information retrieval in Section 3. In Section 4, we introduce the subtask of query selection discussing candidate term selection and several methods for scoring queries. Section 5 describes the datasets used

116

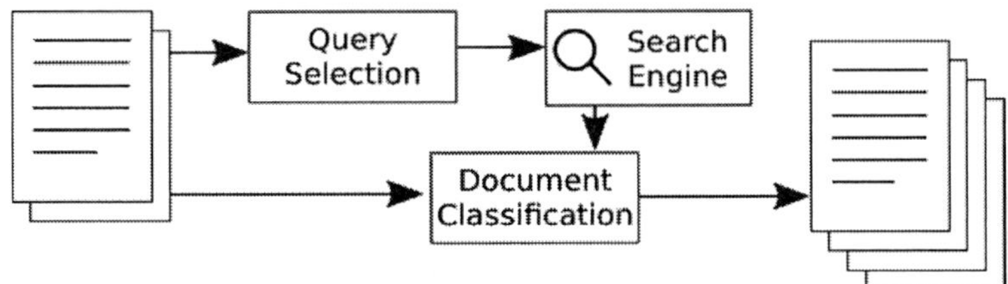

Figure 2: Workflow for automated corpus construction using a collection of sample documents and a search engine

to evaluate our approach for automatically constructing a corpus for FDIs and Section 6 presents the results of an empirical evaluation. Then we provide an overview of related work for this task in Section 7 and we discuss a formal definition for the problem at hand in Section 2. We conclude this work in Section 8.

2 Problem definition

We address the problem of automatically constructing a domain-specific corpus by making use of a discriminative model for information retrieval that defines the problem of document search as a problem of binary classification of relevance (Nallapati, 2004). This allows us to automatically extract queries making use of a sample of relevant documents and then to use an existing search engine as a black box, as can be seen in Figure 2. Sample documents provided as input are used as positives examples to train a binary classifier that can filter retrieved documents based on their relevance.

The problem of query selection for corpora construction is formally defined following the notation introduced in (Bordea et al., 2018) as follows. Given a test collection C of size n where each document c_i is associated with a vector of index terms v_i of a variable size from a set V of size n defined as follows:

$$v_i = \{t_1, ..., t_k\}$$

where t_j is a term from a controlled vocabulary that describes the contents of document c_i, and k is the number of index terms used to annotate the document. We assume that a subset D of size m of relevant documents known to report FDIs is also given, where $m < n$. The subset of index vectors associated with relevant documents is the set V' of size m and each relevant document d_i is annotated with a vector v' of index terms. We also assume that there is a fixed retrieving function S, where $S(q, d)$ gives the score for document d with respect to query q.

We define query selection as the problem of finding a query scoring function R, that gives the score $R(D, q)$ for query q with respect to the collection of relevant documents D. A desired query scoring function would rank higher the queries that perform best when selecting relevant documents.

3 Document classification

In this section, we give an overview of the features and algorithms used to classify scientific articles based on their relevance for the task of FDI discovery, proposing a supervised method to select relevant documents. Classification models are trained using relevant documents as positive examples and irrelevant documents as negative examples.

Preprocessing. Documents are represented as a bag of words that are normalised by replacing numbers by the '#' character. Additionally, other special characters are removed and each word is lowercased.

Word features. Word features are constructed using 1-grams, 1-grams + 2-grams and 1-grams + 2-grams + 3-grams of words. Take for example a document containing the following expression *Food and drug interactions*. The 1-gram features are *food, and, drug, interactions*; 2-grams features are *food and, and drug, drug interactions*; 3-grams are *food and drug, and drug interactions*. In our task, features are constructed from words contained in all documents.

Feature representation. To train classification models, the dataset is transformed into a matrix of size $N \times M$ where N is the number of documents in the dataset and M is the number of features. For each word feature, three types of feature representation approaches are investigated for representing input data:

- **One-hot encoding.** Raw binary occurrence (RBO) matrices. Each document d is represented as a binary feature-document occurrence vector $Rbo = [rbo_0, rbo_1, ...rbo_m]$ of size M where $rbo_i = 1$ if the feature i is in the document d, 0 otherwise.

- **Term frequency.** Count occurrence matrices. Each document d is represented by a vector of counts of term-document occurrences $Tf = [tf_0, tf_1, ...tf_m]$ of size M where tf_i is the number of occurrences of the feature i in the document d.

- **TF-IDF.** Term frequency-inverse document frequency. Each document d is represented by a vector of products of term frequency (TF) and inverse document frequency (IDF).

Index terms features. There is a large number of infrequent index terms that are used to annotate a small number of training documents. To reduce the feature space, we consider as features only index terms that are used to annotate a minimum number of documents. Additionally, we take into account the IDF of each index term in the full collection, that is the number of documents that are annotated with an index term.

Generalised index terms. Index terms are provided from a vocabulary that is hierarchically structured. We exploit this hierarchy to identify terms related to foods and drugs and we introduce three features called *Foods*, *Drugs*, and *Foods and Drugs* that identify documents annotated with one or both types of concepts of interest for our domain. Table 1 gives several examples of nodes from the MeSH hierarchy that are useful for identifying food and drug related concepts.

Classification algorithms. We compare the performance of five classification algorithms with default parameters provided by Scikit-Learn[3]: (1) a decision tree classifier (DTree), (2) a linear SVM classifier (LSVC), (3) a multinomial Naive Bayes classifier (MNB), (4) a logistic regression classifier (LogReg), and (5) a RandomForest classifier (RFC).

4 Query selection

In this section, we discuss the query selection approach presenting first several methods for selecting candidate terms and then proposing different approaches for scoring candidate terms to select the best queries for automatically constructing a domain-specific corpus.

---

[3]`http://scikit-learn.org/stable/`

Food concepts	Node	Drug concepts	Node
Plants	B01.650	Pharmacologic actions	D27.505
Food and beverages	J02	Pharmaceutical preparations	D26
Diet, food, and nutrition	G07.203	Heterocyclic compounds	D03
Fungi	B01.300	Polycyclic compounds	D04
Nutrition therapy	E02.642	Inorganic chemicals	D01
Carbohydrates	D09	Organic chemicals	D02
Plant structures	A18	Amino acids, peptides, and proteins	D12

Table 1: Nodes from the MeSH hierarchy used to identify food and drug related index terms

4.1 Candidate terms for query selection

A first step in automatically selecting queries for constructing a domain-specific corpus is to identify candidate terms that are likely to describe and retrieve relevant documents for the given domain. In our experiments, we consider as candidate queries single terms but more complex queries that combine multiple index terms can also be envisaged.

Index terms. Scientific articles are often annotated with high quality index terms from a controlled vocabulary that can be used as queries to retrieve relevant documents. The controlled vocabulary typically provides in addition hierarchical relations between terms that could be further used to identify more general or abstract concepts. One of the limitations of this approach is that index terms summarise the main topics of an article but might miss some of the more fine-grained information.

Document n-grams. All the sequences of words from a document could be considered as candidate terms for query selection but compared to index terms, this approach is more noisy and increases the ambiguity of terms.

Background knowledge. There are several sources of background knowledge that can be considered to identify terms of interest to retrieve documents that describe FDIs. Queries that mention drugs and a food name are likely to retrieve relevant documents for our domain. There are multiple vocabularies and ontologies that partially cover the food domain from different perspectives, but currently the most complete list of foods can be found by exploiting the DBpedia[4] category structure. DBpedia entities linked to the *Foods* category with the properties *skos:broader* and *dct:subject of* are considered as candidate food terms. Further filtering is required because categories are not necessarily used to identify the type of a DBpedia entity but rather a more loosely defined relatedness relation that often leads to semantic drift when iteratively exploring narrower categories.

Entities are filtered based on their RDF type, based on words but also by excluding categories that are related to foods but are not of interest for FDIs, as can be seen in Table 2. This table is not meant to give an exhaustive list of filters but just a few illustrative examples. We use leaf categories to refer to categories that are taken into consideration as candidate terms but that are not further explored to identify more narrow terms. We identified 15,686 foods from DBpedia and we evaluated the precision of a random sample that is 88%. The recall of this approach was also estimated using a list of 57 foods mentioned in the Stockley 2008 compendium and is 65%.

This is because some of the foods such as *green tea* or *tonic water* can only be found in broader DBpedia categories such as *Food and drink*, *Drinks* or *Diets*, which are more noisy and hence more difficult to filter by hand. The relatively low recall is also due to name variations (e.g., *edible clay* vs. *medicinal clay* in DBpedia), to missing food categories in DBpedia (e.g., *xanthine-containing beverages* and *tyramine-rich foods*), and to errors in the RDF types assigned by DBpedia (e.g., *Brussels sprouts*[5] have the type *Person*).

4.2 Query selection approaches

We consider two types of scoring functions, first based on simple frequency counts of index terms and a second type of scoring functions inspired by existing approaches for feature selection used in supervised classification. The most basic query scoring function is frequency, denoted as the count $c(V', q)$ of query q with respect to the set V' of index vectors associated with relevant documents. The TF-IDF scoring function $tfidf(V', V, q)$ of query q with respect to the set of index vectors associated with relevant documents V' discrimi-

RDF types	Words	Categories	Leaf categories
Book	bakeries	Alcoholic drink brands	Beer
Building	books	Carnivory	Ducks
Company	campaigns	Cherry blossom	Geese
Location	disease	Decorative fruits and seeds	Onions
Organisation	history	Forages	Quails
Person	people	Halophiles	Rubus
Place	pizzerias		Swans
Restaurant	science		Whisky
Software	vineyards		Wine

Table 2: Filters used for selecting candidate foods under the DBpedia *Foods* category

nated against the full set of index terms V is defined as:

$$tfidf(V', V, q) = c(V', q)/ln(c(V, q))$$

For the second category of scoring functions, we consider a binary classifier that distinguishes between relevant documents D and an equal number m of randomly selected documents from the test collection C. Assuming that the size of the test collection is much larger than the number of documents known to be relevant, there is a high probability that randomly selected documents are irrelevant. The first scoring function is the information gain that measures the decrease in entropy when the feature is given vs. absent (Forman, 2003) and is defined as follows:

$$InfoGain(Class, t) = H(Class) - H(Class|t)$$

where the entropy H of a class with two possible values (i.e., relevant *pos* and irrelevant *neg*) is defined based on their probability p as:

$$H(Class) = -p(pos) * log(p(pos)) \\ - p(neg) * log(p(neg))$$

The gain ratio is further defined as the information gain divided by the entropy of the term t:

$$GainR(Class, t) = InfoGain(Class, t)/H(t)$$

Finally, we also consider the Pearson's correlation as a query scoring function for the same binary classifier.

5 Experimental setting

The corpus used in our experiments is manually constructed through a bibliographic analysis of the

---

[4] https://wiki.dbpedia.org/

[5] Brussels sprouts: http://dbpedia.org/page/Brussels_sprout

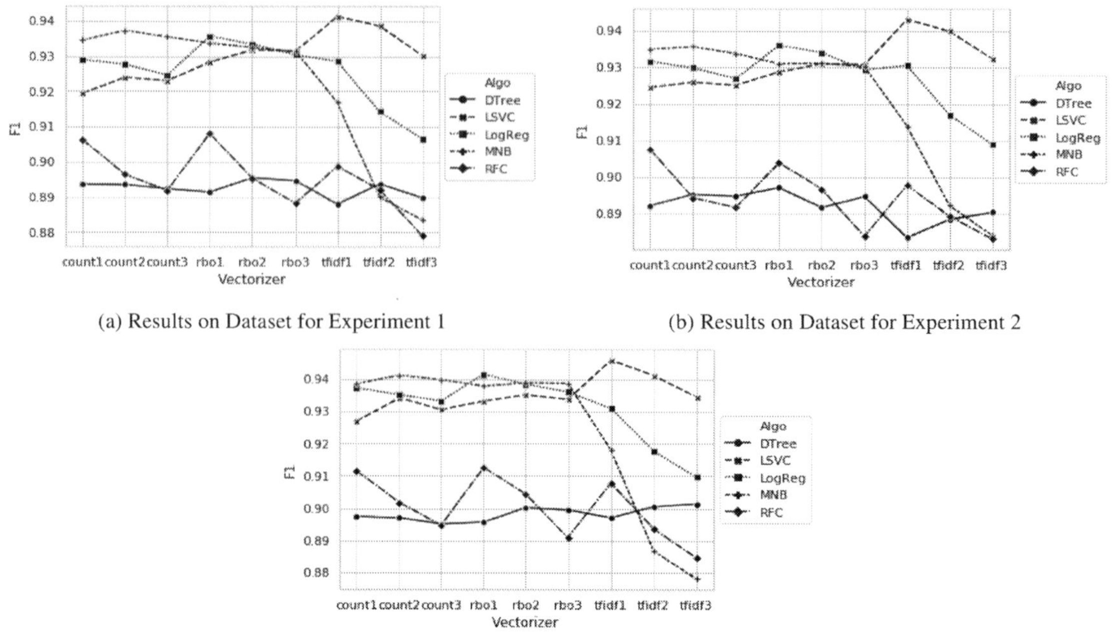

(a) Results on Dataset for Experiment 1   (b) Results on Dataset for Experiment 2

(c) Results on Dataset for Experiment 3

Figure 3: Results of 10-fold cross-validation obtained on each dataset with different classifiers (i.e., decision tree (DTree), linear SVM (LSVC), multinomial Naive Bayes (MNB), logistic regression (LogReg), and RandomForest (RFC)) and vectorizers (i.e., term frequency (count), raw binary occurrence (rbo), and tfidf)

references provided in the Stockley compendium on drug interactions in relation to food. These are considered as positives examples that are used to train a discriminative classifier. The problem of finding negative examples is more challenging because of the problem of unbalanced data and because we aim to train a classifier that is sensitive enough to distinguish between scientific articles that are closely related in topic (i.e., published in the same journals) but that do not describe FDIs.

We manually identify references from pages listed in the index under individual foodstuffs and *Foods*, for a total of 912 references and 460 references, respectively. Using the title and the year of each reference, we retrieve 802 unique PubMed identifiers for references that are available in MEDLINE. In our experiments, we make use of corpora built from MEDLINE abstracts published before 2008 since the version of the Stockley compendium that is available to us was published at this date.

Starting from this collection, several subsets of abstracts are constructed as follows:

(i) references cited in Stockley 2008 (subset *Stockley2008*),

(ii) results of *Food-Drug Interaction* and *Herb-Drug Interaction* MeSH term queries (subset *FDI-HDI*),

(iii) results of the queries *drug and [food name]* where *food name* is one of the 15,686 food names collected from DBpedia as described in Subsection 4.1 (subset *DRUGFOOD*),

(iv) all the MEDLINE abstracts published before 2008 (subset *MEDLINE2008*).

From the first and third subsets, we analyse the list of journals where the articles have been published and all the abstracts published in those journals. In that respect, we have two additional abstract subsets *jrnlAbstracts()* from *Stockley2008* and *jrnlAbstracts()* from *DRUGFOOD* respectively. In our experiments, the set of positive abstracts is the union of Stockley's references with the results of the *FDI-HDI* queries. Table 3 presents the size of the subsets.

The problem of constructing a domain-specific corpus for FDIs is characterised by unbalanced training sets with the non-relevant class representing a large portion of all the examples, while the relevant class has only a small percent of the examples. Dealing with unbalanced class distributions is inherently challenging for discriminative

	Abstracts	Jrnls	jrnlAbstracts()
Stockley2008	895	339	3,344,842
FDI-HDI	3593		
DRUGFOOD	309,327	7421	23,383,538
MEDLINE2008	16,733,485		

Table 3: Overview of different corpora used in our experiments and their size in number of documents

algorithms resulting in trivial classifiers that completely ignore the minority class. We deal with the problem of unbalanced data by under-sampling the majority class such that the training examples in both classes are equal. We define three sets of 4,500 randomly sampled abstracts as negative training examples that successively contain an increasing number of restrictions based on document relevance, publication venue and year of publication:

Experiment 1: abstracts in *jrnlAbstracts()* from *DRUGFOOD* subset that are not cited in Stockley, *FDI-HDI* and *DRUGFOOD* abstracts;

Experiment 2: abstracts in *jrnlAbstracts()* from *DRUGFOOD* subset that are not cited in Stockley, *FDI-HDI* and *jrnlAbstracts()* from *Stockley2008* abstracts;

Experiment 3: MEDLINE abstracts published before 2008 in *jrnlAbstracts()* from *DRUGFOOD* subset which are not cited in Stockley, *FDI-HDI*, *jrnlAbstracts()* from *Stockley2008* and *jrnlAbstracts()* from *DRUGFOOD* abstracts.

6   Results

In this section, we give an overview of the results obtained under different settings. We begin by discussing the results obtained for document classification and we continue with a discussion of the results obtained for the subtask of query selection. In both cases, the classical measures of precision, recall and F-score are used, but in the case of query selection, we adapt these measures to reflect our interest in discovering unseen documents.

6.1   Document classification evaluation

For the purpose of selecting relevant documents regarding food-drug interactions, we evaluate several configurations to construct an efficient classification model. Three sets of experiments are designed around the three training datasets described in the previous section. For each case, we evaluate the models using average of Precision (P), Recall (R) and F1-score (F1) using 10-fold cross-validation. Figure 3 shows the cross-validation re-

sults for different word-based features described in Section 3. The best results in terms of F1-score are obtained across all datasets for TF-IDF features with an SVM classifier. TF-IDF of unigram features combined with SVM classifier produce the best F1-score on all datasets. Focusing on these configurations, results are detailed in Table 4 where we can notice that the recall is higher for the third dataset. The best F1-score presents a low standard deviation, which shows that the obtained model is relatively stable. We conclude that results are better on datasets that use a more restrictive filter for selecting the negative examples (Experiment 3). This demonstrates that the random sampling approach for the majority class can benefit from using a more informed strategy than selecting documents from the full collection.

Exp.	Precision	Recall	F1-score + Std
1	0.962	0.921	0.941 ± 0.010
2	0.965	0.922	0.943 ± 0.007
3	0.964	**0.928**	**0.946*** ± 0.004

Table 4: Results of 10-fold cross-validation on the three datasets using an SVM classifier and 1-gram TF-IDF features. The best result is marked with a star

The next set of experiments is focused on evaluating the performance of features based on index terms as can be seen in Table 5. All the index terms that are used to annotate at least 10 documents from our collection are considered as features, ignoring the less frequent index terms. In general, the results are comparable or better than the best results using word features in terms of F1-score. In the case of index terms features, the best results are obtained for the decision tree classifier that outperforms the linear SVM classifier on all three datasets. The same conclusion can be drawn from these experiments in relation to the random sampling approach as the best results are obtained again for the third experiment.

6.2   Query selection evaluation

The challenge for evaluating queries is that it is preferable to rely on the training examples alone for evaluation. But each selected query will retrieve documents that might be relevant but that are not contained in the provided dataset. To address this issue, we use the best performing classification approach described in the previous section to predict the relevance of retrieved documents instead of computing precision based on the docu-

Exp.	Algorithm	Precision	Recall	F1-score
1	DTree	**0.963**	**0.961**	**0.962**
	LSVC	0.947	0.942	0.944
	LogReg	0.960	0.954	0.957
	MNB	0.941	0.941	0.941
	RFC	0.959	0.955	0.957
2	DTree	0.962	0.958	0.960
	LSVC	0.954	0.950	0.952
	LogReg	**0.964**	0.959	**0.961**
	MNB	0.944	0.943	0.943
	RFC	0.963	**0.960**	0.961
3	DTree	**0.967***	**0.965***	**0.966***
	LSVC	0.959	0.956	0.957
	LogReg	0.965	0.961	0.963
	MNB	0.946	0.946	0.946
	RFC	0.963	0.961	0.962

Table 5: Results of 10-fold cross-validation using different classifiers: decision tree (DTree), linear SVM (LSVC), multinomial Naive Bayes (MNB), logistic regression (LogReg), and RandomForest (RFC) with index terms features. The overall best results are marked with a star

ments known to be relevant alone. Our assumption is that the high performance achieved by the classifier allows us to compute a reliable estimate of precision. Although not perfect, this evaluation strategy allows us to avoid the need for further manual annotation or relevant documents. Recall is calculated for a limited number of retrieved documents as some of the MeSH index terms such as *Humans* and *Animals* are broad enough to be used for annotating most of the documents in the test collection.

Word-based query candidates are not further considered at this stage because the best classification performance is achieved for 1-gram features which are deemed to be too ambiguous for our purposes. Table 6 gives an overview of the top 30 1-gram features selected using the SVM classifier. Several names of drugs such as *aminophylline*, *cyclosporine*, and *ephedrine* that are known to have interactions with foods are among the highest ranked features. Foods such as *caffeine, coffee, cola* and *grapefruit* are also known for their high potential of interactions with drugs. Among these features, names of plants with drug interactions are present including *biloba* and *kava*. Although interesting on their own, we conclude that these features are too generic to be used as queries to extract articles about FDIs without further combining them with other features or index terms.

On the other hand, index term candidates are much more precise, including many terms that refer to food-drug interaction mechanisms such

absorption	cyclosporine	interaction
alcohol	diet	kava
aminophylline	drug	lithium
anticoagulation	effects	medication
biloba	ephedrine	milk
bioavailability	ergotism	monograph
caffeine	food	nutrition
cheese	grapefruit	oral
coffee	herb	pharmacokinetic
cola	ingestion	phytotherapy

Table 6: Top 30 1-gram features selected using the SVM classifier

as *Biological Availability* and *Cytochrome P-450 CYP3A*. Also included in this list are chemical compounds such as *Flavanones* and *Furocoumarins* that are contained in certain foods such as *grapefruit* and that interact with many drugs.

Table 7 gives an overview of the results obtained by each scoring function discussed in the previous section. Performance is computed for the top 20 ranked queries for each method. All the methods score high the *Food-Drug interactions* MeSH term but we remove this term from the results because it was used to construct the FDIs corpus. Overall, the best performance is obtained by the Gain ratio scoring function. Selected queries using this approach include: *Biological Availability*, *Drug Interactions*, and *Intestinal Absorption*. Gain ratio outperforms other approaches because it penalizes high frequency terms that are too broad, such as *Adult, Aged*, and *Female*.

Scoring function	Predicted P@100	Recall @16k	Predicted F1-score
Frequency	0.2020	0.0032	0.0584
TF-IDF	0.2590	0.0084	0.0784
Info gain	0.2755	0.0084	0.0812
Gain ratio	**0.3755**	**0.0557**	**0.0970**
Correlation	0.2590	0.0081	0.0770

Table 7: Scoring functions evaluated for the top 20 MeSH terms using predicted precision at top 100, recall at top 16k and the combined predicted F1-score

7 Related work

Hand-crafted queries based on MeSH terms are often used for retrieving documents related to adverse drug effects (Gurulingappa et al., 2012), but there is a much smaller number of documents available for specific types of adverse effects such as FDIs and herb-drug interactions. The problem of building queries for finding documents related to drug interactions has been recently tackled for herb-drug interactions (Lin et al., 2016). This work addresses a less challenging usage scenario where users have in mind a pair of herbs and drugs and are interested in finding evidences of interaction. Queries are manually constructed by a domain expert using MeSH synonyms for herbs and drugs together with the following MeSH qualifiers: *adverse effects*, *pharmacokinetics*, and *chemistry*. Two additional heuristics rank higher retrieved articles that are annotated with the MeSH terms *Drug Interactions* and *Plant Extracts/pharmacology*. Another limitation of this work is the size of the evaluation dataset that is based on a single review paper (Izzo and Ernst, 2009) that provides about 100 references. In contrast, we propose an automated approach for query selection and we make use of a considerably larger dataset of relevant publications for training and evaluation.

The food-drug interaction discovery task proposed here is similar in setting with the subtask on prior art candidates search from the intellectual property domain (Piroi et al., 2011). In the CLEF-IP datasets, topics are constructed using a patent application and the task is to identify previously published patents that potentially invalidate this application. Keyphrase extraction approaches were successfully applied to generate queries from patent applications (Lopez and Romary, 2010; Verma and Varma, 2011). The input is much larger for our task, that is a corpus of scientific articles describing FDIs manually annotated with index terms from the MeSH thesaurus. A main difference between our work and the CLEF-IP task is that we mainly focus on evaluating different methods for query selection by relying on the PubMed search engine. This makes our task more similar to the term extraction task (Aubin and Hamon, 2006), as we aim to identify relevant terms for a broad domain rather than for a specific document, as done in keyphrase extraction.

The dataset used in (Jovanovik et al., 2015) to infer interactions between drugs and world cuisine is based on textual information from DrugBank about food-drug interactions and optimum drug intake time with respect to food. But this information was manually extracted and structured. The most closely related work to ours is (Bordea et al., 2018) where the authors propose an approach for query selection based on index terms. We extend this work by considering multiple types of classification algorithms and by analysing different query candidates beyond index terms.

8 Conclusion and future work

In this paper, we introduced a large dataset of articles that describe food-drug interactions annotated with index terms to investigate an approach for query selection that allows us to discover other food-drug interactions using an existing search engine. We investigated different strategies for addressing the problem of unbalanced data and we showed that a more informed approach that takes into consideration publication venue and year gives better results than a naive approach for random sampling. We proposed an automatic evaluation of retrieved results using a high-performance classifier and we showed that feature selection approaches outperform frequency-based approaches for this task, with an approach based on gain ratio achieving the best results in terms of predicted F1-score.

In our experiments mainly focused on queries constructed using a single index term, therefore a first direction for future work is to investigate more complex queries that combine multiple terms. The number of queries that have to be evaluated would increase considerably especially for combinations with word-based features. Another improvement would be to compare our results with keyphrase extraction approaches instead of analysing all the n-grams and to generate queries using background knowledge about drugs and foods. Finally, the datasets proposed here are based on an older version of the Stockley compendium from 2008. The results presented in this work could be more relevant if a more recent version is considered as this is a highly dynamic field of research.

9 Acknowledgments

This work was supported by the MIAM project and Agence Nationale de la Recherche through the grant ANR-16-CE23-0012 France and by the KaNNa project and the European Commission through grant H2020 MSCA-IF-217 number 800578.

References

Sophie Aubin and Thierry Hamon. 2006. Improving term extraction with terminological resources. In *Advances in Natural Language Processing*, pages 380–387. Springer.

Karen Baxter and CL Preston. 2010. *Stockley's drug interactions*, volume 495. Pharmaceutical Press London.

Georgeta Bordea, Frantz Thiessard, Thierry Hamon, and Fleur Mougin. 2018. Automatic query selection for acquisition and discovery of food-drug interactions. In *International Conference of the Cross-Language Evaluation Forum for European Languages*, pages 115–120. Springer.

Arik Dahan and Hamutal Altman. 2004. Food–drug interaction: grapefruit juice augments drug bioavailability-mechanism, extent and relevance. *European journal of clinical nutrition*, 58(1):1.

George Forman. 2003. An extensive empirical study of feature selection metrics for text classification. *Journal of machine learning research*, 3(Mar):1289–1305.

Harsha Gurulingappa, Abdul Mateen Rajput, Angus Roberts, Juliane Fluck, Martin Hofmann-Apitius, and Luca Toldo. 2012. Development of a benchmark corpus to support the automatic extraction of drug-related adverse effects from medical case reports. *Journal of biomedical informatics*, 45(5):885–892.

Thierry Hamon, Vincent Tabanou, Fleur Mougin, Natalia Grabar, and Frantz Thiessard. 2017. Pomelo: Medline corpus with manually annotated food-drug interactions. In *Recent Advances in Natural Language Processing (RANLP)*, pages 73–80.

Angelo A Izzo and Edzard Ernst. 2009. Interactions between herbal medicines and prescribed drugs. *Drugs*, 69(13):1777–1798.

Milos Jovanovik, Aleksandra Bogojeska, Dimitar Trajanov, and Ljupco Kocarev. 2015. Inferring cuisine-drug interactions using the linked data approach. *Scientific reports*, 5:9346.

Kuo Lin, Carol Friedman, and Joseph Finkelstein. 2016. An automated system for retrieving herb-drug interaction related articles from medline. *AMIA Summits on Translational Science Proceedings*, 2016:140–149.

Patrice Lopez and Laurent Romary. 2010. Experiments with citation mining and key-term extraction for prior art search. In *CLEF 2010-Conference on Multilingual and Multimodal Information Access Evaluation*.

Ramesh Nallapati. 2004. Discriminative models for information retrieval. In *Proceedings of the 27th annual international ACM SIGIR conference on Research and development in information retrieval*, pages 64–71. ACM.

Florina Piroi, Mihai Lupu, Allan Hanbury, and Veronika Zenz. 2011. Clef-ip 2011: Retrieval in the intellectual property domain. In *CLEF (notebook papers/labs/workshop)*.

Isabel Segura-Bedmar, Paloma Martínez, and María Herrero Zazo. 2013. Semeval-2013 task 9: Extraction of drug-drug interactions from biomedical texts (ddiextraction 2013). In *Second Joint Conference on Lexical and Computational Semantics (*SEM), Volume 2: Proceedings of the Seventh International Workshop on Semantic Evaluation (SemEval 2013)*, volume 2, pages 341–350.

Manisha Verma and Vasudeva Varma. 2011. Exploring keyphrase extraction and ipc classification vectors for prior art search. In *CLEF (Notebook Papers/Labs/Workshop)*.

Enhancing biomedical word embeddings by retrofitting to verb clusters

Billy Chiu[1] **Simon Baker**[1] **Martha Palmer**[2] **Anna Korhonen**[1]

[1] Language Technology Lab, University of Cambridge, 9 West Road, Cambridge, CB3 9DB, UK
[2] Department of Lingusitics, University of Colorado at Boulder, Colorado, 80309–0295, USA
{hwc25,sb895,alk23}@cam.ac.uk
mpalmer@colorado.edu

Abstract

Verbs play a fundamental role in many biomedical tasks and applications such as relation and event extraction. We hypothesize that performance on many downstream tasks can be improved by aligning the input pretrained embeddings according to semantic verb classes. In this work, we show that by using semantic clusters for verbs, a large lexicon of verb classes derived from biomedical literature, we are able to improve the performance of common pretrained embeddings in downstream tasks by retrofitting them to verb classes. We present a simple and computationally efficient approach using a widely-available "off-the-shelf" retrofitting algorithm to align pretrained embeddings according to semantic verb clusters. We achieve state-of-the-art results on text classification and relation extraction tasks.

1 Introduction

Core tasks in biomedical natural language processing (BioNLP) such as relation and event extraction, text classification, syntactic and semantic parsing, natural language inference, and entailment can all benefit from rich computational lexicons containing information about the behaviour and meaning of words in biomedical texts. Verbs are especially important in many of these tasks (Cohen et al., 2008); for example, describing protein–protein interactions in biomedical text can often rely on a wide range of verbs, such as "bind," "activate," "carry," "facilitate," "interact," *etc.* in order to determine the specific type of interaction.

Lexical semantic classes for verbs can be used to abstract away from individual words, or to build a lexical structure (taxonomy) which predicts much of the behaviour of a new word by associating it with an appropriate class (Levin, 1993; Kipper et al., 2008). For example, the verbs "assess," "evaluate," "estimate," "explore," and "analyze"

belong to the class *examine*, while the verbs "utilize," "employ," and "exploit" belong to the class *use*. In addition to simple synonyms of verbs, semantic classes capture similarity in their use and behaviour in text by analysing their contexts (Levin, 1993).

In the past, lexical verb classes have been successfully shown to improve the performance classifiers in a variety of tasks and down stream applications in the biomedical domain; such as relation extraction (Sharma et al., 2010), biomedical fact extraction (Rupp et al., 2010), text classification for cancer (Baker et al., 2015), biomedical discourse analysis (Cox et al., 2017), and biomedical information retrieval (Mahalakshmi, 2015).

Lexical classes are useful for their ability to capture generalizations about a range of linguistic properties (Kipper et al., 2000); our hypothesis is therefore that by retrofitting embedded word representations to semantic verb classes, semantically-similar verbs (*i.e.* member verbs within the same lexical class) like "suppress" and "inhibit" will be pulled together in vector space, whereas verbs like "collect" and "examine" will not. Consequently, this allows NLP systems to generalize away from individual verbs, alleviating the data sparseness problem of representing each verb in the corpus individually.

Retrofitting is a graph-based learning technique for using lexical relational resources to obtain higher quality semantic vectors (Faruqui et al., 2015). It is applied as a post-processing step by running belief propagation on a graph constructed from lexicon-derived relational information to update word vectors. It can be applied to any pretrained word embedding vectors. The intuition behind retrofitting is to encourage the retrofitted vectors to be similar to the vectors of related word

125

Proceedings of the BioNLP 2019 workshop, pages 125–134
Florence, Italy, August 1, 2019. ©2019 Association for Computational Linguistics

types and similar to their original distributional representations.

Using a standard "off-the-shelf" retrofitting algorithm, we apply the idea of retrofitting to verb clusters to two sets of widely-used pretrained embedding vectors in BioNLP (those by Pyysalo et al. (2013a) and by Chiu et al. (2016)) to obtain improved embeddings. We show that by doing nothing more than using this simple approach, we achieve state-of-the-art results on two text classification tasks (both tasks evaluated on document and sentence level classification), and a relation extraction task. We make our retrofitted embeddings freely available to the BioNLP community along with our code.[1]

The main contribution of this work is to be the first of its kind to apply verb-based retrofitting in the biomedical domain. Retrofitting has thus far only been applied for aligning vectors to Medical Subject Headings (MeSH) (Yu et al., 2016), and been validated only in an extrinsic setting. We show that with very little effort, we can achieve state-of-the art results on various downstream tasks in a range of biomedical subdomains.

This paper will first describe relevant work on retrofitting to lexical resources in BioNLP; we then briefly give an overview of two verb cluster and lexicons that we use in our methodology, and then our task-based evaluation. We end with a discussion of the evaluation results.

2 Related work

Lexical resources can be used to enrich representation models by providing them other sources of linguistics information beyond the distributional statistics obtained from corpora. In recent literature, various methods to leverage knowledge available in human- and automatically-constructed lexical resources have been proposed.

One such method involves modifying the objectives in the original representation learning procedures so that they can jointly learn both distributional and lexical information—for example, Yu and Dredze (2014) modify the CBOW objective function by introducing semantic constraints as obtained from the paraphrase database (Ganitkevitch et al., 2013) to train word representations which focus on word similarity over word relatedness.

Another class of methods incorporates lexical information into the vector representations as a post-processing procedure. The method fine-tunes the pretrained word vectors to satisfy linguistic constraints from the external resources. The method can be applied to any off-the-shelf models without requiring large corpora for (re-)training as the joint-learning models do. Among these methods, *retrofitting* (Faruqui et al., 2015) is widely used.

Given any (pretrained) vector-space representations, the goal of retrofitting is to bring closer words which are connected *via* a relation (*e.g.* synonyms) in a given semantic network or lexical resource (*i.e.* linguistic constraints). For example, Yu et al. (2016, 2017) retrofit word vector spaces of MeSH terms by using additional linkage information from the UMNSRS hierarchy to improve the representations of biomedical concepts. Building on retrofitting, Lengerich et al. (2018) generalize retrofitting methods by explicitly modelling individual linguistic constraints that are commonly found in health and clinical-related lexicons (*e.g.* causal-relations between diseases and drugs).

In theory, the joint-learning models could be as effective (or better) as those produced by fine-tuning distributional vectors. However, the performance of joint-learning models has not surpassed that of fine-tuning methods.[2] Furthermore, the joint-learning objectives are usually model-specific and are tailored to a particular model, making it difficult to use them with other methods. In this work, we will use retrofitting to incorporate our lexical features into the word representations.

3 Verb clusters

In this work, we investigate retrofitting popular word embeddings to two publicly available[3] lexicons for verb clusters. The first is composed of 192 relatively frequent verbs from a corpus of 2230 biomedical journal articles which have been hierarchically classified into three levels: 16, 34, and 50 verb classes. The three levels reflect different granularity in the semantics of the verb classes as illustrated in Figure 1. These clusters were annotated by 4 domain experts and 2 linguists, were used to create the gold standard (Korhonen

---

[2]The SimLex-999 home page (`www.cl.cam.ac.uk/~fh295/simlex.html`) lists state-of-the-art performance models, none of which have learned representations jointly

[3]`https://github.com/cambridgeltl/bio-verbnet`

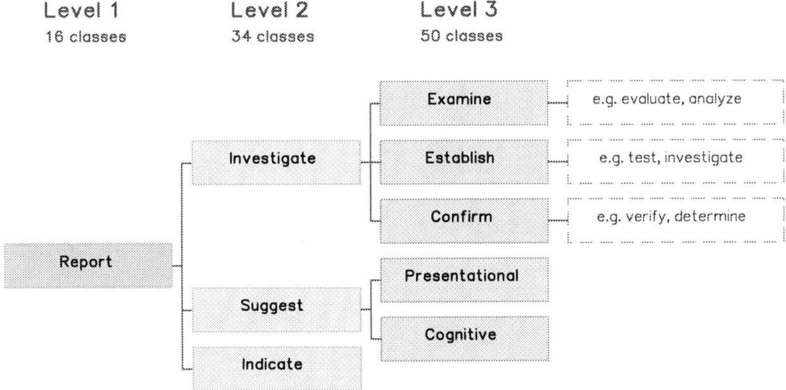

Figure 1: Examples of the verb classes introduced by Korhonen et al. (2006).

et al., 2006). We will refer to this lexicon for the remainder of this paper as the *annotated clusters*.

Chiu et al. (2019) developed a methodology to further extended the annotated clusters automatically using text from PubMed abstracts and full articles with the goal of facilitating the future creation of a BioVerbNet resource, a specialized resource similar to VerbNet (Schuler, 2005). We will refer to this lexicon for the remainder of this paper as the *expanded clusters*.

Chiu et al. (2019) use a two-step method. In the first step, the best contexts for learning biomedical verb representations are identified using a model based on skip-gram with negative sampling (SGNS). It involves first creating a context configuration space based on dependency relations between words, followed by applying an adapted beam search algorithm to search this space for the class-specific contexts, and finally using these contexts to create class-specific representations.

In the second step, the optimized representation is used to provide word features for building a verb classification. This is obtained by expanding the verbs in the annotated clusters, where the candidate verbs are selected from BioSimVerb (Chiu et al., 2018) based on their frequent occurrence in biomedical journals across 120 subdomains of biomedicine. A Nearest Centroid classifier is then used to connect the new candidates to an appropriate class. The resulting classification provides 1149 verbs assigned to the 50 classes in the original annotated clusters. For each verb, the expanded clusters lists the most frequent dependency contexts that reflect their syntactic behaviour along with example sentences.

For the rest of the work, we will investigate the use of both the annotated and expanded clusters

4 Methodology

We apply retrofitting to our default pretrained embeddings[4] The goal is to change the vector-space of the pretrained word embeddings to better capture the semantics represented by the verb classes in both the annotated and expanded clusters. These verb classes provide different levels of generalization to support various tasks, from the coarse-grained level of 16 classes to a fine-grained one of 50 classes.

We base our retrofitting method on that proposed by Faruqui et al. (2015). Given any pretrained vector-space representation, the main idea of retrofitting is to pull words which are connected in relation to the provided semantic lexicon closer together in the vector space. The main objective function to minimize in the retrofitting model is expressed as

$$\sum_{i=1}^{|V|} \left(\alpha_i \left\| \vec{v}_i - \vec{\hat{v}}_i \right\| + \sum_{(i,j) \in S} \beta_{ij} \left\| \vec{v}_i - \vec{v}_j \right\| \right) \quad (1)$$

where $|V|$ represents the size of the vocabulary, $\vec{v}_i$ and $\vec{v}_j$ corresponds to word vectors in a pretrained representation, and $\vec{\hat{v}}_i$ represents the output word vector. S is the input lexicon represented as a set of linguistic constraints—in our case, they are pairs of word indices, denoting the pairwise relations between member verbs in each class. For example,

---

[4]For our default embeddings, we use the embeddings by Chiu et al. (2016) for our text classification tasks and Pyysalo et al. (2013a) for relation extraction.

	Number of verb pairs	
	Annotated clusters	Expanded clusters
16-classes	1,774	96,998
34-classes	638	54,063
50-classes	376	50,104

Table 1: Linguistic constraint counts under each class as obtained from the Korhonen's resource and our automatically-created lexicon.

a pair (i, j) in S implies that the ith and jth words in the vocabulary V belong to the same verb class.

The values of α_i and β_{ij} are predefined and control the relative strength of associations between members. We follow the default settings for these values as stated in the authors' work by setting $\alpha = 1$ and $\beta = 0.05$ in all of the experiments. To minimize the objective function for a set of starting vectors $\vec{v}$ and produce retrofitted vectors $\hat{\vec{v}}$, we run stochastic gradient descent (SGD) for 20 epochs. An implementation of this algorithm has been published online by the authors;[5] we used this implementation in the present work.

Table 1 shows the linguistic constraint counts under each class as derived from the two lexicons. When retrofitted against the three top levels, the member verbs at each subclass are merged with its upper class, as in the work of Faruqui et al. (2015).

5 Evaluation

We apply retrofitting to incorporate the lexical information into word representations. Then we evaluate the quality of the retrofitted-representation as features for two NLP tasks: text classification and relation classification.

5.1 Task 1: Text classification

We evaluate our word representations using two established biomedical datasets for text classification: the Hallmarks of Cancer (**HOC**) (Baker et al., 2015, 2017) and the Exposure taxonomy (**EXP**) (Larsson et al., 2017). We evaluate each based on their document-level and sentence-level classifications.

The Hallmarks of Cancer depicts a set of interrelated biological factors and behaviours that enable cancer to thrive in the body. Introduced by Weinberg and Hanahan (2000), it has been widely used in biomedical NLP, including as part of

[5]`https://github.com/mfaruqui/retrofitting`

the BioNLP Shared Task 2013, "Cancer Genetics task" (Pyysalo et al., 2013b). Baker et al. (2015, 2017) have released an expert-annotated dataset of cancer hallmark classifications for both sentences and documents in PubMed. The data consists of multi-labelled documents and sentences using a taxonomy of 37 classes.

The Exposure taxonomy, introduced by Larsson et al. (2017), is an annotated dataset for the classification of text (documents or sentences) concerning chemical risk assessments. The taxonomy of 32 classes is divided into two branches: one relates to assessment of exposure routes (ingestion, inhalation, dermal absorption, *etc.*) and the second to the measurement of exposure bio-markers (biomonitoring). Table 2 shows basic statistics for each dataset.

	HOC		EXP	
	Document	Sentence	Document	Sentence
Train	1,303	12,279	2,555	25,307
Dev	183	1,775	384	3,770
Test	366	3,410	722	7,100
Total	1,852	17,464	3,661	36,177

Table 2: Summary statistics of the Hallmarks of Cancer (HOC) and the Chemical Exposure Assessment (EXP) datasets.

The model follows the convolutional neural network (CNN) model proposed by Kim (2014). An implementation of this algorithm on HOC and EXP has been published by Baker and Korhonen (2017); we use this implementation in our experiment. The input to the model is an initial word embedding layer that maps input texts into matrices, which is then followed by convolutions of different filter sizes, 1-max pooling, and finally a fully-connected layer leading to an output Softmax layer predicting labels for text. Model hyperparameters and the training setup are summarized in Table 3.

Parameters	Values
Vector dimension	200
Filter sizes	3,4 and 5
Number of filters	300
Dropout probability	0.5
Minibatch size	50
Input size (in tokens)	500 (documents), 100 (sentences)

Table 3: Hyper-parameters used in (Baker and Korhonen, 2017).

For both tasks, we use the embeddings[6] by Chiu et al. (2016). Performance is evaluated using the standard precision, recall, and F_1-score metrics of the labels in the model using the one-*vs.*-rest setup: we train and evaluate K independent binary CNN classifiers (*i.e.* a single classifier per class with the instances of that class as positive samples and all other instances as negatives). Due to their random initialization, we repeat each CNN experiment 20 times and report the mean of the evaluation results to account for variances in neural networks. To address overfitting in the CNN, we follow the authors' early stopping approach, testing only the model that achieved the highest results on the development dataset.

5.2 Task 2: Relation classification

We evaluate our retrofitted representations on the Bio-Creative VI Chemical–Protein relation extraction dataset (**CHEMPROT**) (Krallinger et al., 2017). The corpus provides mention and relation annotations for complex events related to chemical–protein interaction in molecular biology. The goal of this task is to predict whether a given chemical–protein pair is related or not, and to then verify its corresponding relation type. There are five types of relations: *Up-regulator*, *Down-regulator*, *Agonist*, *Antagonist*, and *Substrate*. The corpus is provided in the Turku Event Extraction System (TEES) XML format and are installed with the Turku Extraction System (Björne, 2014). It is parsed with the the BLLIP parser (Charniak and Johnson, 2005) with the McClosky bio-model (Mcclosky, 2010), followed by conversion of the constituency parses into dependency parses using the Stanford Tools (MacCartney et al., 2006). Table 4 summarizes key statistics for the dataset.

	Documents	Entities	Relations
Train	1,020	25,769	4,157
Dev	612	15,571	2,416
Test	800	20,829	3,458
Total	2,432	62,169	10,031

Table 4: Summary statistics of the Chemical-Protein interaction dataset (CHEMPROT).

The model follows the CNN model proposed by Björne and Salakoski (2018). We directly use their published implementation. The model input is an initial word embedding layer that maps input texts into matrices, followed by convolutions of different filter sizes and 1-max pooling, and finally a fully connected layer, leading to an output Softmax layer for predicting labels. Performance is evaluated using the standard precision, recall, and F_1-score metrics of the labels in the model. Classification is performed as multilabel classification where each example may have 0 to n positive labels. Model hyperparameters and the training setup are summarized in Table 5.

Parameters	Values
Vector dimension	200
Filter sizes	1, 3, 5 and 7
Number of filters	400 (100 of each size)
Dropout probability	0.5
Learning rate	0.001
Minibatch size	50

Table 5: Hyperparameters used by Björne and Salakoski (2018).

To account for variance in neural networks due to their random initialization, we adopt the ensemble settings used by Björne and Salakoski (2018). We train 20 models and take the n best ones ($n = 5$), ranked with their F_1-score on the development set, and use their averaged predictions. The ensemble predictions are calculated for each label as the average predicted confidence scores from all the models. We also incorporate the authors' early stopping approach where the model is trained until the development loss no longer decreases. We train for up to 500 epochs, stopping once validation loss has no longer decreased for 10 consecutive epochs. To focus on the effect of verb classes on biomedical representations, we experiment with word representations induced on biomedical texts; this diverges from the authors who use the embeddings[7] by Pyysalo et al. (2013a), induced on a combination of biomedical and general-domain data (PubMed, PMC and Wikipedia texts).

6 Results

We compare the performance of the baseline with the retrofitted embeddings models by measuring their precision (P), recall (R), and F_1-scores in text classification and relation extraction when used as input features.

For the text classification tasks, Tables 6 and 7 show the micro-averaged scores for the HOC and the EXP tasks respectively. Each table shows the performance on document- and sentence-level classification (as columns) with different semantic lexicons (as rows).

For the relation classification task (CHEM-PROT), Table 8 shows the micro-averaged scores. The best results are shown in bold and statistically significant scores are shown with an asterisk. All statistical tests are performed using a two-tailed t-test with $\alpha = 0.05$.

We first describe experiments measuring improvements from the retrofitting method, followed by comparisons against using different sets of lexicons during retrofitting.

6.1 Retrofitting

We use Equation 1 to retrofit word representations using linguistic constraints derived from verb lexicons. Overall, the retrofitted models show improvements in most tasks.

For text classification, the scores have improved in three out of the four cases. For the HOC task (Table 6) all retrofitted models outperform the baseline in F_1-score, which is largely attributed to a substantial improvement in recall (particularly for document-level classification, where there is a 15 point increase over the baseline). In total, five out of the twelve improved scores reported are also statistically significant.

The results for the EXP task (Table 7) are more mixed. At the document level, all retrofitted models achieve a slight F_1-score gain and half of the scores are significant. There is an improvement in recall at the cost of lower precision when compared to the baseline.

However, we can see that sentence-level classification is more difficult, due to the smaller amount of context information available. On the sentence level, the baseline seems to outperform all others, and only two out of six cases are significant. It indicates that the lexicons did not aid sentence-level classification in this particular task.

In relation classification, the word representation achieves the state-of-the-art result after incorporating our lexical information (34 classes). From Table 8, there is approximately a 1.5 point F_1-score increase over the baseline, and half of the improvements reported are significant. The results from both tasks suggest that the class-features provided by verb lexicons improve performance over the raw verb features.

6.2 Semantic lexicons

We compare the performance of our retro-fitted embeddings using both expanded clusters and the manually annotated clusters lexicon. The expanded clusters retrofitted embeddings outperform the original annotated clusters retrofitted embeddings in all evaluated tasks. This is likely due to the larger size of the expanded clusters in comparison to annotated clusters (Table 1), thus providing features for more verbs.

Lexical resources can be useful for NLP tasks for their abilities to capture generalizations about a range of linguistic properties; however, the degree of generalization needed may vary from task to task. When experimenting with retrofitting with different levels of verb classes, we observe a notable difference (1–2 points in F_1-score) between models retrofitted with the coarse-grained level of 16 classes and the fine-grained level of 50 classes.

For document-level text classification in both datasets (Tables 6 and 7), models appear to benefit from a finer-grained classification of 50 classes; on the sentence level a medium level of generalization (34 classes) seems optimal. The best result for relation classification (Table 8) is also obtained with 34 classes.

7 Discussion

The task-based evaluations suggest that verb clusters and a verb-optimized representation, can be a useful resource to support biomedical NLP tasks. In text classification, it has been observed that the occurrence patterns of verbs can be "topic-related" and certain set of verbs frequently appear within a specific topic of documents (Doan et al., 2009; Hatzivassiloglou and Weng, 2002; Sekimizu et al., 1998). Regarding this, expanded clusters appears to have captured some of these topic-related properties. On the HOC dataset, we note that some frequent verbs (such as "proliferate" and "grow") appearing in documents relating to the topic *Sustaining proliferative signaling* also share the same classes in our automatically-created lexicon. Similarly, for exposure assessment documents describing air monitoring data in EXP, we can frequently see member verbs such as "inhale" and "breathe" in the *proceed* class.

Entities–relations described in the biomedical literature are often expressed in a predicative form where a trigger word (most commonly a verb) connects two or more entities; here a range of

Lexicon	Document classification			Sentence classification		
	P	R	F_1	P	R	F_1
No lexicon & SOTA	77.8	51.7	62.1	56.8	30.7	39.9
Annotated clusters						
16-classes	75.1	56.4	64.8	47.1	34.6	39.9
34-classes	74.2	56.6	64.3	48.4	35.5	41.0
50-classes	74.9	59.2	66.2	48.4	35.2	*40.7
Expanded clusters						
16-classes	75.5	64.4	*69.5	45.2	36.5	*40.4
34-classes	74.3	63.5	*68.5	52.7	35.6	**42.5**
50-classes	73.9	66.1	***69.8**	50.9	34.7	41.3

Table 6: Performance results for the Hallmarks of Cancer task (HOC) when different sets of lexicons are used for retrofitting the baseline model. Baseline denotes a skip-gram model generated with our optimized training settings. Scores are adopted from Baker and Korhonen (2017). All figures are micro-averages expressed as percentages (Bold: the best score, *: statistically significant).

Lexicon	Document classification			Sentence classification		
	P	R	F_1	P	R	F_1
No lexicon & SOTA	89.5	87.1	88.3	66.2	62.8	**64.5**
Annotated clusters						
16-classes	88.9	87.7	*88.3	67.1	58.9	62.7
34-classes	89.4	87.8	*88.6	67.2	58.2	*62.4
50-classes	88.9	88.7	88.8	65.6	55.7	60.3
Expanded clusters						
16-classes	89.2	87.9	88.5	66.7	60.0	63.2
34-classes	88.7	88.9	*88.8	67.3	58.7	62.7
50-classes	88.6	89.1	**88.9**	67.5	58.6	*62.7

Table 7: Performance results for the Chemical Exposure Assessment task (EXP). Baseline denotes a skip-gram model generated with our optimized training settings. The "No lexicon" scores are from Baker and Korhonen (2017). All figures are micro-averages expressed as percentages. (Bold: the best score, *: statistically significant).

verbs can be used to describe similar relations. Understanding the commonalities shared among individual verbs helps NLP systems to identify the particular type of relation the text is describing. Consider as an example the *suppress* class in our verb lexicons. It captures the fact that its members are similar in terms of syntax and semantics, and they can be used to make similar statements which describe similar events. In CHEMPROT, member verbs in the *suppress* class such as "suppress" and "inhibit" can often be found in sentences depicting the *down-regulation* relation between chemicals and proteins.

For many NLP applications, lexical classes are useful for their ability to capture generalizations about a range of linguistic properties: by retrofitting word representations to lexical resources, semantically similar verbs (*i.e.* member verbs within the same lexical class) like "suppress" and "inhibit" will be pulled together in the vector space, whereas verbs like "collect" and "examine" will not. Consequently, this allows NLP systems to generalize away from individual verbs, alleviating the data sparseness problem of representing each verb in the corpus individually. The lexical classes provide different levels of generalization to support tasks of various needs, from the coarse-grained level of 16 classes to a fine-grained level of 50. A notable performance difference is observed when we evaluate models retrofitted with different levels of verb classes. Among all three classes, we observe a larger improvement over models at the finer-grained levels of 34 or 50 classes, which reveal that finer-grained levels of verb semantic distinction seem more contributive in our assessed tasks.

Lexicon	P	R	F_1
No lexicon	76.9	63.5	*69.5
SOTA	75.1	65.1	69.7
Annotated clusters			
16-classes	76.5	64.6	70.1
34-classes	78.2	63.8	*70.3
50-classes	76.5	65.0	*70.3
Expanded clusters			
16-classes	76.3	65.2	70.3
34-classes	77.5	65.6	**71.0**
50-classes	76.2	65.9	*70.7

Table 8: Performance results for the Chemical-Protein Interaction (CHEMPROT) when different sets of lexicons are used for retrofitting the baseline model. Baseline denotes a skip-gram model generated with our optimized training settings. SOTA denotes the state-of-the-art result reported by Björne and Salakoski (2018) using the embeddings by Pyysalo et al. (2013a). All figures are micro-averages expressed as percentages. (Bold: the best score for the task, *: statistically significant).

8 Conclusions

Many core NLP tasks and applications in the biomedical domain such as relation and event extraction, text classification, and text mining may benefit from accurate embedded representation of verbs.

Verb semantic classes capture generalizations about a range of linguistic properties, by retrofitting embedded word representations to semantic verb classes, semantically similar verbs (*i.e.* verbs that are members of the same lexical class) are pulled together in the vector space. Consequently, this allows NLP systems to generalize away from individual verbs, reducing the problem of data sparseness in representing less frequent verbs.

The key contribution of this work is to show that by using semantic classes for verbs (such as those provided by both the annotated and expanded clusters) we can improve the downstream performance on several tasks in the biomedical domain by aligning word embeddings according to semantic verb classes.

This is achieved by a post-processing retrofitting procedure, using a standard "off-the-shelf" method, by running belief propagation on a graph constructed from lexicon-derived relational information to update word vectors. It can be applied to any pretrained word embedding vectors.

We applied two lexicons of semantic verb clusters to two sets of widely used pretrained em-bedding vectors in BioNLP on several downstream tasks: two text classification tasks (the Hallmarks of Cancer, and Chemical Exposure Assessment) with both document and sentence classification, as well as a relation extraction task (CHEMPROT). We used a standard "off-the-shelf" retrofitting algorithm to obtain improved embeddings, and we feed the retrofitted representation to the current state-of-the-art models for their respective tasks. We controlled the experimental setup by using the same model implementation, as well as the same training, development and test data folds.

The results show that using verb clusters to retrofit embeddings, we achieved new state-of-the-art performance in the evaluated downstream tasks (with statistically significant scores); the only exception being sentence level classification for the Chemical Exposure Assessment task (however we do improve SOTA in document level classification for the same task). We also note a performance difference when retrofitting with different levels of verb classes, where we see a larger improvement when using finer-grained levels of verb semantic classes (30 or 50 classes), which seem more contributive.

For future work, we will further investigate the possibility of using verb lexicons for retrofitting new generations of word representation models such as contextualized embeddings; we will further evaluate on other downstream biomedical tasks, for instance event and pathway extraction and medical question answering.

Acknowledgement

This work is supported by the Medical Research Council [grant number MR/M013049/1], the ERC Consolidator Grant LEXICAL [grant number: 648909], the ESRC Doctoral Fellowship [grant number: ES/J500033/1] and the Defense Advanced Research Projects Agency [DARPA 15-18-CwC-FP-032].

We would like to thank our reviewers for their constructive feedback. We are very grateful to Tyler Griffiths for helping with proofreading and typesetting this paper.

References

Simon Baker, Imran Ali, Ilona Silins, Sampo Pyysalo, Yufan Guo, Johan Högberg, Ulla Stenius, and Anna Korhonen. 2017. Cancer hallmarks analytics tool (chat): a text mining approach to organize and evaluate scientific literature on cancer. *Bioinformatics*, 33(24):3973–3981.

Simon Baker and Anna Korhonen. 2017. Initializing neural networks for hierarchical multi-label text classification. *BioNLP 2017*, pages 307–315.

Simon Baker, Ilona Silins, Yufan Guo, Imran Ali, Johan Högberg, Ulla Stenius, and Anna Korhonen. 2015. Automatic semantic classification of scientific literature according to the hallmarks of cancer. *Bioinformatics*, 32(3):432–440.

Jari Björne. 2014. *Biomedical Event Extraction with Machine Learning*. Ph.D. thesis, University of Turku.

Jari Björne and Tapio Salakoski. 2018. Biomedical event extraction using convolutional neural networks and dependency parsing. In *Proceedings of the BioNLP 2018 workshop*, pages 98–108.

Eugene Charniak and Mark Johnson. 2005. Coarse-to-fine n-best parsing and maxent discriminative reranking. In *Proceedings of the 43rd annual meeting on association for computational linguistics*, pages 173–180. Association for Computational Linguistics.

Billy Chiu, Gamal Crichton, Anna Korhonen, and Sampo Pyysalo. 2016. How to train good word embeddings for biomedical nlp. In *Proceedings of the 15th Workshop on Biomedical Natural Language Processing*, pages 166–174.

Billy Chiu, Olga Majewska, Sampo Pyysalo, Laura Wey, Ulla Stenius, Anna Korhonen, and Martha Palmer. 2019. A neural classification method for supporting the creation of bioverbnet. *Journal of Biomedical Semantics*, 10(1):2.

Billy Chiu, Sampo Pyysalo, Ivan Vulić, and Anna Korhonen. 2018. Bio-simverb and bio-simlex: wide-coverage evaluation sets of word similarity in biomedicine. *BMC bioinformatics*, 19(1):33.

K Bretonnel Cohen, Martha Palmer, and Lawrence Hunter. 2008. Nominalization and alternations in biomedical language. *PloS one*, 3(9):e3158.

Jessica Cox, Corey A Harper, and Anita de Waard. 2017. Optimized machine learning methods predict discourse segment type in biological research articles. In *Semantics, Analytics, Visualization*, pages 95–109. Springer.

Son Doan, Ai Kawazoe, Mike Conway, and Nigel Collier. 2009. Towards role-based filtering of disease outbreak reports. *Journal of Biomedical Informatics*, 42(5):773–780.

Manaal Faruqui, Jesse Dodge, Sujay K. Jauhar, Chris Dyer, Eduard Hovy, and Noah A. Smith. 2015. Retrofitting word vectors to semantic lexicons. In *Proc. of NAACL*.

Juri Ganitkevitch, Benjamin Van Durme, and Chris Callison-Burch. 2013. Ppdb: The paraphrase database. In *Proceedings of the 2013 Conference of the North American Chapter of the Association for Computational Linguistics: Human Language Technologies*, pages 758–764.

Vasileios Hatzivassiloglou and Wubin Weng. 2002. Learning anchor verbs for biological interaction patterns from published text articles. *International Journal of Medical Informatics*, 67(1-3):19–32.

Yoon Kim. 2014. Convolutional neural networks for sentence classification. In *Proceedings of the 2014 Conference on Empirical Methods in Natural Language Processing (EMNLP)*, pages 1746–1751.

Karin Kipper, Hoa Trang Dang, Martha Palmer, et al. 2000. Class-based construction of a verb lexicon.

Karin Kipper, Anna Korhonen, Neville Ryant, and Martha Palmer. 2008. A large-scale classification of english verbs. *Language Resources and Evaluation*, 42(1):21–40.

Anna Korhonen, Yuval Krymolowski, and Nigel Collier. 2006. Automatic classification of verbs in biomedical texts. In *Proceedings of the 21st International Conference on Computational Linguistics and the 44th annual meeting of the Association for Computational Linguistics*, pages 345–352. Association for Computational Linguistics.

Martin Krallinger, Obdulia Rabal, Saber A Akhondi, et al. 2017. Overview of the biocreative vi chemical-protein interaction track. In *Proceedings of the sixth BioCreative challenge evaluation workshop*, volume 1, pages 141–146.

Kristin Larsson, Simon Baker, Ilona Silins, Yufan Guo, Ulla Stenius, Anna Korhonen, and Marika Berglund. 2017. Text mining for improved exposure assessment. *PloS one*, 12(3):e0173132.

Ben Lengerich, Andrew Maas, and Christopher Potts. 2018. Retrofitting distributional embeddings to knowledge graphs with functional relations. In *Proceedings of the 27th International Conference on Computational Linguistics*, pages 2423–2436. Association for Computational Linguistics.

Beth Levin. 1993. *English verb classes and alternations: A preliminary investigation*. University of Chicago press.

Bill MacCartney, Christopher D Manning, and MC de Marneffe. 2006. Generating typed dependency parses from phrase structure parses. In *Proceedings LREC*.

G Suryanarayanan Mahalakshmi. 2015. Content-based information retrieval by named entity recognition and verb semantic role labelling. *Journal of universal computer science*, 21(13):1830.

David Mcclosky. 2010. *Any Domain Parsing: Automatic Domain Adaptation for Natural Language Parsing*. Ph.D. thesis, Brown University, Providence, RI, USA. AAI3430199.

Sampo Pyysalo, Filip Ginter, Hans Moen, Tapio Salakoski, and Sophia Ananiadou. 2013a. Distributional semantics resources for biomedical text processing. *Proceedings of LBM*.

Sampo Pyysalo, Tomoko Ohta, and Sophia Ananiadou. 2013b. Overview of the cancer genetics (cg) task of bionlp shared task 2013. In *Proceedings of the BioNLP Shared Task 2013 Workshop*, pages 58–66.

CJ Rupp, Paul Thompson, William Black, John Mc-Naught, and Sophia Ananiadou. 2010. A specialised verb lexicon as the basis of fact extraction in the biomedical domain. *Proceedings of Verb 2010*, page 188.

Karin Kipper Schuler. 2005. Verbnet: A broad-coverage, comprehensive verb lexicon.

Takeshi Sekimizu, Hyun S Park, and Jun'ichi Tsujii. 1998. Identifying the interaction between genes and gene products based on frequently seen verbs in medline abstracts. *Genome informatics*, 9:62–71.

Abhishek Sharma, Rajesh Swaminathan, and Hui Yang. 2010. A verb-centric approach for relationship extraction in biomedical text. In *2010 IEEE Fourth International Conference on Semantic Computing*, pages 377–385. IEEE.

RA Weinberg and Douglas Hanahan. 2000. The hallmarks of cancer. *Cell*, 100(1):57–70.

Mo Yu and Mark Dredze. 2014. Improving lexical embeddings with semantic knowledge. In *Proceedings of the 52nd Annual Meeting of the Association for Computational Linguistics (Volume 2: Short Papers)*, volume 2, pages 545–550.

Zhiguo Yu, Trevor Cohen, Byron Wallace, Elmer Bernstam, and Todd Johnson. 2016. Retrofitting word vectors of mesh terms to improve semantic similarity measures. In *Proceedings of the Seventh International Workshop on Health Text Mining and Information Analysis*, pages 43–51.

Zhiguo Yu, Byron C Wallace, Todd Johnson, and Trevor Cohen. 2017. Retrofitting concept vector representations of medical concepts to improve estimates of semantic similarity and relatedness. *arXiv preprint arXiv:1709.07357*.

A Comparison of Word-based and Context-based Representations for Classification Problems in Health Informatics

Aditya Joshi♠, Sarvnaz Karimi♠, Ross Sparks♠, Cécile Paris♠, C Raina MacIntyre♣
♠CSIRO Data61, Sydney, Australia
♣Kirby Institute, University of New South Wales, Sydney, Australia
{firstname.lastname}@csiro.au , r.macintyre@unsw.edu.au

Abstract

Distributed representations of text can be used as features when training a statistical classifier. These representations may be created as a composition of word vectors or as context-based sentence vectors. We compare the two kinds of representations (word versus context) for three classification problems: influenza infection classification, drug usage classification and personal health mention classification. For statistical classifiers trained for each of these problems, context-based representations based on ELMo, Universal Sentence Encoder, Neural Net Language Model and FLAIR are better than Word2Vec, GloVe and the two adapted using the MESH ontology. There is an improvement of 2-4% in the accuracy when these context-based representations are used instead of word-based representations.

1 Introduction

Distributed representations (also known as 'embeddings') are dense, real-valued vectors that capture semantics of concepts (Mikolov et al., 2013). When learned from a large corpus, embeddings of related words are expected to be closer than those of unrelated words. When a statistical classifier is trained, distributed representations of textual units (such as sentences or documents) in the training set can be used as feature representations of the textual unit. This technique of statistical classification that uses embeddings as features has been shown to be useful for many Natural Language Processing (NLP) problems (Zhang et al., 2015; Joshi et al., 2016; Chou et al., 2016; Simova and Uszkoreit, 2017; Fu et al., 2016; Buscaldi and Priego, 2017) and biomedical NLP problems (Yadav et al., 2017; Kholghi et al., 2016). In this paper, we experiment with three classification problems in health informatics: influenza infection classification, drug usage classification and

personal health mention classification. We use statistical classifiers trained on tweet vectors as features. To compute a tweet vector, *i.e.*, a distributed representation for tweets, typical alternatives are: (a) tweet vector as a function of word embeddings of the content words[1] in the tweet; or, (b) a contextualised representation that computes sentence vectors using language models. The former considers meanings of words in isolation, while the latter takes into account the order of these words in addition to their meaning. We compare word-based and context-based representations for the three classification problems. This paper investigates the question:

> '*When statistical classifiers are trained on vectors of tweets for health informatics, how should the vector be computed: using word-based representations that consider words in isolation or context-based representations that account for word order using language models?*'

For these classification problems, we compare five approaches that use word-based representations with four approaches that use context-based representations.

2 Related Work

Distributed representations as features for statistical classification have been used for many NLP problems: semantic relation extraction (Hashimoto et al., 2015), sarcasm detection (Joshi et al., 2016), sentiment analysis (Zhang et al., 2015; Tkachenko et al., 2018), co-reference resolution (Simova and Uszkoreit, 2017), grammatical error correction (Chou et al., 2016), emotion intensity determination (Buscaldi

---

[1] *Content words* refers to all words except stop words.

Proceedings of the BioNLP 2019 workshop, pages 135–141
Florence, Italy, August 1, 2019. ©2019 Association for Computational Linguistics

	Representation	Details
Word-based	A tweet vector is the average of the vectors of the content words in the tweet.	
	Word2Vec_PreTrain, GloVe_PreTrain	Vectors of the content words are obtained from pre-trained embeddings from Word2Vec & GloVe respectively.
	Word2Vec_SelfTrain	Vectors of the content words are based on embeddings learned from the training set, separately for each fold.
	Word2Vec_WithMeSH, Glove_WithMeSH	Vectors of the content words are pre-trained word embeddings from Word2Vec & GloVe (respectively) retrofitted using MeSH ontology.
Context-based	A tweet vector is obtained from a pre-trained language model that uses context.	
	ELMo, USE, NNLM, FLAIR	Context-based representations of tweets are obtained from pre-trained models of ELMo, USE, NNLM and FLAIR respectively. They account for relationship between words using language models.

Table 1: Summary of the representations used in our experiments.

and Priego, 2017) and sentence similarity detection (Fu et al., 2016). In terms of the biomedical domain, word embedding-based features have been used for entity extraction in biomedical corpora (Yadav et al., 2017) or clinical information extraction (Kholghi et al., 2016). Several approaches for personal health mention classification have been reported (Aramaki et al., 2011; Lamb et al., 2013a; Yin et al., 2015). Aramaki et al. (2011) use bag-of-words as features for personal health mention classification. Lamb et al. (2013a) use linguistic features including coarse topic-based features, while Yin et al. (2015) use features based on parts-of-speech and dependencies for a statistical classifier. Feng et al. (2018) compare statistical classifiers with deep learning-based classifiers for personal health mention detection. In terms of detecting drug-related content in text, there has been work on detecting adverse drug reactions (Karimi et al., 2015). Nikfarjam et al. (2015) use word embedding clusters as features for adverse drug reaction detection.

3 Representations

A tweet vector is a distributed representation of a tweet, and is computed for every tweet in the training set. The tweet vector along with the output label is then used to train the statistical classification model. The intuition is that the tweet vector captures the semantics of the tweet and, as a result, can be effectively used for classification. To obtain tweet vectors, we experiment with two alternatives that have been used for several text classification problems in NLP: word-based representations and context-based representations. They are summarised in Table 1, and described in the following subsections.

3.1 Word-based Representations

A word-based representation of a tweet combines word embeddings of the content words in the tweet. We use the average of the word embeddings of content words in the tweet. Average of word embeddings have been used for different NLP tasks (De Boom et al., 2016; Yoon et al., 2018; Orasan, 2018; Komatsu et al., 2015; Ettinger et al., 2018). As in past work, words that were not learned in the embeddings are dropped during the computation of the tweet vector. We experiment with three kinds of word embeddings:

1. **Pre-trained Embeddings**: Denoted as *Word2Vec_PreTrained* and *GloVe_PreTrained* in Table 1, we use pre-trained embeddings of words learned from large text corpora: (A) Word2Vec by Mikolov et al. (2013): This has been pre-trained on a corpus of news articles with 300 million tokens, resulting in 300-dimensional vectors; (B) GloVe by Pennington et al. (2014): This has been pre-trained on a corpus of tweets with 27 billion tokens, resulting in 200-dimensional vectors.

2. **Embeddings Trained on The Training Split**: It may be argued that, since the pre-trained embeddings are learned from a cor-

Classification	# tweets (# true tweets)
IIC	9,006 (2,306)
DUC	13,409 (3,167)
PHMC	2,661 (1,304)

Table 2: Dataset statistics.

pus from an unrelated domain (news and general, in the case of Word2Vec and GloVe respectively), they may not capture the semantics of the domain of the specific classification problem. Therefore, we also use the Word2Vec Model available in the gensim library (Řehůřek and Sojka, 2010) to learn word embeddings from the documents. For each split, the corresponding training set is used to learn the embeddings. The embeddings are then used to compute the tweet vectors and train the classifier. We refer to these as *Word2Vec_SelfTrain*.

3. **Pre-trained embeddings retrofitted with medical ontologies**: Another alternative to adapt word embeddings for a classification problem is to use structured resources (such as ontologies) from a domain same as that of the classification problem. Faruqui et al. (2015) show that word embeddings can be retrofitted to capture relationships in an ontology. We use the Medical Subject Headings (MeSH) ontology (Nelson et al., 2001), maintained by the U.S. National Library of Medicine, which provides a hierarchically-organised terminology of medical concepts. Using the algorithm by Faruqui et al. (2015), we retrofit pre-trained embeddings from Word2Vec and GloVe, with the MeSH ontology. The retrofitted embeddings for Word2Vec and GloVe are referred to as *Word2Vec_WithMeSH*, and *GloVe_WithMeSH* respectively.

The three kinds of word-based representations result in five configurations: *Word2Vec_PreTrained*, *GloVe_PreTrained*, *Word2Vec_SelfTrain*, *Word2Vec_WithMeSH*, and *GloVe_WithMeSH*.

3.2 Context-based Representations

Context-based representations may use language models to generate vectors of sentences. Therefore, instead of learning vectors for individual words in the sentence, they compute a vector for sentences on the whole, by taking into account the order of words and the set of co-occurring words.

We experiment with four deep contextualised vectors: (A) **Embeddings from Language Models (ELMo)** by Peters et al. (2018): ELMo uses character-based word representations and bidirectional LSTMs. The pre-trained model computes a contextualised vector of 1024 dimensions. ELMo is available in the Tensorflow Hub[2], a repository of machine learning modules; (B) **Universal Sentence Encoder (USE)** by Cer et al. (2018): The encoder uses a Transformer architecture that uses attention mechanism to incorporate information about the order and the collection of words (Vaswani et al., 2017). The pre-trained model of USE that returns a vector of 512 dimensions is also available on Tensorflow Hub; (C) **Neural-Net Language Model (NNLM)** by Bengio et al. (2003): The model simultaneously learns representations of words and probability functions for word sequences, allowing it to capture semantics of a sentence. We use a pre-trained model available on Tensorflow Hub, that is trained on the English Google News 200B corpus, and computes a vector of 128 dimensions; (D) **FLAIR** by Akbik et al. (2018): This library by Zalando research[3] uses character-level language models to learn contextualised representations. We use the pooling option to create sentence vectors. This is a concatenation of GloVe embeddings and the forward/backward language model. The resultant is a vector of 4196 dimensions.

Table 1 refers to the four configurations as *ELMo*, *USE*, *NNLM* and *FLAIR* respectively.

4 Experiment Setup

We conduct our experiments on three boolean classification problems in health informatics: (A) **Influenza Infection Classification (IIC)**: The goal is to predict if a tweet reports an influenza infection (*'I have been coughing all day'*, for example) or describes information about influenza (*'flu outbreaks are common in this month of the year'*, for example). We use the dataset presented in Lamb et al. (2013b); (B) **Drug Usage Classification (DUC)**: The objective here is to

[2]https://www.tensorflow.org/hub/; Accessed on 3rd June, 2019.
[3]https://github.com/zalandoresearch/flair; Accessed on 3rd June, 2019.

	# dim.	**IIC**	**DUC**	**PHMC**
(A) *Word-based Representations*				
Word2Vec_PreTrain	300	0.8106 (σ: 0.024)	0.7417 (σ: 0.153)	0.7632 (σ: 0.037)
GloVe_PreTrain	200	0.7996 (σ: 0.015)	0.7549 (σ: 0.120)	0.7765 (σ: 0.033)
Word2Vec_SelfTrain	300	0.5099 (σ: 0.001)	0.7450 (σ: 0.028)	0.7418 (σ: 0.003)
Word2Vec_WithMeSH	300	0.6944 (σ: 0.021)	0.7450 (σ: 0.046)	0.7427 (σ: 0.050)
GloVe_WithMeSH	200	0.7264 (σ: 0.017)	0.7635 (σ: 0.030)	0.7425 (σ: 0.010)
(B) *Context-based Representations*				
ELMo	1024	0.8010 (σ: 0.021)	0.7724 (σ: 0.090)	0.7814 (σ: 0.02)
USE	512	0.8164 (σ: 0.008)	**0.7790** (σ: 0.100)	**0.8155** (σ: 0.030)
NNLM	128	**0.8520** (σ: 0.006)	0.7610 (σ: 0.070)	0.7495 (σ: 0.020)
FLAIR	4196	0.8000 (σ: 0.021)	0.7667 (σ: 0.116)	0.7896 (σ: 0.031)

Table 3: Comparison of five word-based representations with four context-based representations; Average accuracy with standard deviation (σ) indicated in brackets.

detect whether or not a tweet describes the usage of a medicinal drug ('*I took some painkillers this morning*', for example). We use the dataset provided by Jiang et al. (2016); (C) **Personal Health Mention classification (PHMC)**: A personal health mention is a person's report about their illness. We use the dataset provided by Robinson et al. (2015). For example '*I have been sick for a week now*' is a personal health mention while '*Rollercoasters can make you sick*' is not. It must be noted that IIC involves influenza while the PHMC dataset covers a set of illnesses as described later.

The datasets for each of the classification problems consist of tweets that have been manually annotated as reported in the corresponding papers. The statistics of these datasets are shown in Table 2. The values in brackets indicate the number of true tweets (*i.e.*, tweets that have been labeled as true), since these are boolean classification problems. For details on inter-annotator agreement and the annotation techniques, we refer the reader to the original papers. Based on sentence vectors obtained using either word-based or context-based representations, we train logistic regression with default parameters available as a part of the Liblinear package (Fan et al., 2008). We report five-fold cross-validation results for our experiments. Each fold is created using stratified k-fold sampling available in scikit-learn[4].

---

[4]`https://scikit-learn.org/stable/`; Accessed on 3rd June, 2019.

5 Results

We first present a quantitative evaluation to compare the two types of representations. Following that, we analyse sources of errors.

5.1 Quantitative Evaluation

We compare word-based and context-based representations for the three classification problems in Table 3. Accuracy is computed as the proportion of correctly classified instances. The table contains the average accuracy values with standard deviation values shown in parentheses. The table is divided into two parts. Part (A) corresponds to experiments using word-based representations, while Part (B) corresponds to those using context-based representations. In general, context-based representations result in an improvement in the three classification problems as compared to word-based representations. For IIC, the best word-based representation is when pre-trained Word2Vec embeddings ($Word2Vec\_PreTrain$) of content words are averaged to generate the tweet vector. The accuracy in this case is 0.8106. In contrast, the best performing context-based representation is NNLM (0.8520). This is an improvement of 4% points. Similarly, tweet vectors created using USE result in an accuracy of 0.7790 for DUC and 0.8155 for PHMC. This is an improvement of 2-4% points each over the word-based representations for these two classification problems as well. In addition, for pre-trained embeddings (Word2Vec and GloVe) retrofitted with a medical ontology (MeSH), we observe a degrada-

	1st-person mentions		Present Participle	
	Word	**Context**	**Word**	**Context**
IIC	58.2	**41.0**	79.6	**72.5**
DUC	66.4	**54.75**	**33.0**	40.75
PHMC	64.8	**37.5**	61.6	**40.0**

Table 4: Average number of instances (out of 100 randomly sampled mis-classified instances) containing first-person mentions and present participle form for the three classification problems and two types of representations.

tion in the accuracy for IIC and PHMC, as compared to without retrofitting. There is an improvement of 1% point in the case of DUC. Similarly, learning the embeddings on the specific training corpus does not work well. It leads to a degradation as compared to pre-trained embeddings. This could happen because pre-trained embeddings are trained on much larger corpora than our training datasets, thereby capturing semantics more effectively than the *Word2Vec_SelfTrain* variant.

5.2 Qualitative Evaluation

For a qualitative comparison of the two representations, we analyse 100 randomly sampled instances that are mis-classified by each classifier. While these instances need not be the same for each classifier, the trends in the errors show where one kind of representation scores over the other. We compared linguistic properties of these mis-classified instances, such as the person, tense and number. Table 4 shows two linguistic properties where we observed the most variation: first-person mentions and the use of present participles. The two properties are important in terms of the semantics of the three classification problems. First-person mentions are useful indicators to identify if the speaker has influenza, took a drug or reported a personal health mention. Similarly, present participle forms of verbs appear in situations where a person has had an infection or taken a drug. For 'Word', the average is over the five representations, while for 'Context', the average is over the four context-based representations. In the case of IIC, an average of 58.2 mis-classified instances from word-based representations contained first person mentions. The corresponding number for context-based representations was 41. For PHMC, the averages are 64.8 (word-based) and 37.5 (context-

based). The difference is not as high in the case of DUC (66.4 and 54.75 respectively). Differences are observed in the case of present participle in mis-classified instances. However, in the case of DUC, errors from context-based representations contain more average number of present participles (40.75) than word-based representations (33).

6 Conclusions

In this paper, we show that context-based representations are a better choice than word-based representations to create tweet vectors for classification problems in health informatics. We experiment with three such problems: influenza infection classification, drug usage classification and personal health mention classification, and compare word-based representations with context-based representations as features for a statistical classifier. For word-based representations, we consider pre-trained embeddings of Word2Vec and GloVe, embeddings trained on the training split, and the pre-trained embeddings of Word2Vec and GloVe retrofitted to a medical ontology. For context-based representations, we consider ELMo, USE, NNLM and FLAIR. For the three problems, the highest accuracy is obtained using context-based representations. In comparison with pre-trained embeddings, the improvement in classification is approximately 4% for influenza infection classification, 2% for drug usage classification and 4% for personal health mention classification. Embeddings trained on the training corpus or retrofitted on the ontology perform worse than those pre-trained on a large corpus.

While these observations are based on statistical classifiers, the corresponding benefit of context-based representations on neural architectures can be validated as a future work. In addition, while we average the word vectors to obtain tweet vectors, other options for tweet vector computation can be considered for word-based representations. In terms of the dataset, the comparison should be validated for text forms other than tweets, such as medical records. Medical records are expected to have typical challenges such as the use of abbreviations and domain-specific phrases that may not have been learned in pre-trained embeddings.

Acknowledgment

The authors would like to thank the anonymous reviewers for their helpful comments.

References

Alan Akbik, Duncan Blythe, and Roland Vollgraf. 2018. Contextual string embeddings for sequence labeling. In *Proceedings of the 27th International Conference on Computational Linguistics*, pages 1638–1649.

Eiji Aramaki, Sachiko Maskawa, and Mizuki Morita. 2011. Twitter catches the flu: Detecting influenza epidemics using Twitter. In *Empirical Methods in Natural Language Processing*, pages 1568–1576, Edinburgh, UK. Association for Computational Linguistics.

Yoshua Bengio, Réjean Ducharme, Pascal Vincent, and Christian Jauvin. 2003. A neural probabilistic language model. *Journal of machine learning research*, 3(Feb):1137–1155.

Davide Buscaldi and Belem Priego. 2017. Lipn-uam at emoint-2017:combination of lexicon-based features and sentence-level vector representations for emotion intensity determination. In *8th Workshop on Computational Approaches to Subjectivity, Sentiment and Social Media Analysis*, pages 255–258, Copenhagen, Denmark.

Daniel Cer, Yinfei Yang, Sheng-yi Kong, Nan Hua, Nicole Limtiaco, Rhomni St John, Noah Constant, Mario Guajardo-Cespedes, Steve Yuan, Chris Tar, et al. 2018. Universal sentence encoder. *arXiv preprint arXiv:1803.11175*.

Wei-Chieh Chou, Chin-Kui Lin, Yuan-Fu Liao, and Yih-Ru Wang. 2016. Word order sensitive embedding features/conditional random field-based chinese grammatical error detection. In *3rd Workshop on Natural Language Processing Techniques for Educational Applications*, pages 73–81, Osaka, Japan. The COLING 2016 Organizing Committee.

Cedric De Boom, Steven Van Canneyt, Thomas Demeester, and Bart Dhoedt. 2016. Representation learning for very short texts using weighted word embedding aggregation. *Pattern Recognition Letters*, 80:150–156.

Allyson Ettinger, Ahmed Elgohary, Colin Phillips, and Philip Resnik. 2018. Assessing composition in sentence vector representations. In *27th International Conference on Computational Linguistics*, pages 1790–1801, Santa Fe, New Mexico.

Rong-En Fan, Kai-Wei Chang, Cho-Jui Hsieh, Xiang-Rui Wang, and Chih-Jen Lin. 2008. Liblinear: A library for large linear classification. *Journal of machine learning research*, 9:1871–1874.

Manaal Faruqui, Jesse Dodge, Sujay Kumar Jauhar, Chris Dyer, Eduard Hovy, and Noah Smith. 2015. Retrofitting word vectors to semantic lexicons. In *Conference of the North American Chapter of the Association for Computational Linguistics: Human Language Technologies*, pages 1606–1615.

Shichao Feng, Keyuan Jiang, Jiyun Li, Ricardo A Calix, and Matrika Gupta. 2018. Detecting personal experience tweets for health surveillance using unsupervised feature learning and recurrent neural networks. In *Workshops at the 32nd AAAI Conference on Artificial Intelligence*, pages 425–430, New Orleans, Louisiana.

Cheng Fu, Bo An, Xianpei Han, and Le Sun. 2016. Iscas_nlp at semeval-2016 task 1: Sentence similarity based on support vector regression using multiple features. In *10th International Workshop on Semantic Evaluation*, pages 645–649, San Diego, California.

Kazuma Hashimoto, Pontus Stenetorp, Makoto Miwa, and Yoshimasa Tsuruoka. 2015. Task-oriented learning of word embeddings for semantic relation classification. *arXiv preprint arXiv:1503.00095*.

Keyuan Jiang, Ricardo Calix, and Matrika Gupta. 2016. Construction of a personal experience tweet corpus for health surveillance. In *ACL Workshop on biomedical natural language processing*, pages 128–135, Berlin, Germany.

Aditya Joshi, Vaibhav Tripathi, Kevin Patel, Pushpak Bhattacharyya, and Mark Carman. 2016. Are word embedding-based features useful for sarcasm detection? In *Empirical Methods in Natural Language Processing*, pages 1006–1011, Austin, Texas.

Sarvnaz Karimi, Chen Wang, Alejandro Metke-Jimenez, Raj Gaire, and Cecile Paris. 2015. Text and data mining techniques in adverse drug reaction detection. *ACM Computing Surveys*, 47(4):56.

Mahnoosh Kholghi, Lance De Vine, Laurianne Sitbon, Guido Zuccon, and Anthony Nguyen. 2016. The benefits of word embeddings features for active learning in clinical information extraction. In *Australasian Language Technology Association Workshop*, pages 25–34, Melbourne, Australia.

Hiroya Komatsu, Ran Tian, Naoaki Okazaki, and Kentaro Inui. 2015. Reducing lexical features in parsing by word embeddings. In *Pacific Asia Conference on Language, Information and Computation*, pages 106–113, Shanghai, China.

Alex Lamb, Michael J Paul, and Mark Dredze. 2013a. Separating fact from fear: Tracking flu infections on twitter. In *North American Chapter of the Association for Computational Linguistics: Human Language Technologies*, pages 789–795, Atlanta, Georgia.

Alex Lamb, Michael J. Paul, and Mark Dredze. 2013b. Separating fact from fear: Tracking flu infections on twitter. In *North American Chapter of the Association for Computational Linguistics: Human Language Technologies*.

Tomas Mikolov, Ilya Sutskever, Kai Chen, Greg S Corrado, and Jeff Dean. 2013. Distributed representations of words and phrases and their compositionality. In *Advances in neural information processing systems*, pages 3111–3119.

Stuart J Nelson, W Douglas Johnston, and Betsy L Humphreys. 2001. Relationships in medical subject headings (mesh). In *Relationships in the Organization of Knowledge*, pages 171–184.

Azadeh Nikfarjam, Abeed Sarker, Karen Oconnor, Rachel Ginn, and Graciela Gonzalez. 2015. Pharmacovigilance from social media: mining adverse drug reaction mentions using sequence labeling with word embedding cluster features. *Journal of the American Medical Informatics Association*, 22(3):671–681.

Constantin Orasan. 2018. Aggressive language identification using word embeddings and sentiment features. In *1st Workshop on Trolling, Aggression and Cyberbullying*, pages 113–119, Santa Fe, New Mexico.

Jeffrey Pennington, Richard Socher, and Christopher D. Manning. 2014. Glove: Global vectors for word representation. In *Empirical Methods in Natural Language Processing*, pages 1532–1543.

Matthew Peters, Mark Neumann, Mohit Iyyer, Matt Gardner, Christopher Clark, Kenton Lee, and Luke Zettlemoyer. 2018. Deep contextualized word representations. In *North American Chapter of the Association for Computational Linguistics: Human Language Technologies*, pages 2227–2237, New Orleans, Louisiana.

Radim Řehůřek and Petr Sojka. 2010. Software Framework for Topic Modelling with Large Corpora. In *LREC Workshop on New Challenges for NLP Frameworks*, pages 45–50, Valletta, Malta. `http://is.muni.cz/publication/884893/en`.

Bella Robinson, Ross Sparks, Robert Power, and Mark Cameron. 2015. Social media monitoring for health indicators. In *International Congress on Modelling and Simulation*, Gold Coast, Australia.

Iliana Simova and Hans Uszkoreit. 2017. Word embeddings as features for supervised coreference resolution. In *Recent Advances in Natural Language Processing*, pages 686–693, Varna, Bulgaria.

Maksim Tkachenko, Chong Cher Chia, and Hady Lauw. 2018. Searching for the x-factor: Exploring corpus subjectivity for word embeddings. In *Annual Meeting of the Association for Computational Linguistics*, pages 1212–1221, Melbourne, Australia.

Ashish Vaswani, Noam Shazeer, Niki Parmar, Jakob Uszkoreit, Llion Jones, Aidan N Gomez, Łukasz Kaiser, and Illia Polosukhin. 2017. Attention is all you need. In *Advances in Neural Information Processing Systems*, pages 5998–6008.

Shweta Yadav, Asif Ekbal, Sriparna Saha, and Pushpak Bhattacharyya. 2017. Entity extraction in biomedical corpora: An approach to evaluate word embedding features with pso based feature selection. In *European Chapter of the Association for Computational Linguistics*, pages 1159–1170.

Zhijun Yin, Daniel Fabbri, S Trent Rosenbloom, and Bradley Malin. 2015. A scalable framework to detect personal health mentions on twitter. *Journal of medical Internet research*, 17(6).

Su-Youn Yoon, Anastassia Loukina, Chong Min Lee, Matthew Mulholland, Xinhao Wang, and Ikkyu Choi. 2018. Word-embedding based content features for automated oral proficiency scoring. In *3rd Workshop on Semantic Deep Learning*, pages 12–22, Santa Fe, New Mexico.

Zhihua Zhang, Guoshun Wu, and Man Lan. 2015. Ecnu: Multi-level sentiment analysis on twitter using traditional linguistic features and word embedding features. In *9th International Workshop on Semantic Evaluation*, pages 561–567, Denver, Colorado.

Constructing large scale biomedical knowledge bases from scratch with rapid annotation of interpretable patterns

Julien Fauqueur[*]
BenevolentAI
4-8 Maple St, London
W1T 5HD
julien@benevolent.ai

Ashok Thillaisundaram[*]
BenevolentAI
4-8 Maple St, London
W1T 5HD
ashok@benevolent.ai

Theodosia Togia[*]
BenevolentAI
4-8 Maple St, London
W1T 5HD
sia@benevolent.ai

Abstract

Knowledge base construction is crucial for summarising, understanding and inferring relationships between biomedical entities. However, for many practical applications such as drug discovery, the scarcity of relevant facts (e.g. *gene X is therapeutic target for disease Y*) severely limits a domain expert's ability to create a usable knowledge base, either directly or by training a relation extraction model. In this paper, we present a simple and effective method of extracting new facts with a pre-specified binary relationship type from the biomedical literature, without requiring any training data or hand-crafted rules. Our system discovers, ranks and presents the most salient patterns to domain experts in an interpretable form. By marking patterns as compatible with the desired relationship type, experts indirectly batch-annotate candidate pairs whose relationship is expressed with such patterns in the literature. Even with a complete absence of seed data, experts are able to discover thousands of high-quality pairs with the desired relationship within minutes. When a small number of relevant pairs do exist - even when their relationship is more general (e.g. *gene X is biologically associated with disease Y*) than the relationship of interest - our system leverages them in order to i) learn a better ranking of the patterns to be annotated or ii) generate weakly labelled pairs in a fully automated manner. We evaluate our method both intrinsically and via a downstream knowledge base completion task, and show that it is an effective way of constructing knowledge bases when few or no relevant facts are already available.

1 Introduction

In many important biomedical applications, experts seek to extract facts that are often complex and tied to particular tasks, hence data that are truly fit for purpose are scarce or simply nonexistent. Even when only binary relations are sought, useful facts tend to be more specific (e.g. *mutation of gene X has a causal effect on disease Y in an animal model*) than associations typically found in widely available knowledge bases. Extracting facts with a pre-specified relationship type from the literature in the absence of training data often relies on handcrafted rules, which are laborious, ad-hoc and hardly reusable for other types of relations. Recent attempts to create relational data from scratch by denoising the output of multiple hand-written rules (Ratner et al., 2016) or by augmenting existing data through the induction of new black-box heuristics (Varma and Ré, 2018) are still dependent on ad-hoc human effort or pre-existing data. Our approach involves discovering and recommending, rather than prescribing, rules. Importantly, our rules are presented as text-like patterns whose meaning is transparent to human annotators, enabling integration of an automatic data generation (or augmentation) system with a domain expert feedback loop.

In this work, we make the following contributions:

- We propose a number of methods for extracting patterns from a sentence in which two eligible entities co-occur; different types of patterns have different trade-offs between expressive power and coverage.

[*] Equal contribution. Listing order is alphabetical. Theodosia proposed and co-ordinated the research project, built the early prototypes and contributed the different methods for extracting and lexicalising patterns. Ashok provided conceptual work on the metrics for ranking simplifications and for the intrinsic evaluation, developed the simplification extraction module, ran the experiments for the automated workflow (with all the parameter variations) and performed all the extrinsic evaluations. Julien was mainly responsible for the system architecture and workflow, the intrinsic evaluation (including interacting with the experts), handling negation and speculation and the clustering algorithm.

Proceedings of the BioNLP 2019 workshop, pages 142–151
Florence, Italy, August 1, 2019. ©2019 Association for Computational Linguistics

- We propose a simple method for presenting patterns in a readable way, enabling faster, more reliable human annotation

- For cases where a small number of seed pairs are already available, we propose a method which utilises these seed pairs to rank newly discovered patterns in terms of their compatibility with the existing data. The resulting patterns can be used with or without a human in the loop.

The rest of the paper is organised as follows. Section 2 describes some related work. Section 3 explains the relationship between patterns and labelling rules and presents some pattern types along with techniques for rendering them interpretable. Section 4 provides a high-level overview of the system and covers details of our different workflows (with and without seed data; with and without human feedback). Section 5 explains how we measure the system's performance both intrinsically and via a downstream knowledge base completion task. In section 6, we report the details of our main experiments while in sections 7 and 8 we present some analysis along with further experiments. The paper ends with conclusions and proposals for further work in section 9.

2   Related work

The idea of extracting entity pairs by discovering textual patterns dates back to early work on bootstrapping for relation extraction with the DIPRE system (Brin, 1999). This system was designed to find co-occurrences of seed entity pairs of a known relationship type inside unlabelled text, then extract simple patterns (exact string matches) from these occurrences and use them to discover new entity pairs. Agichtein et al. (2000) introduced a pattern evaluation methodology based on the precision of a pattern on the set of entity pairs which had already been discovered; they also used the dot product between word vectors instead of an exact string match to allow for slight variations in text. Later work (Greenwood and Stevenson, 2006; Xu et al., 2007; Alfonseca et al., 2012) has proposed more sophisticated pattern extraction methods (based on dependency graphs or kernel methods on word vectors) and different pattern evaluation frameworks (document relevance scores).

Two recent weak supervision techniques, Data Programming (Ratner et al., 2016) and the method underlying the Snuba system (Varma and Ré, 2018) have attempted to combine the results of handcrafted rules and weak base classifiers respectively. Data Programming involves modelling the accuracy of ideally uncorrelated rules devised by domain experts, then combining their output into weak labels. Although this approach does not require any seed data, it does rely on handwritten rules, which are both time consuming and ad-hoc due to the lack of a data-driven mechanism for exploring the space of possible rules. Snuba learns black-box heuristics (parameters for different classifiers) given seed pairs of the desired relationship. This method avoids the need for manually composing rules, however, the rules it learns are not interpretable, which makes the pipeline harder to combine with an active learning step. Second, the system requires gold standard pairs. In contrast, while our system can leverage gold standard annotations, if available, in order to reduce the space of discovered rules, as well as tune the ranking of newly discovered patterns, it is entirely capable of starting without any gold data if ranking is heuristics-based (e.g. prioritisation by frequency) and a human assesses the quality of the highest coverage rules suggested. Our method does not preclude use within a data programming setup as a way of discovering labelling functions or within a system like Snuba, as a way of generating seed pairs. Another body of work, distant supervision (Verga et al., 2018; Lin et al., 2016) has been a recent popular way to extract relationships from weak labels, but does not give the user any control on the model performance.

A well known body of work, OpenIE (Banko et al., 2007; Fader et al., 2011; Mausam et al., 2012; Angeli et al., 2015) aims to extract patterns between entity mentions in sentences, thereby discovering new surface forms which can be clustered (Mohamed et al., 2011; Nakashole et al., 2012) in order to reveal new meaningful relationship types. In the biomedical domain, Percha and Altman (2018) attempt something similar by extracting and clustering dependency patterns between pairs of biomedical entities (e.g. chemical-gene, chemical-disease, gene-disease). Our work differs from these approaches in that we extract pairs for a pre-specified relationship type (either from scratch or by augmenting existing data written with specific guidelines), which is not guaranteed to correspond to a cluster of discovered surface forms.

3 Extracting interpretable patterns

In a rule-based system, a rule, whether handwritten or discovered, can be described as a hypothetical proposition *"if P then Q"*, where P (the antecedent) is a set of conditions that may be true or false of the system's input and Q (the consequent) is the system's output. For instance, a standard rule-based relation extraction system can **i)** take as input a pair of entities (e.g. `TNF-GeneID:7124` and `Melanoma-MESH:D008545`) that are mentioned in the same piece of text, **ii)** test whether certain conditions are met (e.g. presence of lexical or syntactic features) and **iii)** output a label (e.g. `1: Therapeutic target, 0: Not therapeutic target.`)

In this work, patterns are seen as the antecedents of rules that determine which label (consequent) should be assigned to some input (e.g. candidate pair + text that mentions it.) We aim to extract patterns that are expressive enough to allow a system or a domain expert to discriminate between the different labels available for an input but also generic enough to apply to a wide range of inputs. In this work, we have made the following simplifying assumptions:

1. Relationships are binary (i.e. hold between exactly two entities).

2. A pair of entities are candidates for relation extraction if they are mentioned simultaneously in the same sentence.

3. There is a one-to-many relationship between patterns and inputs. An input (i.e. sentence + entity pair) is described by a single pattern (although this pattern can be a boolean combination of other patterns) but one pattern can correspond to multiple inputs.

4. We can select patterns which are expressive enough to represent the relationship, so it is possible to classify the input from which a given pattern has been extracted by examining the pattern alone. However, the omitted part of the sentence may contain contextual information which specifies the condition when or where the relationship holds. Modeling such contextual information would be useful but is beyond the scope of this work. A consequence of this assumption is that it is possible to batch-annotate a group of inputs that correspond to the same pattern by annotating the pattern itself.

Pattern interpretability An important consideration in this research is pattern interpretability, which could assist domain experts (who are not NLP experts) in exploring the space of labelling rule antecedents for a given relationship type in a given corpus. Hence, for each pattern, we construct what we call a pattern *lexicalisation*, that is converting a pattern to a readable text-like sequence.

Pattern types Simple patterns, which can potentially be combined with boolean operators, can be of different types. We illustrate some types of patterns used in our experiments through the following example sentences that include mentions of a gene-disease pair:

(1) "We investigate the hypothesis that the knockdown of *BRAF* may affect *melanoma* progression."

(2) "The study did not record higher *NF-kb* activity in *cancer* patients."

Below are some types of patterns, as well as their lexicalisations:

- **KEYWORDS**: words (e.g. 'inhibiting') or lemmas (e.g. 'inhibit') in the entire sentence or in the text between the entities. This pattern's lexicalisation is, trivially, the word itself.

- **PATH**: shortest path between the two entity mentions in the dependency graph of the sentence. For instance, in example (1), the path could be `BRAF <-pobj- of <-prep- knockdown <-nsubj- affect -dobj-> progression -compound-> melanoma`; in example (2), the path could be `NF-kb <-compound- activity -prep-> in -pobj-> patients -compound-> cancer`). To lexicalise patterns of this type, we extract the nodes (i.e. words) from the path, arrange them as per their order in the sentence and replace the entity mentions by a symbol denoting simply their entity types. For instance, the first pattern becomes `"knockdown of GENE affect DISEASE progression"`. This pattern is used extensively in our experiments because it strikes a good balance between expressive power and coverage.

We call its lexicalisation a **simplification** because it is a text-like piece that simplifies a sentence by discarding all but the most essential information.

- **PATH_ROOT**: the root (word with no incoming edges) of the shortest path between the two entities (e.g. 'affect' and 'activity' in examples (1) and (2) respectively). The lexicalisation could be trivial (i.e. the root itself) or, alternatively, if this pattern is used in an AND boolean combination with the PATH pattern, the root can simply be highlighted (e.g. `"knockdown of GENE `**`affect`**` DISEASE progression"`)

- **SENTENCE_ROOT**: the root of the dependency graph of the entire sentence (e.g. 'investigate' and 'record' in the examples above), which is often not the same as root of the path connecting the two entities. It can be lexicalised similarly to the pattern above.

- **PATH_BETWEEN_ROOTS**: the path between the root of the entire sentence and the root of the path between the two entities (e.g. `investigate -dobj-> hypothesis -acl-> affect` and `record -dobj-> activity` for examples (1) and (2) respectively). The pattern can be lexicalised as what we have called "simplification" (e.g. `investigate hypothesis affect`, or, if AND-ed with the PATH pattern, all the words from both patterns can be merged and arranged as per their original order in the sentence, potentially with some highlighting to differentiate the two simpler patterns (e.g. `"`_`investigate hypothesis`_` knockdown of GENE affect DISEASE progression"`)

- **SENTENCE_ROOT_DESCENDANTS**: the direct descendants of the SENTENCE_ROOT, for instance, 'did', 'not' and 'activity' in the example (1), because of the edges `did <-aux- record, not <-neg- record` and `record -dobj-> activity`. To lexicalise this pattern, we can extract the words and merge them with words of other patterns. Alternatively, we can devise some simpler sub-patterns, for instance, descendants with `aux`, that is auxiliary, edges, such as 'may', or descendants with `neg` edges such as 'not' and place them outside any simplification: `"`_`investigate hypothesis`_` knockdown

`of GENE affect DISEASE progression `**`+`**` hedging:[may]"`

- **PATH_ROOT_DESCENDANTS**: the direct descendants of the root of the path between the entities (e.g. 'may' and 'progression' in example (1) because `may <-aux- affect` and `affect -dobj-> progression`; 'higher' and 'in' in the example (2) because `higher <-amod- activity` and `activity -prep-> in`). Its lexicalisation can be the same as that of the previous pattern type.

Other examples of patterns could be regular expressions or rules informed by an external biomedical ontology (e.g. `GENE is a Rhodopsin-like receptor`) or with lexical information from databases like WordNet (Miller, 1995) (e.g. for increasing pattern coverage leveraging synonyms or hypernyms of words in a pattern.)

It should be obvious that the more expressive a pattern becomes (for instance by AND-ing multiple other patterns), the less capable it is of subsuming many sentences. It is important to discover patterns with this trade-off in mind.

4 System overview

In this section, we will describe each step of our system, outlined in Figure 1.

4.1 Data preparation

Extracting named entities and patterns The first step is performing named entity recognition (NER) on the sentences in the corpus to enable us to identify all the sentences which contain entity pairs of interest. Our experiments are focused on gene-therapeutic target pairs, however, the system is designed to be agnostic to different types of entities and relationships between them. We then extract the desired patterns from each of these sentences, as described in section 3. For simplicity, we limited our experiments to sentences that contain exactly one gene-disease pair, however, extending the system to handle multiple pairs is straightforward.

We index each sentence in a database along with the lexicalisation for its pattern (e.g. the 'simplification' for PATH or PATH_BETWEEN_ROOTS patterns) and the entity pair found. This allows us to easily query this database i) for all entity pairs that correspond to a pattern (which is now lexicalised

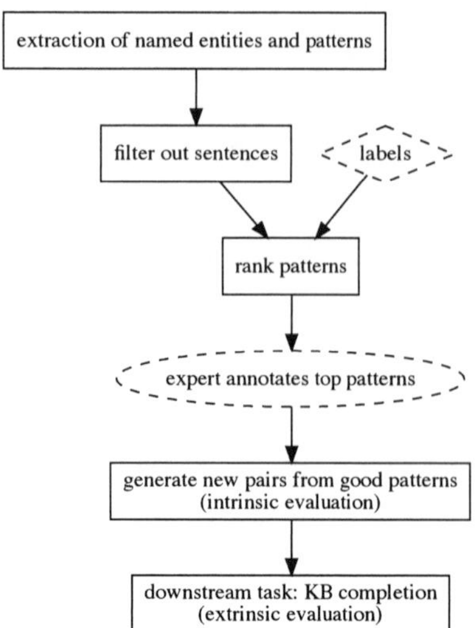

Figure 1: System overview. Diamond box is present only in workflows with seed labels available (i.e. "no expert but labels" and "expert with labels"), elliptical box is only present in workflows involving an expert (i.e. "expert - no labels" and "expert with labels") and rectangular boxes are always present.

and stored as a string) or ii) for all patterns that correspond to an entity pair.

Filtering out sentences with negation and hedging. Since we are interested in inputs which unambiguously encode affirmations of facts about entities, we filter out any sentences which contain negation, speculation, or other forms of hedging. We adopt a conservative approach by excluding sentences which match specific instantiations of these pattern types: **i)** KEYWORDS (e.g. presence of terms such as "no", "didn't", "doubt", "speculate" etc. in the sentence); our list is modified from NegEx (Chapman et al., 2013), **ii)** SENTENCE_ROOT AND SENTENCE_ROOT_DESCENDANTS (e.g. "study we investigated", which makes no statement of results), **iii)** PATH_ROOT AND PATH_ROOT_DESCENDANTS (e.g. "was used", "was performed"), **iv)** path_between_roots (e.g. "found associated") This filtering is applied at all stages in our system where sentences are used.

4.2 Ranking patterns

Below we describe methods for ranking and selecting top patterns in the presence or absence of domain expertise or labelled training data.

Baseline workflow: "no expert - no labels" In this workflow, we simply extract new pairs using simplifications (from the PATH pattern type, but other types are also described in our experiments) that have a high enough ($>= 5$) count of entity pairs.

Manual curation in the absence of any labelled training data: "expert - no labels" In this workflow, we have a domain expert (a biologist) available for manual curation but there is no labelled training data. It is not possible for a domain expert to annotate all simplifications; this would be too time-consuming. In such cases, active learning can be helpful in deciding which simplifications should be shown to the domain experts for manual curation to best improve the output of our system. The approach that we adopt here is simple but the system could be extended with more sophisticated active learning strategies. We rank the simplifications by their count of entity pairs; by this we mean the number of unique pairs contained in the sentences in our corpus which correspond to a given simplification (similar to section 4.2). We then show the top ranked simplifications (i.e. those with the greatest pair count) to our domain expert with a fixed number of random example sentences who then decides if a given simplification is an appropriate heuristic to extract new entity pairs from the corpus, by selecting one of three options "Yes", "No", "Maybe".

Automated workflow: "no expert but labels" For this workflow, a set of gold standard pairs exists as training data but we have no domain experts available. The sentence simplifications can be ranked using various metrics calculated against the gold standard training data. Each simplification is considered as a classifier: A given pair is 'classified' by the simplification as positive if the pair can be discovered using the simplification's underlying rule in the corpus. Otherwise, it is classified as negative. The metrics we use to rank the simplifications are precision and recall. The gold standard pairs will form the positive pairs in our training data. To obtain negative pairs, we operate under the closed world assumption: any entity pair found in our corpus of sentences not present in our gold standard set is taken to be negative. This results in an imbalance in the sizes of positive and negative training data which skews the value of precision. To address this, we use a precision metric where the number of true positives and false positives are normalised by the total number of positive and negative pairs respectively in our training data. For each simplification S we define true positives (TP_S) and false positives (FP_S) as the sets of correctly and incorrectly positive-labelled entity pairs respectively. Our variant of precision for a simplification S is then, $precision_S = \frac{|TP_S|/N_P}{|TP_S|/N_P + |FP_S|/N_N}$ where N_P and N_N are the number of positive and negative pairs respectively in the training data. With this metric, if a simplification classifies 10% of the positive pairs as true positives and 10% of the negative pairs as false positives then $precision_S = \frac{0.1}{0.1+0.1} = 0.5$. The metric utilises the percentage of each class instead of the absolute number of pairs, as would be the case for the standard precision metric. The definition of recall for a given simplification S is with respect to just the positive training data and is thus unaffected by an imbalance in the sizes.

Manual curation with labelled data: "expert with labels" For this workflow, both domain experts and labelled training data are available to us. We improve on our methodology in the "expert - no labels" workflow by making use of the metrics discussed in the "no expert but labels" workflow which are calculated using the labelled training data. As we want to maximise the number and precision of new pairs extracted, we keep only simplifications with recall and precision above certain respective thresholds and present them to domain experts ranked by pair count to ensure they see the most impactful simplifications first.

4.3 Generating new pairs

All previous stages aim at generating a list of good simplifications. We now have a collection of rules which can be used to extract new entity pairs from the corpus. Any simplification selected as useful implies that all entity pairs recovered from the corpus using this rule can be added as positive examples to the dataset. With the selected simplifications, we can batch-annotate thousands of sentences, and hence pairs, with minimal effort. We simply query our database for all new pairs which are found in a sentence expressing any of our selected simplifications.

Clustering simplifications We found that many simplifications can be very similar up to a few characters. We create clusters of quasi-identical simplifications, and use them i) to enforce diversity in the selection of simplifications for the user to annotate, by picking only one simplification per cluster and, ii) to safely extend the selection of positive simplification to other simplification in the cluster. We create clusters of simplifications by detecting connected components in a graph where the nodes are the simplifications and the edges are between simplifications which are at a maximum Levenshtein distance of 2. This allows us to be invariant to plural forms, upper/lower case, to short words like in/of etc. Note that some (not all) of these variations could be captured with a lemmatiser. Example of a distance 2 cluster:

```
{GENE effects on DISEASE,
GENE effect on DISEASE,
GENE effects in DISEASE}
```

With a distance of 2, we typically increase the number of positive simplifications by 50%, which significantly increases the recall on new pairs.

5 Evaluation

We implement two evaluation frameworks. The first is an intrinsic evaluation of the quality of the new extracted pairs. The second is extrinsic; we consider how the inclusion of the new pairs discovered by our system affects the performance of a downstream knowledge base completion task.[*]

5.1 Intrinsic evaluation

Pair-level Our aim in this subsection is to construct an intrinsic evaluation framework which can directly measure the quality of the discovered pairs. We do this by holding out a fraction of the gold standard positive pairs and the negative pairs (under the closed world assumption) to be used as a test set. The remaining fraction is used as training data. We evaluate our system by measuring its recall, specificity (true negative rate), precision, and F-score against this test set. In more detail, the new pairs discovered by our selected simplifications are taken to be the positive pairs predicted by our system. The overlap between these new pairs and the positive test set are the true positives (TP) while the overlap with the negative test set are the false positives (FP). Recall and specificity take their standard definitions. Again, we consider a precision score which is normalised to correct

---

[*]We consider the second type of evaluation extrinsic because knowledge base completion aims to recover latent relationships, whereas knowledge base construction, which the system is built for, is limited to extracting pairs from the literature.

for the imbalance in numbers between positive and negative pairs, $precision = \frac{|TP|/N_P}{|TP|/N_P + |FP|/N_N}$, where N_P and N_N are the numbers of positive and negative pairs respectively present in the test set (as described in section 4.2). We take the F-score to be the harmonic mean of this precision variant and recall.

Simplification-level For the manual workflows, we also consider the expert annotations while assessing the quality of the simplifications. We report MSP, the manual simplification precision, based on N_{Yes}, N_{No} and N_{Maybe}, the number of simplifications that the expert has annotated as "Yes", "No" and "Maybe". $MSP = \frac{N_{Yes}}{N_{Yes}+N_{No}+N_{Maybe}}$. We expect MSP to be as high as possible.

Extrinsic evaluation via knowledge base completion The setup for our extrinsic evaluation framework is straightforward and intuitive. The initial gold standard set of positive pairs is split into training and test data. A graph completion model is then trained using the training data and evaluated to determine whether it can predict the existence of the pairs in the test data. To determine whether our knowledge base construction system can add value, we use the new pairs found from our system to augment the training data for the graph completion model, and observe whether this improves its performance against the test set. We use ComplEx (Trouillon et al., 2017), a well-established tensor factorisation model, as our knowledge base completion model. We provide standard information retrieval metrics to quantify the performance of the graph completion model. These are the precision, $P(k)$, and recall, $R(k)$, calculated for the top k predictions along with the mean average precision (mAP). For gene-disease entity pairs, for example, $mAP = \frac{1}{N_d} \sum_d AveP$, where the sum is over the diseases d with N_d being the total number of diseases, and $AveP = \sum_k P(k)\,(R(k) - R(k-1))$ with $P(k)$ and $R(k)$ as defined above.

6 Main experiments and results

6.1 Datasets

For all the following experiments, our data was drawn from the following datasets: DisGeNET (Pinero et al., 2016) and Comparative Toxicogenomics Database (CTD) (Davis et al., 2018). CTD contains two relation types: 'marker/mechanism' and 'therapeutic'. We use both the entire CTD dataset and the subset of therapeutic gene-disease pairs which we refer to as CTD therapeutic.

The datasets above are first restricted to human genes and then to the gene-disease pairs which appear in our corpus of sentences; this corpus consists of sentences from PubMed articles which have been restricted, for simplicity, to sentences which contain just one gene-disease pair each. With these restrictions in effect, the CTD dataset has 8828 gene-disease pairs, CTD therapeutic has 169 pairs, and Disgenet has 33844 pairs.

6.2 Intrinsic evaluation results

In table 1, we report the pair-level metrics (see section 5.1) for our three proposed workflows and a baseline (see section 4). We also report the expert-based metric MSP (see section 5.1) for the two manual workflows. The CTD therapeutic dataset was the most suitable dataset for this evaluation because **i)** it is very relevant to crucial domains of application such as drug discovery, and **ii)** its small size makes it a good candidate for expansion. In each session, the expert annotated 200 simplifications accompanied by 20 sentences. It took the expert about 3 hours to annotate the first session, which is a rapid way to generate thousands of new pairs from scratch.

We find that our three main proposed workflows ('expert - with labels', 'expert - no labels', and the fully automated 'no expert but labels') all discover a significant number of new gene-disease therapeutic pairs. As confirmed by both pair-level and user-based metrics, incorporating the use of domain expert's time and the use of labelled data results in higher precision at the expense of recall.

6.3 Extrinsic evaluation results

In table 2, we list the results of the downstream knowledge base completion task for the fully automated workflow and the baseline. We compare the performance of our knowledge base completion model when trained with just the initial seed training data versus the seed training data augmented with the new pairs discovered by our fully automated workflow (and baseline workflow).

The addition of new pairs from the fully automated workflow gives us a higher mean average precision (mAP) than with just the seed dataset. We obtain a higher precision (for the top 100 and top 1000 predictions) while maintaining the same

Selection method	MSP	New pairs	Recall	Specificity	Precision	F-score
expert with labels	0.315	8875	0.286	0.976	0.923	0.436
expert - no labels	0.265	9560	0.250	0.975	0.908	0.392
no expert but labels	-	30006	0.679	0.920	0.894	0.772
no expert - no label (baseline)	-	59913	0.774	0.842	0.830	0.801

Table 1: Intrinsic evaluation results for the main experiments on the CTD therapeutic dataset. This was carried out with a train/valid/test split of 0.4/0.1/0.5, and precision threshold of 0.6 for the 'expert with labels' and 'no expert but labels' workflows. MSP is our "manual simplification precision" metric. The precision and F-scores reported here are normalised as described in the section 4.2.

level of recall. For the baseline workflow, mAP is higher but with lower precision (for the top 100 and 1000 predictions respectively).

7 Top simplifications

In table 3, we show the simplifications with the highest count of Disease-Gene pairs in our whole corpus (after the sentence filtering), which have been annotated by the expert as "Yes" or "No", for the CTD therapeutic dataset. While "Yes" and "No" patterns look similar, we can clearly see differences in language. The "No" annotations look unspecific while the "Yes" ones express the target has a therapeutic effect on the disease.

8 Further experiments

We performed several other experiments using our fully automated workflow to evaluate the quality of the new pairs discovered as we varied our experiment parameters.

We consider three dimensions of variation: varying the precision threshold for selecting simplifications, varying the size of the seed training set, and varying the expressiveness of the simplification (for example, by including the SENTENCE_ROOT or restricting to simplifications with at least a specified number of words).

The intrinsic evaluation results for these experiments are listed in tables 4, 5, and 6. In all cases, as we make our system more selective either by raising the precision threshold, by starting with fewer seeds pairs, or by restricting to more informative simplifications, we unsurprisingly obtain higher precision at the expense of lower recall.

The extrinsic evaluation framework is less sensitive to these changes but improvements were observed (without any noticeable trend) for all these parameter changes.

9 Conclusions and further work

We have presented a simple and effective method for knowledge base construction when the desired relational data are scarce or absent. We have demonstrated its effectiveness via **i)** classification metrics on a held-out test set, **ii)** human evaluation and **iii)** performance on a downstream knowledge base completion task. We further show that in the presence of a small set of data, it is possible to control the quality of the pairs discovered, by introducing stricter precision thresholds when ranking patterns. Our method could in principle be extended in order to: **1)** handle higher-order (e.g. ternary) relations between tuples, as opposed to pairs (for instance using dependency subgraphs that connect more than two entities cooccurring in a sentence), **2)** discover explicit negative examples of a binary relation instead of simply positive examples, **3)** train sentence-level relation extraction systems, **4)** collect and utilise continuous, rather than discrete annotations for each pattern (e.g. annotators could indicate the percentage of correct example sentences that correspond to a pattern displayed) as part of a more sophisticated active learning strategy, **5)** extract patterns from a semantic representation (Banarescu et al., 2013) and, finally, **6)** map patterns to a vector space using a distributional representation (e.g. defined by their neighbouring words in sentences) and cluster them for an optimal balance between expressive power and coverage.

Acknowledgments

We are very grateful to Nathan Patel for his engineering support and to Alex de Giorgio for his thorough feedback and domain expertise. We would also like to thank many of our colleagues working on drug discovery and link prediction for insightful conversations, as well as Felix Kruger for proofreading the final version of this paper.

Selection method	New pairs	MAP	Precision	Recall
seed dataset only	-	0.0414	0.0179 / 0.0179	1.0 / 1.0
no expert but labels	30006	0.0545	0.0192 / 0.0192	1.0 / 1.0
no expert - no label (baseline)	59913	0.1885	0.01 / 0.0015	0.6019 / 0.9208

Table 2: Extrinsic evaluation results for the CTD therapeutic dataset. The experiment parameters are the same as those given in table 1. Precision figures are given as 'top 100 / top 1000' and similarly for recall.

Pairs	"Yes" simplif.	Pairs	"No" simplif.
3345	role of GENE in DISEASE	6629	GENE DISEASE
839	GENE plays in DISEASE	4110	DISEASE GENE
648	GENE involved in DISEASE	3350	GENE and DISEASE
321	GENE target in DISEASE	2370	GENE in DISEASE
318	GENE target for DISEASE	2333	DISEASE and GENE
289	GENE mice develop DISEASE	1228	GENE DISEASE cells
279	DISEASE caused by mutations in GENE	904	DISEASE of GENE
		879	DISEASE in GENE
276	GENE gene for DISEASE	638	DISEASE in GENE mice
273	role of GENE in development of DISEASE	572	role for GENE in DISEASE
237	GENE promotes DISEASE	528	GENE in DISEASE patients

Table 3: Top 10 simplifications for CTD Therapeutic annotated "Yes" (left) and "No" (right) by the expert.

Dataset	Thres.	New pairs	R	S	P	F
CTD	0.8	29592	0.297	0.918	0.783	0.430
CTD	0.4	50329	0.379	0.863	0.735	0.500
DG	0.8	17441	0.180	0.947	0.773	0.292
DG	0.4	45446	0.314	0.867	0.703	0.434

Table 4: Intrinsic evaluation results (Recall, Specificity, Precision and F-score) on CTD and DisGeNET (DG) as we vary the precision threshold for the 'no expert but labels' workflow. Experiments are done with a train/valid/test split of 0.8/0.1/0.1 and we restrict to simplifications with at least 5 words to ensure that they are reasonably expressive.

Train/val/test	New pairs	R	S	P	F
0.8/0.1/0.1	29592	0.297	0.918	0.783	0.430
0.5/0.1/0.4	25539	0.274	0.930	0.798	0.408
0.2/0.1/0.7	18268	0.225	0.950	0.818	0.352

Table 5: Intrinsic evaluation results (Recall, Specificity, Precision and F-score) for CTD as we vary size of the seed training data for the 'no expert but labels' workflow. Experiments are done with a precision threshold of 0.8 and we restrict to simplifications with at least 5 words to ensure that they are reasonably specific.

Min length	SENTENCE ROOT	New pairs	R	S	P	F
5	No	29592	0.297	0.918	0.783	0.430
5	Yes	13284	0.174	0.963	0.824	0.288
7	No	5313	0.066	0.986	0.824	0.122
7	Yes	1814	0.028	0.995	0.856	0.055

Table 6: Intrinsic evaluation results (Recall, Specificity, Precision and F-score) for CTD as we vary simplification expressive power (minimum length) for the 'no expert but labels' workflow. Experiments are done with a train/valid/test split of 0.8/0.1/0.1 and a precision threshold of 0.8.

References

Eugene Agichtein and Luis Gravano. 2000. Snowball: Extracting relations from large plain-text collections. In *Proceedings of the Fifth ACM Conference on Digital Libraries*, DL '00, pages 85–94, New York, NY, USA. ACM.

Enrique Alfonseca, Katja Filippova, Jean-Yves Delort, and Guillermo Garrido. 2012. Pattern learning for relation extraction with a hierarchical topic model. In *Proceedings of the 50th Annual Meeting of the Association for Computational Linguistics: Short Papers - Volume 2*, ACL '12, pages 54–59, Stroudsburg, PA, USA. Association for Computational Linguistics.

Gabor Angeli, Melvin Jose Johnson Premkumar, and Christopher D. Manning. 2015. Leveraging linguistic structure for open domain information extraction. In *ACL (1)*, pages 344–354. The Association for Computer Linguistics.

Laura Banarescu, Claire Bonial, Shu Cai, Madalina Georgescu, Kira Griffitt, Ulf Hermjakob, Kevin Knight, Philipp Koehn, Martha Palmer, and Nathan Schneider. 2013. Abstract meaning representation for sembanking. In *Proceedings of the 7th Linguistic Annotation Workshop and Interoperability with Discourse*, pages 178–186, Sofia, Bulgaria. Association for Computational Linguistics.

Michele Banko, Michael J. Cafarella, Stephen Soderland, Matt Broadhead, and Oren Etzioni. 2007. Open information extraction from the web. In *Proceedings of the 20th International Joint Conference on Artifical Intelligence*, IJCAI'07, pages 2670–2676, San Francisco, CA, USA. Morgan Kaufmann Publishers Inc.

Sergey Brin. 1999. Extracting patterns and relations from the world wide web. In *Selected Papers from the International Workshop on The World Wide Web and Databases*, WebDB '98, pages 172–183, London, UK, UK. Springer-Verlag.

WW Chapman, D Hillert, S Velupillai, M Kvist, M Skeppstedt, BE Chapman, M Conway, M Tharp, DL Mowery, and L. Deleger. 2013. Extending the negex lexicon for multiple languages. In *Proceedings of the 14th World Congress on Medical & Health Informatics (MEDINFO)*.

AP Davis, CJ Grondin, RJ Johnson, D Sciaky, R McMorran, J Wiegers, TC Wiegers, and CJ Mattingly.

2018. The comparative toxicogenomics database: update 2019. *Nucleic Acids Res.*

Anthony Fader, Stephen Soderland, and Oren Etzioni. 2011. Identifying relations for open information extraction. In *Proceedings of the Conference on Empirical Methods in Natural Language Processing*, EMNLP '11, pages 1535–1545, Stroudsburg, PA, USA. Association for Computational Linguistics.

MA Greenwood and Mark Stevenson. 2006. Improving semi-supervised acquisition of relation extraction patterns. In *Proceedings of the Workshop on Information Extraction Beyond The Document*, IEBeyondDoc '06, pages 29–35, Stroudsburg, PA, USA. Association for Computational Linguistics.

Yankai Lin, Zhiyuan Liu, Shiqi Shen, Huanbo Luan, and Maosong Sun. 2016. Neural relation extraction with selective attention over instances. In *Proceedings of the 54th Annual Meeting of the Association for Computational Linguistics (Volume 1: Long Papers)*, pages 2124–2133. Association for Computational Linguistics.

Mausam, Michael Schmitz, Robert Bart, Stephen Soderland, and Oren Etzioni. 2012. Open language learning for information extraction. In *Proceedings of the 2012 Joint Conference on Empirical Methods in Natural Language Processing and Computational Natural Language Learning*, EMNLP-CoNLL '12, pages 523–534, Stroudsburg, PA, USA. Association for Computational Linguistics.

George Miller. 1995. Wordnet A lexical database for English. *Communications of ACM*, 38(11):39–41.

Thahir P Mohamed, Estevam R Hruschka Jr, and Tom M Mitchell. 2011. Discovering relations between noun categories. In *Proceedings of the Conference on Empirical Methods in Natural Language Processing*, pages 1447–1455. Association for Computational Linguistics.

Ndapandula Nakashole, Gerhard Weikum, and Fabian M. Suchanek. 2012. Discovering and exploring relations on the web. *PVLDB*, 5(12):1982–1985.

Bethany Percha and Russ B. Altman. 2018. A global network of biomedical relationships derived from text. *Bioinformatics*, 34(15):2614–2624.

Janet Pinero, Alex Bravo, Nuria Queralt-Rosinach, Alba Gutierrez-Sacristan, Jordi Deu-Pons, Emilio Centeno, Javier Garcia-Garcia, Ferran Sanz, and Laura I. Furlong. 2016. Disgenet: a comprehensive platform integrating information on human disease-associated genes and variants. *Nucl. Acids Res.*

Alexander Ratner, Christopher De Sa, Sen Wu, Daniel Selsam, and Christopher Ré. 2016. Data Programming: Creating Large Training Sets, Quickly. *arXiv e-prints*, page arXiv:1605.07723.

Theo Trouillon, Christopher R Dance, Johannes Welbl, Sebastian Riedel, Eric Gaussier, and Guillaume Bouchard. 2017. Knowledge graph completion via complex tensor factorization. *JMLR*.

Paroma Varma and Christopher Ré. 2018. Snuba: Automating weak supervision to label training data. *Proc. VLDB Endow.*, 12(3):223–236.

P Verga, E Strubell, and A McCallum. 2018. Simultaneously self-attending to all mentions for full-abstract biological relation extraction. In *North American Chapter of the Association for Computational Linguistics (NAACL)*.

Feiyu Xu, Hans Uszkoreit, and Hong Li. 2007. A seed-driven bottom-up machine learning framework for extracting relations of various complexity. In *ACL*, pages 584–591.

First Steps towards a Medical Lexicon for Spanish
with Linguistic and Semantic Information

Leonardo Campillos-Llanos

Computational Linguistics Laboratory, Universidad Autónoma de Madrid

`leonardo.campillos@uam.es`

Abstract

We report the work-in-progress of collecting MedLexSp, an unified medical lexicon for the Spanish language, featuring terms and inflected word forms mapped to Unified Medical Language System (UMLS) Concept Unique Identifiers (CUIs), semantic types and groups. First, we leveraged a list of term lemmas and forms from a previous project, and mapped them to UMLS terms and CUIs. To enrich the lexicon, we used both domain-corpora (e.g. Summaries of Product Characteristics and MedlinePlus) and natural language processing techniques such as string distance methods or generation of syntactic variants of multi-word terms. We also added term variants by mapping their CUIs to missing items available in the Spanish versions of standard thesauri (e.g. Medical Subject Headings and World Health Organization Adverse Drug Reactions terminology). We enhanced the vocabulary coverage by gathering missing terms from resources such as the Anatomical Therapeutical Classification, the National Cancer Institute (NCI) Dictionary of Cancer Terms, OrphaData, or the Nomenclátor de Prescripción for drug names. Part-of-Speech information is being included in the lexicon, and the current version amounts up to 76 454 lemmas and 203 043 inflected forms (including conjugated verbs, number and gender variants), corresponding to 30 647 UMLS CUIs. MedLexSp is distributed freely for research purposes.

1   Introduction

Current machine-learning and deep-learning-based methods are *data-intensive*; however, in domains such as Medicine, sufficient data are not always available—due to ethical concerns or privacy issues, especially when dealing with Patient Protected Information. Moreover, some tasks demand high precision outcomes, which either need supervised approaches with annotated data or hybrid methods (e.g. rule-based and dictionary-based). In order to overcome the data bottleneck, richly-structured terminological thesauri enhance the annotation and concept normalization of domain corpora to be used subsequently in supervised models. More importantly, to achieve comparable benchmarks, domain resources should integrate standard terminologies and coding schemes.

In this context, we aim at providing a computational lexicon to be used in the pre-processing of text data used in more complex Natural Language Processing (NLP) tasks. The work here presented reports the first steps towards building the Medical Lexicon for Spanish (MedLexSp). MedLexSp is conceived as an unified resource with linguistic information (lemmas, inflected forms and part-of-speech), concepts mapped to Unified Medical Language System® (hereafter, UMLS) (Bodenreider, 2004) Concept Unique Identifiers (CUIs), and semantic information (UMLS types and groups). Figure 1 is a sample of the lexicon. MedLexSp is firstly aimed at named entity recognition (NER), and it can be used in the pre-annotation step of an NER pipeline. It can also help lemmatization and feed general-purpose Part-of-Speech taggers applied to medical texts—as done in previous works (Oronoz et al., 2013).[1] Because it gathers semantic data of terms, it can ease relation extraction tasks.

Our work makes several contributions. We provide a resource to be distributed for research purposes in the BioNLP community. MedLexSp includes inflected forms (singular/plural, masculine/feminine) and conjugated verb forms of term lemmas, which are mapped to UMLS Concept Unique Identifiers. Verb terms are also mapped to Concept Unique Identifiers; this is the line of current works for expanding terminologies by in-

---

[1] `https://zenodo.org/record/2621286`

Proceedings of the BioNLP 2019 workshop, pages 152–164
Florence, Italy, August 1, 2019. ©2019 Association for Computational Linguistics

C0007102|cáncer colónico|cáncer colónico; cánceres colónicos|N|Neoplastic Process|DISO
C0007102|cáncer de colon|cáncer de colon; cáncer del colon; cánceres de colon; cánceres del colon|N|Neoplastic Process|DISO
C0007102|neoplasia maligna de colon|neoplasia maligna de colon; neoplasias malignas de colon|N|Neoplastic Process|DISO
C0007102|tumor maligno del colon|tumor maligno del colon; tumores malignos del colon|N|Neoplastic Process|DISO
C0018787|cardiaco|cardiaca; cardiacas; cardiaco; cardiacos; cardíaca; cardíacas; cardíaco; cardíacos|ADJ|Body Part, Organ, or Organ Component|ANAT
C0018787|corazón|corazón; corazones|N|Body Part, Organ, or Organ Component|ANAT
C0018787|cardio-|card-; cardi-; cardia-; cardio-; cardió-; cardí-; cardío-; cárdi-; cárdio-|AFF|Body Part, Organ, or Organ Component|ANAT
C0023884|hepático|hepático; hepáticos; hepática; hepáticas|ADJ|Body Part, Organ, or Organ Component|ANAT
C0023884|hígado|hígado; hígados|N|Body Part, Organ, or Organ Component|ANAT
C0346647|cáncer de páncreas|cáncer de páncreas; cáncer del páncreas; cánceres del páncreas; cánceres de páncreas|N|Neoplastic Process|DISO
C0346647|cáncer pancreático|cáncer pancreático; cánceres pancreáticos|N|Neoplastic Process|DISO

Figure 1: Sample of the MedLexSp lexicon. In each entry, field 1 is the UMLS CUI of the entity; field 2, the lemma; field 3, the variant forms; field 4, the Part-of-Speech; field 5, the semantic types(s); and field 6, the semantic group.

cluding verb terms (Thompson et al., 2011; Chiu et al., 2019). We also added inflected terms from MedlinePlus terms, OrphaData (INSERM, 2019), the National Cancer Institute (NCI) Dictionary of Cancer Terms, or the Nomenclator de prescripción (AEMPS, 2019), a knowledge base of medical drugs prescribed in Spain.

Section 2 gives an overview of medical thesauri, and Section 3 describes the methods used to gather terms (both corpora and NLP techniques), map them to UMLS CUIs, and enrich the lexicon. Section 4 reports descriptive statistics of the current version, and Section 5, the results of an evaluation conducted during development. We discuss some limitations and conclude in Section 6.

2 Background and Context

2.1 Health thesauri and taxonomies

Medical thesauri and controlled vocabularies aggregate listings of domain terms, and also gather information about the type of term (e.g. synonym or preferred term), a semantic descriptor (e.g. DRUG or FINDING), an unique concept identifier, and very often a term definition or hierarchical relations between concepts (e.g. IS_A). Thesauri are essential for indexing and populating databases, domain-specific information retrieval, and standardized codification (Cimino, 1996).

Medical thesauri vary according to the application (we only give examples related to our work). The Systematized Nomenclature of Medicine Clinical Terms (SNOMED-CT) (Donnelly, 2006) aims at encoding verbatim mentions in clinical texts, and gathers ontological relations between concepts. To report drug reactions in pharmacovigilance, the World Health Organization created the Adverse Reactions Terminology (WHO ART), although the Medical Dictionary for Regulatory Activities (MedDRA) (Brown et al., 1999) is now preferred. The Medical Subject Headings (MeSH) are developed by the National Library of Medicine for indexing biomedical articles. Lastly, the World Organization of Family Doctors produced the International Classification of Primary Care (ICPC) to classify data aimed at family and primary care physicians (WONCA, 1998).

Medical taxonomies or classifications gather essential domain knowledge.Some examples are the International Classification of Diseases vs. 10 (ICD-10) (WHO, 2004), or the Anatomical Therapeutical Chemical (ATC) classification of pharmacological substances (WHO, 2019).

2.2 Medical Lexicons

Medical lexicons provide a structured representation of terms and their linguistic information (lemmas, inflection, or surface variants); hence, they are essential for NLP tasks. Unlike medical thesauri or classifications, they do not register term hierarchies, classifications nor ontological relations, but they can encode semantic information and, occasionally, argument structure and corpus-based frequency data (Thompson et al., 2011).

Initiatives to collect medical lexicons have been conducted for English (McCray et al., 1994; Johnson, 1999; Davis et al., 2012), German (Weske-Heck et al., 2002), French (Zweigenbaum et al., 2005) or Swedish, even in multilingual initiatives (Markó et al., 2006). For Spanish, some efforts were sparked when a team at the National Library of Medicine (Divita et al., 2007) started to build an equivalent of the MetaMap tool (Aronson, 2001). Other teams conducted experiments to automate the creation of a Spanish MetaMap by applying machine translation and domain ontologies (Carrero et al., 2008). These initiatives, to the best of our knowledge, did not achieve a Spanish lexicon available for medical NLP.

Besides medical lexicons, domain-specific vocabularies were collected for Biology (Thompson et al., 2011). With a different perspective and goal, Consumer Health Vocabularies have been collected to bridge the gap between patients' expressions and healthcare professionals' jargon (Zeng and Tse, 2006; Keselman et al., 2007).

2.3 The Unified Medical Language System

The Unified Medical Language System® (UMLS) (Bodenreider, 2004) MetaThesaurus includes thesauri. The version we used (2018AB) gathers 210 sources and over 3.82 millions of concepts in 23 languages. Synonym terms are encoded with Concept Unique Identifiers (CUIs); and concepts are assigned a semantic type and group (McCray et al., 2001).

2.4 Methods for Creating Medical Lexicons

We will restrict us here to a shallow overview of approaches and will not consider taxonomy nor ontology building. Methods for widening medical vocabularies range from generating syntactic-level variants of multi-word terms (Jacquemin, 1999), inferring derivation rules from string similarity matches and morphological relations between derivational variants (Grabar and Zweigenbaum, 2000), gathering inflected variants semi-automatically (Cartoni and Zweigenbaum, 2010), or deriving terms from corpora (more below).

Graeco-Latin components are very productive for coining medical terms; thus, several BioNLP systems integrate morphology-based lexical resources. For example, for decomposing terms morphosemantically and deriving their definitions (Namer and Zweigenbaum, 2004), or mapping queries to concepts and indexing documents in cross-lingual information retrieval, based on a subword-based morpheme thesaurus (Markó et al., 2005). In this line, generating paraphrase equivalents of neoclassical compounds (e.g. *thyromegalia → enlarged thyroid*) is an approach with potential for deriving new terms, and concept normalization systems (Thompson and Ananiadou, 2018) already implement it. Because string similarity measures and edit distance patterns are used for normalization—e.g (Tsuruoka et al., 2007; Kate, 2015)—and terminology mapping (Dziadek et al., 2017), these approaches are also powerful for expanding medical lexicons from a set of reference terms. Decomposition of multi-word terms and synonym expansion of their components are also alternative strategies applied in normalization systems (Tseytlin et al., 2016).

Corpus-derived medical terminology construction requires collecting domain texts and applying term extraction methods, among others: computing graphs of relations between parse trees and word dependency similarities (Nazarenko et al., 2001), using parallel corpora to map cognates or aligned words (Sbrissia et al., 2004; Deléger et al., 2009), linking terms or abbreviations to their definitions or expanded word forms in the text where they occur (Yu and Agichtein, 2003; McCrae and Collier, 2008), using dictionary features to identify polysemy (Pezik et al., 2008), combining text mining techniques with databases (Thompson et al., 2011), or having experts review terms, a method which has been used to build disease-specific vocabularies (Wang et al., 2016).

Approaches based on the Firthian *Distributional hypothesis* exploit distributional similarity metrics (Carroll et al., 2012). Among them, more recent distributional semantics methods represent terms in the vector space, or calculate word-embeddings to compute similarity measures between vectors, thus allowing the unsupervised expansion of domain terms (Pyysalo et al., 2013; Skeppstedt et al., 2013; Henriksson et al., 2014; Wang et al., 2015; Ahltorp et al., 2016; Segura-Bedmar and Martínez, 2017) or concept normalization (Limsopatham and Collier, 2016).

Lastly, to develop Consumer Health Vocabularies (CHV), a variety of techniques have been used: analysis by experts of Medline queries (Zeng and Tse, 2006), term recognition methods and collaborative review of user logs in medical sites (Zeng et al., 2007), hybrid methods combining n-grams extraction, the C-value, and dictionary look-up (Doing-Harris and Zeng-Treitler, 2011), co-occurrence analysis of terms and seed words (Jiang and Yang, 2013), or approaches based of similarity measures between CHV lexicons and reference lexicons (Seedorff et al., 2013).

3 Methods

Figure 2 depicts the methods used to collect the MedLexSp lexicon. In a first step (left part of Figure 2), we leveraged the lemmas and word forms obtained from a Spanish medical lexicon, mostly corpus-derived; we will refer to it as the *base list*. We only used the subset of lemmas and forms that could be mapped authomatically to UMLS CUIs (exact string match). In a second step, we added missing variants of terms using different methods:

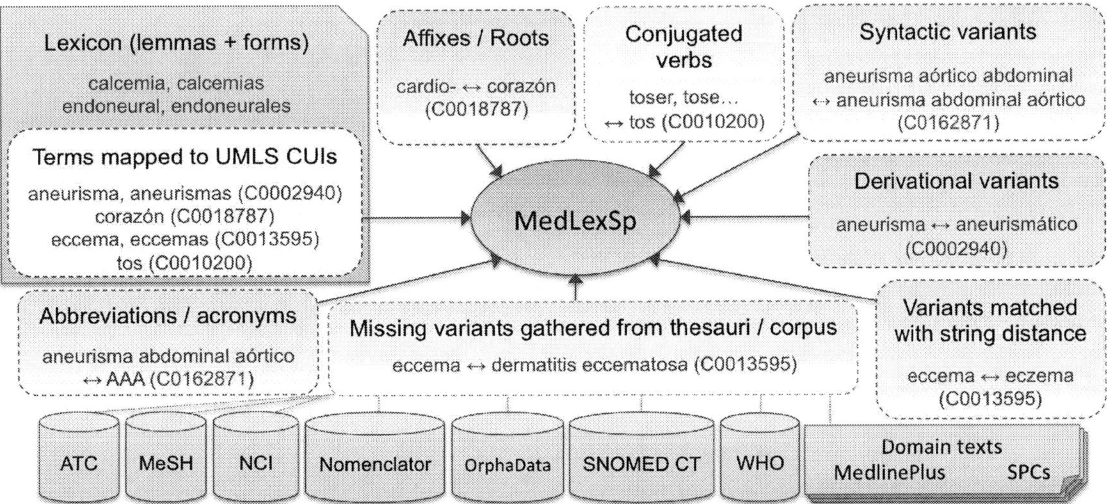

Figure 2: Methods to collect the MedLexSp lexicon.

- Testing **string distance metrics** to match terms in the base list to variants that remained unmatched: e.g. *eccema* ↔ *eczema* ('eczema', C0013595).

- Incorporating **derivational variants** to the base list: e.g. *aneurisma* ('aneurysm') ↔ *aneurismático* ('aneurysmatic', C0002940).

- Including **conjugated verbs** corresponding to the noun terms with CUIs selected in the base list: e.g. *tos* ('cough', C0010200) → *toser, tosiendo...* ('to cough', 'coughing'...).

- Matching **affixes and roots** to those terms in the base list with CUIs: e.g. *corazón* ('heart, C0018787) → *cardio-* ('cardio-').

- Adding **syntactic variants** of the multi-word terms in the base list: e.g. *aneurisma aórtico abdominal* ('aortic abdominal aneurysm') ↔ *aneurisma abdominal aórtico* ('abdominal aortic aneurysm', C0162871).

- Adding **acronyms and abbreviations** of the terms included in the base list: e.g. *aneurisma abdominal aórtico* ('abdominal aortic aneurysm', C0162871) → *AAA*.

- Extending the base list by mapping the CUIs of the terms in the subset to gather **missing variants of synonymous terms**: e.g. *eccema* ('eczema', C0013595) ↔ *dermatitis eccematosa* ('eczematous dermatitis', C0013595). We considered several sources from the UMLS—e.g. Spanish Medical Subject Headings (MeSH), SNOMED CT or the WHO ART terminology—and external sources such as the Anatomical Therapeutical Classification, the National Cancer Institute (NCI) Dictionary of Cancer Terms,[2] the Nomenclator de prescripción (AEMPS, 2019), OrphaData (INSERM, 2019), or the Spanish Drug Effect database (SD-Edb) (Segura-Bedmar et al., 2015).

- Including subsets of **missing terms from thesauri if attested in domain texts**. And vice versa, extracting **corpus-derived terms** from domain texts: synonymous terms from MedlinePlus,[3] and terms from Summaries of Product Characteristics (Segura-Bedmar and Martínez, 2017).

The next subsections explain each method.

3.1 Leveraging an Inflected Lexicon

We started using a list of medical terms collected in a previous project on Spanish medical terminology;[4] we will refer to it as the *base list*. We collected this resource by combining different methods (Moreno Sandoval and Campillos Llanos, 2015) applied on a corpus of 4204 Spanish medical texts (around 4 million tokens) (Moreno-Sandoval and Campillos-Llanos, 2013). To extract candidate medical

---

[2] https://www.cancer.gov/espanol/publicaciones/diccionario

[3] https://medlineplus.gov/spanish/

[4] http://labda.inf.uc3m.es/multimedica/

terms for the base list, we combined rule-based techniques (Part-of-Speech tagging and filtering through medical affixes), corpus-based methods (comparing word forms from a general corpus and from the domain corpus), and statistical methods, namely the Log-Likelihood ratio (Dunning, 1993). We checked in medical sources—e.g. the dictionary published by the Spanish Royal Academy of Medicine (RANME, 2011)—the terms selected by means of those three methods, before being included in the list. This base list was used to build an automatic term extractor (Campillos Llanos et al., 2013), and amounted to 38 354 entries.

Because one of the goals of MedLexSp is concept normalization by using standard domain terminologies, we did not include the full base list. We only used terms that could be assigned UMLS Concept Unique Identifiers (CUIs) in the UMLS MetaThesaurus version 2018AB, namely from those terminologies of special biomedical or clinical interest (e.g. SNOMED CT, WHO ART or Medical Subject Headings) with available Spanish translations. We mapped 18 263 lemmas to CUIs, which means 47.61% entries of the original lexicon. CUIs were assigned according to an *exact match* criterion. For example, *donación* ('donation') is not matched with *donación de tejido* ('Tissue Donation', C0080231), because the latter makes reference to a donation subtype. Note that the current version of MedLexSp does not include the full list of terms from MeSH or SNOMED CT, but only those which were originally mapped from the base list to UMLS terms with CUIs.

3.2 Enriching the Lexicon

String distance metrics We tested mapping terms from the subset of entities with CUIs to terms in the UMLS by applying distance metrics (Levenshtein, 1966) of less than 2. This allowed us mapping hyphenated variants to terms without hyphen (e.g. *creatina-cinasa* ↔ *creatina cinasa*, 'creatine kinase', C0010287), compound terms that are often written as single-words (*dietil éter* ↔ *dietiléter*, 'diethyl ether', C0014994), or matching terms with minimal morphological variation (*eccema* ↔ *eczema*, 'eczema', C0013595). A total of 1463 terms with CUIs were matched to the original base list.

Derivational variants In line with previous work (Grabar and Zweigenbaum, 2000), we collected a list of equivalent derivational variants of terms. Using this list, we assigned a CUI to the corresponding derivational variant: e.g. the CUI of *páncreas* (C0030274) was also ascribed to *pancreático* ('pancreatic'). The current version gathers a total of 801 derivational variants with CUIs.

Conjugated verbs Most terms in the UMLS or standard terminologies are noun or adjective phrases. This limits the named entity recognition of medical concepts expressed with verbs in free text; given a context such as *el paciente tose* ('the patient coughs'), the concept of 'coughing' would not be identified. To widen the scope of concept normalization, verb terms were mapped to CUIs from derived nouns: e.g. *tos* ('coughing', C0010200) → *toser* ('to cough', C0010200). We again used a list of correspondences between verbs and deverbal nouns. We included the conjugated forms of verb lemmas in each verb entry of the lexicon. We used a python script that relies on the lexicon of a Spanish Part-of-Speech tagger (Moreno Sandoval and Guirao, 2006) to generate all conjugated forms of verb terms: e.g. *toser* ('to cough') → *tose* ('he/she coughs'), *tosiendo* ('coughing'), etc. The current version includes a total of 295 single- or multi-word verb items.

Affixes and lexical roots In a first step, we collected affixes and roots from several sources. Firstly, we leveraged a list used in a previous experiment (Sandoval et al., 2013). This list amounts to 1719 forms and considers morphological variants of affixes (e.g. prefix *cardio-* may have accented variant forms in Spanish, such as *cardió-*). Secondly, we translated to Spanish several affixes and roots from the Specialist Lexicon® (McCray et al., 1994) and then added variant forms. In a second step, we assigned UMLS CUIs to affixes and roots in the list. The current list gathers a total of 161 entries (82 prefixes and 79 suffixes) with 134 different CUIs and 386 variant forms. Note that many affixes and roots were not included because they are too underspecified to be assigned to a CUI, or are not restricted to the medical domain (e.g. *kilo-* expresses a quantitative concept).

Abbreviations and acronyms Firstly, we gathered a list of equivalences between full forms and abbreviations and acronyms; we used three sources: 1) the collection of Spanish abbreviations and acronyms used in hospitals, collected by medical doctors (Yetano and Alberola, 2003); 2) abbreviations and acronyms used in the 2nd IberEval Challenge 2018 on Biomedical Abbreviation Recognition and Resolution (Intxaurrondo et al., 2018); and 3) Spanish abbreviations and acronyms from Wikipedia.[5] Secondly, we matched the resulting list of equivalent terms (acronyms and full forms) to UMLS terms, adding the corresponding CUIs to those missing acronyms. For example, the full term *virus de Epstein-Barr* ('Epstein-Barr virus') has CUI C0014644, and we also assigned this code to the corresponding acronym in Spanish (*VEB*). With this method, we assigned CUIs to 1225 items.

Syntactic variants of terms   To widen the coverage of terms mapped to CUIs, we generated variants of multiword entities by swapping the word order of their components. Then, we tried to match each new variant to entities with CUIs. For example, *aneurisma aórtico abdominal* ('aortic abdominal aneurysm') has CUI C0162871, and we assigned the same CUI to the generated variant *aneurisma abdominal aórtico* ('abdominal aortic aneurysm'). With this method, we gathered a total of 154 variants of terms with CUIs in the base list.

Mapping UMLS term variants through CUIs
We gathered synonymous variants referring to each corresponding concept by using the UMLS CUIs from the terms included in the base list. To avoid including noisy terms adequate for biomedical natural language processing, we first cleaned the terms from the terminologies we used. To do so, we applied methods for cleaning term strings (Aronson et al., 2008; Hettne et al., 2010; Névéol et al., 2012; Hellrich et al., 2015). We deleted paraphrastic terms that include a description or specification of the entity type in the term string. These terms commonly come from Spanish SNOMED CT. For example, we deleted *tos (hallazgo)*, 'cough (finding)' (CUI C0010200) and kept the term (*cough*, 'cough'). Likewise, we removed most anatomic terms beginning with *estructura de* ('structure of'): e.g. regarding term *estructura del ojo* ('structure of eyeball', C0015392), we only kept the synonym *ojo* ('eyeball'). Lastly, terms in the WHO ART terminology needed to be accented and reversed regarding word order: e.g. *disociativa, reaccion* → *reacción disociativa* ('dissociative reaction', C0012746).

We also applied an exact-match mapping of Spanish terms from the base list to the English component of the UMLS. This method allowed us to obtain the CUIs of terms unavailable in Spanish terminologies, which remain unchanged in the Spanish language. Namely, Latin scientific names (e.g. *Campylobacter fetus*, C0006814), compound terms with Graeco-Latin roots (e.g. *abdominalgia*, C0000737), English acronyms that are broadly used in the medical discourse without Spanish translation (e.g. *GABA*, 'gamma-aminobutyric acid', C0016904), or international brand drug names (e.g. *abilify*®). In these cases, the same word is used in both English and Spanish. We manually revised the list of mapped terms to discard homonymous terms with a different meaning in English (e.g. *TIP*® is a brand name of a medical drug, but it also means 'point' or 'suggestion' in English).

We extended the list of terms by extracting the information related to rare diseases from OrphaData (INSERM, 2019).[6] We also added terms of pharmacological substances and international non-proprietary names from the Spanish Drug Effect database (SDEdb) (Segura-Bedmar et al., 2015) and the Nomenclator de prescripción (AEMPS, 2019), a resource published and updated regularly by the Spanish Agency of Drugs and Food Products.[7]

For all these procedures and sources, we applied semiautomatic methods to generate the singular and plural inflected forms of the missing terms that were mapped through CUIs. We used the Pattern python library (Smedt and Daelemans, 2012) to create plural forms of terms, which were revised manually before being included in MedLexSp.

Corpus-derived terms   When we started adding variant terms from thesauri, the question of where to stop adding terms came up. In the first version, we decided not to include all terms available in MeSH or SNOMED CT terminologies, given that these thesauri contain terms that are often not necessary in clinical or biomedical NER tasks (e.g. names of trees, wild animals, professions or abstract concepts). On the other hand, to make the

---

[5]https://es.wikipedia.org/wiki/Anexo:
Acrnimos_en_medicina

[6]http://www.orphadata.org/data/
xml/es_product1.xml   We   make   available
the script used to extract terms from OrphaData:
https://github.com/lcampillos/bionlp2019
The code can be adapted to process OrphaData in other languages (e.g. English, French, Italian or Portuguese).

[7]http://listadomedicamentos.aemps.gob.
es/prescripcion.zip.

resource comprehensive, we needed to complement the base list with supplementary terms from thesauri. Hence, in order to decide which items to include in a first version, we computed term frequencies using a medical corpus from a previous project (4 million tokens) (Moreno-Sandoval and Campillos-Llanos, 2013). We currently include terms from the Spanish MeSH and SNOMED CT that were missing in the base list, if they were documented in that corpus. By limiting the inclusion of such subset of terms, we aim at providing quality enriched data (i.e. with revised inflected forms) in a reasonable time and manner.

In a different vein, and similarly to former work (Calleja et al., 2017), we extracted terms from Summaries of Product Characteristics (SPCs). We used Easy Drug Package Leaflets (EasyDPL), a corpus of 306 texts annotated with medical drugs and pathological entities (1400 drug effects) (Segura-Bedmar and Martínez, 2017). We annotated these texts and compared our output annotation with regard to this dataset. We used a purely dictionary-based named-entity recogniser with modules for normalization (e.g. lowercasing), tokenization and lemmatization, implemented in spaCy;[8] then, the MedLexSp lexicon was used for exact string matching. We did not use pre- or post-processing rules in the current version (e.g. rules of term composition).

In several iterative rounds, we annotated the texts, identified the unannotated entities, and added them to the lexicon. We did not add (although annotated in the corpus) entities without a CUI, e.g. coordinated entities (e.g. *pies y manos frías*, 'cold hands and feet') or too specific, post-modified terms (e.g. *dolor de cabeza intenso*, 'intense headache'; only 'headache' has CUI C0018681). By using SPCs, we added 837 term entries to MedLexSp, and we ensure that it includes common terms referring to adverse drug reactions and medical drugs.

Lastly, for Consumer Health Vocabulary terms, we extracted synonyms in MedlinePlus Spanish. This resource provides terms in patient language that were missing: e.g. *ojo vago* ('lazy eye') is a synonym of *ambliopía* ('amblyopia', C0002418). We added 783 term entries from this resource. In addition, we collected 6110 cancer-related terms from the Spanish version of the National Cancer Institute Dictionary.

---

[8]https://spacy.io/

3.3 Semantic and linguistic information

We added to each CUI and lexical entry the corresponding semantic type(s) and group from the UMLS. To avoid noise when annotating biomedical texts semantically, we disfavoured semantic types of the semantic group Concepts and Ideas (CONC, e.g. Quantitative Concept, Functional Concept or Qualitative Concept), which are rather unspecified. We only included terms from that group if no other semantic label was available. If a concept or term can be assigned to two different groups, the element labelled with CONC is not included in our lexicon. For example, the term *inhalación* ('inhalation') can be related to concept C0004048 (semantic type Organism Function, and group PHYS) and also to concept C4521689 (semantic type Intellectual Product, and group CONC). In this case, we only preserve the lexical entry of concept C0004048 and we rule out the entry of concept C4521689.

We have also started adding the Part-of-Speech (PoS) category of each entry in the lexicon. For multiword terms, the category of the head term is selected; e.g. *enfermedad de Crohn* ('Crohn's disease') is categorized as N ('noun'). We are currently testing different techniques to predict the PoS and automate the assignment of categories to each entry, which is still not fully satisfactory.

4 Statistics

Table 1 shows the count of entries in the lexicon according to each source or procedure applied to map terms to UMLS CUIs. Note that the full count exceeds the count of term entries in the current version of MedLexSp, given that some terms were gathered through different methods simultaneously. Table 2 shows the descriptive statistics of the lexicon: counts of lemmas and word forms, and total number of CUIs. Lastly, Table 3 shows a preliminary count of PoS categories in the current version of the lexicon. Note that most entries are nouns or need revision (UNKN stands for 'unknown'); this task is currently being undertaken.

Finally, Figure 3 depicts the distribution of semantic groups. Of note, some groups are underrepresented, due to the corpora and thesauri used to collect terms. For example, few entities belong to the GENE group, which implies that the coverage of the current version of MedLexSp is not adequate for tasks in the Genomics domain.

The amount of lemmas/word forms is lower

Method	# entries
Abbreviations / acronyms	1225
Affixes / roots	161
Derived adjectives	801
Conjugated verb forms	295
Base list mapped to UMLS CUIs:	
Exact match to Spanish UMLS	18 263
Exact match to English UMLS	2534
String distance method	1463
Syntactic variants	134
Terms from thesauri and corpora:	
ATC + Nomenclátor + SDEdb	2931
ICD-10	1299
ICPC	55
MedDRA	5015
MedlinePlus	783
MeSH	6831
NCI	6110
OrphaData	10 741
SNOMED CT	23 096
SPCs (EasyDLP corpus)	837

Table 1: Count of lexical entries according to each source or procedure to map terms to UMLS CUIs.

	Lemmas	Forms	CUIs
Single-words	23 572	23 592	-
Multi-words	52 882	179 451	-
Total	76 454	203 043	30 647

Table 2: Descriptive statistics of the lexicon

PoS	Example	Count
N	*pancreas*	58 830
UNKN	-	13 618
ADJ	*abdominal*	2283
ADJ/N	*gemelo* ('twin')	700
NPR	*Filoviridae*	549
V	*toser* ('to cough')	295
AFF	*cardio-*	161
ADV	*gravemente* ('severely')	20

Table 3: Preliminary counts of Part-of-Speech (PoS) categories. N: 'noun'; UNKN: 'unknown'; ADJ: 'adjective'; ADJ/N: a term that can be an adjective or a noun (depending on the context); NPR: 'proper name'; AFF: 'affix'; ADV: 'adverb'

than in other UMLS-based resources because: 1) we did not include the full thesauri, but only terms from the original base list that were mapped to

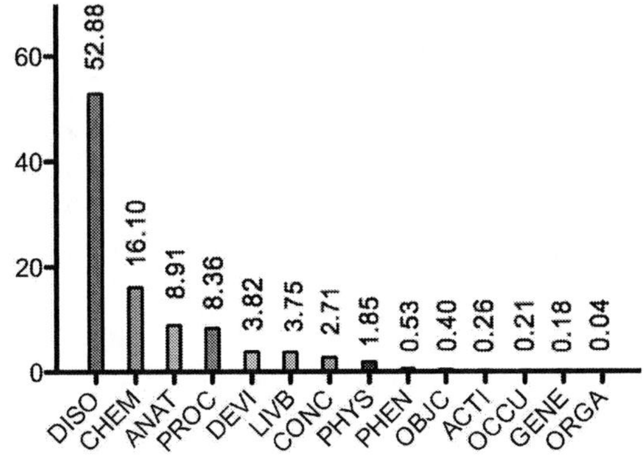

Figure 3: Distribution of semantic groups in the lexicon

UMLS CUIs; and 2) we cleaned noisy terms. As explained, descriptors and qualifiers were removed: e.g. SNOMED CT term *fiebre (hallazgo)* ('fever (finding)', C0015967) was shortened to *fiebre*. We also ruled out some concepts belonging to semantic groups that we can be noisy for clinical or medical NER tasks, such as CONC or GEOG; e.g. *hierro* is related to concept C0302583, 'iron', CHEM; or to concept C0454671, 'Island of Hierro', GEOG (the latter concept was discarded).

5 Development Evaluation

We analysed the coverage of the lexicon with regard to UMLS semantic groups. We applied the dictionary-based NER tool explained below to a gold standard available in the community. We focused on analysing the annotation of few UMLS groups (DISO, CHEM, PROC and ANAT) and assessed how well the lexicon annotated them with regard to the gold standard. We quantified the matched annotations in terms of precision, recall and F1-measure by using the BRAT-Eval script (Verspoor et al., 2013).

A first version of MedLexSp was evaluated with the Spanish texts from the MANTRA corpus (Kors et al., 2015), which gathers 100 texts from the European Medicines Agency (1961 tokens) and 100 texts from Medline (1087 tokens). These texts are available in BRAT format and were annotated with UMLS CUIs, semantic types and groups. We preprocessed the annotated texts for mapping reference annotations to UMLS semantic groups.

With this dataset, we achieved an overall F-measure of 0.83 (exact match) and of 0.87 (approximate match), although the performance var-

	Exact match			Approx. match		
	P	**R**	**F1**	**P**	**R**	**F1**
ANAT	0.64	0.91	0.75	0.67	0.98	0.80
CHEM	0.85	0.96	0.90	0.88	0.97	0.92
DISO	0.83	0.83	0.83	0.89	0.92	0.90
PROC	0.73	0.84	0.78	0.76	0.88	0.82
OA	0.79	0.87	0.83	0.83	0.93	0.87

Table 4: Evaluation of the lexicon; *P*: precision; *R*: recall; *F1*: F-measure; *OA*: overall

ied across semantic groups (Table 4). In our error analysis, we observed that unmatched entities were misspellings (e.g. **detección* instead of *detección*, 'detection'), discontinuous entities (e.g. *hinchazón de la piel* in *hinchazón y hormigueo de la piel*, 'swelling and tingling of skin'), or entities whose scope was wrongly annotated.

6 Discussion and Conclusions

The lexicon is being developed by means of hybrid NLP methods and corpus-derived terms. We combine the mapping of corpus terms to available thesauri, and viceversa, terms missing in the lexicon were attested in domain texts, so that only a subset of attested terms be included in a first version. Interestingly, searching terms from thesauri in a corpus showed us that many of those terms show low frequencies. From a subset of 56 813 MeSH terms missing in the base list, only 6 676 (11.75%) occurred in the corpus we used (Moreno-Sandoval and Campillos-Llanos, 2013). Although this is due to the influence of the text types, it also reflects the difference betweem terms from thesauri and in real usage. This is another argument that stands for the need for dedicated lexicons combined with NLP methods to achieve successful NER results.

A limitation of our evaluation procedure is the restriction to a very small set of texts; hence, results are not comparable to other tasks or text types. To provide more generalizable results, we need to evaluate the MedLexSp lexicon with another annotated medical corpus in Spanish, but such resource is not freely available to date.

We assume the lexicon is not task-independent. To avoid ambiguity, terms would need to be filtered according to the semantic types needed. For example, terms from the Occupation or Discipline group could be removed for most NER tasks. We are also aware of the limits of a purely lexicon-based approach. Contexts of variation occur in multiwords with coordinated terms (e.g. *cáncer de mama y ovario*, 'breast and ovarian cancer') and adjective modifiers. For example, MedLexSp includes the term *cáncer de mama* ('breast cancer'), but not common variants such as *cáncer de mama derecha* ('right breast cancer') or *cáncer de una mama* ('cancer of one breast'). Both phenomena need specific processing techniques.

Mapping concepts to terms differing across varieties of the Spanish language was not exhaustive. As we departed mainly from a set of corpus-derived terms, most terms belong to the variety used in the texts (i.e. Peninsular Spanish). However, since we used other terminological sources, terms from other varieties were included: e.g. *virus sincitial respiratorio* ('respiratory syncytial virus', C0035236) is a term preferred in Spain or Colombia, but we have the variant *virus sincicial respiratorio* (most frequent in Chile or Argentina). These aspects need nonetheless improvement in future versions, in the same way as the coverage of terms from Consumer Health Vocabularies.

Lastly, we are interested in exploring embedding-based methods for term expansion, and in evaluating the lexicon with a broader set of domain texts.

Acknowledgments

This work has been done under the NLPMedTerm project,[9] funded by the European Union's Horizon 2020 research programme under the Marie Skodowska-Curie grant agreement no. 713366 (InterTalentum UAM). We greatly thank the institutions who gave permission to include their data in MedLexSp, and also thank the anonymous reviewers for their valuable comments.

References

AEMPS. 2019. Nomenclátor de prescripción. *www.aemps.gob.es [accessed 2019-03-09]*.

Magnus Ahltorp, Maria Skeppstedt, Shiho Kitajima, Aron Henriksson, Rafal Rzepka, and Kenji Araki. 2016. Expansion of medical vocabularies using distributional semantics on Japanese patient blogs. *Journal of Biomedical Semantics*, 7(1):58.

Alan R Aronson. 2001. Effective mapping of biomedical text to the UMLS Metathesaurus: the MetaMap

---

[9] `http://www.lllf.uam.es/ESP/nlpmedterm_en.html`

program. In *Proc. of the AMIA Symposium*, pages 17–21. American Medical Informatics Association.

Alan R Aronson, James G Mork, Aurélie Névéol, Sonya E Shooshan, and Dina Demner-Fushman. 2008. Methodology for creating UMLS content views appropriate for biomedical natural language processing. In *Proc. of the AMIA Annual Symposium*, pages 21–25. American Medical Informatics Association.

Olivier Bodenreider. 2004. The Unified Medical Language System (UMLS): integrating biomedical terminology. *Nucleic acids research*, 32(suppl 1):D267–D270.

Elliot G Brown, Louise Wood, and Sue Wood. 1999. The Medical Dictionary for Regulatory Activities (MedDRA). *Drug safety*, 20(2):109–117.

Pablo Calleja, Raúl García-Castro, Guadalupe Aguado-de Cea, and Asunción Gómez-Pérez. 2017. Expanding SNOMED-CT through Spanish Drug Summaries of Product Characteristics. In *Proc. of the Knowledge Capture Conference*, pages 29–37. ACM.

L Campillos Llanos, A Moreno Sandoval, and JM Guirao. 2013. An automatic term extractor for biomedical terms in Spanish. In *Proc. of the 5th Int. Symposium on Languages in Biology and Medicine*, Tokyo, Japan.

Francisco Carrero, José Carlos Cortizo, and José María Gómez. 2008. Building a Spanish MMTx by using automatic translation and biomedical ontologies. In *International Conference on Intelligent Data Engineering and Automated Learning*, pages 346–353. Springer.

John Carroll, Rob Koeling, and Shivani Puri. 2012. Lexical acquisition for clinical text mining using distributional similarity. In *International Conference on Intelligent Text Processing and Computational Linguistics*, pages 232–246. Springer.

Bruno Cartoni and Pierre Zweigenbaum. 2010. Semi-Automated Extension of a Specialized Medical Lexicon for French. In *Proc. of LREC*, Valletta, Malta.

Billy Chiu, Olga Majewska, Sampo Pyysalo, Laura Wey, Ulla Stenius, Anna Korhonen, and Martha Palmer. 2019. A neural classification method for supporting the creation of BioVerbNet. *Journal of Biomedical Semantics*, 10(1):2:1–2:12.

James J Cimino. 1996. Coding systems in health care. *Methods of information in medicine*, 35(04/05):273–284.

Allan Peter Davis, Thomas C Wiegers, Michael C Rosenstein, and Carolyn J Mattingly. 2012. MEDIC: a practical disease vocabulary used at the Comparative Toxicogenomics Database. *Database*, 2012.

Louise Deléger, Magnus Merkel, and Pierre Zweigenbaum. 2009. Translating medical terminologies through word alignment in parallel text corpora. *Journal of Biomedical Informatics*, 42(4):692–701.

Guy Divita, Graciela Rosemblat, and Allen C Browne. 2007. Building a Medical Spanish Lexicon. In *Proc. of the AMIA Symposium*, page 941.

Kristina M Doing-Harris and Qing Zeng-Treitler. 2011. Computer-assisted update of a consumer health vocabulary through mining of social network data. *Journal of Medical Internet Research*, 13(2):e37.

Kevin Donnelly. 2006. SNOMED-CT: The advanced terminology and coding system for eHealth. *Studies in health technology and informatics*, 121:279–290.

Ted Dunning. 1993. Accurate methods for the statistics of surprise and coincidence. *Computational linguistics*, 19(1):61–74.

Juliusz Dziadek, Aron Henriksson, and Martin Duneld. 2017. Improving terminology mapping in clinical text with context-sensitive spelling correction. *Informatics for Health: Connected Citizen-Led Wellness and Population Health*, 235:241.

Natalia Grabar and Pierre Zweigenbaum. 2000. A general method for sifting linguistic knowledge from structured terminologics. In *Proc. of the AMIA Symposium*, pages 310–314. American Medical Informatics Association.

Johannes Hellrich, Stefan Schulz, Sven Buechel, and Udo Hahn. 2015. Jufit: A configurable rule engine for filtering and generating new multilingual UMLS terms. In *Proc. of the AMIA Symposium*, pages 604–610. American Medical Informatics Association.

Aron Henriksson, Hans Moen, Maria Skeppstedt, Vidas Daudaravičius, and Martin Duneld. 2014. Synonym extraction and abbreviation expansion with ensembles of semantic spaces. *Journal of Biomedical Semantics*, 5(1):6–.

Kristina M Hettne, Erik M van Mulligen, Martijn J Schuemie, Bob JA Schijvenaars, and Jan A Kors. 2010. Rewriting and suppressing UMLS terms for improved biomedical term identification. *Journal of Biomedical Semantics*, 1(1):5.

INSERM. 2019. Orphadata: Free access data from Orphanet. Data version (XML data version). *http://www.orphadata.org [accessed 2019-05-10]*.

Ander Intxaurrondo, Montserrat Marimón, Aitor González-Agirre, José Antonio López-Martín, H Rodríguez Betanco, J Santamaría, Marta Villegas, and Martin Krallinger. 2018. Finding mentions of abbreviations and their definitions in Spanish Clinical Cases: the BARR2 shared task evaluation results. In *Proc. of IberEval@SEPLN 2018*. SEPLN.

Christian Jacquemin. 1999. Syntagmatic and paradigmatic representations of term variation. In *Proc. of the 37th Annual Meeting of the Association for Computational Linguistics*, pages 341–348.

Ling Jiang and Christopher C Yang. 2013. Using co-occurrence analysis to expand consumer health vocabularies from social media data. In *2013 IEEE International Conference on Healthcare Informatics*, pages 74–81. IEEE.

Stephen B Johnson. 1999. A semantic lexicon for medical language processing. *Journal of the American Medical Informatics Association*, 6(3):205–218.

Rohit J Kate. 2015. Normalizing clinical terms using learned edit distance patterns. *Journal of the American Medical Informatics Association*, 23(2):380–386.

Alla Keselman, Tony Tse, Jon Crowell, Allen Browne, Long Ngo, and Qing Zeng. 2007. Assessing consumer health vocabulary familiarity: an exploratory study. *Journal of Medical Internet Research*, 9(1):e5.

Jan A Kors, Simon Clematide, Saber A Akhondi, Erik M van Mulligen, and Dietrich Rebholz-Schuhmann. 2015. A multilingual gold-standard corpus for biomedical concept recognition: the Mantra GSC. *Journal of the American Medical Informatics Association*, 22(5):948–956.

Vladimir Levenshtein. 1966. Binary codes capable of correcting deletions, insertions, and reversals. In *Soviet Physics Doklady*, volume 10, pages 707–710.

Nut Limsopatham and Nigel Collier. 2016. Normalising medical concepts in social media texts by learning semantic representation. In *Proc. of the 54th Annual Meeting of the Association for Computational Linguistics (Volume 1: Long Papers)*, volume 1, pages 1014–1023.

Kornél Markó, Robert Baud, Pierre Zweigenbaum, Lars Borin, Magnus Merkel, and Stefan Schulz. 2006. Towards a multilingual medical lexicon. In *Proc. of the AMIA Annual Symposium*, pages 534–538. American Medical Informatics Association.

Kornél Markó, Stefan Schulz, and Udo Hahn. 2005. MorphoSaurus. Design and evaluation of an interlingua-based, cross-language document retrieval engine for the medical domain. *Methods of information in medicine*, 44(04):537–545.

John McCrae and Nigel Collier. 2008. Synonym set extraction from the biomedical literature by lexical pattern discovery. *BMC Bioinformatics*, 9(1):159.

Alexa T McCray, Anita Burgun, and Olivier Bodenreider. 2001. Aggregating UMLS semantic types for reducing conceptual complexity. *Studies in health technology and informatics*, 84(0 1):216–220.

Alexa T McCray, Suresh Srinivasan, and Allen C Browne. 1994. Lexical methods for managing variation in biomedical terminologies. In *Proc. of the Annual Symposium on Computer Application in Medical Care*, pages 235–239. American Medical Informatics Association.

Antonio Moreno-Sandoval and Leonardo Campillos-Llanos. 2013. Design and Annotation of MultiMedica–A Multilingual Text Corpus of the Biomedical Domain. *Procedia-Social and Behavioral Sciences*, 95:33–39.

Antonio Moreno Sandoval and Leonardo Campillos Llanos. 2015. Combinación de estrategias léxicas y estadísticas para el reconocimiento automático de términos: aplicación a un corpus de medicina. *Lingüística Española Actual*, 37:173–197.

Antonio Moreno Sandoval and José María Guirao. 2006. Morphosyntactic tagging of the Spanish C-ORAL-ROM corpus: Methodology, tools and evaluation. *Spoken language corpus and linguistic informatics*, 5:199–218.

Fiammetta Namer and Pierre Zweigenbaum. 2004. Acquiring meaning for French medical terminology: contribution of morphosemantics. In *Proc. of MedInfo*, pages 535–539.

Adeline Nazarenko, Pierre Zweigenbaum, Benoît Habert, and Jacques Bouaud. 2001. Corpus-based extension of a terminological semantic lexicon. *Recent Advances in Computational Terminology*, pages 327–351.

Aurélie Névéol, Jiao Li, and Zhiyong Lu. 2012. Linking multiple disease-related resources through UMLS. In *Proc. of the 2nd ACM SIGHIT international health informatics symposium*, pages 767–772. ACM.

Maite Oronoz, Arantza Casillas, Koldo Gojenola, and Alicia Perez. 2013. Automatic annotation of medical records in Spanish with disease, drug and substance names. In *Iberoamerican Congress on Pattern Recognition*, pages 536–543. Springer.

Piotr Pezik, A Jimeno-Yepes, V Lee, and D Rebholz-Schuhmann. 2008. Static dictionary features for term polysemy identification. In *Proc. of Building and Evaluating Resources for Biomedical Text Mining LREC Workshop*, Marrakech, Morocco.

Sampo Pyysalo, Filip Ginter, Hans Moen, Salakoski Tapio, and Sophia Ananiadou. 2013. Distributional semantics resources for biomedical text processing. *Proc. of Languages in Biology and Medicine*, pages 39–44.

RANME. 2011. *Diccionario de Términos Médicos*. Editorial Panamericana.

A Moreno Sandoval, L Campillos Llanos, A Gonzlez Martnez, and JM Guirao. 2013. An affix-based method for automatic term recognition from a medical corpus of Spanish. In *Proc. of the 7th Corpus Linguistics Conference 2013*, Lancaster University.

Eduardo Sbrissia, Percy Nohama, Stefan Schulz, and Kornél Markó. 2004. Semi-Supervised Acquisition of a Spanish Lexicon from a Portuguese Seed Lexicon. *Proc. of COLING*.

Michael Seedorff, Kevin J Peterson, Laurie A Nelsen, Cristian Cocos, Jennifer B McCormick, Christopher G Chute, and Jyotishman Pathak. 2013. Incorporating expert terminology and disease risk factors into consumer health vocabularies. In *Biocomputing 2013*, pages 421–432. World Scientific.

Isabel Segura-Bedmar and Paloma Martínez. 2017. Simplifying drug package leaflets written in Spanish by using word embedding. *Journal of Biomedical Semantics*, 8(1):45.

Isabel Segura-Bedmar, Paloma Martínez, Ricardo Revert, and Julián Moreno-Schneider. 2015. Exploring Spanish health social media for detecting drug effects. In *BMC medical informatics and decision making*, volume 15, page S6. BioMed Central.

Maria Skeppstedt, Magnus Ahltorp, and Aron Henriksson. 2013. Vocabulary expansion by semantic extraction of medical terms. *Proc. of Languages in Biology and Medicine*, pages 63–67.

Tom De Smedt and Walter Daelemans. 2012. Pattern for python. *Journal of Machine Learning Research*, 13(Jun):2063–2067.

Paul Thompson and Sophia Ananiadou. 2018. Hyphen. *Terminology.*, 24(1):91–121.

Paul Thompson, John McNaught, Simonetta Montemagni, Nicoletta Calzolari, Riccardo Del Gratta, Vivian Lee, Simone Marchi, Monica Monachini, Piotr Pezik, Valeria Quochi, et al. 2011. The BioLexicon: a large-scale terminological resource for biomedical text mining. *BMC Bioinformatics*, 12(1):397.

Eugene Tseytlin, Kevin Mitchell, Elizabeth Legowski, Julia Corrigan, Girish Chavan, and Rebecca S Jacobson. 2016. Noble–flexible concept recognition for large-scale biomedical natural language processing. *BMC Bioinformatics*, 17(1):32.

Yoshimasa Tsuruoka, John McNaught, Junichi Tsujii, and Sophia Ananiadou. 2007. Learning string similarity measures for gene/protein name dictionary look-up using logistic regression. *Bioinformatics*, 23(20):2768–2774.

Karin Verspoor, Antonio Jimeno Yepes, Lawrence Cavedon, Tara McIntosh, Asha Herten-Crabb, Zoë Thomas, and John-Paul Plazzer. 2013. Annotating the biomedical literature for the human variome. *Database*, 2013.

Chang Wang, Liangliang Cao, and Bowen Zhou. 2015. Medical synonym extraction with concept space models. In *Proc. of the Twenty-Fourth International Joint Conference on Artificial Intelligence*, pages 989–995.

Liqin Wang, Bruce E Bray, Jianlin Shi, Guilherme Del Fiol, and Peter J Haug. 2016. A method for the development of disease-specific reference standards vocabularies from textual biomedical literature resources. *Artificial intelligence in medicine*, 68:47–57.

Gesa Weske-Heck, Albrecht Zaiss, Matthias Zabel, Stefan Schulz, Wolfgang Giere, Michael Schopen, and Rüdiger Klar. 2002. The German specialist lexicon. In *Proceedings of the AMIA Symposium*, pages 884–888. American Medical Informatics Association.

WHO. 2004. *International Statistical Classification of Diseases and Related Health Problems*. World Health Organization.

WHO. 2019. *Anatomical Therapeutical Chemical classification*. Uppsala: Nordic Council on Medicines.

WONCA. 1998. *International Classification of Primary Care 2nd ed.* Oxford: Oxford University Press, 1998.

Javier Yetano and Vincent Alberola. 2003. *Diccionario de siglas médicas y otras abreviaturas, epónimos y términos médicos relacionados con la codificación de las altas hospitalarias*. SEDOM.

Hong Yu and Eugene Agichtein. 2003. Extracting synonymous gene and protein terms from biological literature. *Bioinformatics*, 19(suppl_1):i340–i349.

Qing Zeng and Tony Tse. 2006. Exploring and developing consumer health vocabularies. *Journal of the American Medical Informatics Association*, 13(1):24–29.

Qing Zeng, Tony Tse, Guy Divita, Alla Keselman, Jonathan Crowell, Allen Browne, Sergey Goryachev, and Long Ngo. 2007. Term identification methods for consumer health vocabulary development. *Journal of Medical Internet Research*, 9(1):e4.

Pierre Zweigenbaum, Robert Baud, Anita Burgun, Fiammetta Namer, Éric Jarrousse, Natalia Grabar, Patrick Ruch, Franck Le Duff, Jean-Franois Forget, Magaly Douyère, and Stéfan Darmoni. 2005. A unified medical lexicon for French. *International Journal of Medical Informatics*, 74(2–4):119–124.

A   Appendix - Copyright and Usage

MedLexSp is distributed freely for research purposes; contact for a license at the email address provided or through the project page. Some thesauri included in MedLexSp were obtained through a distribution and usage agreement from the corresponding institutions who develop them. In addition, some material in the UMLS Metathesaurus is from copyrighted sources of the respective copyright holders. Users of the UMLS Metathesaurus are solely responsible for compliance with any copyright, patent or trademark restrictions and are referred to the copyright, patent or trademark notices appearing in the original sources, all of which are hereby incorporated by reference.

The version of MedLexSp freely available for research does not include terms nor coding data from terminological sources with copyright rights; only the subset of data in MedLexSp without usage restrictions is accessible.

We acknowledge the intellectual property rights of the institutions who develop the sources from which we extracted subsets of terms to compile the lexicon, and who gave permission (or provide a licence to reuse their data) to distribute these subsets of terms: the National Library of Medicine maintains the MedLinePlus resource and the Medical Subject Headings, and BIREME/OPS (Latin-American and Caribbean Center on Health Sciences Information) is in charge of the Spanish translation (Descriptores en Ciencias de la Salud, DeCS); the National Cancer Institute publishes the Dictionary of Cancer Terms; the French National Institute of Health and Medical Research (INSERM) supports OrphaNet and gathers the information provided in OrphaData; the World Health Organization produces the Adverse Drug Reactions terminology, the International Classification of Diseases vs. 10, and the Anatomical Therapeutical Classification; the Spanish translation of the International Classification of Primary Care (ICPC) is supported by the World Organization of Family Doctors; and the Spanish Agency of Drugs and Food Products (AEMPS) publishes the Nomenclátor de prescripción. MedLexSp also gathers some terms from the Spanish version of the Medical Dictionary for Regulatory Activities (MedDRA), which is maintained by the Maintenance and Support Services Organization (MSSO). However, the distributed version of MedLexSp does not include terms coming solely from the MedDRA sources, because of copyright restrictions. In addition, MedLexSp includes a subset of the Spanish version of SNOMED Clinical Terms®, which is used by permission of the International Health Terminology Standards Development Organization (IHTSDO; all rights reserved). SNOMED CT® was originally created by The College of American Pathologists.

Incorporating Figure Captions and Descriptive Text in MeSH Term Indexing

Xindi Wang
Department of Computer Science
The University of Western Ontario
London, Ontario, Canada
xwang842@uwo.ca

Robert E. Mercer
Department of Computer Science
The University of Western Ontario
London, Ontario, Canada
mercer@csd.uwo.ca

Abstract

The goal of text classification is to automatically assign categories to documents. Deep learning automatically learns effective features from data instead of adopting human-designed features. In this paper, we focus specifically on biomedical document classification using a deep learning approach. We present a novel multichannel TextCNN model for MeSH term indexing. Beyond the normal use of the text from the abstract and title for model training, we also consider figure and table captions, as well as paragraphs associated with the figures and tables. We demonstrate that these latter text sources are important feature sources for our method. A new dataset consisting of these text segments curated from 257,590 full text articles together with the articles' MEDLINE/PubMed MeSH terms is publicly available.

1 Introduction

Text classification is a process that assigns labels or tags to text according to its contents. It can be done manually or automatically. Most text classification tasks were done by human annotators prior to the information age. A human annotator reads and interprets the content of the text and then classifies it into certain categories. Traditional text classification is time consuming and expensive, especially when dealing with a large number of documents.

Currently, there is a trend to support text classification through automatic tools as it does the same job as human annotators, but accomplishes it in more accurate and efficient ways. Automatic text classification is an important application and research topic in natural language processing because of the exponentially increasing number of online documents. It saves time and money in general, leading to its continued and enthusiastic usage in both business and research.

MEDLINE[1] and PubMed[2] are databases that can access publications of life sciences and biomedical topics. They are maintained by the United States National Library of Medicine (NLM).

The MEDLINE database includes bibliographic information for articles in various disciplines of life sciences and biomedicine, such as medicine, health care, biology, biochemistry and molecular evolution. The database contains more than 25 million records in over 5,200 worldwide journals. More than 800,000 citations were added to MEDLINE in 2017, which is more than 2,000 updates daily[1].

PubMed has a web server that can freely access the MEDLINE database of references and abstracts. Some PubMed records have full text articles available on PubMed Central[3]. Journal articles in MEDLINE are indexed according to Medical subject headings (MeSH)[4], which are the NLM's controlled vocabulary thesaurus.

MeSH is a hierarchically-organized terminology indexing system that categorizes biomedical documents in the NLM databases. It is updated annually. The 2018 version of MeSH contains 28,939 headings[5]. Among these MeSH terms, there are 29 check tags which are a special group of MeSH terms describing subjects of research (human or animal; mice or rats, etc.). MeSH terms are distinctive features of MEDLINE, which are great tools for indexers and searchers. Indexers from NLM use MeSH terms to classify documents based on the contents of journal articles in the MEDLINE database. Searchers and researchers use MeSH terms to assist subject searching in MEDLINE, PubMed and other databases.

---

[1]https://www.nlm.nih.gov/bsd/medline.html
[2]https://www.nlm.nih.gov/bsd/pubmed.html
[3]https://en.wikipedia.org/wiki/PubMed_Central
[4]https://www.nlm.nih.gov/mesh/meshhome.html
[5]https://www.nlm.nih.gov/pubs/techbull/nd17/nd17_mesh.html

Proceedings of the BioNLP 2019 workshop, pages 165–175
Florence, Italy, August 1, 2019. ©2019 Association for Computational Linguistics

Currently MeSH term indexing is performed by a large number of human annotators, who review full text documents and assign suitable MeSH terms to each article. Human annotation is time consuming and costly. Research shows that the average cost of annotation per document is around $9.40 (Mork et al., 2013), which translates into a huge cost for indexing a large number of documents. Meanwhile, there is a large number of documents uploaded to MEDLINE and PubMed databases every day (approximately 2,000–4,000 on a daily basis)[2]. It is challenging to annotate all new incoming documents in a relatively short time. Therefore, a computational system that can assist the indexing of a large number of biomedical articles is highly desired.

In this paper, we focus on the task of automatic MeSH indexing. We propose a novel deep learning based discriminative method, multichannel TextCNN, which uses convolutional neural network based feature selection to extract important information from the article to be indexed. In addition to extracting information from the title and abstract of the article, our innovation integrates figure and table captions, as well as relevant paragraphs into the indexing process. We summarize the most major contributions as follows:

- We explore the use of multichannel deep learning architectures for the automatic MeSH indexing task.

- Experimental results show that incorporating figure and table information improves the performance of automatic MeSH indexing.

- We make available a labeled full text biomedical document dataset (including title, abstract, figure and table captions, as well as paragraphs related to the figures and tables) to the research community.

2   Related Work

Due to the growth in the number of documents in MEDLINE, and the increasing number of MeSH terms every year, automatic MeSH indexing is a difficult challenge. The Medical Text Indexer (MTI) (Aronson et al., 2004) produced by the U.S. National Library of Medicine (NLM), is the first program that automatically produces MeSH indexing recommendations. Given the title and abstract for an article in MEDLINE, MTI will provide a ranked list of MeSH terms. The initial MTI system was developed in 2002, and has been continuously improved over the years. There are two main components in MTI, namely, MetaMap (Aronson and Lang, 2010), and PubMed Related Citations (PRC) (Lin and Wilbur, 2007). MetaMap analyzes documents and annotates them using the Unified Medical Language System (UMLS)[6]. The PRC algorithm[7] with k-nearest neighbours (k-NN) uses document similarity to find MeSH terms. MTI is an important tool in MeSH indexing, and indexers can use MTI suggestions for documents that they are annotating. Another method, Restrict-To-Mesh (Kin-Wah Fung, 2007) also maps from UMLS to MeSH terms.

BioASQ[8], a European Union-funded project, has organized challenges on automatic MeSH indexing since 2013. Participants are required to annotate unlabelled PubMed citations with abstracts and titles using their models before these articles are indexed by human annotators. The winning system in 2013, for example, used the MetaLabeler algorithm (Tang et al., 2009) to learn two models, one for ranking and the other for predicting the number of related labels. MeSHLabeler (Liu et al., 2015) won first place in 2014. It also has two components: MeSHRanker and MeSHNumber. MeSHRanker returns a ranked list of candidate MeSH terms. MeSHNumber predicts the number of output MeSH terms. DeepMeSH (Peng et al., 2016) was the best system in 2017. It incorporates deep semantic information into MeSHLabeler using a dense semantic representation for documents, namely document to vectors (D2V). In addition, DeepMeSH has a second classifier to find the number of MeSH terms returned. AttentionMeSH (Jin et al., 2018), also proposed in 2017, uses a bi-direction recurrent gated unit (Bi-GRU) architecture to capture contextual features, and attention mechanisms to select MeSH terms from the candidate list.

Rios and Kavuluru (2015) used a convolutional neural network (CNN) to classify the 29 most frequent MeSH terms on a small dataset comprised of 9,000 citations. Gargiulo et al. (2018) applied deep CNN on the abstracts and titles of 1,115,090 articles. Besides deep learning approaches, other

---

[6]https://www.nlm.nih.gov/research/umls/
[7]https://ii.nlm.nih.gov/MTI/Details/related.shtml
[8]http://bioasq.org

machine learning algorithms have also been explored in the hopes of solving MeSH indexing tasks. A few examples are: Naïve Bayes (NB), support vector machines (SVM), linear regression, and AdaBoost (Jimeno-Yepes et al., 2012, 2013).

3 Proposed Model

3.1 Problem Statement

Multi-label classification studies the problem where each document is associated with a set of labels (Zhang and Zhou, 2014). In the MeSH indexing problem, each MeSH term can be treated as a class label and each biomedical article can have multiple MeSH terms. Because of the large number of MeSH terms we regard automatic MeSH term indexing as an extreme multi-label classification problem.

The learning framework is defined as follows. Suppose $\mathcal{X}$ is a set of biomedical documents (at this point we won't prejudice how these documents are represented, these representational details are discussed below) and $\mathcal{Y}$ is the set of MeSH terms. Multi-label classification studies the learning function $f : \mathcal{X} \rightarrow 2^{\mathcal{Y}}$ using the training set $\mathcal{D} = (x_i, Y_i), i = 1 \ldots D$, where D is the number of documents in the set $\mathcal{X}$. Each instance x_i is an n-dimensional vector, where n is the number of words in document x_i, and $Y_i \subseteq \mathcal{Y}$ is the set of labels associated with instance x_i. The objective of multi-label classification is to predict the proper label set Y_k for any unseen instance x_k (Zhang and Zhou, 2014).

Two challenges should be considered when solving automatic MeSH indexing tasks (Zhai et al., 2015). First, the number of MeSH terms is large and they have widely varying occurrence frequencies. There are around 29,000 MeSH term and they are updated annually. The frequency of each MeSH term appearing as a document label is quite biased. For instance, of the 29,000 MeSH terms, the most frequent term "Humans", appears in 8,152,852 citations; and "Pandanaceae", on the other hand, only appears in 31 documents (Zhai et al., 2015). Second, the number of MeSH terms assigned to each document varies. Some documents have more than 30 MeSH terms and some have fewer than 5. In this paper, we have used the 2018 version of MeSH which contains 28,939 headings in total.

3.2 Model Overview

We propose multichannel TextCNN, a novel deep learning approach to assign proper MeSH terms to given documents. To make use of multimodal features, our model has two input channels:

- Channel 1: word embeddings from abstract and title

- Channel 2: word embeddings from figure and table captions and corresponding paragraphs that mention the figures and tables

As promised above, we now discuss the representational details of a document. A document is composed of n words. We use d-dimensional word embeddings to represent the words. The word embedding matrix e for each document is then $e \in \mathbb{R}^{d \times n}$. For each document, we have two texts: the abstract and title, and the captions and paragraphs. These two texts are represented by two embedding matrices, namely e_{AT}, the word embedding matrix for the abstract and title text, and e_{CP}, the word embedding matrix for the captions and paragraphs.

The model structure, shown in Figure 1, is a variant convolutional neural network (CNN) with multichannel inputs, which is inspired by TextCNN (Kim, 2014). We have chosen the CNN-based model because it has been successful in various text classification tasks. For each channel, the architecture is similar to TextCNN. The representation of abstract and title, e_{AT}, is input to one channel. The representation of captions and paragraphs, e_{CP}, is input to the other. We also use a single channel architecture by concatenating these two representations as input to one of the channels.

The model learns feature representations by passing embedded documents to the convolutional layer. The entire input document in each channel can be represented as $e_{1:n} = [e_1, e_2, \ldots, e_n] \in \mathbb{R}^{d \times n}$, where n is the length of the document and $e_i \in \mathbb{R}^d$, where e_i represents the i-th word in the document. The convolutional layer is composed of 128 convolutional filters each with sizes 3, 4, and 5. Recalling, we have d-dimensional word embedding vectors. So, the convolutional windows are $m \times d$, where $m \in \{3, 4, 5\}$.

In the convolutional layer, we have 128 feature maps for each filter size. The feature maps are then passed to a pooling layer which takes the maximum value for each associated feature map. After pooling, we get the feature map for each chan-

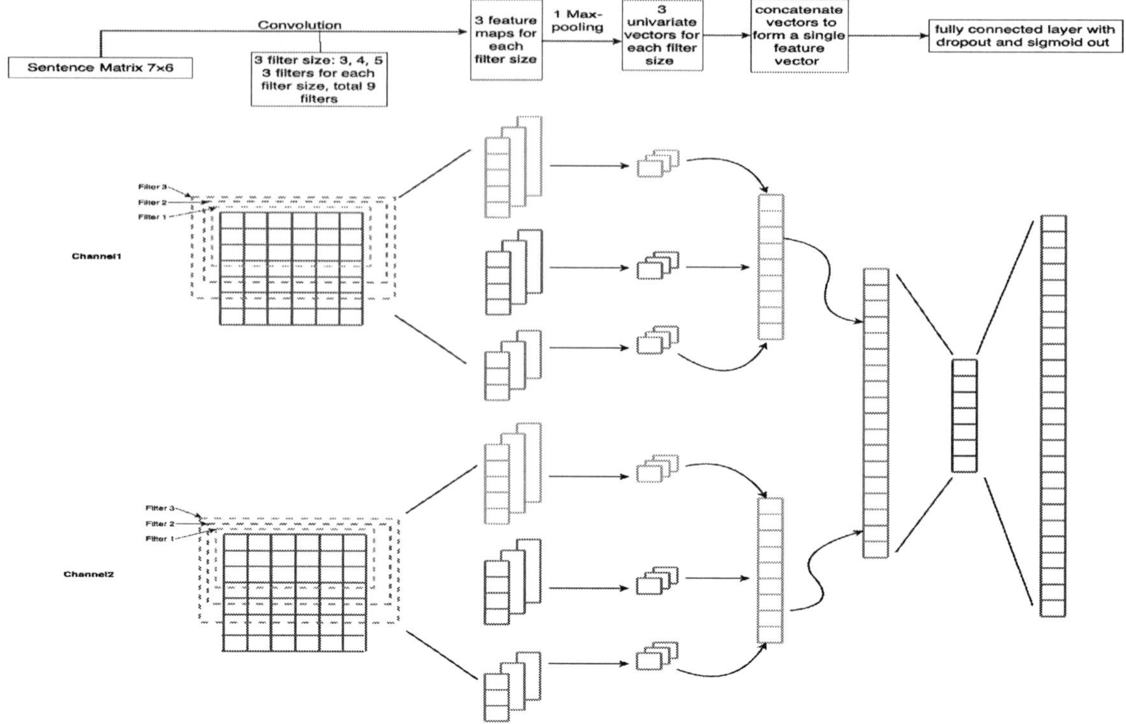

Figure 1: Multichannel TextCNN Architecture - filter 1, filter 2, and filter 3 indicate convolutional filters of size 3, 4, and 5, respectively. In this figure, we characterized our model with a 7×6 input document, where the number of words in the document n is 7, and the dimensionality of the word embedding d is 6.

nel and we concatenate these two feature maps to form a single feature vector. This feature vector is then passed to a fully connected bottleneck layer with 512 hidden units followed by a sigmoid classifier that returns a probability value for each of the 28,939 MeSH terms.

The training of our proposed methods uses binary cross-entropy as the loss function on the sigmoid classifier. We use the sigmoid function to return the probability score of each class. The sigmoid function is defined as:

$$\sigma(x) = \frac{1}{1 + e^{-x}}$$

Binary cross-entropy is formulated as:

$$H(q) = -\frac{1}{L} \sum_{i=1}^{L} y_i \cdot log(\sigma(y_i)) +$$

$$(1 - y_i) \cdot log(1 - \sigma(y_i))$$

where σ is the sigmoid function, L is the total number of labels, y_i is the original label of document i, and $\sigma(y_i)$ is the predicted probability of label y for document i. The sigmoid binary cross-entropy optimizes a label one-versus-all loss based on max-entropy.

We have also experimented with multichannel XMLCNN, which is inspired by XMLCNN (Liu et al., 2017), a variant CNN model developed for extreme multi-label classification; multichannel biLSTM, a bidirectional long short term memory neural network (Schuster and Paliwal, 1997) with multichannel inputs; and multichannel attention based convLSTM, which is a stacked CNN and LSTM followed by an attention layer. The experimental results indicate that the multichannel TextCNN model performed best among all of the models mentioned above in both execution time and evaluation metrics. Details of these experiments are available (Wang, 2019).

3.3 Setup and Model Hyper-parameters

In our proposed multichannel TextCNN model, we used rectified linear units (ReLU) as the activation function, convolutional filter windows of size 3, 4, and 5 with 128 filters each, dropout rate of 0.5, 512 hidden units in the bottleneck layer, batch size of 10, and learning rate of 0.001. The number of epochs in training is 20. These hyper-parameters are fixed across the different datasets. In each dataset, we used 90% of the data as the training set and 10% to test the performance of the model. We reserved 20% of the training data, chosen randomly, as the validation set, and the remaining 80% is used for training the model. All experiments are performed on the Nvidia GeForce 1080Ti GPU. Models for Fulltext (Large) are trained on 2 GPUs and training with the other datasets is performed on a single GPU.

For word embeddings in our proposed model we used the pre-trained 200-dimensional BioASQ word embedding vectors (Pavlopoulos et al., 2019) to represent the words in our vocabulary. These pre-trained word vectors are trained on 10,876,004 English biomedical abstracts from PubMed, and represent 1,701,632 distinct words.

4 Experiments

4.1 Datasets

Most existing approaches in automatic MeSH indexing are performed on datasets with abstracts and titles only. In this paper, we created a full text dataset which is composed of table and figure captions as well as associated paragraphs, as we believe figures and tables might provide important MeSH features for classification. The two datasets that were used to build our four datasets are described below:

- **2015 Subject Extraction Test Collection (SETC2015)**: SETC2015 contains 14,828 PMC full text articles used by Demner-Fushman and Mork (2015). We used this dataset to create the following two Small (S) datasets:

    - **AT (S)**: labelled documents from SETC2015 which contain abstract and title only
    - **Full (S)**: labelled documents from SETC2015 which contain abstract, title, figure and table captions, and associated paragraphs

- **PMC Full Text Collection[9] (PMC Collection)**: We used a downloaded dataset of 257,590 PMC full text documents in XML format, and used this dataset to create the following two Large (L) datasets:

---

[9]https://www.ncbi.nlm.nih.gov/pmc/tools/ftp/

- **AT (L)**: labelled documents from PMC Collection which are composed of abstract and title only
- **Full (L)**: labelled documents from PMC Collection which are composed of abstract, title, figure and table captions, and associated paragraphs

Datasets	D	F	L	$\bar{L}$	$\tilde{L}$
AT (S)	14828	63004	14365	13.15	13.5
Full (S)	14828	148330	14365	13.15	13.5
AT (L)	257590	188693	22881	13.34	150
Full (L)	257590	669999	22881	13.34	150

Table 1: Statistics of Datasets: D is the total number of documents (90% training, 10% testing); F represents the number of unique tokens contained in all of the documents; L is the number of class labels; $\bar{L}$ is the average number of labels per document; $\tilde{L}$ is the average number of documents per label

Table 1 provides statistical information for the described datasets. Our labeled datasets are using 28,939 MeSH terms in total. To assist in our understanding of the hierarchical evaluation, we explored the MeSH hierarchical structure and split them into 5 levels to see how many MeSH terms exist at each level (it should be noted that there is some overlap of MeSH terms between levels). The number of MeSH terms in the first, the second, the third, the fourth and the fifth level, are: 16, 120, 1903, 6,808, and 11,127, respectively.

4.2 Data pre-processing

The full text source files from PMC are in XML format. We extracted article information (including PMID, abstract, title, captions, and figure and table related paragraphs) from these downloaded XML files. Paragraphs are considered related to figures or tables if they contain the words "Figure" or "Table". MeSH terms for each article were scraped from its citation on PubMed by locating the citation using its PMID, the unique article identifier number used in PubMed.

In pre-processing, we first did word level tokenization of our input documents, to split the documents into a list of words. Then we prepared our data by using the following process: set all characters to lowercase; convert numbers to "NUM", percentage sign "%" to "PERCENTAGE", chemical notations (i.e., H_2O) to "CHEM", and relation symbols, namely "$=$", "$<$", "$>$", "$\leq$", "$\geq$", to "EQUAL", "LESS", "GREATER", "LessAndEqual", "GreaterAndEqual"; remove punctuation.

After the above process, we utilized the Keras (Chollet et al., 2015) Tokenizer API to vectorize our data into a sequence of integers. Each integer represents the index of a token in the dictionary generated from the dataset.

4.3 Evaluation Metrics

There is generally no accepted standard for the evaluation of multi-label classifications. Evaluation metrics adopted from multi-class classification and binary classification are used to measure multi-label classification in an effective way. In automatic MeSH indexing, even if the label space is very large, only relatively few MeSH terms match each document. To evaluate the performance of our proposed model, we present three groups of measurements suggested by Tsoumakas et al. (2010) and Kosmopoulos et al. (2015), namely bipartition-based, ranking-based and hierarchy-based evaluation.

To set the stage to discuss the three metrics, we define a test set of N document-label pairs $\{x_i, y_i\}_{i=1}^{N}$ taken from the dataset, where x_i is the document text and $y_i \in \{0, 1\}^L$. The vector y_i denotes the set of true labels (i.e., MeSH terms) for each document i (0 meaning the label is not in the set, 1 meaning it is in the set), N denotes the number of test examples, and L is the total number of labels. Given a document x_i, the set of labels predicted by the classifiers is denoted as $\{\hat{y}_i\}_{i=1}^{N}$, where $\hat{y}_i \in \{0, 1\}^L$, and the ranking indexes of predicted labels among the top k is denoted as $r_k(\hat{y})$, where $\hat{y} = \{\hat{y}_i\}_{i=1}^{N}$.

Bipartition evaluation is further divided into example-based and label-based metrics. Example-based measurements calculate precision, recall, and F-score over (in our evaluation) the top 5, top 10, and top 15 ranked labels over all of the documents of the test set. The measurements include example-based precision (*EBP*), example-based recall (*EBR*) and example-based F-score (*EBF*). The metrics are defined as:

$$EBP = \frac{1}{N} \sum_{i=1}^{N} \frac{|y_i \cap \hat{y}_i|}{|\hat{y}_i|}$$

$$EBR = \frac{1}{N} \sum_{i=1}^{N} \frac{|y_i \cap \hat{y}_i|}{|y_i|}$$

$$EBF = \frac{1}{N} \sum_{i=1}^{N} \frac{2 \times |y_i \cap \hat{y}_i|}{|y_i| + |\hat{y}_i|}$$

Label-based evaluation is calculated for each label in the label set. The measurements include macro- and micro-average precision (*MaP, MiP*), macro- and micro-average recall (*MaR, MiR*),and macro- and micro-average F-score (*MaF, MiF*). The metrics are defined as:

$$MaP = \frac{1}{L} \sum_{j=1}^{L} \frac{TP_j}{TP_j + FP_j}$$

$$MiP = \frac{\sum_{j=1}^{L} TP_j}{\sum_{j=1}^{L} TP_j + \sum_{j=1}^{L} FP_j}$$

$$MaR = \frac{1}{L} \sum_{j=1}^{L} \frac{TP_j}{TP_j + FN_j}$$

$$MiR = \frac{\sum_{j=1}^{L} TP_j}{\sum_{j=1}^{L} TP_j + \sum_{j=1}^{L} FN_j}$$

$$MaF = \frac{2 \times MaR \times MaP}{MaR + MaP}$$

$$MiF = \frac{2 \times MiR \times MiP}{MiR + MiP}$$

where TP_j, FP_j and FN_j as true positives, false positives, and false negatives respectively for each label l_j in the set of total labels L.

Ranking-based evaluation, including precision at k ($p@k$), and normalized discounted cumulative gain (*nDCG*), ranks the predicted labels and aims to rank the relevant labels higher than the irrelevant ones. The metrics are defined as follows:

$$p@k = \frac{1}{k} \sum_{l \in r_k(\hat{y})} y_l$$

$$DCG@k = \sum_{l \in r_k(\hat{y})} \frac{y_l}{log(l+1)}$$

$$IDCG = \sum_{l=1}^{min(k, \|y\|_0)} \frac{1}{log(l+1)}$$

$$nDCG@k = \frac{DCG@k}{IDCG}$$

Hierarchy-based evaluation, including hierarchical precision (*HP*) and hierarchical recall (*HR*), is used to measure a hierarchical classification that classifies elements into a hierarchy of classes. It measures performance based on the gold standard labels and the predicted labels augmented with their ancestors and descendants within distances 1 and 2. The augmented gold standard labels Y_{aug} and predicted labels $\hat{Y}_{aug}$ are used in the hierarchical evaluation. *HP* and *HR* are defined as follows:

$$HP = \frac{1}{N} \sum_{i=1}^{N} \frac{|\hat{Y}_{aug} \cap Y_{aug}|}{|\hat{Y}_{aug}|}$$

$$HR = \frac{1}{N} \sum_{i=1}^{N} \frac{|\hat{Y}_{aug} \cap Y_{aug}|}{|Y_{aug}|}$$

In ranking-based evaluation, for $p@k$, $k \in \{1, 3, 5, 10, 15\}$, and $k \in \{1, 3, 5\}$ for $nDCG@k$. In example-based, label-based, and hierarchy-based evaluations, the calculation is done with the top 5, 10, and 15 predicted labels. In hierarchical evaluation, we used distances 1 and 2 for *HP* and *HR*. The example-based, ranking-based, and hierarchical evaluation metrics are calculated for each document. An average score over all documents in the test set is returned. Likewise, the label-based evaluation is calculated for each label and averaged over all labels in the test set.

5 Results

We first conducted our experiments on datasets with titles and abstracts only (designated AT), passing the appropriate word embeddings to the single channel TextCNN. Next, we did our experiments on the full text datasets (designated Full), passing the word embeddings for titles, abstracts, captions and paragraphs to the single channel TextCNN. Finally, we conducted our experiments on the full text datasets using the multichannel model: we passed word embeddings for titles and abstracts to the first channel, and word embeddings for captions and paragraphs to the second channel. Four datasets have been used: two Small datasets (comprised of text from SETC2015)—AT (S) and Full (S)—and two Large datasets (comprised of text from PMC Collection)—AT (L) and Full (L).

The $p@k$ and $nDCG@k$ performance of the single channel TextCNN and the multichannel TextCNN on all four datasets is summarized in Table 2. Each row in the table compares all datasets on a specific metric, where the best score for each metric (the Small and Large datasets being observed separately) is in boldface. The results clearly indicate that when dealing with the same dataset, multi-channel TextCNN performs the best, which indicates that integrating captions and paragraphs indeed helps to improve the performance of classification. Also, the multi-channel TextCNN outperforms the single channel TextCNN, suggesting that the multi-channel TextCNN architecture has an advantage over the single channel TextCNN architecture. The reason for this could be that the single channel model misses some important features in the captions and related paragraphs in the convolutional and pooling layers. To be more explicit, the single channel model may be extracting insignificant features in the convolutional layer from which the max-pooling layer can take only one value in each filter. Further observation indicates that for the Small dataset, the Fulltext single channel TextCNN outperforms the AT TextCNN model (with only the $p@10$ and $p@15$ results differing from this general trend), while interestingly for the Large dataset, the AT TextCNN model outperforms the Fulltext single channel TextCNN model by a wide margin.

Now looking only at the multichannel TextCNN model when comparing the results, using data that comes from abstracts and titles only (AT) with the Fulltext data, the Small dataset shows the

Metrics		Datasets					
		AT (S)	Full (S) Single_Channel	Full (S) Multichannel	AT (L)	Full (L) Single_Channel	Full (L) Multichannel
$p@k$	$p@1$	0.76197	0.78220	**0.80512**	0.87600	0.72305	**0.87907**
	$p@3$	0.58283	0.59699	**0.62980**	0.70951	0.51016	**0.72139**
	$p@5$	0.47633	0.48901	**0.52057**	0.60532	0.41908	**0.61479**
	$p@10$	0.39641	0.38815	**0.41958**	0.51000	0.32631	**0.51793**
	$p@15$	0.37281	0.35910	**0.39587**	0.47127	0.28318	**0.48009**
$nDCG@k$	$nDCG@1$	0.76197	0.78219	**0.80512**	0.87600	0.72305	**0.87907**
	$nDCG@3$	0.62306	0.63744	**0.66982**	0.74737	0.55327	**0.75737**
	$nDCG@5$	0.53918	0.55227	**0.58409**	0.66640	0.48009	**0.67521**

Table 2: Results for TextCNN in $p@k$ and $nDCG@k$. Boldface indicates the best result on the each dataset.

Datasets	top_k	Metrics										
		EBP	*EBR*	*EBF*	*MiP*	*MaP*	*MiF*	*MaF*	HP_1	HR_1	HP_2	HR_2
AT (S)	@5	0.477	0.187	0.261	0.498	0.499	0.456	0.478	0.521	0.135	0.605	0.144
	@10	0.349	0.253	0.288	0.473	0.498	0.456	0.478	0.381	0.208	0.456	0.225
	@15	0.301	0.278	0.288	0.458	0.497	0.453	0.478	0.325	0.241	0.399	0.262
Full (S) Single Channel	@5	0.490	0.185	0.261	0.499	0.500	0.456	0.478	0.521	0.126	0.578	0.129
	@10	0.345	0.245	0.281	0.472	0.499	0.454	0.478	0.338	0.193	0.377	0.212
	@15	0.291	0.267	0.277	0.455	0.498	0.449	0.478	0.281	0.221	0.329	0.250
Full (S) Multi-channel	@5	0.521	0.200	0.282	0.503	0.498	0.460	0.478	0.539	0.161	0.608	0.176
	@10	0.377	0.270	0.309	0.478	0.495	0.460	0.477	0.386	0.237	0.442	0.265
	@15	0.325	0.298	0.310	0.463	0.494	0.457	0.477	0.326	0.268	0.380	0.304
AT (L)	@5	0.606	0.239	0.332	0.575	0.502	0.364	0.398	0.632	0.196	0.685	0.197
	@10	0.462	0.334	0.380	0.479	0.500	0.407	0.401	0.478	0.304	0.532	0.315
	@15	0.404	0.372	0.386	0.434	0.497	0.414	0.403	0.413	0.352	0.468	0.371
Full (L) Single Channel	@5	0.420	0.162	0.227	0.446	0.500	0.282	0.396	0.394	0.100	0.405	0.091
	@10	0.290	0.206	0.236	0.338	0.500	0.287	0.396	0.313	0.119	0.336	0.105
	@15	0.240	0.220	0.228	0.290	0.500	0.277	0.396	0.223	0.135	0.262	0.137
Full (L) Multi-channel	@5	0.616	0.243	0.338	0.581	0.503	0.369	0.400	0.640	0.199	0.702	0.199
	@10	0.468	0.339	0.386	0.484	0.500	0.413	0.403	0.492	0.307	0.558	0.319
	@15	0.411	0.379	0.392	0.440	0.497	0.420	0.405	0.426	0.357	0.491	0.378

Table 3: Flat and Hierarchical Measures for TextCNN on Different Datasets. top_k indicates the top k labels returned by the classifier; *EBP*, *EBR*, *EBF* are example based precision, recall, and F-score, respectively; *MiP* and *MiF* are micro precision and F-score; *MaP* and *MaF* are macro precision and F-score; HP_m and HR_m are hierarchical precision, where m denotes the maximum distance from the original label to its ancestors and descendants.

greater improvement, approximately 2-5 percentage points for each $p@k$ and each $nDCG@k$ value. The improvement for the Large dataset is typically closer to 1 percentage point. It should be noted that the Large dataset has a somewhat higher, thus more difficult to improve upon, abstract and title baseline for each metric (10 percentage points or more than the Small dataset). Another reason for this difference could be that more training examples simply gives better models, so the extra information provided by the new data sources does not have as significant an effect as the increase in the number of training examples. Comparing AT (L) to AT (S) and Full (L) to Full (S) shows an approximately 7-13 percentage point improvement for each $p@k$ and $nDCG@k$. Another possibility could be that the Small and Large datasets were generated from documents with different attributes. We have not investigated this possibility.

Table 3 reports the performance of flat and hierarchical evaluations on all datasets giving a further assessment of introducing the extra information sources. When comparing AT to Full Multichannel in the Small and Large datasets, we see an approximate .5-5 percentage point improvement in all of the measures except *MaP*. Most importantly, there is improvement in precision without a decrease in recall. The obtained results further suggest that our hypothesis that adding captions and paragraphs indeed provides valuable information in automatic MeSH indexing. Comparing *EBP*, which is the same as HP_0, with the *HP* values, an approximate 1-5 percentage point improvement in all cases at HP_1 and an approximate 6-13 percentage point improvement in all cases at HP_2 can be seen. These observations indicate that some of the predicted MeSH terms are not exactly the same as the gold standard labels, but the model has suggested MeSH terms that are in the correct branch of the MeSH term hierarchy. With this latter observation we have investigated how the predicted results correspond to the gold standard results. To do this investigation, we look at the parents above and the children below the predicted labels.

An in-depth analysis of the hierarchical evaluation on the AT (L) and Full (L) datasets are reported in Table 4. We have computed the average number of gold standard MeSH term labels in common with the predicted labels including m levels up and n levels down over all documents, where $m \in \{0, 1, 2\}$ and $n \in \{0, 1, 2\}$. Each row in the table compares model performance at a certain MeSH hierarchy, where C_m indicates the predicted label augmented with children with dis-

	top_5_predicted			top_10_predicted			top_15_predicted		
	AT (L)	Full (L) Single Channel	Full (L) Multi-channel	AT (L)	Full (L) Single Channel	Full (L) Multi-channel	AT (L)	Full (L) Single Channel	Full (L) Multi-channel
$C_0\_P_1$	0.0256	0.0021	0.0748	0.0236	0.0008	0.1107	0.0290	0.0067	0.1424
$C_1\_P_0$	0.1119	0.4154	0.4551	0.2420	0.5040	0.8545	0.2791	0.5763	1.0379
$C_0\_P_2$	0.2387	0.1988	0.4904	0.2933	0.1536	0.6983	0.3405	0.2741	0.8269
$C_2\_P_0$	0.1591	0.4860	0.6528	0.3202	0.6480	1.1166	0.3681	0.7296	1.3542
$C_2\_P_1$	0.1847	0.4881	0.7276	0.3438	0.6489	1.2267	0.3971	0.7363	1.4954
$C_1\_P_2$	0.3506	0.6142	0.9455	0.5354	0.6576	1.5528	0.6196	0.8504	1.8649

Table 4: Hierarchical Analysis on TextCNN - top_k_selected indicates the top k labels return by the classifier

tance m, and P_n is predicted label augmented with parents with distance n. As an example: $C_0\_P_1$ on AT (L) with the top 5 predicted labels indicates that if the predicted labels are augmented with their parents with distance 1, the number of common labels between true labels and predicted ones will increase on average by *0.0256* over all documents in the test set. For each top_k_predicted labels returned by the TextCNN model, comparisons within the same dataset but expanded augmentations show that the number of common MeSH terms between the gold standard and predicted ones increase in all but four cases: two instances of an increased window size for the multichannel TextCNN, the single channel TextCNN augmented with two parent labels, and the AT (L) dataset for top_5_predicted. Observing each column, this can a ten-fold increase or more. Comparing AT with Full multichannel TextCNN the increase is approximately three times when adding captions and related texts. This observation gives us confidence in concluding that the multichannel TextCNN model gives MeSH terms that are in the correct branch of the MeSH hierarchy and adding figure captions and related texts does provide valuable improvement in automatic MeSH indexing.

6 Conclusions and Future Work

This paper has presented a novel multichannel TextCNN model for MeSH term indexing. In addition, this paper has included figure and table information for the automatic MeSH indexing task. Notably, our deep learning model introduced a variety of features obtained from different parts of the document. The experimental results indicate that adding more features obtained from captions and related paragraphs indeed improve the performance of our proposed multi-channel TextCNN

model, supporting the initial hypothesis that figure and table captions as well as associated paragraphs provide valuable evidence in automatic MeSH indexing. In addition, introducing the extra information in a separate channel appears to have a positive effect compared with presenting all of the information in one channel.

We have contributed a labeled text-enhanced biomedical document dataset for the research community. It includes title, abstract, figure and table captions, and paragraphs related to figures and tables. This dataset and our software is available at https://github.com/xdwang0726/Mesh.

In the future, we first intend to extend our experiments on different optimizers, learning rates and classifiers in order to improve the performance of our models. Secondly, in this paper, we focused on finding a classifier to capture important features in the document. We manually set the number of MeSH terms returned from the model, i.e., in this work, we asked our model to return the top k predicted MeSH terms, where $k \in \{5, 10, 15\}$. We plan to improve our model by implementing a ranking system module which can be added right after the classifier. The ranking module would automatically suggest the number of labels returned for each document, which could help the indexing system to return more accurate MeSH terms. Thirdly, we also aim to develop a tool which could help human annotators locate the places in the document that has text important for determining MeSH terms in order to improve the efficiency of computer assisted human MeSH indexing.

Acknowledgements This research is partially funded by The Natural Sciences and Engineering Research Council of Canada through a Discovery Grant to Robert E. Mercer. We also acknowledge the helpful comments provided by the reviewers.

References

Alan R Aronson and François-Michel Lang. 2010. An overview of MetaMap: Historical perspective and recent advances. *Journal of the American Medical Informatics Association*, 17(3):229–236.

Alan R. Aronson, James G. Mork, Clifford W. Gay, Susanne M. Humphrey, and Willie J. Rogers. 2004. The NLM Indexing Initiative's Medical Text Indexer. *Studies in Health Technology and Informatics*, 107 Pt 1:268–272.

François Chollet et al. 2015. Keras. https://github.com/fchollet/keras.

Dina Demner-Fushman and James G. Mork. 2015. Extracting characteristics of the study subjects from full-text articles. In *Proceedings of the American Medical Informatics Association (AMIA) Annual Symposium*, pages 484–491.

Francesco Gargiulo, Stefano Silvestri, and Mario Ciampi. 2018. Deep convolution neural network for extreme multi-label text classification. In *Proceedings of the 11th International Joint Conference on Biomedical Engineering Systems and Technologies – Volume 5: AI4Health*, pages 641–650.

Antonio Jimeno-Yepes, James G. Mork, Dina Demner-Fushman, and Alan R. Aronson. 2012. A one-size-fits-all indexing method does not exist: Automatic selection based on meta-learning. *Journal of Computing Science and Engineering*, 6(2):151–160.

Antonio Jimeno-Yepes, James G. Mork, Dina Demner-Fushman, and Alan R. Aronson. 2013. Comparison and combination of several MeSH indexing approaches. In *Proceedings of the American Medical Informatics Association (AMIA) Annual Symposium*, pages 709–718.

Qiao Jin, Bhuwan Dhingra, and William W. Cohen. 2018. AttentionMeSH: Simple, effective and interpretable automatic MeSH indexer. In *Proceedings of the 2018 EMNLP Workshop BioASQ: Large-scale Biomedical Semantic Indexing and Question Answering*, pages 47–56.

Yoon Kim. 2014. Convolutional neural networks for sentence classification. In *Proceedings of the 2014 Conference on Empirical Methods in Natural Language Processing (EMNLP)*, pages 1746–1751.

Olivier Bodenreider KinWah Fung. 2007. Utilizing the umls for semantic mapping between terminologies. In *Proceedings of the American Medical Informatics Association (AMIA) Annual Symposium*, pages 266–270.

Aris Kosmopoulos, Ioannis Partalas, Eric Gaussier, Georgios Paliouras, and Ion Androutsopoulos. 2015. Evaluation measures for hierarchical classification: A unified view and novel approaches. *Data Mining and Knowledge Discovery*, 29(3):820–865.

Jimmy Lin and W. John Wilbur. 2007. Pubmed related articles: A probabilistic topic-based model for content similarity. *BMC Bioinformatics*, 8(1):423.

Jingzhou Liu, Wei-Cheng Chang, Yuexin Wu, and Yiming Yang. 2017. Deep learning for extreme multi-label text classification. In *Proceedings of the 40th International ACM SIGIR Conference on Research and Development in Information Retrieval*, pages 115–124.

Ke Liu, Shengwen Peng, Junqiu Wu, ChengXiang Zhai, Hiroshi Mamitsuka, and Shanfeng Zhu. 2015. MeSHLabeler: Improving the accuracy of large-scale mesh indexing by integrating diverse evidence. *Bioinformatics*, 31(12):i339–i347.

James G. Mork, Antonio Jimeno-Yepes, and Alan R. Aronson. 2013. The NLM Medical Text Indexer system for indexing biomedical literature. In *Proceedings of the first Workshop on Bio-Medical Semantic Indexing and Question Answering (BioASQ)*. CEUR-WS.org, online http://CEUR-WS.org/Vol-1094/bioasq2013_submission_3.pdf.

Ioannis Pavlopoulos, Aris Kosmopoulos, and Ion Androutsopoulos. 2019. Continuous space word vectors obtained by applying word2vec to abstracts of biomedical articles. Retrieved from http://bioasq.lip6.fr/info/BioASQword2vec/.

Shengwen Peng, Ronghui You, Hongning Wang, ChengXiang Zhai, Hiroshi Mamitsuka, and Shanfeng Zhu. 2016. Deepmesh: deep semantic representation for improving large-scale mesh indexing. *Bioinformatics*, 32(12):i70–i79.

Anthony Rios and Ramakanth Kavuluru. 2015. Convolutional neural networks for biomedical text classification: Application in indexing biomedical articles. In *Proceedings of the 6th ACM Conference on Bioinformatics, Computational Biology and Health Informatics*, pages 258–267.

Mike Schuster and Kuldip K. Paliwal. 1997. Bidirectional recurrent neural networks. *IEEE Transactions on Signal Processing*, 45(11):2673–2681.

Lei Tang, Suju Rajan, and Vijay K. Narayanan. 2009. Large scale multi-label classification via MetaLabeler. In *Proceedings of the 18th International World Wide Web Conference (WWW)*, pages 211–220.

Grigorios Tsoumakas, Ioannis Katakis, and Ioannis P. Vlahavas. 2010. Mining multi-label data. In *Data Mining and Knowledge Discovery Handbook (2nd ed.)*, pages 667–685. Springer, Boston, MA.

Xindi Wang. 2019. Incorporating figure captions and descriptive text in mesh term indexing: A deep learning approach. Master's thesis, The University of Western Ontario.

Chengxiang Zhai, Hiroshi Mamitsuka, Junqiu Wu, Ke Liu, Shanfeng Zhu, and Shengwen Peng. 2015. MeSHLabeler: Improving the accuracy of large-scale MeSH indexing by integrating diverse evidence. *Bioinformatics*, 31(12):i339–i347.

M. Zhang and Z. Zhou. 2014. A review on multi-label learning algorithms. *IEEE Transactions on Knowledge and Data Engineering*, 26(8):1819–1837.

BioRelEx 1.0: Biological Relation Extraction Benchmark

Hrant Khachatrian[1,2], Lilit Nersisyan[3], Karen Hambardzumyan[1,2], Tigran Galstyan[1,2],
Anna Hakobyan[3], Arsen Arakelyan[3], Andrey Rzhetsky[4], and Aram Galstyan[5]

[1]YerevaNN, Yerevan, Armenia
[2]Department of Informatics and Applied Mathematics, Yerevan State University, Yerevan, Armenia
[3]Bioinformatics Group, Institute of Molecular Biology, NAS RA, Yerevan, Armenia
[4]Institute for Genomics and Systems Biology, Departments of Medicine and
Human Genetics, University of Chicago, Chicago, Illinois, USA
[5]USC Information Sciences Institute, Marina Del Rey, California, USA
hrant@yerevann.com

Abstract

Automatic extraction of relations and interactions between biological entities from scientific literature remains an extremely challenging problem in biomedical information extraction and natural language processing in general. One of the reasons for slow progress is the relative scarcity of standardized and publicly available benchmarks. In this paper we introduce BioRelEx, a new dataset of fully annotated sentences from biomedical literature that capture *binding* interactions between proteins and/or biomolecules. To foster reproducible research on the interaction extraction task, we define a precise and transparent evaluation process, tools for error analysis and significance tests. Finally, we conduct extensive experiments to evaluate several baselines, including SciIE, a recently introduced neural multi-task architecture that has demonstrated state-of-the-art performance on several tasks.

1 Introduction

Biological interaction databases capture a small portion of knowledge depicted in biomedical papers, due to time consuming nature of manual information extraction. As experimental methodologies to identify such interactions tend to increase in scale and throughput, the problem stands to rapidly update these databases for relevant applications (Oughtred et al., 2018). The long-term aim of our efforts is to provide bases for filling this gap automatically.

Despite significant progress in recent years, extracting relationships and interactions between different biological entities is still an extremely chellenging problem. Some of those challenges are due to objective reasons such as lack of very large annotated datasets for training complex models, or wide variability in biomedical literature which can lead to domain mismatch and poor generalization. Another important challenge, which is the main focus of the present paper, is the scarcity of publicly available datasets. Indeed, with despite some notable exceptions (Kim et al., 2003; Dogan et al., 2017), there is a relative lack of adequate, high-quality benchmark datasets which would facilitate reproducible research and allow for robust comparative evaluation of existing approaches.

Here we have processed biological texts to annotate biological entities and interaction pairs. In contrast to other related databases, our efforts were focused on delineation of biological entities from experimental ones, and on distinguishing between indirect regulatory interactions and direct physical interactions. Furthermore, we have performed grounding via cross-reference of annotated entities with external databases. This allows for merging interactions from different sources into a single network of biomolecular interactions.

The main contributions of this work are:

1. We publish a dataset of 2010 sentences with complete annotations of biological entities and binding interactions between the entities,

2. We propose a benchmark task with a well-defined evaluation system, which follows the best practices of machine learning research,

3. We perform extensive evaluation of several competing methods on the dataset and report the results.

Proceedings of the BioNLP 2019 workshop, pages 176–190
Florence, Italy, August 1, 2019. ©2019 Association for Computational Linguistics

2   Related work

In this section we briefly summarize prior work on relation extraction from unstructured text.

Since 2009, NIST has organized Knowledge Base Population evaluations as part of Text Analysis Conferences (TAC KPB). Thousands of sentences from newswire and informal web pages were annotated for training and evaluation purposes (Getman et al., 2018). In 2017, a team from Stanford released TACRED (Zhang et al., 2017), a dataset of 106 264 sentences with 42 relation types. The relations are mainly between people, places and organizations.

A large number of papers focused on biological relation extraction. (Bunescu et al., 2005) built a manually annotated corpus of 225 abstracts to evaluate various extraction methods. This dataset is referred as AIMed in subsequent papers. Later, (Pyysalo et al., 2007) developed a smaller dataset called BioInfer with more detailed annotations. In particular, the authors developed large ontologies for biological entities and relations between them and attempted to classify each entity and relation according to these ontologies. The small number of sentences and interactions is 1100 and 2662, respectively, so for many types of relations there were too few samples. Because of that, almost all subsequent papers that applied machine learning techniques on BioInfer discarded the detailed labels and used it as a dataset of binary relations. In 2008, (Pyysalo et al., 2008a) presented a detailed comparison of AIMed, BioInfer and three other datasets (IEPA, HPRD50 and LLL) and found significant differences in the data collection and evaluation procedures.

In (Pyysalo et al., 2008b), the authors concluded that the results on the five datasets reported in different papers are incomparable and suggested to unify the datasets in a common format with a precise evaluation procedure. This proved to be successful as a large number of subsequent papers use the unified versions of the datasets. On the other hand, these datasets are currently used only for binary relation classification, as the unified versions keep the lowest common level of annotations (only entity locations and binary labels between the pairs). It means that the models trained on these datasets cannot be used for end-to-end relation extraction from text. Moreover, many recent papers violate evaluation strategies (e.g. perform cross-validation on splits that do not respect doc-

ument boundaries) and report unrealistically high scores (Hsieh et al., 2017; Ahmed et al., 2019).

One of the highest quality datasets is developed as part of GENIA project (Kim et al., 2003). It involves annotations of entities, syntactic features, wide variety of events, including around 2500 binding interactions (Thompson et al., 2017). GENIA does not have a training/test split, but various subsets of it have been used as training and test sets of BioNLP Shared Tasks in 2009 (Kim et al., 2009), 2011 (Kim et al., 2011) and 2013 (Nédellec et al., 2013). Several protein-protein interaction (PPI) datasets appeared in BioCreative series of shared tasks. There was a track on PPI extraction in BioCreative II, including a binary relation extraction subtask from full texts and another subtask for finding evidence sentences for the given interaction (Krallinger et al., 2008). BioCreative V Track 4 included a subtask on extraction of more complex data structures called Biological Expression Language (BEL) statements (Rinaldi et al., 2016).

Other biological relation extraction datasets include ADE (Gurulingappa et al., 2012), a dataset of adverse drug effects; BB3 (Deléger et al., 2016), a dataset of relations between bacteria and their habitats, which was used in BioNLP Shared Task 2016; SeeDev, a dataset of sentences about seed development of plants; AGAC, a dataset on gene mutations and diseases. The latter three datasets are included in BioNLP 2019 Shared Tasks. Precision Medicine Track of BioCreative VI (Dogan et al., 2017) introduced a large dataset of protein-protein interactions that are affected by mutations.

SemEval 2017 Task 10 (Augenstein et al., 2017) was about extracting relations from scientific paper abstracts (physics, computer science and materials science). SemEval 2018 Task 7 focused on sentences from computational linguistics papers. SciERC (Luan et al., 2018) is a dataset consisting of 500 research paper abstracts from major AI conferences with annotated entities, coreference links and relations between entities.

3   Dataset description

3.1   The choice of sentences

We have annotated 2010 sentences for binding interactions between biological entities. Those sentences came from a much larger set of 40,000 sentences that were automatically extracted from various biomedical journals and underwent minimal manual post-processing (Rzhetsky et. al., 2019). While the original set contained numerous interaction types, here our focus is on binding interactions only. The text of the sentences are mostly copied from the journal websites and can include uncommon Unicode symbols. In rare cases we had to copy the sentences from PDF versions of the papers and manually fix incorrect characters.

As stated above, the current version of the dataset is focused on binding interactions. All sentences in the dataset contain one of the following words: "bind", "binds", "binding", "bound". This will potentially limit the applicability of the models trained on this dataset on other sentences that contain information about binding interactions.

3.2 Entities

3.2.1 Entity definition

Every annotated entity is a continuous span of characters in the sentence surrounded by non-alphanumeric symbols (can include spaces, hyphens etc.).

Tokenization of biomedical texts can be a challenging task. To ensure consistency, we have verified that all annotated entities in the dataset are surrounded by the symbols described in Table 3 of Appendix A.3. Note that all these symbols can also appear inside an entity name.

3.2.2 Entity types

We have annotated 33 types of entities. For classification of entities we were governed both, by biological function and by chemical structure. More specifically, we distinguish between biological and experimental entities. For example, if the sentence refers to an oligonucleotide in an experiment, we do not annotate it as DNA, but as an experimental-construct. Furthermore, we define main organic entity types as protein, protein-family, protein-complex, DNA and RNA, while refer to the rest of organic compounds as chemicals. The complete list of entity types is listed in Appendix A.4.

These decisions were motivated by two main reasons: (a) only biological entities should be annotated and cross-referenced in order to arrive at biologically meaningful interaction networks; (b) a higher level of annotation that disregards details (e.g. chemicals) significantly reduces annotation resources with no loss to our targetted aim. This

contrasts to the Genia ontology, where entity annotation was only based on chemical structure of substances (Thompson et al., 2017).

Note that while the majority of entities are annotated to a single type, two entities with the same name may be annotated to different types (e.g. protein or protein-family) depending on the context, and sometimes these cases may co-occur in the same sentence (e.g. protein and gene (`1.0.train.166`)).

3.2.3 Coreference

Pairs of entities may be in *is_a* or *part_of* relationships. We have undertaken two approaches to mark such relationships for unambigous placement of entities when merging relations from one or many sentences.

3.2.4 Links between entities

1. Sometimes the same entity appears in multiple forms in the sentence. We annotate them with a "synonym" link. Sometimes, one of the forms is just an acronym for another form, in which case we use "abbreviation" link.

2. Biologically nested entities are linked with a *part_of* link. For example, protein-domains and protein-regions are part of proteins, while protein subunits are part of complexes. These links correspond to the substrate chemical structure ontology presented in Genia dataset (Thompson et al., 2017).

3.2.5 Grounding

Entities of types gene, protein, protein-family and chemical have been cross-referenced with external database identifiers. The aim of grounding is to introduce unique naming/identification of entities. This is particularly useful for unambiguous identification of entities in the process of merging relations derived from different sentences into a single network.

Notably, as a side effect, the process of grounding increased the quality of entity annotation for the specified entity types.

3.2.6 Ambiguities

Entity annotation is not a straightforward task, as entities usually appear in a variety of grammatical and biological forms. Therefore, we have developed the following guidelines for standardized annotations. Formation of these guidelines was a result of iterative annotations followed by resolution of inter-annotator conflicts.

1. *Entity modifications*

   Sometimes the text contains an entity which is a mutated form of another entity, or it is an entity in an unusual state. In these cases we tag the entity with "mutant" and/or "state" labels (Appendix A.1, example 1).

2. *Spanned and nested entities*

   If an entity contains multiple tokens, those may be separated by other words in the text, or may themselves contain nested simpler entities. In cases when the same token is shared between multiple complex entities, we annotate the shared tokens only as part of the first entity (Appendix A.1, example 2). A better solution to these cases would be to annotate the shared tokens in all the entities that they are part of and use a text-span notation to mark those cases. However, considering the small number of such cases, we didn't find this worthwhile. Sometimes a complex entity name contains a name of another entity. We annotate both, and both can appear in interactions. In extreme cases, the second entity can be a single digit. In contrast to our approach, entity recognizer systems that do not support nested entities are not be able to find these cases. In evaluation, we have a separate score that reports performance on the nested entities (Appendix A.1, examples 3-5).

3. *A/B syntax*

   In many cases A/B means a complex of the proteins A and B. In other cases it refers to separate proteins A and B, and the interaction with A/B means interactions with both of them. In both cases, we annotate A and B as individual entities. In case of complexes, we also annotate A/B as a complex. If A/B is involved in an interaction with a protein C, we annotate an interaction between A/B and C only if A/B is a complex. If A/B is not a complex, we annotate two interactions between A and C, and B and C. (Appendix A.1, example 6)

4. *Hidden entity names and implicit coreferences*

Sometimes the sentence is about an entity which is not explicitly mentioned, but there are words that refer to it. We do not annotate these words as entities and do not annotate corresponding interactions (Appendix A.1, example 7).

3.3 Interactions

We annotate binding interactions between several types of entities.

3.3.1 Interaction types

We use three labels: 1 if the interaction exists, 0 for speculations (if the sentence does not conclude whether the interaction exists or not), and -1 for negations (if the sentence concludes that there is no binding interaction between the entities).

We conclude that an interaction exists (1) if we find explicit triggers describing direct physical interactions, such as *A binds/ associates with/ interacts with /recuits /phosphorylates B*, and their grammatical varieties.

Speculative interactions (Appendix A.2, examples 1-2) arise either due to lack of experimental evidence or due to the sentence not reaching the conlusion yet. We mark such cases with a "hypothesis" label. Other cases may be sentences that are actually titles of the sections or even the papers. In practice, title of the paper might be extracted both from the title section of the paper and from the reference sections of other papers. We tag the sentences extracted from paper, section or figure titles by "title" label (Appendix A.2, examples 3-4).

3.3.2 Ambiguities

1. *Entity polymorphisms*

   When an entity participating in an interaction appears in multiple forms in the sentence (e.g. plural forms, synonyms, etc.), we annotate the one which is the most obvious from the sentence. In evaluation, we do not penalize the predictions with another form of the same entity (Appendix A.2, example 5).

2. *Static interactions: protein complexes and domains*

   Static or implicit interactions refer to cases where an interaction is inferred from the context, but is not mentioned with any explicit trigger.

When the sentence contains a complex of two or more proteins, and the components of the complex are present in the sentence, we annotate a binding interaction between them and tag it with a "complex" label. In rare cases, the same sentence contains another explicit mention of the interaction between two proteins. In this cases we do not tag the interaction with "complex" label (Appendix A.2, examples 6-7). In evaluation, we additionally report the performance on such implicit binding interactions inside complexes.

Sometimes we annotate a (positive) binding interaction between entities A and B, where B is a region (*part_of*) of another entity C. The most common scenario is when B is a protein domain and A and C are proteins. In this case, we annotate another interaction between A and C and tag it with an "implicit" label. The full list of entity types that can get involved in similar implicit interactions is presented in Appendix A.4. We have automatically verified that all such implicit interactions are annotated (Appendix A.2, examples 8-9).

3. *Self interactions*

There are cases when an entity binds to itself, especially when the entity is a protein-family and the binding can refer to different members of the same family (Appendix A.2, example 10).

In rare cases, the sentence talks about homodimers or oligomerization, which implies that there is a protein which binds to itself. We tag these cases with an "implicit" label (Appendix A.2, example 11-12).

4. *Interactions with implicit entities*

Sometimes the sentences contain interactions with entities without naming them. We exclude these interactions from the dataset (A.2, example 13).

3.4 Dataset statistics

The lengths of sentences vary from 3 to 138. The median length is 29, the mean is around 30. 95% of all sentences have less than 50. The average number of entity clusters per sentence is 3.92, while the average number of entity mentions per sentence is 4.91. On average, there are 1.61 interaction per sentence.

We used Cytoscape (Shannon et al., 2003) to construct a graph based on positive interactions annotated from our dataset. It has 2248 nodes (entities) and 3235 edges (interactions) (see Figure 2 in Appendix A.5). The graph had a large connected component, containing 65% (1475) of nodes and 81% (2635) of edges. Many interactions were annotated multiple times, with 67% (2177) of unique interactions, and up to 11 duplications per entity pair. The graph showed small-world properties, with average shortest path between any pairs of nodes being 5, and with very few hub nodes. Degrees range from 1 to 83 with median 1.

3.5 Comparison with other datasets

Table 1 compares BioRelEx 1.0 with the popular related datasets. The original version of AIMed has similar number of sentences to BioRelEx, but the number of annotated relations is significantly lower due to different annotation guidelines and choice of sentences. BioInfer contains fewer sentences with a lot more detailed annotations, which is not suitable for the current machine learning techniques, hence most of the models designed for BioInfer simply ignore the details of annotations. Both datasets do not have corresponding well-defined benchmarks. The five datasets in a unified format from (Pyysalo et al., 2008a) suit better for machine learning research, but they are limited to relation classification tasks.

The dataset for BioCreative VI Precision Medicine Track has 6.5 times more sentences than BioRelEx 1.0, but has two times less relations, as it is focused on a more rare kind of interactions.

GENIA corpus is the closest in spirit to ours. It has more detailed annotations and covers more relation types. As a result, the density of binding interactions in GENIA is much lower (only 2448 binding interactions in 9372 sentences). Also, there is a slight difference in the goals of GENIA and BioRelEx. GENIA is best suited for functional annotation and biomedical search optimization. We however, had a different aim in mind - to retrieve interactions in a way to make them useful for interaction network generation. This difference affected the way we have designed the annotation guidelines, as described in the previous subsections. Because of these differences we did not use the ontologies developed in GENIA.

In contrast to all mentioned datasets, BioRelEx includes grounding information for most of the labeled entities.

4 Benchmark

We propose a relation extraction benchmark on top of our dataset. The task is to take the raw text input and produce clusters of entity mentions along with binding interactions between the clusters. We define two main evaluation metrics, one for entity recognition and one for relation extraction. In addition to these, we define several other evaluation metrics that can be helpful in error analysis.

The main evaluation metrics are:

- Entity recognition performance in terms of micro-averaged precision, recall and F-score. In this metric we count each occurence of an entity as a separate item, and measure if the system could find all mentions in the sentence.

- Relation extraction performance in terms of micro-averaged precision, recall and F-score. Relation extraction is measured between entity clusters. Each cluster can be represented by multiple entity names in the sentence. We consider a relation between two entity clusters correctly detected, if the system predicts a relation between all pairs of entity names from the two clusters.

Two common problems of experimental setups used in relation extraction literature, as described in (Pyysalo et al., 2008b), are the inconsistent training/dev/test splits and hyperparameter tuning on the test set. To prevent these issues, we enforce a precise evaluation procedure. Following (Luan et al., 2018), we randomly split the dataset into training/dev/test sets with 70%/10%/20% ratio. The training, dev and test sets contain 1405, 201 and 404 sentences, respectively. Training and dev parts are publicly available as JSON files. We will set up a publicly available evaluation server to ensure having a truly blind test set. Additionally, we have released the evaluation script used in the server[1]. We encourage everyone to use the dev set for model selection only.

---

[1] The dataset files along with the description of the JSON structure and the evaluation scripts are available at `https://github.com/YerevaNN/BioRelEx/`

4.1 Error analysis

To help with error analysis, we propose few more evaluation metrics.

Entity names: Each entity name can be mentioned multiple times in the sentence. If a model finds only one of the mentions, it is considered as a match for this score. This metric helps to verify the consistency of entity recognition in different parts of the sentence.

Flat entities: Many relation extraction systems do not support recognition of nested entities. This score acts as if there are no flat entities. More precisely, we do two modifications before calculating precision and recall:

1. If an entity mention was found by a system, we remove all entity mentions that intersect with that one from the prediction and ground truth.

2. For the remaining entity mentions we keep only the ones which do not contain another mention (e.g., only shortest mentions).

Entity coreferences: Sometimes, several entity names refer to the same actual entity. For each sentence we construct a graph, where entity names are the vertices, and two vertices are joined with an edge if they refer to the same underlying entity (are synonyms or abbreviations). This graph consists of one or more connected components, where each component is a clique and refers to a single unique entity. We measure precision, recall and f-score of the edges of the abovementioned graph. This metric helps to measure the impact of synonym or abbreviation detection.

Relation extraction (any): This metric measures relation extraction in a weaker form. We consider a relation between two entity clusters correctly detected, if the system predicts a relation between any pair of entity names from the two clusters.

Relation extraction (positive): Annotated relations have one of the three labels: 1 if the sentence confirms there is an interaction, -1 if the sentence confirms there is no interaction, and 0 if the sentence is inconclusive. We report scores that do not penalize if relations with labels 0 or -1 are not detected.

Relation extraction (non-implicit): Some of the interactions are marked as "implicit" by the annotators. These are the interactions which can be

	Task	Split	Relation Types	Sentences	Entities	Relations
AIMed (Bunescu et al., 2005)	Relation extraction	No	No	1978	4141	816
BioInfer (Pyysalo et al., 2007)	Relation extraction	No	Ontology	1100	6349	2662
AIMed* (Bunescu et al., 2005)	Classification	Yes	No	1955	4301	978
BioInfer* (Pyysalo et al., 2007)	Classification	Yes	No	1100	6349	2662
HPRD50* (Fundel et al., 2006)	Classification	Yes	No	145	406	160
IEPA* (Ding et al., 2001)	Classification	Yes	No	486	1118	340
LLL* (Nédellec, 2005)	Classification	Yes	No	77	239	162
BioC V BEL (Rinaldi et al., 2016)	BEL extraction	Yes	Yes	6353	N/A	11066
BioC VI PM (Dogan et al., 2017)	Relation Extraction	Yes	No	12751	10325	1629
BioNLP GE (Kim et al., 2003)	Classification+Coref	Yes	Ontology	9372	93293	36114
BioRelEx 1.0	Relation Extraction	Yes	Only binding	2010	9871	3235

Table 1: Comparison of BioRelEx 1.0 with the most popular protein-protein interaction datasets. The ones mentioned by asterisk are the unified versions from (Pyysalo et al., 2008a)

hard to detect, as they require relatively complex reasoning. We report scores that do not penalize if an implicit interaction is not detected.

All our evaluation scripts use test set bootstrapping to compute confidence intervals for the scores and to test whether the difference between two models is significant.

5 Experiments

5.1 Baselines

We provide several baselines for the benchmark described in the previous section. First, we report several trivial baselines with gold standard entities, as well as using an off-the-shelf named entity recognizer. Next, we evaluate REACH, an end-to-end biological relation extraction system, which does not require re-training. Finally, we train SciIE, an end-to-end neural network which is known to produce state-of-the-art results on similar tasks.

5.1.1 Trivial baselines

Following (Pyysalo et al., 2008a), we report scores produced by co-occurence baselines. First, we take all gold entities from the dataset and assume that there are binding interactions between all of them. This baseline gives a perfect recall and is called "Co-occur (gold)". Then, we pass the sentences to a biomedical named entity recognition system `SciSpacy` (Neumann et al., 2019) (trained on JNLPBA corpus) and assume that there are binding interactions between all pairs. This baseline is called "Co-occur (SciSpacy)".

5.1.2 REACH

REACH (Valenzuela-Escárcega et al., 2018) is a rule-based relation extraction system The authors

host a web-based service for extracting relations from biomedical texts. We did not train or tune the system. The technical details on how we evaluated REACH system on our dataset is presented in Appendix A.6.

5.1.3 SciIE model

SciIE (Luan et al., 2018) is a complex multi-task neural architecture developed by University of Washington for relation extraction from computer science paper abstracts. The model produces candidate spans of tokens, and then attempts to jointly predict entities, coreferences and relations between entities based on the spans. SciIE supports multi-word and nested entities. The technical details about adapting our data for SciIE architecture are available in Appendix A.7.

5.2 Results

The results of the four baselines on the test set of BioRelEx 1.0 are presented in Table 2. If the entity names are known, getting 35% F-score for relation extraction is trivial. Recall for relation extraction of the co-occurrence baseline is less than 100% because of the self interactions in the dataset. On the other hand, entity recognition is not easy. SciSpacy's named entity recognizer trained on the famous JNLPBA dataset (derived from GENIA corpus) gets 67% precision and less than 53% recall. Part of the low recall is because SciSpacy's NER cannot produce nested entities. The co-occurrence baseline with these entities gets less than 20% F-score for relation extraction.

SciIE model has a large number of hyperparameters. We kept the values mentioned in the official repository for SciERC dataset with one exception: we have changed `max_arg_width` to 5, as

		Entity Recognition	Relation Extraction	Co-occur (SciSpacy)	Co-occur (Gold)	REACH	SciIE
Co-occur (SciSpacy)	P	$67.3 \pm 1.4 \,(64.6 - 69.8)$	$12.6 \pm 1.3 \,(10.3 - 15.2)$		0.0%	0.2%	0.0%
	R	$52.6 \pm 1.5 \,(49.8 - 55.5)$	$45.1 \pm 3.7 \,(38.5 - 52.3)$				
	F_1	$59.0 \pm 1.3 \,(56.4 - 61.6)$	$19.6 \pm 1.9 \,(16.3 - 23.5)$				
Co-occur (Gold)	P	$100.0 \pm 0.0 \,(100 - 100)$	$21.5 \pm 1.3 \,(19.2 - 24.2)$	100.0%		64.8%	0.0%
	R	$100.0 \pm 0.0 \,(100 - 100)$	$99.2 \pm 0.5 \,(98.1 - 99.9)$				
	F_1	$100.0 \pm 0.0 \,(100 - 100)$	$35.3 \pm 1.8 \,(32.2 - 38.9)$				
REACH	P	$70.6 \pm 1.4 \,(68.1 - 73.1)$	$63.2 \pm 3.9 \,(55.6 - 70.7)$	99.8%	35.2%		0.0%
	R	$65.9 \pm 1.3 \,(63.4 - 68.3)$	$23.2 \pm 2.3 \,(19.1 - 27.6)$				
	F_1	$68.2 \pm 1.1 \,(65.9 - 70.3)$	$33.9 \pm 2.8 \,(28.6 - 39.2)$				
SciIE	P	$87.7 \pm 1.0 \,(85.8 - 89.6)$	$53.2 \pm 2.3 \,(48.9 - 57.9)$	100.0%	100.0%	100.0%	
	R	$63.3 \pm 1.6 \,(60.2 - 66.3)$	$47.4 \pm 3.1 \,(41.1 - 53.1)$				
	F_1	$73.5 \pm 1.3 \,(71.0 - 75.8)$	$50.1 \pm 2.3 \,(45.5 - 54.3)$				

Table 2: Results of the four baselines on the test set of BioRelEx 1.0. We report precision (P), recall (R) and F-score (F_1) for entity recognition and relation extraction. Every metric is calculated $n = 1000$ times by bootstrapping on the test set. The table shows mean, standard deviation and 95% confidence interval of 1000 runs. The right part of the table shows how often one baseline beats the other ones in 1000 evaluations according to F-score of relation extraction. We consider the difference between two models to be significant if one performs better than the other in 95% of cases.

there are very few entities with more than five tokens. We did several experiments with different weights for the NER and coreference branches of the model and picked the combination which performed best on the dev set of our dataset.

SciIE model significantly outperforms REACH system on the F-score of relation extraction: 50.1% vs 33.9%. On the other hand, REACH has a better precision for relation extraction. The difference between REACH and co-occurrence baseline with gold entities is not significant.

5.3 Error analysis

To measure the impact of nested entities on entity prediction performance we calculate **Flat entities** metric and compare it with the main entity recognition metrics. Recall jumps from 65.8% to 71.2% for REACH and from 63.3% to 68.9% for SciIE.

Our error analysis tools measure coreference detection performance. Both REACH and SciIE baselines do not output coreferences. SciIE is capable of producing coreference clusters, but the best performance on the dev set.

The relaxed versions of relation extraction evaluation do not change the results significantly. In particular, **Relation extraction (any)** metric gives 35.5% (vs. 33.9%) for REACH and 51.0% (vs. 50.1%) for SciIE.

To understand the impact of sentence lengths on the performance of the models we calculate our main metrics on the top and bottom halves of the list of sentences from dev set sorted by length.

For REACH, F-score on longer sentences is worse by 1.2 and 0.8 percentage points for entity recognition and relation extraction, respectively. For SciIE, the differences are much larger, 7.4 and 9.9 percentage points respectively.

5.4 Qualitative analysis

To understand how the SciIE baseline model performs in real-world settings, we did the following experiment. We took a figure from a paper that describes MAPK-ERK signaling pathway. Figure 1a shows the schematic representation of the pathway, as described in the paper (Dantonio et al., 2018). The caption of the figure in the original paper reads: "In regular conditions, ligands such as growth factors or mitogens bind to the RTK, which is activated by autophosphorylation. Phosphotyrosine residues recruit adaptor protein Grb2 and Sos, promoting Ras:GTP association. Activated by GAPs such as NF1, Ras hydrolyzes GTP and activates Raf, the first effector kinase in the MAPK pathway. Raf then phosphorylates MEK, which in turn phosphorylates ERK. p-ERK activates cytoplasmic and nuclear substrates".

Figure 1b shows the network extracted by our SciIE model from the original caption with no modifications. The original scheme is depicted as an underlay with light gray shades. The true positive entities and interactions are highlighted in red.

Our dataset is biased towards sentences with the verb "bind". To see how it affects the performance of our model, we have replaced three triggers in

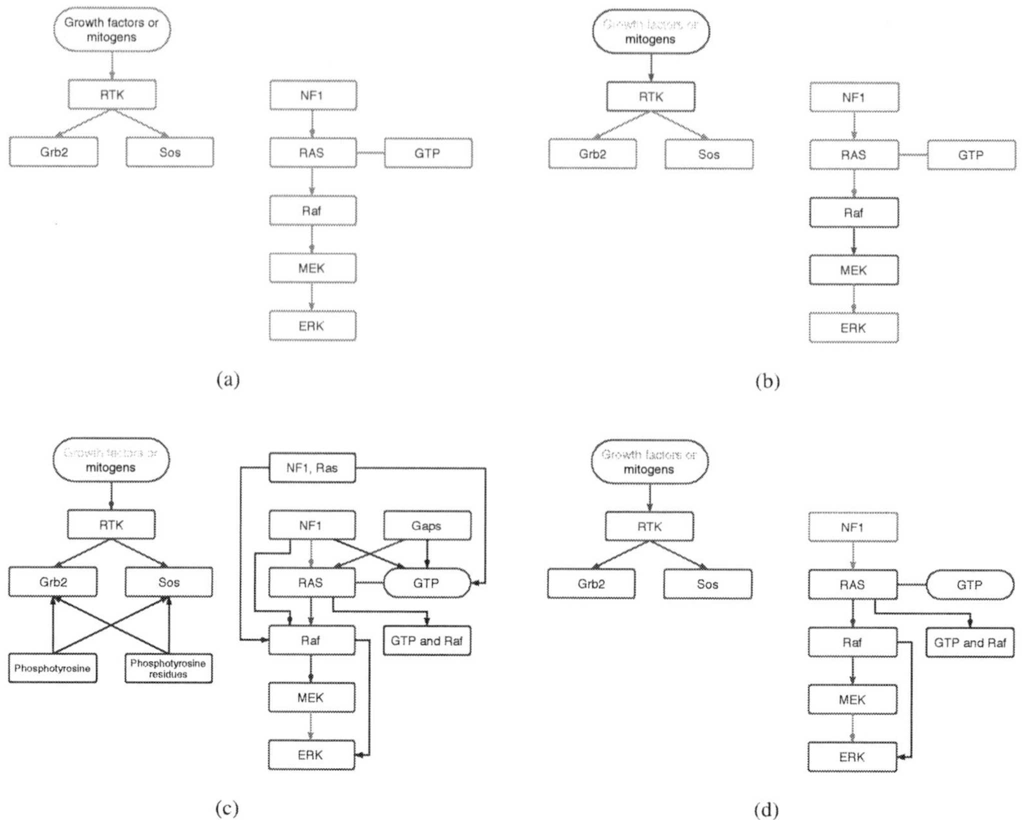

Figure 1: A network extracted by SciIE model. Refer to Section 5.4 for the details.

the original caption with "binding". The resulting network produced by SciIE is presented in Figure 1c. True positives are highlighted with red, while false positives - with blue. Note that many false entities, such as "NF1, Ras" are extracted in this case.

Finally, we removed the sentence containing the misleading "NF1", and replaced the "which in turn" coreference with "MEK". Additionally, the "phospohorylated residues" were replaced by "phosphorylated RTK" to hint the model that these residues belong to RTK. The network produced by SciIE on this version is shown in Figure 1d. The full captions used in these experiments are shown in Appendix A.8.

The results demonstrate that our SciIE baseline works much better when the interactions are expressed with the verb "bind". Additionally, we see that the lack of coreference resolution between sentences severely limits the applications of this model.

6 Conclusion

In this paper we have introduced BioRelEx 1.0, a manually annotated corpus for interaction extraction from biomedical literature. We have developed detailed guidelines for annotating binding interactions between various biological entities. The dataset is publicly available at https://github.com/YerevaNN/BioRelEx/. Based on the dataset we have designed a benchmark and evaluated several baselines on it. Finally, we have demonstrated the quality of a neural relation extraction model trained on the dataset in a real-world setting. We hope this benchmark will help to develop more accurate methods for relation extraction from unstructured text.

7 Acknowledgments

We would like to thank Sahil Garg and Martin Mirakyan for their help in the project. We would like to thank NVIDIA for donating Titan V GPUs used in the experiments.

References

Mahtab Ahmed, Jumayel Islam, Muhammad Rifayat Samee, and Robert E Mercer. 2019. Identifying protein-protein interaction using tree lstm and structured attention. In *2019 IEEE 13th International Conference on Semantic Computing (ICSC)*, pages 224–231. IEEE.

Andrey Rzhetsky et. al. 2019. *in preparation*.

Isabelle Augenstein, Mrinal Das, Sebastian Riedel, Lakshmi Vikraman, and Andrew McCallum. 2017. Semeval 2017 task 10: Scienceie - extracting keyphrases and relations from scientific publications. *CoRR*, abs/1704.02853.

Razvan Bunescu, Ruifang Ge, Rohit J Kate, Edward M Marcotte, Raymond J Mooney, Arun K Ramani, and Yuk Wah Wong. 2005. Comparative experiments on learning information extractors for proteins and their interactions. *Artificial intelligence in medicine*, 33(2):139–155.

Paola M. Dantonio, Marianne O. Klein, Maria Renata V.B. Freire, Camila N. Araujo, Ana Carolina Chiacetti, and Ricardo G. Correa. 2018. Exploring major signaling cascades in melanomagenesis: a rationale route for targetted skin cancer therapy. *Bioscience Reports*, 38(5).

Louise Deléger, Robert Bossy, Estelle Chaix, Mouhamadou Ba, Arnaud Ferré, Philippe Bessières, and Claire Nédellec. 2016. Overview of the bacteria biotope task at bionlp shared task 2016. In *Proceedings of the 4th BioNLP Shared Task Workshop*, pages 12–22, Berlin, Germany. Association for Computational Linguistics.

Jing Ding, Daniel Berleant, Dan Nettleton, and Eve Wurtele. 2001. Mining medline: abstracts, sentences, or phrases? In *Biocomputing 2002*, pages 326–337. World Scientific.

Rezarta Islamaj Dogan, Andrew Chatr-aryamontri, Sun Kim, Chih-Hsuan Wei, Yifan Peng, Donald Comeau, and Zhiyong Lu. 2017. Biocreative vi precision medicine track: creating a training corpus for mining protein-protein interactions affected by mutations. In *BioNLP 2017*, pages 171–175.

Katrin Fundel, Robert Küffner, and Ralf Zimmer. 2006. Relexrelation extraction using dependency parse trees. *Bioinformatics*, 23(3):365–371.

Jeremy Getman, Joe Ellis, Stephanie Strassel, Zhiyi Song, and Jennifer Tracey. 2018. Laying the groundwork for knowledge base population: Nine years of linguistic resources for tac kbp. In *Proceedings of the Eleventh International Conference on Language Resources and Evaluation (LREC-2018)*.

Harsha Gurulingappa, Abdul Mateen Rajput, Angus Roberts, Juliane Fluck, Martin Hofmann-Apitius, and Luca Toldo. 2012. Development of a benchmark corpus to support the automatic extraction of drug-related adverse effects from medical case reports. *Journal of Biomedical Informatics*, 45(5):885–892.

Yu Lun Hsieh, Yung-Chun Chang, Nai Wen Chang, and Wen Lian Hsu. 2017. Identifying protein-protein interactions in biomedical literature using recurrent neural networks with long short-term memory. In *Proceedings of the Eighth International Joint Conference on Natural Language Processing*, pages 240–245. Asian Federation of Natural Language Processing.

J.-D. Kim, T. Ohta, Y. Tateisi, and J. Tsujii. 2003. GENIA corpusa semantically annotated corpus for biotextmining. *Bioinformatics*, 19:i180–i182.

Jin-Dong Kim, Tomoko Ohta, Sampo Pyysalo, Yoshinobu Kano, and Jun'ichi Tsujii. 2009. Overview of BioNLP'09 shared task on event extraction. In *Proceedings of the BioNLP 2009 Workshop Companion Volume for Shared Task*, pages 1–9, Boulder, Colorado. Association for Computational Linguistics.

Jin-Dong Kim, Sampo Pyysalo, Tomoko Ohta, Robert Bossy, Ngan Nguyen, and Jun'ichi Tsujii. 2011. Overview of BioNLP shared task 2011. In *Proceedings of BioNLP Shared Task 2011 Workshop*, pages 1–6, Portland, Oregon, USA. Association for Computational Linguistics.

Martin Krallinger, Florian Leitner, Carlos Rodriguez-Penagos, and Alfonso Valencia. 2008. Overview of the protein-protein interaction annotation extraction task of biocreative ii. *Genome biology*, 9(2):S4.

Yi Luan, Luheng He, Mari Ostendorf, and Hannaneh Hajishirzi. 2018. Multi-task identification of entities, relations, and coreference for scientific knowledge graph construction. In *Proceedings of the 2018 Conference on Empirical Methods in Natural Language Processing*, pages 3219–3232. Association for Computational Linguistics.

Claire Nédellec. 2005. Learning language in logicgenic interaction extraction challenge. In *Proceedings of the 4th Learning Language in Logic Workshop (LLL05)*, volume 7, pages 1–7. Citeseer.

Claire Nédellec, Robert Bossy, Jin-Dong Kim, Jung-Jae Kim, Tomoko Ohta, Sampo Pyysalo, and Pierre Zweigenbaum. 2013. Overview of bionlp shared task 2013. In *Proceedings of the BioNLP Shared Task 2013 Workshop*, pages 1–7.

Mark Neumann, Daniel King, Iz Beltagy, and Waleed Ammar. 2019. Scispacy: Fast and robust models for biomedical natural language processing.

Rose Oughtred, Chris Stark, Bobby-Joe Breitkreutz, Jennifer Rust, Lorrie Boucher, Christie Chang, Nadine Kolas, Lara ODonnell, Genie Leung, Rochelle McAdam, Frederick Zhang, Sonam Dolma, Andrew Willems, Jasmin Coulombe-Huntington, Andrew Chatr-aryamontri, Kara Dolinski, and Mike Tyers. 2018. The BioGRID interaction database: 2019 update. *Nucleic Acids Research*, 47(D1):D529–D541.

Sampo Pyysalo, Antti Airola, Juho Heimonen, Jari Björne, Filip Ginter, and Tapio Salakoski. 2008a. Comparative analysis of five protein-protein interaction corpora. *BMC Bioinformatics*, 9(3):S6.

Sampo Pyysalo, Filip Ginter, Juho Heimonen, Jari Björne, Jorma Boberg, Jouni Järvinen, and Tapio Salakoski. 2007. Bioinfer: a corpus for information extraction in the biomedical domain. *BMC bioinformatics*, 8(1):50.

Sampo Pyysalo, Rune Sætre, Junichi Tsujii, and Tapio Salakoski. 2008b. Why biomedical relation extraction results are incomparable and what to do about it. In *Proceedings of the Third International Symposium on Semantic Mining in Biomedicine (SMBM 2008). Turku*, pages 149–152. Citeseer.

Fabio Rinaldi, Tilia Renate Ellendorff, Sumit Madan, Simon Clematide, Adrian Van der Lek, Theo Mevissen, and Juliane Fluck. 2016. Biocreative v track 4: a shared task for the extraction of causal network information using the biological expression language. *Database*, 2016.

Paul Shannon, Andrew Markiel, Owen Ozier, Nitin S. Baliga, Jonathan T. Wang, Daniel Ramage, Nada Amin, Benno Schwikowski, and Trey Ideker. 2003. Cytoscape: A software environment for integrated models of biomolecular interaction networks. *Genome Research*, 13(11):2498–2504.

Paul Thompson, Sophia Ananiadou, and Junichi Tsujii. 2017. The genia corpus: Annotation levels and applications. In *Handbook of Linguistic Annotation*, pages 1395–1432. Springer.

Marco A Valenzuela-Escárcega, Özgün Babur, Gus Hahn-Powell, Dane Bell, Thomas Hicks, Enrique Noriega-Atala, Xia Wang, Mihai Surdeanu, Emek Demir, and Clayton T Morrison. 2018. Large-scale automated machine reading discovers new cancer driving mechanisms. *Database: The Journal of Biological Databases and Curation*.

Yuhao Zhang, Victor Zhong, Danqi Chen, Gabor Angeli, and Christopher D. Manning. 2017. Position-aware attention and supervised data improve slot filling. In *Proceedings of the 2017 Conference on Empirical Methods in Natural Language Processing (EMNLP 2017)*, pages 35–45.

A Appendices

A.1 Examples of entity annotation ambiguities

1. (`1.0.train.104`) "The inability of tyrosine-phosphorylated SLP-76 to interact with nck(SH2*)". We annotate "nck" as a protein and "nck(SH2*)" as a protein with label "mutant".

2. (`1.0.train.45`) "... equal amounts of REGs α and β bound to the proteasome ...". We annotate "REGs α" as a protein and link it to the implicit "REG α", and annotate "β" as a protein and link it to the implicit "REG β".

3. (`1.0.dev.118`) "NF-Y binds the HSP70 promoter in vivo.". We annotate three entities in this sentence: "NF-Y" is a protein, "HSP70" is a gene, and "HSP70 promoter" is a DNA. There is a binding interaction between "NF-Y" and "HSP70 promoter", but not with "HSP70".

4. (`1.0.train.964`) "...the binding of cortactin to the Arp2/3 complex....". Here, "Arp2/3" is annotated as a complex, "Arp2" is a protein, while "3" is annotated as a protein and is linked to an implicit entity "Arp3".

5. (`1.0.train.430`) "A18 hnRNP Binds Specifically to RPA2 and Thioredoxin 3'-UTRs". Here, "RPA2 3'-UTR" is a region of "RPA2" RNA. But it is not a continuous span of characters, so we are forced to annotate only "RPA2". As a result, the same sequence of characters "RPA" is annotated both as an RNA and as an RNA-region.

6. (`1.0.train.121`) "JNK/SAPK Binds and Phosphorylates a MEKK1 Fragment In Vitro". Here JNK and SAPK are separate entities. We annotate binding interaction between "JNK" and "MEKK1" and between "SAPK" and "MEKK".

7. (`1.0.train.50`) "... Apaf-1 binds cytochrome c and dATP, and this complex recruits caspase-9 ...". "This complex" refers to an implicit complex with three entities. We do not annotate the complex and its interactions.

A.2 Examples of interaction annotation ambiguities

1. (`1.0.train.540`) "We also attempted to examine the actin-binding ability of partially phosphorylated F-rad." This sentence motivates the performed experiment, but does not talk about the outcome.

Space, full stop	"S. cerevisiae" (cell), "S-1.MgADP.Pi" (protein-domain)
Question mark	No examples
Comma, colon, semicolon	"PI(4,5)P2" (chemical), "f:TFIID" (fusion-protein)
Round brackets	"NAD(H)" (chemical), "HMG-1(A-B)" (protein region)
Square brackets	"DB[a,l]PDE" (chemical), "[3H]LY341495" (drug)
Hyphen-like symbols	"IGF-II promoter" (DNA), "hTcf-4-(180)" (protein-region)
Apostrophe	"3'UTR" (RNA), "3'dE5" (chemical)
Asterisk	"Rh*" (protein), "C2A* mutant" (protein-domain)
Plus	"Ca2+" (chemical), "Na+,K+-ATPase" (protein-complex)
Dot-like symbols	"DBAD" (protein-region), "actinφ" (protein-family)

Table 3: All entities in the dataset are surrounded by any of the symbols described in the first column. On the other hand, most of these symbols can appear inside entity names. The second column of the table shows examples of entities which contain these symbols.

2. (1.0.train.755) "We expect that in the intact BAF complex, the actin monomer is bound to Brg1 at both of these sites." This sentence does not confirm the existence of a binding interaction.

3. (1.0.train.1397) "Binding of Hairy derivatives to Gro in vitro.". This is a title that uses an indefinite verb, and the contents of the following paragraphs might imply both existence and non-existence of the binding interaction. We annotate the binding interaction between "Hairy derivatives" and "Gro" with label 0.

4. (1.0.train.1234) "Phosphorylation of L1 Y1176 inhibits L1 binding to AP-2." This is a subsection title, but it clearly implies that "L1" binds to "AP-2" (which is inhibited by phosphorylation), so we annotate this interaction with label 1.

5. (1.0.train.1154) "... the ORC-Cdc6p complex (and perhaps other proteins) recruits the six minichromosome maintenance (MCM) proteins ...". Here "minichromosome maintenance" and "MCM" refer to the same protein family and are annotated as synonyms. We annotate binding interaction between "MCM" and "ORC-Cdc6p", and the evaluation script does not penalize the model if it predicts an interaction between "minichromosome maintenance" and "ORC-Cdc6p".

6. (1.0.train.785) "... TR/RXR binds to the TRE ...". Here we annotate a binding interaction between "TR" and "RXR" and tag it as "complex".

7. (1.0.train.1154) "... Cdc6p most likely binds to ORC and then the ORC-Cdc6p complex ...". Here the binding interaction between "ORC" and "Cdc6p" can be inferred explicitly from the first part of the sentence and implicitly from the name of the complex. In these cases we do not tag the interaction with "complex" label.

8. (1.0.train.630) "hTcf-4-(180) interacts directly with the Armadillo repeats of β-catenin". Here "hTcf-4-(180)" is annotated as a domain of "hTcf-4" protein, and "Armadillo repeats" is annotated as a region of "β-catenin" protein. We annotate the interaction between "hTcf-4-(180)" and "Armadillo repeats". Additionally, we annotate three other interactions: "hTcf-4-(180)" and "β-catenin", "hTcf-4" and "Armadillo repeats", "hTcf-4" and "β-catenin", and tag them with an "implicit" label.

9. (1.0.train.758) "Synaptotagmin binds β-SNAP, but not α-SNAP...". Here "Synaptotagmin" and "SNAP" are annotated as proteins, while "α-SNAP" and "β-SNAP" are annotated as isoforms of "SNAP". We annotate a negative binding interaction between "α-SNAP" and "Synaptotagmin", but it does not imply that "Synaptotagmin" does not bind "SNAP". This shows that the implicit "transfer" of an interaction does not hold if the interaction is negative.

10. (1.0.test.171) "Myozenin binds to both

α-actinin-2 and -3 but not to itself, whereas α-actinin-2 and -3 both bind to myozenin as well as to themselves." In this sentence we annotate a negative interaction between "My-ozenin" and "Myozenin", and another positive interaction between "α-actinin-2" and "α-actinin-2".

11. (1.0.dev.95) "... thereby inhibiting the binding of c-Jun homodimer to TRE." Here c-Jun homodimer implies that there is a binding interaction between "c-Jun" proteins.

12. (1.0.train.69) "... Shs1 can bind to Gin4 and induce Gin4 oligomerization ..." Here oligomerization implies a binding interaction between "Gin4" and "Gin4".

13. (1.0.train.783) "Binding of IL-1 and TNF-alpha to their receptors activates several signaling pathways, including the NFkappaB and AP-1 pathways.". We do not annotate any binding interactions in this sentence, as "IL-1 receptor" is not an explicitly mentioned entity.

A.3 Tokenization rules

Table 3 describes the tokenization rules used in BioRelEx 1.0.

A.4 Entity types

Table 5 lists all entity types with descriptions used in BioRelEx 1.0 and some useful statistics[2].

Table 4 lists the pairs of entity types that are in *part_of* relationship for which we automatically add interactions to the dataset.

A.5 BioRelEx 1.0 graph

We have constructed a graph that represents the whole annotated dataset (Fig. 2) using Cytoscape tool (Shannon et al., 2003). We use grounding information to match entities from different sentences. If grounding information is not available, we fall back to entity names.

A.6 REACH baseline

We use two API calls to get information from REACH system[3]:

Child	Parent
protein-domain	protein
protein-region	protein
protein-state	protein
protein-isoform	protein

Table 4: If the sentence contains a positive binding interaction between entities A and B, where A is of a "child" type listed in this table, and it belongs to another entity C of a corresponding "parent" type, then we additionally annotate an implicit binding interaction between B and C.

- In `fries` mode, the server outputs information about entities. Each object corresponds to one entity mention in the text. Each mention has a text, location in the text, type of the entity and grounding information. In rare cases, the same entity name has different grounding information for different locations in the text. Our system does not support this scenario, so we keep the grounding information from the first mention.

- In `indexcard` mode, the server outputs information about interactions between entities. Entities have grounding identifiers which can be matched to the output of the `fries` mode. We only take the interactions which have `binds` type. In one case this API returned an interaction, where the second participant was a list of two entities. In these cases we take the first one only.

We group multiple mentions of the same entity name by matching the string. Then we group multiple entity names into an entity cluster (`unique_entity` object) by taking into account the grounding information (the concatenation of `namespace` and `ID` from REACH output).

REACH attempts to detect many entity types. We keep only the following entity types: `celline`, `family`, `protein`, `simple-chemical`, `site`. Including other types (e.g. `bioprocess`, `organ`, etc.) decreases precision of entity recongition (as these are not annotated in the dataset).

The implementation of our pipeline based on REACH is available on GitHub[4].

---

[2]We originally annotated DNA-motifs and DNA-regions as separate entity types, but after some analysis we have seen inconsistencies: sometimes DNA-motifs were annotated as DNA. We made a decision to merge all these entity types into a single cluster with name "DNA".

[3] http://agathon.sista.arizona.edu:8080/odinweb/api/text

[4] https://github.com/YerevaNN/Relation-extraction-pipeline/

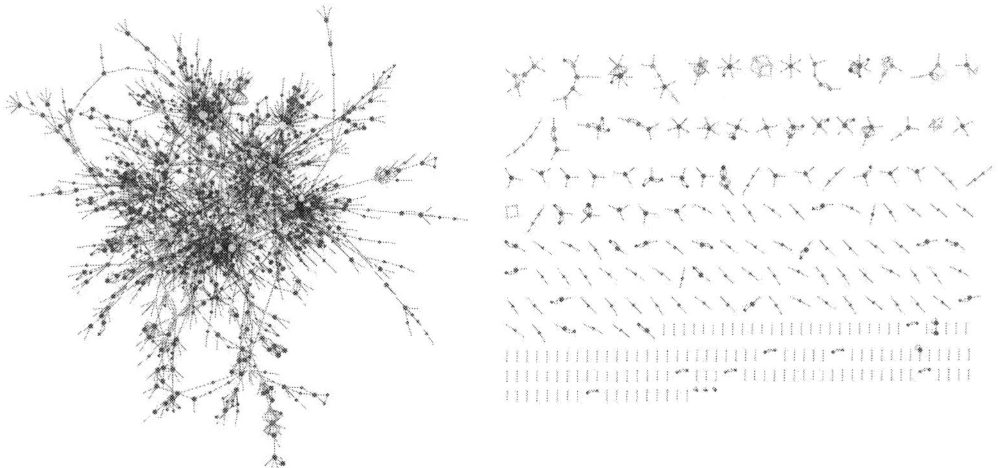

Figure 2: The network of the interactions annotated in BioRelEx 1.0

A.7 SciIE baseline

To use the SciIE model for our dataset we had to convert our data to the format the model can accept. We used a tokenzier from `scispacy` (Neumann et al., 2019), matched the tokens with our annotated entities, and added entity type, cluster (coreference) and relation information. The model supports multiple relation types. We have only one type: `bind`. Additionally, we have converted SciIE code to Python 3, and the converted version was made available on GitHub[5].

Unfortunately, the set of entities produced by the entity recognizer submodule is not syncronized with the entities that appear in the predicted coreference clusters and relations. We have developed another script to convert the output of SciIE to a JSON format that our evaluation script can handle. For entities, we used the output of SciIE entity recognizer submodule (along with the predicted entity types) and concatenated the entities that were produced by coreference and relation extraction submodules (with a label `other`). In our JSON, we specify a relation between entity clusters, although SciIE produces relations between individual entity mentions.

Our preprocessing and post-processing scripts are available on GitHub[6].

A.8 Captions for Figure 1 of Section 5.4

C In regular conditions, ligands such as growth factors or mitogens bind to the RTK, which is activated by autophosphorylation. Phosphotyrosine residues bind to adaptor protein Grb2 and Sos, promoting Ras·GTP association. Activated by GAPs such as NF1, Ras binds GTP and Raf, the first effector kinase in the MAPK pathway. Raf then phosphorylates MEK, which in turn binds ERK. p-ERK activates cytoplasmic and nuclear substrates.

D In regular conditions, ligands such as growth factors or mitogens bind to the RTK, which is activated by autophosphorylation. Phosphotyrosine residues bind to adaptor protein Grb2 and Sos, promoting Ras:GTP association. Ras binds GTP and Raf, the first effector kinase in the MAPK pathway. Raf then phosphorylates MEK, afterwards MEK binds ERK. p-ERK activates cytoplasmic and nuclear substrates.

---

[5]`https://github.com/YerevaNN/SciERC/`
[6]`https://github.com/YerevaNN/Relation-extraction-pipeline/`

Entity type	Statistics	Description
protein	3640 / 3777 / 82 / 147	Entities either represented with protein names; or with gene names (X) but factually standing as actual proteins in the sentence (either explicitly: X protein; or implicitly X binds the promoter)
protein-family	1086 / 1017 / 0 / 0	Entities represented with protein-family names (e.g. actin) or representing a group of protein with common properties (e.g. globular proteins; x-domain containing proteins, etc)
chemical	532 / 295 / 0 / 0	Any chemical compound other than protein or DNA or RNA, excluding experimental reagents/antibodies.
DNA	506 / 468 / 2 / 0	Any entity type that represents a region of or full DNA molecule, except for gene names. These include explicit 'DNA' mentions; DNA-regions, such as gene promoters, DNA elements; DNA sequences represented with nucleotides and DNA-motifs represented with names; chromosomes and plastids.
protein-complex	419 / 294 / 1 / 0	Protein complexes are either explicitlly mentioned with name followed by 'complex' suffix, or with name containing subunits seperated with slashes or dashes, or with names that do not contain the members, but are known to be complexes.
protein-domain	318 / 134 / 2 / 1	Domains may or may not be explicitly annotated with the suffix 'domain'. They may be specific domains of proteins present in the sentence, or general domain names without reference to the proteins they belong to.
cell	152 / 1 / 3 / 2	Explicit mentions of a cell or entities representing cell names, cell-line, bacterium, as well as viruses.
experimental-construct	141 / 60 / 0 / 0	Entities refering to artificially merged molecules, including tagged proteins, tagged RNA and DNA and chemically modified proteins/RNA/DNA.
RNA	137 / 105 / 0 / 0	All the entities representing physical RNA molecules (mRNA, tRNA, rRNA, etc.), or RNA-motifs (represented by RNA sequence or motif name) or RNA regions (represented by region names). mRNAs presented in text with corresponding gene names are also annotated as RNA.
experiment-tag	128 / 35 / 0 / 0	Chemicals or proteins experimentally added to proteins (e.g. GST tag).
reagent	128 / 43 / 0 / 0	Chemicals/biomolecules used in experimental settings (e.g. antibody)
protein-motif	122 / 43 / 0 / 0	Amino acid sequence patterns represented either by motif names or amino acid sequences, which may or may not be followed by explicit 'motif' mention.
gene	109 / 6 / 2 / 0	Entities represented with gene names.
amino-acid	69 / 2 / 0 / 0	Amino acids represented by amino acid names or explicit amino acid mentions.
protein-region	66 / 37 / 0 / 0	Protein regions are entities refering to amino-acid sequences (motif names or actual sequence representations); or regions on the protein not refering to whole domains.
assay	55 / 0 / 0 / 0	Entities refering to exprimental method names or assays or procedures.
organelle	51 / 20 / 0 / 0	Subcellular entities represented with their names (e.g. ribosome).
peptide	37 / 24 / 0 / 0	Short amino-acid polymers represented by their names, which may or may not be followed by explict 'peptide' mentions.
fusion-protein	32 / 25 / 0 / 0	Fusion-proteins
protein-isoform	32 / 33 / 0 / 0	Protein sub-types encoded by the same gene, but resulting from its differential post-processing. These entities may or may not appear in a sentence with explicit isoform mentions.
process	31 / 0 / 0 / 0	Entities refering to sequences of events at molecular, cellular or organismal levels. These may be pathway names (represented either by member gene names or target process names, with or without explicit 'pathway' mentions); process names/descriptions (e.g. autophagy); disorders and biological phenotypes.
mutation	20 / 0 / 0 / 0	Specifications of mutations in the form of nucleotide-to-nucleotide (A55G) or amino acid-to-amino acid transitions (Ala55Ser) or sequence to sequence transitions (ACGT to AGGT).
protein-RNA-complex	20 / 11 / 0 / 0	Complexes composed of proteins and RNA, mentioned either with component names or the complex alias, with or without explicit 'protein-RNA' mention.
drug	18 / 8 / 0 / 0	Drug names
organism	7 / 0 / 0 / 0	Multi-cellular organisms (i.e. excluding cells, bacteria and viruses)
disease	6 / 0 / 0 / 0	Entities representing disease names.
protein-DNA-complex	5 / 7 / 0 / 0	Complexes composed of proteins and DNA, mentioned either with component names or the complex alias, with or without explicit 'protein-DNA' mention.
brand	4 / 0 / 0 / 0	Entities representing company names or reagent/drug brands.
tissue	2 / 0 / 0 / 0	Entities representing tissues.
RNA-family	2 / 1 / 0 / 0	Entities representing groups of RNA with common properties.
gene-family	2 / 0 / 0 / 0	Entities representing sets of genes encoding for protein-families or combined by a common characteristic. Usually mentioned with name followed by 'gene family'.
fusion-gene	1 / 1 / 0 / 0	Entities representing fusion products of two genes. Usually represented by gene names separated with dashes followed (or not) by 'fusion' suffix.

Table 5: Entity types annotated in the dataset. The second column shows the number of mentions of those entities in the sentences, number of binding interactions involving those entities, number of mutated entities and number of entities that appear in a special state (e.g. phosphorylated).

Extraction of Lactation Frames from Drug Labels and LactMed

Heath Goodrum, Meghana Gudala, Ankita Misra, Kirk Roberts
School of Biomedical Informatics
University of Texas Health Science Center at Houston
{heath.goodrum,kirk.roberts}@uth.tmc.edu

Abstract

This paper describes a natural language processing (NLP) approach to extracting lactation-specific drug information from two sources: FDA-mandated drug labels and the NLM Drugs and Lactation Database (LactMed). A frame semantic approach is utilized, and the paper describes the selected frames, their annotation on a set of 900 sections from drug labels and LactMed articles, and the NLP system to extract such frame instances automatically. The ultimate goal of the project is to use such a system to identify discrepancies in lactation-related drug information between these resources.

1 Introduction

Medical information about prescription drugs is publicly available in a variety of sources, including the biomedical literature, consumer-focused websites, and the drug labels mandated by the U.S. Food & Drug Administration (FDA). But the rapid advances in biomedicine—especially recently-approved drugs—threatens to make these sources discordant. Synchronizing such sources is difficult due to their unstructured nature and the wide variety of ways in which they are organized. This paper presents initial work in an effort to align two such sources—drug labels and a single consumer health website—for information particular to a single sub-population—nursing mothers.

This is a critical sub-population for providing validated health information to, especially for prescription drugs. Notably, randomized trials, the gold standard in drug evaluation, contain few if any nursing mothers in their trial populations. Thus, the information that a pharmaceutical substance has on such mothers and their children is scarce and of poor evidence quality, which only serves to promote misinformation and discourage mothers from taking needed medications. Authoritative guidance is critical in regards to what is supported or contradicted by the limited evidence, as well as what is simply unknown. Several public sources attempt to provide such authoritative information. Here, two such sources are studied: a section of drug labels specific to nursing mothers and a government website specific to drugs and lactation, LactMed. By identifying discrepancies in the free text narratives of these sources, further review can pinpoint information gaps, conflicting opinions, and out-of-date guidance.

The general strategy proposed in this paper involves (a) identifying seven key information types of drug information specific to nursing mothers, (b) utilizing linguistically-motivated frame semantic representations for these information types, (c) annotating instances of these frames on both lactation information sources, and (d) developing natural language processing (NLP) methods to extract this information automatically from these sources.

Our specific contributions include:

1. The first NLP method to focus specifically on drug information for nursing mothers.

2. Development of frame representations for lactation-specific drug information.

3. Application of a deep learning-based system on two separate lactation information sources, drug labels and LactMed.

4. Evaluation of cross-corpus similarity in terms of important lactation information.

While this paper's scope is quite narrow, just lactation information from two sources, we posit the techniques described here are generalizable to other lactation information sources (with minimal annotation/training) as well as to other important pharmaceutical sub-populations (with, albeit, considerable annotation effort).

Proceedings of the BioNLP 2019 workshop, pages 191–200
Florence, Italy, August 1, 2019. ©2019 Association for Computational Linguistics

2 Related Work

The existing work related to that proposed here is broken down into information extraction efforts on drug labels (§2.1), maternal health in particular (§2.2), and frame semantics in biomedicine (§2.3).

2.1 Drug Label Information Extraction

Drug labels contain a wealth of unstructured information relating to FDA-approved pharmaceuticals, and thus have proven to be a consistent target of NLP-based systems interested in automatically creating knowledge bases (KBs) (Harpaz et al., 2014). For instance, SIDER (Kuhn et al., 2010, 2016) is a well-used KB, constructed from drug labels, for adverse drug reaction (ADR) information (i.e., side effects). The 2017 TAC ADR task (Roberts et al., 2017) utilized a corpus of 200 drug labels with sections specific to ADR information (Demner-Fushman et al., 2018b). On the other hand, Duke et al. (2013) demonstrated the dangers of using drug labels as an ADR KB by identifying numerous inconsistencies between the labels for bioequivalent drugs. Meanwhile, drug indications (i.e., the medical condition the drug is intended to treat) have also been well-studied (Névéol and Lu, 2010; Fung et al., 2013; Li et al., 2013; Khare et al., 2014), as have drug interactions (Demner-Fushman et al., 2018a). All of these focus on general aspects of a drug, while hardly any work has focused on the information in drug labels related to specific populations, though both the TAC task as well as Culbertson et al. (2014) identified ADR-population relations.

2.2 Maternal Health Information Extraction

A few NLP methods have been applied to support maternal health. This includes processing biomedical literature to support evidence-based review of maternal mortality (de Groot et al., 2015) and identifying genes associated with placenta-mediated maternal diseases (Rodriguez et al., 2017). Electronic health record (EHR) data has been used to identify important maternal health information (Borra et al., 2013; Abhyankar and Demner-Fushman, 2013) and screen for suicide (Zhong et al., 2018, 2019). Social media has been used to identify pregnant women (Chandrashekar et al., 2017; Sarker et al., 2017). Finally, only one known work focuses drug labels for maternal health, focusing on the identification of pregnancy risk categories (Rodriguez and Fushman, 2015).

2.3 Frame Semantics in Biomedicine

Frame semantics (Fillmore, 1976, 1982) is a linguistic theory that postulates the meaning of most words is understood in relation to a conceptual frame in which entities take part. E.g., the meaning of sell in the "*Jerry sold a car to Chuck*" evokes a frame related to COMMERCE, which includes four elements: BUYER, SELLER, MONEY, and GOODS, though not all elements are required (as with MONEY here). Frames also include a lexical unit that triggers the frame ("*sold*" in the example). Frames provide a good connection between an abstract information representation and the actual text that specifies that information, and is thus a natural choice for a task such as identifying detailed lactation information in drug labels. Most notably, frame semantics have been operationalized in the large-scale resource FrameNet (Baker et al., 1998, 2003), though this resource is not specific to biomedicine.

Several works have explicitly extended FrameNet for biomedical tasks. This includes frame for molecular biology information (Dolbey et al., 2006; Dolbey, 2009; Tan, 2014), cancer information from EHRs (Roberts et al., 2018; Si and Roberts, 2018; Datta et al., 2017), and general medical information for Swedish (Kokkinakis, 2013). Many other works have implicitly used representations that are similar to frames, including the TAC ADR task data on drug labels (Roberts et al., 2017; Demner-Fushman et al., 2018b).

3 Data

Two different datasets were used to create the text corpus for frame annotation. Section 3.1 describes the drug labels dataset and Section 3.2 describes the LactMed dataset.

3.1 Lactation Information in Drug Labels

Drug labels were downloaded in August 2018 from the full release collection made available by DailyMed[1]. DailyMed is a public website operated by the National Library of Medicine (NLM) and is the official provider of FDA label information. These labels are maintained in a document markup standard approved by Health Level Seven (HL7) referred to as Structured Product Labeling (SPL), which specifies various drug label sections.

---

[1] https://dailymed.nlm.nih.gov/dailymed/

Figure 1 appears within a boxed region:

> **8.2 Lactation**
>
> *Risk Summary*
>
> LIPITOR use is contraindicated during breastfeeding *[see CONTRAINDICATIONS (4)]*. There is no available information on the effects of the drug on the breastfed infant or the effects of the drug on milk production. It is not known whether atorvastatin is present in human milk, but it has been shown that another drug in this class passes into human milk and atorvastatin is present in rat milk. Because of the potential for serious adverse reactions in a breastfed infant, advise women that breastfeeding is not recommended during treatment with LIPITOR.

Figure 1: Example "Lactation" section of a drug label

Figure 2 appears within a boxed region:

> **Summary of Use during Lactation:**
>
> The consensus opinion is that women taking a statin should not breastfeed because of a concern with disruption of infant lipid metabolism. However, others have argued that children homozygous for familial hypercholesterolemia are treated with statins beginning at 1 year of age, that statins have low oral bioavailability, and risks to the breastfed infant are low, especially with rosuvastatin and pravastatin.[1] Some evidence indicates that atorvastatin can be taken by nursing mothers with no obvious developmental problems in their infants. Until more data become available, an alternate drug may be preferred, especially while nursing a newborn or preterm infant.

Figure 2: Example "Use During Lactation" section from LactMed

For this work, only the lactation section was extracted. An example of this section is shown in Figure 1.

3.2 LactMed

LactMed[2] is a database created by the National Library of Medicine under the collection of TOXNET databases. LactMed provides information about various drugs and chemicals that nursing mothers may be exposed to that may then be passed to their infant through breast feeding. Information provided in LactMed includes the amount of a substance that may be excreted into breast milk, the absorption rate of an infant, and any potential adverse effects to a nursing infant. Data in LactMed is derived from reviews of the scientific literature, with each entry including references. Additionally, all records are peer-reviewed by a panel of experts. For this work, only the "Summary of Use During Lactation" section was extracted for each LactMed article. An example of this section is shown in Figure 2.

3.3 Preprocessing

Each individual drug label is stored in the Daily-Med collection as a zip compressed folder that includes the drug label as an XML file and scanned images of the label. We extracted the folders and parsed each XML document to identify the relevant lactation information. While the drug labels provide additional information regarding the use of the drug, only section 8.2, "Lactation", was extracted into individual documents for each label. Prior to a specification change in June 2015,

---

[2] https://toxnet.nlm.nih.gov/newtoxnet/lactmed.htm

this section was labeled as section 8.3, "Nursing Mothers". There were 37,005 separate drug labels parsed. Of those, lactation information was identified in 31,309 drug labels. Additionally, since many drug labels exist for the same drug, due to multiple manufactures and dosage amounts, only the lactation information from the most recent label for a drug was extracted. After this process of selecting only unique drug labels based on name, a dataset of 4,486 documents was created.

The entirety of LactMed is made available as a single XML document. This file was parsed to identify the drug name and the Summary of Use During Lactation section. Each LactMed article was already unique, therefore no de-duplication process is required. In total, 1,151 documents were created from LactMed.

4 Lactation Frames

Section 4.1 describes the frames annotated for both the drug labels and LactMed. Section 4.2 describes the annotation process.

4.1 Frame Descriptions

Since one of the primary purposes of annotating these two datasets is to compare information between them, a standard set of frames was chosen that would be applicable to both datasets. Seven lactation-related frames were chosen based on an initial review of sample drug labels and LactMed entries. These frames, detailed in Table 1, are: INFORMATION_AVAILABILITY, EFFECT_ON_MILK_SUPPLY, EXCRETION_INTO_MILK, ABSORPTION, ADVERSE_REACTION, ALTERNATIVES, and VERDICT. For each of these frames, elements

Table 1: Frames and Frame Elements

Element	Description
Non-Core Elements - Elements that are common across all/most frames.	
ANIMAL	Marks non-humans to which the frame applies. Frequently information is only available in animals studies and as not been verified/observed in human studies.
CONDITION	A condition (specific circumstance) under which the rest of the frame applies.
DRUG	The name of the drug or the class of drugs to which the frame applies.
INFORMATION	Any reference to how the information was obtained/published or the information quality that results in the frame's information.
LIKELIHOOD	Any expression that suggests the frame is less than 100% positive, including hedging ("possible"), infrequency ("sometimes"), and negation ("no evidence").
???	Marks any span that the annotator feels is important but does not currently have an annotation to match.
INFORMATION_AVAILABILITY - The quantity/quality of lactation information for the drug.	
QUALITY	A reference to the quality of information available. (e.g., observational studies, randomized controlled trials)
QUANTITY	A reference to the quantity of information available (e.g., a large number of studies, minimal information)
SOURCE	The source of information (e.g., journal article, post marketing surveillance)
EFFECT_ON_MILK_SUPPLY - The impact the drug has on the overall milk supply.	
QUALITY	A reference to the change in quality of the breast milk due to the drug.
QUANTITY	A quantitative expression of the impact of the drug on the milk supply.
TREND	The generalized trend (e.g., increases, decreases) in milk supply due to the drug.
EXCRETION_INTO_MILK - Information that the drug is excreted into the breast milk.	
QUANTITY	A quantitative expression of how much of the drug (or other substance) is excreted into the breast milk.
TIMEFRAME	Either the span of time from taking the medication till initial excretion (e.g., "2 hours after taking") or the span of time (possibly half-life) until the drug will no longer be excreted (e.g., "within 4 days")
ABSORPTION - Information that the nursing infant absorbs the drug from the breast milk.	
QUANTITY	A quantitative expression of how much of the drug (or other substance) is actually absorbed by the infant from the breast milk.
TIMEFRAME	Some span of time related to the absorption of the drug/substance by the infant.
ADVERSE_REACTION - Reactions the infant may have from being exposed to the drug.	
REACTION	The adverse reaction resulting from the drug.
ALTERNATIVES - Alternative drug options for breastfeeding mothers.	
ALTERNATIVE	The name of the alternative drug, drug class, or agent.
PREFERENCE	A statement about the preference for the alternative, which can be positive ("preferred") or negative ("not recommended").
VERDICT - Recommendations for nursing mothers using the drug.	
POLARITY	Positive or negative verdict.
DECISION	What the nursing mother taking the drug should do (or not do).
MONITOR	Statement that the mother/child should be monitored (e.g., for adverse reactions).
REASON	The particular reason leading to the verdict decision.

	DL	LM	Total	Length
FRAMES				
EXCRETION_INTO_MILK	631	222	853	1.24
VERDICT	492	360	852	1.10
ADVERSE_REACTION	376	351	727	1.84
EFFECT_ON_MILK_SUPPLY	132	52	184	1.70
INFORMATION_AVAILABILITY	15	113	128	1.09
ABSORPTION	30	96	126	1.09
ALTERNATIVES	3	111	114	1.33
ELEMENTS				
DRUG	1452	901	2353	1.31
CONDITION	603	639	1242	4.66
REACTION	606	554	1160	1.92
DECISION	669	357	1026	2.84
INFORMATION	563	341	904	2.53
REASON	310	283	593	5.57
QUANTITY	202	315	517	2.15
MONITOR	37	118	155	1.97
ANIMAL	128	3	131	1.09
PREFERENCE	2	101	103	1.62
TREND	45	42	87	1.31
TIMEFRAME	21	12	33	4.79
POLARITY	3	26	29	1.41
QUALITY	16	1	17	1.08
SOURCE	3	14	17	1.50
ALTERNATIVE	0	17	17	2.67

Table 2: Frame and Element Frequency and average number of tokens. DL: drug labels, LM: LactMed, Length: Average length of each lexical unit or frame element (in tokens)

	LM	DL	DL + LM
FRAMES			
INFORMATION_AVAILABILITY	73.64	28.57	64.24
EFFECT_ON_MILK_SUPPLY	62.30	92.00	83.41
EXCRETION_INTO_MILK	85.14	72.47	76.39
ABSORPTION	74.84	63.64	72.36
ADVERSE_REACTION	66.24	63.10	64.84
ALTERNATIVES	30.25	0.00	27.69
VERDICT	56.77	64.56	61.09
ELEMENTS			
DRUG	88.22	81.92	84.79
ANIMAL	66.67	77.91	77.71
QUANTITY	72.66	35.16	60.67
INFORMATION	56.77	62.92	60.40
TREND	60.38	57.63	58.93
MONITOR	63.77	40.00	58.43
LIKELIHOOD	67.52	31.19	53.52
REASON	29.31	61.84	47.76
DECISION	41.78	48.63	46.30
QUALITY	0.00	50.00	46.15
REACTION	41.94	39.31	45.77
ALTERNATIVE	38.71	0.00	36.36
CONDITION	43.29	24.62	32.62
TIMEFRAME	47.62	12.90	26.92
PREFERENCE	29.21	0.00	26.53
POLARITY	40.00	0.00	24.49
SOURCE	31.58	13.33	23.53

Table 3: Inter-Annotator Agreement, measured by F1.

were selected that describe the individual attributes and relations for the frame. These elements are where the detailed semantic information is located. Certain elements were selected that exist across all frames, these are referred to as Non-Core Elements.

4.2 Annotation

A random subset of equal amounts of documents from the drug labels dataset and LactMed dataset were selected for manual annotation. Example annotations from LactMed articles and the drug labels are shown in Figure 3.

The annotation process was completed by three individuals using BRAT (Stenetorp et al., 2013). Documents were double-annotated with a pair of individuals first annotating a collection of documents independently and then meeting to reconcile any differences. In cases where annotations could not be easily reconciled, the case was presented to two other individuals to help establish rules which could be used in similar situations moving forward. Annotation guidelines, which included frequently occurring lexical units for a given frame and example annotations, were developed in order to identify and ensure consistency.

After annotation of each subset of documents was completed and reconciled, a final review was performed by one of the annotators to ensure that any newly-established guidelines were consistent throughout all documents.

In total, 900 documents were double-annotated, 450 drug labels and 450 LactMed entries. Within these 900 documents a total of 2,984 frames and 8,384 frame elements were annotated. The frequency breakdown for each frame and frame element type is shown in Table 2. The most frequently identified frames were EXCRETION_INTO_MILK with 853 frames, VERDICT with 852 frames, and ADVERSE_REACTION with 727 frames.

Table 3 shows the inter-annotator agreement for each frame and frame element. When determining the inter-annotator agreement, only exact matches are considered, though partial disagreements were quite common. For example if one annotator choose the lexical unit "breast milk" for an EXCRETION_INTO_MILK frame and the second annotator choose "milk", this would be considered a mismatch. (The annotation guidelines specify that "breast milk" is the correct lexical unit in such a case.)

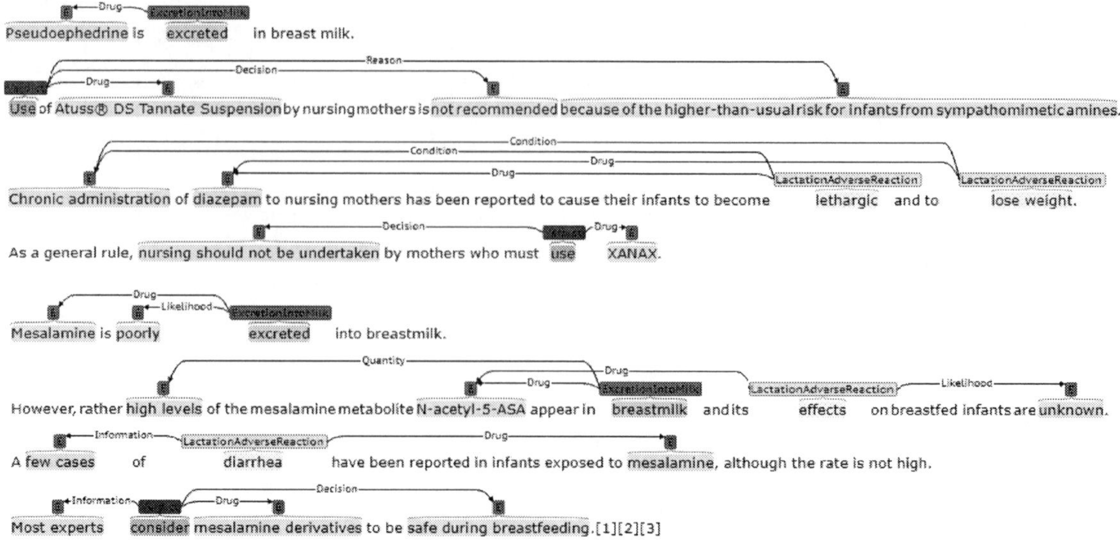

Figure 3: Annotation Examples

5   Extraction

A standard bi-directional Long Short-Term Memory (Bi-LSTM) Conditional Random Field (CRF) was utilized to extract lactation frames. The Bi-LSTM utilizes both character embeddings (dynamic) and word embeddings (static, described below). Specifically, a pipeline approach was used that extracts frames and frame elements are identified in two separate steps. The first step identifies lexical units in a sentence for all potential frames, essentially equivalent to a named entity recognition approach. The second step performs relation extraction for each identified lexical unit, identifying frame elements associated for the frames identified in the first step.

We experimented with four different embedding corpora: (i) pre-built 300-dimension GloVe (Pennington et al., 2014) embeddings built from Wikipedia; (ii) pre-built 100-dimension word2vec (Mikolov et al., 2013) embeddings built from MIMIC-III (Johnson et al., 2016); (iii) pre-built 200-dimension word2vec embeddings built from PubMed, PMC, and Wikipedia (Pyysalo et al., 2013); and (iv) specially-built 300-dimension GloVe embeddings on the lactation data (all drug labels and LactMed articles combined). Section 7 describes experiments with how these embedding combinations perform.

6   Evaluation

For our evaluation we created three separate collections of training, test, and validation sets by

Embeddings	Frame	FE
MIMIC + GloVe (Wikipedia)	84.53	71.33
MIMIC + GloVe (DL + LM)	83.89	70.93
W2V (PubMed + PMC + Wikipeida)	82.57	73.31
GloVe (DL + LM)	80.78	68.67

Table 4: Experiment with different word embeddings, measured by F1. DL: drug labels; LM: LactMed; FE: frame elements.

splitting the documents of the drug labels (DL), LactMed (LM), and LactMed and drug label combined (DL+LM). 80 percent of each dataset was used for training, 10 percent for testing, and 10 percent for validation. We trained and tested on various combinations of datasets.

We also experimented with training and testing on the different combinations of datasets, for example training on drug labels and LactMed (DL+LM) and testing on LactMed (LM), or training on LactMed (LM) and testing on the drug labels (DL).

Finally, to determine the effect that creating more manual annotations may improve the results of our model we generated a learning curve, using the full LactMed and drug label combined datasets. For generation of the learning curve, the same testing set was maintained and documents were added to the training set 50 documents at a time to generate a new model and evaluate against the test set.

Train	Test	Frame	FE
DL + LM	DL + LM	84.53	71.33
DL + LM	DL	90.07	78.32
DL + LM	LM	77.18	65.49
DL	DL	88.52	76.7
DL	LM	51.49	12.15
DL	DL + LM	68.9	56.24
LM	DL	57.41	27.9
LM	LM	71.54	52.03
LM	DL + LM	67.81	30.81

Table 5: Experiment with different train/test combinations, measured by F1, using best system from Table 4. DL: drug label; LM: LactMed; FE: frame elements

7 Results

Table 4 shows the results of our different embedding experiments. The GloVe embeddings generated from the drug labels and LactMed articles combined with the MIMIC embeddings, which had an F1 measure for the frames of 83.89 and the elements of 70.93, performed slightly worse than the top performing embeddings. The GloVe (Wikipedia) embeddings combined with MIMIC embeddings had an F1 of 84.53 for frames and 71.33 for frame elements. The W2V (PubMed + PMC + Wikipeida) embeddings performed best on frame element extraction with an F1 of 73.31.

Table 5 shows the different combinations of training and testing on various datasets. This data shows that training on the drug labels and LactMed together does improve the prediction performance on a single dataset opposed to just training on one dataset alone. For example, the model that was trained on drug labels and LactMed performed better on the LactMed test set (frame F1 of 77.18) than the model that was trained only on the LactMed dataset (frame F1 of 71.54). This effect is likely caused by an increase in training data overcoming the differences between the datasets.

Table 6 shows the breakdown of the results by each frame and frame element for the model that was created using the embeddings that performed best on frame identification (MIMIC + GloVe (Wikipedia)) and the combined drug labels and LactMed dataset for training and testing.

Figure 4 shows the learning curve as additional documents were added to the training set and the effect it has on the overall F1-measure. For both the frame and frame element the curve is beginning to level off, however it does seem to show that additional training data may continue to have a positive effect on the overall F1 for both cases.

	P	R	F1
FRAME			
OVERALL	85.52	83.55	84.53
ABSORPTION	71.43	35.71	47.62
ALTERNATIVES	86.67	86.67	86.67
EFFECT_ON_MILK_ SUPPLY	100.0	85.71	92.31
EXCRETION_INTO_MILK	86.21	83.33	84.75
INFORMATION_ AVAILABILITY	100.0	91.67	95.65
ADVERSE REACTION	89.71	85.92	87.77
VERDICT	78.02	87.65	82.56
ELEMENT			
OVERALL	69.74	72.99	71.33
ALTERNATIVE	100.0	50.00	66.67
ANIMAL	76.47	92.86	83.87
CONDITION	54.19	67.20	60.00
DECISION	76.84	71.57	74.11
DRUG	74.52	83.55	78.78
INFORMATION	92.59	82.42	87.21
LIKELIHOOD	68.12	74.60	71.21
MONITOR	44.19	76.00	55.88
POLARITY	100.0	75.00	85.71
PREFERENCE	81.25	100.0	89.66
QUALITY	100.0	100.0	100.0
QUANTITY	62.75	55.17	58.72
REACTION	92.86	50.00	65.00
REASON	54.39	58.49	56.36
SOURCE	100.0	100.0	100.0
TIMEFRAME	33.33	16.67	22.22
TREND	66.67	76.92	71.43

Table 6: Frame and Element Breakdown

8 Discussion

This paper addresses a critical component for assessing the consistency of drug information for nursing mothers, namely the information extraction techniques to extract semi-structured information from two drug information sources: manufacturer-supplied drug labels and expert-sourced LactMed. A frame-based approach was devised utilizing seven frames dealing with the availability/quality of lactation information, the effects the drug has on a mother's milk supply, the degree to which the drug is excreted into the milk, the degree to which that drug is absorbed into the child's body, any potential adverse reactions the child may experience due to breastfeeding, recommended alternative drugs while nursing, and any general statements or verdicts on what nursing mothers should do as it relates to the particular drug. Each of these seven frames was double-annotated on a corpus of 450 drug label sections and 450 LactMed article summaries. A standard Bi-LSTM-CRF combining character and word embeddings is trained to extract these frames automatically. Experiments were performed to assess the best set of embeddings to use, the transferability of drug label and LactMed annotations, and

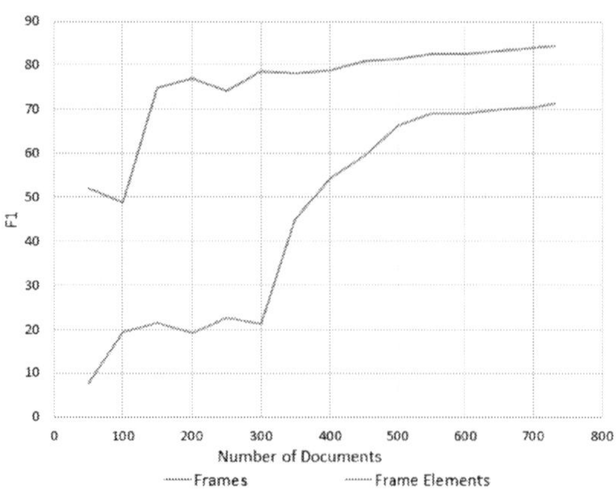

Figure 4: Learning curve of test performance with increasing amounts of training data.

whether sufficient annotated data exists to maximize frame extraction performance. These experiments yield several observations that have implications on further development of such a frame extraction system.

First, the fact that open-domain embeddings outperformed embeddings trained on the drug labels and LactMed (see Table 4) can be considered a negative, but not entirely conclusive, result. In our initial error analysis it was clear that the lack of embedding information for common terms in the dataset (such as particular drug names) resulted in numerous errors. We did not experiment with additional embedding combinations, such as concatenating separate embeddings for Wikipedia, MIMIC, and drug label/LactMed, though this concatenation strategy has shown promise in other biomedical tasks (Roberts, 2016).

Second, the experiments demonstrated that training on both drug labels and LactMed improves performance over training on each individually (Table 5). This improvement is despite the fact that the drug labels and LactMed data appears to be quite different, as can be seen by comparing the result of training on one and testing on the other. For instance, LactMed results are quite poor when only training on drug labels (51.49 F1) and improve significantly when training on LactMed (71.54), but improve further still when training on both drug labels and LactMed (77.18). This would suggest that there is sufficient similarity to train on both, but perhaps domain adaptation methods could be employed to gain the benefits of larger training datasets while still identifying source-specific differences in the data.

Third, the amount of data available for training (Figure 4) suggest small gains are still likely to be expected given more data. Our error analysis, however, suggested that many of the "errors" could in fact be considered legitimate frame instances. This is typically a result of inconsistent frame annotation, which is of course quite common in complex semantic annotation tasks. However, it is clear that further quality control on the existing annotations will likely be a more promising effort prior to adding further annotations.

Beyond the work described in this paper, there is still a good distance to go before an automatic method exists for detecting inconsistencies in lactation information sources. Notably, this work only extracts the basic frame instances from each of these sources, but does nothing to compare frames. Future work will thus be necessary to compare frame instances from a drug label and its corresponding LactMed article (not to mention that there are often multiple labels per drug). Comparing frames is certainly much easier than comparing full documents, but not without its challenges. Comparing individual frames requires both frame element-specific comparisons (e.g., is "*safe during breastfeeding*" equivalent to "*no major concerns*") as well as comparing to null frame elements (e.g., if one frame has a QUANTITY of "*high levels*" but the other frame has no QUANTITY at all). It is unlikely simple rule-based procedures can be used to identify equivalent or contradictory frames with high accuracy. However, this need not be the goal. Instead of providing a complete list of all inconsistent labels/articles, a likely application for the use of such a system is to provide a ranked list of labels/articles that are most likely to be incongruous. This approach may have greater robustness to errors in frame matching, and is a likely direction of future work.

9 Conclusion

This paper described a frame-based approach for lactation information extraction from drug labels and LactMed. Seven lactation-related frames were identified, manually annotated, and automatically extracted using a standard NLP approach. Future work will involve utilizing this system in order to identify discordant information present in drug information sources for nursing mothers.

Acknowledgements This work was supported in part by the U.S. National Library of Medicine under award R00LM012104.

References

Swapna Abhyankar and Dina Demner-Fushman. 2013. A simple method to extract key maternal data from neonatal clinical notes. In *Proceedings of the AMIA Annual Symposium*, pages 2–9.

Collin F Baker, Charles J Fillmore, and Beau Cronin. 2003. The Structure of the Framenet Database. *International Journal of Lexicography*, 16(3):281–296.

Collin F. Baker, Charles J. Fillmore, and John B. Lowe. 1998. The Berkeley FrameNet project.

Allan Borra, Karla Delos Santos, Daniel Gonzales, and Alden Reyes. 2013. Maternal Medication Information Extraction (MaMIE) System. In *Proceedings of the 9th National Natural Language Processing Research Symposium*, pages 25–29.

Pramod Bharadwaj Chandrashekar, Arjun Magge, Abeed Sarker, and Graciela Gonzalez. 2017. Social media mining for identification and exploration of health-related information from pregnant women. *CoRR*, abs/1702.02261.

Adam Culbertson, Marcelo Fiszman, Dongwook Shin, and Thomas C. Rindflesch. 2014. Semantic Processing to Identify Adverse Drug Event Information from Black Box Warnings. In *Proceedings of the AMIA Annual Symposium*, pages 442–448.

Surabhi Datta, Elmer V Bernstam, and Kirk Roberts. 2017. A frame semantic overview of NLP-based information extraction for cancer-related EHR notes. *CoRR*, abs/1904.01655.

Christianne J. M. de Groot, Thed van Leeuwen, Ben Willem J. Mol, and Ludo Waltman. 2015. A Longitudinal Analysis of Publications on Maternal Mortality. *Paediatric and Perinatal Epidemiology*, 29(6):481–489.

Dina Demner-Fushman, Kin Wah Fung, Phong Do, Richard D. Boyce, and Travis R. Goodwin. 2018a. Overview of the TAC 2018 Drug-Drug Interaction Extraction from Drug Labels Track. In *Proceedings of the Text Analysis Conference*.

Dina Demner-Fushman, Sonya E. Shooshan, Laritza Rodriguez, Alan R. Aronson, Francois Lang, Willie Rogers, Kirk Roberts, and Joseph Tonning. 2018b. A dataset of 200 structured product labels annotated for adverse drug reactions. *Scientific Data*, 5:180001.

A. Dolbey. 2009. *BioFrameNet: A FrameNet Extension to the Domain of Molecular Biology*. Ph.D. thesis, UC Berkeley.

A. Dolbey, M. Ellsworth, and J. Scheffzyk. 2006. BioFrameNet: A Domain-specific FrameNet Extension with Links to Biomedical Ontologies. In *Proceedings of the Biomedical Ontology in Action Workshop at KR-MED*, pages 86–94.

Jon Duke, Jeff Friedlin, and Xiaochun Li. 2013. Consistency in the safety labeling of bioequivalent medications. *Pharmacoepidemiology Drug Safety*, 22(3):294–301.

Charles J Fillmore. 1976. Frame Semantics and the Nature of Language. *Annals of the New York Academy of Sciences*, 280(1):20–32.

Charles J Fillmore. 1982. Frame semantics. *Linguistics in the Morning Calm*, pages 111–137.

Kin Wah Fung, Chiang S Jao, and Dina Demner-Fushman. 2013. Extracting drug indication information from structured product labels using natural language processing. *Journal of the American Medical Informatics Association*, 20(3):482–488.

Rave Harpaz, Alison Callahan, Suzanne Tamang, Yen Low, David Odgers, Sam Finlayson, Kenneth Jung, Paea LePendu, and Nigam H. Shah. 2014. Text Mining for Adverse Drug Events: the Promise, Challenges, and State of the Art. *Drug Safety*, 37(10):777–790.

Alistair E.W. Johnson, Tom J. Pollard, Lu Shen, Li wei H. Lehman, Mengling Feng, Mohammad Ghassemi, Benjamin Moody, Peter Szolovits, Leo Anthony Celi, , and Roger G. Mark. 2016. MIMIC-III, a freely accessible critical care database. *Scientific Data*, 3:160035.

Ritu Khare, Jiao Li, and Zhiyong Lu. 2014. LabeledIn: Cataloging Labeled Indications for Human Drugs. *Journal of Biomedical Informatics*, 52:448–456.

Dimitrios Kokkinakis. 2013. Medical Event Extraction using Frame Semantics – Challenges and Opportunities. *International Journal of Computational Linguistics and Applications*, 4(2):121–133.

Michael Kuhn, Monica Campillos, Ivica Letunic, Lars Juhl Jensen, and Peer Bork. 2010. A side effect resource to capture phenotypic effects of drugs. *Molecular Systems Biology*, 6:343.

Michael Kuhn, Ivica Letunic, Lars Juhl Jensen, and Peer Bork. 2016. The SIDER database of drugs and side effects. *Nucleic Acids Research*, 44(D1):D1075–D1079.

Qi Li, Louise Deleger, Todd Lingren, Haijun Zhai, Megan Kaiser, Laura Stoutenborough, Anil G Jegga, Kevin Bretonnel Cohen, and Imre Solti. 2013. Mining FDA drug labels for medical conditions. *BMC Medical Informatics and Decision Making*, 13(53).

Tomas Mikolov, Ilya Sutskever, Kai Chen, Greg Corrado, and Jeffrey Dean. 2013. Distributed Representations of Words and Phrases and their Compositionality. In *Proceedings of the 26th International Conference on Neural Information Processing Systems*, pages 3111–3119.

Aurélie Névéol and Zhiyong Lu. 2010. Automatic Integration of Drug Indications from Multiple Health Resources. In *Proceedings of the 1st ACM International Health Informatics Symposium*, pages 666–673.

Jeffrey Pennington, Richard Socher, and Christopher D Manning. 2014. GloVe: Global Vectors for Word Representation. In *Proceedings of the 2014 Conference on Empirical Methods in Natural Language Processing*, pages 1532–1543.

Sampo Pyysalo, Filip Ginter, Hans Moen, Tapio Salakoski, and Sophia Ananiadou. 2013. Distributional Semantics Resources for Biomedical Text Processing. In *Proceedings of Languages in Biology and Medicine*.

Kirk Roberts. 2016. Assessing the Corpus Size vs Similarity Trade-off for Word Embeddings in Clinical NLP. In *Proceedings of the COLING Workshop on Clinical Natural Language Processing*.

Kirk Roberts, Dina Demner-Fushman, and Joseph Tonning. 2017. Overview of the TAC 2017 Adverse Reaction Extraction from Drug Labels Track. In *Proceedings of the Text Analysis Conference (TAC)*.

Kirk Roberts, Yuqi Si, Anshul Gandhi, and Elmer Bernstam. 2018. A FrameNet for Cancer Information in Clinical Narratives: Schema and Annotation. In *Proceedings of the Language Resources and Evaluation Conference*, pages 272–279.

Laritza M. Rodriguez and Dina Demner Fushman. 2015. Automatic Classification of Structured Product Labels for Pregnancy Risk Drug Categories, a Machine Learning Approach. In *Proceedings of the AMIA Annual Symposium*, pages 1093–1102.

Laritza M. Rodriguez, Stephanie M. Morrison, Kathleen Greenberg, and Dina Demner Fushman. 2017. Mining the literature for genes associated with placenta-mediated maternal diseases. In *Proceedings of the AMIA Annual Symposium*, pages 1498–1506.

Abeed Sarker, Pramod Chandrashekar, Arjun Magge, Haitao Cai, Ari Klein, and Graciela Gonzalez. 2017. Discovering Cohorts of Pregnant Women From Social Media for Safety Surveillance and Analysis. *Journal of Medical Internet Research*, 19(10):e361.

Yuqi Si and Kirk Roberts. 2018. A Frame-Based NLP System for Cancer-Related Information Extraction. In *Proceedings of the AMIA Annual Symposium*.

Pontus Stenetorp, Sampo Pyysalo, Goran Topic, Tomoko Ohta, Sophia Ananiadou, and Junichi Tsujii. 2013. BRAT: a Web-based Tool for NLP-Assisted Text Annotation. In *Proceedings of the 13th Conference of the European Chapter of the Association for Computational Linguistics*, pages 102–107.

He Tan. 2014. A System for Building FrameNet-like Corpus for the Biomedical Domain. In *Proceedings of the 5th International Workshop on Health Text Mining and Information Analysis (Louhi)*, pages 46–53.

Qiu-Yue Zhong, Elizabeth W. Karlson, Bizu Gelaye, Sean Finan, Paul Avillach, Jordan W. Smoller, Tianxi Cai, and Michelle A. Williams. 2018. Screening pregnant women for suicidal behavior in electronic medical records: diagnostic codes vs. clinical notes processed by natural language processing. *BMC Medical Informatics and Decision Making*, 18:30.

Qiu-Yue Zhong, Leena P. Mittal Margo D. Nathan, Kara M. Brown, Deborah Knudson Gonzlez, Tianrun Cai, Sean Finan, Bizu Gelaye, Paul Avillach, Jordan W. Smoller, Elizabeth W. Karlson, Tianxi Cai, and Michelle A. Williams. 2019. Use of natural language processing in electronic medical records to identify pregnant women with suicidal behavior: towards a solution to the complex classification problem. *European Journal of Epidemiology*, 34(2):153–162.

Annotating Temporal Information in Clinical Notes for Timeline Reconstruction: Towards the Definition of Calendar Expressions

Natalia Viani
IoPPN, King's College London
London, UK

Hegler Tissot
University College London
London, UK

Ariane Bernardino
IoPPN, King's College London
London, UK

Sumithra Velupillai
IoPPN, King's College London
London, UK
EECS, KTH, Stockholm, Sweden

Abstract

To automatically analyse complex trajectory information enclosed in clinical text (e.g. timing of symptoms, duration of treatment), it is important to understand the related temporal aspects, anchoring each event on an absolute point in time. In the clinical domain, few temporally annotated corpora are currently available. Moreover, underlying annotation schemas - which mainly rely on the TimeML standard - are not necessarily easily applicable for applications such as patient timeline reconstruction. In this work, we investigated how temporal information is documented in clinical text by annotating a corpus of medical reports with time expressions (TIMEXes), based on TimeML. The developed corpus is available to the NLP community. Starting from our annotations, we analysed the suitability of the TimeML TIMEX schema for capturing timeline information, identifying challenges and possible solutions. As a result, we propose a novel annotation schema that could be useful for timeline reconstruction: CALendar EXpression (CALEX).

1 Introduction and Background

When applying natural language processing (NLP) methods to the analysis of clinical notes, understanding the temporal aspects of narratives is crucial (e.g. *when* the patient experienced a certain symptom, or *when* a particular drug was prescribed). To model and extract the temporal information enclosed in free text, the development of suitable annotation schemas and reliably annotated corpora is essential.

The TimeML specification language was developed to enable the recognition of events and their temporal ordering in general-domain texts (Pustejovsky et al., 2003a). In the original schema, four major elements are modelled: time expressions (TIMEXes), events, signals, and their relations. Signals are function words (e.g. "during", "before") that indicate how temporal objects can be related to each other. Relations are represented by either temporal links (e.g. "before", "simultaneous"), subordination links (e.g. "intensional", "factive"), or aspectual links (e.g. "initiates, "continues"). The TimeML schema was used to develop the TimeBank corpus, consisting of 183 news articles (Pustejovsky et al., 2003b). Gold annotations were reused in the TempEval tasks on temporal information extraction (Verhagen et al., 2007; Pustejovsky and Verhagen, 2009), where a simplified TimeML annotation was applied.

The TimeML specification language provides a standard model for the mark-up of time expressions (with type Date, Time, Duration, or Set), events (mostly verbs or noun phrases), and their temporal ordering (Pustejovsky et al., 2010), and it can be in principle applied to any type of text. In the clinical domain, two reference corpora based on TimeML are available. The 2012 i2b2 corpus (310 discharge summaries) includes annotations for time expressions, clinical events, and eight types of temporal relations (Sun et al., 2013a). In addition, a section time (SECTIME) is used to keep track of section creation dates. The THYME corpus (1,254 oncology notes) contains annotations for events, time expressions (with two additional types), and 5 types of temporal relations (Styler IV et al., 2014). In this corpus, narrative containers were introduced, representing temporal buckets (mostly dates) containing a set of events. Figure 1 provides a graphical representation of the main changes introduced by i2b2 2012 and THYME on the original TimeML model.

Most clinical NLP development based on available corpora have focused on the three separate main tasks: detecting and classifying 1) events and

Proceedings of the BioNLP 2019 workshop, pages 201–210
Florence, Italy, August 1, 2019. ©2019 Association for Computational Linguistics

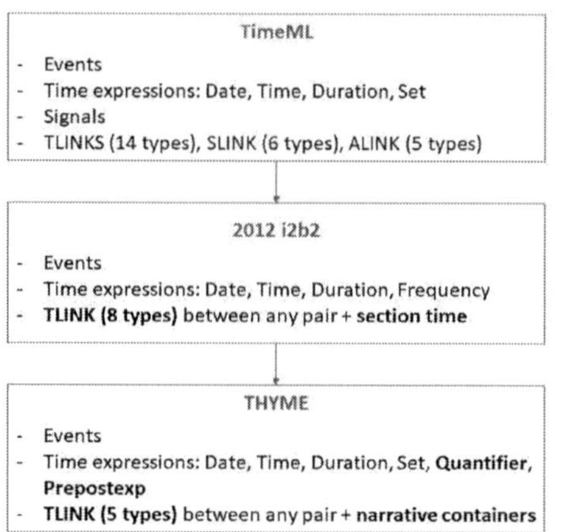

Figure 1: The TimeML model and how it has been adapted to, and implemented in, the clinical domain in the 2012 i2b2 and THYME corpora.

2) time expressions (and their attributes), and 3) classifying temporal links (TLINKs) (Sun et al., 2013b; Bethard et al., 2016, 2017). However, these separate tasks do not directly address the problem of anchoring events on an absolute timeline, which would be important for an improved understanding of patient trajectories. Moreover, mainly due to the inherent complexity of identifying temporal links (which can exist between any pair of entities and with different types), temporal annotation becomes a challenging task.

A few studies have proposed alternate approaches or extensions to the TimeML model for temporal ordering of events in various domains (Chambers et al., 2014; Jeblee and Hirst, 2018; Kolomiyets et al., 2012; Raghavan et al., 2014), but without addressing *timeline reconstruction*, i.e. anchoring events in absolute time. An approach that aimed to anchor events in time and simplify the annotation task was proposed by Reimers et al. (2016, 2018), where the event time is modelled as an argument of the event mention. However, the emphasis lies on the events, not on time expressions. Another approach proposed by Zhao et al. (2012) focuses on an alternative way to normalise time expressions using time intervals, allowing for more efficient temporal reasoning.

The existing temporally annotated corpora for the clinical domain have not, to our knowledge, been studied in great detail with respect to timeline reconstruction, particularly as regards TIMEX annotations. Tissot et al. (2015) found a surprising number of repeated inconsistencies between the guidelines and the manually created corpora for certain regular and unambiguous temporal language constructs. These were mainly related to inconsistencies in span and class assignments, incorrect annotations (false positives) or missing annotations (false negatives). This evidences how hard it is to create coherent annotated resources, and how NLP development and evaluation can be affected by the quality of the underlying data.

To further support the growth of clinical-temporal NLP and the development of translational applications, the release of additional annotated corpora (including different clinical specialties) is needed. Moreover, despite the efforts put into creating resources like the i2b2 2012 and THYME corpora, the suitability of the underlying annotation schemas to support clinical timeline reconstruction has not been widely investigated. While some types of TIMEXes are definitely useful to anchor clinical events on a timeline (e.g. explicit references to the calendar, like *in February 2009*), the importance of capturing other entities (e.g. the frequency of medication intake) remains unclear. Moreover, there might be TIMEX types that could be relevant in the clinical domain, but are not currently considered by temporal models based on TimeML, e.g. age-related expressions. As another important point, the way in which TIMEXes are typically normalised is not necessarily optimal for timeline reconstruction (e.g. duration values such as `P4M` cannot be immediately placed on a timeline).

Our intuition is that, by changing the way time expressions are defined and normalised, temporal relation annotation could be much simplified. More specifically, adding timeline information at the time expression level could help to temporally anchor entities without the need for multiple types of temporal links. To assess this, we applied a TimeML-based TIMEX annotation schema on a corpus of clinical texts, investigating how time expression information is documented, and then analysed how it can be reused for timeline reconstruction. As a result, we created a corpus of clinical texts (for four different clinical specialties), annotated with TIMEXes which mostly relies on the TimeML schema. This corpus is publicly available, as an additional resource that could be reused for temporal NLP.[1] In this paper, we analyse the suitability of these TIMEXes for timeline recon-

---

[1] https://github.com/medesto/timeline-reconstruction

struction, and from this analysis propose an innovative way to annotate temporal information in free text: CALendar EXpressions (CALEX).

2 Materials and Methods

2.1 Dataset

We downloaded and extracted documents from MTSamples,[2] a collection of Medical Transcription Sample reports for multiple clinical specialties (where the same document can belong to different groups), created for educational purposes and for working transcriptionists.

In the MTSamples resource, there is no availability of document creation times (DCTs). Most documents follow a semi-structured format, including different section headings/textual content depending on the specific clinical specialty. We selected the following specialties for manual annotation and analysis: discharge summaries (108 documents), psychiatry-psychology (53 documents), paediatrics (70 documents), and emergency (75 documents).[3]

2.2 Manual Annotation

All MTSamples subsets were double-annotated for five types of time expressions: Date, Time, Duration, Frequency (from TimeML (Pustejovsky et al., 2010)), and Age-related (*at the age of 16, in his teens* (Viani et al., 2018)). Identified expressions were also normalised to a standard temporal value (e.g. `"2011-05"` for *May 2011*). Manual annotations were performed by two annotators: a native English-speaker undergraduate student and a non-native English-speaker researcher. To guide the annotation process, we created specific annotation guidelines, which we refined by adding relevant examples from the text. Resulting guidelines are available on our repository.

Besides including an Age-related time expression type, our annotation task differed from TimeML in two ways. First, we included as TIMEXes domain-specific (and temporally-anchored) concepts, e.g. *On the day of admission*. This is similar to the Prepostexp type used in the THYME corpus, but instead of creating a different category, we included these expressions among existing types (e.g. Date). Depending on the clinical specialty (and more generally, on the domain), different concepts could be considered as

temporal anchors within temporal annotation (e.g. *discharge* for discharge summaries, *pregnancy* for paediatrics notes). Second, we allowed annotators to use relative values in the normalisation phase, if needed. These relative values, or formulas, can either refer to the document creation date (e.g. `"DCT-P2D"` for *Two days ago*) or to the domain-specific concepts (e.g. `"OP+P2D"` for *postoperative day #2*).

To compute inter-annotator agreement (IAA) on textual spans, we used the F1 score (allowing overlapping annotations). Expressions identified by both annotators (*overlap*) were used to compute IAA on types and normalised values (accuracy). For IAA on types, we also report the Cohen's Kappa measure (K), to take the possibility of chance agreement into account.

2.3 Annotation Analysis for Timeline Reconstruction

To analyse the suitability of using the TimeML-based TIMEX annotations for timeline reconstruction, we based our analysis on the following: 1) timeline properties of the TIMEX type Frequency, 2) properties of normalised values for Date annotations, and 3) properties of common annotation disagreements.

Our hypothesis regarding Frequency annotations was that these would not be necessarily useful as temporal references on an absolute timeline, as they would be mostly related to drug prescriptions. To assess this, we applied the MedEx-UIMA tool (Jiang et al., 2014) on the text surrounding each Frequency expression (the *context*),[4] and quantified the proportion of annotations close to drug mentions.

Normalised values for Date can represent a specific point on a timeline, e.g. `"YYYY-MM-DD"`, `"YYYY-MM"` or `"YYYY"`. However, they can also represent other less straightforward points in time, e.g. DCT-related (*yesterday*), vague references (*in the past*), incomplete dates (*on the 13th*), and concept-related (*on the day of admission*). To better understand these latter types of normalised Date values and their relation to timeline reconstruction, we analysed annotations marked as such by at least one annotator.

Finally, to inform the development of a new annotation schema for timeline reconstruction based

---

[2]`www.mtsamples.com`
[3]The number of unique documents is 286.

[4]We considered a window of 50 characters before and after the annotation.

on calendar expressions, we counted all annotation disagreements and analysed the most common type. We manually reviewed the documents containing these expressions, assessing whether: a) they could be placed on a timeline; and b) how they could be normalised in a non-ambiguous way. During this review, we also added new types of expressions that we believe would be crucial elements to anchor on a timeline, thus forming a proposal for a novel annotation schema.

3 Results

We report results for the new TimeML-based TIMEX annotated corpus in terms of IAA, and a breakdown of the number of documents, tokens and TIMEXes for each clinical specialty (Table 1). Furthermore, we report the results for the analysis on different aspects of the suitability of these annotations for timeline reconstruction that was used to inform the development of a novel annotation schema (further outlined in Section 4).

3.1 Manual Annotation

For each clinical specialty, Table 1 reports the number of documents (with total number of tokens), the number of time expressions marked by at least one annotator (*merged*), and those marked by both annotators (*overlap*). For IAA, we report F1 score for text spans (allowing overlapping annotations) and type/value agreement measures (on *overlap* annotations). We also report the prevalence of time expression types (only looking at overlap annotations with type agreement).

IAA results for text spans are encouraging, 76-84%. We observe that the distribution of TIMEX types is similar across clinical specialties, where Date is most common (28-36%) and Time is least common (3-9%), with the exception of discharge summaries, where Frequency is most common (39%) – probably due to the abundance of drugs prescribed after discharge. Agreement for normalised values measured by accuracy is slightly lower, overall (72-75%).

3.2 Annotation Analysis for Timeline Reconstruction

As shown in Table 1, Frequency expressions are common across all MTSamples subsets. By applying MedEx-UIMA and extracting the related contexts, we found that most frequencies occurred close to a drug mention (94%, 82%, 59%,

and 80%, in discharge summaries, psychiatry-psychology, paediatrics, and emergency, respectively). By manually reviewing a sample of the remaining expressions, we noticed that some of them referred to alcohol/smoking (*he drank one bottle of wine everyday*) or recommendations (*continue bathing twice a week*), and would therefore not be placed on a timeline. In other cases, examples were still related to a drug mention which, however, did not fall in the selected context or was not extracted by MedEx.

As regards the analysis of Date normalised values, we noticed that most "non-standard" values were given by DCT-related formulas (e.g. `"DCT-P2Y"`). For discharge summaries, the second most frequent type was concept-related (e.g. `"ADM+P2Y"`, where ADM stands for ADMission day). Vague values were used across all subsets to mark time references that were not explicitly written (*at that time, on the following Tuesday*).

In all MTSamples subsets, the most frequent type of disagreement was Duration-vs-Date, with a proportion of 47% over all other types of disagreements (41/86, 51/98, 21/59, 34/69, in discharge summaries, psychiatry-psychology, paediatrics, and emergency, respectively). In Table 2, we report the most common types of disagreement across all subsets.

By manually analysing documents where these disagreements were present, and taking into considerations our findings on the (TimeML-based TIMEX) annotated corpus, we propose a new annotation schema for capturing time expressions that are actually useful for timeline reconstruction: CALendar Expressions (CALEX).

4 CALEX

CALEX refers to a temporal annotation schema restricted to time expressions and concepts that can be (directly or not) connected to an absolute timeline. The key novelty of this model is to better utilise time expression properties that are relevant for anchoring points on a timeline, including the introduction of certain timeline-relevant concepts.

In relation to TimeML-based TIMEX definitions, CALEX *excludes* the following, because they cannot be directly used for timeline reconstruction:

- FREQUENCY/SET/QUANTIFIER, e.g. *once a week, two units of blood*;

	dis. summaries	psych.	paediatrics	emergency
Documents	108	53	70	75
Tokens	55,433 (513/doc)	67,569 (1275/doc)	36,675 (524/doc)	52,041 (694/doc)
TIMEXes (merged)	1,378	1,227	566	801
TIMEXes (overlap)	994	840	360	496
TIMEXes (same type)	908	742	301	427
Date	326 (36%)	234 (32%)	85 (28%)	154 (36%)
Duration	110 (12%)	122 (16%)	49 (16%)	44 (10%)
Time	29 (3%)	31 (4%)	23 (8%)	39 (9%)
Frequency	355 (39%)	216 (29%)	61 (20%)	88 (21%)
Age_related	88 (10%)	139 (19%)	83 (28%)	102 (24%)
IAA F1	0.84	0.81	0.78	0.76
type acc.	0.91	0.88	0.84	0.86
type K	0.89	0.85	0.79	0.82
value acc.	0.74	0.72	0.75	0.74

Table 1: Manual annotation results - time expressions (TIMEXes) on documents from MTSamples: discharge summaries (dis. summaries), psychiatry/psychology (psych.), paediatrics and emergency department documents.

Type	dis. summ.	psych.	paediatrics	emergency
Duration-vs-Date	41	51	21	34
Duration-vs-Time	11	5	8	16
Duration-vs-Frequency	16	5	7	6
Age_related-vs-Date	1	10	11	1
Age_related-vs-Duration	1	7	8	2
Date-vs-Time	2	6	2	6
Frequency-vs-Time	6	7	1	1

Table 2: TIMEX type disagreement counts on the MTSamples subsets - discharge summaries (dis. summ.), psychiatry/psychology (psych.), paediatrics and emergency department documents.

- DURATION when it is a temporal *attribute* describing other events, e.g. "the procedure usually takes *15 minutes*";

- TIME when it refers to temporal *attributes* describing other events, e.g. "to be always taken *around 9am*".

There are three main elements in the proposed CALEX annotation schema: TYPE, METADATA, and VALUE.

TYPE

TYPE defines the type of a calendar expression. The possible types within the CALEX schema are described as follows:

- CALENDAR: this type covers all calendar expressions that do not require any metadata in order to provide the final normalised VALUE, including:

  - explicit calendar references in different temporal granularities such as date, month, year
  - timestamps
  - explicit ranges
  - when time is described as a *period* of time (duration) but the connection with the timeline is not clear or explicit - this type refers to the original DURATION type as part of the TimeML annotation guidelines (e.g. "he took this medication *for one month*")

- AGE: age-related expressions can either define the current age of a patient (e.g. "a *56 year old* woman"), or be a reference to a certain point in time in which the patient had a given age (*at the age of 17*).

- DOMAIN: expressions that either explicitly define the value of a domain-specific concept (*admitted on 2010 Jun 6th*), or are references to a given domain-specific concept (*on post-operative day #4*).

- DCT: expressions that require information about the document creation time in order to be normalised (*last month*).

- TENSE: imprecise expressions that refer to conditions in the past, present or future (*recently*).

- CONTEXT: expressions that refer to a *temporal context*, represented by either the last mentioned temporal reference or the most recent temporal reference available within the document. [5] This type includes times/periods of the day where the connection to the timeline is not clear and relies on the temporal context (e.g. *the previous night*). However, times/periods of the day representing frequencies (e.g. "one tablet *at night*") are NOT considered as CALEXes.

METADATA

We introduce METADATA as a feature to allow for a computationally more efficient way of calculating a particular time reference for CALEXes that are not explicitly anchored in time. An essential aspect of this feature is that it can include concepts in its definition. These can also be explicitly set within the METADATA feature, to ensure the original values are used in order to normalise the final calendar expression.

Document-related concepts include the document creation time {doc.DCT} and contextualised references to the last or more recent temporal mentions within the text ({doc.LAST} and {doc.RECENT}). Patient-related concepts are used to describe patient demographic features, such as {patient.AGE}, {patient.DOB}, {patient.DOD}, the later possibly useful when analysing death certificates.[6] Patient-stay-related concepts will basically refer to the period within admission and discharge ({patient.ADMISSION} and {patient.DISCHARGE}).

One important concept that may require some disambiguation is related to the pregnancy period.

The terms *Pregnancy* and *Prenatal* are generally interchangeably used when referring either to the mother or the child. We formalise *Pregnancy* as being the period of time used when referring to the mother as a patient, whereas *Prenatal* refers to the period of time (usually 40 weeks) before {patient.DOB}, which refers to the child as a patient. This way, {patient.PREGNANCY} can occur at any time in the patient's life, whereas {patient.PRENATAL} is the period of 40 weeks preceding the patient's date of birth.

Finally, some social- and family-related concepts can be used in order to refer to some temporal values regarding the patient's relatives, such as {mother.AGE} or {father.DOB}.

Besides making use of timeline-relevant concepts, METADATA also contains functions that are used to derive values:

- .set(): for explicitly defining the value of domain-specific concepts;

- .add(): adds a period of time to a given point in the calendar, moving to a later point in time;

- .sub(): subtracts a period of time from a given point in the calendar, moving to an earlier point in time;

- .next(): finds the next occurrence of a temporal granularity based on an anchor calendar expression;

- .prev(): finds the previous occurrence of a temporal granularity based on an anchor calendar expression.

For example, a reference to DCT cannot be properly normalised when DCT is unknown. However, the metadata can keep the definition for a calendar expression, to be converted to an actual value when DCT is given: instead of parsing the entire document, only the metadata has to be re-evaluated – e.g. metadata for the expression "yesterday" is "{doc.DCT}.sub(P1D)".

VALUE

This component gives a normalised value of a calendar expression, mostly following the previous TimeML notation, with an extension: *range* values are used to normalise periods of time in the form of [begin,end].[7]

---

[5] Other specific contextual references can be required for documents in different domains.

[6] {patient.AGE} and {patient.DOB} represent complementary concepts.

[7] To indicate included endpoints, we use standard square brackets: [A,B]. To indicate excluded endpoints, we use re-

Example	Type	Metadata	Value
dated June 15, 2007	CALENDAR	null	2007-06-15
on June 15, 2007 at 10:00	CALENDAR	null	2007-06-15T10:00
in 2009	CALENDAR	null	2009
between 2007 and 2009	CALENDAR	null	[2007,2009]
since 2007	CALENDAR	null	[2007,]
after 2007	CALENDAR	null	]2007,]
for one month	CALENDAR	null	[,P1M]
a 20-year-old male patient ...	AGE	{patient.AGE}.set(P20Y)	P20Y
at age 15, when...	AGE	{patient.DOB}.add(P15Y)	XXXX (unknown DOB)
at age 15, when...	AGE	{patient.DOB}.add(P15Y)	[2002-04,2003-03] (known DOB)
since age 25 ...	AGE	[{patient.DOB}.add(P25Y),]	[XXXX,] (unknown DOB)
admitted on 05-27-2009	DOMAIN	{patient.ADMISSION}.set(2009-05-27)	2009-05-27
born in 07/2007	DOMAIN	{patient.DOB}.set(2007-07)	2007-07
discharged on 01/21/10	DOMAIN	{patient.DISCHARGE}.set(2010-01-21)	2010-01-21
upon discharge	DOMAIN	{patient.DISCHARGE}	XXXX-XX-XX (unknown)
18 hours prior to admission	DOMAIN	{patient.ADMISSION}.sub(PT18H)	2010-06-25T02:00
tomorrow	DCT	{doc.DCT}.add(P1D)	2010-07-02
11 years ago	DCT	{doc.DCT}.sub(P11Y)	1999
for the next 2 weeks	DCT	[{doc.DCT},P2W]	[2010-07-01,2010-07-15]
next Tuesday	DCT	{doc.DCT}.next(WD,3)	2010-07-06
in july of next year	DCT	{doc.DCT}.add(P1Y).next(M,7)	2011-07
in the past	TENSE	[,{doc.DCT}[	[,2010-07-01[
recently	TENSE	],{doc.DCT}[	],2010-07-01[
at this time	TENSE	]{doc.DCT}[	]2010-07-01[
in the future	TENSE	]{doc.DCT},]	]2010-07-01,]
at that time	CONTEXT	{doc.LAST}	2010-03-15
3 days prior	CONTEXT	{doc.LAST}.sub(P3D)	2010-03-12
10am	CONTEXT	{doc.LAST}.next(TH,10)	2010-03-15T10:00
was...on Tuesday	CONTEXT	{doc.LAST}.prev(WD,3)	2010-03-09

* doc.DCT = "2010-07-01" for all the examples

Table 3: Calendar Expression — CALEX — examples.

Table 3 presents some examples on how the CALEX annotation schema works in terms of normalising the main features.[8]

As shown in the examples, a key element of the CALEX schema is the handling of domain-specific concepts in the METADATA element.

In Table 4, we show how different expressions would be represented within CALEX and TimeML, highlighting the types to be added to capture timeline-related expressions in CALEX format ("N/A" values in the *TimeML type* column).

Figure 2 provides an example of timeline creation using CALEX instead of TimeML (for the psychiatry domain). First, to temporally anchor the first emergence of auditory hallucinations, an age-related time expression is added (*since the age of 14*). Second, to capture the admission date, a specific domain concept is used (*On admission*, abbreviated as {patient.ADM}). For these expressions, the METADATA feature allows identifying a specific point in the timeline without the need for temporal links. As another difference, the medication frequency (*twice a day*), which cannot be represented at the timeline level, is removed.

---

verse square brackets:]A,B[. Open ranges/periods of time are indicated by [A,] or [,B].

[8]Note that relevant prepositions are included in the expression textual span.

5 Discussion

In this paper, we investigated how temporal information is documented in clinical text by focusing on time expressions (TIMEXes), using clinical notes from MTSamples for four different specialties (discharge summaries, psychiatry and psychology, paediatrics, and emergency). Our goal was to assess whether TIMEX annotation schemas based on TimeML would be suitable to capture the information needed to reconstruct patient timelines. First, we annotated documents using TimeML-inspired TIMEX types. Then, we analysed which of these expressions actually indicate a connection to the timeline, thus proposing a new annotation schema based on calendar expressions: CALEX.

Annotating MTSamples documents with a TimeML-based TIMEX model was helpful to investigate how temporal information is reported across different clinical specialties. Despite the use of sample reports, which might be more structured as compared to real clinical records, the distribution of time expression types (Table 1) is similar to those found in i2b2 2012 and THYME, where Date represents the most common TIMEX type and Time the least common. By analysing our manually annotated time expressions, we identified some key points to be taken into account to simplify timeline reconstruction. First, we ob-

CALEX type	Example	Definition	TimeML type
CALENDAR	{on 02/12/2009}	directly connected to calendar	Date
DCT	{tomorrow}	relative to the DCT	Date
TENSE	{in the past}	imprecise reference	Date
DCT	{two years ago}	relative to the DCT	Duration
DOMAIN	{18 hours prior to admission}	related to a domain concept	Duration
CONTEXT	{three days before}	related to another expression	Duration
—	the procedure usually takes {15 minutes}	not directly connected to calendar	Duration
CALENDAR	on 02/12/2009 {at 9am}	directly connected to calendar	Time
—	{twice a day}	any re-occurring expression	Frequency/Set/Quantifier
AGE	a {56 years old} woman	age of the patient	N/A
AGE	{when she was 17}	reference to age	N/A
DOMAIN	{admitted on Oct 12th}	domain-concept definition	N/A
DOMAIN	{the day before admission}	reference to domain	N/A

Table 4: Time expression examples as represented within CALEX and TimeML.

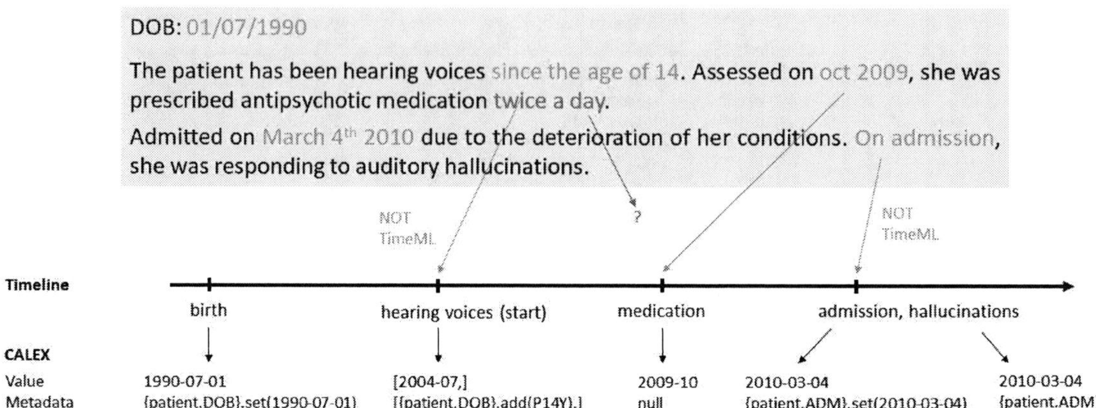

Figure 2: Example of timeline creation using CALEX instead of TimeML: an age-related and a domain-specific time expressions are added, while the medication frequency is removed.

served that most Frequency annotations are not helpful to anchor clinical events on an absolute timeline. Our suggestion is to remove such elements from time expression annotations, and to capture them as entity attributes instead (e.g. drug prescriptions). Moreover, the TimeML TIMEX normalisation step is not always directly useful for timeline reconstruction, as some expressions would still require different types of temporal links to be connected to the calendar.

To address these points, our proposed model, CALEX, integrates timeline information at the time expression level, specifying three different components: TYPE, METADATA, and VALUE. The new TYPE classification allows distinguishing between expressions directly connected to the calendar (e.g. full explicit dates) and relative/contextual expressions. Besides facilitating timeline reconstruction, this should also reduce ambiguity when assigning expression types, as the different type categories are more clearly separated. The METADATA feature, in combination with TYPE, allows storing the information needed for calendar normalisation, making use of functions and timeline-relevant concepts (Table 3). While functions are general and reusable across different document types, timeline-relevant concepts are specific to each domain or use-case, capturing the most appropriate anchor points within a finite set (e.g. {Admission, Discharge} for discharge summaries). The METADATA feature is useful to automatically derive or evaluate the normalised VALUE, especially for concept-related/contextual expressions where manually assigning values might be not straightforward.

Compared to TimeML, the CALEX model removes the Frequency/Quantifier/Set type and introduces new types and normalised values. In particular, Date- and Duration- like expressions are assigned different CALEX types depending on how they can be linked to the timeline (Table 4), which will be useful to reduce manual annotation for temporal links. Within the CALEX model, instead, a greater annotation effort is required for the normalisation task. Especially for relative expressions, assigning standardized values to METADATA is likely to be hard for non-technical annotators. Therefore decisions on what to manually annotate in the CALEX model will need to be defined. As a first step, we would require manual annotations mainly for calendar expression VALUEs, specifying the METADATA feature only if necessary (e.g. when no is DCT available) and using a simplified notation (e.g. "DCT-P2Y"). In most cases, this feature would be derived programmatically, and its derivative value used for evaluating the manual VALUE.

This study has some limitations and directions for future work. The TimeML-based TIMEX annotations have not been adjudicated. However, we have released the corpus as it is, so that NLP researchers can integrate/reuse annotations for analysis and system development. In particular, we have made available all annotations (merged), specifying which ones are overlapping (and could therefore be considered as more reliable). Our study has been heavily focused on analysing time expressions: we have not systematically also analysed how existing annotation schemas can capture calendar information by other annotation elements. For example, i2b2 2012 and THYME include annotations for admission and discharge, but they are classified as *events* to be linked to other temporal entities. In other studies, it has been proposed to add timeline information directly as event attributes, e.g. Reimers et al. (2018).

When normalising time expressions, another aspect to be considered is the presence of imprecise temporal references, which are abundant in the clinical domain (Tissot et al. (2019)). As part of the CALEX model, TENSE expressions are included, which are used to refer to the past, present or future. At the moment, we are also evaluating how to incorporate other types of imprecise temporal references. More generally, we are designing a CALEX annotation guideline which is focused on both manual (e.g. VALUE) and potentially automatic (e.g. METADATA) tasks. As future work, we plan to create a reference corpus annotated with CALEXes, and design a shared task for further evaluation. Creating a CALEX annotated corpus will be crucial to assess the utility of our model, as well as to highlight potential issues and areas for improvement (with a specific focus on the proposed types and the METADATA feature).

6   Conclusion

In this paper we developed a corpus of medical reports annotated with TimeML-based time expressions and systematically analysed their usefulness for timeline reconstruction. As a result, we proposed a new annotation schema, CALEX, which will be used to design and develop new resources.

Acknowledgments

We thank the reviewers for valuable comments on our manuscript. SV has received support from the Swedish Research Council (2015-00359), and the Marie Skłodowska Curie Actions, Cofund, Project INCA 600398. HT is funded and supported by the Health Data Research UK (grant No. LOND1).

References

Steven Bethard, Guergana Savova, Wei-Te Chen, Leon Derczynski, James Pustejovsky, and Marc Verhagen. 2016. Semeval-2016 task 12: Clinical tempeval. In *Proceedings of the 10th International Workshop on Semantic Evaluation (SemEval-2016)*, pages 1052–1062, San Diego, California. Association for Computational Linguistics.

Steven Bethard, Guergana Savova, Martha Palmer, and James Pustejovsky. 2017. Semeval-2017 task 12: Clinical tempeval. In *Proceedings of the 11th International Workshop on Semantic Evaluation (SemEval-2017)*, pages 565–572, Vancouver, Canada. Association for Computational Linguistics.

Nathanael Chambers, Taylor Cassidy, Bill McDowell, and Steven Bethard. 2014. Dense event ordering with a multi-pass architecture. *Transactions of the Association for Computational Linguistics*, 2(1):273–284.

Serena Jeblee and Graeme Hirst. 2018. Listwise temporal ordering of events in clinical notes. In *Proceedings of the Ninth International Workshop on Health Text Mining and Information Analysis*, pages 177–182, Brussels, Belgium. Association for Computational Linguistics.

Min Jiang, Yonghui Wu, Anushi Shah, Priyanka Priyanka, Joshua C Denny, and Hua Xu. 2014. Extracting and standardizing medication information in clinical text–the medex-uima system. *AMIA Summits on Translational Science Proceedings*, 2014:37.

Oleksandr Kolomiyets, Steven Bethard, and Marie-Francine Moens. 2012. Extracting narrative timelines as temporal dependency structures. In *Proceedings of the 50th Annual Meeting of the Association for Computational Linguistics: Long Papers - Volume 1*, ACL '12, pages 88–97, Stroudsburg, PA, USA. Association for Computational Linguistics.

James Pustejovsky, José M Castano, Robert Ingria, Robert J Gaizauskas, Andrea Setzer, Graham Katz, and Dragomir R Radev. 2003a. Timeml: Robust specification of event and temporal expressions in text.

James Pustejovsky, Patrick Hanks, Roser Sauri, Andrew See, Robert Gaizauskas, Andrea Setzer, Dragomir Radev, Beth Sundheim, David Day, Lisa Ferro, et al. 2003b. The timebank corpus. In *Corpus linguistics*, volume 2003, page 40. Lancaster, UK.

James Pustejovsky, Kiyong Lee, Harry Bunt, and Laurent Romary. 2010. Iso-timeml: An international standard for semantic annotation. In *LREC*, volume 10, pages 394–397.

James Pustejovsky and Marc Verhagen. 2009. Semeval-2010 task 13: evaluating events, time expressions, and temporal relations (tempeval-2). In *Proceedings of the Workshop on Semantic Evaluations: Recent Achievements and Future Directions (SEW-2009)*, pages 112–116.

Preethi Raghavan, Eric Fosler-Lussier, Noémie Elhadad, and Albert M. Lai. 2014. Cross-narrative temporal ordering of medical events. In *Proceedings of the 52nd Annual Meeting of the Association for Computational Linguistics (Volume 1: Long Papers)*, pages 998–1008, Baltimore, Maryland. Association for Computational Linguistics.

Nils Reimers, Nazanin Dehghani, and Iryna Gurevych. 2016. Temporal anchoring of events for the timebank corpus. In *Proceedings of the 54th Annual Meeting of the Association for Computational Linguistics (Volume 1: Long Papers)*, pages 2195–2204, Berlin, Germany. Association for Computational Linguistics.

Nils Reimers, Nazanin Dehghani, and Iryna Gurevych. 2018. Event time extraction with a decision tree of neural classifiers. *Transactions of the Association for Computational Linguistics*, 6:77–89.

William F Styler IV, Steven Bethard, Sean Finan, Martha Palmer, Sameer Pradhan, Piet C de Groen, Brad Erickson, Timothy Miller, Chen Lin, Guergana Savova, and James Pustejovsky. 2014. Temporal annotation in the clinical domain. *Transactions of the Association for Computational Linguistics*, 2:143–154.

Weiyi Sun, Anna Rumshisky, and Ozlem Uzuner. 2013a. Annotating temporal information in clinical narratives. *Journal of biomedical informatics*, 46:S5–S12.

Weiyi Sun, Anna Rumshisky, and Ozlem Uzuner. 2013b. Evaluating temporal relations in clinical text: 2012 i2b2 Challenge. *Journal of the American Medical Informatics Association*, 20(5):806–813.

Hegler Tissot, Marcos Didonet Del Fabro, Leon Derczynski, and Angus Roberts. 2019. Normalisation of imprecise temporal expressions extracted from text. *Knowledge and Information Systems*.

Hegler Tissot, Angus Roberts, Leon Derczynski, Genevieve Gorrell, and Marcos Didonet Del Fabro. 2015. Analysis of temporal expressions annotated in clinical notes. In *Proceedings of 11th Joint ACL-ISO Workshop on Interoperable Semantic Annotation*, pages 93–102, London, UK. ACL.

Marc Verhagen, Robert Gaizauskas, Frank Schilder, Mark Hepple, Graham Katz, and James Pustejovsky. 2007. Semeval-2007 task 15: Tempeval temporal relation identification. In *Proceedings of the fourth international workshop on semantic evaluations (SemEval-2007)*, pages 75–80.

Natalia Viani, Lucia Yin, Joyce Kam, Ayunni Alawi, André Bittar, Rina Dutta, Rashmi Patel, Robert Stewart, and Sumithra Velupillai. 2018. Time expressions in mental health records for symptom onset extraction. In *Proceedings of the Ninth International Workshop on Health Text Mining and Information Analysis*, pages 183–192, Brussels, Belgium. Association for Computational Linguistics.

Ran Zhao, Quang Do, and Dan Roth. 2012. A robust shallow temporal reasoning system. In *Proceedings of the Demonstration Session at the Conference of the North American Chapter of the Association for Computational Linguistics: Human Language Technologies*, pages 29–32, Montréal, Canada. Association for Computational Linguistics.

Leveraging Sublanguage Features for the
Semantic Categorization of Clinical Terms

Leonie Grön
Quantitative Linguistics and
Lexical Variation (QLVL)
KU Leuven
leonie.gron@kuleuven.be

Ann Bertels
Leuven Language Institute (ILT)
Quantitative Linguistics and
Lexical Variation (QLVL)
KU Leuven
ann.bertels@kuleuven.be

Kris Heylen
Quantitative Linguistics and
Lexical Variation (QLVL)
KU Leuven
kris.heylen@kuleuven.be

Abstract

The automatic processing of clinical documents, such as Electronic Health Records (EHRs), could benefit substantially from the enrichment of medical terminologies with terms encountered in clinical practice. To integrate such terms into existing knowledge sources, they must be linked to corresponding concepts. We present a method for the semantic categorization of clinical terms based on their surface form. We find that features based on sublanguage properties can provide valuable cues for the classification of term variants.

1 Background

Structured terminologies and ontologies play a pivotal role in the automatic processing of health data, as they provide the framework for mapping unstructured information into a machine-readable format. Moreover, the term bases themselves can serve as input for the identification of medical entities in free text. Even though methods from machine learning are gaining popularity, many state-of-the-art systems rely strongly on pre-compiled terminologies (e.g. Savova et al. 2010). The performance of such applications thus relies crucially on the lexical coverage of the term base. However, the major biomedical terminologies, such as the

Systematic Nomenclature of Medicine – Clinical Terms (SNOMED CT)[1] and the Unified Medical Language System (UMLS)[2] do not adequately reflect the range of term variants encountered in clinical practice. Especially in languages other than English, where the available terminologies are less comprehensive, this discrepancy can harm performance (Henriksson et al. 2014; Skeppstedt et al. 2014). One strategy to overcome this bottleneck is to enrich the available terminologies with additional variants acquired from domain corpora. Concretely, this involves the recognition of variants in text, and their association with the semantic classes or concepts provided by the respective terminology.

The focus of this paper is on the second task, i.e. the semantic categorization of term variants. In particular, we investigate whether the features of a given sublanguage can be leveraged to associate individual variants with semantic classes. According to sublanguage theory, specialized languages can be characterized by semantic constraints, as well as stylistic preferences and distinctive syntactic patterns (Friedman, Kra, and Rzhetsky 2002; Harris 1982, 1991; 2002). In the medical domain, such differences manifest themselves at fine-grained levels, e.g. between clinical specialties and different document types (Feldman, Hazekamp,

---

[1] https://browser.ihtsdotools.org/?

[2] https://uts.nlm.nih.gov/home.html

Proceedings of the BioNLP 2019 workshop, pages 211–216
Florence, Italy, August 1, 2019. ©2019 Association for Computational Linguistics

Section	Function	Stylistic properties
Anamnesis	Assess environmental and behavioral factors that could influence the patient's condition.	Narrative; high proportion of abbreviations
Comments	Inform colleagues about the current state and further course of treatment.	Telegraphic; high proportion of abbreviations and non-standard variants
Complaints	Summarize the current mental and physical state as experienced by the patient himself.	Narrative; high proportion of lay terms
Conclusion	Inform the patient's GP about the outcome of the consultation and the course of therapy.	Narrative; well-formed syntax; standard terms
Examination	Report on procedures carried out during the consultation.	Telegraphic; high proportion of abbreviations
History	Enumerate prior conditions and procedures that the patient underwent.	List-style; mostly nominal forms; standard terms
Medication	List the pharmaceutical substances administered to the patient.	List-style; mostly nominal forms
Therapy	Document further therapeutic measures.	List-style; mostly nominal forms

Table 1: Overview of the sections of the EHRs in the corpus, their communicative function and style.

and Chawla 2016). We capitalize on this phenomenon for the semantic classification of clinical terms: Drawing on the observation that, even within one clinical document, there are fundamental semantic and stylistic differences between the individual sections, we consider the languages found in different parts of the EHR sublanguages of their own. Based on the assumption that, within the context of a sublanguage, certain variation processes pattern with conceptual properties, we use properties of the surface form as predictors for the semantic classification of the term.

The remainder of this paper is structured as follows: In Section 2, we give an overview of related research. In Section 3, we describe our materials and methods. After the presentation of the results (Section 4) and their discussion (Section 5), we conclude in Section 6.

2   Related research

Especially in emerging domains and under-resourced languages, domain corpora are a valuable resource for terminology development. Automatic Term Recognition (ATR) from biomedical and clinical text is thus a well-studied field (cf. e.g. Spasić et al. 2013; Carroll, Koeling, and Puri 2012; Doing-Harris, Livnat, and Meystre 2015; Zhang et al. 2017 for state-of-the-art systems).

To leverage the acquired terms for NLP, they are typically organized according to their semantic properties. If the target categories are not yet defined, clustering can be used to group semantically related terms and infer taxonomical relations

(Siklósi 2015). However, the more common scenario is that the newly acquired variants need to be integrated into an existing knowledge source. To associate terms with pre-defined semantic categories, both external and internal features of the terms have been used. Most approaches rely on external context. In particular, they draw on the core assumption of distributional semantics, which is that semantically similar words tend to occur in similar lexical contexts and syntactic constellations (Sibanda et al. 2006; Weeds et al. 2014). A number of studies showed, though, that term-internal properties can inform the task as well: Medical terms contain a high number of descriptive elements, such as neoclassical affixes or roots associated with a semantic type. Such features have been successfully employed to classify biological concept names and validate the assignment of semantic types in biomedical knowledge sources (Torii, Kamboj, and Vijay-Shanker 2004; Fan, Xu, and Friedman 2007). Morpho-semantic decomposition has also been employed for the semantic grouping of medical compounds in a multilingual setting (Namer and Baud 2007).

However, these approaches only work for a very confined group of terms, namely specialized terms that are based on neoclassical roots, spelled out in their full form, and adhere to grammatical and orthographic conventions. While these conditions might be met in the biomedical genre, they are unrealistic when dealing with input from the clinical domain: In clinical practice, medical staff use both specialized terms and lay variants, which do not contain neoclassical elements. Moreover, clinical

Feature	Criteria	Example term from corpus
REGISTER	Standard term as in SNOMED CT	hypotensie "hypotension"
REDUCTION	Abbreviation or acronym	asp "aspirine"
MORPHO-SYNTACTIC VARIANT	Derivation, paraphrase or compound	thoraxwand "thorax wall"

Table 2: Formal term features.

Semantic class	Example concepts (SNOMED CT term)
ANATOMY	Thoracic structure
CHEMICALS & DRUGS	Human insulin analog product
CONCEPTS & IDEAS	Chronic persistent
DISORDERS	Hypotension
PROCEDURES	Thyroid panel

Table 3: Semantic classes and example concepts.

records are composed in a hectic environment and primarily intended for peer-to-peer communication. They are thus known to contain a high proportion of irregular or intransparent forms, such as misspellings and abbreviations. Therefore, in this paper, we investigate whether the approach can be taken to a more abstract level. Instead of using the words themselves as predictors, we employ a set of non-lexical features reflecting formal properties of the surface form.

3 Materials and Methods

3.1 Corpus Characteristics

We evaluate the approach on a set of terms extracted from a clinical corpus written in Belgian Dutch. This corpus consists of 4,426 EHRs, which were provided by a Belgian hospital. All of them relate to patients diagnosed with diabetes, who visit the hospital in regular intervals for routine checkups. The EHRs were exported from the clinical data warehouse and de-identified by the ICT team of the hospital. In particular, all personal information concerning the patients themselves, their families, or members of clinical staff was removed. In addition, all researchers that had insight into the data signed confidentiality agreements with the hospital.

All EHRs relate to individual clinical encounters. They were composed with a semi-structured template, which contains different sections relating to the individual stages of a consultation. These sections differ with regard to their thematic scope and communicative function, resulting in characteristic semantic structures and stylistic properties. They can thus be considered distinct sublanguages.

For example, the section *complaints* serves to assess the current mental and physical condition. This section is composed in interaction with the patient, which manifests itself in the narrative style and a high proportion of lay terms. By contrast, the *comments* are used for the informal exchange among colleagues. This section is composed in a telegraphic style, containing a high proportion of ungrammatical constructions and jargon expressions. Table 1 gives an overview of the sections and their characteristics.

3.2 Semantic and Formal Annotation

In an earlier project, all EHRs in the corpus were manually annotated with concept codes from SNOMED CT. After manual validation of the term-concept association, a total of 15,025 unique terms, relating to 7,687 different concepts, remain. All concepts were mapped to the semantic groups of the UMLS (McCray, Burgun, and Bodenreider 2001). In a second pass, the terms obtained in the earlier stage were also annotated at the formal level. To this end, the unique terms were manually annotated with a set of binary features reflecting the term's register, morpho-syntactical alternations and reduction processes. Table 2 gives an overview of the formal term features. [3]

Each term was inspected individually. For those features that applied to the term, a positive value was assigned; for the remaining features, the values remained negative by default. For example, the term *hypotens* 'hypotensive' would be assigned the following features: REGISTER – *positive*; REDUCTION – *negative*; MORPHO-SYNTACTIC VARIANT – *positive*.

---

[3] While the original set of features was more extensive, we used a reduced version for the present study to create more realistic conditions. In a real-life scenario, it is unlikely that resources would be available for the manual coding of term features. Therefore, we only included those features that could be assigned automatically, e.g. by dictionary lookup or morphological analysis.

Section	Number of terms	Number of target classes	F1-score	REGISTER	REDUCTION	MORPHO-SYNTAX
Anamnesis	1081	5	0.73	0.08	0.23	**0.69**
Comments	3105	5	0.64	0.14	**0.54**	0.31
Complaints	1592	5	0.5	0.16	0.02	**0.82**
Conclusion	8214	5	0.48	**0.49**	0.03	0.48
Examination	804	3	0.85	0.25	0.31	**0.44**
History	4202	5	0.45	**0.6**	0.15	0.26
Medication	3529	3	0.99	0.24	0.11	**0.65**
Therapy	508	4	0.86	0.27	0.03	**0.7**
	23035		**0.69**	**0.28**	**0.18**	**0.55**

Table 4: Details of the terms and the results of the classification by section. The last three columns specify the mean importance of the different predictor types; for each section, the highest value is printed in bold. The last row provides the sum of the second column and the mean values of the last four columns.

3.3 Composition of the Concept and Term Sample

For the classification task, we focused on the five most frequently occurring semantic groups, namely DISORDERS, PROCEDURES, CONCEPTS & IDEAS, CHEMICALS & DRUGS and ANATOMY (cf. Table 3). For each group, the associated concepts were ranked by absolute frequency and the number of associated variants. Five concepts per group were chosen for the classification task. The final selection of concepts was based on the diversity of formal alternations observed in the associated variants. For instance, a concept whose terms showed variation in both morpho-syntax and reduction (e.g. a noun phrase and a paraphrase, and an abbreviation and a full form) would be preferred over a concept whose terms only vary at the morpho-syntactical level. Moreover, we aimed to compose the sample such that the full spectrum of the semantic class would be covered. For instance, for ANATOMY, we chose concepts relating to visible body parts (e.g. *leg*) as well as internal organs (e.g. *thyroid*) The final sample consisted of 25 concepts. For each concept, the annotated terms were retrieved from our corpus and sorted by the section of occurrence. Concepts occurring with a frequency of less than 500 within a section were excluded. Consequently, the number of semantic classes varies across sections.

3.4 Experimental Setup

We approached the categorization task as a multiclass classification problem with multiple predictors: *Given the observation of a term in a particular section, predict the semantic category based on the formal features.* Our hypothesis is that the sublanguage features of each section influence the informativity of the formal predictors. For example, in a narrative section like the *complaints*, MORPHO-SYNTACTICAL features should be better predictors than in the *medication*, which contains few full sentences, but merely enumerates drugs and dosage instructions. On the other hand, the REDUCTION feature is likely more insightful in the *comments*, which are dominated by informal expressions, than in the *conclusion*, where well-formed expressions prevail.

For the classification experiment, we used a Python implementation[4] of the Random Forest Classifier (Breiman 2001). For each section, the list of annotated terms is split into a training and test set, containing 70% and 30% of all terms respectively. One model is trained and tested per section. To evaluate the results, we calculate the F1-score as well as the mean importance of the different predictor types.

4 Results

Overall, the best results were achieved in those sections that only contain a small number of target classes, namely the *medication, therapy* and *examination*. By contrast, the F1-values tend to be lower in those sections that are more diverse. On average, the MORPHO-SYNTACTIC features are the most important predictors, followed by the REGISTER feature. The REDUCTION feature, on the other hand, seems less informative overall.

At the same time, the relative contribution of the feature types varies considerably across the sections: In the *conclusion* and *history*, REGISTER is

---

[4] https://scikit-learn.org/stable/modules/generated/sklearn.ensemble.RandomForestClassifier.html

the strongest predictor; however, in the *conclusion,* the MORPHO-SYNTACTIC features are almost on par with REGISTER. While REDUCTION is most important in the *comments,* it also has a substantial effect in the *examination.* The MORPHO-SYNTACTIC features make the strongest contribution in the *complaints, therapy, anamnesis* and *medication*; they are also strongest, but not quite as dominant, in the *examination.* Table 4 provides the full results.

5   Discussion

The results show that the semantic complexity of the respective sublanguage influences classification performance. The best F1-scores were achieved in those sections devoted to a very confined topic, while the values were lower in the more heterogeneous ones. This tendency corroborates the findings of previous work studying the effect of sublanguage properties on NLP in the clinical domain (Doing-Harris et al. 2013).

However, we found striking differences in the relative importance of the predictor types. On the whole, the contribution of the predictors patterns with the stylistic properties of the respective sublanguages: For instance, MORPHO-SYNTACTIC features are most informative in those sections composed in a narrative style; REDUCTION is strongest in the informal parts of the document. This finding confirms our initial hypothesis. At a closer look, though, another effect emerges: In semantically homogeneous sections, infrequent features can serve to identify conceptual outliers. For instance, in the therapy-centered sections, which are dominated by nouns relating to pharmaceutical substances, the presence of non-nominal morphological properties, such as an adjective ending, is a strong predictor for a term belonging to another semantic class, such as a temporal modifier.

Our study has its limitations, as it only considers a very small sample of highly frequent concepts. Possibly, for low-frequency concepts, the formal features would not be informative enough to allow a reliable classification. Therefore, in future work, we plan to replicate the experiment at a larger scale, including a more diverse concept sample. Besides, in order to test the generalizability of the method, it would be interesting to evaluate the performance on data from different clinical specialties, and from multiple clinical institutions.

6   Conclusion

We presented a first attempt for the classification of clinical terms by formal features alone. While there is much variation in the results, our experiment demonstrates that sublanguage properties can be exploited to associate terms acquired from domain corpora with semantic categories. This approach could be integrated with other systems to support the enrichment of medical terminologies. In further research, we plan to replicate the study at a larger scale.

Acknowledgements

This work was supported by Internal Funds KU Leuven.

References

Breiman, Leo. 2001. "Random Forests." *Machine Learning* 45: 5–32. https://doi.org/10.1023/A:1010933404324.

Carroll, John, Rob Koeling, and Shivani Puri. 2012. "Lexical Acquisition for Clinical Text Mining Using Distributional Similarity." *Lexical Computational Linguistics and Intelligent Text Processing. CICLing 2012,* 232–4. https://doi.org/10.1007/978-3-642-28601-8_20.

Doing-Harris, Kristina, Olga Patterson, Sean Igo, and John Hurdle. 2013. "Document Sublanguage Clustering to Detect Medical Specialty in Cross-Institutional Clinical Texts." *Proc ACM Int Workshop Data Text Min Biomed Inform,* 9–12. https://doi.org/10.1145/2512089.2512101.

Doing-Harris, Kristina, Yarden Livnat, and Stephane Meystre. 2015. "Automated Concept and Relationship Extraction for the Semi-Automated Ontology Management (SEAM) System." *J Biomed Sem* 6: 15. https://doi.org/10.1186/s13326-015-0011-7.

Fan, Jung Wei, Hua Xu, and Carol Friedman. 2007. "Using Contextual and Lexical Features to Restructure and Validate the Classification of Biomedical Concepts." *BMC Bioinform* 8: 264. https://doi.org/10.1186/1471-2105-8-264.

Feldman, Keith, Nicholas Hazekamp, and Nitesh V. Chawla. 2016. "Mining the Clinical Narrative: All Text Are Not Equal." *2016 IEEE International Conference on Healthcare Informatics (ICHI),* 271–80. https://doi.org/10.1109/ICHI.2016.37.

Friedman, Carol, Pauline Kra, and Andrey Rzhetsky. 2002. "Two Biomedical Sublanguages: A Description Based on the Theories of Zellig Harris." *J Biomed Inform* 35 (4): 222–35. https://doi.org/10.1016/S1532-0464(03)00012-1.

Harris, Zellig. 1982. "Discourse and Sublanguage." In *Sublanguage. Studies of Language in Restricted Semantic Domains*, edited by Richard Kittredge and John Lehrberger, 231–36. Berlin/New York: De Gruyter.

Harris, Zellig. 1991. *A Theory of Language and Information. A Mathematical Approach.* Oxford: Clarendon Press.

Harris, Zellig. 2002. "The Structure of Science Information." *J Biomed Inform* 35 (4): 215–21. https://doi.org/10.1016/S1532-0464(03)00011-X.

Henriksson, Aron, Hans Moen, Maria Skeppstedt, Vidas Daudaravičius, and Martin Duneld. 2014. "Synonym Extraction and Abbreviation Expansion with Ensembles of Semantic Spaces." *J Biomed Sem* 5: 6. https://doi.org/10.1186/2041-1480-5-6.

McCray, Alexa T., Anita Burgun, and Olivier Bodenreider. 2001. "Aggregating UMLS Semantic Types for Reducing Conceptual Complexity." Stud Health Technol Inform 84 (0 1): 216–20. https://doi.org/10.3233/978-1-60750-928-8-216.

Namer, Fiammetta, and Robert Baud. 2007. "Defining and Relating Biomedical Terms: Towards a Cross-Language Morphosemantics-Based System." Int J Med Inform 76 (2–3): 226–33. https://doi.org/10.1016/j.ijmedinf.2006.05.001.

Savova, Guergana K, James J Masanz, Philip V Ogren, Jiaping Zheng, Sunghwan Sohn, Karin C Kipper-Schuler, Christopher G Chute. 2010. "Mayo Clinical Text Analysis and Knowledge Extraction System (CTAKES): Architecture, Component Evaluation and Applications." *J Am Med Inform Assoc* 17 (5): 507–13. https://doi:10.1136/jamia.2009.001560 .

Sibanda, Tawanda, Tian He, Peter Szolovits, and Ozlem Uzuner. 2006. "Syntactically-Informed Semantic Category Recognizer for Discharge Summaries." *AMIA Annu Symp Proc 2006*, 714–18.

Siklósi, Borbála. 2015. "Clustering Relevant Terms and Identifying Types of Statements in Clinical Records." *Computational Linguistics and Intelligent Text Processing. CICLing 2015.* 619–30. https://doi.org/10.1007/978-3-319-18117-2_46.

Skeppstedt, Maria, Maria Kvist, Gunnar H. Nilsson, and Hercules Dalianis. 2014. "Automatic Recognition of Disorders, Findings, Pharmaceuticals and Body Structures from Clinical Text: An Annotation and Machine Learning Study." *J Biomed Inform* 49: 148–58. https://doi.org/10.1016/j.jbi.2014.01.012.

Spasić, Irena, Mark Greenwood, Alun Preece, Nick Francis, and Glyn Elwyn. 2013. "FlexiTerm: A Flexible Term Recognition Method." *J Biomed Sem* 4: 27. https://doi.org/10.1186/2041-1480-4-27.

Torii, Manabu, Sachin Kamboj, and K. Vijay-Shanker. 2004. "Using Name-Internal and Contextual Features to Classify Biological Terms." *J Biomed Inform* 37: 498–511. https://doi.org/10.1016/j.jbi.2004.08.007.

Weeds, Julie, James Dowdall, Gerold Schneider, Bill Keller, and David J. Weir. 2014. "Using Distributional Similarity to Organise BioMedical Terminology." *Terminology* 11 (1): 107–41. https://doi.org/10.1075/bct.2.07wee.

Zhang, Rui, Jialin Liu, Yong Huang, Miye Wang, Qingke Shi, Jun Chen, and Zhi Zeng. 2017. "Enriching the International Clinical Nomenclature with Chinese Daily Used Synonyms and Concept Recognition in Physician Notes." *BMC Med Inform Decis Mak* 17: 54. https://doi.org/10.1186/s12911-017-0455-z.

Enhancing PIO Element Detection in Medical Text Using Contextualized Embedding

Hichem Mezaoui
IMRSV Data Labs
Ottawa, Canada
hichem@imrsv.ai

Aleksandr Gontcharov
IMRSV Data Labs
Ottawa, Canada
aleksandr.gontcharov@imrsv.ai

Isuru Gunasekara
IMRSV Data Labs
Ottawa, Canada
isuru@imrsv.ai

Abstract

In this paper, we investigate a new approach to Population, Intervention and Outcome (PIO) element detection, a common task in Evidence Based Medicine (EBM). The purpose of this study is two-fold: to build a training dataset for PIO element detection with minimum redundancy and ambiguity and to investigate possible options in utilizing state of the art embedding methods for the task of PIO element detection. For the former purpose, we build a new and improved dataset by investigating the shortcomings of previously released datasets. For the latter purpose, we leverage the state of the art text embedding, Bidirectional Encoder Representations from Transformers (BERT), and build a multi-label classifier. We show that choosing a domain specific pre-trained embedding further optimizes the performance of the classifier. Furthermore, we show that the model could be enhanced by using ensemble methods and boosting techniques provided that features are adequately chosen.

1 Introduction

Evidence-based medicine (EBM) is of primary importance in the medical field. Its goal is to present statistical analyses of issues of clinical focus based on retrieving and analyzing numerous papers in the medical literature (Haynes et al., 1997). The PubMed database is one of the most commonly used databases in EBM (Sackett et al., 1996).

Biomedical papers, describing randomized controlled trials in medical intervention, are published at a high rate every year. The volume of these publications makes it very challenging for physicians to find the best medical intervention for a given patient group and condition (Borah et al., 2017). Computational methods and natural language processing (NLP) could be adopted in order to expedite the process of biomedical evidence synthesis. Specifically, NLP tasks applied to well structured documents and queries can help physicians extract appropriate information to identify the best available evidence in the context of medical treatment.

Clinical questions are formed using the PIO framework, where clinical issues are broken down into four components: Population/Problem (P), Intervention (I), Comparator (C), and Outcome (O). We will refer to these categories as PIO elements, by using the common practice of merging the C and I categories. In (Rathbone et al., 2017) a literature screening performed in 10 systematic reviews was studied. It was found that using the PIO framework can significantly improve literature screening efficacy. Therefore, efficient extraction of PIO elements is a key feature of many EBM applications and could be thought of as a multi-label sentence classification problem.

Previous works on PIO element extraction focused on classical NLP methods, such as Naive Bayes (NB), Support Vector Machines (SVM) and Conditional Random Fields (CRF) (Chung, 2009; Boudin et al., 2010). These models are shallow and limited in terms of modeling capacity. Furthermore, most of these classifiers are trained to extract PIO elements one by one which is suboptimal since this approach does not allow the use of shared structure among the individual classifiers.

Deep neural network models have increased in popularity in the field of NLP. They have pushed the state of the art of text representation and information retrieval. More specifically, these techniques enhanced NLP algorithms through the use of contextualized text embeddings at word, sentence, and paragraph levels (Mikolov et al., 2013; Le and Mikolov, 2014; Peters et al., 2017; Devlin et al., 2018; Logeswaran and Lee, 2018; Radford et al., 2018).

More recently, Jin and Szolovits (2018) proposed a bidirectional long short term memory (LSTM) model to simultaneously extract PIO

Proceedings of the BioNLP 2019 workshop, pages 217–222
Florence, Italy, August 1, 2019. ©2019 Association for Computational Linguistics

components from PubMed abstracts. To our knowledge, that study was the first in which a deep learning framework was used to extract PIO elements from PubMed abstracts.

In the present paper, we build a dataset of PIO elements by improving the methodology found in (Jin and Szolovits, 2018). Furthermore, we built a multi-label PIO classifier, along with a boosting framework, based on the state of the art text embedding, BERT. This embedding model has been proven to offer a better contextualization compared to a bidirectional LSTM model (Devlin et al., 2018).

2 Datasets

In this study, we introduce PICONET, a multi-label dataset consisting of sequences with labels Population/Problem (P), Intervention (I), and Outcome (O). This dataset was created by collecting structured abstracts from PubMed and carefully choosing abstract headings representative of the desired categories. The present approach is an improvement over a similar approach used in (Jin and Szolovits, 2018).

Our aim was to perform automatic labeling while removing as much ambiguity as possible. We performed a search on April 11, 2019 on PubMed for 363,078 structured abstracts with the following filters: Article Types (Clinical Trial), Species (Humans), and Languages (English). Structured abstract sections from PubMed have labels such as introduction, goals, study design, findings, or discussion; however, the majority of these labels are not useful for P, I, and O extraction since most are general (e.g. *methods*) and do not isolate a specific P, I, O sequence. Therefore, in order to narrow down abstract sections that correspond to the P label, for example, we needed to find a subset of labels such as, but not limited to *population*, *patients*, and *subjects*. We performed a lemmatization of the abstract section labels in order to cluster similar categories such as *subject* and *subjects*. Using this approach, we carefully chose candidate labels for each P, I, and O, and manually looked at a small number of samples for each label to determine if text was representative.

Since our goal was to collect sequences that are uniquely representative of a description of Population, Intervention, and Outcome, we avoided a keyword-based approach such as in (Jin and Szolovits, 2018). For example, using a keyword-

Category	Sentences
I	22818
I O	7
I P	337
O	10994
P	30106
P O	13
NEGATIVE	32053

Table 1: Number of occurrences of each category P, I and O in abstracts.

based approach would yield a sequence labeled *population and methods* with the label P, but such abstract sections were not purely about the population and contained information about the interventions and study design making them poor candidates for a P label. Thus, we were able to extract portions of abstracts pertaining to P, I, and O categories while minimizing ambiguity and redundancy. Moreover, in the dataset from (Jin and Szolovits, 2018), a section labeled as P that contained more than one sentence would be split into multiple P sentences to be included in the dataset. We avoided this approach and kept the full abstract sections. The full abstracts were kept in conjunction with our belief that keeping the full section retains more feature-rich sequences for each sequence, and that individual sentences from long abstract sections can be poor candidates for the corresponding label.

For sections with labels such as *population and intervention*, we created a mutli-label. We also included negative examples by taking sentences from sections with headings such as *aim*. Furthermore, we cleaned the remaining data with various approaches including, but not limited to, language identification, removal of missing values, cleaning unicode characters, and filtering for sequences between 5 and 200 words, inclusive.

3 BERT-Based Classification Model

3.1 Background

BERT (Bidirectional Encoder Representations from Transformers) is a deep bidirectional attention text embedding model. The idea behind this model is to pre-train a bidirectional representation by jointly conditioning on both left and right contexts in all layers using a transformer (Vaswani et al., 2017; Devlin et al., 2018). Like any other language model, BERT can be pre-trained on different contexts. A contextualized representation is generally optimized for downstream NLP tasks.

Since its release, BERT has been pre-trained on a multitude of corpora. In the following, we describe different BERT embedding versions used for our classification problem. The first version is based on the original BERT release (Devlin et al., 2018). This model is pre-trained on the BooksCorpus (800M words) (Zhu et al., 2015) and English Wikipedia (2,500M words). For Wikipedia, text passages were extracted while lists were ignored. The second version is BioBERT (Lee et al., 2019), which was trained on biomedical corpora: PubMed (4.5B words) and PMC (13.5B words).

3.2 The Model

The classification model is built on top of the BERT representation by adding a dense layer corresponding to the multi-label classifier with three output neurons corresponding to PIO labels. In order to insure that independent probabilities are assigned to the labels, as a loss function we have chosen the binary cross entropy with logits (BCEWithLogits) defined by

$$E = -\sum_{i=1}^{n}(t_i log(y_i) + (1 - t_i)log(1 - y_i)); \quad (1)$$

where **t** and **y** are the target and output vectors, respectively; **n** is the number of independent targets (n=3). The outputs are computed by applying the logistic function to the weighted sums of the last hidden layer activations, s,

$$y_i = \frac{1}{1 + e^{-s_i}}, \quad (2)$$

$$s_i = \sum_{j=1} h_j w_{ji}. \quad (3)$$

For the original BERT model, we have chosen the smallest uncased model, Bert-Base. The model has 12 attention layers and all texts are converted to lowercase by the tokenizer (Devlin et al., 2018). The architecture of the model is illustrated in Figure 1.

Using this framework, we trained the model using the two pretrained embedding models described in the previous section. It is worth to mention that the embedding is contextualized during the training phase. For both models, the pretrained embedding layer is frozen during the first epoch

(the embedding vectors are not updated). After the first epoch, the embedding layer is unfrozen and the vectors are fine-tuned for the classification task during training. The advantage of this approach is that few parameters need to be learned from scratch (Howard and Ruder, 2018; Radford et al., 2018; Devlin et al., 2018).

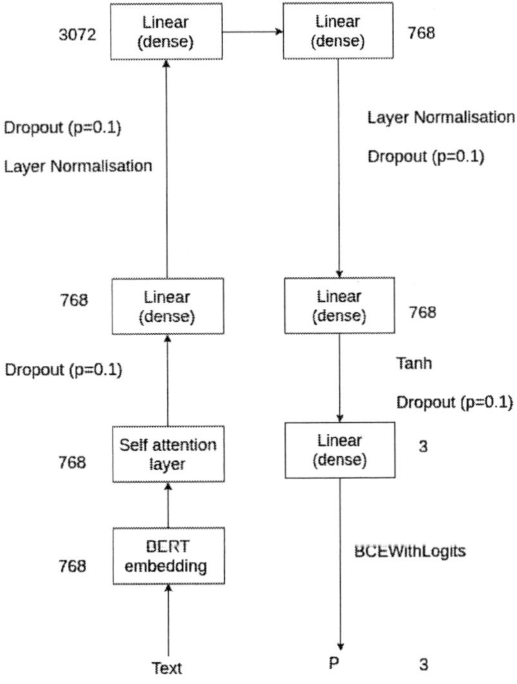

Figure 1: Structure of the classifier.

4 Results

4.1 Performance Comparison

In order to quantify the performance of the classification model, we computed the precision and recall scores. On average, it was found that the model leads to better results when trained using the BioBERT embedding. In addition, the performance of the PIO classifier was measured by averaging the three Area Under Receiver Operating Characteristic Curve (ROC_AUC) scores for P, I, and O. The ROC_AUC score of 0.9951 was obtained by the model using the general BERT embedding. This score was improved to 0.9971 when using the BioBERT model pre-trained on medical context. The results are illustrated in Figure 2.

4.2 Model Boosting

We further applied ensemble methods to enhance the model. This approach consists of combin-

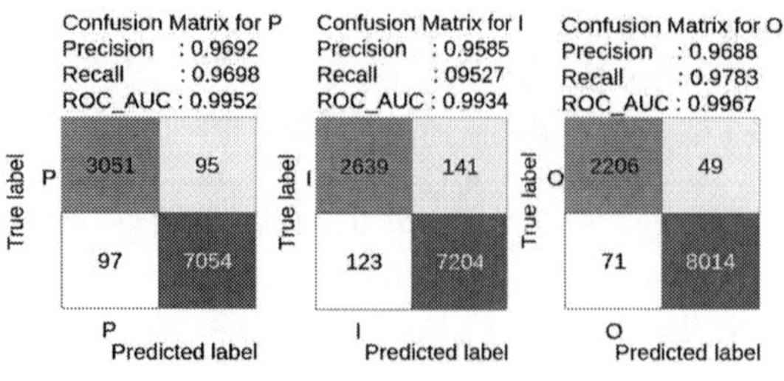

(a) BERT (ROC_AUC: 0.9951)

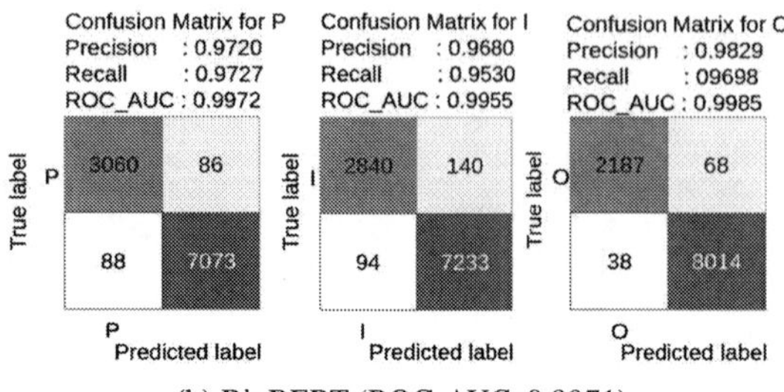

(b) BioBERT (ROC_AUC: 0.9971)

Figure 2: ROC_AUC scores and confusion matrices.

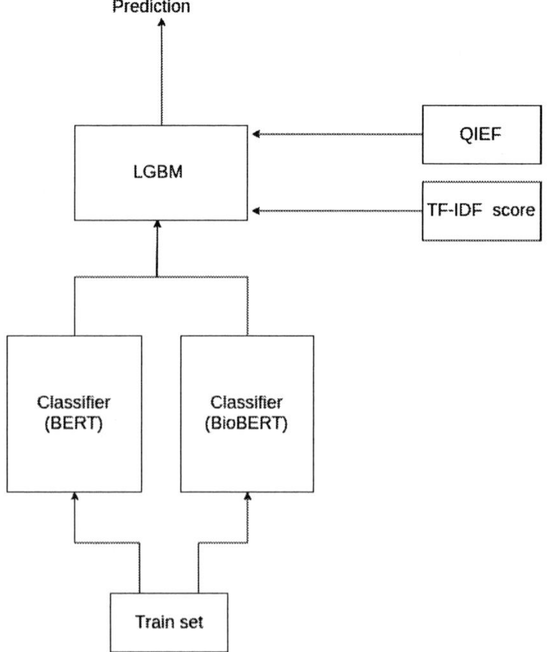

Figure 3: An illustration of the LGBM framework: : combining the two base models and the TF-IDF and QIEF features.

ing predictions from base classifiers with features of the input data to increase the accuracy of the model (Merz, 1999).

We investigate an important family of ensemble methods known as boosting, and more specifically

Model	ROC_AUC	F1
BERT	0.9951	0.9666
BioBERT	0.9971	0.9697
BERT + TF-IDF + QIEF	0.9981	0.9784
BioBERT + TF-IDF + QIEF	0.9996	0.9793
BERT + BioBERT + TF-IDF + QIEF	0.9998	0.9866

Table 2: Performance of the classifiers in terms of ROC_AUC and F1 scores.

a Light Gradient Boosting Machine (LGBM) algorithm, which consists of an implementation of fast gradient boosting on decision trees. In this study, we use a library implemented by Microsoft (Ke et al., 2017). In our model, we learn a linear combination of the prediction given by the base classifiers and the input text features to predict the labels. As features, we consider the average term frequency-inverse document frequency (TF-IDF) score for each instance and the frequency of occurrence of quantitative information elements (QIEF) (e.g. percentage, population, dose of medicine). Finally, the output of the binary cross entropy with logits layer (predicted probabilities for the three classes) and the feature information are fed to the LGBM.

We train the base classifier using the original training dataset, using 60% of the whole data as training dataset, and use a five-fold cross-validation framework to train the LGBM on the remaining 40% of the data to avoid any information leakage. We train the LGBM on four folds and test on the excluded one and repeat the process for all five folds.

The results of the LGBM classifier for the different boosting frameworks and the scores from the base classifiers are illustrated in Table 2. The highest average ROC_AUC score of 0.9998 is obtained in the case of combining the two base learners along with the TF-IDF and QIEF features.

5   Discussion and Conclusion

In this paper, we presented an improved methodology to extract PIO elements, with reduced ambiguity, from abstracts of medical papers. The proposed technique was used to build a dataset of PIO elements that we call PICONET. We further proposed a model of PIO elements classification using state of the art BERT embedding. It has been shown that using the contextualized BioBERT embedding improved the accuracy of the classifier. This result reinforces the idea of the importance of embedding contextualization in subsequent classification tasks in this specific context.

In order to enhance the accuracy of the model, we investigated an ensemble method based on the LGBM algorithm. We trained the LGBM model, with the above models as base learners, to optimize the classification by learning a linear combination of the predicted probabilities, for the three classes, with the TF-IDF and QIEF scores. The results indicate that these text features were adequate for boosting the contextualized classification model. We compared the performance of the classifier when using the features with one of the base learners and the case where we combine the base learners along with the features. We obtained the best performance in the latter case.

The present work resulted in the creation of a PIO elements dataset, PICONET, and a classification tool. These constitute an important component of our system of automatic mining of medical abstracts. We intend to extend the dataset to full medical articles. The model will be modified to take into account the higher complexity of full text data and more efficient features for model boosting will be investigated.

References

Rohit Borah, Andrew W Brown, Patrice L Capers, and Kathryn A Kaiser. 2017. Analysis of the time and workers needed to conduct systematic reviews of medical interventions using data from the prospero registry. *BMJ open*, 7(2):e012545.

Florian Boudin, Jian-Yun Nie, Joan C Bartlett, Roland Grad, Pierre Pluye, and Martin Dawes. 2010. Combining classifiers for robust pico element detection. *BMC medical informatics and decision making*, 10(1):29.

Grace Y Chung. 2009. Sentence retrieval for abstracts of randomized controlled trials. *BMC medical informatics and decision making*, 9(1):10.

Jacob Devlin, Ming-Wei Chang, Kenton Lee, and Kristina Toutanova. 2018. Bert: Pre-training of deep bidirectional transformers for language understanding. *arXiv preprint arXiv:1810.04805*.

R Brian Haynes, David L Sackett, W Scott Richardson, William Rosenberg, and G Ross Langley. 1997. Evidence-based medicine: How to practice & teach ebm. *Canadian Medical Association. Journal*, 157(6):788.

Jeremy Howard and Sebastian Ruder. 2018. Universal language model fine-tuning for text classification. *arXiv preprint arXiv:1801.06146*.

Di Jin and Peter Szolovits. 2018. Pico element detection in medical text via long short-term memory neural networks. In *Proceedings of the BioNLP 2018 workshop*, pages 67–75.

Guolin Ke, Qi Meng, Thomas Finley, Taifeng Wang, Wei Chen, Weidong Ma, Qiwei Ye, and Tie-Yan Liu. 2017. Lightgbm: A highly efficient gradient boosting decision tree. In *Advances in Neural Information Processing Systems*, pages 3146–3154.

Quoc Le and Tomas Mikolov. 2014. Distributed representations of sentences and documents. In *International conference on machine learning*, pages 1188–1196.

Jinhyuk Lee, Wonjin Yoon, Sungdong Kim, Donghyeon Kim, Sunkyu Kim, Chan Ho So, and Jaewoo Kang. 2019. Biobert: pre-trained biomedical language representation model for biomedical text mining. *arXiv preprint arXiv:1901.08746*.

Lajanugen Logeswaran and Honglak Lee. 2018. An efficient framework for learning sentence representations. *arXiv preprint arXiv:1803.02893*.

Christopher J Merz. 1999. Using correspondence analysis to combine classifiers. *Machine Learning*, 36(1-2):33–58.

Tomas Mikolov, Ilya Sutskever, Kai Chen, Greg S Corrado, and Jeff Dean. 2013. Distributed representations of words and phrases and their compositionality. In *Advances in neural information processing systems*, pages 3111–3119.

Matthew E Peters, Waleed Ammar, Chandra Bhagavatula, and Russell Power. 2017. Semi-supervised sequence tagging with bidirectional language models. *arXiv preprint arXiv:1705.00108*.

Alec Radford, Karthik Narasimhan, Time Salimans, and Ilya Sutskever. 2018. Improving language understanding with unsupervised learning. Technical report, Technical report, OpenAI.

John Rathbone, Loai Albarqouni, Mina Bakhit, Elaine Beller, Oyungerel Byambasuren, Tammy Hoffmann, Anna Mae Scott, and Paul Glasziou. 2017. Expediting citation screening using pico-based title-only screening for identifying studies in scoping searches and rapid reviews. *Systematic reviews*, 6(1):233.

David L Sackett, William MC Rosenberg, JA Muir Gray, R Brian Haynes, and W Scott Richardson. 1996. Evidence based medicine: what it is and what it isn't.

Ashish Vaswani, Noam Shazeer, Niki Parmar, Jakob Uszkoreit, Llion Jones, Aidan N Gomez, Łukasz Kaiser, and Illia Polosukhin. 2017. Attention is all you need. In *Advances in neural information processing systems*, pages 5998–6008.

Yukun Zhu, Ryan Kiros, Rich Zemel, Ruslan Salakhutdinov, Raquel Urtasun, Antonio Torralba, and Sanja Fidler. 2015. Aligning books and movies: Towards story-like visual explanations by watching movies and reading books. In *Proceedings of the IEEE international conference on computer vision*, pages 19–27.

Contributions to Clinical Named Entity Recognition in Portuguese

Fábio Lopes
CISUC, DEI
University of Coimbra
Portugal
fadcl@student.dei.uc.pt

César Teixeira
CISUC, DEI
University of Coimbra
Portugal
cteixei@dei.uc.pt

Hugo Gonçalo Oliveira
CISUC, DEI
University of Coimbra
Portugal
hroliv@dei.uc.pt

Abstract

Having in mind that different languages might present different challenges, this paper presents the following contributions to the area of Information Extraction from clinical text, targeting the Portuguese language: a collection of 281 clinical texts in this language, with manually-annotated named entities; word embeddings trained in a larger collection of similar texts; results of using BiLSTM-CRF neural networks for named entity recognition on the annotated collection, including a comparison of using in-domain or out-of-domain word embeddings in this task. Although learned with much less data, performance is higher when using in-domain embeddings. When tested in 20 independent clinical texts, this model achieved better results than a model using larger out-of-domain embeddings.

1 Introduction

In recent years, much data has been produced on different areas, including healthcare, which, besides its general relation to well-being, is also economically-relevant (Folland et al., 2017). We focus on the clinical field, where valuable information is hidden on produced admission notes, diagnostic test reports, patient discharge letters or clinical case reports. The latter contain information about patient clinical histories, such as their condition; diagnostic tests and respective results; or treatments and how they were administered. Such data is very useful for clinical professionals in their future decisions about what diagnostic tests or therapies a patient has to do, based on past clinical information. However, manually processing all available texts and looking for important information is impractical for humans. To make it more tractable, Natural Language Processing (NLP) tools have been developed for automating tasks such as Information Extraction (IE), including Named Entity Recognition (NER), and ultimately store acquired information in relational databases, where queries should be more efficient.

Similarly to many other NLP-related tasks, the field of clinical NLP has been growing. This is both reflected in the organization of shared tasks (Uzuner et al., 2011; Stubbs and Uzuner, 2015; Doğan et al., 2014; Pestian et al., 2007; Elhadad et al., 2015; Bethard et al., 2016; Kelly et al., 2016), which made available several datasets, such as Informatics for Integrating Biology & the Bedside (i2b2); or in the adoption of deep neural network architectures that lead to state-of-the-art results, namely Bidirectional Long Short Term Memory with a stacked Conditional Random Fields layer (BiLSTM-CRF) (Xu et al., 2017; Unanue et al., 2017). However, most of the work going on targets text written in English. When it comes to other languages, such as Portuguese, the number of studies on this field is much lower (Névéol et al., 2018).

This work aims to boost clinical NLP in Portuguese with three main contributions: (i) A collection of Portuguese clinical texts with manually-labelled named entities; (ii) A model of word embeddings learned from a larger collection of Portuguese clinical text (i.e., Neurology clinical case descriptions); (iii) An analysis of the performance of state-of-the-art models in Portuguese clinical NER, namely BiLSTM-CRF neural networks (Lample et al., 2016), tested on the labelled collection, either using the previous word embeddings or general-language word embeddings.

In the next section, we introduce deep learning architectures and word embedding (WE) models that have been used in NER. Section 3 describes how texts were labelled and provides some figures on the resulting dataset and its revision. Furthermore, we explain how the in-domain WE model was trained and its qualitative difference towards

Proceedings of the BioNLP 2019 workshop, pages 223–233
Florence, Italy, August 1, 2019. ©2019 Association for Computational Linguistics

the pre-trained out-of-domain WE model used. Finally, we explain the architecture of our deep learning model. Section 4 reports the results for hyperparameters grid search. After choosing the best model for both in-domain and out-of-domain WEs, we tested it on an independent test set. We report micro-averaged relaxed F1-score and strict F1-score of 70.41% and 62.71%, respectively. We conclude with a brief discussion.

2   Related Work

Training a model for clinical NER requires access to much clinical textual data. Although much text of this kind is produced everyday, its availability is highly limited due to strict ethical regulations that constrain using data with personal information, as in clinical case or diagnostic test reports. Still, when available, such texts constitute valuable sources of data, and may be used in the development of models for Information Extraction, including Named Entity Recognition (NER).

In order to create machine learning models that identify and classify named entities (NEs), the latter have to be annotated on a collection of texts, which can be used as training and/or testing data. That is generally done manually, as several authors did. For instance, Uzuner et al. (2011) annotated 871 medical records with Medical Problems, Treatments and Tests, in order to provide a dataset for the 2010 i2b2/VA concept extraction shared task; and Stubbs and Uzuner (2015) labelled 1,304 individual longitudinal records with heart-risk NEs (e.g. Diabetes references or Hypertension) with 0.95 agreement ratio. Beyond English, some studies involved the creation of datasets in other languages. Skeppstedt et al. (2014) annotated Disorders, Findings, Body Structures and Pharmaceutical Drugs, in 1,104 clinical notes in Swedish, with agreement ratios of 0.79, 0.66, 0.80 and 0.90, respectively. Mykowiecka et al. (2009) annotated 700 mammography reports and 100 diabetic discharge documents, in Polish, with NEs that carry information about Pathological Findings, Breast Tissue, and Crucial Health information about diabetic patients. Ferreira et al. (2010) manually labelled 90 clinical notes in Portuguese with NEs such as Condition, Anatomical Site and Finding. Although in Portuguese, the previous dataset is not publicly available due to ethical regulations, but the annotation guidelines followed are published (Ferreira, 2011).

In recent years, deep learning approaches have been used for NER, leading to state-of-the-art results. Clinical NER is not an exception, with such models used for extracting data from Electronic Medical Records (EMR). Adopted architectures include Recurrent Neural Networks (RNN), with simple RNN layers, LSTM layers, BiLSTM layers or Gated Recurrent Unit (GRU) layers; Convolutional Neural Networks (CNN); and also Feed-Forward Networks (FFN). Luu et al. (2018) showed that a vanilla RNN outperforms a FNN using the same features on clinical texts provided in the CLEF eHealth 2016 task (Kelly et al., 2016) on the extraction of relevant information from nursing shift changes notes. This was expected because FNNs do not consider past information.

Chokwijitkul et al. (2018) assessed the performance of CNN, RNN, LSTM, BiLSTM and GRU networks for identifying heart risk factors in EMRs and found that BiLSTM networks achieved the best F-measure. They further show that such models perform near the rule-based and shallow machine learning models, but do not resort to gazetteers or knowledge bases. Wu et al. (2018) compared different classifiers (CRF, CNN and BiLSTM) for NER, using the dataset of the 2010 i2b2 NLP challenge. They also compared their models with the best models at the time (Structured SVM) and trained during the competition (Semi-Markov model), and used pre-trained word embeddings (WEs) as features for the BiLSTM network and the CNN. For the CRF, they used three different feature sets: only word and n-gram features; the previous plus linguistic features and document level features, such as section names; and all the previous plus features from general clinical NLP systems (MedLEE, MetaMap, KnowledgeMap) and gazetteer features from the UMLS terminology. Similarly to Chokwijitkul et al. (2018), they report that the BiLSTM network outperformed all the others.

Others developed a BiLSTM network with a character embedding layer, a WE layer and a CRF layer. Xu et al. (2017) evaluated their architecture on the NCBI Disease Corpus (793 PubMed medical literature abstracts), while Unanue et al. (2017) evaluated their models with three different datasets (2010 i2b2/VA dataset, DrugBank and MedLine). Both showed that the CRF layer and the character embedding feature have great importance on the performance of a BiLSTM network.

Although these models became the trend in NER, they rely heavily on the quality of the WE models for converting each word to its embedding vector. On the clinical domain, Newman-Griffis and Zirikly (2018) compared WEs using in-domain and out-of-domain corpora. In-domain corpora consisted of two different datasets, one with 154,967 Electronic Health Records (EHR) and a subset with 17,952 EHR documents focused on Physical Therapy (PT) and Occupational Therapy (OT). Out-of-domain corpora were constituted by 14.7 million abstracts from the 2016 PubMed baseline and two million free-text documents released as part of the MIMIC-III critical care DB. Besides those, they used a FastText model, pre-trained on Wikipedia 2017 documents. They reported that, with WEs trained with small in-domain corpora, results were similar to those achieved with the large out-of-domain corpora. Unanue et al. (2017) additionally showed that re-training WE models with domain-specific texts improves the performance of the model.

Although not on the clinical domain, there is some related work on Portuguese. On general NER, de Castro et al. (2018) recently achieved state-of-art results using a BiLSTM-CRF model. On distributional similarity, Hartmann et al. (2017) compared Portuguese word WEs, learned with different methods, in both intrinsic (syntactic and semantic analogies) and extrinsic (PoS tagging and sentence similarity) tasks. There are also studies suggesting that, in tasks such as PoS tagging and NER, combining character embedding with pre-trained WE outperforms approaches that use only WEs (Santos and Zadrozny, 2014; dos Santos and Guimarães, 2015).

3 Experimental Set-up

This section presents the textual data used, the guidelines followed for its annotation and characterizes the resulting dataset with some numbers on its contents and revision. It further explains how the WE models used were learned and the architecture of the NER model, including how its hyperparameters grid search was made.

3.1 Dataset

Three different datasets were used in different stages of this work:

- For training and validation, 281 clinical case texts collected from the numbers 1 and 2

of volume 17 of the clinical journal Sinapse (Sinapse, 2017a,b), published by the Portuguese Society of Neurology. Neurology texts were used because the testing texts, that originally motivated this work, were obtained from the Neurology service.

- For testing, a small set of 20 clinical texts obtained from the Neurology service of the Coimbra University Hospital Centre (CHUC), in Coimbra, Portugal. These include admission notes, diagnostic test reports and patient discharge letters and were originally used in the development of the European Epilepsy Database (Klatt et al., 2012).

- For training the in-domain WE model, a total of 3,377 clinical texts were collected from all the volumes of the Sinapse journal, published between 2001 and 2018[1]. Although the journal contains clinical cases and experimental reports we just collected the clinical cases.

As all the texts used for training, validation and test were in a raw format, they were preprocessed with tools in NLPPort (Rodrigues et al., 2018), a NLP toolkit for Portuguese, based on OpenNLP – each text was tokenized with TokPort, PoS-tagged with TagPort, and lemmas for each token-PoS pair obtained with LemPort. After preprocessing, manual NE annotation was based on the guidelines described in Ferreira's PhD Thesis (Ferreira, 2011), originally developed with the help of physicians and linguists and used in the annotation of Ferreira's dataset. All the NEs in the guidelines were considered, with the exception of Location, because it represents geographical locations, e.g, "Coimbra" (a city) or "domicílio" (home, in Portuguese), which does not represent important clinical information. Although DateTime does not represent clinical information as well, it is important to know what temporal information is related to diseases or therapies, e.g., their frequency or duration. Furthermore, two new NE classes were introduced, namely Genetics and Additional Observations. The former was used for information about genes related to diseases (e.g., "...o estudo do *gene PMP22* identificou..." (*...study of the gene PMP22 identified...*)), and the latter for all clinically-relevant information that did not suit any of the other classes (e.g. "...medicada e

---

[1]http://www.sinapse.pt/archive.php

ex-fumador, refere..." (*...medicated and ex-smoker, states...*). The dataset thus considers 14 different tags, one for each NE class, plus the Out tag, for tokens not belonging to a NE. For annotation, we adopted the Inside-Outside-Beginning (IOB) format, which allows to distinguish between tokens in the beginning and inside a NE. This is essential to sequential classifiers and allows for better rules, which do not enable to tag a token as inside-NE before the beginning of the same NE. Table 1 illustrates the annotated data.

Tables 2 and 4 provide a quantitative analysis of the training and validation datasets, while tables 3 and 5 a quantitative analysis of the independent test set. Tables 2 and 3 quantify the tokens for each IOB tag (NT), the number of distinct tokens (NDT), and their ratios (NTR, NDTR). Finally, tables 4 and 5 show the number of NE occurrences (O), the number of distinct NE occurrences (DO) and their ratios (OR, DOR). As the test set has only reports related to epilepsy, it does not have occurrences of the Genetics NE.

The entire dataset was annotated by the first author of this paper, a last-year student of the MSc in Biomedical Engineering. After that, to validate the annotation, 30% of the dataset was revised by two MSc students in Biomedical Engineering, two PhD students in Data Science, one Computer Science Professor working on NLP and NER, and one Physiotherapist. Each of the previous revised 15 texts. Based on the revised subset, we calculated the agreement ratios as the ratio between the number of tokens which were annotated with the same tag as our annotation and the total number of tokens for each NE. Although there were some tokens annotated with different tags, we did not change dataset labels. Agreement ratios (ARs) for each NE, as well as the number of agreed (AT) and of not-agreed tags (NAT) are in table 6.

The lowest ARs are for Additional Observations, Characterization and Results. They were also the classes whose original labelling raised more doubts. Additional Observations is a general class which may include other NEs, for instance, in case it does not relate to the patient but to their family — e.g., "...diagnóstico de doença neoplástica no marido..." (*...diagnosis of neoplastic disease in her husband...*) — , or information about the patient that is important but does not suit any other class — e.g. "...abandono do acompanhamento médico..." (*...abandonment of medi-*

cal assistance...). Characterization may have tokens from the Condition or Evolution classes, depending on the perspective of the reader — e.g., "possível" (*possible*) in "possível processo vascular" (*possible vascular process*) or "hipótese" (*hypothesis*) in "hipótese de metástase" (*hypothesis of metastasis*), for Condition, and "progressivo" (*progressive*) in "declínio cognitivo progressivo" (*progressive cognitive decline*) for Evolution. Depending on their interpretation, results may also have tokens from Condition — e.g. "nova lesão" (*new injury*) in "...RM-CE que documentou nova lesão..." (*...RM-CE which documents a new injury...*), or "hematoma" in "...TAC-CE que mostrou aumento do hematoma..." (*...TAC-CE which shown an increase of the hematoma...*). Overall, the agreement for all the NE classes is above 90%, except for Characterization. This is high, especially considering the number of classes covered and that the used documents are not always easy to interpret, due to the high presence of medical terminology. We recall that these numbers apply for only 30% of the dataset. Due to lack of time, the remaining documents were not revised.

Token	POS Tag	Lemma	IOB Tag
de	prp	de	O
66	num	66	O
anos	n	ano	O
,	punc	,	O
com	prp	com	O
antecedentes	n	antecedente	B-DT
de	prp	de	O
dislipidemia	n	dislipidemia	B-C
e	conj-c	e	O
síndrome	n	síndrome	B-C
depressiva	adj	depressivo	I-C
,	punc	,	O
começou	v-fin	começar	O
por	prp	por	O

Table 1: Example of dataset annotation. Sentence: "...de 66 anos, com antecedentes de dislipidemia e síndrome depressiva, começou por..."

3.2 Word Embeddings

In-domain WE models were trained with 3,377 clinical texts collected from the Sinapse journal, comprising 686,762 tokens all together. For training the model, we used the FastText algorithm (Bojanowski et al., 2017), available in the Gensim library (Rehurek and Sojka, 2010). FastText learns embeddings for characters and represents each word by the sum of its characters. It was used instead of word2vec (Mikolov et al., 2013b) because, while word2vec would consider unseen

IOB Tags	NT	NTR (%)	NDT	NDTR (%)	Examples	Examples (English)
B-AS	2,491	4.272	770	6.794	seio (B-AS)	venous
I-AS	2,510	4.305	599	5.285	venoso (I-AS)	sinous
B-C	3,884	6.662	1,074	9.476	paramnésia (B-C)	reduplicative
I-C	3,634	6.233	1,269	11.196	reduplicativa (I-C)	paramnesia
B-CH	1,043	1.789	503	4.438	mais (B-CH)	more
I-CH	576	0.988	358	3.159	marcado (I-CH)	marked
B-DT	1,516	2.600	280	2.470	18 (B-DT)	18
I-DT	2,495	4.279	378	3.335	semanas (I-DT)	weeks
B-EV	794	1.362	184	1.623	desenvolveu (B-EV)	gradually
I-EV	452	0.775	120	1.059	gradualmente (I-EV)	developed
B-G	61	0.105	15	0.132	gene (B-G)	EGFR
I-G	62	0.106	47	0.415	EGFR (I-G)	gene
B-N	768	1.317	46	0.406	não (B-N)	not
I-N	2	0.003	2	0.018	impedindo (I-N)	hindering
B-OBS	217	0.372	153	1.350	restantes (B-OBS)	remaining
I-OBS	227	0.389	144	1.271	irmãos (I-OBS)	siblings
B-R	1,767	3.031	589	5.197	VS (B-R)	increased
I-R	2,520	4.322	922	8.135	aumentada (I-R)	ESR
B-RA	71	0.122	14	0.124	intravenoso (B-RA)	intravenous
I-RA	0	0.000	0	0.000		
B-T	2,041	3.501	490	4.323	estudo (B-T)	cytogenetic
I-T	2,113	3.624	677	5.973	citogénico (I-T)	study
B-THER	894	1.533	384	3.388	correção (B-THER)	correction
I-THER	709	1.216	332	2.929	de (I-THER)	of
B-V	410	0.703	276	2.435	0.8 (B-V)	0.8
I-V	584	1.002	112	0.988	células (I-V)	cells
O	26,463	45.388	1,596	14.082	-	-
Total	58,304	100.000	11,334	100.000	-	-

Table 2: Quantitative analysis of the training/validation dataset.

Reference: CH: Characterization; T: Test; EV: Evolution; G: Genetics; AS: Anatomical Site; N: Negation; OBS: Additional Observations; C: Condition; R: Results; DT: DateTime; THER: Therapeutics; V: Value; RA: Route of Administration; O: Out

IOB Tag	NT	NTR (%)	NDT	NDTR (%)
B-AS	17	0.628	13	1.343
I-AS	12	0.444	8	0.826
B-C	99	3.660	48	4.959
I-C	109	4.030	58	5.992
B-CH	51	1.885	42	4.339
I-CH	48	1.774	33	3.409
B-DT	130	4.806	67	6.921
I-DT	194	7.172	96	9.917
B-EV	52	1.922	30	3.099
I-EV	12	0.444	10	1.033
B-G	0	0.000	0	0.000
I-G	0	0.000	0	0.000
B-N	33	1.220	7	0.723
I-N	0	0.000	0	0.000
B-OBS	47	1.738	26	2.686
I-OBS	58	2.144	35	3.616
B-R	19	0.702	16	1.653
I-R	14	0.518	13	1.343
B-RA	3	0.111	3	0.310
I-RA	0	0.000	0	0.000
B-T	66	2.440	36	3.719
I-T	36	1.331	28	2.893
B-THER	88	3.253	62	6.405
I-THER	59	2.181	37	3.822
B-V	38	1.405	29	2.996
I-V	62	2.292	18	1.860
O	1,458	53.900	253	26.136
Total	2,705	100	968	100

Table 3: Quantitative analysis of the test dataset

NE	O	OR (%)	DO	DOR (%)
AS	2,488	15.59	1,412	16.14
C	3,887	24.35	2,203	25.18
CH	1,044	6.54	632	7.22
DT	1,519	9.52	883	10.09
EV	793	4.97	331	3.78
G	63	0.39	50	0.57
OBS	217	1.36	166	1.90
N	768	4.81	48	0.55
R	1,766	11.06	1,090	12.46
RA	71	0.45	14	0.16
T	2,041	12.79	1,012	11.57
THER	894	5.60	563	6.44
V	411	2.57	344	3.93
Total	15,962	100.00	8,748	100.00

Table 4: NE Training/Validation Dataset Description

words as out-of-vocabulary, FastText may represent some of them, based on their characters.

For training the FastText model, the following parameters were used: 300 dimensions, skip-gram with negative sampling, minimum count of 5 words, minimum char-gram length of 1, and default settings for the remaining hyperparameters. The skip-gram algorithm (Mikolov et al., 2013a) predicts the surrounding context given the input word, which allows to relate words to their neigh-

NE	O	OR (%)	DO	DOR (%)
AS	17	2.644	14	2.960
C	99	15.397	66	13.953
CH	51	7.932	45	9.514
DT	130	20.218	102	21.564
EV	52	8.087	34	7.188
G	0	0.000	0	0.000
N	33	5.132	7	1.480
OBS	47	7.309	34	7.188
R	19	2.955	17	3.594
RA	3	0.467	3	0.634
T	66	10.264	44	9.302
THER	88	13.686	73	15.433
V	38	5.910	34	7.188
Total	643	100	473	100

Table 5: NE Test Dataset Description

NE	AR (%)	AT	NAT	Total
AS	98.01	1,821	37	1,858
C	94.16	2,323	144	2,467
CH	86.29	428	68	496
DT	93.79	1,193	79	1,272
EV	97.15	375	11	386
G	100.00	27	0	27
N	97.74	259	6	265
OBS	91.11	164	16	180
R	91.68	1,322	120	1,442
RA	91.30	21	2	23
T	96.81	1,273	42	1,315
THER	95.13	605	31	636
V	96.78	331	11	342
O	96.91	8,941	285	9,226
Total	95.73	19,083	852	19,935

Table 6: Agreement Ratios for all NEs and Non-Entity

bors, an important characteristic for NER. The number of dimensions (300) and minimum word count (5) were the same as in the out-of-domain WE model. Minimum char-grams length (1) was used for training the model with all the characters, thus enabling to recognize unknown words. Finally, all the words in the dataset starting with an uppercase character were converted to lowercase, since they represent the same word but in the beginning of a sentence. After preprocessing, only 7,312 tokens occur more than 5 times.

For the out-of-domain WEs, we used a general Portuguese WE model downloaded from the FastText website[2], trained with billions of tokens from Wikipedia and Common Crawl (Grave et al., 2018). As it was trained with a character window of 5 characters, a total of 27 words and 80 lemmas in our dataset do not have an embedding vector in this model. For them, we assign the embedding of the word 'UNK', meaning unknown, but not a Portuguese word, thus not introducing much

noise to the embedding datasets. This strategy was followed because simply putting out these words could influence the labelling of the network, as the classification of each word depends on the classification of the others around.

3.3 Model Architecture

Given the current trend on NER and its state-of-the-art results, we adopted a BiLSTM-CRF neural network as our model for this purpose. The architecture used is presented in figure 1. The word embedding step is where all the tokens are converted to their embedding vectors. Lemmas are also converted to their WE vectors and concatenated to the previous vectors. PoS tags, orthographic and morphological features, e.g. first character is uppercase, all characters are uppercase, digit/non-digit were added as well. Afterwards, the embedding vectors are inserted in a BiLSTM layer with one backward layer and one forward layer. The former enables the network to preserve the information from the past to the future, since it analyses the information from the left to the right. The forward layer enables the network to do the inverse of the backward. Together, these types of LSTM improve the prediction of the network, which, this way, understands better the context of each token.

Finally, the output of the BiLSTM layer is inserted in the CRF layer, which enables the network to consider the neighbor tags. In other words, it allows the network to create tag relations, e.g., if a token is tagged with a beginning of NE, the following token is probably the continuation of such NE. This layer is also responsible for not allowing a token to be tagged with an in-NE tag without this NE being started previously.

Adam optimization function (Kingma and Ba, 2014) was used with a learning rate of 0.001. A grid search was not performed here because this study does not focus on the architecture, but on the application of these models to Portuguese.

In order to get the best number of hidden units and dropout percentages for our model, we performed a grid search using 50 training epochs with 10-fold cross validation. As the dataset has a low number of instances, we used a small set of values for the grid search of the number of hidden units $[2^3, 2^7]$. Keeping the network with a low number of parameters prevents overfitting to the data (Zhang et al., 2016). Furthermore, we used an interval of dropout percentage values from 10%

to 50%. This hyperparameter allows the network to prevent both overfitting and under-learning (Srivastava et al., 2014). An independent grid search was run for each WE model, because they had been trained in different types of text.

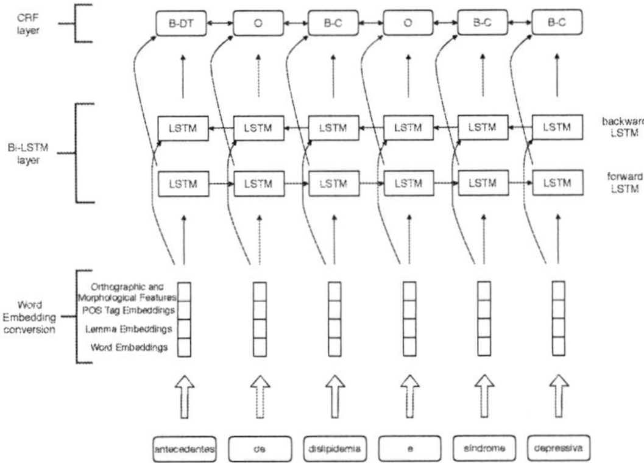

Figure 1: BiLSTM-CRF Neural Network Architecture on the sentence: "antecedentes de dislipidemia e síndrome depressiva" (*history of dyslipidemia and depressive syndrome*)

4   Results and Discussion

According to grid search, the best number of hidden units is 2^6 and 2^5, respectively for the network that uses the in-domain WEs and for the one that uses out-of-domain WEs. The best dropout percentage is 50% for both. This confirms that, for small datasets, the value of each parameter should also be small. Furthermore, the results corroborate that dropout regularization helps avoiding overfitting, since the best results were obtained for high dropout percentage. Validation results for both models and all NE classes are in table 7.

Besides looking at recall and precision, we focus our discussion on the F1-score. Table 7 shows relaxed and strict results. Relaxed or one-point performance measures the performance of the model for each token, while the strict performance considers all occurrences, i.e., one occurrence is well predicted if all its tokens are well predicted too. For example, with the relaxed evaluation, "síndrome depressiva" (*depressive syndrome*) counts as two tokens, i.e, each token's tag is independently compared to its golden tag. With the strict evaluation, if the model fails on a single token's tag, all NE occurrence is considered incorrect.

Results show that the in-domain WE model performs better than the out-of-domain, which is in line with Newman-Griffis and Zirikly (2018). An important reason for this is that the out-of-domain model was not trained with unigrams, leading to the representation of some tokens with the 'UNK' vector, instead of the original token, thus introducing bias. A second reason is that the out-of-domain model was not trained specifically for the clinical domain. Although trained in a much larger collection of text, the out-of-domain model fails to learn clinical relations between different diseases or diagnostic tests, as the in-domain model does. Table 8 shows examples that confirm this fact, e.g. in the in-domain model the word "ECG" is related to three other cardiac diagnostic tests, beyond its extended form, while in the out-of-domain model, it is only related to one more ("eco-cardiograma"); or the neighbors of "diabetes" in the in-domain model, which include related diseases (e.g., "dislipidemia" and arterial hypertension ("HTA"), while, in the out-of-domain model, the neighbors of the same word are words that contain it (e.g., "pré-diabetes" and "diabetes.O"). Furthermore, in the out-of-domain model, several words are not related with the clinical domain, as "hemiparasita" (*hemiparasite*) in the "hemiparésia" (*hemiparesis*) example, or words are not related with anything understandable, as in the "poliangeíte" example.

Table 9 has the results for both WE models on the independent test set, and for a CRF model used as a baseline. The CRF was trained in the same dataset, using the same features as the deep learning model, but raw tokens and lemmas, instead of their embeddings. The best hyperparameters of the validation dataset were used for both WEs. This experiment aims to analyze how well the models trained in text from the journal perform on text collected directly from the hospital.

Once again, the in-domain WE model outperformed the out-of-domain model. Average results for this independent dataset are about 10% lower than for the validation dataset. A possible reason for this is that the test set contains some admission notes and patient discharge letters, structured on items (e.g., origin, admission motive) and their description, which is different from the clinical cases in the validation dataset, described in a full paragraph that covers all related information. Furthermore, since they were not published, these texts were written less carefully, and therefore have some orthographic errors.

WE	NE	Recall		Precision		F1-Score	
		Relaxed	Strict	Relaxed	Strict	Relaxed	Strict
In-Domain	mic Avg	82.34±1.97	74.48±2.37	82.77±1.72	75.25±2.36	82.54±1.61	74.86±2.17
Out-of-Domain		81.63±2.07	73.35±1.57	82.31±1.48	75.06±1.62	81.96±1.50	74.19±1.44
In-Domain	mac Avg	79.04±1.99	73.08±3.00	81.06±2.12	75.59±2.77	79.54±1.89	73.87±2.66
Out-of-Domain		77.75±2.84	70.87±3.07	79.71±2.87	73.73±3.42	78.02±2.76	71.58±2.91
In-Domain	Weighted Avg	82.34±1.97	74.48±2.37	82.84±1.49	75.23±2.39	82.44±1.59	74.73±2.15
Out-of-Domain		81.63±2.07	73.35±1.57	82.35±1.54	74.82±1.65	81.76±1.59	73.90±1.42

Table 7: 10-fold Cross Validation Results with both WEs

WE	Word	Top-5 Nearest Neighbors
In-Domain	ECG	ECG-Holter; electrocardiograma; ecodoppler; ecocardiograma ecocardiogramas
Out-of-Domain	ECG	eletrocardiograma; Electrocardiograma; electrocardiograma; ecocardiograma; Ecocardiograma
In-Domain	diabetes	mellitus; dislipidemia; dislipidémia; HTA; diabética
Out-of-Domain	diabetes	diabete; pré-diabetes; Diabetes; Pré-diabetes; diabetes.O
In-Domain	paramnésia	amnésia; amnésico; mnésico; mnésica; desorientação
Out-of-Domain	paramnésia	paramécia; param3; paranóia.; alucinatória; articulatória
In-Domain	polineuropatia	neuropatia; mononeuropatia; axonal; sensitivo-motora; miopatia
Out-of-Domain	polineuropatia	Polineuropatia; polineuropatias; mononeuropatia; polineurite; neuropatia
In-Domain	poliangeíte	ganglonopatia; citopatia; mielopatia; linfoproliferativa; granulomatosa
Out-of-Domain	poliangeíte	CH12CH14CH15CH18CH26CH30CH4DH5DH6DH8DH9DH10DH12DH15DH20DH30DH; estômagoCarbosymagDulcolaxGavisconImodiumIpraaloxLansoylLubentylMaaloxMicrolaxRennieSmectaSpasfon; XIII787980818283848586878889909192Colóquio; AnguloSimulacrosVeículosABCIABSCABTDABTMBRTPBRTSBSRPBSRSLTRGVAMEVAPAVCOCVCOTVEVE-CIVETAVFCIVGEOVLCIVOPEVPVPMEVPMTVRCIVSAEVSAMVSATVTGCVTPGVTPTVTTFVTTRVTTUVUCIA1; biológicoCaméfitoLigações
In-Domain	hemiparésia	hemiparesia; hemiplegia; hemianopsia; hemianópsia; biparésia
Out-of-Domain	hemiparésia	hemiparéticos; hemiparesia; hemiparasita; hemiplegia; hemiparasitas
In-Domain	artralgias	poliartralgias; algias; mialgias; cervicalgias; lombalgias
Out-of-Domain	artralgias	Artralgias; artralgia; mialgias; Mialgias; Nevralgias

Table 8: Top-5 Nearest Neighbors for both WE models

Average results for the CRF are lower than the average results for both BiLSTM-CRF models. This difference is in line with the results obtained by Chokwijitkul et al. (2018) and Wu et al. (2018). In general, the results of table 9 follow the agreement ratios presented in table 6. Additional Observations and Characterization present the lowest results because they carry too general information easily labelled by the model as a more specific NE (e.g. Condition or Evolution) as explained in section 3.1. Results show low results as well, due to their similarity with Condition, also shown in the examples of section 3.1. Value, Negation, DateTime, Evolution and Anatomical Site show the highest results because they are very specific. Value is related to numbers of therapeutic doses or to the results of diagnostic texts, Negation and Evolution are NEs with many repeated tokens (see tables 2 and 3) and they are highly related to Condition and Results, a characteristic caught by the CRF layer. DateTime is related with time, usually written using the same words and not depending on the author of the text (e.g. training texts contain "aos 60 anos" (*at 60 years old*) and "durante 21 dias" (*during 21 days*) and test texts have "aos 14 anos" (*at 14 years old*) and "durante o período da manhã" (*during the morning*)). Although Anatomical Site has few tokens on the test texts, they are frequent on the training data, which is why results for this NE are high. We were expecting better results for Condition, Test and Therapeutics because they are too specific. This did not happen, and a possible explanation is the different style of writing in the training and testing texts.

Finally, it is important to recall that the Genetics NE is not in the test set, and that the same set has only one token for Negation and Route of Administration, which explains the same relaxed and strict results for these NEs.

5 Conclusion

With this study, we achieved our the three main goals: we gathered and annotated a new dataset for Portuguese clinical text; we applied a BiLSTM-CRF neural network for NER on the previous dataset; we learned a WE model of Portuguese clinical text and compared the performance of the previous approach when using this model and when using general language WEs. The datasets and the learned WE model are publicly available

Algorithm	WE	NE	Recall		Precision		F1-Score	
			Relaxed	Strict	Relaxed	Strict	Relaxed	Strict
BiLSTM-CRF	In-Domain	AS	100.00	88.24	80.56	68.18	89.23	76.92
	Out-of-Domain		93.10	88.24	75.00	65.22	83.08	75.00
CRF	-		86.21	70.59	42.37	40.00	56.82	51.06
BiLSTM-CRF	In-Domain	C	70.19	70.71	59.11	54.26	64.18	61.40
	Out-of-Domain		72.12	68.69	67.87	59.13	69.93	63.55
CRF	-		72.12	61.62	52.63	42.07	60.85	50.00
BiLSTM-CRF	In-Domain	CH	24.24	23.53	42.11	38.71	30.77	29.27
	Out-of-Domain		21.21	21.57	47.73	45.83	29.37	29.33
CRF	-		15.15	21.57	50.00	44.00	23.26	28.95
BiLSTM-CRF	In-Domain	DT	85.80	66.15	84.50	71.07	85.15	68.53
	Out-of-Domain		87.64	61.54	82.08	68.38	84.78	64.78
CRF	-		82.41	48.46	76.95	64.29	79.58	55.26
BiLSTM-CRF	In-Domain	EV	81.25	75.00	82.54	81.25	81.89	78.00
	Out-of-Domain		64.06	53.85	78.85	80.00	70.69	64.37
CRF	-		60.94	51.92	92.86	90.00	73.58	65.85
BiLSTM-CRF	In-Domain	N	96.97	96.97	88.89	88.89	92.75	92.75
	Out-of-Domain		96.97	96.97	91.43	91.43	94.12	94.12
CRF	-		93.94	93.94	91.18	91.18	92.54	92.54
BiLSTM-CRF	In-Domain	OBS	17.14	12.77	64.29	40.00	27.07	19.35
	Out-of-Domain		0.95	0.00	33.33	0.00	1.85	0.00
CRF	-		4.76	6.38	100.00	75.00	9.09	11.76
BiLSTM-CRF	In-Domain	R	63.64	68.42	38.18	44.83	47.73	54.17
	Out-of-Domain		57.58	47.37	45.24	37.50	50.67	41.86
CRF	-		54.55	42.11	19.78	22.22	29.03	29.09
BiLSTM-CRF	In-Domain	RA	33.33	33.33	50.00	50.00	40.00	40.00
	Out-of-Domain		33.33	33.33	50.00	50.00	40.00	40.00
CRF	-		33.33	33.33	100.00	100.00	50.00	50.00
BiLSTM-CRF	In-Domain	T	62.75	54.55	68.82	59.02	65.64	56.69
	Out-of-Domain		60.78	48.48	57.41	44.44	59.05	46.38
CRF	-		50.98	34.85	43.70	33.33	47.06	34.07
BiLSTM-CRF	In-Domain	THER	84.35	67.05	58.49	57.84	69.08	62.11
	Out-of-Domain		79.59	64.77	68.42	62.64	73.58	63.69
CRF	-		69.39	61.36	82.93	80.60	75.56	69.68
BiLSTM-CRF	In-Domain	V	96.00	84.21	88.07	80.00	91.87	82.05
	Out-of-Domain		89.00	73.68	83.18	66.67	85.99	70.00
CRF	-		86.00	63.16	82.69	63.16	84.31	63.16
BiLSTM-CRF	In-Domain	mic Avg	70.97	62.36	69.85	63.05	70.41	62.71
	Out-of-Domain		67.68	56.14	72.32	62.03	69.93	58.94
CRF	-		63.43	49.46	63.79	55.11	63.61	52.13
BiLSTM CRF	In-Domain	mac Avg	67.97	61.74	67.13	61.17	65.45	60.10
	Out-of-Domain		63.03	54.87	65.04	55.94	61.93	54.42
CRF	-		59.15	49.11	69.59	62.15	56.81	50.12
BiLSTM-CRF	In-Domain	Weighted Avg	70.97	62.36	69.75	61.91	68.52	61.10
	Out-of-Domain		67.68	56.14	68.20	57.87	66.07	56.26
CRF	-		63.43	49.46	70.07	60.77	61.39	51.31

Table 9: Results of BiLSTM-CRF model using both WEs and of baseline CRF model on independent test set

in our GitHub repository[3]. We hope that making all these resources available for everyone has a positive impact on IE from text written in Portuguese, namely on clinical text.

In-domain WEs were trained with much less text, but lead to higher performance in NER. Although in a different language, this is in line with Newman-Griffis and Zirikly (2018), and confirms that, in the clinical domain, it should be better to train WE models exclusively with clinical texts, even if there is substantially more in-domain text.

The performance of the model in the independent test confirms that it is possible to train models for extracting information from hospital clinical texts without having direct access to them. In other words, IE models trained with public clinical cases extracted from journals are able to extract information from texts never seen before by the model. This is important, given the difficulty to access clinical texts from hospitals.

In order to improve the current results, we plan to make a better parameter optimization and to explore other deep learning architectures, such as those using residual learning (Tran et al., 2017). Furthermore, we aim to increase the datasets used and tackle relation extraction between NEs (Sahu et al., 2016), which would make it easier to summarize clinical reports.

6 Acknowledgements

We would like to thank the Neurology service of the Coimbra University Hospital Centre (CHUC) for providing us access to the testing dataset; and to those responsible for the revision of 30% of the training/validation dataset, namely Paulo Francisco Valente, João Marques, Adriana Leal, Mauro Pinto, Fellipe Allevato and Ricardo Rodrigues. We also acknowledge the financial support of Fundação para a Ciência e a Tecnologia through CISUC (UID/CEC/00326/2019).

[3]https://github.com/fabioacl/
PortugueseClinicalNER

References

Steven Bethard, Guergana Savova, Wei-Te Chen, Leon Derczynski, James Pustejovsky, and Marc Verhagen. 2016. Semeval-2016 task 12: Clinical TempEval. In *Proceedings of the 10th International Workshop on Semantic Evaluation (SemEval-2016)*, pages 1052–1062.

Piotr Bojanowski, Edouard Grave, Armand Joulin, and Tomas Mikolov. 2017. Enriching Word Vectors with Subword Information. *Transactions of the Association for Computational Linguistics*, 5:135–146.

Pedro Vitor Quinta de Castro, Nádia Félix Felipe da Silva, and Anderson da Silva Soares. 2018. Portuguese named entity recognition using LSTM-CRF. In *Computational Processing of the Portuguese Language - 13th International Conference, PROPOR 2018, Canela, Brazil, September 24-26, 2018, Proceedings*, pages 83–92.

Thanat Chokwijitkul, Anthony Nguyen, Hamed Hassanzadeh, and Siegfried Perez. 2018. Identifying risk factors for heart disease in electronic medical records: A deep learning approach. In *Proceedings of the BioNLP 2018 workshop*, pages 18–27, Melbourne, Australia. Association for Computational Linguistics.

Rezarta Islamaj Doğan, Robert Leaman, and Zhiyong Lu. 2014. NCBI disease corpus: a resource for disease name recognition and concept normalization. *Journal of biomedical informatics*, 47:1–10.

Noémie Elhadad, Sameer Pradhan, Sharon Gorman, Suresh Manandhar, Wendy Chapman, and Guergana Savova. 2015. SemEval-2015 task 14: Analysis of clinical text. In *Proceedings of the 9th International Workshop on Semantic Evaluation (SemEval 2015)*, pages 303–310.

Liliana Ferreira, António J. S. Teixeira, and João Paulo Cunha. 2010. Information Extraction from Portuguese Hospital Discharge Letters. *VI Jornadas en Technologia del Habla and II Iberian SL Tech Workshop*, (January):39–42.

Liliana da Silva Ferreira. 2011. *Medical Information Extraction in European Portuguese*. Ph.D. thesis, Universidade de Aveiro.

Sherman Folland, Allen C Goodman, and Miron Stano. 2017. Introduction. In *The Economics of Health and Health Care*, 8th edition, chapter 1, pages 29–54. Pearson Prentice Hall Upper Saddle River, NJ.

Edouard Grave, Piotr Bojanowski, Prakhar Gupta, Armand Joulin, and Tomas Mikolov. 2018. Learning Word Vectors for 157 Languages. In *Proceedings of the International Conference on Language Resources and Evaluation (LREC 2018)*.

Nathan S. Hartmann, Erick R. Fonseca, Christopher D. Shulby, Marcos V. Treviso, Jéssica S. Rodrigues, and Sandra M. Aluísio. 2017. Portuguese word embeddings: Evaluating on word analogies and natural language tasks. In *Proceedings the 11th Brazilian Symposium in Information and Human Language Technology*, STIL 2017.

Liadh Kelly, Lorraine Goeuriot, Hanna Suominen, Aurélie Névéol, Joao Palotti, and Guido Zuccon. 2016. Overview of the CLEF eHealth evaluation lab 2016. In *International Conference of the Cross-Language Evaluation Forum for European Languages*, pages 255–266. Springer.

Diederik P Kingma and Jimmy Ba. 2014. Adam: A method for stochastic optimization. *arXiv preprint arXiv:1412.6980*, page 13.

Juliane Klatt, Hinnerk Feldwisch-Drentrup, Matthias Ihle, Vincent Navarro, Markus Neufang, Cesar Teixeira, Claude Adam, Mario Valderrama, Catalina Alvarado-Rojas, Adrien Witon, et al. 2012. The epilepsiae database: An extensive electroencephalography database of epilepsy patients. *Epilepsia*, 53(9):1669–1676.

Guillaume Lample, Miguel Ballesteros, Sandeep Subramanian, Kazuya Kawakami, and Chris Dyer. 2016. Neural architectures for named entity recognition. In *Proceedings of the 2016 Conference of the North American Chapter of the Association for Computational Linguistics: Human Language Technologies*, pages 260–270, San Diego, California. Association for Computational Linguistics.

Thoai Man Luu, Robert Phan, and Rachel Davey. 2018. Clinical Name Entity Recognition Based on Recurrent Neural Networks. *2018 18th International Conference on Computational Science and Applications (ICCSA)*, pages 1–9.

Tomas Mikolov, Kai Chen, Greg Corrado, and Jeffrey Dean. 2013a. Efficient estimation of word representations in vector space. *arXiv preprint arXiv:1301.3781*.

Tomas Mikolov, Ilya Sutskever, Kai Chen, Greg Corrado, and Jeffrey Dean. 2013b. Distributed representations of words and phrases and their compositionality. In *Proceedings of the 26th International Conference on Neural Information Processing Systems - Volume 2*, NIPS'13, pages 3111–3119, USA. Curran Associates Inc.

Agnieszka Mykowiecka, Małgorzata Marciniak, and Anna Kupść. 2009. Rule-based information extraction from patients' clinical data. *Journal of Biomedical Informatics*, 42(5):923–936.

Aurélie Névéol, Hercules Dalianis, Sumithra Velupillai, Guergana Savova, and Pierre Zweigenbaum. 2018. Clinical natural language processing in languages other than english: opportunities and challenges. *Journal of biomedical semantics*, 9(1):12.

Denis Newman-Griffis and Ayah Zirikly. 2018. Embedding transfer for low-resource medical named entity recognition: A case study on patient mobility. In *Proceedings of the BioNLP 2018 workshop*, pages 1–11, Melbourne, Australia. Association for Computational Linguistics.

John P Pestian, Christopher Brew, Dj J Matykiewicz Pawełand Hovermale, Neil Johnson, K Bretonnel Cohen, and Włodzisław Duch. 2007. A shared task involving multi-label classification of clinical free text. In *Proceedings of the Workshop on BioNLP 2007: Biological, Translational, and Clinical Language Processing*, pages 97–104. Association for Computational Linguistics.

Radim Rehurek and Petr Sojka. 2010. Software Framework for Topic Modelling with Large Corpora. In *Proceedings of the LREC 2010 Workshop on New Challenges for NLP Frameworks*, pages 45–50, Valletta, Malta. ELRA.

Ricardo Rodrigues, Hugo Gonçalo Oliveira, and Paulo Gomes. 2018. NLPPort: A Pipeline for Portuguese NLP (Short Paper). In *7th Symposium on Languages, Applications and Technologies (SLATE 2018)*, volume 62 of *OpenAccess Series in Informatics (OASIcs)*, pages 18:1—-18:9, Dagstuhl, Germany. Schloss Dagstuhl–Leibniz-Zentrum fuer Informatik.

Sunil Sahu, Ashish Anand, Krishnadev Oruganty, and Mahanandeeshwar Gattu. 2016. Relation extraction from clinical texts using domain invariant convolutional neural network. In *Proceedings of the 15th Workshop on Biomedical Natural Language Processing*, pages 206–215, Berlin, Germany. Association for Computational Linguistics.

Cicero dos Santos and Victor Guimarães. 2015. Boosting named entity recognition with neural character embeddings. In *Proceedings of the Fifth Named Entity Workshop*, pages 25–33, Beijing, China. Association for Computational Linguistics.

Cicero D Santos and Bianca Zadrozny. 2014. Learning character-level representations for part-of-speech tagging. In *Proceedings of the 31st International Conference on Machine Learning (ICML-14)*, pages 1818–1826.

Sinapse. 2017a. *Publicações da Sociedade Portuguesa de Neurologia*, volume 17:1. Sociedade Portuguesa de Neurologia, Lisbon.

Sinapse. 2017b. *Publicações da Sociedade Portuguesa de Neurologia*, volume 17:2. Sociedade Portuguesa de Neurologia, Lisbon.

Maria Skeppstedt, Maria Kvist, Gunnar H. Nilsson, and Hercules Dalianis. 2014. Automatic recognition of disorders, findings, pharmaceuticals and body structures from clinical text: An annotation and machine learning study. *Journal of Biomedical Informatics*, 49:148–158.

Nitish Srivastava, Geoffrey Hinton, Alex Krizhevsky, Ilya Sutskever, and Ruslan Salakhutdinov. 2014. Dropout: a simple way to prevent neural networks from overfitting. *The Journal of Machine Learning Research*, 15(1):1929–1958.

Amber Stubbs and Özlem Uzuner. 2015. Annotating risk factors for heart disease in clinical narratives for diabetic patients. *Journal of biomedical informatics*, 58:S78–S91.

Quan Tran, Andrew MacKinlay, and Antonio Jimeno Yepes. 2017. Named Entity Recognition with Stack Residual LSTM and Trainable Bias Decoding. In *Proceedings of the Eighth International Joint Conference on Natural Language Processing (Volume 1: Long Papers)*, pages 566–575, Taipei, Taiwan. Asian Federation of Natural Language Processing.

Inigo Jauregi Unanue, Ehsan Zare Borzeshi, and Massimo Piccardi. 2017. Recurrent neural networks with specialized word embeddings for health-domain named-entity recognition. *Journal of Biomedical Informatics*, 76(June):102–109.

Özlem Uzuner, Scott L DuVall, Brett R South, and Shuying Shen. 2011. 2010 i2b2/VA challenge on concepts, assertions, and relations in clinical text. *Journal of the American Medical Informatics Association*, 18(5):552–556.

Yonghui Wu, Min Jiang, Jun Xu, Degui Zhi, and Hua Xu. 2018. Clinical Named Entity Recognition Using Deep Learning Models. *AMIA Annual Symposium Proceedings*, pages 1812–1819.

Kai Xu, Zhanfan Zhou, Tianyong Hao, and Wenyin Liu. 2017. A Bidirectional LSTM and Conditional Random Fields Approach to Medical Named Entity Recognition. *International Conference on Advanced Intelligent Systems and Informatics*, pages 355–365.

Chiyuan Zhang, Samy Bengio, Moritz Hardt, Benjamin Recht, and Oriol Vinyals. 2016. Understanding deep learning requires rethinking generalization. *arXiv preprint arXiv:1611.03530*.

Can Character Embeddings Improve Cause-of-Death Classification for Verbal Autopsy Narratives?

Zhaodong Yan
Dept of Electrical and
Computer Engineering
University of Toronto
Toronto, Ontario, Canada
zhaodong.yan@mail.utoronto.ca

Serena Jeblee
Dept of Computer Science
University of Toronto
Toronto, Ontario, Canada
sjeblee@cs.toronto.edu

Graeme Hirst
Dept of Computer Science
University of Toronto
Toronto, Ontario, Canada
gh@cs.toronto.edu

Abstract

We present two models for combining word and character embeddings for cause-of-death classification of verbal autopsy reports using the text of the narratives. We find that for smaller datasets (500 to 1000 records), adding character information to the model improves classification, making character-based CNNs a promising method for automated verbal autopsy coding.

1 Introduction

1.1 Verbal autopsies

Each year, two-thirds of the 60 million deaths in low-and-middle-income countries do not have a known cause of death (CoD), usually because they occurred outside of health facilities and no physical autopsy was performed (United Nations, 2013). Verbal autopsy (VA) surveys are one method of assessing the true distribution of CoDs in these regions. These surveys are conducted by lay interviewers and typically include demographic data, multiple-choice questions, and a free-text narrative, which details the events leading up to the person's death. These records are later coded by physicians for cause of death.

Although several attempts have been made to automate this coding process, including systems such as InterVA (Byass et al., 2012), InSilicoVA (McCormick et al., 2016), the Tariff method (Serina et al., 2015), and others (Miasnikof et al., 2015), the results have not been adequate, in part because they have focused only on the multiple-choice questions and not at all, or only to a limited extent, on the narrative text. However, using the narrative is more convenient because it does not require a specific questionnaire format, and also because it takes less time to collect a short questionnaire and narrative than a long, very detailed survey. Although the narratives present some text

processing problems, they allow for more detail and explanation than the structured data alone.

Only a few methods have used the full text of the narrative for CoD classification. Danso et al. (2013) used term frequency and TF-IDF (term frequency–inverse document frequency) features to classify CoD from VA narratives of neonatal deaths. The Tariff method (Serina et al., 2015) uses a small set of word occurrence features from the narrative, but both of these methods ignore word order. Jeblee et al. (2018) used VA narrative text to jointly predict CoD and a list of keywords for each record using a neural network model with word embeddings.

In our work, we therefore focus on the narrative text. However, the models that have been developed to date for VA classification using the narrative, including SVMs (Danso et al., 2013) and neural networks (Jeblee et al., 2018), have used only word-level information. However, recent research has shown that character-level information can improve text classification models, especially in cases where there are many spelling errors and variations, which is the case with VA narratives. Therefore, we investigate here the use of character embeddings for the VA CoD classification task.

1.2 Character embedding models

Instead of representing each word as a vector, as is typically done with word embeddings, we can represent each character in the text as a vector. With traditional word embeddings, any word that is not found in the vocabulary is represented as a vector of zeros, essentially losing all the information from that word. The character-based model does not have this limitation, and therefore can represent unseen words as well as misspelled words.

Another benefit to character-based models is that because of the much smaller vocabulary size, they result in less variation in the input representa-

234

Proceedings of the BioNLP 2019 workshop, pages 234–239
Florence, Italy, August 1, 2019. ©2019 Association for Computational Linguistics

tion, which can be especially useful for very small datasets such as our verbal autopsy records.

Zhang et al. (2015) used a character-level convolutional neural network (CNN) for text classification tasks on a dataset of news articles and internet reviews, demonstrating that the character-level model could outperform word-level models. Verwimp et al. (2017) combined character-level and word-level embeddings by concatenation with padding, and used them with a Long Short-Term Memory (LSTM) language model, achieving better perplexity than similar word-based models.

Si and Roberts (2018) used an LSTM model to learn character embeddings, which were then concatenated with pre-trained word embeddings to extract cancer-related information such as diagnosis, showing that combined character and word-based models can be used successfully for tasks in the medical domain.

2   Data: The Million Death Study

Our dataset of informal medical narratives consists of verbal autopsy reports from the Million Death Study (MDS) (Westly, 2013), a program that collects VAs in India that cover adult, child, and neonatal deaths. We currently have a dataset of 12,045 adult records, 1851 child records, and 572 neonatal records with English narratives (transcribed from handwritten forms). The records are classified into several broad CoD categories: 18 for adult deaths, 9 for child deaths, and 5 for neonatal deaths. (See Table 4 in the Appendix for the list of CoD categories.)

The process of translating the local languages and converting handwritten texts to digital format creates many errors. Many narratives have frequent spelling and grammar errors, such as inconsistent pronouns, sentence fragments, incorrect punctuation, and transcription errors, in addition to many local terms. See Table 1 for an example narrative. The nature of the text means that purely word-based models, especially ones trained on other corpora, are likely to miss key information. In order to address this issue, we add character embedding representations to the classification model to see whether it will improve the results. We also compare this model to the word-only model.

Narrative

Heart failure. The patient death due to breathlessness. The person sufering paralysis and stroke lost on year with chest pain very pressure after then person was head.

CoD category: Other cardiovascular diseases

Table 1: A verbal autopsy narrative with spelling and grammar errors, and the associated CoD category.

3   Models

3.1   Pre-processing

All text is lowercased before being passed to the model, and punctuation is separated from words. Spelling is corrected using PyEnchant's English dictionary (Kelly, 2015) and a 5-gram language model for scoring candidate replacements, using KenLM (Heafield et al., 2013). However, many instances remain where misspellings result in another valid word (such as *dead* being mistyped as *head*) or are too badly misspelled to be corrected. Moreover, many local terms are not handled properly by our automated spelling correction, so while the spelling correction model fixes some of the more apparent errors, many misspellings persist even after this step.

3.2   Word–based model

For the word-based model, we represent each word in the narrative as a 100-dimensional word embedding. The embeddings are trained using the word2vec CBOW algorithm (Mikolov et al., 2013) on the training set of the VA narratives, as well as data from the ICE corpus of Indian English[1] and about 1M posts from MedHelp, an online medical advice forum for patients[2]. The maximum length of input is 200 words, and shorter narratives are padded with zeros.

The classification model is a convolutional neural network (CNN) implemented in PyTorch (Paszke et al., 2017), with windows of 1 to 5 words, max-pooling, and 0.1 dropout.

3.3   Character–based model

For the character-based model we use publicly available pre-trained character embeddings[3] de-

---

[1] http://ice-corpora.net/ice/avail.htm
[2] http://www.medhelp.org
[3] https://github.com/minimaxir/
char-embeddings

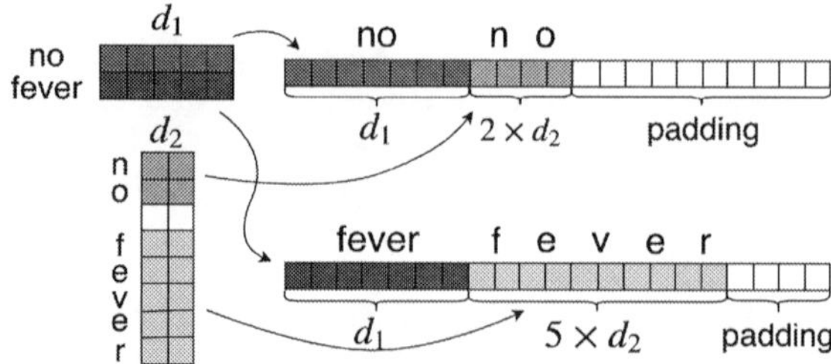

Figure 1: Embedding concatenation model architecture. d_1 is the dimensionality of the word embedding (100), and d_2 is the dimensionality of the character embedding (24).

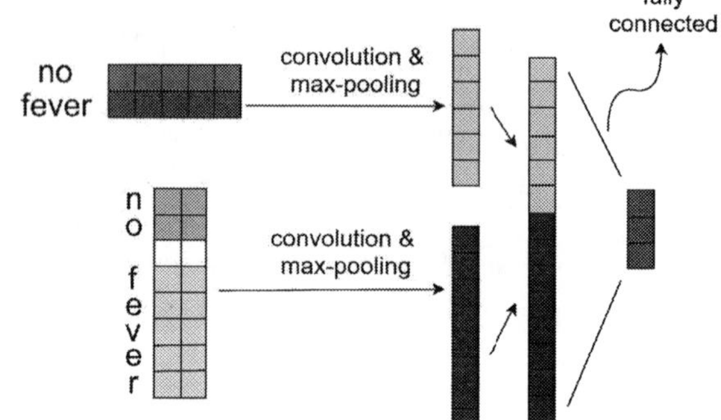

Figure 2: Model combination architecture.

rived from GloVe vectors (Pennington et al., 2014) trained on Common Crawl. The dimensionality of the character embeddings is reduced from 300 to 24 with principal component analysis (PCA).

We also tried learning the embeddings directly as a first layer in the model, but the model was unable to learn useful embeddings, likely because our training set is too small.

The character-based classification model is also a CNN, with a maximum of 1000 characters for each narrative. We also remove punctuation for the character-based model.

3.4 Combined models

We use two different methods of combining the word and character embeddings: embedding concatenation and model combination.

For embedding concatenation, we simply concatenate the word embedding for each word with the ordered character embeddings for the characters in the word. Since words have different numbers of characters, we keep only the first 7 characters of the word, and if the word is shorter than 7 characters we pad the embedding with zeros. In the dataset, 87% of words have 7 characters or fewer, and no improvement was seen by using thresholds of 5, 6, 8, 9, or 10 characters. See Figure 1 for a diagram of the embedding concatenation.

For the model combination, we use all but the final layer of both the word-based CNN and the character-based CNN in parallel, which each produce a feature vector. Before the final classification layer, we concatenate the output vectors from these two networks, and use the combined vector as input to the final feed-forward layer that produces the classification probabilities. See Figure 2

for the diagram of the model architecture[4]. This model allows us to combine the full information from both the word-level and character-level models. However, it also requires the model to learn almost twice as many parameters.

4 Results

We evaluate the four different models using precision, recall, and F_1 score. We also report *cause-specific mortality fraction accuracy* (Murray et al., 2011, 2014), which measures how similar the predicted CoD distribution is to the true distribution. A *cause-specific mortality fraction* (CSMF) is the fraction of a population whose death is attributable to a specific cause. *CSMF accuracy* (CSMFA) is then defined in terms of the difference between the true and predicted fraction for each of k causes:

$$CSMFA = 1 - \frac{\sum_{j=1}^{k} |CSMF_j^{true} - CSMF_j^{pred}|}{2(1 - min(CSMF_j^{true}))}$$

The results of CoD classification using 10-fold cross-validation are presented in Table 2.

Since we hypothesized that the character information would improve results particularly for smaller datasets, we also evaluated the models on a subset of the adult data, which consists of 10% of the original adult dataset, evaluated with 10-fold cross-validation (about 137 records in each test set). We call this dataset "Adult small".

5 Discussion

Overall, the embedding concatenation model performs the best across all individual-level metrics, except on the full adult dataset, where the word

---

[4]The model code is available at: `https://github.com/sjeblee/verbal-autopsy`

Model	Precision	Recall	F_1	CSMFA
Adult (18 categories)				
Word embedding	**.759**	**.755**	**.751**	**.933**
Char. embedding	.690	.684	.680	.922
Emb. concatenation	.716	.699	.699	.912
Model combination	.629	.620	.609	.872
Adult small (18 categories)				
Word embedding	.453	.500	.456	.773
Char. embedding	.609	.603	.589	.837
Emb. concatenation	**.691**	**.669**	**.660**	**.861**
Model combination	.590	.596	.571	.827
Child (11 categories)				
Word embedding	.713	.707	.697	**.902**
Char. embedding	.655	.638	.623	.851
Emb. concatenation	**.740**	**.718**	**.712**	.890
Model combination	.640	.638	.627	.890
Neonate (5 categories)				
Word embedding	.515	.556	.515	.795
Char. embedding	.504	.502	.482	.795
Emb. concatenation	**.562**	**.585**	**.556**	**.819**
Model combination	.502	.530	.495	.807

Table 2: Results from 10-fold cross-validation for each age group in the MDS dataset.

Cat	1	2	3	4	5	n
1	0.870	0.043	0.000	0.043	0.043	23
2	0.294	0.588	0.118	0	0	17
3	0.818	0.091	0.091	0	0	11
4	0.500	0.250	0	0.250	0	4
5	0.500	0.333	0.167	0	0	6

Cat	1	2	3	4	5	n
1	0.826	0.043	0.043	0	0.0869	23
2	0.235	0.588	0.176	0	0	17
3	0.545	0.182	0.273	0	0	11
4	0.500	0	0.250	0	0	4
5	0.500	0	0.333	0	0.167	6

Table 3: Confusion matrices for the neonatal test set (iteration 0). **Top:** results from the word embedding model. **Bottom:** results from the embedding concatenation model. Rows are the correct CoD categories and columns are the predicted categories. n is the number of records belonging to that category in the test set.

embedding model performs the best. For the child dataset, the word-based model performs the best in terms of CSMF accuracy, which means that it best captures the distribution of CoD categories, but the character-based model achieves better accuracy on classifying individual records.

For the adult data, reducing the dataset size to 10% of the original size causes a sharp decrease in accuracy for the word-based model, but only a smaller decrease for the character-based and combined models, showing that the character embeddings are more robust to data size.

Table 3 shows the confusion matrix for the five classes of the neonatal test set from the word embedding model versus the embedding concatenation model. We can see that both models have a heavy preference for the most frequent class (1 *Prematurity and low birthweight*). The embedding concatenation model achieves better accuracy on class 3 (*Birth asphyxia and birth trauma*) and class 5 (*Ill-defined*), but performs worse on class 4 (*Congenital anomalies*), which is the smallest class.

For the child data, the embedding concatenation performs much better on class 1 (*Pneumonia*) (68% accuracy vs. 44%) and class 6 (*Noncommunicable diseases*) (83% vs.78%), and class 10 (*Ill-defined*) (33% vs. 11%), while the word-based model performs better on class 4 (*Other infections*) (76% with the embedding concatenation model vs. 84% with the word model).

The best performing classes for the adult dataset are class 5 (*Maternal*), 15 (*Road traffic incidents*), and 16 (*Suicide*), which are also the categories which have the highest physician agreement. For the small adult dataset, the embedding concatenation model performs noticeably better on classes 4 (*Unspecified infection*), 8 (*Neoplasms*), 16 (*Suicide*), and 18 (*Ill-defined*).

Overall the character information seems to improve accuracy with the smaller datasets, due to its much smaller vocabulary size and its ability to handle spelling variations and unknown words. The combined model performs the best on all of the small datasets, regardless of the number of categories, and especially seems to perform better on more ambiguous categories like *Ill-defined* and *Unspecified infections*.

6 Conclusion and future work

We have shown that character information can improve classification of CoD for verbal autopsies, for smaller datasets, which are very common in the case of VAs. To our knowledge, this is the first application of character-based neural network models to VA narratives.

Due to differences in the datasets, we cannot make direct comparisons to other automated methods. However, since they typically have recall scores around 0.6, our method is competitive. In addition, this method can be applied to any VA dataset with narratives, regardless of the country of origin or the specific survey form.

Future work may include using a language model with character information, such as ELMo (Peters et al., 2018), but we would have to rely on out-of-domain data since the VA dataset is too small to effectively train ELMo or a similar model. The paucity of VA data is one of the biggest obstacles to automated coding.

In the future we also plan to expand these models to other languages, as there are larger VA datasets in languages such as Portuguese and Hindi. We will also investigate using the structured data in addition to the narrative to improve performance.

Acknowledgments

We thank Prabhat Jha of the Centre for Global Health Research for providing the dataset. Our work is supported by funding from the Natural Sciences and Engineering Research Council of Canada and by a Google Faculty Research Award.

References

Peter Byass, Daniel Chandramohan, Samuel Clark, Lucia D'Ambruoso, Edward Fottrell, Wendy Graham, Abraham Herbst, Abraham Hodgson, Sennen Hounton, Kathleen Kahn, Anand Krishnan, Jordana Leitao, Frank Odhiambo, Osman Sankoh, and Stephen Tollman. 2012. Strengthening standardised interpretation of verbal autopsy data: The new InterVA-4 tool. *Global Health Action*, 5:19281.

Samuel Danso, Eric Atwell, and Owen Johnson. 2013. A comparative study of machine learning methods for verbal autopsy text classification. *International Journal of Computer Science Issues*, 10(6).

Kenneth Heafield, Ivan Pouzyrevsky, Jonathan H. Clark, and Philipp Koehn. 2013. Scalable modified Kneser-Ney language model estimation. In *Proceedings of the 51st Annual Meeting of the Association for Computational Linguistics*, pages 690–696, Sofia, Bulgaria.

Serena Jeblee, Mireille Gomes, and Graeme Hirst. 2018. Multi-task learning for interpretable cause of death classification using key phrase prediction. In *Proceedings of the BioNLP 2018 Workshop*, pages 12–17, Melbourne, Australia. Association for Computational Linguistics.

Ryan Kelly. 2015. Pyenchant. http:// pythonhosted.org/pyenchant/.

Tyler H McCormick, Zehang Richard Li, Clara Calvert, Amelia C Crampin, Kathleen Kahn, and Samuel Clark. 2016. Probabilistic cause-of-death assignment using verbal autopsies. *Journal of the American Statistical Association*, 111(15):1036–1049.

Pierre Miasnikof, Vasily Giannakeas, Mireille Gomes, Lukasz Aleksandrowicz, Alexander Y Shestopaloff, Dewan Alam, Stephen Tollman, Akram Samarikhalaj, and Prabhat Jha. 2015. Naïve Bayes classifiers for verbal autopsies: Comparison to physician-based classification for 21,000 child and adult deaths. *BMC Medicine*, 13(1):286–294.

Tomas Mikolov, Ilya Sutskever, Kai Chen, Greg Corrado, and Jeffrey Dean. 2013. Distributed representations of words and phrases and their compositionality. In *Proceedings of the 26th International Conference on Neural Information Processing Systems - Volume 2*, NIPS'13, pages 3111–3119.

Christopher JL Murray, Rafael Lozano, Abraham D Flaxman, Alireza Vahdatpour, and Alan D Lopez. 2011. Robust metrics for assessing the performance of different verbal autopsy cause assignment methods in validation studies. *Population Health Metrics*, 9:28. Erratum (Murray et al., 2014).

Christopher JL Murray, Rafael Lozano, Abraham D Flaxman, Alireza Vahdatpour, and Alan D Lopez. 2014. Erratum to: Robust metrics for assessing the performance of different verbal autopsy cause assignment methods in validation studies. *Population Health Metrics*, 12:7.

Adam Paszke, Sam Gross, Soumith Chintala, Gregory Chanan, Edward Yang, Zachary DeVito, Zeming Lin, Alban Desmaison, Luca Antiga, and Adam Lerer. 2017. Automatic differentiation in PyTorch. In *NIPS 2017 Autodiff Workshop*, pages 1–4.

Jeffrey Pennington, Richard Socher, and Christopher D. Manning. 2014. GloVe: global vectors for word representation. In *Proceedings of the Conference on Empirical Methods in Natural Language Processing (EMNLP)*, pages 1532–1543.

Matthew E. Peters, Mark Neumann, Mohit Iyyer, Matt Gardner, Christopher Clark, Kenton Lee, and Luke Zettlemoyer. 2018. Deep contextualized word representations. In *Proceedings of NAACL-HLT 2018*, pages 2227–2237.

Peter Serina et al. 2015. Improving performance of the Tariff method for assigning causes of death to verbal autopsies. *BMC Medicine*, 13(1):291.

Yuqi Si and Kirk Roberts. 2018. A frame-based NLP system for cancer-related information extraction. *AMIA Annual Symposium Proceedings*, 2018:1524–1533.

Department of Economic and Social Affairs, Population Division, United Nations. 2013. *World Population Prospects: The 2012 revision.* ST/ESA/SER.A/334.

Lyan Verwimp, Joris Pelemans, Hugo Van hamme, and Patrick Wambacq. 2017. Character-word LSTM language models. In *Proceedings of the 15th Conference of the European Chapter of the Association for Computational Linguistics: Volume 1, Long Papers*, pages 417–427, Valencia, Spain. Association for Computational Linguistics.

Erica Westly. 2013. Global health: One million deaths. *Nature*, 504(7478):22–23.

Xiang Zhang, Junbo Zhao, and Yann LeCun. 2015. Character-level convolutional networks for text classification. In *Proceedings of the 28th International Conference on Neural Information Processing Systems - Volume 1*, NIPS '15, pages 649–657.

A   Appendix

Num	Category
	Adult
1	Acute respiratory infections
2	Tuberculosis
3	Diarrhoeal
4	Unspecified infections
5	Maternal
6	Nutrition
7	Chronic respiratory diseases
8	Neoplasms
9	Ischemic heart disease
10	Stroke
11	Diabetes
12	Other cardiovascular diseases
13	Liver and alcohol
14	Other non-communicable diseases
15	Road traffic incidents
16	Suicide
17	Other injuries
18	Ill-defined
	Child
1	Pneumonia
2	Diarrhoea
3	Malaria
4	Other infections
5	Congenital anomalies
6	Non-communicable diseases
7	Injuries
8	Nutritional
9	Other
10	Ill-defined
11	Cancer
	Neonate
1	Prematurity and low birthweight
2	Neonatal infections
3	Birth asphyxia and birth trauma
4	Congenital anomalies
5	Ill-defined

Table 4: Cause of death categories used for the MDS data.

Is artificial data useful for biomedical Natural Language Processing algorithms?

Zixu Wang[1], **Julia Ive**[2], **Sumithra Velupillai**[3] and **Lucia Specia**[1]

[1]Department of Computing, Imperial College London, UK
[2]DCS, University of Sheffield, UK
[3]IoPPN, King's College London, UK and KTH, Sweden

`zixu.wang18@imperial.ac.uk`
`j.ive@sheffield.ac.uk`
`sumithra.velupillai@kcl.ac.uk`
`l.specia@imperial.ac.uk`

Abstract

A major obstacle to the development of Natural Language Processing (NLP) methods in the biomedical domain is data accessibility. This problem can be addressed by generating medical data artificially. Most previous studies have focused on the generation of short clinical text, and evaluation of the data utility has been limited. We propose a generic methodology to guide the generation of clinical text with key phrases. We use the artificial data as additional training data in two key biomedical NLP tasks: text classification and temporal relation extraction. We show that artificially generated training data used in conjunction with real training data can lead to performance boosts for data-greedy neural network algorithms. We also demonstrate the usefulness of the generated data for NLP setups where it fully replaces real training data.

1 Introduction

Data availability is a major obstacle in the development of more powerful Natural Language Processing (NLP) methods in the biomedical domain. In particular, current state-of-the-art (SOTA) neural techniques used for NLP rely on substantial amounts of training data.

In the NLP community, this low-resource problem is typically addressed by generating complementary data artificially (Poncelas et al., 2018; Edunov et al., 2018). This approach is also gaining attention in biomedical NLP. Most of these studies present work on the generation of short text (typically under 20 tokens), given structural information to guide this generation (e.g., chief complaints using basic patient and diagnosis information (Lee, 2018)). Evaluation scenarios for the utility of the artificial text usually involve a single downstream NLP task (typically, text classification).

SOTA approaches tackle other language generation tasks by applying neural models: variations of the encoder-decoder architecture (ED) model (Sutskever et al., 2014; Bahdanau et al., 2015), a.k.a sequence to sequence (seq2seq), e.g., the `Transformer` model (Vaswani et al., 2017). In this work, we follow these approaches and guide the generation process with key phrases in the `Transformer` model.

Our main contribution is thus twofold: (1) a single methodology to generate medical text for a series of downstream NLP tasks; (2) an assessment of the utility of the generated data as complementary training data in two important biomedical NLP tasks: text classification (phenotype classification) and temporal relation evaluation. Additionally, we thoroughly study the usefulness of the generated data in a set of scenarios where it fully replaces real training data.

2 Related Work

Natural Language Generation. Natural language generation is an NLP area with a range of applications such as dialogue generation, question-answering, machine translation (MT), summarisation, simplification, storytelling, etc.

SOTA approaches attempt to solve these tasks by using neural models. One of the most widely used models is the encoder-decoder architecture (ED) (Sutskever et al., 2014; Bahdanau et al., 2015). In this architecture, the decoder is a conditional language model. It generates a new word at a timestep taking into account the previously generated words, as well as the information provided by the encoder (a sequence of hidden states, roughly speaking, a set of automatically learned features).

For different tasks, the input to the encoder may be different: questions for question-answering,

Proceedings of the BioNLP 2019 workshop, pages 240–249
Florence, Italy, August 1, 2019. ©2019 Association for Computational Linguistics

source text for MT, story prompts for story generation, etc.

Long text generation. One of the main challenges of the `ED` architecture remains the generation of long coherent text. In this work, we consider paragraphs as long text. Other NLP tasks may target documents, or even group of documents (e.g., multi-document summarisation systems).

Existing vanilla `ED` models mainly focus on local lexical decisions which limits their ability to model the global integrity of the text. This issue can be tackled by varying the generation conditions: e.g., guiding the generation with prompts (Fan et al., 2018), with named entities (Clark et al., 2018) or template-based generation (Wiseman et al., 2018). All these conditions serve as binding elements to relate generated sentences and ensure the cohesion of the resulting text.

In this work, we follow the approach of Peng et al. (2018) and guide the generation of Electronic Health Record (EHR) notes with the help of key phrases (phrases composed of frequent content words often co-occurring with other content words). These key phrases are sense-bearing elements extracted at the paragraph level. Using them as guidance ensures semantic integrity and relevance of the generated text. We experiment with the SOTA `ED Transformer` model. The model is based on multi-head attention mechanisms. Such mechanisms decide which parts of input and previously generated output are relevant for the next generation decision. Heads are designed to attend to information from different representation subspaces. Recent studies show that their roles are potentially linguistically intepretable: e.g., attending to syntactic dependencies or rare words (Voita et al., 2019).

Usage of artificial data in NLP In MT, artificial data has been successfully used in addition to real data for training `ED` models. There have also been attempts to build MT models in low-resource conditions only with artificial data (Poncelas et al., 2018). In this work, we investigate the usefulness of the generated data both in the complementary setting and in the full replacement setting.

Medical text generation. The generation of medical data destined to help clinicians has been addressed e.g. through generation of imaging reports by Jing et al. (2018); Liu (2018).

However, to our knowledge, there have been very few attempts to create artificial medical data to help NLP. One attempt to create such data can be found in (Suominen et al., 2015), where nursing handover data is generated in a very costly way with the help of a clinical professional who wrote imaginary text.

The attempt closest to ours is the one of Lee (2018). They generate short-length (under 20 tokens) chief complaints using diagnosis and patient- and admission-related information as conditions in the conditional LM. The authors investigate the clinical validity of the generated text by using it as test data for NLP models built with real data. But they do not look into the utility of the generated data for building NLP models.

3 Methodology

As mentioned before, in our attempt to find an optimal way to generate synthetic EHRs we experiment with the `Transformer` architecture. We extract key phrases at the paragraph level, match them at the sentence level and further use them as inputs into our generation model (see Figure 1). Thus, each paragraph is generated sentence by sentence but taking the information ensuring its integrity into account.

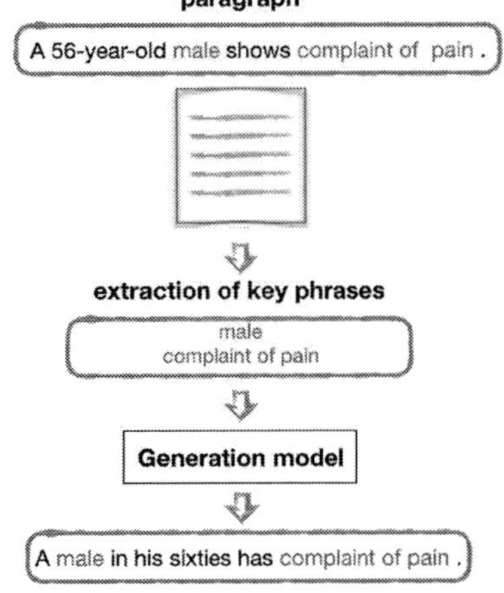

Figure 1: Our generation methodology to guide the generation with key phrases.

The intrinsic evaluation of the generated data is performed with a set of metrics standard for text generation tasks: `ROUGE-L` (Lin, 2004) and `BLEU` (Papineni et al., 2002). `ROUGE-L` measures the n-gram recall, `BLEU`– the n-gram precision. We also assess the length of the generated text.

At the extrinsic evaluation step, we use *generated data* as training data in a phenotype classification task and a temporal relation extraction task. For each task, we experiment with neural models. We compare performance of three models: one trained with real data, one trained using upsampled real data (the real dataset repeated twice) and one built using real data augmented with generated data for *real test sets* (see Figure 2). Development sets are also real across setups. By upsampling the real data twice we create a baseline mimicking a very bad generation model simply reproducing the original data without adding any variation to it.

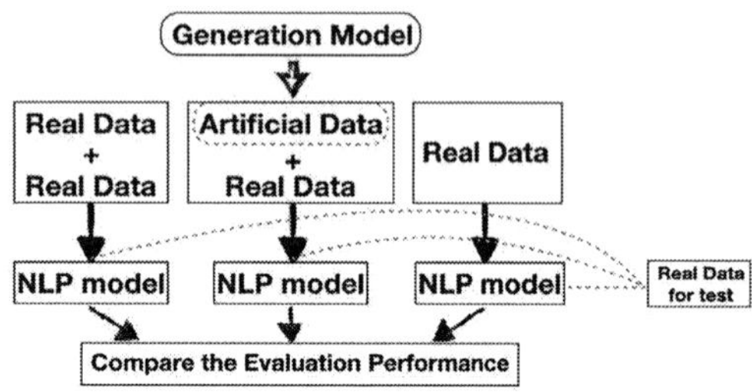

Figure 2: Our extrinsic evaluation procedure with real test data.

We further investigate the actual contribution of the artificial data to the classification process in experiments where we fully replace the real training data with the artificial training data for both neural and non-neural algorithms. Useful artificial data models should demonstrate similar performance results to real models. And, most importantly, those artificial data models should correctly preserve any differences between classification algorithms trained using the real data.

4 Experimental Setup

In what follows, we describe the data used in experiments (Subsection 4.1), details of generation models (Subsection 4.2) and classification models (Subsection 4.3) we use.

4.1 Data

In our study we use EHRs from the publicly available MIMIC-III database (Johnson et al., 2016; Johnson and Pollard, 2016). MIMIC-III contains de-identified clinical data of around 50K adult patients to the intensive care units (ICU) at the Beth Israel Deaconess Medical Center from 2001 to 2012. The dataset comprises several types of clinical notes, including discharge summaries, nursing notes, radiology and ECG reports.

Text generation dataset. For the text generation experiments, we extract all the MIMIC-III

discharge summaries of the patients with the 3 first diagnoses (ordered by their priority, represented by 2 first characters of each respective ICD-9 code) matching at least one sequence of the 3 first diagnoses for the patients from our phenotyping dataset (used later in our phenotype classification experiments). Thus, our text generation dataset do not contain the patients from the phenotyping dataset.

From all the extracted data we randomly select records of 126 patients for development purposes. This results in two subsets: `train-gen` and `val-gen` (see Table 1). As our test sets we used parts of the phenotyping dataset (`test-gen-pheno`) and of the temporal relations dataset (`test-gen-temp`) described below.

set	#, patient ID	#, admission ID	#, lines	#, tok.
`train-gen`	9767	10926	1.2M	20M
`val-gen`	126	132	13K	224K

Table 1: Statistics over `train-gen`, and `val-gen`. # denotes number.

Our preprocessing pipeline including sentence detection uses the spaCy-2.0.18 toolkit.[1] We lowercase all texts. In addition, we replace dates with a placeholder `date`. We discard all the sentences with length under 5 words.

Phenotyping dataset. In our text classification experiments we use the phenotyping dataset from MIMIC-III database released by Gehrmann et al. (2018). Phenotyping is the task of determining whether a patient has a medical condition or is at risk for developing one. The dataset includes discharge summaries annotated with 13 phenotypes (e.g., advanced cancer, advanced heart disease, advanced lung disease, etc.)[2]

The phenotyping dataset used in our experiments contains 1,600 discharge summaries of 1,561 patients (around 180K sentences). We follow Gehrmann et al. (2018) and randomly select 10% and 20% of this data for development and test purposes respectively (`dev-pheno` and `test-pheno`). The rest 70% is used as the test set for the generation experiments and as the training set for the phenotype classification experiments (`test-gen-pheno`).[3]

---

[1]https://spacy.io
[2]https://github.com/sebastianGehrmann/phenotyping
[3]Because of structural differences between MIMIC-III

Temporal relations dataset. In the temporal relation classification experiments, we use the data set from the 2012 i2b2 temporal relations shared task (Sun et al., 2013b). The task focuses on determining the relative ordering of the events in medical history with respect to each other and to time expressions. The dataset contains texts of discharge summaries from MIMIC-II. Various textual segments in these summaries are manually annotated for events (EVENT), time expressions (TIMEX3) and eight temporal relations between them (TLINK). In this study we focus only on detecting the presence of the most frequent OVERLAP temporal relation between events (33% of the annotated relations). OVERLAP indicates that two related events happen almost the same time, but not exactly (Sun et al., 2013a) (see Figure 3).

Figure 3: Example of an OVERLAP temporal relation (paraphrased).

The original training set includes 190 discharge summaries. We experiment with this dataset to demonstrate the transferability of our generation methodology. Hence, we do not modify our generation model but instead filter out the discharge summaries in the 2012 i2b2 dataset that overlap in their content with train-gen (according to the ≥ 10 sentences criteria).

For the 2012 i2b2 data, we condition the generation using the textual segments annotated as EVENT. These could also be seen as binding elements of parts of longer text. Moreover, textual segments given in the input are mostly preserved in the generated output. The advantage of this approach is that in most of the cases we do not need to redo human annotation in the generated text because they are preserved if given in the input. Table 2 reports the statistics of the original (all) data versus the data (reduced) for which the annotations are preserved.

10% of the data is randomly selected for devel-

opment purposes (dev-temp). The rest of the data is again used as the test data for the generation task and as the training data for the temporal classification models (test-gen-temp). The test set provided with the 2012 i2b2 temporal relations shared task was used as is for temporal classification models (test-temp).

	#, docs	#, lines	#, tok.	%, OVERLAP.
all	190	7447	97K	33.0
reduced	175	6762	89K	33.6

Table 2: Statistics over test-gen-temp. The all dataset corresponds to the one provided by the organisers. The reduced dataset is the one for which the annotations are preserved by the generation model. # denotes number.

4.2 Text Generation Models

In our text generation experiments we use the Transformer model, which generates text sentence by sentence. To ensure the semantic integrity of paragraphs resulting from the concatenation of generated sentences, we guide the generation with key phrases. Key phrases are extracted from each original paragraph of train-gen. For this, we use the Rake algorithm (Rose et al., 2010)[4] and take the highest scored 50% per paragraph. We further generate a paragraph sentence by sentence using as inputs only those extracted key phrases that are present in each particular sentence. This results in approximately 2.4 key phrases with an average length of 1.7 words per sentence (as computed for train-gen).[5] Boundaries of key phrases in the input to models are fixed by a reserved token. During training, the model is learned to restore real text from key phrases, basically by filling the gaps between those key phrases.

We trained our Transformer models as provided by the OpenNMT toolkit (Klein et al., 2017) with default parameters. In train-gen we replaced all the words with frequency 1 with a placeholder. This resulted in a vocabulary of around 50K words. Each model was trained for 30K epochs.[6] Outputs are produced with the standard beam decoding procedure with a beam size of 5.

---

and MIMIC-II database that was initially used to collect the phenotyping dataset, we could not correctly identify text fields for records with duplicated admission IDs. We simply merged those records together giving preferences to annotations with a higher rate of positive labels. This resulted in a small reduction of the initial dataset (less than 1%).

[4]The algorithm selects phrases composed of frequent content words co-occurring with other content words.

[5]We used the implementation available at https://github.com/csurfer/rake-nltk

[6]We noticed that this quantity of epochs is necessary for

test-gen-pheno		
1	gen	a ct was obtained which revealed a very poor study but <u>no evidence of</u> a brain injury .
	real	ct was a poor study but <u>did not reveal</u> a brain injury .
2	gen	he had <u>a walk of losing blood</u> .
	real	she is unable to <u>walk without losing blood</u> .
test-gen-temp		
3	gen	he was treated with increasing doses of <u>rosuvastatin and atorvastatin</u> .
	real	he has been on increasing doses of <u>rosuvastatin receiving atorvastatin in addition</u> on a basis .
4	gen	he was started on ibuprofen and <u>his wife back pain was improved</u> .
	real	the patient was initially treated with ibuprofen which was stopped after <u>his back pain</u> improved .

Table 3: Examples of real and generated text. The underlined text highlights "good" (examples 1 and 3) or "bad" (examples 2 and 4) modifications. All sentences have been paraphrased.

4.3 Models for Phenotype Classification

For the phenotype classification task, we train two standard NLP models:

1. **Convolutional Neural Network (CNNs) model** inspired by (Kim, 2014). The CNN model is built with 3 convolutional layers with window sizes of 3, 4 and 8 respectively. The word embedding dimensionality is 300, both convolution layers have 100 filters. The size of the hidden units of the dense layer is 100. We also use a dropout layer with a probability of 0.5. The network is implemented using the Pytorch toolkit[7] with the pre-trained GloVe word embeddings (Pennington et al., 2014).

2. **Word-level bag-of-words (BoW) model** trained with the Naive Bayes (NB) algorithm. We applied the MultinomialNB implementation from `Scikit-learn` (Pedregosa et al., 2011).

We cast the task as a binary classification task and evaluate the detection of each phenotype computing the F1-score of the positive class.

4.4 Models for Temporal Relations Extraction

Inspired by the SOTA approaches for the task (Tourille et al., 2017), we build a Bidirectional Long Short-Term Memory (BiLSTM) classifier (Hochreiter and Schmidhuber, 1997). The BiLSTM model is constructed with two hidden layers of opposite directions. The size of hidden LSTM units is 512. We use a dropout layer before the output layer with a probability of 0.2 and the

---

stabilization of the model perplexity.

[7]`https://pytorch.org`

	ROUGE-L	BLEU	avg. sent. l (gen./real)
test-gen-pheno	67.74	40.62	13.27 / 17.50
test-gen-temp	48.47	20.91	18.61 / 16.81

Table 4: Qualitative evaluation and average sentence lengths.

concatenation of the last hidden states of both layers goes into the ouput layer. We train our network with the Adam (Kingma and Ba, 2014) optimization algorithm with a batch size of 64 and a learning rate of 0.001. We use again the pre-trained GloVe word embeddings. The classifier is implemented using Pytorch. As for a non-neural model, we use again the NB model as for the phenotype classification task.

We cast the task as a binary classification task (for each event-event pair, classify as OVERLAP or not) and evaluate the result by computing the F1-score of the positive decision.

5 Experimental Results

In this section we present results of our experiments, first of the intrinsic evaluation of the quality of generated text (Section 5.1) and then of the extrinsic evaluation of its utility for NLP (text classification and temporal relation extraction tasks, Section 5.2).

5.1 Intrinsic Evaluation

Table 4 shows the intrinsic evaluation results for both generated test-gen-pheno and test-gen-temp. The BLEU and ROUGE-L are computed between the original text (the one used to extract key phrases) and the generated text. We also compare the average lengths sentences for those two texts.

	Obesity	Non Adherence	Developmental.Delay.Retardation	Adv. Heart Disease	Adv. Lung Disease	Schizo and other Psych. Disorders	Alcohol Abuse	Other Substance Abuse	Chr. Pain Fibromyalgia	Chr. Neurological Dystrophies	Adv. Cancer	Depression	Dementia	avg.
freq, %	8	9	3	17	10	18	12	10	20	23	10	29	7	
CNN														
real + gen	0.3257	0.3394	**0.3636**	**0.6384**	0.5333	**0.3664**	0.7428	**0.5714**	0.3846	**0.5574**	**0.6173**	0.4373	0.5714	**0.4961**
real	**0.3789**	**0.3589**	0.2500	0.6019	0.5085	0.2909	0.7200	0.4912	0.4000	0.4782	0.5567	**0.4623**	0.5667	0.4665
2 × real	0.3636	0.3333	0.2857	0.5347	**0.5758**	0.3057	**0.7435**	0.4789	**0.4040**	0.4580	0.5667	0.4162	**0.6341**	0.4692
gen	0.2500	0.3656	0.2000	0.4667	0.5574	0.3221	0.7297	0.4478	0.3978	0.4564	0.6575	0.4598	0.3273	0.4337
gen-key	0.1365	0.2443	0.0252	0.5200	0.1429	0.2978	0.2581	0.1914	0.3781	0.3740	0.3778	0.4262	0.0800	0.2656
NB														
real	0.2000	0.4722	0.0000	0.5812	0.4838	0.5614	0.6756	0.5000	0.4109	0.5270	0.6779	0.5700	0.3846	0.4650
gen	0.2424	0.4719	0.0000	0.5893	0.4687	0.5000	0.6506	0.4594	0.4022	0.5122	0.6562	0.5391	0.3125	0.4465
gen-key	0.1407	0.1984	0.0447	0.3022	0.2108	0.2857	0.2367	0.1723	0.3284	0.3815	0.2032	0.4398	0.1039	0.2345

Table 5: Phenotyping results for CNN and Naive Bayes (NB), `test-pheno`. Best performing models for CNN data augmentation experiments are highlighted in bold. We report results for the models trained with: `real` data augmented with generated `gen` data, `real` data only, 2 × `real` data upsampled twice, `gen` data only, `gen-key` data without traces of the input real data.

As expected, automatic evaluation scores show that for both test sets our model generates context preserving pieces of the real text from the input (e.g., ROUGE-L = 67.74 for `test-gen-pheno`, ROUGE-L = 48.47 for `test-gen-temp`). The proximity of average lengths of sentences for the generated text and the real text supports this statement.

As automatic metrics perform only a shallow comparison, we also manually reviewed a sample of texts. In general, most of the generated text preserves the main meaning of the original text adding or dropping some details. Incomprehensible generated sentences are rare.

Table 3 shows examples of the generated text for both datasets. In examples 1 and 3, `Transformer` generates text with a meaning very close to the original one (e.g., *no evidence of* ≈ *did not reveal*, for `test-gen-pheno`). Examples 2 and 4 are "bad" modifications. In general, such examples are infrequent. For instance in Example 2, the real phrase *unable to walk without losing blood* is incorrectly modified into *a walk of losing blood*. However, the main sense of losing blood is preserved.

Overall, our observations indicate that the generation methodology successfully adapts to changes in generation conditions.

5.2 Extrinsic Evaluation

Phenotype Classification. Table 5 shows results of our text classification experiments. They indicate that the artificial training data used as complimentary to the real training data is in general beneficial for the CNN model (e.g., av. F-score=0.50 for `real + gen` > 0.47 for `real`). `real +gen` setup also outperforms the model trained using larger volume data, where the training data was repeated two times (2 × `real`). Overall, `real +gen` outperforms `real` for 9 phenotypes out of 13 with an average ΔF-score=0.06, while 2 × `real` for 6 phenotypes with an average ΔF-score=0.04 only.

To get further insights into the actual informativeness of the generated data, we study the performance of both CNN and NB in a series of setups where the artificial training data fully replace the real training data. To be more precise, we study: (a) `gen` setup, where the full generated data with traces of input key phrases are used as the training data; and (b) `gen-key` setup, where the generated text without traces of input data is used as the training data (see Figure 4). The results of these experiments are in Table 5, lower part. They show that average performances of `gen` and `real` tend to be comparable for each algorithm (e.g., Δ avg. F-score=0.03 for both CNN and NB). The `gen-key` setup results in a signif-

	F-score	Features (words)
`real`	0.5614	**chest**, 20, given, 11, hours, time, history, admission, continued, capsule, needed, 25, disease, refills, follow, negative, started, status, disp, days, release, discharge, ml, stable, hct, prior, dr, showed, 40, fax, neg, telephone, likely, 15, glucose, wbc, home, renal, care, seen, iv, 24, acute, urine, post, noted, artery, 14, year, unit, tube, inr, bid, 50, **edema**, units, plt, insulin, known, course, **pulmonary**, mild, did, dose
`gen`	0.5000	follow, 12, fax, renal, admission, care, telephone, prior, artery, bid, acute, dr, unit, known, time, post, likely, seen, neg, discharge, iv, insulin, tube, units, admitted, placed, year, 11, 25, 13, **pulmonary**, urine, dose, delayed, mild, chronic, transferred, **edema**, lower, pressure, heart, course, fluid, failure, ventricular, aortic, abdominal, 50, discharged, medications, valve, evidence, noted, increased
`gen-key`	0.2857	blood, day, mg, 10, 07, date, pt, 10pm, refills, 100, 20, tablet, needed, started, **ct**, plt, 12, 30, inr, 11, 25, 13, dr, times, 50, sig, 213, 24, patient, daily, 40, 500, telephone, release, transferred, negative, discharged, 81, follow, final, admitted, 15, 30pm, time, fax, hours, delayed, normal, placed, history, 20am, seen, **breath**, 00, did, 18, 15pm, evidence, 80, admission, consulted, home, wbc, po, hct, bedtime, **shortness**

Table 6: Top-30 words contributing the most to the `Advanced Lung Disease` phenotype detection using Naive Bayes.

icant performance drop (of F-score=0.2 on average). However, the `gen-key` text still potentially bears some relevant information that allows both CNN and NB have comparable performance for this setup.

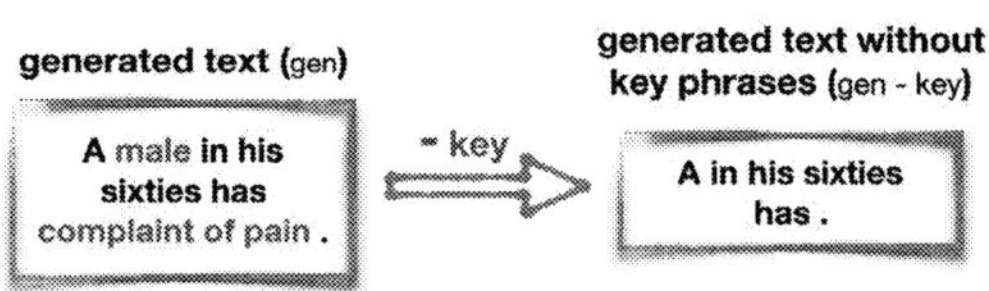

Figure 4: Example of creating `gen gen-key` data – the generated text without traces of input data (paraphrased)

Taking advantage of the easy interpretability of the NB model, we analyse the words that contribute the most to classification decisions (highest likelihoods given the positive class) for the `Adv. Lung Disease` as a an example of a phenotype with an average frequency for the dataset. Table 6 displays those words in order of importance for `real`, `gen` and `gen-key`. As expected, for `real` and `gen` with higher F-score values, there are more relevant medical terms: e.g., *pulmonary* and *chest*. For `gen-key`, there are words more distantly related to the phenotype: e.g., *ct* and *breath*.

Temporal Evaluation. For the i2b2 dataset, we focus only on the evaluation for the OVERLAP temporal relation between events as the most well-represented group. Inspired by the SOTA solutions for the temporal relations extraction task (Tourille et al., 2017), we provide only the text spans that link the two events as inputs to our models. This setup is particularly beneficial to assess the utility of the generated text (see Figure 5).

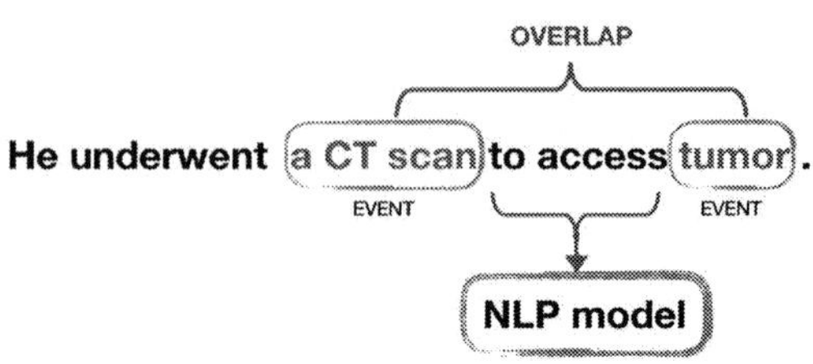

Figure 5: Example of an input to our models for temporal relations extraction – a text span that links the two events (paraphrased).

As mentioned earlier, for this dataset we guide the text generation with event text spans. Thus, for this setup, we take only the text between those real text spans essentially copied from the input. This allows us to better assess the utility only of what was generated.[8]

Table 7 reports results for our experiments with the i2b2 dataset. They are similar to the ones performed for the phenotyping dataset. Note that we reduce the initial training set provided by the task due to particularities of our generation procedure. In our data augmentation experiments we add this reduced generated data to all the provided real training data (`real all`).

The results show that `real all + gen` (F-score=0.62) outperforms the `real` setup (F-score=0.59), as well as the upsampled setup (2 × `real all`, F-score=0.58). This confirms the utility of our data augmentation procedure for the BiLSTM model. Results for `gen` and `real reduced` are again comparable for BiL-

---

[8]However, it should be noted here that the generated text between two events may still contain other event spans copied from the input, especially for the cases when events are in different sentences.

	BiLSTM
`real all + gen`	**0.6217**
`real all`	0.5896
`2 × real all`	0.5803
`gen`	0.5138
`real reduced`	0.5312
	NB
`gen`	0.5769
`real reduced`	0.5024

Table 7: Temporal relations extraction for OVERLAP for CNN and Naive Bayes (NB), `test-temp`. Only the real/generated text between events serves as input. Best performing models for data augmentation experiments are highlighted in bold. We report results for the models trained using: `real all` training data from the i2b2 task augmented with the generated `gen` data, `real all` data only, `2 × real all` data upsampled twice, `real reduced` data only, `gen` data only.

STM. For NB, we even observe an improvement of ΔF-score=0.08 for `gen` as compared to `real reduced` for NB. This may be explained by a stronger semantic signal in the generated data. Overall, our results demonstrate the potential of developing a model that would generate artificial medical data for a series of NLP tasks.

6 Discussion

Our study is designed as a proof-of-concept and the main objective of this work is to study the utility of using SOTA approaches for generating artificial EHR data and to evaluate the impact of using this to augment real data for common NLP tasks in the clinical domain. Our results are promising. From a preliminary manual analysis, most meaning is preserved in the generated texts. For both extrinsic evaluation tasks (phenotype classification, and temporal relation classification), using generated text to augment real data in the training phase improved results. Moreover, for both tasks, results using only generated data was comparable to those using only real data, further indicating usefulness.

To our knowledge, this is the first study looking at the problem of generating longer clinical text, and that is extrinsically evaluated on two downstream NLP tasks. Although the MIMIC data is comprehensive, it represents a particular type of clinical documentation from an ICU setting, in further work we plan to extend to other clinical domains.

If artificial data was to be used for further downstream tasks, particularly those that are intended to support secondary uses in a clinical research setting, further analysis is needed to assess the clinical validity of the generated text. This would require domain expertise. For instance, the temporal relation classification problem imposes different constraints as compared with the document classification task, which might require other approaches for designing the text generation models. Moreover, other temporal information representation models have been proposed in other studies, for other use-cases, such as the CONTAINS relation in the THYME corpus (Styler IV et al., 2014). In future studies, we will invite clinicians to review the generated text with a focus on clinical validity aspects, as well as study further downstream NLP tasks. We will also study additional alternative metrics for intrinsic evaluation, such as the modified CIDEr metric proposed by Lee (2018).

7 Conclusion

In this work, we attempt to generate artificial training data for two downstream clinical NLP tasks: text classification and temporal relation extraction. We propose a generic methodology to guide the generation in both cases. Our experiments show the utility of artificial data for neural NLP models in data augmentation setups. Our generation methodology holds promise for the development of a more universal approach that will allow medical text generation for an even wider range of biomedical NLP tasks. We also plan to further investigate the validity and utility of artificial data. We think thus, that artificial data generation is an approach that has the potential to solve current data accessibility issues associated with biomedical NLP.

Acknowledgments

This work was partly funded by EPSRC Healtex Feasibility Funding (Towards Shareable Data in Clinical Natural Language Processing: Generating Synthetic Electronic Health Records). The third author has received support from the Swedish Research Council (2015-00359), and the Marie Skłodowska Curie Actions, Cofund, Project INCA 600398. We would like to thank the anonymous reviewers for their helpful comments.

References

Dzmitry Bahdanau, Kyunghyun Cho, and Yoshua Bengio. 2015. Neural machine translation by jointly learning to align and translate. In *Proceedings of International Conference on Learning Representations (ICLR)*.

Elizabeth Clark, Yangfeng Ji, and Noah A. Smith. 2018. Neural text generation in stories using entity representations as context. In *Proceedings of the 2018 Conference of the North American Chapter of the Association for Computational Linguistics: Human Language Technologies, Volume 1 (Long Papers)*, pages 2250–2260. Association for Computational Linguistics.

Sergey Edunov, Myle Ott, Michael Auli, and David Grangier. 2018. Understanding back-translation at scale. *arXiv preprint arXiv:1808.09381*.

Angela Fan, Mike Lewis, and Yann Dauphin. 2018. Hierarchical neural story generation. In *Proceedings of the 56th Annual Meeting of the Association for Computational Linguistics (Volume 1: Long Papers)*, pages 889–898. Association for Computational Linguistics.

Sebastian Gehrmann, Franck Dernoncourt, Yeran Li, Eric T Carlson, Joy T Wu, Jonathan Welt, John Foote, Edward T Moseley, David W Grant, Patrick D Tyler, and Leo A Celi. 2018. Comparing deep learning and concept extraction based methods for patient phenotyping from clinical narratives. *PLoS ONE*, 13(2):e0192360.

S Hochreiter and J Schmidhuber. 1997. Long short-term memory. *Neural Computation*, 9(8):1735–1780.

Baoyu Jing, Pengtao Xie, and Eric Xing. 2018. On the Automatic Generation of Medical Imaging Reports. In *Proceedings of the 56th Annual Meeting of the Association for Computational Linguistics (Volume 1: Long Papers)*, pages 2577–2586. Association for Computational Linguistics.

Alistair E W Johnson and Tom J Pollard. 2016. The MIMIC-III Clinical Database.

Alistair E W Johnson, Tom J Pollard, Lu Shen, Liwei H Lehman, Mengling Feng, Mohammad Ghassemi, Benjamin Moody, Peter Szolovits, Leo Anthony Celi, and Roger G Mark. 2016. MIMIC-III, a freely accessible critical care database. *Scientific Data*, 3:160035.

Yoon Kim. 2014. Convolutional neural networks for sentence classification. *CoRR*, abs/1408.5882.

Diederik P Kingma and Jimmy Ba. 2014. Adam: A method for stochastic optimization. *arXiv preprint arXiv:1412.6980*.

Guillaume Klein, Yoon Kim, Yuntian Deng, Jean Senellart, and Alexander Rush. 2017. OpenNMT: Open-source toolkit for neural machine translation. In *Proceedings of ACL 2017, System Demonstrations*, pages 67–72. Association for Computational Linguistics.

Scott H. Lee. 2018. Natural language generation for electronic health records. *npj Digital Medicine*, 1(1):63.

Chin-Yew Lin. 2004. ROUGE: A package for automatic evaluation of summaries. In *ACL workshop on Text Summarization Branches Out*.

Peter J. Liu. 2018. Learning to write notes in electronic health records. *CoRR*, abs/1808.02622.

Kishore Papineni, Salim Roukos, Todd Ward, and Wei-Jing Zhu. 2002. Bleu: a method for automatic evaluation of machine translation. In *Proceedings of 40th Annual Meeting of the Association for Computational Linguistics*, pages 311–318.

Fabian Pedregosa, Gaël Varoquaux, Vincent Michel Alexandre Gramfort, Bertrand Thirion, Olivier Grisel, Peter Prettenhofer Mathieu Blondel, Ron Weiss, Vincent Dubourg, Jake Vanderplas, Alexandre Passos, David Cournapeau, Matthieu Brucher, Matthieu Perrot, and Édouard Duchesnay. 2011. Scikit-learn: Machine learning in Python. *Journal of Machine Learning Research*, 12:2825–2830.

Nanyun Peng, Marjan Ghazvininejad, Jonathan May, and Kevin Knight. 2018. Towards controllable story generation. In *Proceedings of the First Workshop on Storytelling*, pages 43–49. Association for Computational Linguistics.

Jeffrey Pennington, Richard Socher, and Christopher Manning. 2014. Glove: Global vectors for word representation. In *Proceedings of the 2014 conference on empirical methods in natural language processing (EMNLP)*, pages 1532–1543.

Alberto Poncelas, Dimitar Shterionov, Andy Way, Gideon Maillette de Buy Wenniger, and Peyman Passban. 2018. Investigating backtranslation in neural machine translation. In *Proceedings of th 21st International Conference of the European Association for Machine Translation (EAMT)*.

Stuart Rose, Dave Engel, Nick Cramer, and Wendy Cowley. 2010. Automatic keyword extraction from individual documents. *Text Mining: Applications and Theory*, pages 1 – 20.

William F Styler IV, Steven Bethard, Sean Finan, Martha Palmer, Sameer Pradhan, Piet C de Groen, Brad Erickson, Timothy Miller, Chen Lin, Guergana Savova, and James Pustejovsky. 2014. Temporal annotation in the clinical domain. *Transactions of the Association for Computational Linguistics*, 2:143–154.

Weiyi Sun, Anna Rumshisky, and Ozlem Uzuner. 2013a. Annotating temporal information in clinical narratives. *Journal of biomedical informatics*, 46 Suppl(0):S5–S12.

Weiyi Sun, Anna Rumshisky, and Ozlem Uzuner. 2013b. Evaluating temporal relations in clinical text: 2012 i2b2 challenge. *Journal of the American Medical Informatics Association*, 20(5):806–813.

Hanna Suominen, Liyuan Zhou, Leif Hanlen, and Gabriela Ferraro. 2015. Benchmarking Clinical Speech Recognition and Information Extraction: New Data, Methods, and Evaluations. *JMIR Medical Informatics*, 3(2):e19.

Ilya Sutskever, Oriol Vinyals, and Quoc V. V Le. 2014. Sequence to sequence learning with neural networks. In *Advances in Neural Information Processing Systems 27*, pages 3104–3112.

Julien Tourille, Olivier Ferret, Aurélie Névéol, and Xavier Tannier. 2017. Neural architecture for temporal relation extraction: A bi-lstm approach for detecting narrative containers. In *Proceedings of the 55th Annual Meeting of the Association for Computational Linguistics (Volume 2: Short Papers)*, pages 224–230, Vancouver, Canada. Association for Computational Linguistics.

Ashish Vaswani, Noam Shazeer, Niki Parmar, Jakob Uszkoreit, Llion Jones, Aidan N Gomez, Łukasz Kaiser, and Illia Polosukhin. 2017. Attention is all you need. In I. Guyon, U. V. Luxburg, S. Bengio, H. Wallach, R. Fergus, S. Vishwanathan, and R. Garnett, editors, *Advances in Neural Information Processing Systems 30*, pages 5998–6008. Curran Associates, Inc.

Elena Voita, David Talbot, Fedor Moiseev, Rico Sennrich, and Ivan Titov. 2019. Analyzing Multi-Head Self-Attention: Specialized Heads Do the Heavy Lifting, the Rest Can Be Pruned. In *Proceedings of the 57th Annual Meeting of the Association for Computational Linguistics*, Florence, Italy. Association for Computational Linguistics.

Sam Wiseman, Stuart M Shieber, and Alexander M Rush. 2018. Learning neural templates for text generation. *arXiv preprint arXiv:1808.10122*.

ChiMed: A Chinese Medical Corpus for Question Answering

Yuanhe Tian
Department of Linguistics
University of Washington
yhtian@uw.edu

Weicheng Ma
Computer Science Department
New York University
wm724@nyu.edu

Fei Xia
Department of Linguistics
University of Washington
fxia@uw.edu

Yan Song
Tencent AI Lab
clksong@gmail.com

Abstract

Question answering (QA) is a challenging task in natural language processing (NLP), especially when it is applied to specific domains. While models trained in the general domain can be adapted to a new target domain, their performance often degrades significantly due to domain mismatch. Alternatively, one can require a large amount of domain-specific QA data, but such data are rare, especially for the medical domain. In this study, we first collect a large-scale Chinese medical QA corpus called *ChiMed*; second we annotate a small fraction of the corpus to check the quality of the answers; third, we extract two datasets from the corpus and use them for the relevancy prediction task and the adoption prediction task. Several benchmark models are applied to the datasets, producing good results for both tasks.

1 Introduction

In the big data era, it is often challenging to locate the most helpful information in many real-world applications, such as search engine, customer service, personal assistant, etc. A series of NLP tasks, such as text representation, text classification, summarization, keyphrase extraction, and answer ranking, are able to help QA systems in finding relevant information (Siddiqi and Sharan, 2015; Allahyari et al., 2017; Yang et al., 2016; Joulin et al., 2016; Song et al., 2017, 2018).

Currently, most QA corpora are built for the general domain focusing on extracting/generating answers from articles, such as CNN/Daily Mail (Hermann et al., 2015), SQuAD (Rajpurkar et al., 2016), Dureader (He et al., 2017), SearchQA (Dunn et al., 2017), CoQA (Reddy et al., 2018), etc., with few others from community QA forums, such as TrecQA (Wang et al., 2007), WikiQA (Yang et al., 2015), and SemEval-2015 (Nakov et al., 2015).

In the medical domain, most medial QA corpora consist of scientific articles, such as BioASQ (Tsatsaronis et al., 2012), emrQA (Pampari et al., 2018), and CliCR (Šuster and Daelemans, 2018). Although some studies were done for conversational datasets (Wang et al., 2018a,b), corpora designed for community QA are extremely rare. Meanwhile, given that many online medical service forums have emerged (e.g. MedHelp[1]), there are increasing demands from users to search for answers for their medical concerns. One might be tempted to build QA corpora from such forums. However, in doing so, one must address a series of challenges such as how to ensure the quality of the derived corpus despite the noise in the original forum data.

In this paper, we introduce our work on building a Chinese medical QA corpus named *ChiMed* by crawling data from a big Chinese medical forum[2]. In the forum, the questions are asked by web users and all the answers are provided by accredited physicians. In addition to (Q, A) pairs, the corpus contains rich information such as the title of the question, key phrases, age and gender of the user, the name and affiliation of the accredited physicians who answer the question, and so on. As a result, the corpus can be used for many NLP tasks. In this study, we focus on two tasks: relevancy prediction (whether an answer is relevant to a question) and adoption prediction (whether an answer will be adopted).

---

[1] https://www.medhelp.org

[2] The code for constructing the corpus and the datasets used in this study are available at https://github.com/yuanheTian/ChiMed.

Proceedings of the BioNLP 2019 workshop, pages 250–260
Florence, Italy, August 1, 2019. ©2019 Association for Computational Linguistics

# of As per Q	# of Qs	% of Qs
1	5,517	11.8
2	39,098	83.7
≥ 3	2,116	4.5
Total	46,731	100.0

Table 1: Statistics of *ChiMed* with respect to the number of answers (As) per question (Q).

2 The *ChiMed* Corpus

To benefit NLP research in the medical domain, we create a Chinese medical corpus (*ChiMed*). This section describes how the corpus was constructed, the main content of the corpus, and and its potential usage.

2.1 Data Collection

Ask39[3] is a large Chinese medical forum where web users (to avoid confusion, we will call them *patients*) can post medical questions and receive answers provided by licensed physicians. Each question, together with its answers and other related information (e.g., the names of physicians and similar questions), is displayed on a page (aka a QA page) with a unique URL. Currently, approximately 145 thousand forum-verified physicians have joined the forum to answer questions and there are 17.6 million QA pages. We started with fifty thousand URLs from the URL pool and downloaded the pages using the selenium package[4]. After removing duplicates or pages with no answers, 46,731 pages remain and most of the questions (83.7%) have two answers (See Table 1).

2.2 QA Records

From each QA page, we extract the question, the answers and other related information, and together they form a *QA record*. Table 2 displays the main part of a QA record, which has five fields that are most relevant to this study: (1) "*Department*" indicates which medical department the question is directed to;[5] (2) "*Title*" is a brief description of disease/symptoms (5-20 characters); (3) "*Question*" is a health question with a more detailed description of symptoms (at least 20 characters); (4) "*Keyphrases*" is a list of phrases related to the question and the answer(s); (5) The

last field is a list of *answers*, and each answer has an *Adopted* flag indicating whether it has been adopted. Among the five fields, *Title* and *Question* are entered by patients; *Answers* are provided by physicians; *Department* is determined by the forum engine automatically when the question is submitted. As for the *Keyphrases* field and the *Adopted* flag, it is not clear to us whether they are created manually (if so, by whom) or generated automatically.[6] In addition to these fields, a QA record also contains other information such as the name and affiliation of the physicians who answer the question, the patient's gender and age, etc.

Table 3 shows the statistics of *ChiMed* in terms of QA records. On average, each QA record contains one question, 1.96 answers, and 4.48 keyphrases. Overall, 69.1% of the answers in the corpus have an adopted flag.

2.3 Potential Usage of the Corpus

Given the rich content of the QA record, ChiMed can be used in many NLP tasks. For instance, one can use the corpus for **text classification** (to predict the medical department that a Q should be directed to), **text summarization** (to generate a title given a Q), **keyphrase generation** (to generate keyphrases given a Q and/or its As), **answer ranking** (to rank As for the same Q, if adopted As are indeed better than unadopted As), and **question answering** (retrieve/generate As given a Q).

Because the content of the corpus comes from an online forum, before we use the corpus for any NLP task, it is important to check the quality of the corpus with respect to that task. As a case study, for the rest of the paper, we will focus on three closely related tasks, all taking a question and an answer (or a set of answers) as the input: The first one determines whether the answer is relevant to the question; the second determines whether the answer will be adopted for the question (as indicated by the *Adopted* flag in the corpus); the third one ranks all the answers for the question if there are more than one answer. We name them the *relevancy task*, the *adoption prediction task*, and the *answer ranking task*, respectively. The first two are binary classification tasks, while the last one is a ranking task. In the next section, we will manually check a small fraction of the corpus to determine whether its quality is high for those tasks.

---

[3]http://ask.39.net
[4]https://github.com/SeleniumHQ/selenium
[5]There are 13 departments such as pediatrics, infectious diseases, and internal medicine.

[6]We have made many attempts to no avail to contact the forum about those and other questions.

Department	内科 > 淋巴增生 Internal Medicine > Lymphocytosis
Title	胃部淋巴增生会癌变吗？ Will lymphatic hyperplasia in the stomach cause cancer?
Question	我最近检查出患有胃部淋巴增生的疾病，非常担心，请问它会癌变吗？ I recently checked out the disease of lymphoid hyperplasia in the stomach. I am very worried. Will it cause cancer?
Keypharses	慢性浅表性胃炎，幽门螺旋杆菌感染，淋巴增生，胃，消化 Chronic superficial gastritis, Helicobacter pylori infection, lymphatic hyperplasia, stomach, digestion
Answer 1	这一般是幽门螺旋杆菌感染造成的，一般不会造成癌变，所以不必惊慌。建议饮食规律，吃易消化的食物，细嚼慢咽，少量多餐，禁食刺激性食物。 In general, this is caused by Helicobacter pylori infection and does not cause cancer. So do not panic. It is recommended to have a regular diet, eat digest friendly food and chew slowly. Do not eat much in one meal and no spicy food is allowed.
Adopted	True
Answer 2	这是普通的慢性胃粘膜炎症，与幽门螺旋杆菌感染有关。可用阿莫西林治疗。 This is a common chronic gastric mucosal inflammation and has a relationship with Helicobacter pylori infection. You can choose amoxicillin for treatment.
Adopted	False

Table 2: An example of QA record in *ChiMed*. The English translation is not part of the corpus.

# of Questions	46,731
# of Answers	91,416
Avg. # of Answers per Question	1.96
# (%) of Answers Adopted	63,153 (69.1%)
# of Keyphrases	209,261
# of Keyphrases per Q	4.48
# of Unique Keyphrases	10,360

Table 3: Statistics of *ChiMed*.

# of Adopted As	# of Qs	% of Qs
0	30	0.08%
1	25,594	65.46%
2	13,474	34.46%
Total	39,098	100%

Table 4: QA records with exactly two answers.

3 Relevancy, Answer Ranking, and Answer Adoption

Given *ChiMed*, it is easy to synthesize a "labeled" dataset for the relevancy task. E.g., given a question, we can treat answers in the same QA record as relevant, and answers in other QA records as irrelevant. The quality of such a synthesized dataset will depend on how often answers in a QA record are truly relevant to the question in the same record. For the adoption prediction task, we can directly use the *Adopted* flag in the QA records.

For the answer ranking task, the answers in a QA record are not ranked. However, if adopted answers are often better than unadopted answers, the former can be considered to rank higher than the latter if both answers come from the same QA record. Table 4 shows among the QA records with exactly two answers, 65.46% of them have exactly one adopted answer and 34.46% have two adopted answers. We can use these 65.46% of QA records as a labeled dataset for the answer ranking task. However, the quality of such a dataset will depend on the correlation between the *Adopted* flag and the high quality of an answer.

To evaluate whether the answers are relevant to the question in the same QA record, and whether adopted answers are better than unadopted ones, we randomly sampled QA records containing exactly two questions, and picked 60 records with exactly one adopted and one unadopted answers (called **Subset-60**) and 40 records with both answers adopted (called **Subset-40**). The union of subset-60 and subset-40 is called **Full-100**, and it contains 100 questions, 200 answers (140 answers are adopted and 60 are not).

3.1 Annotating Relevancy and Answer Ranking

To determine the quality of *ChiMed*, we manually added two types of labels to each QA record in

Possible Relevancy Labels for a (Q, A) pair:

1: The A fully answers the Q

2: The A partially answers the Q

3: The A does not answer the Q

4: Cannot tell whether the A is relevant to Q

Possible Ranking Labels for one Q and two As:

1: The first A is better

2: The second A is better

3: The two As are equally good

4: Neither of As is good (fully answers the Q)

5: Cannot tell which A is better

Properties of Good As:

1: Answer more sub-questions

2: Analyze symptoms or causes of disease

3: Offer advice on treatments or examinations

4: Offer instructions for drug usage

5: Soothe patients' emotions

Properties of Bad As:

1: Answer the Q indirectly

2: The A has grammatical errors

3: Offer irrelevant information

Table 5: Labels and part of annotation Guidelines for relevancy and ranking annotation.

the Full-100 set. The first is *relevancy* label, indicating whether an answer is relevant to a question (i.e., whether the answer field provides a satisfactory answer to the question). There are four possible values as shown in the top part of Table 5.

The second type of labels ranks the two answers for a question. Sometimes, determining which answer is better can be challenging especially when both answers are relevant. Intuitively, people tend to prefer answers that address the question directly, that are easy to understand while supported by evidence, etc. Based on such intuition, we create a set of annotation guidelines, parts of which are shown in the second half of Table 5. Because both types of annotation may require medical expertise, we include a *Cannot tell* label (label "4" for relevancy annotation and label "5" for ranking annotation) for non-expert annotators to annotate different cases.

3.2 Inter-annotator Agreement on Relevancy and Answer Ranking

We hired two annotators without medical background to first annotate the Full-100 set independently and then resolve any disagreement via discussion. The results in terms of percentage and

	Relevancy		Ranking	
	%	κ	%	κ
I vs. II	90.5	55.6	62.0	43.0
I vs. Agreed	97.0	83.7	79.0	69.2
II vs. Agreed	93.5	70.4	76.0	64.4

Table 6: Inter-annotator agreement for relevancy and ranking labeling on the Full-100 set in terms of percentage (%) and Cohen's Kappa (κ). I and II refer to the annotations by the two annotators before any discussion, and *Agreed* is the annotation after the annotators have resolved their disagreement.

I \ II	1	2	3	4	Total
1	170	10	0	1	181
2	2	9	1	0	12
3	0	4	2	0	6
4	0	1	0	0	1
Total	172	24	3	1	200

(a) Confusion matrix of two annotators on relevancy labels on the Full-100 set. The agreement is 90.5% (55.6% in Cohen's Kappa) and the four labels are explained in Table 5.

I \ II	1	2	3	4	5	Total
1	25	6	4	1	0	36
2	7	25	5	0	0	37
3	9	4	11	0	0	24
4	0	1	0	1	0	2
5	1	0	0	0	0	1
Total	42	36	20	2	0	100

(b) Confusion matrix of two annotators on ranking labels on the Full-100 set. The agreement is 62.0% (43.0% in Cohen's Kappa) and the five labels are explained in Table 5.

Table 7: The confusion matrices of two annotators on relevancy labels and ranking labels on the Full-100 set.

Cohen's Kappa are in Table 6. Inter-annotator agreement on the relevancy label is quite high (90.5% in percentage and 55.6% in kappa), while the agreement on the ranking label is much lower (62.0% in percentage and 43.0% in kappa).

Table 7a and Table 7b show the confusion matrices of the two annotators on the relevancy annotation and ranking annotation, respectively. Out of four relevancy labels and five ranking labels, relevancy label "3" and ranking label "4" are rare as most answers in the corpus are relevant; relevancy label "4" and ranking label "5" are also rare, but they do occur as sometimes choosing the relevant/better answer requires medical expertise.

Q	我一直卷发，拉直也没用，这是卷毛性综合症吗？怎么治疗呢？ I have curly hair and straightening is useless. Is this a curl syndrome? How to treat it?
A1	卷发有两个原因，一是先天的自然卷发；另一种是后天的不慎引起，如烫发或染发。先天矫直或化学矫直只能是暂时的。除了洗头和护发产品要调整外，避免使用热吹风机。梳头时要小心。不要用头绳或橡皮筋发夹，防止头发拉伤。 There are two reasons for curly hair: one is congenital natural curly hair; the other is caused by inadvertently acquired, such as perming or dyeing hair. Congenital straightening or chemical straightening can only be temporary. In addition to the shampoo and hair care products need to be adjusted, avoid using a hot hair dryer. Be careful when combing your hair. Do not use a headband or rubber band hairpin to prevent hair strain.
A2	自然卷是一种受遗传因素影响的发型。头发自然卷成一卷。形成的原因是由于人类基因的不同。卷发并不是一件坏事。这种自卷曲的类型是药物无法改变的。如果拉直用的是直板，离子是热的，经过熨烫，一段时间后，它就会回到原来的状态。 Natural rolls are a type of hair that is affected by genetic factors. The hair is naturally rolled into a roll. The reason for the formation is due to differences in human genes. Curly hair is not a bad thing. This type of self-curling is that the drug cannot be changed. If the way of straightening is straight, the ions are hot and after ironing, after a while, it will return to its original state.

Table 8: An example where one annotator thinks the two answers are equally good because they both answer the question informatively. The other annotator thinks A1 is better because it tells the patient how to take care of his/her hair in daily life, although A1 provides less analysis of the causes of the symptom. After discussion, the two annotators reach an agreement that advice on daily care is very important and thus A1 is better than A2.

For ranking annotation, disagreement tends to occur when the two answers are very similar. That is why the majority of disagreed annotations (22 out of 38) occur when one annotator chooses one answer to be better while the other annotator considers the two answers to be equally good (an example is given in Table 8). There are 13 examples where annotators have completely opposite annotation (e.g., one annotates "1" while the other annotates "2"), which further shows the difficulty in identifying which answer is better.

3.3 The *Adopted* flag in *ChiMed*

As is mentioned above, each answer in *ChiMed* has a flag indicating whether or not the answer has been adopted. While we do not know the exact meaning of the flag and whether the flag is set manually (e.g., by the staff at the forum) or automatically (e.g., according to factors such as the physicians' past performance or seniority), we would like to know whether the flag is a good indicator of relevant or better answers.

Among four relevancy labels, we regard answers with label "1" or "2" as relevant answers because they fully or partially answer the question, and answers with label "3" or "4" as irrelevant answers. Table 9 shows that 98.0% of the answers in the Full-100 set are considered to be relevant, according to the *Agreed* relevancy annotation. In

	# of As	# (%) of Relevant As
Adopted	140	137(97.9%)
Unadopted	60	59(98.3%)
Total	200	196(98.0%)

Table 9: The *Adopted* flag vs. relevancy label on the Full-100 set. Here, answers with relevancy label "1" or "2" are regarded as relevant answers.

other words, approximately 98% of (Q, A) pairs in the corpus are good question-answer pairs. On the other hand, the adopted answers are not more likely to be relevant to the question than the unadopted ones. Therefore, the *Adopted* flag is not a good indicator of an answer's relevancy.

The next question is whether adopted answers tend to be better answers than unadopted ones. If so, we can use the *Adopted* flag to infer ranking labels as follows: if a QA record in the Full-100 set has exactly one adopted answer, we rank that answer higher than the unadopted one in the same record; if both answers in a QA record are adopted, they are considered to be equally good. Table 10 shows such inferred labels do not correlate well with human annotation. In fact, the correlation between inferred labels and the *Agreed* human annotation is only 0.068, when we use the 97 QA records with ranking label "1", "2", or "3". Therefore, the *Adopted* flag is not a good indicator

	Subset-60	Full-100
Adopted vs. I	43.3%	34.0%
Adopted vs. II	46.7%	32.0%
Adopted vs. Agreed	43.3%	36.0%

(a) Agreements between the ranking labels from annotators (I, II, and Agreed) and the labels induced from the adopted flag (Adopted). The Subset-60 is the subset of the Full-100 set where each question has exactly one adopted answer and one unadopted answer (See Section 3).

Agreed / Adopted	1	2	3	4	5	Total
1	17	6	9	0	1	33
2	7	9	10	1	0	27
3	14	15	10	1	0	40
Total	38	30	29	2	1	100

(b) Confusion matrix between the agreed human annotation and ranking labels induced from the adopted flag. The meaning of the five labels are explained in Table 5.

Table 10: The adopted flags vs. the ranking labels from annotators on the Full-100 set.

for better answers.

So far we have demonstrated that the *Adopted* flag is not a good indicator for relevant or better answers. So what does the *Adopted* flag really indicate? While we are waiting for responses from the Ask39 forum, there are two possibilities. One is that the flag is intended to mean something totally different from relevant or better answers. The other possibility is that the flag intends to mark relevant or better answers but their criteria for relevant or better answers are very different from ours. Table 11 shows a (Q, A) pair, where the answer is adopted. On the one hand, the answer does not directly answer the question. On the other hand, it does provide some useful information about gallstone, and one can argue that the adopted flag in the original corpus is plausible.

3.4 Two Datasets from *ChiMed*

As shown in Table 9, the majority of answers in *ChiMed* are relevant to the questions in the same QA records. To create a dataset for the relevancy task, we start with the 25,594 QA records which have exactly one adopted and one unadopted answer (see Table 4), Next, we filter out QA records whose questions or answers are too long or too short,[7] because very short questions or answers

---

[7]We will remove a QA record if it contains a question/answer that is ranked either top 1% or bottom 1% of all questions/answers according to their character-based length.

Q	请问为什么胆结石总是晚上发作？ Why does gallstone always occur at night?
A	有些人会出现过度劳累、腹胀、打鼾症状。可能是胆结石的原因，且通常晚上疼痛更严重。可以选择药物治疗。手术复发的可能性很大。建议平时多运动。 Some people have symptoms of fatigue, bloating and snoring. They may be caused by gallstones, and usually the pain is more severe at night. You can choose medication. There is a high probability of recurrence of surgery. It is recommended to exercise more usually.

Table 11: The answer does not directly answer the question, but it has an adopted flag.

	Train	Dev	Test
# of Qs	19,952	2,494	2,494
# of As	39,904	4,988	4,988
Avg. Length of Qs	63.5	63.8	63.3
Avg. Length of As in *ChiMed-QA1*	118.7	118.6	118.0
Avg. Length of As in *ChiMed-QA2*	128.0	127.6	127.1

Table 12: Statistics of the two *ChiMed-QA* Datasets. Average lengths of Qs and As are in characters.

tend to be lack of crucial information, whereas very long ones tend to include much redundant or irrelevant information. The remaining dataset contains 24,940 QA records. We divide it into training/development/testing set with portions of 80%/10%/10% and call the dataset *ChiMed-QA1*. Since each QA record has one adopted and one unadopted answer, we will use the dataset to train an adoption predictor.

For the relevancy task, we need both positive and negative examples. We start with *ChiMed-QA1*, and for each QA record, we keep the adopted answer as a positive instance, and replace the unadopted answer with an adopted answer from another QA record randomly selected from the same training/dev/testing subsets to distinguish relevant vs. irrelevant answers. We call this synthesized dataset *ChiMed-QA2*. We will use those two datasets for the adoption prediction task and the relevancy task (see the next section). We are not able to use the corpus for the answer ranking task as we cannot infer the ranking label from the *Adopted* flag.

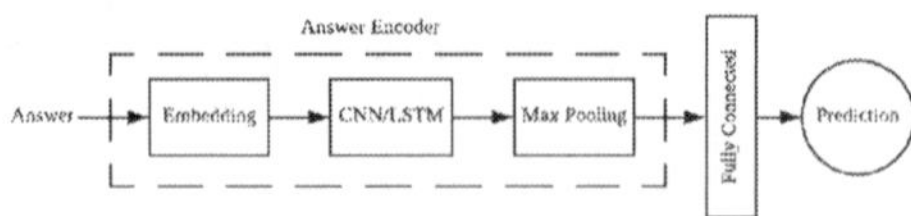

Figure 1: The architecture of CNN- and LSTM-based systems under A-Only setting.

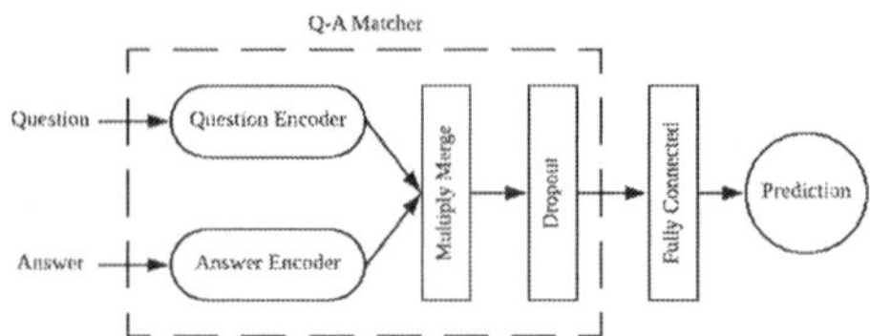

Figure 3: The architecture of our systems under Q-A setting. The architecture of question and answer encoders are identical with the architecture in Figure 1.

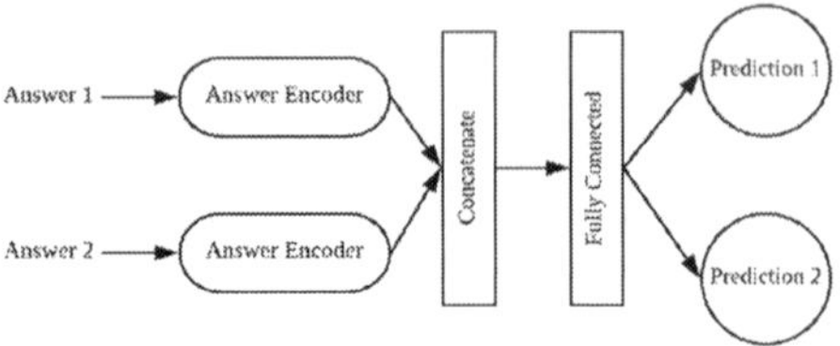

Figure 2: The architecture of our systems under A-A setting. The architecture of answer encoder is identical with the one in Figure 1. Prediction 1 and 2 means the prediction for answer 1 and 2, respectively.

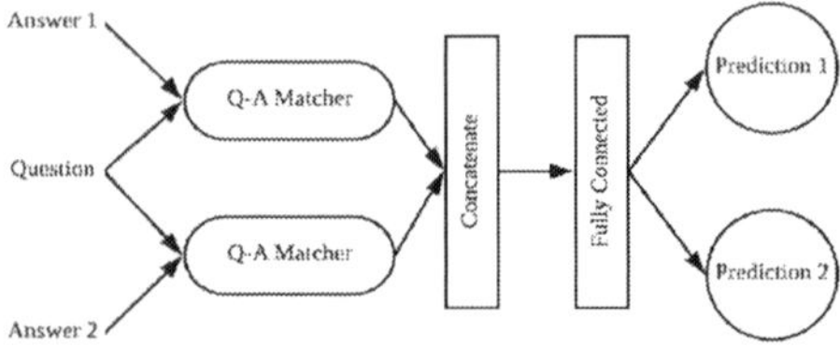

Figure 4: The architecture of our systems under Q-As setting. The architecture of Q-A matcher is shown in Figure 3. We use five Q-A matchers in our experiment: CNN, LSTM, ARC-I, DUET, and DRMM.

Table 12 shows the statistics of the two datasets. The first three rows are the same for the two datasets; the average length of As in *ChiMed-QA2* is slightly longer than that in *ChiMed-QA1* because adopted answers tend to be longer than un-adopted ones.

4   Experiments on Two Prediction Tasks

In this section, we use *ChiMed-QA1* and *ChiMed-QA2* (See Table 12) to build NLP systems for the adoption prediction task and the relevancy prediction task, respectively. Both tasks are binary classification tasks with the same type of input; the only difference is the meaning of class labels (relevancy vs. adopted flag). Therefore, we build a set of NLP systems and apply them to both tasks.

4.1   Systems and Settings

We implemented both CNN- and LSTM-based systems, and applied three state-of-the-art sentence matching systems to the two tasks. The three existing systems are: (1) **ARC-I** (Hu et al., 2014) matches questions and answers by directly concatenating their embeddings; (2) **DUET** (Mitra et al., 2017) computes the Q-A similarity by matching exact terms and high-level sentence embeddings (Hadamard production) simultaneously; (3) **DRMM** (Guo et al., 2016) makes its final prediction based on the similarity matrix of each pair

of word embeddings in a question and an answer.

We run our CNN- and LSTM-based systems under four different settings: (1) **A-Only** where an answer is the only input (See Figure 1); (2) **A-A** where both answers are input (See Figure 2); (3) **Q-A** where a question and one of its answers are input (See Figure 3); (4) **Q-As** where a question and both of its answers are input (See Figure 4). ARC-I, DUET, and DRMM are run under the settings of Q-A and Q-As, because the systems require a question to be one of the input. The reason we apply the A-Only and A-A settings to the adoption prediction task is that it helps identify whether features from an answer itself will contribute to its adopted flag assignment without knowing its question. To compare the relevancy task and the adoption prediction task, we also apply these two settings to the former task although they are not common settings in previous studies (Lai et al., 2018).

Word segmentation has always been a challenge in Chinese NLP especially when it is applied to a particular domain (Song et al., 2012; Song and Xia, 2012, 2013). Therefore, instead of word embeddings (Song et al., 2018), we use Chinese-character-based embeddings to avoid word segmentation errors. We set the embedding size to 150. We use 155 and 245 as the lengths of questions and answers respectively. Short texts are padded with blank characters. We use 32 filters

Sys ID	Input Setting	NLP System	Relevancy Prediction		Adoption Prediction	
			-CR	+CR	-CR	+CR
1	A-Only	CNN	50.80	51.64	74.10	81.64
2		LSTM	50.66	50.72	74.24	82.00
3	A-A	CNN	49.40	-	84.20	-
4		LSTM	50.28	-	85.00	-
5	Q-A	CNN	74.32	81.84	74.84	81.07
6		LSTM	80.19	87.09	75.28	83.64
7		ARC-I	50.34	50.60	75.20	82.64
8		DUET	81.03	91.74	75.28	82.48
9		DRMM	93.60	98.16	71.49	83.88
10	Q-As	CNN	76.98	-	83.52	-
11		LSTM	88.41	-	84.24	-
12		ARC-I	48.84	-	83.88	-
13		DUET	87.17	-	83.36	-
14		DRMM	98.32	-	83.28	-

Table 13: Results of all systems under different settings with respect to (Q, A) pair prediction accuracy with (+CR) and without (-CR) conflict resolution. We do not present results of +CR in A-A and Q-As settings because they are equivalent to the results of -CR.

with the kernel size 3 for every CNN layer and we set the LSTM hidden size to 32. We apply a pooling size of 2 to all max pooling layers. Besides, the activation function of the output layers under A-Only and Q-A settings is *sigmoid*, that of output layers under A-A and Q-As settings is *softmax*, and that of all other layers is *tanh*.

In addition, noting that the two answers for the same question have opposite labels in both tasks, we evaluate all systems in terms of (Q, A) pair predication accuracy with and without **conflict resolution** (CR), with which the model resolves conflicts when either two relevant/adopted answers or two irrelevant/unadopted answers are predicted. Because the activation function of the output layers under A-A and Q-As settings is *softmax* and because there are always two answers for each question, systems under these two settings never generate conflict predictions. We do not apply MAP (Mean Average Precision) (Lai et al., 2018) to the tasks because the number of candidate answers of each question in the datasets is limited to 2.

4.2 Experimental Results

Table 13 shows the experimental results of running the five predictors on the testing set under four different settings. There are a few observations.

First, for the relevancy task, by designing only half of the (Q, A) pairs in *ChiMed-QA2* come from the same QA records. When Q is not given as part of the input (System 1-4), it is impossible for the predictors to determine whether an answer is relevant; therefore, the system performances are no better than random guesses. In contrast, for the adoption prediction task, by designing all the (Q, A) pairs in *ChiMed-QA1* come from the same QA records, and according to Table 9 we also know that about 98% of the answers, regardless of whether they are adopted or not, are relevant. Therefore, the absence of Qs in System 1-4 does not affect system performance a lot.

Second, when both Q and A are present (System 5-9), the accuracy of relevancy prediction is higher than that of adoption prediction, because the former is an easier task (at least for humans). The only exception is ARC-I (System 7), whose results on relevancy is close to random guess (50.34% and 50.60%) while the result on adoption is comparable with other systems. This is due to the way that ARC-I matches questions and answers. Because embeddings of a question and an answer are directly concatenated in ARC-I, Q-A similarity are not fully captured, leading to low performance on relevancy. On the contrary, the adoption prediction does not rely much on the Q-A similarity (as explained above).

Third, for the relevancy task, systems that capture more features of Q-A similarity tend to have a better result. For example, under the Q-A setting, DUET (System 8) outperforms CNN, LSTM and ARC-I (System 5-7) because DUET has an additional model of exact phrase matching between questions and answers. DRMM (System 9) performs better than DUET (System 8) because DRMM uses word embedding instead of exact phrase when matching pairs of phrases between a question and an answer. In contrast, the performances of the five systems on the adoption task are very similar.

In addition, except for the relevancy task evaluated with CR, the contrast between System 10-14 vs. System 5-9 indicates comparing two As always helps predictors in both tasks because intuitively knowing both answers would help us to decide which one is relevant/adopted. On the contrary, the comparison between the same two groups of systems with CR in the relevancy task indicates comparing two As may hurt the relevancy predictors (System 5, 7, 8) because the relevancy is really between Q and A, which might be affected by the existence of other As.

Finally, all the systems under A-Only and Q-A settings (Systems 1-2 and 5-9) benefit from CR. It is also worth noting that running the models under Q-A setting and to evaluate them without CR in previous studies (Lai et al., 2018) is much more common. Under this setting, the highest performance achieved is 93.60% (System 9). The score is not as high as our expectation and there still exist room for improvement.

4.3 Error Analysis for Relevancy Prediction

We go though errors of system 9 in the relevancy prediction task without CR and find three main types of errors. Note that we artificially build *ChiMed-QA2* for the relevancy prediction task by keeping the adopted answer a of a question q and replacing the unadopted answer of q with an adopted answer a' from another question q'. And we therefore regard a as a relevant answer of q and a' as an irrelevant answer of q (See Section 3.4).

The first type of error is that the answer a is actually irrelevant to the question q. In other words, the gold standard is wrong; system 9 does make a correct prediction. This is not surprising as there are around 2% irrelevant answers in the dataset according to our annotation (See Table 9).

Second, the system fails to capture the relationship between a disease and a corresponding treatment. E.g., a patient describes his/her symptoms and asks for treatment. The doctor offers a drug directly without analyzing the symptoms and causes of disease. In that case, the overlap between the question and the answer is relatively low. The system therefore cannot predict the answer to be relevant without the help of a knowledge base.

Finally, it is quite common that a patient describes his/her symptoms at the beginning of the question q and asks something else at the end (e.g. whether drug X will help with his/her illness). In this case, if q' (the original question of the irrelevant answer a') describes similar symptoms, the system may fail to capture what exactly q wants to ask and therefore mistakes a' for a relevant answer. Table 14 gives an error in this type where q and q' describe similar diseases but they are in fact expecting totally different answers.

Given the three types of errors, we find out the latter two are relatively challenging. This therefore requires further exploration on the way of modeling (Q, A) pairs in the relevancy prediction task. In addition, because current irrelevant answers are randomly sampled from the entire dataset, the current dataset does not include many

q	我上周感冒咳嗽，现在感冒好了，但咳嗽更加厉害了。蜂蜜可以治疗咳嗽吗？ I had a cold and cough last week. Now, the cold has gone, but the cough is even worse. Can honey treat cough?
q'	我是支气管扩张患者，最近感冒病情加重。支气管扩张病人感冒怎么治疗？ I am a patient with bronchiectasis. I have recently become worse with a cold. How to treat a cold for a bronchiectasis patient?
a'	正常的情况下，支气管病人如果感冒，就应该立即到医院就医，并在医生的指导下用药物治疗。如果耽误治疗的话病情会加重，而且会出现一些并发症。 Normally, if a bronchial patient has a cold, he should go to the hospital immediately and take medication under the guidance of a doctor. If the treatment is delayed, the condition will worsen and complications will occur.

Table 14: An example where system 9 mistakes irrelevant answer a' for a relevant answer. Both questions q and q' are talking about cold and cough, but they are totally different because q is asking whether honey is helpful for cough while q' is looking for treatment.

challenging examples. This makes relevancy prediction task appear easier than what it could be. For future work, we plan to balance the easy and hard instances in the dataset by adding more challenging examples to *ChiMed-QA2*.

5 Conclusion and Future Work

In this paper, we present *ChiMed*, a Chinese medical QA corpus collected from an online medical forum. Our annotation on a small fraction of the corpus shows that the corpus is of high quality as approximately 98% of the answers successfully address the questions raised by the forum users. To demonstrate the usage of the corpus, we extract two datasets and use them for two prediction tasks. A few benchmark systems yield good performance on both tasks.

For the future work, we are collecting data to expand the corpus and plan to add more challenging samples to the datasets. In addition, we plan to use *ChiMed* for other NLP tasks such as automatic answer generation, keyphrase generation, summarization, and question classification. We also plan to explore various methods of adding more annotations (e.g., answer ranking) to the corpus.

References

Mehdi Allahyari, Seyedamin Pouriyeh, Mehdi Assefi, Saeid Safaei, Elizabeth D Trippe, Juan B Gutierrez, and Krys Kochut. 2017. Text summarization techniques: a brief survey. *arXiv preprint arXiv:1707.02268*.

Matthew Dunn, Levent Sagun, Mike Higgins, V Ugur Guney, Volkan Cirik, and Kyunghyun Cho. 2017. Searchqa: A new q&a dataset augmented with context from a search engine. *arXiv preprint arXiv:1704.05179*.

Jiafeng Guo, Yixing Fan, Qingyao Ai, and W Bruce Croft. 2016. A deep relevance matching model for ad-hoc retrieval. In *Proceedings of the 25th ACM International on Conference on Information and Knowledge Management*, pages 55–64. ACM.

Wei He, Kai Liu, Jing Liu, Yajuan Lyu, Shiqi Zhao, Xinyan Xiao, Yuan Liu, Yizhong Wang, Hua Wu, Qiaoqiao She, et al. 2017. Dureader: a chinese machine reading comprehension dataset from real-world applications. *arXiv preprint arXiv:1711.05073*.

Karl Moritz Hermann, Tomas Kocisky, Edward Grefenstette, Lasse Espeholt, Will Kay, Mustafa Suleyman, and Phil Blunsom. 2015. Teaching machines to read and comprehend. In *Advances in neural information processing systems*, pages 1693–1701.

Baotian Hu, Zhengdong Lu, Hang Li, and Qingcai Chen. 2014. Convolutional neural network architectures for matching natural language sentences. In *Advances in neural information processing systems*, pages 2042–2050.

Armand Joulin, Edouard Grave, Piotr Bojanowski, and Tomas Mikolov. 2016. Bag of tricks for efficient text classification. *arXiv preprint arXiv:1607.01759*.

Tuan Manh Lai, Trung Bui, and Sheng Li. 2018. A review on deep learning techniques applied to answer selection. In *Proceedings of the 27th International Conference on Computational Linguistics*, pages 2132–2144.

Bhaskar Mitra, Fernando Diaz, and Nick Craswell. 2017. Learning to match using local and distributed representations of text for web search. In *Proceedings of the 26th International Conference on World Wide Web*, pages 1291–1299. International World Wide Web Conferences Steering Committee.

Preslav Nakov, Lluís Màrquez, Walid Magdy, Alessandro Moschitti, Jim Glass, and Bilal Randeree. 2015. Semeval-2015 task 3: Answer selection in community question answering. In *Proceedings of the 9th International Workshop on Semantic Evaluation (SemEval 2015)*, pages 269–281.

Anusri Pampari, Preethi Raghavan, Jennifer Liang, and Jian Peng. 2018. emrqa: A large corpus for question answering on electronic medical records. *arXiv preprint arXiv:1809.00732*.

Pranav Rajpurkar, Jian Zhang, Konstantin Lopyrev, and Percy Liang. 2016. Squad: 100,000+ questions for machine comprehension of text. *arXiv preprint arXiv:1606.05250*.

Siva Reddy, Danqi Chen, and Christopher D Manning. 2018. Coqa: A conversational question answering challenge. *arXiv preprint arXiv:1808.07042*.

Sifatullah Siddiqi and Aditi Sharan. 2015. Keyword and keyphrase extraction techniques: a literature review. *International Journal of Computer Applications*, 109(2).

Yan Song, Prescott Klassen, Fei Xia, and Chunyu Kit. 2012. Entropy-based training data selection for domain adaptation. In *Proceedings of COLING 2012: Posters*, pages 1191–1200, Mumbai, India.

Yan Song, Chia-Jung Lee, and Fei Xia. 2017. Learning word representations with regularization from prior knowledge. In *Proceedings of the 21st Conference on Computational Natural Language Learning (CoNLL 2017)*, pages 143–152, Vancouver, Canada.

Yan Song, Shuming Shi, Jing Li, and Haisong Zhang. 2018. Directional skip-gram: Explicitly distinguishing left and right context for word embeddings. In *Proceedings of the 2018 Conference of the North American Chapter of the Association for Computational Linguistics: Human Language Technologies, Volume 2*, pages 175–180, New Orleans, Louisiana.

Yan Song and Fei Xia. 2012. Using a goodness measurement for domain adaptation: A case study on Chinese word segmentation. In *Proceedings of the Eighth International Conference on Language Resources and Evaluation (LREC-2012)*, pages 3853–3860, Istanbul, Turkey.

Yan Song and Fei Xia. 2013. A common case of jekyll and hyde: The synergistic effect of using divided source training data for feature augmentation. In *Proceedings of the Sixth International Joint Conference on Natural Language Processing*, pages 623–631, Nagoya, Japan.

Simon Šuster and Walter Daelemans. 2018. Clicr: a dataset of clinical case reports for machine reading comprehension. *arXiv preprint arXiv:1803.09720*.

George Tsatsaronis, Michael Schroeder, Georgios Paliouras, Yannis Almirantis, Ion Androutsopoulos, Eric Gaussier, Patrick Gallinari, Thierry Artieres, Michael R Alvers, Matthias Zschunke, et al. 2012. Bioasq: A challenge on large-scale biomedical semantic indexing and question answering. In *2012 AAAI Fall Symposium Series*.

Mengqiu Wang, Noah A Smith, and Teruko Mitamura. 2007. What is the jeopardy model? a quasi-
synchronous grammar for qa. In *Proceedings of
the 2007 Joint Conference on Empirical Methods
in Natural Language Processing and Computational
Natural Language Learning (EMNLP-CoNLL)*.

Nan Wang, Yan Song, and Fei Xia. 2018a. Coding structures and actions with the COSTA scheme
in medical conversations. In *Proceedings of the
BioNLP 2018 workshop*, pages 76–86, Melbourne,
Australia.

Nan Wang, Yan Song, and Fei Xia. 2018b. Constructing a Chinese medical conversation corpus annotated with conversational structures and actions. In
*Proceedings of the 11th Language Resources and
Evaluation Conference*, Miyazaki, Japan.

Liu Yang, Qingyao Ai, Jiafeng Guo, and W Bruce
Croft. 2016. anmm: Ranking short answer texts
with attention-based neural matching model. In *Proceedings of the 25th ACM international on conference on information and knowledge management*,
pages 287–296. ACM.

Yi Yang, Wen-tau Yih, and Christopher Meek. 2015.
Wikiqa: A challenge dataset for open-domain question answering. In *Proceedings of the 2015 Conference on Empirical Methods in Natural Language
Processing*, pages 2013–2018.

Clinical Concept Extraction for Document-Level Coding

Sarah Wiegreffe[1], Edward Choi[1]*, Sherry Yan[2], Jimeng Sun[1], Jacob Eisenstein[1]
[1]Georgia Institute of Technology
[2]Sutter Health
* Current Affiliation: Google Inc
saw@gatech.edu, edwardchoi@google.com, yansx@sutterhealth.org,
jsun@cc.gatech.edu, jacobe@gatech.edu

Abstract

The text of clinical notes can be a valuable source of patient information and clinical assessments. Historically, the primary approach for exploiting clinical notes has been information extraction: linking spans of text to concepts in a detailed domain ontology. However, recent work has demonstrated the potential of supervised machine learning to extract document-level codes directly from the raw text of clinical notes. We propose to bridge the gap between the two approaches with two novel syntheses: (1) treating extracted concepts as *features*, which are used to supplement or replace the text of the note; (2) treating extracted concepts as *labels*, which are used to learn a better representation of the text. Unfortunately, the resulting concepts do not yield performance gains on the document-level clinical coding task. We explore possible explanations and future research directions.

1 Introduction

Clinical decision support from raw-text notes taken by clinicians about patients has proven to be a valuable alternative to state-of-the-art models built from structured EHRs. Clinical notes contain valuable information that the structured part of the EHR does not provide, and do not rely on expensive and time-consuming human annotation (Torres et al., 2017; American Academy of Professional Coders, 2019). Impressive advances using deep learning have allowed for modeling on the raw text alone (Mullenbach et al., 2018; Rios and Kavuluru, 2018a; Baumel et al., 2018). However, there exist some shortcomings to these approaches: clinical text is noisy, and often contains heavy amounts of abbreviations and acronyms, a challenge for machine reading (Nguyen and Patrick, 2016). Additionally, rare words replaced with "UNK" tokens for better generalization may be crucial for predicting rare labels.

Clinical concept extraction tools abstract over the noise inherent in surface representations of clinical text by linking raw text to standardized concepts in clinical ontologies. The Apache clinical Text Analysis Knowledge Extraction System (cTAKES, Savova et al., 2010) is the most widely-used such tool, with over 1000 citations. Based on rules and non-neural machine learning methods and engineered for almost a decade, cTAKES provides an easily-obtainable source of human-encoded domain knowledge, although it cannot leverage deep learning to make document-level predictions.

Our goal in this paper is to maximize the predictive power of clinical notes by bridging the gap between information extraction and deep learning models. We address the following research questions: how can we best leverage tools such as cTAKES on clinical text? Can we show the value of these tools in linking unstructured data to structured codes in an existing ontology for downstream prediction?

We explore two novel hybrids of these methods: data augmentation (augmenting text with extracted concepts) and multi-task learning (learning to predict the output of cTAKES). Unfortunately, in neither case does cTAKES improve downstream performance on the document-level clinical coding task. We probe this negative result through an extensive series of ablations, and suggest possible explanations, such as the lack of word variation captured through concept assignment.

2 Related Work

Clinical Ontologies Clinical concept ontologies facilitate the maintenance of EHR systems with standardized and comprehensive code sets, allowing consistency across healthcare institutions and practitioners. The Unified Medical Language System (UMLS) (Lindberg et al., 1993) maintains

Proceedings of the BioNLP 2019 workshop, pages 261–272
Florence, Italy, August 1, 2019. ©2019 Association for Computational Linguistics

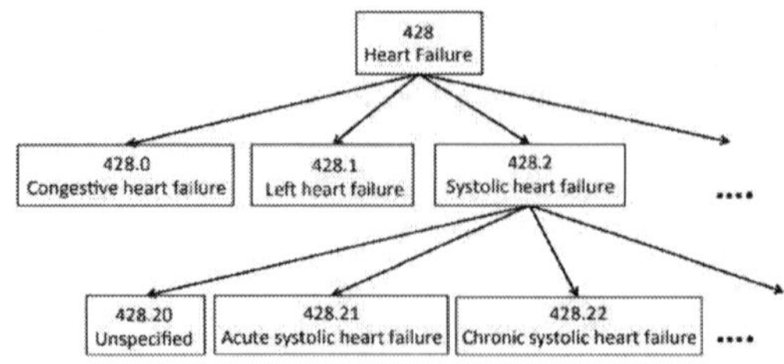

Figure 1: A subtree of the ICD ontology (figure from Singh et al., 2014).

a standardized vocabulary of clinical concepts, each of which is assigned a concept unique identifier (CUI). The Systematized Nomenclature of Medicine- Clinical Terms (SNOMED-CT) (Donnelly, 2006) and the International Classification of Diseases (ICD) (National Center for Health Statistics, 1991) build off of the UMLS and provide structure by linking concepts based on their relationships. The SNOMED ontology has over 340,000 active concepts, ranging from fine-grained ("Adenylosuccinate lyase deficiency") to extremely general ("patient"). The ICD ontology is narrower in scope, with around 13,000 diagnosis and procedure codes used for insurance billing. Unlike SNOMED, which has an unconstrained graph structure, ICD9 is organized into a top-down hierarchy of specificity (see Figure 1).

Clinical Information Extraction Tools There are several tools for extracting structured information from clinical text. Popular types of information extraction include *named-entity recognition*, identifying words or phrases in the text which align with clinical concepts, and *ontology mapping*, labelling the identified words and phrases with their respective clinical codes from an existing ontology.[1] Of the tools which perform both of these tasks, the open-source Apache cTAKES is used in over 50% of recent work (Wang et al., 2017), outpacing competitors such as MetaMap (Aronson, 2001) and MedLEE (Friedman, 2000).

cTAKES utilizes a rule-based system for performing ontology mapping, via a UMLS dictionary lookup on the noun phrases inferred by a part-of-speech tagger. Taking raw text as input, the software outputs a set of UMLS concepts identified in

the text and their positions, with functionality to map them to other ontologies such as SNOMED and ICD9. It is highly scalable, and can be deployed locally to avoid compromising identifiable patient data. Figure 2 shows an example cTAKES annotation on a clinical record.

Clinical Named-Entity Recognition (NER) Recent work has focused on developing tools to replace cTAKES in favor of modern neural architectures such as Bi-LSTM CRFs (Boag et al., 2018; Tao et al., 2018; Xu et al., 2018; Greenberg et al., 2018), varying in task definition and evaluation. Newer approaches leverage contextualized word embeddings such as ELMo (Zhu et al., 2018; Si et al., 2019). In contrast, we focus on maximizing the power of existing tools such as cTAKES. This approach is more practical in the near-term, because the adoption of new NER systems in the clinical domain is inhibited by the amount of computational power, data, and gold-label annotations needed to build and train such token-level models, as well as considerations for the effectiveness of domain transfer and a necessity to perform annotations locally in order to protect patient data. Newer models do not provide these capabilities.

NER in Text-based Models Prior works use the output of cTAKES as features for disease- and drug-specific tasks, but either concatenate them as shallow features, or substitute them for the text itself (see Wang et al. (2017) for a literature review). Weng et al. (2017) incorporate the output of cTAKES into their input feature vectors for the task of predicting the medical subdomain of clinical notes. However, they use them as shallow features in a non-neural setting, and combine cTAKES annotations with the text representations by concatenating the two into one larger feature vector. In contrast, we propose to learn dense neural concept embedding representations, and integrate the concepts in a learnable fashion to guide the representation learning process, rather than simply concatenating them or using them as a text replacement. We additionally focus on a more challenging task setting.

Boag and Kané (2017) augment a Word2Vec training objective to predict clinical concepts. This work is orthogonal to ours as it is an unsupervised "embedding pretraining" approach rather than an end-to-end supervised model.

---

[1]Ontology mapping also serves as a form of text normalization.

[2]Figure from `https://healthnlp.github.io/examples/`.

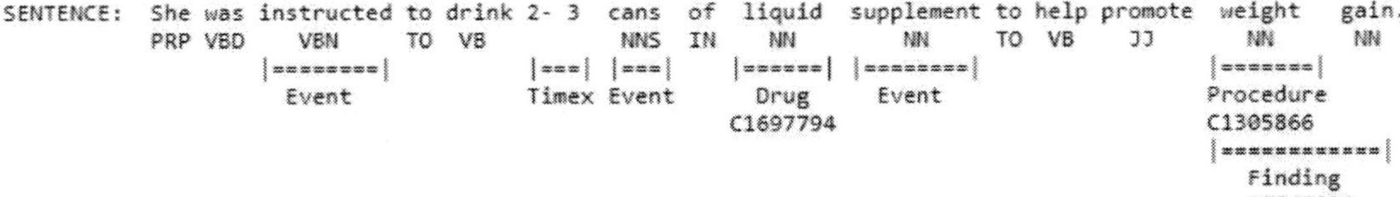

Figure 2: An example of cTAKES annotation output with part-of-speech tags and UMLS CUIs for named entities.[2]

Automated Clinical Coding The automated clinical coding task is to predict from the raw text of a hospital discharge summary describing a patient encounter all of the possible ICD9 (diagnosis and procedure) codes which a human annotator would assign to the visit. Because these annotators are trained professionals, the ICD codes assigned serve as a natural label set for describing a patient record, and the task can be seen as a proxy for a general patient outcome or treatment prediction task. State-of-the-art methods such as CAML (Mullenbach et al., 2018) treat each label prediction as a separate task, performing many binary classifications over the many-thousand-dimensional label space. The model is described in more detail in the next section.

The label space is very large (tens of thousands of possible codes) and frequency is long-tailed. Rios and Kavuluru (2018b) find that CAML performs weakly on rare labels.

3 Problem Setup

Task Notation A given discharge summary is represented as a matrix $X \in \mathbb{R}^{d_e \times N}$.[3] The set of diagnosis and procedure codes assigned to the visit is represented as the one-hot vector $y \in \{0,1\}^L$. The task can be framed as $L = |\mathcal{L}|$ binary classifications: predict $y_l \in \{0,1\}$ for code l in labelspace $\mathcal{L}$.

Data We use the publically-available MIMIC-III dataset, a collection of deidentified discharge summaries describing patient stays in the Beth Israel Deaconess Medical Center ICU between 2001 and 2012 (Johnson et al., 2016; Pollard and Johnson, 2016). Each discharge summary has been tagged with a set of ICD9 codes. See Figure 3 for an example of a record, and Appendix A for a description of the dataset and preprocessing.

Concept Annotation We run cTAKES on the discharge summaries (described in Appendix B).

Results on the extracted concepts are presented in Table 1. Note the difference in number of annotations provided by using the SNOMED ontology compared to ICD9.[4]

ICD9	
Total concepts extracted	1,005,756
Mean # extracted concepts per document	19.10
Mean % words assigned a concept per document	1.26%
SNOMED	
Total concepts extracted	28,090,075
Mean # extracted concepts per document	532.76
Mean % words assigned a concept per document	35.21%
Mean # tokens per document	1513.00

Table 1: Descriptive Statistics on concept extraction for the MIMIC-III corpus.

Base model We evaluate against CAML (Mullenbach et al., 2018), a state-of-the-art text-based model for the clinical coding task. The model leverages a convolutional neural network (CNN) with per-label attention to predict the combination of codes to assign to a diven discharge summary. Applying convolution over X results in a convolved input representation $H \in \mathbb{R}^{d_c \times N}$ (with $d_c < d_e$) in which the column-dimensionality N is preserved. H is then used to predict y, by attentional pooling over the columns.

We include implementation details of all methods, including hyperparameters and training, in Appendix A.

4 Approach 1: Augmentation Model

One limitation of learning-based models is their tendency to lose uncommon words to "UNK" tokens, or to suffer from poor representation learning for them. We hypothesize that rare words are important for predicting rare labels, and that text-based

---

[3]We use notation for a single instance throughout.

[4]Preliminary experiments with sparser ontologies (RXNORM) were not promising, leading us to choose these two ontologies based on their annotation richness (SNOMED) and direct relation to the prediction task (ICD9).

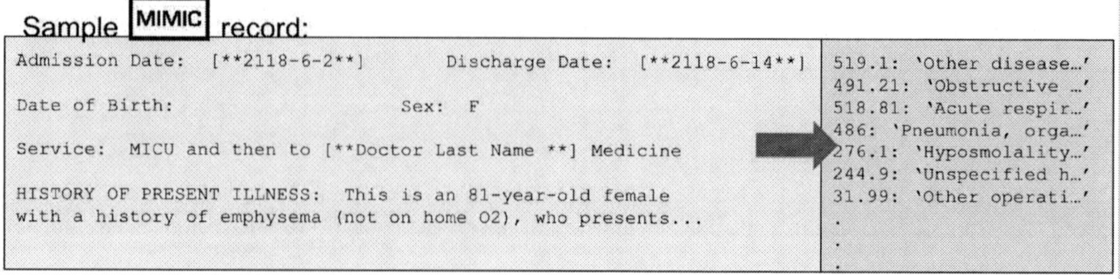

Figure 3: An example clinical discharge summary and associated ICD codes.

models may be improved by augmenting word embeddings with concept embeddings as a means to strengthen representations of rare or unseen words. We additionally hypothesize that linking multiple words to a shared concept via cTAKES annotation will reduce textual noise by grouping word variants to a shared representation in a smaller and more frequently updated parameter space.

4.1 Method

Given a discharge summary containing words $w_1, w_2, ..., w_N \in \mathcal{W}^*$ and an embedding function $\gamma : \mathcal{W} \to \mathbb{R}^{d_e}$, we construct input matrix $\boldsymbol{X} = [\boldsymbol{x}_1^T, \boldsymbol{x}_2^T, ..., \boldsymbol{x}_N^T] \in \mathbb{R}^{d_e \times N}$ as column-stacked word embeddings, where $\boldsymbol{x}_n = \gamma(w_n)$.

We additionally assume a code embedding function $\phi : \mathcal{C} \to \mathbb{R}^{d_e}$ and a set of annotated codes for a given document $c_1, c_2, ..., c_N \in \mathcal{C}^*$, where $\mathcal{C}$ is the full codeset for the ontology used to annotate the document, and c_n is the code annotated for word token w_n, if one exists (else $c_n = \varnothing$, by abuse of notation). We construct a representation for each document, $\boldsymbol{D}$, of the same dimensionality as $\boldsymbol{X}$, by learning one representation leveraging both the concept and word embedding at each position:

For token n,

$$\boldsymbol{d}_n = \beta_{w_n, c_n} \phi(c_n) + (1 - \beta_{w_n, c_n}) \boldsymbol{x}_n, \quad (1)$$

$\beta_{w_n, c_n} \in [0, 1]$ is a learned parameter specific to each observed word+concept pair, including UNK tokens.[5] Intuitively, if there is a concept associated with index n, a concept embedding $\phi(c_n)$ is generated and a *linear combination* of the word and concept embedding is learned, using a learned

parameter specific to that word+concept pair.[6] We fix $\beta_{w_n, c_n = \varnothing} = 0$, which reverts to the word embedding when there is no concept assigned.

We additionally propose a simpler version of this method, *full replace*, in which word embeddings are completely replaced with concept embeddings if they exist (i.e. $\beta_{w_n, c_n} = 1, \forall w_n, c_n \neq \varnothing$). In this formulation, if a concept spans multiple words, all of those words are represented by the same vector. Conversely, the CAML baseline corresponds to a model in which $\beta_{w_n, c_n} = 0, \forall w_n, c_n$.

4.2 Evaluation Setup

Metrics In addition to the metrics reported in prior work, we report average precision score (AP), which is a preferred metric to AUC for imbalanced classes (Saito and Rehmsmeier, 2015; Davis and Goadrich, 2006). We report both macro- and micro- metrics, with the former being more favorable toward rare labels by weighting all classes equally. We additionally focus on the precision-at-k (P@k) metric, representing the fraction of the k highest-scored predicted labels that are present in the ground truth. Both macro-metrics and P@k are useful in a computer-assisted coding use-case, where the desired outcome is to correctly identify needle-in-the-haystack labels as opposed to more frequent ones, and to accurately suggest a small subset of codes with the highest confidence as annotation suggestions (Mullenbach et al., 2018).

Baselines Along with CAML, we evaluate on a *raw codes* baseline where the ICD9 annotations generated by cTAKES $c_1, c_2, ..., c_N$ are used directly as the document-level predictions. Formally,

---

[5]We experimented with models in which this gate was computed element-wise and shared by all word+concept pairs (e.g. by passing $\boldsymbol{x}_n$ and $\phi(c_n)$ through a linear layer or simple multi-layer perceptron to compute $\boldsymbol{d}_n$), but this did not improve performance.

[6]A single token may have multiple concept annotations associated with it. We experiment with an attention mechanism for this case (see Appendix C), but find a heuristic of arbitrarily selecting the first concept assigned to each word performs just as well.

$\hat{y}_{c_n} = 1$ when $c_n \in \mathcal{L}$ and $c_n \neq \varnothing$, for all n in integers 1 to N.

4.3 Results

We present results on the test set in Table 2. Overall, the concept-augmented models are indistinguishable from the baseline, and there is no significant difference between annotation type or recombination method, although the linear combination method with ICD9 annotations is the best performing and rivals the baseline.

Following the negative results for our initial attempt to augment word embeddings with concept embeddings, we tried two alternative strategies:

- We concatenated the ICD9 annotations with two other ontologies: RXNORM and SNOMED. While this led to greater coverage over the text (with slightly more than one third of the tokens in the text receiving corresponding concept annotations), it did not improve downstream performance.

- Prior work has demonstrated that leveraging clinical ontological structure can allow models to learn more effective code embeddings in fully structured data models (Singh et al., 2014; Choi et al., 2017). We applied the methodology of Choi et al. (2017) on both the ICD9 and SNOMED annotations, but this did not improve performance. For more details, see Appendix D.

4.4 Error Analysis

Error analysis of the word-to-concept mapping produced by cTAKES exposes limitations of our initial hypothesis that cTAKES mitigates word-level variation by assigning multiple distinct word phrases to shared concepts. Figure 4 demonstrates that the vast majority of the ICD9 concepts in the corpus are assigned to only one distinct word phrase, and the same results are observed for SNOMED concepts. This may explain the virtually indistinguishable performance of the augmentation models from the baseline, because randomly-initialized word and concept embeddings which are observed in strictly identical contexts should theoretically converge to the same representation.[7]

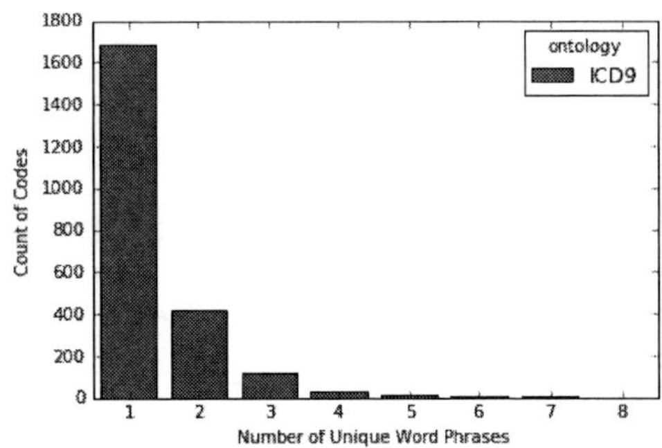

Figure 4: A histogram showing the distribution of ICD9 concepts in $\mathcal{C}$ grouped according to the number of unique word phrases in the MIMIC-III corpus associated with each. We observe the same trend when plotting SNOMED annotations.

The raw codes baseline performs poorly, which aligns with the observation that cTAKES codes assigned to a discharge summary often do not have appropriate or proportional levels of specificity (for example, the top-level ICD9 code '428 Heart Failure' may be assigned by cTAKES, but the gold-label code is '428.21 Acute Systolic Heart Failure'). This may also contribute to the negative result of the proposed model.

Figure 6 (included in the Appendix) illustrates prediction performance as a function of code frequency in the training set, showing that the proposed model does not improve upon the baseline for rare or semi-rare codes.[8]

4.5 Ablations

We separate and analyze the two distinct components of cTAKES' annotation ability for further analysis: 1) how well cTAKES recognizes the location of concepts in the text (*NER*), and 2) how accurately cTAKES maps the recognized positions to the correct clinical concepts (*ontology mapping*). Annotation sparsity (NER) and/or cTAKES mapping error may lend the raw text on its own equally useful, as observed in Table 2. We investigate these hypotheses here. We evaluate performance of ablations relative to the augmentation model and baseline to determine whether each component individ-

---

*These metrics were computed by randomly selecting k elements from those predicted, since there are no sorted probabilities associated with this baseline. For the same reasons we cannot report AUC or AP metrics.

[7]Simulations of the augmentation method under a contrived setting with more concept annotations per note as well as more unique word phrases mapping to a single concept demonstrate solid performance increases over the baseline. This provides supporting evidence that the findings presented in this section may be the cause of the negative result rather than our proposed architecture.

[8]We use the following grouping criteria: rare codes have 50 or fewer occurrences in the training data, semi-rare have between 50 and 1000, and common have more than 1000.

Model	AUC		AP		F1		R@k		P@k	
	Macro	Micro	Macro	Micro	Macro	Micro	8	15	8	15
Baseline (Mullenbach et al., 2018)	0.8892	0.9846	**0.2492**	0.5426	**0.0796**	**0.5421**	**0.3731**	**0.5251**	0.7120	0.5616
Baseline (raw codes)	n/a*	n/a*	n/a*	n/a*	0.0189	0.0877	0.0534*	0.0640*	0.1132*	0.0747*
Augmentation with ICD9										
full replace	0.8846	0.9838	0.2242	0.5329	0.0691	0.5363	0.3688	0.5189	0.7048	0.5564
linear combination	**0.8914**	**0.9849**	0.2467	**0.5427**	0.0763	0.5419	**0.3732**	**0.5267**	**0.7121**	**0.5634**
Augmentation with SNOMED										
full replace	0.8744	0.9830	0.2221	0.5271	0.0724	0.5326	0.3675	0.5177	0.7022	0.5547
linear combination	0.8781	0.9835	0.2238	0.533	0.0692	0.5357	0.3687	0.5194	0.7042	0.5563

Table 2: Test set results using the augmentation methods.

ually adds value. The ablations are:

1. *Dummy Concepts* We replace all word embeddings annotated by cTAKES with 0-vectors, and only use remaining embeddings for prediction. If this alternative shows similar performance to the baseline, then we conclude that the positions in the text annotated by cTAKES (NER) are not valuable for prediction performance.

2. *Concepts Only* We test the complement by replacing all word embeddings *not* annotated by cTAKES with a 0-vector. In contrast to Dummy Concepts, strong performance of this approach relative to the baseline will allow us to conclude that the positions in the text annotated by cTAKES are valuable for prediction performance.

3. *Concepts Only, Concept Embeddings* We replace all word embeddings not annotated by cTAKES with a 0-vector, and then replace all remaining word embeddings with their concept embedding. If this model performs better than Concepts Only, it will demonstrate the strength of cTAKES' ontology mapping component.

Note that Dummy Concepts and Concepts Only are the decomposition of the baseline CAML. Similarly, Dummy Concepts and Concepts Only, Concept Embeddings are the decomposition of the full-replace augmentation model presented in Section 4.

Results Results are presented in Tables 3 and 4. Results are consistent with previous experiments in that augmentation with concept annotations does not improve performance. For both ontologies, neither the Dummy Concepts nor the Concepts Only models outperform the full-text models (in which both token representations are used). However, there are some interesting findings. Using SNOMED annotations, performance of the Concepts Only model is significantly higher than Dummy Concepts and very close to full-text model performance. This finding is strengthened by considering the concept coverage discussed in Table 1: the Concepts Only model achieves comparable performance receiving only about 35% (1% in the ICD9 setting) of the input tokens which the full-text baseline receives, and the Dummy Concepts Model receives about 65% (99% in the ICD9 setting). Thus, a significant proportion of downstream prediction performance can be attributed a small portion of the text which is recognized by cTAKES in both the SNOMED and ICD9 settings, indicating the strength of cTAKES' NER component.

5 Approach 2: Multi-task Learning

We present an alternative application of cTAKES as a form of distant supervision. Our approach is inspired by recent successes in multi-task learning for NLP which demonstrate that cheaply-obtained labels framed as an auxiliary task can improve performance on downstream tasks (Swayamdipta et al., 2018; Ruder, 2017; Zhang and Weiss, 2016). We propose to predict clinical information extraction system annotations as an auxiliary task, and share lower-level representations with the clinical coding task through a jointly-trained model architecture. We hypothesize that domain-knowledge embedded in cTAKES will guide the shared layers of the model architecture towards a more optimal representation for the clinical coding task.

We formulate the auxiliary task as follows: given each word-embedding or word-embedding span in the input which cTAKES has assigned a code, can the model predict the code assigned to it by cTAKES?

Model	Token Representation		AUC		AP		F1		R@k		P@k	
	Concept Match	No Match	Macro	Micro	Macro	Micro	Macro	Micro	8	15	8	15
Baseline (Mullenbach et al., 2018)	Word	Word	**0.8892**	**0.9846**	**0.2492**	**0.5426**	**0.0796**	**0.5421**	**0.3731**	**0.5251**	**0.7120**	**0.5616**
Dummy Concepts	0	Word	0.8876	0.9839	0.2119	0.5236	0.0732	0.5261	0.3634	0.5141	0.6943	0.5506
Concepts Only	Word	0	0.7549	0.9626	0.0538	0.2487	0.0080	0.1961	0.2063	0.2880	0.4196	0.3197
Concepts Only, Concept Embeddings	Concept	0	0.7534	0.9620	0.0552	0.2464	0.0086	0.1972	0.2058	0.2855	0.4200	0.3166
Augmentation Model (full replace)	Concept	Word	0.8846	0.9838	0.2242	0.5329	0.0691	0.5363	0.3688	0.5189	0.7048	0.5564

Table 3: Test set results of ablation experiments on the MIMIC-III dataset, using ICD9 concept annotations.

Model	Token Representation		AUC		AP		F1		R@k		P@k	
	Concept Match	No Match	Macro	Micro	Macro	Micro	Macro	Micro	8	15	8	15
Baseline (Mullenbach et al., 2018)	Word	Word	**0.8892**	**0.9846**	**0.2492**	**0.5426**	**0.0796**	**0.5421**	**0.3731**	**0.5251**	**0.7120**	**0.5616**
Dummy Concepts	0	Word	0.8472	0.9780	0.1461	0.4375	0.0413	0.4426	0.3202	0.4439	0.6234	0.4804
Concepts Only	Word	0	0.8736	0.9817	0.2059	0.4518	0.0515	0.4295	0.3278	0.4583	0.6300	0.4903
Concepts Only, Concept Embeddings	Concept	0	0.8739	0.9813	0.2019	0.4451	0.0519	0.4258	0.3247	0.4538	0.6254	0.4851
Augmentation Model (full replace)	Concept	Word	0.8744	0.9830	0.2221	0.5271	0.0724	0.5326	0.3675	0.5177	0.7022	0.5547

Table 4: Test set results of ablation experiments on the MIMIC-III dataset, using SNOMED concept annotations.

5.1 Method

We annotate the set of non-null ground-truth codes output by cTAKES for document i in the training data as $\{(a_{i,1}, c_{i,1}), (a_{i,2}, c_{i,2}), \ldots, (a_{i,M}, c_{i,M})\}$, where each anchor $a_{i,m}$ indicates the span of tokens in the text for which concept $c_{i,m}$ is annotated, and $c_{i,m} \neq \varnothing$.

The loss term of the model is augmented to include the multi-class cross-entropy of predicting the correct code for all annotated spans in the training batch:

$$\mathcal{L} = \sum_{i=1}^{I} BCE(\boldsymbol{y}_i, \hat{\boldsymbol{y}}_i)$$
$$+ \lambda \frac{\sum_{i=1}^{I} \sum_{m=1}^{M_i} - \log p(c_{i,m} \mid a_{i,m})}{\sum_{i=1}^{I} M_i}$$

where $BCE(\boldsymbol{y}_i, \hat{\boldsymbol{y}}_i)$ is the standard (binary cross-entropy) loss from the baseline for the clinical coding task, $p(c_{i,m} \mid a_{i,m})$ is the probability assigned by the auxiliary model to the true cTAKES-annotated concept given word span $a_{i,m}$ as input, λ is the hyperparameter to tradeoff between the two objectives, and I is the number of instances in the batch.

Because we use the auxiliary task as a "scaffold" (Swayamdipta et al., 2018) for transferring domain knowledge encoded in cTAKES' rules into the learned representations for the clinical coding task, we must only run cTAKES and compute a forward pass through the auxiliary module at training time. At test-time, we evaluate only on the clinical coding task, so the time complexity of model inference remains the same as the baseline, an advantage of this architecture.

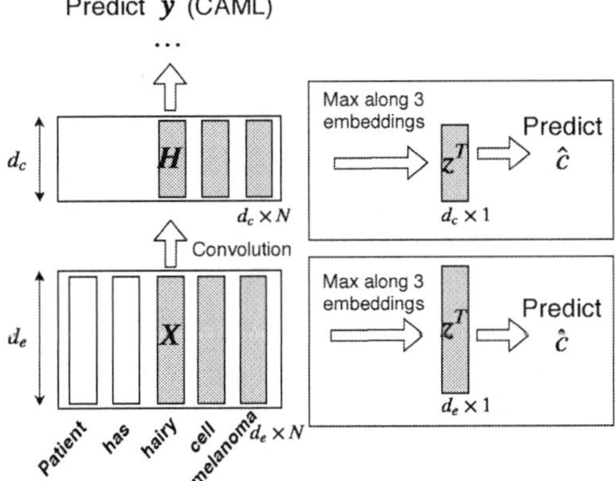

Figure 5: The proposed architecture (for prediction on a single document, i, and auxiliary supervision on a single annotation, m). The bottom box illustrates the pre-convolution model, and the top box post-convolution. The architecture on the left is the baseline.

We model $p(c_{i,m} \mid a_{i,m})$ via a multi-layer perceptron with a Softmax output layer to obtain a distribution over the codeset, $\mathcal{C}$. We additionally experiment with a linear layer variant to combat overfitting on the auxiliary task by reducing the capacity of this module. The input to this module is a single vector, $\boldsymbol{z}_{i,m} \in \mathbb{R}^{d_e}$, constructed by selecting the maximum value over s word embeddings for each dimension, where s is the length of the input span.[9] To facilitate information transfer between the clinical coding and auxiliary task, we experiment with tying both the randomly-initialized embedding layer, $\boldsymbol{X}$, and a higher-level layer of the

---

[9] While this is simple representation, we find that multi-word concept annotations are rather rare, in which case $\boldsymbol{z}_{i,m}$ is equivalent to $\boldsymbol{x}_{i,m}$.

Shared Features	Auxiliary Model	AUC		AP		F1		R@k		P@k	
		Macro	Micro	Macro	Micro	Macro	Micro	8	15	8	15
Baseline (Mullenbach et al., 2018)	n/a	**0.8892**	**0.9846**	**0.2492**	**0.5426**	**0.0796**	**0.5421**	**0.3731**	0.5251	**0.7120**	0.5616
Pre-convolution	MLP	0.8874	0.9839	0.2365	0.5390	0.0734	0.5376	0.3724	0.5235	0.7102	0.5597
Pre-convolution	Linear Layer	0.8834	0.9838	0.2398	0.5412	0.0766	0.5414	**0.3731**	**0.5265**	0.7113	**0.5633**
Post-convolution	MLP	0.7252	0.9619	0.0578	0.3002	0.0159	0.2966	0.2449	0.3417	0.4879	0.3748
Post-convolution	Linear Layer	0.7562	0.9655	0.0606	0.3035	0.0123	0.2934	0.2461	0.3392	0.4900	0.3700

Table 5: Test set performance on the ICD9 coding task for $\lambda = 1$ and using ICD9 annotations.

Shared Features	Auxiliary Model	Tagging Accuracy	
		After one epoch	After last epoch
Pre-convolution	MLP	0.9343	0.9398
Pre-convolution	Linear Layer	0.8940	**0.9400**
Post-convolution	MLP	0.9102	0.9335
Post-convolution	Linear Layer	0.7524	0.9341

Table 6: Dev set performance on the auxiliary task for $\lambda = 1$ and using ICD9 annotations. Relatively high task performance is achieved even after one epoch with a simple model.

network (e.g. the outputs of the document-level convolution layer H described in Section 3). See Figure 5 for the model architecture.

5.2 Experiment and Results

Results are presented in Table 5 and Table 6 for ICD9 annotations. Overall, the cTAKES span-prediction task does more to hurt than help performance on the main task. Tying the model weights at a higher layer (post-convolution as opposed to pre-convolution) results in worse performance, even though the model fits the auxiliary task well. This indicates either that the model may not have enough capacity to adequately fit both tasks, or that the cTAKES prediction task as formulated may actually misguide the clinical coding task slightly in parameter search space.[10]

We additionally remark that increasing the weight of the auxiliary task generally lowers performance on the clinical coding task, and tuning λ on the dev set does not result in more optimal performance (we include results with $\lambda = 1$ here; see Table 9 in the Appendix). Notably, for even very small values of λ, we achieve very high validation accuracy on the auxiliary task. This performance does not change with larger weightings, indicating that the auxiliary task may not be difficult enough to result in effective knowledge transfer.[11]

---

[10]We found similar results using SNOMED annotations.

[11]While the models in Sections 4 did not introduce new hyperparameters to the baseline architecture, hyperparameters for this architecture were selected by human intuition. Room for future work includes more extensive tuning (see Table 8 in Appendix A).

6 Conclusion

Integrating existing clinical information extraction tools with deep learning models is an important direction for bridging the gap between rule-based and learning-based methods. We have provided an analysis of the quality of the widely-used clinical concept annotator cTAKES when integrated into a state-of-the-art text-based prediction model. In two settings, we have shown that cTAKES does not improve performance over raw text alone on the clinical coding task. We additionally demonstrate through error analysis and ablation studies that the amount of word variation captured and the differentiation between the named-entity recognition and ontology-mapping tasks may affect cTAKES' effectiveness.

While automated coding is one application area, the models presented here could easily be extended to other downstream prediction tasks such as patient diagnosis and treatment outcome prediction. Future work will include evaluating newly-developed clinical NER tools with similar functionalities to cTAKES in our framework, which can potentially serve as a means to evaluate the effectiveness of newer systems vis-à-vis cTAKES.

Acknowledgments We thank the Research, Development, and Dissemination group at Sutter Health as well as members of the Georgia Tech Computational Linguistics Lab for helpful discussions. Figure 3 was provided by James Mullenbach.

References

American Academy of Professional Coders. 2019. What is medical coding?

Alan R Aronson. 2001. Effective mapping of biomedical text to the umls metathesaurus: the metamap program. In *Proceedings of the AMIA Symposium*, page 17. American Medical Informatics Association.

Dzmitry Bahdanau, Kyunghyun Cho, and Yoshua Bengio. 2014. Neural machine translation by jointly learning to align and translate. *arXiv preprint arXiv:1409.0473*.

Tal Baumel, Jimena Nassour-Kassis, Raphael Cohen, Michael Elhadad, and Noémie Elhadad. 2018. Multi-label classification of patient notes a case study on icd code assignment. In *Proceedings of the 2018 AAAI Joint Workshop on Health Intelligence*.

Willie Boag and Hassan Kané. 2017. Awe-cm vectors: Augmenting word embeddings with a clinical metathesaurus. *arXiv preprint arXiv:1712.01460*.

Willie Boag, Elena Sergeeva, Saurabh Kulshreshtha, Peter Szolovits, Anna Rumshisky, and Tristan Naumann. 2018. Cliner 2.0: Accessible and accurate clinical concept extraction. *arXiv preprint arXiv:1803.02245*.

Edward Choi, Mohammad Taha Bahadori, Le Song, Walter F Stewart, and Jimeng Sun. 2017. Gram: graph-based attention model for healthcare representation learning. In *Proceedings of the 23rd ACM SIGKDD International Conference on Knowledge Discovery and Data Mining*, pages 787–795. ACM.

Jesse Davis and Mark Goadrich. 2006. The relationship between precision-recall and roc curves. In *Proceedings of the 23rd international conference on Machine learning*, pages 233–240. ACM.

Kevin Donnelly. 2006. Snomed-ct: The advanced terminology and coding system for ehealth. *Studies in health technology and informatics*, 121:279.

Carol Friedman. 2000. A broad-coverage natural language processing system. In *Proceedings of the AMIA Symposium*, page 270. American Medical Informatics Association.

Nathan Greenberg, Trapit Bansal, Patrick Verga, and Andrew McCallum. 2018. Marginal likelihood training of bilstm-crf for biomedical named entity recognition from disjoint label sets. In *Proceedings of the 2018 Conference on Empirical Methods in Natural Language Processing*, pages 2824–2829.

Alistair EW Johnson, Tom J Pollard, Lu Shen, H Lehman Li-wei, Mengling Feng, Mohammad Ghassemi, Benjamin Moody, Peter Szolovits, Leo Anthony Celi, and Roger G Mark. 2016. Mimiciii, a freely accessible critical care database. *Scientific data*, 3:160035.

Donald AB Lindberg, Betsy L Humphreys, and Alexa T McCray. 1993. The unified medical language system. *Methods of information in medicine*, 32(04):281–291.

James Mullenbach, Sarah Wiegreffe, Jon Duke, Jimeng Sun, and Jacob Eisenstein. 2018. Explainable prediction of medical codes from clinical text. In *Proceedings of the 2018 Conference of the North American Chapter of the Association for Computational Linguistics: Human Language Technologies*, volume 1.

National Center for Health Statistics. 1991. *The International Classification of Diseases: 9th Revision, Clinical Modification: ICD-9-CM*.

Hoang Nguyen and Jon Patrick. 2016. Text mining in clinical domain: Dealing with noise. In *Proceedings of the 22nd ACM SIGKDD international conference on knowledge discovery and data mining*, pages 549–558. ACM.

T.J. Pollard and A.E.W. Johnson. 2016. The MIMIC-III clinical database.

Anthony Rios and Ramakanth Kavuluru. 2018a. Emr coding with semi-parametric multi-head matching networks. In *Proceedings of the 2018 Conference of the North American Chapter of the Association for Computational Linguistics: Human Language Technologies*, volume 1, pages 2081–2091.

Anthony Rios and Ramakanth Kavuluru. 2018b. Fewshot and zero-shot multi-label learning for structured label spaces. In *Proceedings of the 2018 Conference on Empirical Methods in Natural Language Processing*, pages 3132–3142.

Sebastian Ruder. 2017. An overview of multi-task learning in deep neural networks. *arXiv preprint arXiv:1706.05098*.

Takaya Saito and Marc Rehmsmeier. 2015. The precision-recall plot is more informative than the roc plot when evaluating binary classifiers on imbalanced datasets. *PloS one*, 10(3):e0118432.

Guergana K Savova, James J Masanz, Philip V Ogren, Jiaping Zheng, Sunghwan Sohn, Karin C Kipper-Schuler, and Christopher G Chute. 2010. Mayo clinical text analysis and knowledge extraction system (ctakes): architecture, component evaluation and applications. *Journal of the American Medical Informatics Association*, 17(5):507–513.

Yuqi Si, Jingqi Wang, Hua Xu, and Kirk Roberts. 2019. Enhancing clinical concept extraction with contextual embedding. *arXiv preprint arXiv:1902.08691*.

Anima Singh, Girish Nadkarni, John Guttag, and Erwin Bottinger. 2014. Leveraging hierarchy in medical codes for predictive modeling. In *Proceedings of the 5th ACM conference on Bioinformatics, Computational Biology, and Health Informatics*, pages 96–103. ACM.

Swabha Swayamdipta, Sam Thomson, Kenton Lee, Luke Zettlemoyer, Chris Dyer, and Noah A. Smith. 2018. Syntactic scaffolds for semantic structures. *CoRR*, abs/1808.10485.

Yifeng Tao, Bruno Godefroy, Guillaume Genthial, and Christopher Potts. 2018. Effective feature representation for clinical text concept extraction. *arXiv preprint arXiv:1811.00070*.

Jacqueline M Torres, John Lawlor, Jeffrey D Colvin, Marion R Sills, Jessica L Bettenhausen, Amber Davidson, Gretchen J Cutler, Matt Hall, and Laura M Gottlieb. 2017. Icd social codes. *Medical care*, 55(9):810–816.

Yanshan Wang, Liwei Wang, Majid Rastegar-Mojarad, Sungrim Moon, Feichen Shen, Naveed Afzal, Sijia Liu, Yuqun Zeng, Saeed Mehrabi, Sunghwan Sohn, et al. 2017. Clinical information extraction applications: a literature review. *Journal of biomedical informatics*.

Wei-Hung Weng, Kavishwar B Wagholikar, Alexa T McCray, Peter Szolovits, and Henry C Chueh. 2017. Medical subdomain classification of clinical notes using a machine learning-based natural language processing approach. *BMC medical informatics and decision making*, 17(1):155.

Kai Xu, Zhanfan Zhou, Tao Gong, Tianyong Hao, and Wenyin Liu. 2018. Sblc: a hybrid model for disease named entity recognition based on semantic bidirectional lstms and conditional random fields. *BMC medical informatics and decision making*, 18(5):114.

Yuan Zhang and David Weiss. 2016. Stack-propagation: Improved representation learning for syntax. *arXiv preprint arXiv:1603.06598*.

Henghui Zhu, Ioannis Ch Paschalidis, and Amir Tahmasebi. 2018. Clinical concept extraction with contextual word embedding. *arXiv preprint arXiv:1810.10566*.

A Experimental Details

Data Following Mullenbach et al. (2018), we use the same train/test/validation splits for the MIMIC-III dataset, and concatenate all supplemental text for a patient discharge summary into one record. We use the authors' provided data processing pipeline[12] to preprocess the corpus. The vocubulary includes all words occurring in at least 3 training documents. See Table 7 for descriptive statistics of the dataset.

We construct a concept vocabulary for embedding initialization following the same specification as the word vocabulary: any concept which does not occur in at least 3 training documents is replaced with an UNK token. Details on the size of the vocabulary can be found in Table 8.

# training documents	47,723
# test documents	3,372
# dev documents	1,631
Mean # tokens per document	1,513.0
Mean # labels per document	16.09
Total # labels (L)	8,921

Table 7: Dataset Descriptive Statistics.

Training We train with the same specifications as Mullenbach et al. (2018) unless otherwise specified, with dropout performed after concept augmentation for the models in Sections 4, and early stopping with a patience of 10 epochs on the precision at 8 metric, for a maximum of 200 epochs (note that in the multi-task learning models the stopping criterion is only a function of performance on the clinical coding task). Unlike previous work, we reduce the batch size to 12 in order to allow each batch to fit on a single GPU, and we do not use pretrained embeddings as we find this improves performance. All models are trained on a single NVIDIA Titan X GPU with 12,189 MiB of RAM.

We port the optimal hyperparameters reported in Mullenbach et al. (2018) to our experiments. With more extensive hyperparameter tuning, we may expect to see a potential increase in the performance of our models over the baseline. See Table 8 for hyperparameters and other details specific to our proposed model architectures. All neural models

are implemented using PyTorch[13], and built on the open-source implementation of CAML.[14]

Parameter	Value
Vocabulary Size	51,917
SNOMED Concept Vocabulary ($\mathcal{C}$) Size	20,775
ICD9 Concept Vocabulary ($\mathcal{C}$) Size	1,529
Embedding Size (d_e)	100
Post-convolution Embedding Size (d_c)	50
Dropout Probability	0.2
Learning Rate	0.0001
Attention Mechanism Hidden State Size	20
Attention Mechanism Activation Function	ReLU
Auxiliary hidden layer size	700
Auxiliary activation function	ReLU

Table 8: Model details.

B Concept Extraction

We build a custom dictionary from the UMLS Metathesaurus that includes mappings from UMLS CUIs to SNOMED-CT and ICD9-CM concepts. We run the cTAKES annotator in advance of training for all 3 dataset splits using the resulting dictionary, allowing us to obtain annotations for each note in the dataset, and the positions of the annotations in the raw text. Note that for the multi-task learning experiments (Section 5), we only require annotations for training data. Annotating the MIMIC-III datafiles using these specifications takes between 4 and 5 hours for 3,000 discharge summaries on a single CPU, and can be parallelized for efficiency.

C Attention for Overlapping Concepts

We implement an attention mechanism (Bahdanau et al., 2014) to compute a single concept embedding $\phi(\mathcal{C}_n) \in \mathbb{R}^{d_e}$ when $\mathcal{C}_n = \{c_1, c_2, \ldots, c_J\}$ represents a set of concepts annotated at position n instead of a single concept. Intuitively, we want to more heavily weight those concepts in the set which have the most similarity to the surrounding text. We define a context vector for position n as:

$$v_n = [x_{n-2}, x_{n-1}, x_{n+1}, x_{n+2}] \in \mathbb{R}^{4d_e}$$

The context is defined as the concatenated word embeddings surrounding position n. We use a context size of $n +/- 2$, where 2 is a hyperparameter. We choose to use a smaller value for computational efficiency.

λ	AUC		AP		F1		R@k		P@k		Auxiliary Tagging Accuracy	
	Macro	Micro	Macro	Micro	Macro	Micro	8	15	8	15	After one epoch	After last epoch
0.001	**0.9002**	**0.9848**	0.3129	0.5470	**0.0704**	**0.5511**	0.3902	0.5447	0.7164	0.5631	0.8888	0.9398
0.01	0.8954	0.9842	0.2885	0.5352	0.0636	0.5425	0.3843	0.5328	0.7088	0.5528	0.8938	**0.9401**
0.1	0.9000	0.9846	**0.3145**	0.5465	0.0689	0.5471	**0.3909**	0.5426	**0.7183**	0.5617	0.8940	0.9400
0.5	0.8934	0.9840	0.2892	0.5362	0.0624	0.5386	0.3844	0.5361	0.7089	0.5546	**0.8941**	0.9400
1	0.8975	0.9840	0.3087	0.5460	0.0668	0.5477	0.3886	0.5439	0.7169	0.5624	0.8940	0.9400
10	0.8979	0.9842	0.3122	**0.5484**	0.0678	0.5474	0.3908	**0.5457**	0.7182	**0.5644**	0.8940	0.9400
50	0.8939	0.9837	0.2982	0.5410	0.0638	0.5427	0.3855	0.5391	0.7111	0.5592	0.8940	**0.9401**
100	0.8913	0.9835	0.2943	0.5383	0.0632	0.5407	0.3849	0.5374	0.7096	0.5577	0.8940	**0.9401**
1000	0.8851	0.9827	0.2750	0.5260	0.0564	0.5309	0.3803	0.5290	0.7016	0.5491	0.8940	**0.9401**

Table 9: The effect of tuning λ on dev set performance on the ICD9 coding task, for the pre-convolution model with a linear auxiliary layer and ICD9 annotations. We select $\lambda = 1$ for reporting test results; there isn't a clear value which produces strictly better performance.

We concatenate the word-context vector and each concept embedding c_j in $\mathcal{C}_n$ as $[\boldsymbol{v}_n, \phi(c_j)] \in \mathbb{R}^{5d_e}$, and pass it through a multi-layer perceptron to compute a similarity score: $f : \mathbb{R}^{5d_e} \to \mathbb{R}^1$. An attention score for each c_j is computed as:

$$\alpha_j = \frac{exp(f(\boldsymbol{v}_n, \phi(c_j))}{\sum_{k=1}^{J} exp(f(\boldsymbol{v}_n, \phi(c_k))}$$

This represents the relevance of the concept to the surrounding word-context, normalized by the other concepts in the set. A final concept embedding $\phi(\mathcal{C}_n) \in \mathbb{R}^{d_e}$ is computed as a linear combination of the concept vectors, weighted by their attention scores:

$$\phi(\mathcal{C}_n) = \sum_{j=1}^{J} \alpha_j \cdot \phi(c_j)$$

D   Leveraging Ontological Graph Structure

Following the methodology of Choi et al. (2017), we experiment with learning higher-quality concept representations using the hierarchical structure of the ICD9 ontology. We replace concept embedding $\phi(c_n)$ with a learned linear combination of itself and its parent concepts' embeddings (see Figure 1). For child concepts which are observed infrequently or have poor representations, prior work has shown that a trained model will learn to weight the parent embeddings more heavily in the linear combination. Because the parent concepts represent more general concepts, they have most often been observed more frequently in the training set and have stronger representations. This also allows for learned representations which capture relationships between concepts. We refer the reader to Choi et al. (2017) for details.

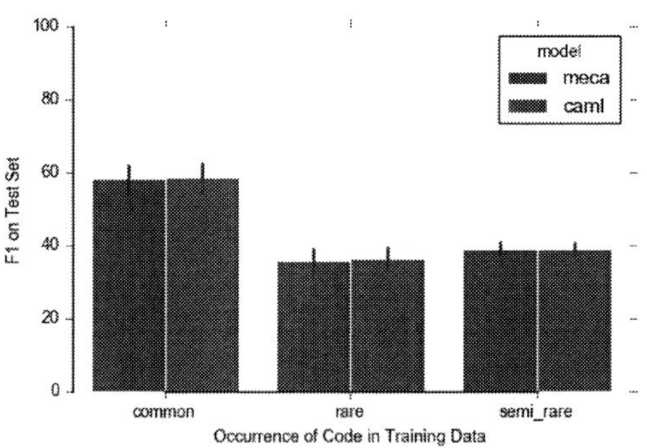

Figure 6: F1 on Test Data based on Frequency of Codes in Training Data, where the metric is defined ('meca' indicates the *linear combination* ICD9 augmentation model).

Clinical Case Reports for NLP

Cyril Grouin
LIMSI, CNRS, Université Paris Saclay
Campus universitaire d'Orsay
91405 Orsay cedex, France
`cyril.grouin@limsi.fr`

Natalia Grabar
STL, CNRS, Université de Lille
Domaine du Pont-de-Bois
59653 Villeneuve-d'Ascq cedex, France
`natalia.grabar@univ-lille.fr`

Vincent Claveau
IRISA, CNRS
Czampus universitaire de Beaulieu
35042 Rennes cedex, France
`vincent.claveau@irisa.fr`

Thierry Hamon
LIMSI, CNRS, Université Paris Saclay
Université Paris 13
99 avenue Jean-Baptiste Clément
93430 Villetaneuse, France
`thierry.hamon@limsi.fr`

Abstract

Textual data are useful to access expert information. Since the texts are representative of distinct language uses, it is necessary to build specific corpora in order to be able to design suitable NLP tools. In some domains, such as medical domain, it may be complicated to access the representative textual data and their semantic annotations, while there exists a real need for providing efficient tools and methods. In this paper, we present a corpus of 717 clinical cases written in French. We manually annotated this corpus into four general categories (age, gender, outcome, and origin) for a total number of 2,835 annotations. The values of age, gender, and outcome have been normalized. We also manually annotated a subset of 70 files into 27 fine-grained categories, for a total number of 5,198 annotations. In addition, we present a few basic experiments made on those annotations in order to highlight their usefulness.

1 Introduction

In Natural Language Processing (NLP), texts are useful to access information, especially expert information. Nevertheless, the linguistic diversity (type of narratives, common or specialized vocabulary, regular or complex syntactic structures, etc.) requires robust tools to access the information present in those texts. In order to build suitable NLP-based tools, to model linguistic elements (machine-learning, word-embeddings), or to produce gold standards for evaluating automatic systems, texts are needed (Nadkarni et al., 2011). However, due to privacy and ethical reasons, documents from specialized domains (e.g., clinical notes or justice decisions) are not easily accessible unless authorization (Chapman et al., 2011).

When such data exist for the research, they are generally limited to English language, such as the MIMIC-III database (Johnson et al., 2016) and derived corpora. For French language, the Quaero medical corpus (Névéol et al., 2014) is composed of a limited number of documents (13 documents from the European Medicines Agency, 25 documents from the European Patent Organization) or very short documents (2,500 Medline titles).

In order to make available documents concerned by privacy issues, de-identification techniques have been widely used to replace nominative data by plausible information (Meystre et al., 2010; Kayaalp, 2017). Despite the recent improvements of these techniques, especially based on artificial neural networks (Dernoncourt et al., 2017), one can not assure that all nominative data have been removed and humans must further check those documents. Another solution relies on the production of synthetic data (Lohr et al., 2018). Originally, they were generated and used to train OCR systems for handwriting recognition (Doermann and Yao, 1995). They are now used when original data are missing or to provide more data, despite their artificial character (Eger et al., 2019). Besides, whether the texts are de-identified or artificially generated, their linguistic specificity will have an impact on further designed NLP rule-based and statistically-based approaches.

In this paper, we present the semantic annotations we made on a corpus of clinical cases written in French by domain experts. Since this corpus is composed of already published and freely accessible clinical cases, our aim is to make this annotated corpus available for the research. In order to present the usefulness of those annotations, we present a few basic experiments we made.

273

Proceedings of the BioNLP 2019 workshop, pages 273–282
Florence, Italy, August 1, 2019. ©2019 Association for Computational Linguistics

2 Corpus and annotation guidelines

2.1 Corpus

In the clinical domain, in order to overcome the privacy and ethical issues when working on electronic health records, one solution consists in using clinical case reports. Indeed, it is quite common to find freely available publications from scientific journals which report in clinical cases of real de-identified or fake patients. Such clinical cases are usually published and discussed to improve medical knowledge (Atkinson, 1992) of colleagues and medical students. One may find scientific journals specifically dedicated to case reports, such as the *Journal of Medical Case Reports* launched in 2006 (Rison et al., 2017). Clinical cases consist of a detailed and hierarchically structured description of history, signs and symptoms, diseases, tests, treatments, follow-up and outcome of a given patient or of a cohort of patients (Rison, 2013). As pinpointed by Lysanets et al. (2017), clinical cases are composed of linguistic particularities which constitute a specific genre of medical texts: active voice sentences, past simple tense, personal pronouns, and modal verbs. Beyond this warning, they represent both an available and useful clinical content, especially for the NLP community for which the access to EHRs is becoming harder and harder.

We assume that this new orientation to tackle the medical data accessibility problem may become popular in the years to come within the biomedical domain. Let's for instance mention the work by Satomura and Amaral (1992), which produced back in 90's an automatic system designed for the indexing of clinical cases with ICD-9 codes. These clinical cases written in English have been extracted from the *New England Journal of Medicine* and permitted the researchers to develop their NLP system and to test it. More recently, Gurulingappa et al. (2012) produced a benchmark corpus composed of 3,000 clinical case reports in English, which has been then annotated into several categories (drug, dosage, and adverse effects), and relationships among them in order to provide mentions of adverse drug reactions.

The corpus we present in this work is composed of 717 clinical case reports written in French (see table 1 for general statistics). These cases have been previously published and are freely accessible. The cases from scientific literature often go with their discussion and keywords. In this work, we only focus on the clinical case description. This set has been manually annotated with general and fine-grained information, which is described in the two following sections. This corpus is part of a larger and yet growing corpus, which currently contains over 4,100 clinical cases (Grabar et al., 2018).

Element	Number
Documents	717
Sentences	1,124
Words (occurrences)	26,787
Words (forms)	5,030

Table 1: General statistics on the corpus annotated in this work

2.2 Annotations of general information

We considered four general categories of information for the annotation. They are related to demographic data (age and gender) and to medical data (the starting medical problem or origin and the outcome). Most of the clinical cases describe the clinical events of one patient. Yet, some clinical cases may be dedicated to the description of several patients, in which case, all relevant information are annotated for each patient. For this reason, the total number of annotations may be higher than the number of clinical cases. For three out of four categories, the values are normalized and taken from finite sets:

- Age $\in \mathbb{N}$: numerical value rounded in years; age in letters is converted into numerical value;

- Gender $\in \{$ feminine, masculine $\}$;

- Outcome $\in \{$ recovery, improvement, stable condition, worsening, death $\}$.

Besides, when several ages are given for the same patient, only the age at the moment of the main clinical event is considered. For the category Origin, the values correspond to text spans describing the initial medical problem.

Two scientists with a biomedical computer science background created the annotations independently, and then elaborated consensual annotations. Hence, all spans of text providing the expected information were annotated. For the category origin, the most inclusive text spans have been chosen.

2.3 Annotations of fine-grained information

The corpus has also been enriched with fine-grained annotations of entities concerning physiology, surgery, diseases, drugs, temporal data, lab and exam results. The annotations are based on the semantic types from the UMLS (Lindberg et al., 1993), on existing annotation guidelines such as the I2B2 NLP Challenges (Uzuner et al., 2010, 2011), and on medical entities from our corpus. We provide those annotations as a basis for several NLP tasks such as information extraction or automatic classification based on clinical entities.

In this section, we present the guidelines we defined. For each category, we give a definition and a few examples from the corpus.

2.3.1 Physiology

Body measurements: weight *(71.8 kg)*, size *(165 cm)*, and body surface area *(1.81 m^2)*

Vital signs: temperature *(38.2 °C)*, and physiological liquid mentions *(blood, urine)*

Biology: anatomical parts *(left lung, thyroid)*, localization of procedures or diseases *(arterial, pulmonary)*, and biological functions *(pregnancy, pulse)*

2.3.2 Surgery

These categories are related to the surgery:

Medical speciality including the types of medical units *(oncology, surgical care units)*.

Tests including names of tested elements *(radiography, biological check-up, blood pressure)*

Surgical treatments: treatments done by physicians *(chemotherapy, resection)*

Surgical approach: access used by the physician *(apical access)*

Medical devices used by patients or by physicians *(drainage, mask, sensor)*

2.3.3 Diseases

We considered four types of disease-related information:

Pathology: mentions of diseases or diseased condition *(acute lymphoblastic leukemia, tumor)*

Signs or symptoms which are not chronic diseases *(cough, fever, headache, hypertension)*

Biological organism: bacteria and infectious organisms *(escherichia coli, group B streptococcus)*

Nature: indication of quality (qualifying adjectives, grade) for diseases, signs and symptoms *(pT2 G1 carcinoma, benign cyst)*

2.3.4 Drugs

Pharmaceutical class or family of drugs *(antibiotic, anticoagulant, anti-vitamin K)*

Substance: commercial and generic drug names or generic substance *(acetaminophen, ferrous sulphate)*

Concentration of molecules in drugs *(10%, 5 mg/ml)*

Mode of administration *(intravenous, oral route, by nebulization)*

Dose: composed of value and unit for drug dose *(0.5 mg, four doses, one to two pills, three million units)* or rates *(5 mg/kg)*. If a dose was changed according to a past condition, the modification is annotated among two normalized values (increase, decrease)

2.3.5 Temporal data

Date: absolute and relative dates *(January 2005)*

Moment: moment of a day for drug intake or surgical intervention *(at bedtime, the morning)* or specific time during the hospital stay *(at D1-D2)*

Duration especially for treatments and diseases *(since 10 years, for four weeks)*

Frequency for intakes, diseases, signs and symptoms *(once a day, if needed, chronic, every two weeks)*

2.3.6 Lab and exam results

This category is related to all numerical values from lab results *(105/80 mm Hg, 68 bpm)* and analysis result from examination (e.g., *normal* for imaging or palpation).

2.4 Additional information

Some categories are annotated with additional information.

2.4.1 Linguistic annotations

Similarly to Uzuner et al. (2011), we added assertion values among the six tags possible: *present, absent, associated to someone else, conditional, hypothetical, possible. Present*: default value; *Absent*: element planned but not realized; *Conditional*: element that can occur under certain circumstances; *Hypothetical*: element that may occur in the future; *Possible*: element that may occur; *Associated to someone else*: element concerning family or acquaintances. Assertions may be used for the annotation of the Pathology, Signs and Symptoms, Tests, and Treatments categories.

2.4.2 Medical information

Linguistic interpretation: With Substances and Weight, if the medication or the weight change according to their previous values, this modification is annotated according to two normalized values: *stop* and *titration* for Substances, and *gain* and *loss* for Weight.

Medical interpretation: For lab results (e.g., blood pressure) and physiological data (temperature), if values can be compared to known ranges (external medical knowledge), three normalized levels are used *(high, normal, low)* in order to provide a better comprehension of those values.

3 Annotated corpus

3.1 Inter-annotator agreement

The inter-annotator agreement is computed with Cohen's κ, and with Precision, Recall and F-measure values (Sebastiani, 2002).

General information We computed inter-annotator agreement scores on the normalized values for general information: Age, Gender and Outcome, and on the annotated text spans for Origin. We achieved excellent agreements for Age and Gender (κ=0.939), differences being due to omissions; poor agreement for Outcome (κ=0.369) due to differences of interpretation between close values (e.g., recovery vs. improvement for long-term diseases); and very low agreement for Origin (κ=-0.762) since spans of text were often distinct between annotators. As stated by Grouin et al. (2011), the κ metric is not well suited for annotations of text since it relies on a random baseline for which the number of units that may be annotated is hard to define. As a consequence, the classical F-measure is often used as an approximation of inter-annotator agreement. In the following experiments, we present the inter-annotator agreements through Precision, Recall, and F-measure.

Outcome The outcome value is complex since differences between recovery and improvement may imply more knowledge than the information presented in the clinical case. As an example, for a patient presenting arterial hypertension at the consultation, do we consider a "recovery" or an "improvement" when clinicians indicate *a complete remission 18 months after the intervention?* Can we consider a recovery for a remission? Is a period of eighteen months sufficient to take a decision? If *no tumor recurrence after fifteen months of decline* is considered, since a tumor may appear again, can we still consider a "recovery"?

At last, we made a difference between cancers or malign tumors ("improvement") and benign tumors or other diseases ("recovery"). For chronic diseases, we only considered an "improvement".

Fine-grained categories In Table 2, we indicate the inter-annotator agreement for the main categories from fine-grained annotations on a subset of 70 clinical cases we annotated in duplicate.

Category	P	R	F
Anatomy	0.5660	0.8511	0.6799
Concentration	0.5714	0.2857	0.3810
Date	0.7042	0.2747	0.3953
Devices	0.3151	0.8519	0.4600
Dose	0.3744	0.8913	0.5273
Duration	0.7500	0.5816	0.6552
Examen.	0.4260	0.8267	0.5623
Function	0.5135	0.2879	0.3689
Frequency	0.5597	0.8824	0.6849
Localisation	0.4328	0.8056	0.5631
Mode	0.5563	0.8778	0.6810
Pathology	0.2596	0.6116	0.3645
SOSY	0.5567	0.6888	0.6157
Specialty	0.3077	0.2051	0.2462
Substance	0.5950	0.7163	0.6500
Treatment	0.5378	0.4054	0.4623
Overall	0.4426	0.6924	0.5400

Table 2: Inter-annotator agreement for the main categories (fine-grained annotations) on the 70 files dataset

We observe that the categories yield better Recall than Precision, which means that similar units are annotated by the two annotators. Yet, Precision values are often lower because the units may correspond to different text spans. The average agreement in terms of F-measure is 0.5400. This first round of fine-grained annotations permitted to elaborate strong annotation guidelines, which

Physiology								Numerical Values
Body measurements			*Vital signs*		*Biology*			
Weight	Size	Surface	Temp.	Liquid	Anatomy	Local.	Function	
8	5	3	5	47	424	603	37	310
Surgery					**Diseases**			
Speciality	Tests	Treatm.	Access	Devices	Pathology	SOSY	Organism	Nature
26	784	251	20	73	285	803	11	192
Drugs					**Temporal**			
Class	Substance	Conc.	Mode	Dose	Date	Moment	Duration	Frequency
44	437	14	142	219	71	174	76	134

Table 3: Number of annotations for each fine-grained category within the subset of 70 files. (Temp.=temperature, Local.=localization, Treatm.=surgical treatments, Access=surgical approach, SOSY=signs or symptoms, Class=pharmaceutical class, Conc.=concentration)

is being applied to the whole set of 717 clinical cases. We expect that the further annotations will provide with better inter-annotator agreement.

3.2 Statistics

Table 3 indicates the number of annotations for each fine-grained category based on a subset of 70 cases. The total number of annotations is 5,198, which gives on average 74.3 annotations per case.

As shown on figure 1, all fine-grained categories have not been used in each file. Six categories are mainly used in the dataset of 70 files: Test (annotations found in 95.7% of all files), Localisation (90.0%), Sign or Symptom (78.6%), Anatomy (75.7%), Pathology (72.9%), and Surgical Treatment (68.6%).

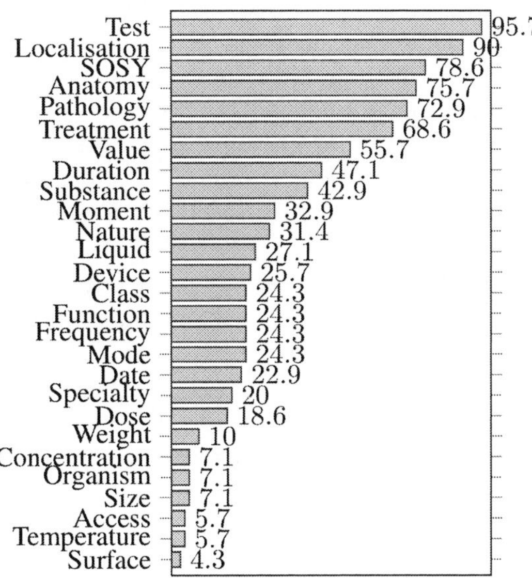

Figure 1: Distribution of fine-grained annotations in the dataset of 70 files (percentage)

Physiological information (body measurements and vital signs) are found in a few number of files (less than 10% of files from the dataset). Since those types of information are useful for a limited number of pathologies or signs or symptoms, they have been found in few documents.

Table 4 presents the final number of annotations on the four general categories and their distribution on the whole dataset of 717 files. Since a few clinical cases describe several patients (either a cohort of patients or a pathology affecting several patients), the total number of annotations may be higher than the total number of files in the corpus. This has been observed for Gender and Origin.

Category	#	Distribution
Age	717	from new born to 98 y.o.
Gender	727	317 feminine, 410 masculine
Outcome	678	227 recovery, 256 improvement, 55 stable, 23 worsening, 117 death
Origin	722	722 distinct spans of text

Table 4: Number of mentions for the general information annotations on the whole dataset of 717 files

Nevertheless, apart from the Gender category, other general annotations are not found in all files: Origin is present in 716 files (99.9% of files), Age in 698 files (97.4%), and Outcome in 675 files (94.1%). Annotations are missing when it was not possible to identify the information.

3.3 Annotated clinical case report

Figure 2 shows the following clinical case: *A 73-year-old woman who had only one child by caesarean section, but had for several years a*

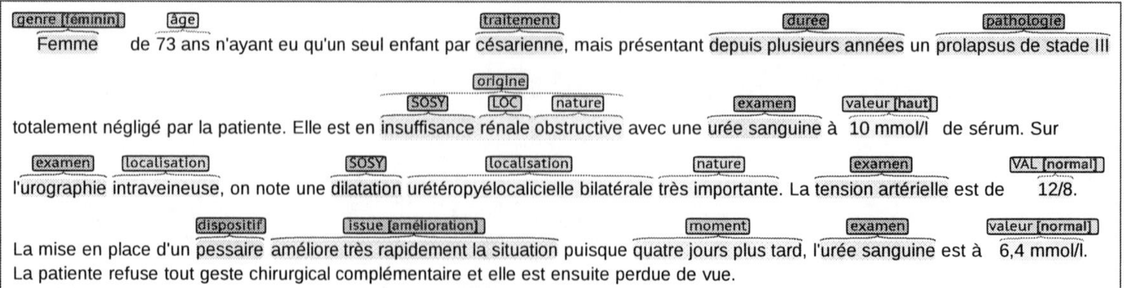

Figure 2: Annotated case report. General information includes the following tags: *genre* (gender), *âge* (age), *origine* (origin), *issue* (outcome). Other tags are related to fine-grained information. Normalized values appear between square brackets (feminine gender, high or normal values, improvement outcome)

stage III prolapse totally neglected by the patient. She is in obstructive renal failure with blood urea at 10 mmol/l serum. On the intravenous urography, we notice a very significant bilateral ureteropyelocaliceal dilation. The blood pressure is 12/8. The pessary placement very quickly improves the situation since four days later, the blood urea is 6.4 mmol/l. The patient refuses any additional surgery and is then lost to follow-up. The case is annotated with general and fine-grained information. Elements in square brackets correspond to normalized tags: feminine *("féminin")* for gender, high *("haut")* and normal for values, and improvement *("amélioration")* for outcome.

3.3.1 Types of information in the typical clinical case report

This case report is composed of several parts, annotated as follows:

- *general description with patient history:* gender (woman, *"femme"*); age (73-year-old, *"73 ans"*); surgical treatment (caesarean, *"césarienne"*); duration (for several years, *"depuis plusieurs années"*); pathology (stage III prolapse, *"prolapsus de stade III"*)

- *origin of consultation, tests and results:* origin (obstructive renal failure, *"insuffisance rénale obstructive"*), composed of three elements: sign or symptom (failure, *"insuffisance"*), localization (renal, *"rénale"*), and nature (obstructive, *"obstructive"*); three tests (blood urea, *"urée sanguine"*, blood pressure, *"tension artérielle"*, urography, *"urographie"*) with lab results (10 mmol/l, 12/8, 6.4 mmol/l) and localization (intravenous, *"intraveineuse"*); sign or symptom (dilation, *"dilatation"*) with localization (bilateral ureteropyelocaliceal, *"urétéropyélocalicielle bilatérale"*) and nature (very significant, *"très importante"*)

- *surgical treatment and issue:* medical device (pessary, *"pessaire"*); outcome (very quickly improves the situation, *"améliore très rapidement la situation"*); moment (four days later, *"quatre jours plus tard"*)

- *follow-up:* no annotation in this clinical case

We observe the types of information contained in clinical case reports are similar to those typically provided by patient health documents in hospitals.

3.3.2 Distribution of annotations

Columns two and three from Table 4 indicate that general information are found in all clinical cases. For gender and origin, the number of annotations if higher than the number of clinical cases because several people are described in some cases (gender), and because several origins of consultation may be indicated (namely, several signs or symptoms).

From Table 3, one can observe a very imbalanced number of annotations per category. The main categories are: signs or symptoms (15.4%), tests (15.1%), localizations (11.6%), substances (8.4%), and anatomical parts (8.2%). The number of signs and symptoms mentions are three times higher than annotations of diseases (5.5%). Small categories are related to specific data (especially body measurements and vital signs) that are indicated in a limited number of cases. This may correspond to the average difference with the clinical patient reports.

4 Experiments and analysis

The annotated corpus has been exploited to perform similar annotations automatically and for their evaluation. Our aim is to verify the adequateness of the annotations for this information extraction task, as well as to serve as baseline for future work. We specify we do not aim to provide new methods, nor to improve existing systems, but to present a few use cases that may be done on the annotations presented in section 2.

4.1 Linguistic analysis

Syntax. Depending on the outcome observed in clinical cases, we studied the distribution of a few verbal tenses based on the POS annotations provided by the TreeTagger system (Schmid, 1994). As presented in Table 5, past perfect is the main tense for death outcome while present is the main tense for both improvement and stable condition outcomes. Conversely, we observe no future tense in case reports concerned by death.

Verbal tense	R	I	S	W	D
future	0.01	0.01	0.01	**0.02**	0.00
imperfect	**0.19**	0.15	0.12	0.13	0.16
past perfect	0.41	0.41	0.41	0.41	**0.45**
present	0.23	**0.29**	**0.29**	0.24	0.26

Table 5: Percentage of verbal tenses use depending on the outcome value (R=recovery, I=improvement, S=stable, W=worsening, D=death)

Table 6 presents the distribution of demonstrative pronouns (PRO:dem) vs. personal pronouns (PRO:per) depending on the outcome. We observe that impersonal linguistic constructions are mainly used for stable condition outcomes (less personal pronouns and more demonstrative pronouns) than in other outcome types, as if the uncertainty of the stable condition (no improvement nor worsening) would prevent from a too much personal representation of the case.

POS tag	R	I	S	W	D
PRO:dem	0.19	0.19	**0.24**	0.21	0.18
PRO:per	**0.52**	0.50	0.47	0.48	0.51

Table 6: Percentage of types of pronoun use (PRO:dem=demonstrative, PRO:per=personal) depending on the outcome value (R=recovery, I=improvement, S=stable, W=worsening, D=death)

Semantics. Table 7 presents the main elements annotated as anatomical parts, pathologies, signs or symptoms, and surgical treatments depending on the gender. The observed differences of medical entities mainly highlight differences due to anatomical parts specific to men or women, or to distinct prevalences of pathologies. We observe less differences in surgical treatments than in other categories.

Category	F/M	Annotated spans
Anatomy	F	kidney, bladder, torso
	M	testicle, bladder, prostate
Pathology	F	acute pyelonephristis, adenocarcinoma, carcinoma, edema, mydriasis, tumor
	M	adenocarcinoma, fistula, rhabdomyosarcoma, tuberculosis, tumor, ulcer
Signs or Symptoms	F	dilation, hematuria, hypersensitivity, lesion, mass, pain, rash, stone, vomiting
	M	fever, infection, lesion, mass, nodule, pain, pneumonia, relapse, retention, trouble
Treatments	F	chemotherapy, curettage, desensitization, exeresis, lumpectomy, nephrectomy
	M	ablation, chemotherapy, clamping, desensitization, exeresis, orchiectomy, plasma exchange, resection

Table 7: Most used anatomical parts, pathologies, signs or symptoms, and surgical treatments depending on the gender (F=feminine, M=masculine)

4.2 Information extraction

The information extraction experiments rely on the Wapiti tool (Lavergne et al., 2010) that implements linear chain CRF (Lafferty et al., 2001). We trained a model on the 16 fine-grained categories presented in Table 2, through a 10 fold cross-validation process, using a $l1$ regularization. We used the following features: unigrams and bigrams of tokens, number of characters, typographic case, presence of punctuation and digit, Soundex code[1] value of each token, relative position of token within the document (beginning, middle, end), POS tags from the TreeTagger system (Schmid, 1994) and syntactic chunks based on those tags, presence of the token in a dictionary of 251k inflected forms for French, and cluster id (120 classes) of each token using the clustering algorithm from Brown et al. (1992) implemented by Liang (2005). The results that we achieved are presented in table 8. Overall, we obtain 0.76 Precision, 0.45 Recall and 0.67 F-measure.

Category	P	R	F
Anatomy	0.7260	0.4823	0.5795
Concentration	0.5000	0.0714	0.1250
Date	1.0000	0.4507	0.6214
Devices	0.3077	0.0548	0.0930
Dose	0.7805	0.5818	0.6667
Duration	0.9545	0.2692	0.4200
Examen.	0.8308	0.6303	0.7168
Function	0.8889	0.2162	0.3478
Frequency	0.9630	0.1955	0.3250
Localisation	0.7812	0.5795	0.6654
Mode	0.8929	0.5245	0.6608
Pathology	0.5918	0.2086	0.3085
SOSY	0.6067	0.3639	0.4549
Specialty	1.0000	0.3846	0.5556
Substance	0.8490	0.3721	0.5175
Treatment	0.8190	0.3785	0.5177
Overall	0.7640	0.4492	0.5658

Table 8: Results achieved using a CRF through a 10 folds cross-validation

5 Discussion

Corpus.   One contribution of this work is related to the availability of the annotated corpus from the medical domain for French. We based our annotation schema on both existing ones (semantic types from the UMLS, i2b2 NLP Challenges) and on types of elements found in our corpus. This annotated corpus will be made available for the research purposes and may be of interest for several NLP tasks related to the biomedical domain: information extraction, relationships identification, classification, discourse analysis, temporality, etc.

Human annotations vs. CRF. We observed that results obtained by the designed CRF system are in line with results obtained by humans when annotating the corpus. More specifically, while humans were producing the gold standard, they had to deal with categories harder to process than others. We also observe that those categories

are generally difficult to retrieve and annotate with the CRF model as well: Concentration (F=0.38 vs. 0.13), Function (F=0.37 vs. 0.35), and Pathology (F=0.36 vs. 0.31). An explanation is the lack of regularity (for the CRF system) and ambiguous content w.r.t. content from other categories.

Yet, two categories considered as hard for humans yielded better results than expected with the CRF model: Specialty (F=0.25 vs. 0.56) and Dates (F=0.40 vs. 0.67). The differences observed between humans which produce those bad results were mainly due to omissions. Conversely, humans outperformed the CRF model on Frequency (F=0.68 vs. 0.33), Duration (F=0.66 vs. 0.42), and Devices (F=0.46 vs. 0.09). Those categories are composed of distinct elements with low frequencies of use which are complex to process for a probability-based system, but basic for humans.

As future work, we plan to continue the fine-grained annotation of the whole corpus. We also plan to define relationships between the existing entities, in order to provide annotations of relations. Despite the absence of relationships annotations, the corpus can still serve to perform unsupervised experiments. Such results may be used for automatic pre-annotation of relationships, in order to make it easier the human annotation work.

6 Conclusion

In this paper, we presented a corpus composed of 717 medical clinical case reports, written in French, with two levels of annotations (general and fine-grained annotations). Our annotation schema is composed of four general categories (age, gender, outcome, origin) for a total of 2,835 annotations, and 27 fine-grained categories dealing with five domains (physiology, surgery, diseases, drugs, temporal) for a total of 5,198 annotations on a subset of 70 files. For certain categories, the annotations are provided under a normalized format (age, gender, outcome) while other categories are associated with additional information based on a human judgement, either of linguistic nature (assertions, change of conditions) or medical nature (lab results compared to known ranges). The corpus and its annotations will be made available for the research. We expect that the availability of this corpus may boost the research on biomedical textual data in French, and provide the domain with more robust and stable tools leading to a better reproducibility of the results.

Acknowledgments

This work has been founded by the French ANR (grant number ANR-17-CE19-0016-01) as part of the project CLEAR (Communication, Literacy, Education, Accessibility, Readability) and by the French government support, granted to the Comin-Labs LabEx, managed by the ANR in Investing for the Future program (grant number ANR-10-LABX-07-01).

References

Dwight Atkinson. 1992. The evolution of medical research writing from 1735 to 1985: The case of the Edinburgh medical journal. *Applied Linguistics*, 13(4):337–74.

Peter F Brown, Vincent J Della Pietra, Peter V de Souza, Jenifer C Lai, and Robert L Mercer. 1992. Class-based n-gram models of natural language. *Computational Linguistics*, 18(4):467–79.

Wendy W Chapman, Prakash M Nadkarni, Lynette Hirschman, Leonardo W D'Avolio, Guergana K Savova, and Özlem Uzuner. 2011. Overcoming barriers to NLP for clinical text: the role of shared tasks and the need for additional creative solutions. *J Am Med Inform Assoc*, 18(5):540–3.

Franck Dernoncourt, Ji Young Lee, Özlem Uzuner, and Peter Szolovits. 2017. De-identification of patient notes with recurrent neural networks. *J Am Med Inform Assoc*, 24(3):596–606.

David Doermann and Shee Yao. 1995. Generating synthetic data for text analysis systems. In *Symposium on Document Analysis and Information Retrieval*, Las Vegas, USA.

Steffen Eger, Gözde Gül Şahin, Andreas Rücklé, Ji-Ung Lee, Claudia Schulz, Mohsen Mesgar, Krishnkant Swarnkar, Edwin Simpson, and Iryna Gurevych. 2019. Text processing like humans do: Visually attacking and shielding NLP systems. ArXiv:1903.11508v1.

Natalia Grabar, Vincent Claveau, and Clément Dalloux. 2018. Cas: French corpus with clinical cases. In *Proc of LOUHI*, pages 122–128, Brussels, Belgium.

Cyril Grouin, Sophie Rosset, Pierre Zweigenbaum, Karën Fort, Olivier Galibert, and Ludovic Quintard. 2011. Proposal for an extension of traditional named entities: From guidelines to evaluation, an overview. In *Proc of Linguistic Annotation Workshop (LAW-V)*, pages 92–100, Portland, OR. Association for Computational Linguistics.

Harsha Gurulingappa, Abdul Mateen-Rajput, Angus Roberts, Juliane Fluck, Martin Hofmann-Apitius, and Luca Toldo. 2012. Development of a benchmark corpus to support the automatic extraction of drug-related adverse effects from medical case reports. *J Biomed Inform*, 45(5):885–92.

Alistair EW Johnson, Tom J Pollard, Lu Shen, Li-Wei H Lehman, Mengling Feng, Mohammad Ghassemi, Benjamin Moody, Peter Szolovits, Leo Anthony Celi, and Roger G Mark. 2016. MIMIC-III, a freely accessible critical care database. *Scientific Data*, 3.

Mehmet Kayaalp. 2017. Modes of de-identification. In *AMIA Annu Symp Proc*, pages 1044–50, San Francisco, USA.

John D. Lafferty, Andrew McCallum, and Fernando C. N. Pereira. 2001. Conditional Random Fields: Probabilistic models for segmenting and labeling sequence data. In *Proc of ICML*, pages 282–9, Williamstown, MA.

Thomas Lavergne, Olivier Cappé, and François Yvon. 2010. Practical very large scale CRFs. In *Proc of ACL*, pages 504–13, Uppsala, Sweden.

Percy Liang. 2005. Semi-supervised learning for natural language. Master's thesis, Massachusetts Institute of Technology.

Donald A. Lindberg, Betsy L. Humphreys, and Alexa T. McRay. 1993. The Unified Medical Language System. *Methods Inf Med*, 32(4):281–91.

Christina Lohr, Sven Buechel, and Udo Hahn. 2018. Sharing copies of synthetic clinical corpora without physical distribution – A case study to get around IPRs and privacy constraints featuring the German JSynCC corpus. In *Proc of LREC*, pages 1259–66, Miyazaki, Japan.

Yuliia Lysanets, Halyna Morokhovets, and Olena Bieliaieva. 2017. Stylistic features of case reports as a genre of medical discourse. *J Med Case Rep*, 11:83.

Stéphane M Meystre, F Jeffrey Friedlin, Brett R South, Shuying Shen, and Matthew H Samore. 2010. Automatic de-identification of textual documents in the electronic health record: a review of recent research. *BMC Med Res Method*, 10(70).

Prakash M Nadkarni, Lucila Ohno-Machado, and Wendy W Chapman. 2011. Natural languague processing: an introduction. *J Am Med Inform Assoc*, 18(5):544–51.

Aurélie Névéol, Cyril Grouin, Jeremy Leixa, Sophie Rosset, and Pierre Zweigenbaum. 2014. The QUAERO French medical corpus: A ressource for medical entity recognition and normalization. In *Proc of BioTextMining Work*, pages 24–30.

Richard A Rison. 2013. A guide to writing case reports for the journal of medical case reports and biomed central research notes. *J Med Case Rep*, 7:239.

Richard A Rison, Jennifer Kelly Shepphird, and Michael R Kidd. 2017. How to choose the best journal for your case report. *J Med Case Rep*, 11:198.

Yoichi Satomura and Marcio Biczyk Do Amaral. 1992. Automated diagnostic indexing by natural language processing. *Med Inform*, 17(3):149–63.

Helmut Schmid. 1994. Probabilistic part-of-speech tagging using decision trees. In *Proc of International Conference on New Methods in Language*.

F Sebastiani. 2002. Machine learning in automated text categorization. *ACM Computing Surveys*, 34(1):1–47.

Özlem Uzuner, Imre Solti, and Eithon Cadag. 2010. Extracting medication information from clinical text. *J Am Med Inform Assoc*, 17(5):514–8.

Özlem Uzuner, Brett R South, Shuying Shen, and Scott L. DuVall. 2011. 2010 i2b2/VA challenge on concepts, assertions, and relations in clinical text. *J Am Med Inform Assoc*, 18(5):552–6.

Two-stage Federated Phenotyping and Patient Representation Learning

Dianbo Liu
CHIP
Boston Children's Hospital
Harvard Medical School
Boston, MA, USA, 02115
Dianbo.liu@childrens.harvard.edu

Dmitriy Dligach
loyola university chicago
Chicago, IL ,USA 60660
ddligach@luc.edu

Timothy Miller
CHIP
Boston Children's Hospital
Harvard Medical School
Boston, MA, USA, 02115
Timothy.Miller@childrens.harvard.edu

Abstract

A large percentage of medical information is in unstructured text format in electronic medical record systems. Manual extraction of information from clinical notes is extremely time consuming. Natural language processing has been widely used in recent years for automatic information extraction from medical texts. However, algorithms trained on data from a single healthcare provider are not generalizable and error-prone due to the heterogeneity and uniqueness of medical documents. We develop a two-stage federated natural language processing method that enables utilization of clinical notes from different hospitals or clinics without moving the data, and demonstrate its performance using obesity and comorbities phenotyping as medical task. This approach not only improves the quality of a specific clinical task but also facilitates knowledge progression in the whole healthcare system, which is an essential part of learning health system. To the best of our knowledge, this is the first application of federated machine learning in clinical NLP.

1 Introduction

Clinical notes and other unstructured data in plain text are valuable resources for medical informatics studies and machine learning applications in healthcare. In clinical settings, more than 70% of information are stored as unstructured text. Converting the unstructured data into useful structured representations will not only help data analysis but also improve efficiency in clinical practice (Jagannathan et al., 2009; Kreimeyer et al., 2017; Ford et al., 2016; Demner-Fushman et al., 2009; Murff et al., 2011; Friedman et al., 2004). Manual extraction of information from the vast volume of notes from electronic health record (EHR) systems is too time consuming.

To automatically retrieve information from unstructured notes, natural language processing (NLP) has been widely used. NLP is a subfield of computer science, that has been developing for more than 50 years, focusing on intelligent processing of human languages (Manning et al., 1999). A combination of hard-coded rules and machine learning methods have been used in the field, with machine learning currently being the dominant paradigm.

Automatic phenotyping is a task in clinical NLP that aims to identify cohorts of patients that match a predefined set of criteria. Supervised machine learning is curently the main approach to phenotyping, but availability of annotated data hinders the progress for this task. In this work, we consider a scenario where multiple instituitions have access to relatively small amounts of annotated data for a particular phenotype and this amount is not sufficient for training an accurate classifier. On the other hand, combining data from these institutions can lead to a high accuracy classifier, but direct data sharing is not possible due to operational and privacy concerns.

Another problem we are considering is learning patient representations that can be used to train accurate phenotyping classifiers. The goal of patient representation learning is mapping the text of notes for a patient to a fixed-length dense vector (embedding). Patient representation learning has been done in a supervised (Dligach and Miller, 2018) and unsupervised (Miotto et al., 2016) setting. In both cases, patient representation learning requires massive amounts of data. As in the scenario we outlined in the previous paragraph, combining data from several institutions can lead to higher quality patient representations, which in turn will improve the accuracy of phenotyping classifiers. However, direct data sharing, again, is difficult or impossible.

Proceedings of the BioNLP 2019 workshop, pages 283–291
Florence, Italy, August 1, 2019. ©2019 Association for Computational Linguistics

To tackle the challenges we mentioned above, we developed a federated machine learning method to utilize clinical notes from multiple sources, both for learning patient representations and phenotype classifiers.

Federated machine learning is a concept that machine learning models are trained in a distributed and collaborative manner without centralised data (Liu et al., 2018a; McMahan et al., 2016; Bonawitz et al., 2019; Konečný et al., 2016; Huang et al., 2018; Huang and Liu, 2019). The strategy of federated learning has been recently adopted in the medical field in structured data-based machine learning tasks (Liu et al., 2018a; Huang et al., 2018; Liu et al., 2018b). However, to the best of our knowledge, this work is the first time a federated learning strategy has been used in medical NLP.

We developed our two-stage federated natural language processing method based on previous work on patient representation (Dligach and Miller, 2018). The first stage of our proposed federated learning scheme is supervised patient representation learning. Machine learning models are trained using medical notes from a large number of hospitals or clinics without moving or aggregating the notes. The notes used in this stage need not be directly relevant to a specific medical task of interest. At the second stage, representations from the clinical notes directly related to the phenotyping task are extracted using the algorithm obtained from stage 1 and a machine learning model specific to the medical task is trained.

Clinicians spend a significant amount of time reviewing clinical notes. This time can be saved or reduced with reasonably designed NLP technologies. One such task is phenotying from medical notes. In this study, we demonstrated, using phenotyping from clinical note as a clinical task (Conway et al., 2011; Dligach and Miller, 2018), that the method we developed will make it possible to utilize notes from a wide range of hospitals without moving the data.

The ability to utilize clinical notes distributed at different healthcare providers not only benefits a specific clinical practice task but also facilitates building a learning healthcare system, in which meaningful use of knowledge in distributed clinical notes will speed up progression of medical knowledge to translational research, tool development, and healthcare quality assessment (Friedman et al., 2010; Blumenthal and Tavenner, 2010). Without the needs of data movement, the speed of information flow can approach real time and make a rapid learning healthcare system possible (Slutsky, 2007; Friedman et al., 2014; Abernethy et al., 2010).

2 Methods

2.1 Study Cohorts

Two datasets were used in this study. The MIMIC-III corpus (Johnson et al., 2016) was used for representation learning. This corpus contains information for more than 58,000 admissions for more than 45,000 patients admitted to Beth Israel Deaconess Medical Center in Boston between 2001 and 2012. Relevant to this study, MIMIC-III includes clinical notes, ICD9 diagnostic codes, ICD9 procedure codes, and CPT codes. The notes were processed with cTAKES[1] to extract UMLS[2] unique concept identifiers (CUIs). Following the cohort selection protocol from (Dligach and Miller, 2018), patients with over 10,000 CUIs were excluded from this study. We obtained a cohort of 44,211 patients in total.

The Informatics for Integrating Biology to the Bedside (i2b2) Obesity challenge dataset was used to train phenotyping models (Uzuner, 2009). The dataset consists of 1237 discharge summaries from Partners HealthCare in Boston. Patients in this cohort were annotated with respect to obesity and its comorbidities. In this study we consider the more challenging *intuitive* version of the task. The discharge summaries were annotated with obesity and its 15 most common comorbidities, the presence, absence or uncertainty (questionable) of which were used as ground truth label in the phenotyping task in this study. Table 1 shows the number of examples of each class for each phenotype. Thus, we build phenotyping models for 16 different diseases.

2.2 Data Extraction and feature choice

At the representation learning stage (stage 1), all notes for a patient were aggregated into a single document. CUIs extracted from the text were used as input features. ICD-9 and CPT codes for the patient were used as labels for supervised representation learning.

---

[1]https://ctakes.apache.org
[2]https://www.nlm.nih.gov/research/umls/

Table 1: i2b2 cohort of obesity comorbidities

Disease	#Absence	#Presence	#Questionable
Asthma	86	596	0
CAD	391	265	5
CHF	308	318	1
Depression	142	555	0
Diabetes	473	205	5
GERD	144	447	1
Gallstones	101	609	0
Gout	94	616	2
Hypercholesterolemia	315	287	1
Hypertension	511	127	0
Hypertriglyceridemia	37	665	0
OA	117	554	1
OSA	99	606	8
Obesity	285	379	1
PVD	110	556	1
Venous Insufficiency	54	577	0

At the phenotyping stage (stage 2), CUIs extracted from the discharge summaries were used as input features. Annotations of being present, absent, or questionable for each of the 16 diagnoses for each patient were used as multi-class classification labels.

2.3 Two-stage federated natural language processing of clinical notes

We envision that clinical textual data can be useful in at least two ways: (1) for pre-training patient representation models, and (2) for training phenotyping models.

In this study, a patient representation refers to a fixed-length vector derived from clinical notes that encodes all essential information about the patient. A patient representation model trained on massive amounts of text data can be useful for a wide range of clinical applications. A phenotyping model, on the other hand, captures the way a specific medical condition works, by learning the function that can predict a disease (e.g., asthma) from the text of the notes.

Until recently, phenotyping models have been trained from scratch, omitting stage (1), but recent work (Dligach and Miller, 2018) included a pre-training step, which derived dense patient representations from data linking large amounts of patient notes to ICD codes. Their work showed that including the pre-training step led to learning patient representations that were more accurate for a number of phenotyping tasks.

Our goal here is to develop methods for federated learning for both (1) pre-training patient representations, and (2) phenotyping tasks. These methods will allow researchers and clinicans to utilize data from multiple health care providers, without the need to share the data directly, obviating issues related to data transfer and privacy.

To achieve this goal, we design a two-stage federated NLP approach (Figure 1). In the first stage, following (Dligach and Miller, 2018), we pre-train a patient representation model by training an artificial neural network (ANN) to predict ICD and CPT codes from the text of the notes. We extend the methods from (Dligach and Miller, 2018) to facilitate federated training.

In the second stage, a phenotyping machine learning model is trained in a federated manner using clinical notes that are distributed across multiple sites for the target phenotype. In this stage, the notes mapped to fixed-length representations from stage (1) are used as input features and whether the patient has a certain disease is used as a label with one of the three classes: presence, absence or questionable.

In the following sections, we first describe a simple notes pre-processing step. We then discuss the method for pre-training patient representations and the method for training phenotyping models. Finally, we describe our framework for performing the latter two steps in a federated manner.

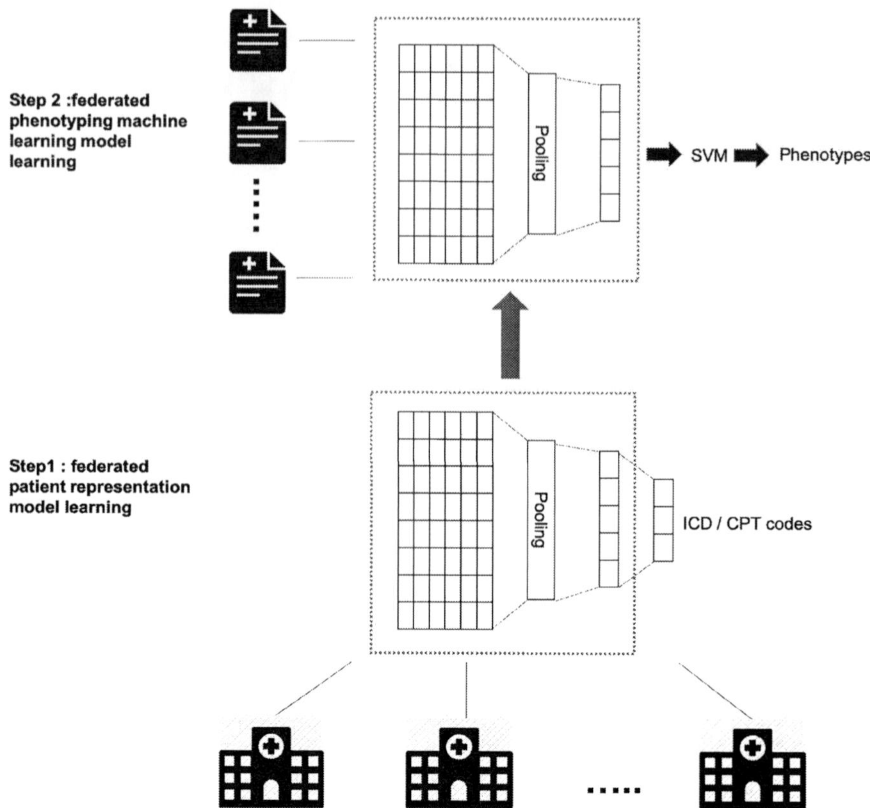

Figure 1: Two stage federated natural language processing for clinical notes phenotyping. In the first stage, a patient representation model was trained using an artificial neural network (ANN) to predict ICD and CPT codes from the text of the notes from a wide range of healthcare providers. The model without output layer was then used as "representation extractor" in the next stage. In the second stage, a phenotyping support vector machine model was trained in a federated manner using clinical notes for the target phenotype distributed across multiple silos.

2.4 Pre-processing

All of our models rely on standardized medical vocabulary automatically extracted from the text of the notes rather than on raw text.

To obtain medically relevant information from clinical notes, Unified Medical Language System (UMLS) concept unique identifiers (CUIs) were extracted from each note using Apache cTAKES (https://ctakes.apache.org). UMLS is a resource that brings together many health and biomedical vocabularies and standardizes them to enable interoperability between computer systems.

The Metathesaurus is a large, multi-purpose, and multi-lingual vocabulary that contains information about biomedical and health related concepts, their various names, and the relationships among them. The Metathesaurus structure has four layers, Concept Unique Identifies(CUIs), Lexical (term) Unique Identifiers (LUI), String Unique Identifiers (SUI) and Atom Unique Iden-

tifiers (AUI). In this study, we focus on CUIs, in which a concept is a medical meaning. Our models use UMLS CUIs as input.

2.5 Representation learning

We adapted the architecture from (Dligach and Miller, 2018) for pre-training patient representations. A deep averaging network (DAN) that consists of an embedding layer, an average pooling layer, a dense layer, and multiple sigmoid outputs, where each output corresponds to an ICD or CPT code being predicted.

This architecture takes CUIs as input and is trained using binary cross-entropy loss function to predict ICD and CPT codes. After the model is trained, the dense layer can be used to represent a patient as follows: the model weights are frozen and the notes of a new patient are fed into the network; the patient representation is collected from the values of the units of the dense layer. Thus, the

Stage 1

Input: MIMIC3 data clinical notes distributed at 10 simulated sites
Representation learning model

Output: 174 ICD or CPT codes

Extract CUIs from each patient's clinical notes using cTAKE.

for $t \in 1$ **to** T **do**

    for $k \in 1$ **to** K *in parallel* **do**
       | Train patient representation learning model f_k
    end

    aggregate models from all sites by $W_{ag}^t = \sum_{k=1}^{K} \frac{n_k}{N} w_k^t$
end

;

Stage 2

Input: i2b2 clinical notes for obesity comorbidities distributed at 3 sites
phenotyping machine learning model

Output: 1 single binary output (one of the comorbidities)

Extract CUIs from each clinical notes using cTAKES.

for $t \in 1$ **to** T' **do**

    for $k \in 1$ **to** K' *in parallel* **do**
       | Train phenotyping model f_k'
    end

    aggregate models from all sites by $W_{ag}'^t = \sum_{k=1}^{K'} \frac{n_k'}{N'} w_k'^t$
end

Algorithm 1: Two-stage federated natural language processing

text of the notes is mapped to a fixed-length vector using a pre-trained deep averaging network.

2.6 Phenotyping

A linear kernel Support Vector Machine (SVM) taking input from representations generated using the pre-trained model from stage 1 was used as the classifer for each phenotype of interest. No regularization was used for the SVM and stochastic gradient descent was used as the optimization algorithm.

2.7 Federated machine learning learning on clinical notes

To train the ANN model in either stage 1 or stage 2, we simulated sending out models with identical initial parameters to all sites such as hospitals or clinics. At each site, a model was trained using only data form that site. Only model parameters of the models were then sent back to the analyzer for aggregation but not the original training data. An updated model is generated by averaging the parameters of models distributively trained, weighted by sample size (Konečný et al., 2016; McMahan et al., 2016). In this study, sample size is defined as the number of patients.

After model aggregation, the updated model was sent out to all sites again to repeat the global training cycle (Algorithm 1). Formally, the weight update is specified by:

$$W_{ag}^t = \sum_{k=1}^{K} \frac{n_k}{N} W_k^t \tag{1}$$

where W_{ag} is the parameter of aggregated model at the analyzer site, K is the number of data sites, in this study the number of simulated healthcare providers or clinics. n_i is the number of samples at the ith site, N is the total number of samples across all sites, and W_i is the parameters learned from the ith data site alone. t is the global cycle number in the range of [1,T]. The algorithm tries to minimize the following objective function:

$$argmin_f(-\sum_{j=1}^{N}\sum_{p=1}^{M}[y_{jp}logf(x_{jp})+$$
$$(1-y_{jp})log(1-f(x_{jp}))])$$

Where x_j is the feature vector of CUIs. and y is the class label. p is the output number and M is the total number of outputs. f is the machine learning model such as artificial neural network or SVM.Codes that accompany this article can be found at our github repository [3].

3   Experiments

To imitate real world medical setting where data are distributed with different healthcare providers, we randomly split patients in MIMIC-III data into 10 sites for stage 1 (federated representation learning). The training data of i2b2 was split into 3 sites for stage 2 (phenotype learning) to mimic obesity related notes distributed with three different healthcare providers. i2b2 notes were not included in the representation learning as in clinic settings information exchange routes for disease-specific records are often not the same as general medical information and ICD/CPT codes were not available for i2b2 dataset.

Experiments were designed to answer three questions:

1. Whether clinical notes distributed in different silos can be utilized for patient representation learning without data sharing

2. Whether utilizing data from a wide range of sources will help improve performance of phenotyping from clinical notes

3. Whether models trained in a two-stage federated manner will have inferior performance to models trained with centralized data.

To answer these questions, two-stage NLP algorithms were trained. Performance of models trained using only i2b2 notes from one of the

three sites were compared with two-stage federated NLP results. Furthermore, performance of machine learning models using distributed or centralized data at patient representation learning stage or phenotyping stage were compared.

4   Results

4.1   Two-stage federated natural language processing improves performance of automatic phenotyping

We looked at the scenarios where no representation learning was performed. In those cases, the standard TF-IDF weighted sparse bag-of-CUIs vectors were used to represent i2b2 notes. The sparse vectors were used as input into the phenotyping SVM model. We also looked at the scenarios where representation learning was performed by predicting ICD codes. For each of these conditions, we trained our phenotyping models using centralized vs. federated learning. Finally, we considered a scenario where the phenotyping model was trained using the notes from a single site (the metrics we report were averaged across three sites).

To summarize, seven experiments were conducted:

1. No representation learning + centralized phenotyping learning

2. No representation learning + federated phenotyping learning where i2b2 training data were randomly split into 3 silos

3. No representation learning + single source phenotyping learning, where i2b2 data were randomly split into 3 silos, but phenotyping algorithm was only trained using data from one of the silos

4. Centralized representation learning + centralized phenotyping learning

5. Centralized representation learning + federated phenotyping learning

6. Federated representation learning + centralized phenotyping learning,where MIMIC-III data were randomly split into 10 silos

7. Federated representation learning + federated phenotyping learning, where MIMIC-III data were randomly split into 10 silos and i2b2 data into 3 silos (Table 2).

Table 2: Performance of different experiments

Experiment	Patient representations	Phenotyping	Precision	Recall	F1
1	Bag-of-CUIs	Centralized	0.649	0.627	0.634
2	Bag-of-CUIs	Federated	0.650	0.623	0.632
3	Bag-of-CUIs	Single source	0.552	0.540	0.542
4	Centralized learned	Centralized	0.749	0.714	0.726
5	Centralized learned	Federated	0.743	0.713	0.723
6	Federated learned	Centralized	0.729	0.716	0.715
7	Federated learned	Federated	**0.753**	**0.715**	**0.724**

Table 3: Performance of two-stage federated NLP in obesity comobidity phenotyping by disease

Disease	Prec	Rec	F1
Asthma	0.941	0.919	0.930
CAD	0.605	0.606	0.605
CHF	0.583	0.588	0.585
Depression	0.844	0.774	0.801
Diabetes	0.879	0.873	0.876
GERD	0.578	0.543	0.558
Gallstones	0.775	0.619	0.650
Gout	0.948	0.929	0.938
Hypercholesterolemia	0.891	0.894	0.892
Hypertension	0.877	0.854	0.865
Hypertriglyceridemia	0.725	0.519	0.524
OA	0.531	0.520	0.525
OSA	0.627	0.594	0.609
Obesity	0.900	0.894	0.897
PVD	0.590	0.604	0.596
Venous Insufficiency	0.763	0.712	0.734
Average	0.753	0.715	0.724

The results of our experiments are shown in Table 3. First of all, we looked at whether phenotyping model training can be conducted in a federated manner without compromising performance. When only i2b2 data from one of three silos was used for phenotyping training (experiment 3), the F1 score of 0.542 was achieved. When data from all three i2b2 sites were used for phenotyping model training (experiment 1) the F1 score improved to 0.634, which suggests that more data did improve the model. If we assume data from the three i2b2 silos can not be moved and aggregated together, the model trained in a federated manner (experiment 2) achieved a comparable F1 score of 0.632. This suggested federated learning worked for phenotyping model training.

Previous work showed that using learned representations from clinical notes from a different source using a transfer learning strategy helps to improve the performance of phenotyping NLP models (Dligach and Miller, 2018). When patient representations learned from centralized MIMIC-III notes were used as features and centralized phenotyping training was conducted (experiment 4), the phenotyping performance increased significantly with F1 score of 0.714, which was consistent with previous findings (Dligach and Miller, 2018).

When a federated approach was applied in both representation learning and phenotyping stages, the algorithm achieved F1 score of 0.724. It is worth pointing out that F1 scores from experiment 7 , where both representation and phenotyping training were conducted in a federated manner, were not statistically different from F1 scores of experiment 4 over multiple rounds of experiment using different data shuffling and initialization. In comparison, when only data from a single simulated silo was used, the average F1 score 0.634. When the centralized approach was taken at both stages, the precision, recall and F1 score were 0.718, 0.711 and 0.714 respectively. These results suggested utilizing clinical notes from different silos in a federated manner did improve accuracy of the phenotyping NLP algorithm, and the performance is comparable to NLP trained on centralized data. The performance of federated NLP on each single obesity commodity were shown in Table 3. It is necessary to point out that it was impractical to conduct federated phenotyping training when the number of "questionable" cases for many diseases are small (Table 1). This is true for many diseases in the i2b2 dataset. In such situation, "questionable" cases were excluded from the training and testing process. Instead of 3-class classification, a 2-class binary classification of "presence" or "absence" were conducted. Therefore, the performance metrics can not be directly compared with results in the original i2b2 challenge, though the scores were similar.

5   Conclusion

In this article, we presented a two-stage method that conducts patient representation learning and obesity comorbidity phenotyping, both in a federated manner. The experimental results suggest that federated training of machine learning models on distributed datasets does improve performance of NLP on clinical notes compared with algorithms trained on data from a single site. In this study, we used CUIs as input features into machine learning models, but the same federated learning strategies can also be applied to raw text.

References

Amy P Abernethy, Lynn M Etheredge, Patricia A Ganz, Paul Wallace, Robert R German, Chalapathy Neti, Peter B Bach, and Sharon B Murphy. 2010. Rapid-learning system for cancer care. *Journal of Clinical Oncology*, 28(27):4268.

David Blumenthal and Marilyn Tavenner. 2010. The meaningful use regulation for electronic health records. *New England Journal of Medicine*, 363(6):501–504.

Keith Bonawitz, Hubert Eichner, Wolfgang Grieskamp, Dzmitry Huba, Alex Ingerman, Vladimir Ivanov, Chloe Kiddon, Jakub Konecny, Stefano Mazzocchi, H Brendan McMahan, et al. 2019. Towards federated learning at scale: System design. *arXiv preprint arXiv:1902.01046*.

Mike Conway, Richard L Berg, David Carrell, Joshua C Denny, Abel N Kho, Iftikhar J Kullo, James G Linneman, Jennifer A Pacheco, Peggy Peissig, Luke Rasmussen, et al. 2011. Analyzing the heterogeneity and complexity of electronic health record oriented phenotyping algorithms. In *AMIA annual symposium proceedings*, volume 2011, page 274. American Medical Informatics Association.

Dina Demner-Fushman, Wendy W Chapman, and Clement J McDonald. 2009. What can natural language processing do for clinical decision support? *Journal of biomedical informatics*, 42(5):760–772.

Dmitriy Dligach and Timothy Miller. 2018. Learning patient representations from text. *arXiv preprint arXiv:1805.02096*.

Elizabeth Ford, John A Carroll, Helen E Smith, Donia Scott, and Jackie A Cassell. 2016. Extracting information from the text of electronic medical records to improve case detection: a systematic review. *Journal of the American Medical Informatics Association*, 23(5):1007–1015.

Carol Friedman, Lyudmila Shagina, Yves Lussier, and George Hripcsak. 2004. Automated encoding of clinical documents based on natural language processing. *Journal of the American Medical Informatics Association*, 11(5):392–402.

Charles Friedman, Joshua Rubin, Jeffrey Brown, Melinda Buntin, Milton Corn, Lynn Etheredge, Carl Gunter, Mark Musen, Richard Platt, William Stead, et al. 2014. Toward a science of learning systems: a research agenda for the high-functioning learning health system. *Journal of the American Medical Informatics Association*, 22(1):43–50.

Charles P Friedman, Adam K Wong, and David Blumenthal. 2010. Achieving a nationwide learning health system. *Science translational medicine*, 2(57):57cm29–57cm29.

Li Huang and Dianbo Liu. 2019. Patient clustering improves efficiency of federated machine learning to predict mortality and hospital stay time using distributed electronic medical records. *arXiv preprint arXiv:1903.09296*.

Li Huang, Yifeng Yin, Zeng Fu, Shifa Zhang, Hao Deng, and Dianbo Liu. 2018. Loadaboost: Loss-based adaboost federated machine learning on medical data. *arXiv preprint arXiv:1811.12629*.

Vasudevan Jagannathan, Charles J Mullett, James G Arbogast, Kevin A Halbritter, Deepthi Yellapragada, Sushmitha Regulapati, and Pavani Bandaru. 2009. Assessment of commercial nlp engines for medication information extraction from dictated clinical notes. *International journal of medical informatics*, 78(4):284–291.

Alistair EW Johnson, Tom J Pollard, Lu Shen, H Lehman Li-wei, Mengling Feng, Mohammad Ghassemi, Benjamin Moody, Peter Szolovits, Leo Anthony Celi, and Roger G Mark. 2016. Mimic-iii, a freely accessible critical care database. *Scientific data*, 3:160035.

Jakub Konečný, H Brendan McMahan, Felix X Yu, Peter Richtárik, Ananda Theertha Suresh, and Dave Bacon. 2016. Federated learning: Strategies for improving communication efficiency. *arXiv preprint arXiv:1610.05492*.

Kory Kreimeyer, Matthew Foster, Abhishek Pandey, Nina Arya, Gwendolyn Halford, Sandra F Jones, Richard Forshee, Mark Walderhaug, and Taxiarchis Botsis. 2017. Natural language processing systems for capturing and standardizing unstructured clinical information: a systematic review. *Journal of biomedical informatics*, 73:14–29.

Dianbo Liu, Timothy Miller, Raheel Sayeed, and Kenneth Mandl. 2018a. Fadl: Federated-autonomous deep learning for distributed electronic health record. *arXiv preprint arXiv:1811.11400*.

Dianbo Liu, Nestor Sepulveda, and Ming Zheng. 2018b. Artificial neural networks condensation: A strategy to facilitate adaption of machine learning in medical settings by reducing computational burden. *arXiv preprint arXiv:1812.09659*.

Christopher D Manning, Christopher D Manning, and Hinrich Schütze. 1999. *Foundations of statistical natural language processing*. MIT press.

H Brendan McMahan, Eider Moore, Daniel Ramage, Seth Hampson, et al. 2016. Communication-efficient learning of deep networks from decentralized data. *arXiv preprint arXiv:1602.05629*.

Riccardo Miotto, Li Li, Brian A Kidd, and Joel T Dudley. 2016. Deep patient: an unsupervised representation to predict the future of patients from the electronic health records. *Scientific reports*, 6:26094.

Harvey J Murff, Fern FitzHenry, Michael E Matheny, Nancy Gentry, Kristen L Kotter, Kimberly Crimin, Robert S Dittus, Amy K Rosen, Peter L Elkin, Steven H Brown, et al. 2011. Automated identification of postoperative complications within an electronic medical record using natural language processing. *Jama*, 306(8):848–855.

Jean R Slutsky. 2007. Moving closer to a rapid-learning health care system. *Health affairs*, 26(2):w122–w124.

Özlem Uzuner. 2009. Recognizing obesity and co-morbidities in sparse data. *Journal of the American Medical Informatics Association*, 16(4):561–570.

Transfer Learning for Causal Sentence Detection

Manolis Kyriakakis[1], Ion Androutsopoulos[2], Joan Ginés i Ametllé[1], Artur Saudabayev [1]
[1]Causaly, London, UK
[2]Department of Informatics, Athens University of Economics and Business, Greece
{m.kyriakakis, joan.g, artur}@causaly.com, ion@aueb.gr

Abstract

We consider the task of detecting sentences that express causality, as a step towards mining causal relations from texts. To bypass the scarcity of causal instances in relation extraction datasets, we exploit transfer learning, namely ELMO and BERT, using a bidirectional GRU with self-attention (BIGRUATT) as a baseline. We experiment with both generic public relation extraction datasets and a new biomedical causal sentence detection dataset, a subset of which we make publicly available. We find that transfer learning helps only in very small datasets. With larger datasets, BIGRU-ATT reaches a performance plateau, then larger datasets and transfer learning do not help.

1 Introduction

A wide range of biomedical questions, from what causes a disease to what drug dosages should be recommended and which side effects might be triggered, center around detecting particular causal relationships between biomedical entities. Causality, therefore, has long been a focus of biomedical research, e.g., in medical diagnostics (Rizzi, 1994), pharmacovigilance (Agbabiaka et al., 2008), and epidemiology (Karhausen, 2000). The most common way to detect causal relationships is by carrying highly controlled randomized controlled trials, but it is also possible to mine evidence from observational studies and meta-analyses (Ward and Johnson, 2008), where information is often expressed in natural language (e.g., journal articles or clinical study reports).

In natural language processing (NLP), causality detection is often viewed as a type of relation extraction, where the goal is to determine which relations (e.g., part-whole, content-container, cause-effect), if any, hold between two entities in a text (Hendrickx et al., 2009), using deep learning in most recent works (Bekoulis et al., 2018; Zhang

et al., 2018). The same view of causality detection is typically adopted in biomedical NLP (Cohen and Demner-Fushman, 2014; Li and Mao, 2019).

Existing relation extraction datasets, however, contain few causal instances, which may not allow relation extraction methods to learn to infer causality reliably. Note that causality can be expressed in many ways, from using explicit lexical markers (e.g., "smoking *causes* cancer") to markers that do not always express causality (e.g., "heavy smoking *led* to cancer" vs. "the nurse *led* the patient to her room") to no explicit markers ("she was infected by a virus and admitted to a hospital"). Also, existing relation extraction datasets contain sentences from generic, not biomedical documents.

In this paper, we focus on *detecting causal sentences*, i.e., sentences conveying at least one causal relation. This is a first step towards mining causal relations from texts. Once causal sentences have been detected, computationally more intensive relation extraction methods can be used to identify the exact entities that participate in the causal relations and their roles (cause, effect). To bypass the scarcity of causal instances in relation extraction datasets, we exploit *transfer learning*, namely ELMO (Peters et al., 2018) and BERT (Devlin et al., 2018), comparing against a bidirectional GRU with self-attention (Cho et al., 2014; Bahdanau et al., 2015). We experiment with *generic* public relation extraction datasets and a new larger *biomedical* causal sentence detection dataset, a subset of which we make publicly available.[1]

Unlike recently reported results in other NLP tasks (Peters et al., 2018; Devlin et al., 2018; Peters et al., 2019), we find that transfer learning

---

[1]We cannot provide the entire biomedical dataset, because it is used to develop commercial products. We report, however, results for both the entire biomedical dataset and the publicly available subset.

Proceedings of the BioNLP 2019 workshop, pages 292–297
Florence, Italy, August 1, 2019. ©2019 Association for Computational Linguistics

helps only in datasets with hundreds of training instances. When a few thousands of training instances are available, BIGRUATT reaches a performance plateau (both in generic and biomedical texts), then increasing the size of the dataset or employing transfer learning does not improve performance. We believe this is the first work to (a) focus on causal sentence detection as a binary classification task, (b) consider causal sentence detection in both generic and biomedical texts, and (c) explore the effect of transfer learning in this task.

2   Methods

BIGRUATT: Our baseline model is a bidirectional GRU (BIGRU) with self-attention (BIGRUATT) (Cho et al., 2014; Bahdanau et al., 2015), a classifier that has been reported to perform well in short text classification (Pavlopoulos et al., 2017; Chalkidis et al., 2019). The model views each sentence as the sequence $\langle e_1, \ldots, e_n \rangle$ of its word embeddings (Fig. 1). We use WORD2VEC embeddings (Mikolov et al., 2013) pre-trained on approx. (a) 3.5 billion tokens from PUBMED texts (McDonald et al., 2018)[2] or (b) 100 billion tokens from Google News.[3] The BIGRU computes two lists H_f, H_b of hidden states, reading the word embeddings left to right and right to left, respectively. The corresponding elements of H_f, H_b are then concatenated to form the output H of the BIGRU:

$$
\begin{aligned}
H^f &= \left\langle h_1^f, \ldots, h_n^f \right\rangle = \text{GRU}^f(e_1, \ldots, e_n) \\
H^b &= \left\langle h_1^b, \ldots, h_n^b \right\rangle = \text{GRU}^b(e_1, \ldots, e_n) \\
H &= \left\langle [h_1^f; h_1^b], \ldots, [h_n^f; h_n^b] \right\rangle
\end{aligned}
$$

where ^{f}, ^{b} indicate the forward and backward directions, $e_i \in \mathbb{R}^{d_e}$, $h_i^f, h_i^b \in \mathbb{R}^{d_h}$, and ';' denotes concatenation.[4] A linear attention computes an attention score $a_i \in \mathbb{R}$ for each element h_i of H:

$$
\widetilde{a}_i = u_{att} \cdot h_i, \quad a_i = \text{softmax}(\widetilde{a}_i; \widetilde{a}_1, \ldots, \widetilde{a}_n)
$$

where $u_{att} \in \mathbb{R}^{2 \times d_h}$ and $\cdot$ is the dot product. A sentence embedding s, representing the entire sentence, is then formed as the weighted (by the attention scores) sum of the elements of H and is

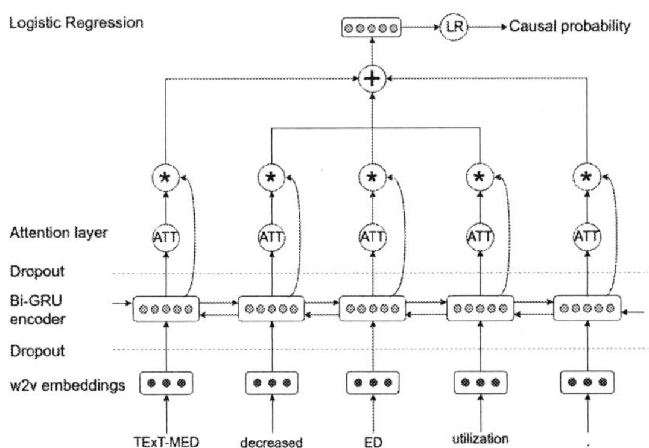

Figure 1: Illustration of the BIGRUATT model.

passed to a logistic regression (LR) layer to estimate the probability p that a sentence is causal:

$$
s = \sum_{i=1}^{n} a_i h_i, \quad p = \sigma(u_p \cdot s + b_p)
$$

where $u_p \in \mathbb{R}^{2 \times d_h}$, $b_p \in \mathbb{R}$, and σ is the sigmoid function. We use cross-entropy loss, the Adam optimizer (Kingma and Ba, 2015), and dropout layers (Srivastava et al., 2014) before and after the BIGRU (Fig. 1). Word embeddings are not updated.

BIGRUATT+ELMO: ELMO (Peters et al., 2018) produces word embeddings by passing the input text (in our case, a sentence) to a pre-trained stacked bidirectional LSTM language model (LM). It then uses a linear combination of the states of the LM (from the different layers of the stacked LSTM) at each word position to produce the corresponding word embedding. Like WORD2VEC, ELMO (its LM) is pre-trained on large corpora. However, ELMO maps occurrences of the same word to possibly different embeddings, depending on context. Furthermore, it uses CNNs (LeCun et al., 1989) to produce the initial word embeddings (that are fed to the LM) from word characters, alleviating the problem of out-of-vocabulary words. BIGRUATT+ELMO is the same as BIGRUATT, except that the embedding of each word is now the concatenation of its WORD2VEC and ELMO embeddings. We do not update the parameters of ELMO and the word embeddings when training BIGRUATT+ELMO. We used the original pre-trained ELMO of Peters et al. (2018).[5] For biomedical sentences, we also experimented with an ELMO model pre-trained on PUBMED texts, but performance was very similar as with the original ELMO.

BERT+LR: BERT (Devlin et al., 2018) is a model based on Transformers (Vaswani et al., 2017), pre-

---

[2]http://nlp.cs.aueb.gr/software.html
[3]https://drive.google.com/file/d/0B7XkCwpI5KDYNlNUTTlSS21pQmM

[4]In our experiments, $d_h = 128$; d_e is 300 for Google News and 200 for biomedical embeddings; d_e increases by 1,024 when ELMO is added. We also tried LSTMs (Hochreiter and Schmidhuber, 1997), but performance was similar.

[5]https://allennlp.org/elmo

trained on large corpora to predict (a) masked words from their left and right contexts, and (b) the next sentence. For a new NLP task, a task-specific layer is added on top of a pre-trained BERT model. The extra layer is trained jointly with BERT on task-specific data (in our case, a causal sentence detection dataset), a process that also fine-tunes the parameters of the pre-trained BERT for the new task. In BERT+LR, we add a logistic regression (LR) layer on top of BERT, which estimates the probability that the input sentence is causal. The LR layer is fed with the embedding of the 'classification' token, which BERT also produces for each sentence. We used the pre-trained 'base' BERT model of Devlin et al. (2018), which we fine-tuned jointly with the LR layer. For biomedical sentences, we also experimented with BIOBERT (Lee et al., 2019), a BERT model pre-trained on biomedical texts, but performance was very similar.

BERT+BIGRUATT: Common practice (Devlin et al., 2018; Peters et al., 2019) is to combine BERT with very shallow task-specific models, usually only an LR layer. To explore if deeper task-specific models can yield improved performance, we replaced the LR layer of BERT+LR with BIGRUATT, leading to BERT+BIGRUATT. This is the same as BIGRUATT, but uses the context-aware word embeddings that BERT produces at its top layer as the input to BIGRUATT, instead of WORD2VEC embeddings. Again, we use the 'base' pre-trained BERT model of (Devlin et al., 2018), and we fine-tune the entire BERT+BIGRUATT network on causal sentence detection datasets.

LR (n-grams): A plain LR classifier with TF-IDF n-gram features (word n-grams, $n = 1, 2, 3$).[6]

3 Datasets

SemEval-2010 (Task 8): This dataset contains 10,674 samples, of which 1,325 causal (Hendrickx et al., 2009). Each sample is a sentence annotated with a pair of entities and the type of their relationship. Since we are only interested in causality, we treat sentences with a *Cause-Effect* relationship as causal, and all the others as non-causal.

Causal-TimeBank (CausalTB): In this dataset (Mirza et al., 2014), we identified causal sentences using C-SIGNAL (causal signal) and CLINK (causal link) tags, discarding causal relationships between entities from different sentences, following Li and Mao (2019). The resulting dataset contains 2,470 sentences, of which 244 are causal.

Event StoryLine (EventSL): In this dataset (Caselli and Vossen, 2017), we detected causal sentences by examining the CAUSES and CAUSED_BY attributes in the PLOT_LINK tags, again following Li and Mao (2019). Again, we discarded causal relationships between entities from different sentences. The resulting dataset contains 4,107 sentences, of which 77 are causal.

BioCausal: The full biomedical causal detection dataset we developed (**BioCausal-Large**) contains 13,342 sentences from PUBMED, of which 7,562 causal. Each sentence was annotated by a single annotator familiar with biomedical texts.[7] The publicly available subset (**BioCausal-Small**) contains 2,000 sentences, of which 1,113 causal.[8]

Li and Mao (2019) report that SemEval-2010 contains a large number of causal samples with explicit causal markers; by contrast, CausalTB and EventSL contain more complex causal relations with no explicit clues. BioCausal includes causal sentences both with and without explicit clues.

SemEval-2010, CausalTB, and EventSL are highly imbalanced, with the vast majority of sentences being non-causal. To prevent a high bias towards the non-causal class, in our experiments we randomly selected 2500, 500, 200 non-causal sentences respectively, discarding the rest. The resulting causal to non-causal ratios (Table 1) are, thus, roughly 1:2 (SemEval, CausalTB) or 1:3 (EventSL). By contrast, the BioCausal (Large and Small) datasets are roughly balanced. All five datasets were then split into train (70%), validation (15%) and test (15%) subsets, maintaining the same ratio between the two classes in the three subsets.

4 Experimental results

Tables 1–2 report our experimental results. For each neural model, we performed 10 repetitions (with different random seeds) and report averages and standard deviations. For completeness, we show precision, recall, F1, and area under the precision-recall curve (AUC), though AUC scores are the main ones to consider, since they examine

---

[6]We used the LR code of SCIKIT-LEARN (`https://scikit-learn.org/`). For all other methods, we used our own PyTorch implementations (`https://pytorch.org/`), with the BERT API of `https://github.com/huggingface/pytorch-pretrained-BERT`.

[7]The average inter-annotator agreement on a sample of 300 sentences was 79.36%. Cohen's Kappa was 0.56.

[8]BioCausal-Small is available at `https://archive.org/details/CausalySmall`.

Dataset (causal:non-causal)	SemEval (1,325 : 2,500)				CausalTB (244 : 500)				EventSL (77 : 200)				BioCausal-Small (1,113 : 887)			
Model	F1	P	R	AUC	F1	P	R	AUC	F1	P	R	AUC	F1	P	R	AUC
LR (n-grams)	76.22	88.67	66.83	87.50	36.36	**100.00**	22.22	65.02	42.86	**100.00**	27.27	73.55	77.49	73.91	81.44	87.65
BIGRUATT	90.64	93.96	87.59	96.57	69.98	67.04	73.89	74.38	63.65	70.09	60.91	70.36	85.97	83.57	88.62	93.91
	±0.70	±1.71	±1.52	±0.32	±3.58	±5.16	±6.60	±4.16	±10.12	±7.47	±17.28	±9.84	±0.90	±2.03	±2.69	±0.88
BIGRUATT+ELMO	**92.81**	**94.45**	91.26	97.03	75.08	81.29	70.28	82.06*	66.55	77.47	59.09	77.31	**87.32**	**89.46**	85.39	**94.95**
	±0.78	±0.94	±1.77	±1.44	±4.20	±5.43	±6.81	±3.59	±7.82	±5.05	±10.17	±4.84	±0.78	±2.34	±2.50	**±0.33**
BERT+LR	91.55	86.62	**97.09**	96.94	**80.55**	71.17	**93.33**	82.26*	72.35	62.44	**87.17**	78.15*	85.64	78.87	**93.71**	90.75
	±0.53	±1.16	±0.67	±2.25	±3.62	±6.02	±3.33	±3.41	±5.36	±8.21	±4.58	±9.48	±0.61	±1.16	±1.54	±3.69
BERT+BIGRUATT	91.45	86.80	96.63	**97.61**	80.06	74.52	86.94	**84.27***	**73.09**	66.15	83.64	**84.17***	85.87	79.43	93.47	93.75
	±0.59	±1.28	±0.60	**±0.29**	±2.94	±4.46	±5.56	**±1.71**	±5.27	±7.86	±10.60	**±4.04**	±0.88	±1.53	±1.08	±0.45

Table 1: Precision, recall, F1, AUC on the four publicly available datasets, averaged over 10 repetitions, with standard deviations (±). Next to each dataset name, we show in brackets the total causal and non-causal sentences that we used. The best results are shown in bold. The best AUC results are also shown in gray background. In the AUC columns, stars indicate statistically significant ($p \leq 0.05$) differences compared to BIGRUATT.

performance at multiple classification thresholds; the other measures are computed only for a particular threshold, which was 0.5 in our experiments.

Model	F1	P	R	AUC
LR (n-grams)	79.21	76.75	81.82	86.54
BIGRUATT	85.84	84.28	87.47	93.71
	±0.36	±0.66	±0.75	±0.15
BIGRUATT+ELMO	86.77	**87.46**	86.12	94.64*
	±0.52	±1.29	±1.49	±0.26
BERT+LR	**87.33**	82.11	**93.27**	92.77
	±0.47	±0.89	±0.67	±1.57
BERT+BIGRUATT	87.09	81.70	93.25	**94.70**
	±0.34	±0.58	±0.40	**±0.24**

Table 2: Results on BioCausal-Large (7,562 : 5,780).

Focusing on AUC scores, BIGRUATT outperforms the simpler LR with n-grams by a wide margin, with the exception of EventSL, which is probably too small for the capacity of BIGRUATT.[9] The precision of LR is perfect on CausalTB and EventSL, at the expense of very low recall, suggesting that LR learned perfectly few high-precision n-grams in those datasets. Transfer learning (ELMO, BERT) improves the AUC of BIGRUATT by a wide margin in the two smallest datasets (CausalTB, EventSL), which contain only hundreds of instances, and the AUC differences from BIGRUATT are statistically significant (stars in Table 1), except for BIGRUATT+ELMO in EventSL.[10] However, in the other three datasets which contain thousands of instances, the AUC differences between transfer learning and plain BIGRUATT are small, with no statistical significance in most cases. Also, the AUC scores of all methods on BioCausal-Large are close to those on BioCausal-Small, despite the fact that BioCausal-Large is approx. seven times larger. Similar observations can be made by looking at the F1 scores.

It seems that causal sentence detection, at least with the neural methods we considered, reaches a plateau with few thousands of training sentences both in generic and biomedical texts; then increasing the dataset size or employing transfer learning does not help. The latter finding is not in line with previously reported results (Peters et al., 2018; Devlin et al., 2018; Peters et al., 2019), where ELMO and BERT were found to improve performance in most NLP tasks without studying, however, the effect of dataset size. Furthermore, BERT+BIGRUATT consistently performed better than BERT+LR in AUC (but not in F1), which casts doubts on the practice of adding only shallow task-specific models to BERT.[11] BIGRUATT+ELMO is competitive in AUC to BERT+BIGRUATT (with the exception of EventSL). Mainly comprised of sentences with simple explicit causal statements (Li and Mao, 2019), SemEval expectedly demonstrated the best classification performance across datasets.

5 Related and Future Work

Recent work on (causal) relation extraction uses LSTMs (Zhang et al., 2017) or CNNs (Li and Mao, 2019), assuming however that the spans of the two entities (cause, effect) are known. A notable exception is the model of Bekoulis et al. (2018), which jointly infers the spans of the entities and their relationships. Such finer relation extraction methods, however, are computationally more expensive than our causal sentence detection methods, especially when they involve parsing (Zhang et al., 2018). We plan to consider pipelines where computationally cheaper causal sentence detection components will first detect sentences likely to express causality, and then finer relation extraction components will pinpoint the entities, the type of

---

[9]Indeed BIGRUATT overfits the training set of EventSL.

[10]We performed two-tailed Approximate Randomization tests (Dror et al., 2018), $p \leq 0.05$, with 10k iterations, randomly swapping in each iteration 50% of the decisions (over all tested sentences) across the two compared methods. When testing statistical significance, for each method we use the repetition (among the 10) with the best validation F1 score.

[11]We note, however, that the AUC difference between BERT+LR and BERT+BIGRUATT is statistically significant ($p \leq 0.05$) only in BioCausal-Large.

causality (e.g., up-regulate), and entity roles.

Appendix

A   Hyper-parameters

Batch sizes of 128, 32, 16, 32 and 256 were used for Semeval, CausalTB, EventSL, BioCausal-Small and BioCausal-Large, respectively, for all neural models. Trainable parameters were initialized using the default PyTorch initialization methods except from self-attention weights where the method of Glorot and Bengio (2010) was used.

BIGRU and BIGRUATT+ELMO were trained for 100 epochs using Adam (Kingma and Ba, 2015) with initial learning rate $2e^{-3}$, $\beta_1/\beta_2 = 0.9/0.999$ and $eps = 1e^{-8}$. The learning rate was decayed linearly every 20 epochs as $lr_{new} \leftarrow lr_{prev} \cdot 0.75$. Gradients were clipped using a clip norm threshold of 0.25. The GRU's hidden size was set to 128 and its depth to 1. A dropout of 0.5 was applied to the input and output connections of the BIGRU encoder. Validation F1 was checked periodically in order to keep the model's checkpoint with the best validation performance.

BERT and BERT+BIGRUATT used the BERT-BASE uncased pre-trained model, which has 12 layers, 768 hidden size, 12 attention heads, and 110M parameters. For both models the entire network was fine-tuned for 10 epochs using Adam with a very small learning rate of $2e^{-5}$, $\beta_1/\beta_2 = 0.9/0.999$, $eps = 1e^{-6}$, L2 weight decay of 0.01 and linear warmup of 0.1. A dropout of 0.1 was applied to all BERT-specific layers. For BERT+BIGRUATT, an additional dropout of 0.5 was applied to the input and output connections of its BIGRU encoder. Similarly to BIGRU and BI-GRUATT+ELMO, the hidden size of the GRU was set to 128 and its depth to 1.

References

Taofikat B. Agbabiaka, Jelena Savović, and Edzard Ernst. 2008. Methods for causality assessment of adverse drug reactions. *Drug Safety*, 31(1):21–37.

Dzmitry Bahdanau, Kyunghyun Cho, and Yoshua Bengio. 2015. Neural machine translation by jointly learning to align and translate. In *3rd International Conference on Learning Representations*, San Diego, California.

Giannis Bekoulis, Johannes Deleu, Thomas Demeester, and Chris Develder. 2018. Adversarial training for multi-context joint entity and relation extraction. In *Proceedings of the Conference on Empirical Methods in Natural Language Processing*, pages 2830–2836, Brussels, Belgium.

Tommaso Caselli and Piek Vossen. 2017. The event storyline corpus: A new benchmark for causal and temporal relation extraction. In *Proceedings of the Events and Stories in the News Workshop*, pages 77–86, Vancouver, Canada.

Ilias Chalkidis, Manos Fergadiotis, Prodromos Malakasiotis, Nikolaos Aletras, and Ion Androutsopoulos. 2019. Extreme multi-label legal text classification:a case study in EU legislation. In *NAACL Workshop in Natural Legal Language Processing*.

Kyunghyun Cho, Bart van Merrienboer, Caglar Gulcehre, Dzmitry Bahdanau, Fethi Bougares, Holger Schwenk, and Yoshua Bengio. 2014. Learning phrase representations using RNN encoder-decoder for statistical machine translation. In *Proceedings of the Conference on Empirical Methods in Natural Language Processing*, pages 1724–1734, Doha, Qatar.

Kevin Bretonnel Cohen and Dina Demner-Fushman. 2014. *Biomedical Natural Language Processing*. John Benjamins.

Jacob Devlin, Ming-Wei Chang, Kenton Lee, and Kristina Toutanova. 2018. BERT: Pre-training of deep bidirectional transformers for language understanding. *arXiv preprint arXiv:1810.04805*.

Rotem Dror, Gili Baumer, Segev Shlomov, and Roi Reichart. 2018. The hitchhiker's guide to testing statistical significance in natural language processing. In *Proceedings of the 56th Annual Meeting of the ACL (Volume 1: Long Papers)*, pages 1383–1392.

Xavier Glorot and Yoshua Bengio. 2010. Understanding the difficulty of training deep feedforward neural networks. In *Proceedings of the Thirteenth International Conference on Artificial Intelligence and Statistics*, pages 249–256.

Iris Hendrickx, Su Nam Kim, Zornitsa Kozareva, Preslav Nakov, Diarmuid Ó Séaghdha, Sebastian Padó, Marco Pennacchiotti, Lorenza Romano, and Stan Szpakowicz. 2009. SemEval-2010 Task 8: Multi-way classification of semantic relations between pairs of nominals. In *Proceedings of the Workshop on Semantic Evaluations: Recent Achievements and Future Directions*, pages 94–99, Boulder, Colorado.

Sepp Hochreiter and Jürgen Schmidhuber. 1997. Long short-term memory. *Neural Computation*, 9:1735–1780.

Lucien R. Karhausen. 2000. Causation: The elusive grail of epidemiology. *Medicine, Health Care and Philosophy*, 3(1):59–67.

Diederik P. Kingma and Jimmy Ba. 2015. Adam: A method for stochastic optimization. *CoRR*, abs/1412.6980.

Yann LeCun, Bernhard E. Boser, John S. Denker, Don Henderson, R. E. Howard, W. Hubbard, and Larry Jackel. 1989. Backpropagation applied to handwritten zip code recognition. *Neural Computation*, 1(4):541–551.

Jinhyuk Lee, Wonjin Yoon, Sungdong Kim, Donghyeon Kim, Sunkyu Kim, Chan Ho So, and Jaewoo Kang. 2019. Biobert: a pre-trained biomedical language representation model for biomedical text mining. *arXiv preprint arXiv:1901.08746*.

Pengfei Li and Kezhi Mao. 2019. Knowledge-oriented convolutional neural network for causal relation extraction from natural language texts. *Expert Systems with Applications*, 115:512–523.

Ryan McDonald, Georgios-Ioannis Brokos, and Ion Androutsopoulos. 2018. Deep relevance ranking using enhanced document-query interactions. In *Proceedings of the Conference on Empirical Methods in Natural Language Processing*, pages 1849–1860, Brussels, Belgium.

Tomas Mikolov, Ilya Sutskever, Kai Chen, Greg S Corrado, and Jeff Dean. 2013. Distributed representations of words and phrases and their compositionality. In C. J. C. Burges, L. Bottou, M. Welling, Z. Ghahramani, and K. Q. Weinberger, editors, *Advances in Neural Information Processing Systems 26*, pages 3111–3119.

Paramita Mirza, Rachele Sprugnoli, Sara Tonelli, and Manuela Speranza. 2014. Annotating causality in the tempeval-3 corpus. In *Proceedings of the EACL Workshop on Computational Approaches to Causality in Language (CAtoCL)*, pages 10–19.

John Pavlopoulos, Prodromos Malakasiotis, and Ion Androutsopoulos. 2017. Deeper attention to abusive user content moderation. In *Proceedings of the Conference on Empirical Methods in Natural Language Processing*, pages 1125–1135, Copenhagen, Denmark.

Matthew Peters, Mark Neumann, Mohit Iyyer, Matt Gardner, Christopher Clark, Kenton Lee, and Luke Zettlemoyer. 2018. Deep contextualized word representations. In *Proceedings of the Conference of the NAACL: Human Language Technologies, Volume 1 (Long Papers)*, pages 2227–2237, New Orleans, Louisiana.

Matthew Peters, Sebastian Ruder, and Noah A. Smith. 2019. To tune or not to tune? adapting pre-trained representations to diverse tasks. *CoRR*, abs/1903.05987.

Dominick A. Rizzi. 1994. Causal reasoning and the diagnostic process. *Theoretical medicine*, 15:315–33.

Nitish Srivastava, Geoffrey Hinton, Alex Krizhevsky, Ilya Sutskever, and Ruslan Salakhutdinov. 2014. Dropout: A simple way to prevent neural networks from overfitting. *Journal of Machine Learning Research*, 15:1929–1958.

Ashish Vaswani, Noam Shazeer, Niki Parmar, Jakob Uszkoreit, Llion Jones, Aidan N Gomez, Ł ukasz Kaiser, and Illia Polosukhin. 2017. Attention is all you need. In I. Guyon, U. V. Luxburg, S. Bengio, H. Wallach, R. Fergus, S. Vishwanathan, and R. Garnett, editors, *Advances in Neural Information Processing Systems 30*, pages 5998–6008.

Andrew Ward and Pamela Jo Johnson. 2008. Addressing confounding errors when using non-experimental, observational data to make causal claims. *Synthese*, 163(3):419–432.

Yuhao Zhang, Peng Qi, and Christopher D. Manning. 2018. Graph convolution over pruned dependency trees improves relation extraction. In *Proceedings of the Conference on Empirical Methods in Natural Language Processing*, pages 2205–2215, Brussels, Belgium.

Yuhao Zhang, Victor Zhong, Danqi Chen, Gabor Angeli, and Christopher D. Manning. 2017. Position-aware attention and supervised data improve slot filling. In *Proceedings of the Conference on Empirical Methods in Natural Language Processing*, pages 35–45, Copenhagen, Denmark.

Embedding Biomedical Ontologies by Jointly Encoding Network Structure and Textual Node Descriptors

Sotiris Kotitsas[1], Dimitris Pappas[1,2], Ion Androutsopoulos[1],
Ryan McDonald[1,3] and Marianna Apidianaki[4]

[1]Department of Informatics, Athens University of Economics and Business, Greece
[2]Institute for Language and Speech Processing, Research Center 'Athena', Greece
[3]Google Research
[4]CNRS, LLF, Univ. Paris Diderot, France

{p3150077, pappasd, ion}@aueb.gr
ryanmcd@google.com, marianna@limsi.fr

Abstract

Network Embedding (NE) methods, which map network nodes to low-dimensional feature vectors, have wide applications in network analysis and bioinformatics. Many existing NE methods rely only on network structure, overlooking other information associated with the nodes, e.g., text describing the nodes. Recent attempts to combine the two sources of information only consider local network structure. We extend NODE2VEC, a well-known NE method that considers broader network structure, to also consider textual node descriptors using recurrent neural encoders. Our method is evaluated on link prediction in two networks derived from UMLS. Experimental results demonstrate the effectiveness of the proposed approach compared to previous work.

1 Introduction

Network Embedding (NE) methods map each node of a network to an embedding, meaning a low-dimensional feature vector. They are highly effective in network analysis tasks involving predictions over nodes and edges, for example link prediction (Lu and Zhou, 2010), and node classification (Sen et al., 2008).

Early NE methods, such as DEEPWALK (Perozzi et al., 2014), LINE (Tang et al., 2015), NODE2VEC (Grover and Leskovec, 2016), GCNs (Kipf and Welling, 2016), leverage information from the network structure to produce embeddings that can reconstruct node neighborhoods. The main advantage of these *structure-oriented* methods is that they encode the network context of the nodes, which can be very informative. The downside is that they typically treat each node as an atomic unit, directly mapped to an embedding in a look-up table (Fig. 1a). There is no attempt to model information other than the network structure, such

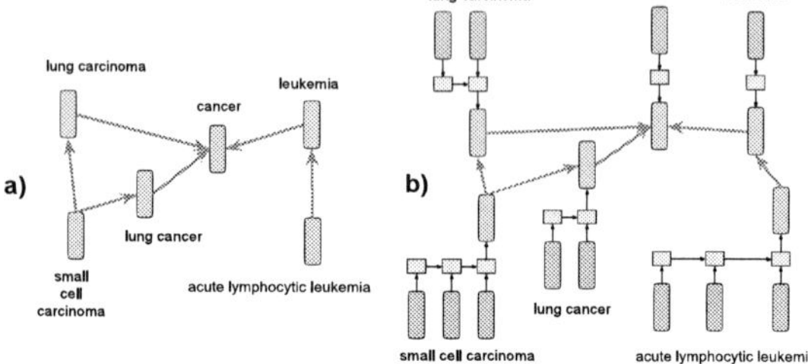

Figure 1: Example network with nodes associated with textual descriptors. a) A model where each node is represented by a vector (node embedding) from a look-up table. b) A model where each node embedding is generated compositionally from the word embeddings of its descriptor via an RNN. The latter model can learn node embeddings from both the network structure and the word sequences of the textual descriptors.

as textual descriptors (labels) or other meta-data associated with the nodes.

More recent NE methods, e.g., CANE (Tu et al., 2017), WANE (Shen et al., 2018), produce embeddings by combining the network structure and the text associated with the nodes. These *content-oriented* methods embed networks whose nodes are rich textual objects (often whole documents). They aim to capture the compositionality and semantic similarities in the text, encoding them with deep learning methods. This approach is illustrated in Fig. 1b. However, previous methods of this kind considered impoverished network contexts when embedding nodes, usually single-edge hops, as opposed to the non-local structure considered by most structure-oriented methods.

When embedding biomedical ontologies, it is important to exploit both wider network contexts and textual node descriptors. The benefit of the latter is evident, for example, in 'acute leukemia' IS-A 'leukemia'. To be able to predict (or reconstruct) this IS-A relation from the embeddings of 'acute leukemia' and 'leukemia' (and the word embeddings of their textual descriptors in Fig. 1b),

Proceedings of the BioNLP 2019 workshop, pages 298–308
Florence, Italy, August 1, 2019. ©2019 Association for Computational Linguistics

a NE method only needs to model the role of 'acute' as a modifier that can be included in the descriptor of a node (e.g., a disease node) to specify a sub-type. This property can be learned (and encoded in the word embedding of 'acute') if several similar IS-A edges, with 'acute' being the only extra word in the descriptor of the sub-type, exist in the network. This strategy would not however be successful in 'p53' (a protein) IS-A 'tumor suppressor', where no word in the descriptors frequently denotes sub-typing. Instead, by considering the broader network context of the nodes (i.e. longer paths that connect them), a NE method can detect that the two nodes have common neighbors and, hence, adjust the two node embeddings (and the word embeddings of their descriptors) to be close in the representation space, making it more likely to predict an IS-A relation between them.

We propose a new NE method that leverages the strengths of both structure and content-oriented approaches. To exploit wide network contexts, we follow NODE2VEC (Grover and Leskovec, 2016) and generate random walks to construct the network neighborhood of each node. The SKIP-GRAM model (Mikolov et al., 2013) is then used to learn node embeddings that successfully predict the nodes in each walk, from the node at the beginning of the walk. To enrich the node embeddings with information from their textual descriptors, we replace the NODE2VEC look-up table with various architectures that operate on the word embeddings of the descriptors. These include simply averaging the word embeddings of a descriptor, and applying recurrent deep learning encoders. The proposed method can be seen as an extension of NODE2VEC that incorporates textual node descriptors. We evaluate several variants of the proposed method on link prediction, a standard evaluation task for NE methods. We use two biomedical networks extracted from UMLS (Bodenreider, 2004), with PART-OF and IS-A relations, respectively. Our method outperforms several existing structure and content-oriented methods on both datasets. We make our datasets and source code available.[1]

2   Related work

Network Embedding (NE) methods, a type of representation learning, are highly effective in network analysis tasks involving predictions over nodes and edges. Link prediction has been extensively studied in social networks (Wang et al., 2015), and is particularly relevant to bioinformatics where it can help, for example, to discover interactions between proteins, diseases, and genes (Lei and Ruan, 2013; Shojaie, 2013; Grover and Leskovec, 2016). Node classification can also help analyze large networks by automatically assigning roles or labels to nodes (Ahmed et al., 2018; Sen et al., 2008). In bioinformatics, this approach has been used to identify proteins whose mutations are linked with particular diseases (Agrawal et al., 2018).

A typical structure-oriented NE method is DEEPWALK (Perozzi et al., 2014), which learns node embeddings by applying WORD2VEC's SKIPGRAM model (Mikolov et al., 2013) to node sequences generated via random walks on the network. NODE2VEC (Grover and Leskovec, 2016) explores different strategies to perform random walks, introducing hyper-parameters to guide them and generate more flexible neighborhoods. LINE (Tang et al., 2015) learns node embeddings by exploiting first- and second-order proximity information in the network. Wang et al. (2016) learn node embeddings that preserve the proximity between 2-hop neighbors using a deep autoencoder. Yu et al. (2018) encode node sequences generated via random walks, by mapping the walks to low dimensional embeddings, through an LSTM autoencoder. To avoid overfitting, they use a generative adversarial training process as regularization. Graph Convolutional Networks (GCNs) are a graph encoding framework that also falls within this paradigm (Kipf and Welling, 2016; Schlichtkrull et al., 2018). Unlike other methods that use random walks or static neighbourhoods, GCNs use iterative neighbourhood averaging strategies to account for non-local graph structure. All the aforementioned methods only encode the structural information into node embeddings, ignoring textual or other information that can be associated with the nodes of the network.

Previous work on biomedical ontologies (e.g., Gene Ontology, GO) suggested that their terms, which are represented through textual descriptors, have compositional structure. By modeling it, we can create richer representations of the data encoded in the ontologies (Mungall, 2004; Ogren et al., 2003, 2004). Ogren et al. (2003) strengthen

---

the argument of compositionality by observing that many GO terms contain other GO terms. Also, they argue that substrings that are not GO terms appear frequently and often indicate semantic relationships. Ogren et al. (2004) use finite state automata to represent GO terms and demonstrate how small conceptual changes can create biologically meaningful candidate terms.

In other work on NE methods, CENE (Sun et al., 2016) treats textual descriptors as a special kind of node, and uses bidirectional recurrent neural networks (RNNs) to encode them. CANE (Tu et al., 2017) learns two embeddings per node, a text-based one and an embedding based on network structure. The text-based one changes when interacting with different neighbors, using a mutual attention mechanism. WANE (Shen et al., 2018) also uses two types of node embeddings, text-based and structure-based. For the text-based embeddings, it matches important words across the textual descriptors of different nodes, and aggregates the resulting alignment features. In spite of performance improvements over structure-oriented approaches, these content-aware methods do not thoroughly explore the network structure, since they consider only direct neighbors.

By contrast, we utilize NODE2VEC to obtain wider network neighborhoods via random walks, a typical approach of structure-oriented methods, but we also use RNNs to encode the textual descriptors, as in some content-oriented approaches. Unlike CENE, however, we do not treat texts as separate nodes; unlike CANE, we do not learn separate embeddings from texts and network structure; and unlike WANE, we do not align the descriptors of different nodes. We generate the embedding of each node from the word embeddings of its descriptor via the RNN (Fig. 1), but the parameters of the RNN, the word embeddings, hence also the node embeddings are updated during training to predict NODE2VEC's neighborhoods.

Although we use NODE2VEC to incorporate network context in the node embeddings, other neighborhood embedding methods, such as GCNs, could easily be used too. Similarly, text encoders other than RNNs could be applied. For example, Mishra et al. (2019) try to detect abusive language in tweets with a semi-supervised learning approach based on GCNs. They exploit the network structure and also the labels associated with the tweets, taking into account the linguistic behavior of the authors.

3   Proposed Node Embedding Approach

Consider a network (graph) $G = \langle V, E, S \rangle$, where V is the set of nodes (vertices); $E \subseteq V \times V$ is the set of edges (links) between nodes; and S is a function that maps each node $v \in V$ to its textual descriptor $S(v) = \langle w_1, w_2, \ldots, w_n \rangle$, where n is the word length of the descriptor, and each word w_i comes from a vocabulary W. We consider only undirected, unweighted networks, where all edges represent instances of the same (single) relationship (e.g., IS-A or PART-OF). Our approach, however, can be extended to directed weighted networks with multiple relationship types. We learn an embedding $f(v) \in \mathbb{R}^d$ for each node $v \in V$. As a side effect, we also learn a word embedding $e(w)$ for each vocabulary word $w \in W$.

To incorporate structural information into the node embeddings, we maximize the *predicted* probabilities $p(u|v)$ of observing the *actual* neighbors $u \in N(v)$ of each 'focus' node $v \in V$, where $N(v)$ is the neighborhood of v, and $p(u|v)$ is predicted from the node embeddings of u and v. The neighbors $N(v)$ of v are not necessarily directly connected to v. In real-world networks, especially biomedical, many nodes have few direct neighbors. We use NODE2VEC (Grover and Leskovec, 2016) to obtain a larger neighborhood for each node v, by generating random walks from v. For every focus node $v \in V$, we compute r random walks (paths) $P_{v,i} = \langle v_{i,1} = v, v_{i,2}, \ldots, v_{i,k} \rangle$ $(i = 1, \ldots, r)$ of fixed length k through the network $(v_{i,j} \in V)$.[2] The predicted probability $p(v_{i,j} = u)$ of observing node u at step j of a walk $P_{v,i}$ that starts at focus node v is taken to depend only on the embeddings of u, v, i.e., $p(v_{i,j} = u) = p(u|v)$, and can be estimated with a softmax as in the SKIPGRAM model (Mikolov et al., 2013):

$$p(u|v) = \frac{\exp(f'(u) \cdot f(v))}{\sum_{u' \in V} \exp(f'(u') \cdot f(v))} \quad (1)$$

where it is assumed that each node v has two different node embeddings, $f(v), f'(v)$, used when

---

[2]Our networks are unweighted, hence we use uniform edge weighting to traverse them. NODE2VEC has two hyper-parameters, p, q, to control the locality of the walk. We set $p = q = 1$ (default values). For efficiency, NODE2VEC actually performs r random walks of length $l \geq k$; then it uses r sub-walks of length k that start at each focus node.

v is the focus node or the predicted neighbor, respectively, and $\cdot$ denotes the dot product. NODE2VEC minimizes the following objective function:

$$L = -\sum_{v \in V} \sum_{i=1}^{r} \sum_{j=2}^{k} \log p(v_{i,j} | v_{i,1} = v) \quad (2)$$

in effect maximizing the likelihood of observing the actual neighbors $v_{i,j}$ of each focus node v that are encountered during the r walks $P_{v,i} = \langle v_{i,1} = v, v_{i,2}, ..., v_{i,k} \rangle$ $(i = 1, \ldots, r)$ from v. Calculating $p(u|v)$ using a softmax (Eq. 1) is computationally inefficient. We apply negative sampling instead, as in WORD2VEC (Mikolov et al., 2013). Thus, NODE2VEC is analogous to SKIP-GRAM WORD2VEC, but using random walks from each focus node, instead of using a context window around each focus word in a corpus.

As already mentioned, the original NODE2VEC does not consider the textual descriptors of the nodes. It treats each node embedding $f(v)$ as a vector representing an atomic unit, the node v; a look-up table directly maps each node v to its embedding $f(v)$. This does not take advantage of the flexibility and richness of natural language (e.g., synonyms, paraphrases), nor of its compositional nature. To address this limitation, we substitute the look-up table where NODE2VEC stores the embedding $f(v)$ of each node v with a neural sequence encoder that produces $f(v)$ from the word embeddings of the descriptor $S(v)$ of v.

More formally, let every word $w \in W$ have two embeddings $e(w)$ and $e'(w)$, used when w occurs in the descriptor of a focus node, and when w occurs in the descriptor of a neighbor of a focus node (in a random walk), respectively. For every node $v \in V$ with descriptor $S(v) = (w_1, \ldots, w_n)$, we create the sequences $T(v) = \langle e(w_1), \ldots, e(w_n) \rangle$ and $T'(v) = \langle e'(w_1), \ldots, e'(w_n) \rangle$. We then set $f(v) = \text{ENC}(T(v))$ and $f'(v) = \text{ENC}(T'(v))$, where ENC is the sequence encoder. We outline below three specific possibilities for ENC, though it can be any neural text encoder. Note that the embeddings $f(v)$ and $f'(v)$ of each node v are constructed from the word embeddings $T(v)$ and $T'(v)$, respectively, of its descriptor $S(v)$ by the encoder ENC. The word embeddings of the descriptor and the parameters of ENC, however, are also optimized during back-propagation, so that the resulting node embeddings will predict (Eq. 1) the actual neighbors of each focus node (Fig. 2).

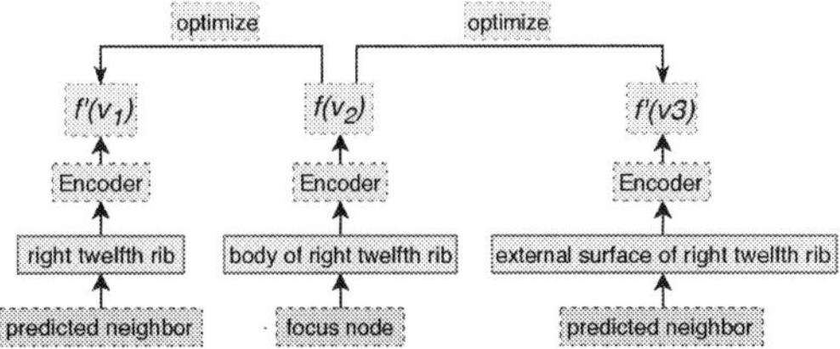

Figure 2: Illustration of the proposed NE approach.

For simplicity, we only mention $f(v)$ and $T(v)$ below, but the same applies to $f'(v)$ and $T'(v)$.

AVG-N2V: For every node $v \in V$, this model constructs the node's embedding $f(v)$ by simply averaging the word embeddings $T(v) = \langle e(w_1), \ldots, e(w_n) \rangle$ of $S(v) = (w_1, w_2, \ldots, w_n)$.

$$f(v) = \frac{1}{n} \sum_{i=1}^{n} e(w_i) \quad (3)$$

GRU-N2V: Although averaging word embeddings is effective in text categorization (Joulin et al., 2016), it ignores word order. To account for order, we apply RNNs with GRU cells (Cho et al., 2014) instead. For each node $v \in V$ with descriptor $S(v) = \langle w_1, \ldots, w_n \rangle$, this method computes n hidden state vectors $H = \langle h_1, \ldots, h_n \rangle = \text{GRU}(e(w_1), \ldots, e(w_n))$. The last hidden state vector h_n is the node embedding $f(v)$.

BIGRU-MAX-RES-N2V: This method uses a bidirectional RNN (Schuster and Paliwal, 1997). For each node v with descriptor $S(v) = \langle w_1, w_2, \ldots, w_n \rangle$, a bidirectional GRU (BIGRU) computes two sets of n hidden state vectors, one for each direction. These two sets are then added to form the output H of the BIGRU:

$$H_f = \text{GRU}_f(e(w_1), \ldots, e(w_n)) \quad (4)$$
$$H_b = \text{GRU}_b(e(w_1), \ldots, e(w_n)) \quad (5)$$
$$H = H_f + H_b \quad (6)$$

where $_f$, $_b$ denote the forward and backward directions, and $+$ indicates component-wise addition. We add residual connections (He et al., 2015) from each word embedding $e(w_t)$ to the corresponding hidden state h_t of H. Instead of using the final forward and backward states of H, we apply max-pooling (Collobert and Weston, 2008; Conneau et al., 2017) over the state vectors h_t of H. The output of the max pooling is the node embedding $f(v)$. Figure 3 illustrates this method.

Additional experiments were conducted with several variants of the last encoder. A unidirectional GRU instead of a BIGRU, and a BIGRU with

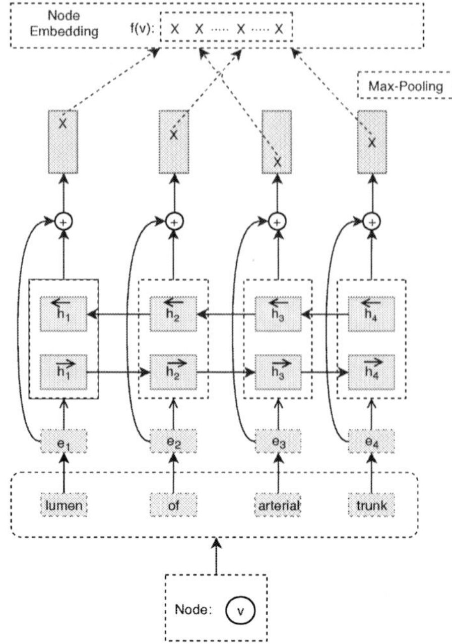

Figure 3: Obtaining the embedding of a node v by applying a BIGRU encoder with max-pooling and residuals to the embeddings of v's textual descriptor.

self-attention (Bahdanau et al., 2015) instead of max-pooling were also tried. To save space, we described only the best performing variant.

4 Experiments

We investigate the effectiveness of our proposed approach by conducting link prediction experiments on two biomedical datasets derived from UMLS. Furthermore, we devise a new approach of generating negative edges for the link prediction evaluation – beyond just random negatives – that makes the problem more difficult and aligns more with real-world use-cases. We also conduct a qualitative analysis, showing that the proposed framework does indeed leverage both the textual descriptors and the network structure.

4.1 Datasets

We created our datasets from the UMLS ontology, which contains approx. 3.8 million biomedical concepts and 54 semantic relationships. The relationships become edges in the networks, and the concepts become nodes. Each concept (node) is associated with a textual descriptor. We extract two types of semantic relationships, creating two networks. The first, and smaller one, consists of PART-OF relationships where each node represents a part of the human body. The second network

Statistics	IS-A	PART-OF
Nodes	294,693	16,894
Edges	356,541	19,436
Training true positive edges	294,692	16,893
Training true negative edges	294,692	16,893
Test true positive edges	61,849	2,543
Test true negative edges	61,849	2,543
Avg. descriptor length	5 words	6 words
Max. descriptor length	31 words	14 words

Table 1: Statistics of the two datasets (IS-A, PART-OF). The true positive and true negative edges are used in the link prediction experiments.

contains IS-A relationships, and the concepts represented by the nodes vary across the spectrum of biomedical entities (diseases, proteins, genes, etc.). To our knowledge, the IS-A network is one of the largest datasets employed for link prediction and learning network embeddings. Statistics for the two datasets are shown in Table 1.

4.2 Baseline Node Embedding Methods

We compare our proposed methods to baselines of two types: *structure-oriented* methods, which solely focus on network structure, and *content-oriented* methods that try to combine the network structure with the textual descriptors of the nodes (albeit using impoverished network neighborhoods so far). For the first type of methods, we employ NODE2VEC (Grover and Leskovec, 2016), which uses a biased random walk algorithm based on DEEPWALK (Perozzi et al., 2014) to explore the structure of the network more efficiently. Our work can be seen as an extension of NODE2VEC that incorporates textual node descriptors, as already discussed, hence it is natural to compare to NODE2VEC. As a *content-oriented* baseline we use CANE (Tu et al., 2017), which learns separate text-based and network-based embeddings, and uses a mutual attention mechanism to dynamically change the text-based embeddings for different neighbors (Section 2). CANE only considers the direct neighbors of each node, unlike NODE2VEC, which considers larger neighborhoods obtained via random walks.

4.3 Link Prediction

In link prediction, we are given a network with a certain fraction of edges removed. We need to infer these missing edges by observing the incomplete network, facilitating the discovery of links (e.g., unobserved protein-protein interactions).

Concretely, given a network we first randomly remove some percentage of edges, ensuring that the network remains connected so that we can perform random walks over it. Each removed edge e connecting nodes v_1, v_2 is treated as a *true positive*, in the sense that a link prediction method should infer that an edge should be added between v_1, v_2. We also use an equal number of *true negatives*, which are pairs of nodes v'_1, v'_2 with no edge between v'_1, v'_2 in the original network. When evaluating NE methods, a link predictor is given true positive and true negative pairs of nodes, and is required to discriminate between the two classes by examining only the node embeddings of each pair. Node embeddings are obtained by applying a NE method to the pruned network, i.e., after removing the true positives. A NE method is considered better than another one, if it leads to better performance of the same link predictor.

We experiment with two approaches to obtain true negatives. In *Random Negative Sampling*, we randomly select pairs of nodes that were not directly connected (by a single edge) in the original network. In *Close Proximity Negative Sampling*, we iterate over the nodes of the original network considering each one as a focus. For each focus node v, we want to find another node u in close proximity that is not an ancestor or descendent (e.g., parent, grandparent, child, grandchild) of v in the IS-A or PART-OF hierarchy, depending on the dataset. We want u to be close to v, to make it more difficult for the link predictor to infer that u and v should not be linked. We do not, however, want u to be an ancestor or descendent of v, because the IS-A and PART-OF relationships of our datasets are transitive. For example, if u is a grandparent of v, it could be argued that inferring that u and v should be linked, is *not* an error. To satisfy these constraints, we first find the ancestors of v that are between 2 to 5 hops away from v in the original network.[3] We randomly select one of these ancestors, and then we randomly select as u one of the ancestor's children in the original network, ensuring that u was not an ancestor or descendent of v in the original network. In both approaches, we randomly select as many true negatives as the true positives, discarding the remaining true negatives.

We experimented with two link predictors:

Cosine similarity link predictor (CS): Given a pair of nodes v_1, v_2 (true positive or true negative edge), CS computes the cosine similarity (ignoring negative scores) between the two node embeddings as $s(v_1, v_2) = \max(0, \cos(f(v_1), f(v_2)))$, and predicts an edge between the two nodes if $s(v_1, v_2) \geq t$, where t is a threshold. We evaluate the predictor on the true positives and true negatives (shown as 'test' true positives and 'test' true negatives in Table 1) by computing AUC (area under ROC curve), in effect considering the precision and recall of the predictor for varying t.[4]

Logistic regression link predictor (LR): Given a pair of nodes v_1, v_2, LR computes the Hadamard (element-wise) product of the two node embeddings $f(v_1) \odot f(v_2)$ and feeds it to a logistic regression classifier to obtain a probability estimate p that the two nodes should be linked. The predictor predicts an edge between v_1, v_2 if $p \geq t$. We compute AUC on a test set by varying t. The test set of this predictor is the same set of true positives and true negatives (with Random or Close Proximity Negative Sampling) that we use when evaluating the CS predictor. The training set of the logistic regression classifier contains as true positives all the other edges of the network that remain after the true positives of the test set have been removed, and an equal number of true negatives (with the same negative sampling method as in the test set) that are not used in the test set.

4.4 Implementation Details

For NODE2VEC and our NE methods, which can be viewed as extensions of NODE2VEC, the dimensionality of the node embeddings is 30. The dimensionality of the word embeddings (in our NE methods) is also 30. In the random walks, we set $r = 5$, $l = 40$, $k = 10$ for IS-A, and $r = 10$, $l = 40$, $k = 10$ for PART-OF; these hyper-parameters were not particularly tuned, and their values were selected mostly to speed up the experiments. We train for one epoch with a batch size of 128, setting the number of SKIPGRAM's negative samples to 2. We use the Adam (Kingma and Ba, 2015) optimizer in our NE methods. We implemented our NE methods and the two link predictors using PyTorch (Paszke et al., 2017) and Scikit-Learn (Pedregosa et al., 2011). For NODE2VEC and CANE, we used the implementations provided.[5]

---

[3]The edges of the resulting datasets are not directed. Hence, looking for descendents would be equivalent.

[4]We do not report precision, recall, F1 scores, because these require selecting a particular threshold t values.

[5]See `https://github.com/aditya-grover/node2vec`, `https://github.com/thunlp/CANE`.

NE Method + Link Predictor	Random Negative Sampling	Close Proximity Sampling
Node2Vec + CS	66.6	54.3
CANE + CS	94.1	69.6
Avg-N2V + CS	95.0	78.6
GRU-N2V + CS	**98.7**	**79.2**
BiGRU-Max-Res-N2V + CS	98.5	79.0
Node2Vec + LR	77.2	56.3
CANE + LR	95.3	70.0
Avg-N2V + LR	97.6	73.9
GRU-N2V + LR	99.0	79.6
BiGRU-Max-Res-N2V + LR	**99.3**	**82.1**

Table 2: AUC scores (%) for the IS-A dataset. Best scores per link predictor (CS, LR) shown in bold.

NE Method + Link Predictor	Random Negative Sampling	Close Proximity Sampling
Node2Vec + CS	76.8	61.8
CANE + CS	93.9	75.3
Avg-N2V + CS	95.9	81.8
GRU-N2V + CS	98.0	83.1
BiGRU-Max-Res-N2V + CS	**98.5**	**83.3**
Node2Vec + LR	85.2	66.5
CANE + LR	94.4	76.3
Avg-N2V + LR	97.6	79.4
GRU-N2V + LR	99.0	85.6
BiGRU-Max-Res-N2V + LR	**99.5**	**88.6**

Table 3: AUC scores (%) for the PART-OF dataset. Best scores per link predictor (CS, LR) shown in bold.

For CANE, we set the dimensionality of the node embeddings to 200, as in the work of Tu et al. (2017). We also tried 30-dimensional node embeddings, as in NODE2VEC and our NE methods, but performance deteriorated significantly.

4.5 Link Prediction Results

Link prediction results for the IS-A and PART-OF networks are reported in Tables 2 and 3. All content-oriented NE methods (CANE and our extensions of NODE2VEC) clearly outperform the structure-oriented method (NODE2VEC) on both datasets in both negative edge sampling settings, showing that modeling the textual descriptors of the nodes is critical. Furthermore, all methods perform much worse with Close Proximity Negative Sampling, confirming that the latter produces more difficult link prediction datasets.

All of our NE methods (content-aware extensions of NODE2VEC) outperform NODE2VEC and CANE in every case, especially with Close Proximity Negative Sampling. We conclude that it is important to model not just the textual descriptor of a node or its direct neighbors, but as much non-local network structure as possible.

For PART-OF relations (Table 3), BIGRU-MAX-RES-N2V obtains the best results with both link predictors (CS, LR) in both negative sampling settings, but the differences from GRU-N2V are very small in most cases. For IS-A (Table 2), BIGRU-MAX-RES-N2V obtains the best results with the LR predictor, and only slightly inferior results than GRU-N2V with the CS predictor. The differences of these two NE methods from AVG-N2V are larger, indicating that recurrent neural encoders of textual descriptors are more effective than simply averaging the word embeddings of the descriptors.

Target Node: Left Eyeball (PART-OF)		
Most Similar Embeddings	**Cos**	**Hops**
equator of left eyeball	99.3	1
episcleral layer of left eyeball	99.2	4
cavity of left eyeball	99.1	1
wall of left eyeball	99.0	1
vascular layer of left eyeball	98.9	1
Target Node: Lung Carcinoma (IS-A)		
Most Similar Embeddings	**Cos**	**Hops**
recurrent lung carcinoma	97.6	1
papillary carcinoma	97.1	2
lung pleomorphic carcinoma	97.0	3
ureter carcinoma	96.6	2
lymphoepithelioma-like lung carcinoma	96.6	3

Table 4: Examples of nodes whose embeddings are closest (cosine similarity, Cos) to the embedding of a target node in the PART-OF (top) and IS-A (bottom) datasets. We also show the distances (number of edges, Hops) between the nodes in the networks.

Finally, we observe that the best results of the LR predictor are better than those of the CS predictor, in both datasets and with both negative edge sampling approaches, with the differences being larger with Close Proximity Sampling. This is as one would expect, because the logistic regression classifier can assign different weights to the dimensions of the node embeddings, depending on their predictive power, whereas cosine similarity assigns the same importance to all dimensions.

- skin of left upper quadrant of right breast
- skin of right breast

(a) Two nodes connected by a PART-OF edge.

- wall of inflow part of right atrium
- inflow part of right atrium

(b) Two nodes connected by a PART-OF edge.

- anterior tongue carcinoma
- malignant anterior tongue neoplasm

(c) Two nodes connected by an IS-A edge.

Figure 4: Visualization of the importance that BIGRU-MAX-RES-N2V assigns to the words of the descriptors of the nodes of three edges. Edges (a) and (b) are from the PART-OF dataset. Edge (c) is from the IS-A dataset.

4.6 Qualitative Analysis

To better understand the benefits of leveraging both network structure and textual descriptors, we present examples from the two datasets.

Most similar embeddings: Table 4 presents the five nearest nodes for two target nodes ('Left Eyeball' and 'Lung Carcinoma'), based on the cosine similarity of the corresponding node embeddings in the PART-OF and IS-A networks, respectively. We observe that all nodes in the PART-OF example are very similar content-wise to our target node. Furthermore, the model captures the semantic relationship between concepts, since most of the returned nodes are actually parts of 'Left Eyeball'. The same pattern is observed in the IS-A example, with the exception of 'ureter carcinoma', which is not directly related with 'lung carcinoma', but is still a form of cancer. Finally, it is clear that the model extracts meaningful information from both the textual content of each node and the network structure, since the returned nodes are closely located in the network (Hops 1–4).

Heatmap visualization: BIGRU-MAX-RES-N2V can be extended to highlight the words in each textual descriptor that mostly influence the corresponding node embedding. Recall that this NE method applies a max-pooling operator (Fig. 3) over the state vectors $h_1, \ldots, h_n$ of the words $w_1, \ldots, w_n$ of the descriptor, keeping the maximum value per dimension across the state vectors. We count how many dimension-values the max-pooling operator keeps from each state vector h_i, and we treat that count (normalized to $[0, 1]$) as the importance score of the corresponding word w_i.[6] We then visualize the importance scores as

Edges/Descriptors	BN2V	CANE	N2V	Hops
(a) bariatric surgery (b) bypass gastrojejunostomy	82.7	38.0	56.2	11
(a) anatomical line (b) anterior malleolar fold	82.3	29.0	50.0	22
(a) zone of biceps brachii (b) short head of biceps brachii	93.0	70.0	61.6	13

Table 5: Examples of true positive edges, showing how structure and textual descriptors affect node embeddings. The first two edges are IS-A, the third one is PART-OF. The NE methods used are BIGRU-MAX-RES-N2V (BN2V), CANE and NODE2VEC (N2V). We report cosine similarities between node embeddings and the distances between the nodes (number of edges, Hops) in the networks after removing true positive edges.

heatmaps of the descriptors. In the first two example edges of Fig. 4, the highest importance scores are assigned to words indicating body parts, which is appropriate given that the edges indicate PART-OF relations. In the third example edge, the highest importance score of the first descriptor is assigned to 'carcinoma', and the highest importance scores of the second descriptor are shared by 'malignant' and 'neoplasm'; again, this is appropriate, since these words indicate an IS-A relation.

Case Study: In Table 5, we present examples that illustrate learning from both the network structure and textual descriptors. All three edges are true positives, i.e., they were initially present in the network and they were removed to test link prediction. In the first two edges, which come from the IS-A network, the node descriptors share no words. Nevertheless, BIGRU-MAX-RES-N2V (BN2V) produces node embeddings with high cosine similarities, much higher than NODE2VEC that uses only network structure, presumably because the word embeddings (and neural encoder) of BN2V correctly capture lexical relations (e.g., near-synonyms). Although CANE also considers the textual descriptors, its similarity scores are much lower, presumably because it uses only local neighborhoods (single-edge hops). The nodes in the third example, which come from the PART-OF network, have a larger word overlap. NODE2VEC

---

[6]We actually obtain two importance scores for each word in the descriptor of a node, since each node v has two embeddings $f(v), f'(v)$, used when v is the focus or a neighbor (Section 3), i.e., there are two results of the max-pooling operator. We average the two importance scores of each word.

is unaware of this overlap and produces the lowest score. The two content-oriented methods (BN2V, CANE) produce higher scores, but again BN2V produces a much higher similarity, presumably because it uses larger neighborhoods. In all three edges, the two nodes are distant (>10 hops), yet BN2V produces high similarity scores.

5 Conclusions and Future Work

We proposed a new method to learn content-aware node embeddings, which extends NODE2VEC by considering the textual descriptors of the nodes. The proposed approach leverages the strengths of both structure- and content-oriented node embedding methods. It exploits non-local network neighborhoods generated by random walks, as in the original NODE2VEC, and allows integrating various neural encoders of the textual descriptors. We evaluated our models on two biomedical networks extracted from UMLS, which consist of PART-OF and IS-A edges. Experimental results with two link predictors, cosine similarity and logistic regression, demonstrated that our approach is effective and outperforms previous methods which rely on structure alone, or model content along with local network context only.

In future work, we plan to experiment with networks extracted from other biomedical ontologies and knowledge bases. We also plan to explore if the word embeddings that our methods generate can improve biomedical question answering systems (McDonald et al., 2018).

Acknowledgements

This work was partly supported by the Research Center of the Athens University of Economics and Business. The work was also supported by the French National Research Agency under project ANR-16-CE33-0013.

References

Monica Agrawal, Marinka Zitnik, and Jure Leskovec. 2018. Large-scale analysis of disease pathways in the human interactome. *Pacific Symposium on Biocomputing. Pacific Symposium on Biocomputing*, 23:111–122.

Nesreen K. Ahmed, Ryan A. Rossi, John Boaz Lee, Xiangnan Kong, Theodore L. Willke, Rong Zhou, and Hoda Eldardiry. 2018. Learning role-based graph embeddings. *CoRR*, abs/1802.02896.

Dzmitry Bahdanau, Kyunghyun Cho, and Yoshua Bengio. 2015. Neural machine translation by jointly learning to align and translate. In *3rd International Conference on Learning Representations, ICLR 2015, San Diego, CA, USA, May 7-9, 2015, Conference Track Proceedings*.

James Bergstra, Daniel L K Yamins, and David D. Cox. 2013. Hyperopt: A python library for optimizing the hyperparameters of machine learning algorithms.

Olivier Bodenreider. 2004. The unified medical language system (UMLS): integrating biomedical terminology. *Nucleic Acids Research*, 32(Database-Issue):267–270.

Kyunghyun Cho, Bart van Merrienboer, Çaglar Gülçehre, Dzmitry Bahdanau, Fethi Bougares, Holger Schwenk, and Yoshua Bengio. 2014. Learning phrase representations using RNN encoder-decoder for statistical machine translation. In *Proceedings of the 2014 Conference on Empirical Methods in Natural Language Processing, EMNLP 2014, October 25-29, 2014, Doha, Qatar, A meeting of SIGDAT, a Special Interest Group of the ACL*, pages 1724–1734.

Ronan Collobert and Jason Weston. 2008. A unified architecture for natural language processing: deep neural networks with multitask learning. In *Machine Learning, Proceedings of the Twenty-Fifth International Conference (ICML 2008), Helsinki, Finland, June 5-9, 2008*, pages 160–167.

Alexis Conneau, Douwe Kiela, Holger Schwenk, Loïc Barrault, and Antoine Bordes. 2017. Supervised learning of universal sentence representations from natural language inference data. In *Proceedings of the 2017 Conference on Empirical Methods in Natural Language Processing, EMNLP 2017, Copenhagen, Denmark, September 9-11, 2017*, pages 670–680.

Aditya Grover and Jure Leskovec. 2016. node2vec: Scalable feature learning for networks. In *KDD*, pages 855–864. ACM.

Kaiming He, Xiangyu Zhang, Shaoqing Ren, and Jian Sun. 2015. Deep residual learning for image recognition. *CoRR*, abs/1512.03385.

Armand Joulin, Edouard Grave, Piotr Bojanowski, and Tomas Mikolov. 2016. Bag of tricks for efficient text classification. *CoRR*, abs/1607.01759.

Diederik P. Kingma and Jimmy Ba. 2015. Adam: A method for stochastic optimization. *CoRR*, abs/1412.6980.

Thomas N. Kipf and Max Welling. 2016. Semi-supervised classification with graph convolutional networks. *CoRR*, abs/1609.02907.

Chengwei Lei and Jianhua Ruan. 2013. A novel link prediction algorithm for reconstructing protein-protein interaction networks by topological similarity. *Bioinformatics*, 29(3):355–364.

Linyuan Lu and Tao Zhou. 2010. Link prediction in complex networks: A survey. *CoRR*, abs/1010.0725.

Ryan McDonald, Georgios-Ioannis Brokos, and Ion Androutsopoulos. 2018. Deep relevance ranking using enhanced document-query interactions. In *Proceedings of the Conference on Empirical Methods in Natural Language Processing*, pages 1849–1860, Brussels, Belgium.

Tomas Mikolov, Ilya Sutskever, Kai Chen, Gregory S. Corrado, and Jeffrey Dean. 2013. Distributed representations of words and phrases and their compositionality. In *NIPS*, pages 3111–3119.

Pushkar Mishra, Marco Del Tredici, Helen Yannakoudakis, and Ekaterina Shutova. 2019. Abusive language detection with graph convolutional networks. *CoRR*, abs/1904.04073.

Christopher J Mungall. 2004. Obol: Integrating language and meaning in bio-ontologies. *Comparative and Functional Genomics*, 5:509 – 520.

Philip V. Ogren, K. Bretonnel Cohen, George K. Acquaah-Mensah, Jens Eberlein, and Lawrence E Hunter. 2003. The compositional structure of gene ontology terms. *Pacific Symposium on Biocomputing. Pacific Symposium on Biocomputing*, pages 214–25.

Philip V. Ogren, K. Bretonnel Cohen, and Lawrence E Hunter. 2004. Implications of compositionality in the gene ontology for its curation and usage. *Pacific Symposium on Biocomputing. Pacific Symposium on Biocomputing*, pages 174–85.

Adam Paszke, Sam Gross, Soumith Chintala, Gregory Chanan, Edward Yang, Zachary DeVito, Zeming Lin, Alban Desmaison, Luca Antiga, and Adam Lerer. 2017. Automatic differentiation in pytorch.

Fabian Pedregosa, Gaël Varoquaux, Alexandre Gramfort, Vincent Michel, Bertrand Thirion, Olivier Grisel, Mathieu Blondel, Peter Prettenhofer, Ron Weiss, Vincent Dubourg, Jacob VanderPlas, Alexandre Passos, David Cournapeau, Matthieu Brucher, Matthieu Perrot, and Edouard Duchesnay. 2011. Scikit-learn: Machine learning in python. *Journal of Machine Learning Research*, 12:2825–2830.

Bryan Perozzi, Rami Al-Rfou, and Steven Skiena. 2014. Deepwalk: online learning of social representations. In *KDD*, pages 701–710. ACM.

Michael Schlichtkrull, Thomas N Kipf, Peter Bloem, Rianne Van Den Berg, Ivan Titov, and Max Welling. 2018. Modeling relational data with graph convolutional networks. In *European Semantic Web Conference*, pages 593–607. Springer.

Mike Schuster and Kuldip K. Paliwal. 1997. Bidirectional recurrent neural networks. *IEEE Trans. Signal Processing*, 45:2673–2681.

Prithviraj Sen, Galileo Namata, Mustafa Bilgic, Lise Getoor, Brian Gallagher, and Tina Eliassi-Rad. 2008. Collective classification in network data. *AI Magazine*, 29(3):93–106.

Dinghan Shen, Xinyuan Zhang, Ricardo Henao, and Lawrence Carin. 2018. Improved semantic-aware network embedding with fine-grained word alignment. In *Proceedings of the 2018 Conference on Empirical Methods in Natural Language Processing, Brussels, Belgium, October 31 - November 4, 2018*, pages 1829–1838.

Ali Shojaie. 2013. Link prediction in biological networks using multi-mode exponential random graph models.

Xiaofei Sun, Jiang Guo, Xiao Ding, and Ting Liu. 2016. A general framework for content-enhanced network representation learning. *CoRR*, abs/1610.02906.

Jian Tang, Meng Qu, Mingzhe Wang, Ming Zhang, Jun Yan, and Qiaozhu Mei. 2015. LINE: large-scale information network embedding. In *Proceedings of the 24th International Conference on World Wide Web, WWW 2015, Florence, Italy, May 18-22, 2015*, pages 1067–1077.

Cunchao Tu, Han Liu, Zhiyuan Liu, and Maosong Sun. 2017. CANE: context-aware network embedding for relation modeling. In *Proceedings of the 55th Annual Meeting of the Association for Computational Linguistics, ACL 2017, Vancouver, Canada, July 30 - August 4, Volume 1: Long Papers*, pages 1722–1731.

Daixin Wang, Peng Cui, and Wenwu Zhu. 2016. Structural deep network embedding. In *Proceedings of the 22nd ACM SIGKDD International Conference on Knowledge Discovery and Data Mining, San Francisco, CA, USA, August 13-17, 2016*, pages 1225–1234.

Peng Wang, Baowen Xu, Yurong Wu, and Xiaoyu Zhou. 2015. Link prediction in social networks: the state-of-the-art. *SCIENCE CHINA Information Sciences*, 58(1):1–38.

Wenchao Yu, Cheng Zheng, Wei Cheng, Charu C. Aggarwal, Dongjin Song, Bo Zong, Haifeng Chen, and Wei Wang. 2018. Learning deep network representations with adversarially regularized autoencoders. In *Proceedings of the 24th ACM SIGKDD International Conference on Knowledge Discovery & Data Mining, KDD 2018, London, UK, August 19-23, 2018*, pages 2663–2671.

Appendix

A   CANE Hyper-parameters

CANE has 3 hyper-parameters, denoted α, β, γ, which control to what extent it uses information from network structure or textual descriptors. We learned these hyper-parameters by employing the HyperOpt (Bergstra et al., 2013) on the validation set.[7] All three hyper-parameters had the same search space: $[0.2, 1, 0]$ with a step of 0.1. The optimization ran for 30 trials for both datasets. Table 6 reports the resulting hyper-parameter values.

Parameters	PART-OF	IS-A
α	0.2	0.7
β	1.0	0.7
γ	1.0	0.7

Table 6: Hyper-parameter values used in CANE.

---

[7] For more information on HyperOpt see: `https://github.com/hyperopt/hyperopt/wiki/FMin`. For a tutorial see: `https://github.com/Vooban/Hyperopt-Keras-CNN-CIFAR-100`.

Simplification-induced transformations: typology and some characteristics

Anaïs Koptient[1]   **Rémi Cardon[2]**   **Natalia Grabar[2]**

1. `{firstname.lastname}.etu@univ-lille.fr`
2. `{firstname.lastname}@univ-lille.fr`
CNRS, Univ. Lille, UMR 8163 STL - Savoirs Textes Langage F-59000 Lille, France

Abstract

The purpose of automatic text simplification is to transform technical or difficult to understand texts into a more friendly version. The semantics must be preserved during this transformation. Automatic text simplification can be done at different levels (lexical, syntactic, semantic, stylistic...) and relies on the corresponding knowledge and resources (lexicon, rules...). Our objective is to propose methods and material for the creation of transformation rules from a small set of parallel sentences differentiated by their technicity. We also proposc a typology of transformations and quantify them. We work with French-language data related to the medical domain, although we assume that the method can be exploited on texts in any language and from any domain.

1 Introduction

The purpose of automatic text simplification is to provide a simplified version for a given text. Simplification can be done at lexical, syntactic, semantic but also pragmatic and stylistic levels. Simplification can be useful in two main contexts: as help provided to human readers, which guarantees better access and understanding of the content of documents (Son et al., 2008; Paetzold and Specia, 2016; Chen et al., 2016; Arya et al., 2011; Leroy et al., 2013), and as a pre-processing step for other NLP tasks and applications, which makes easier the work of other NLP modules and may improve the overall results (Chandrasekar and Srinivas, 1997; Vickrey and Koller, 2008; Blake et al., 2007; Stymne et al., 2013; Wei et al., 2014; Beigman Klebanov et al., 2004). We can see that potentially this task may play an important role.

Three main types of methods are currently exploited in text simplification:

- *Methods based on knowledge and rules.* For instance, the use of WordNet (Miller et al., 1993) may provide equivalent expressions for difficult words (Carroll et al., 1998; Bautista et al., 2009), or help with the choice of synonyms using their frequency (Devlin and Tait, 1998; De Belder and Moens, 2010; Drndarevic et al., 2012) or their length (Bautista et al., 2009). One limitation of such methods is their weak recall performance (De Belder and Moens, 2010) and confusion between difficult and simple words (Shardlow, 2014);

- *Methods based on distribution probabilities*, like word embeddings (Mikolov et al., 2013; Pennington et al., 2014), are used to acquire a lexicon and substitution rules for simplification. When trained on relevant data (Wikipedia, Simple Wikipedia, PubMed Central...), word embeddings can contain simpler equivalents, that can be exploited to perform the simplification (Glavas and Stajner, 2015; Kim et al., 2016). Nonetheless, such methods require consequent filtering to keep only the best candidates. Those methods generally provide good coverage and, when the filtering is efficient, good precision;

- *Methods issued from machine translation* tackle the problem as translation from technical to simple text. A growing number of works propose to exploit this type of method to English texts (Zhao et al., 2010; Zhu et al., 2010; Wubben et al., 2012; Sennrich et al., 2016; Xu et al., 2016; Wang et al., 2016a,b; Zhang and Lapata, 2017; Nisioi et al., 2017). They exploit corpora made of parallel and aligned sentences, that mainly derive from the Simple English Wikipedia - English Wikipedia corpus (SEW-EW). Globally, those methods seem to maintain a balance between the quality of the simplification, good coverage and precision.

Proceedings of the BioNLP 2019 workshop, pages 309–318
Florence, Italy, August 1, 2019. ©2019 Association for Computational Linguistics

Almost all the existing works address text simplification in English, while other languages are poorly described. Yet, whatever the method and language it is necessary to have available suitable resources for making the transformations required by the task. This work is intended as a basis to design a method and to use it for preparing linguistic data for the creation of transformation rules.

2 Linguistic Data

We exploit an existing corpus with comparable documents[1] differentiated by their technicity: technical documents and the corresponding simplified documents. The corpus is composed of documents from three sources: information on drugs, encyclopedia articles and abstracts from systematic reviews. We use *simple* and *simplified* interchangeably in our work. Yet, a *simplified* document is the result of the simplification process of a technical document, like the simplified abstracts from systematic reviews; while a *simple* document is issued from an independently written simple document, like drug information and encyclopedia articles. In the used corpus, the technical part contains over 2.8M occurrences, and the simplified part contains over 1.5M occurrences. A subset of this corpus has been manually aligned at the level of sentences, which provides 663 pairs of parallel sentences exploited in our work. These pairs of sentences show two types of relations:

- *Semantic equivalence:* two sentences of a pair have the same or very close meaning:
  - *les sondes gastriques sont couramment utilisées pour administrer des médicaments ou une alimentation entérale aux personnes ne pouvant plus avaler(feeding tubes are often used to administer medicine or enteral nutrition to people who cannot swallow)*
  - *les sondes gastriques sont couramment utilisées pour administrer des médicaments et de la nourriture directement dans le tractus gastro-intestinal (un tube permettant de digérer les aliments) pour les personnes ne pouvant pas avaler (feeding tubes are often used to administer medicine and food directly into the gastrointestinal tract (a tube that allows to digest food))*

With the semantic equivalence, simplification is mainly performed at lexical level, typ-

ically using lexical substitutions. Simplification can also be done by adding information and, in this case, complex notions are followed by their explanations, like *le tractus gastro-intestinal (un tube permettant de digérer les aliments)*. Often, those two processes (substitution and addition of information) are applied jointly;

- *Semantic inclusion:* the meaning of one sentence is included in the meaning of the other sentence. The inclusion is oriented: the technical sentence as well as the simplified sentence can be inclusive or included. In this example, the technical sentence is inclusive and informs in addition on the number of participants and the evaluation metric:
  - *peu de données (43 participants) étaient disponibles concernant la détection d'un mauvais placement (la spécificité) en raison de la faible incidence des mauvais placements (only a few data (43 participants) were available concerning the detection of a bad placement (specificity), due to the weak incidence of bad placements)*
  - *cependant, peu de données étaient disponibles concernant les sondes placées incorrectement et les complications possibles d'une sonde mal placée (however, only a few data were available concerning the badly placed probes and the potential complications of a bad probe placement)*

With inclusion, simplification is also peformed at the syntactic level, as the example above illustrates. Typically, subordinate and inserted clauses, information between brackets, some adjectives of adverbs are deleted during the simplification, like the information between brackets (*43 participants* and *la spécificité*). Semantic inclusion also involves enumerations: technical sentences with coordination can be segmented into lists with separate items in the simplified versions. Yet, enumerations with comma-separated items can be found in either technical and simplified documents. We should also point out that syntactic and lexical transformations often occur together.

---

[1]`http://natalia.grabar.free.fr/`
`resources.php`

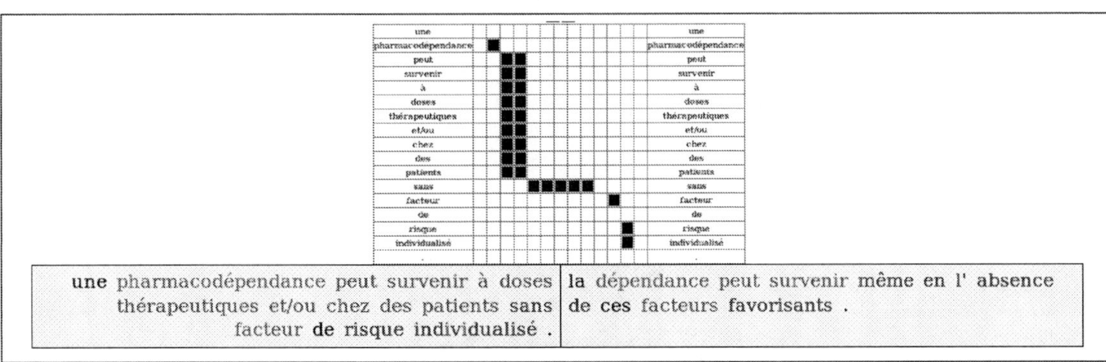

Figure 1: Matrix-based alignment of words within YAWAT

3 Methods

The methods for annotating and preparing the linguistic data for the description of simplification-induced transformations rely on three main dimensions: (1) control of the semantic inclusion relations, when sentences are split or merged during the simplification (Section 3.1); (2) semantic annotation of pairs of sentences to describe more precisely the transformations (Section 3.2); and (3) syntactic tagging and analysis for joining the semantic and syntactic information (Section 3.3).

3.1 Merging and Splitting of Sentences

One typical strategy applied during text simplification consists in merging or splitting the technical sentences when creating simple sentences (Brouwers et al., 2014). When merged, the technical sentences become shorted, which allows their merging into one sentence which yet remains readable in the simplified version. On contrary, when a given technical sentence contains more than one clause, like one main and one secondary, it can be split into two sentences by transforming the secondary clause into the main clause of another sentence. Sometimes, the splitting should be blocked because it can make the understanding of the main clause more difficult (Brunato et al., 2014).

In our corpus, merged and split sentences are detected using their proximity in the corpus and multiple alignments, like in these examples:

T_1 *elle impose l' arrêt du traitement et contre-indique toute nouvelle administration de clindamycine.* (It forces the stopping of the cure and it is contra-indicated to administer once again the clindamycin.)

S_1 *prévenez votre médecin immédiatement car cela impose l' arrêt du traitement.* (tell it to your doctor immediately for it forces the stopping of the cure.)
cette réaction va contre-indiquer toute nouvelle administration de clindamycine. (this response contra-indicates any new administration of the clindamycin.)

T_2 *abcès.* (abscess)
douleurs. (pain)

S_2 *douleurs ou accumulation de pus au niveau du site d' injection* (pain or accumulation of pus at the injection site)

Note that in the case of merging, the complex sentences when they are merged get also other simplifications, such as synonymy for instance.

3.2 Semantic Annotation

The simplification-induced transformations are annotated semantically using YAWAT (Yet Another Word Alignment Tool) (Germann, 2008). YAWAT permits to visualize and manipulate parallel texts. The tool was designed for working with parallel bilingual texts related to mutual translations (Yu et al., 2012). We propose to exploit it with monolingual parallel texts related to simplification. YAWAT displays the two parallel and aligned sentences side by side. The annotator can then align the words using the matrix (Figure 1), and to assign the type of transformation to each pair of text segments considered. The number of squares displayed vertically correspond to the number of words that are counted in the sentence on the left (that is, the technical sentence). The number of squares displayed horizontally correspond to the number of words that are counted in the sentence on the right (that is, the simple sentence). Then, in order to match word/group

311

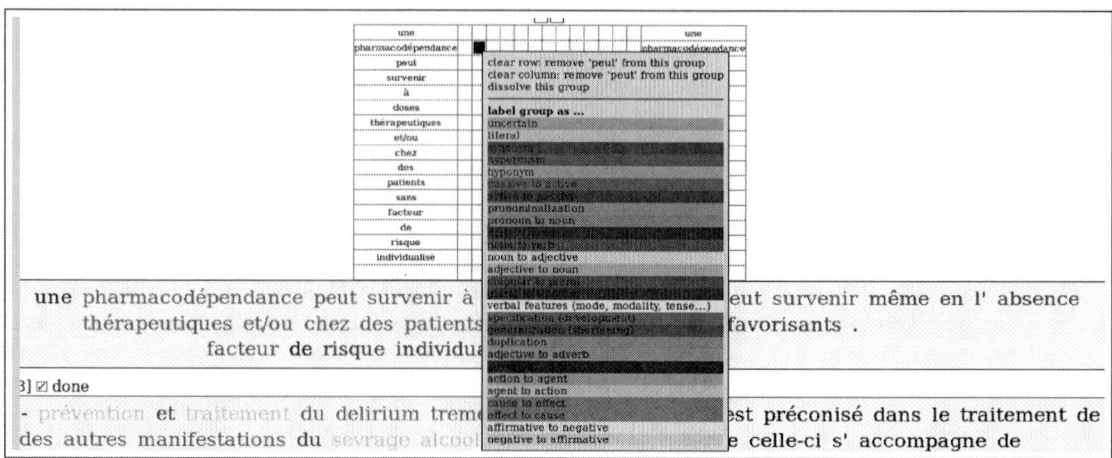

Figure 2: Annotation scheme within YAWAT

of words that correspond to a simplification phenomenon, the user has to click on the square that corresponds in both sentences. The fact that the text in the two columns are the same permits the user to click on the right square easily. The transformation types permits to describe more precisely their semantic nature. We defined a set of transformation types using previous similar work and observations on our corpus (Brunato et al., 2014). The proposed typology contains up to 25 transformations (Figure 2):

- *literal* is the default value which is kept when the words are identical in both sentences,

- *synonym*: substitution of technical term by its synonym {*effets négligeables* ; *effets délétères*} ({*insignificant effects* ; *deleterious effects*}),

- *hyperonym*: technical term is replaced by its hyperonym {*clindamicine* ; *médicament*} ({*clindamycin* ; *drug*}),

- *hyponym*: technical term is replaced by its hyponym {*benzodiazépines* ; *bromazepam*} ({*benzodiazepine* ; *bromazepam*})

- *p2a* (and *a2p*): verb in passive voice in technical sentence is replaced by its active voice {*ne doit jamais être utilisé* ; *ne prenez jamais*} ({*should never be used* ; *never use*}), and the contrary {*n'a aucun* ; *n'est pas attendu*} ({*does not have* ; *is not expected*})

- *pronominalization*: substitution by pronouns {*l'antibioprophylaxie* ; *elle*} ({*the antibiotic prophylaxis* ; *it*}),

- *p2n*: substitution of pronoun by its reference {*elles* ; *ce médicament*} ({*they* ; *this drug*}),

- *v2n* (and *n2v*): substitution of verbs by nouns {*conduire* ; *conduite*} ({*to drive* ; *driving*}), and the contrary {*l'arrêt du traitement* ; *arrêter brutalement*} ({*cessation of treatment* ; *stop at once*})

- *n2a* (and *a2n*): substitution of nouns by their adjectives {*allergies* ; *allergiques*} ({*allergies* ; *allergic*}), and the contrary {*cardiaque* ; *du coeur*} ({*cardiac* ; *of the heart*}),

- *s2p* (and *p2s*): substitution of singular forms by plural forms {*de tout antibiotique* ; *d'antibiotiques*} ({*of any antibiotic* ; *of antibiotics*}), and the contrary {*les enfants* ; *l'enfant*} ({*the children* ; *the child*}),

- *specification*: adding explanation to technical term {*bêta-lactamines* ; *bêta-lactamines (pénicilline, céphalosporine)*} ({*beta-lactam* ; *beta-lactam (penicillin, cephalosporins)*}). The difference with *synonymy* is that, instead of substitution, the technical term remains and its explanation (definition, examples) is added,

- *generalization*: removal of some information {*arrêt du traitement et contre indique toute nouvelle administration du clindamycine* ; *arrêt du traitement*} ({*cessation of treatment and contra-indicate any new administration of clindamycine* ; *cessation of treatment*}),

- *duplication*: two or more occurrences of a given term in simple sentence,

312

- *adj2adv* (and *adv2adj*): substitution of adjectives by adverbs {*récente* ; *récemment*} ({*recent* ; *recently*}), and the contrary {*tardif* ; *tard*} ({*late* ; *late*}),

- *agt2act* (and *act2agt*): substitution of agent by the action {*conducteurs* ; *conduite*} ({*drivers* ; *driving*}), and the contrary {*conduite* ; *conducteurs*} ({*driving* ; *drivers*}),

- *cau2eff* (and *eff2cau*): substitution of cause by its effect {*prescrits* ; *utilisés*} ({*prescribed* ; *used*}), and the contrary {*dans le traitement* ; *chez les patients atteints*} ({*during the treatment* ; *in affected patients*}),

- *aff2neg* (and *neg2aff*): substitution of affirmative form of information by negative form with the same meaning {*présentant une absence complète* ; *n'avez aucune*} ({*indicating total absence* ; *you have no*}), and the contrary {*ne pas* ; *éviter*} ({*should not* ; *avoid*}).

Since it is common that some sequences can be tagged with several concurrent tags, we defined the priority rules, such as *u2n > synonym*, like in {*cardiaque* ; *du coeur*} ({*cardiac* ; *of the heart*}), because it describes the transformation more precisely. Since it is common that some sequences can be tagged with several concurrent tags, we prioritized part-of-speech related tags over synonymy because it is more precize, like in {*cardiaque* ; *du coeur*} ({*cardiac* ; *of the heart*}). Similarly, pronominalization is prioritized over verbal features, and also all the lexical transformations over syntactic transformations.

3.3 Syntactic Analysis

Syntactic analysis permits to linguistically annotate the parallel sentences and to mark within them the syntactic groups. Syntactic processing is done with Cordial (Laurent et al., 2009), which performs tokenization, POS-tagging, lemmatisation and syntactic analysis into constituents. In Table 1, we provide an example of Cordial tagging and analysis for the sentence *dalacine n'a aucun effet ou qu'un effet négligeable sur l'aptitude à conduire des véhicules et à utiliser des machines.* *(dalacine has no effect or the effect is insignificant on the capacity to drive vehicles and to use machines.)* We can see that the sequence *un effet négligeble (insignificant effect)* belongs to the same syntactic group, stated in column *synt. group*. Besides, the syntactic head is

nb.	form	POS tag	synt. group
1	dalacine	NCI	1
2	n'	ADV	3
3	a	VINDP3S	3
4	aucun	ADJIND	5
5	effet	NCMS	5
6	ou	COO	-
7	qu'	ADV	3
8	un	DETIMS	9
9	effet	NCMS	9
10	négligeable	ADJSIG	9
11	sur	PREP	13
12	l'	DETDFS	13
13	aptitude	NCFS	13
14	à	PREP	15
15	conduire	VINF	15
16	des	DETDPIG	17
17	véhicules	NCMP	17
18	et	COO	-
19	à	PREP	20
20	utiliser	VINF	20
21	des	DETDPIG	22
22	machines	NCFP	22
23	.	PCTFORTE	-

Table 1: Example of syntactic annotation by Cordial (word position in the sentence, word form, POS tag, and syntactic group)

effet (effect), which has the same number as the syntactic groupe (9) and, being common noun (*NC*), it characterizes this group as nominal phrase.

4 Results and Discussion

4.1 Merging and Splitting of Sentences

We counted 51 cases in which two of more technical sentences are merged into one simple sentence, and 16 cases in which technical sentences are split into two or more simple sentences. In a previous work, it was noticed that the merging of sentences during the simplification is rare (Brouwers et al., 2014). Yet, in our corpus, we observe the contrary: much more technical sentences are merged than split. We can see several explanations:

- The cited work (Brouwers et al., 2014) is done on articles from Wikipedia and Vikidia. Vikidia is designed for 8-13 year old children and relies on strong guidelines when creating the articles. One of the rules is to use short and clear sentences. In our work, Wikipedia and Vikidia correspond to the encycopedia part of the corpus. The two other subcorpora (drug leaflets and scientific abstracts) do not respect same writing principles.

- Drug leaflets frequently use coordinations with disorders, known adverse effects, functions, etc. Often, they are presented as itemized lists in technical documents, while in simplified documents then occur within coordinated sentences.

- In abstracts of systematic reviews, technical sentences are often shortened during their simplification in order to keep the main information. Then, possibly as consequence of it, the sentences may be merged. Notice also that there is no clear guidelines when writing plain-language abstracts and that each editor may apply its own principles.

4.2 Semantic Annotation

In Figure 3, we present the typology of the simplification-induced transformations. The Figure also contains information on prevalence of each transformation in terms of its frequence and percentage. We distinguish several high-level transformations, which may also be present in the existing typologies (Brunato et al., 2014; Brouwers et al., 2014): lexical substitution, lexical addition, lexical deletion, syntactic substitution, pronominalization and use of affirmative and negated forms. The biggest set of transformations (965 occurrences, 69%) is related to lexical substitutions, within which we differentiate substitutions with semantic shift (hyponymy and hyperonymy) and without semantic shift (synonymy and morphological transformation). We subsequently have lexical additions or specifications (199 occurrences, 14%), when explanations are added to technical terms in simplifed sentences, and lexical deletions or generalizations (132 occurrences, 9%), when some information is shortened and removed during the simplification. Then we consider that the only pure syntactic substitutions correpospond to active and passive voices of verbs. Hence, singular/plural and other verbal features belong to lexical substitutions without semantic shift. Pronominalization, and use of positive and negative equivalent expressions correspond to distinct small types of transformations.

By comparison with the typology from (Brouwers et al., 2014), we separated synonymy from hy-peronymy because they have fundamental differences (semantic equivalence or subsumption) and require specific methods and resources. We differentiate several syntactic and morphological transformations, while in the citetd work, only the passive/active transformation is considered. Another difference is that we do not differentiate betweel lexical and semantic transformations: semantics becomes a feature of lexical substitutions.

By comparison with the typology from (Brunato et al., 2014), the authors differentiate several types of word insertion and deletion, according to the syntactic nature of these words (verb, noun...). We do not make this differentiation because, in most cases, insertions and deletions apply to syntactic clauses. Besides, we considered the shift of grammatical categories as lexical substitution, which we describe with detail according to the POS categories. Unlike in the cited work, we consider separately hyperonymy, hyponymy and synonymy, because they have fundamental differences and require specific methods and resources.

Finally, by comparison with the typology from (Vila et al., 2011), which is dedicated to the general description of paraphrases and does not specifically aim transformations due to the simplification, we notice several similarities. The main difference is that the authors separated lexical substitutions and morphological derivations, which we keep together because they all apply at the word level. Yet, we can differentiate them through the use of syntactic infomartion.

On the whole, we count 1,394 transformations, which gives 2.1 transformations per pair of sentences on average. In Table 2, we indicate the frequency of the most frequent types of transformations according to whether they occur in split or merged sentences, or generally in the corpus (the *total* column). As in Figure 3, the most frequent transformations are related to the use of synonyms, and the specification and generalization of contents. These types are frequent in the whole corpus and, by consequence, in merged and split sentences. There is no real association between sentence splitting or merging and transformations. At the more fine-grained level, we observe that:

- *a2n* (adjective → noun) transformations (53 occ.) may be necessary to replace adjectives, often coined on suppletive bases (*cardiac*), by the corresponding nouns, often coined on

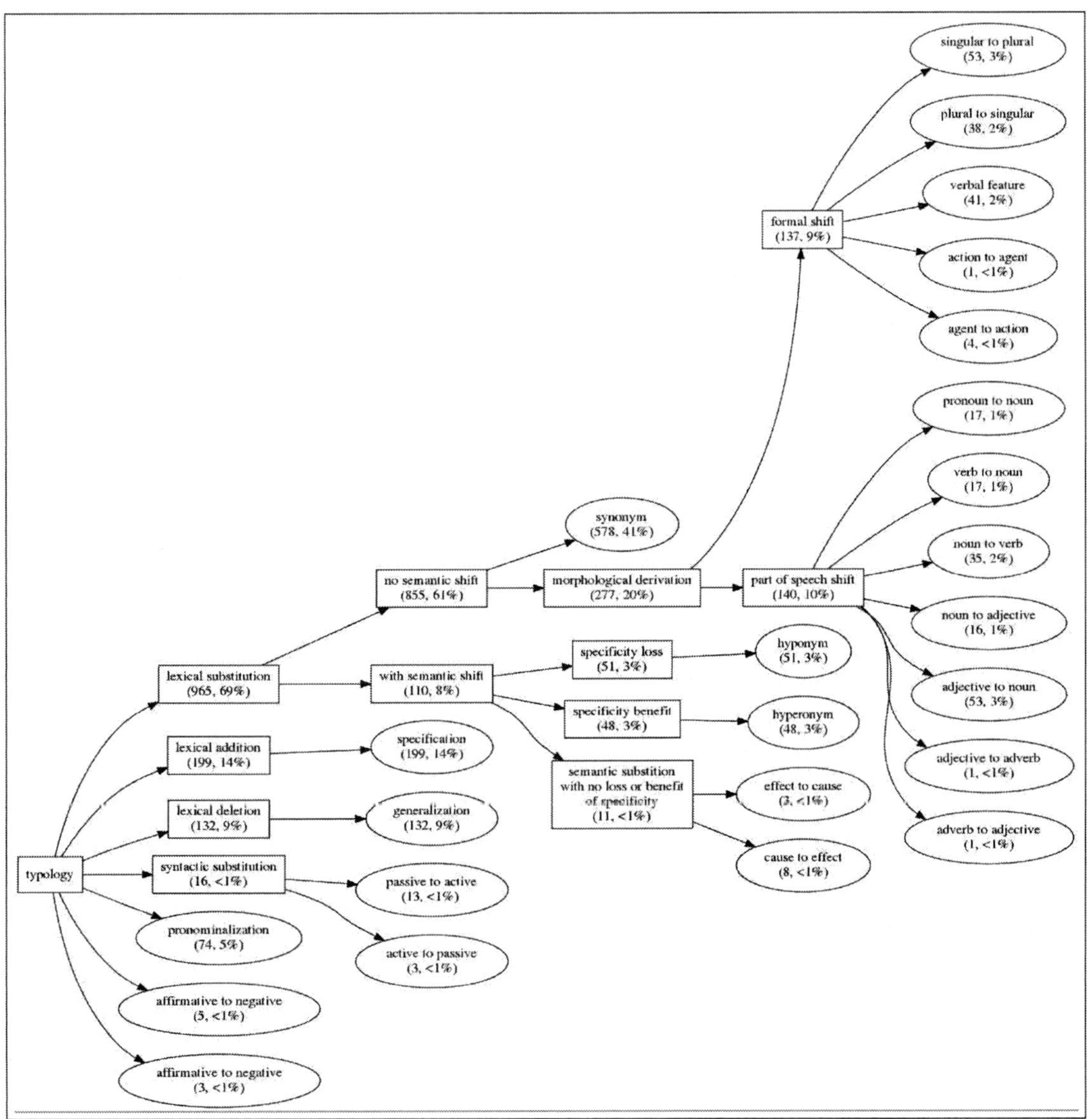

Figure 3: Typology of simplification-induced transformations

alternative native bases (*heart*),

- *hyperonymy* transformations (48 occ.) permit to use words with broader meaning, which may make the understanding easier,

- *hyponymy* transformations (51 occ.) permit to use instanciations and terms with narrower meaning, which may also make the understanding easier,

- *n2v* (noun → verb) transformations (35 occ.) make the sentence less abstract by replacing concept by the action, and hence easier to understand.

It may seem counter-intuitive that there are more cases of hyponymy than hyperonymy in simplification, however, this can be explained. Indeed, in the simple side of the drug corpus, the exact name of the drug is given, when on the technical side of the drug corpus, the name given is the therapeutical class of the drug. For instance, there is a case where we have *IEC* (*ACE inhibitor*) on the technical side and *Moex* (the name of a drug) on the simple side. Since *Moex* is a kind of *IEC*, then *Moex* is a hyponym for *IEC*.

315

tag	split	merge	total
syno	24	112	578
hypero	1	10	48
hypo	0	13	51
pronoun	9	2	74
v2n	1	1	17
n2v	2	5	35
n2a	2	0	16
a2n	2	17	53
s2p	0	6	53
p2s	5	3	38
vfea	0	4	41
specif	12	34	199
gener	14	10	132

Table 2: Frequency of the most frequent transformations in split and merged sentences, and in all the parallel sentences

4.3 Syntactic Analysis

Syntactic analysis permitted to associate semantic and syntactic information. One issue is that, with the substitutions, the POS-tags or syntactic groups remain identical in 221 cases. In several other cases, the original syntactic group is completed with other groups ($GN \rightarrow GP\ GN$, $GN \rightarrow GN\ GAdj$). Besides, up to 531 transformations start from nominal groups, up to 190 from prepositional groups and up to 174 from verbal groups. Overall, this means that: (1) the syntactic analysis may provide important clues for the detection of frontiers of the sequences to transform; (2) words and expressions of various syntactic nature are involved in transformations (nouns, verbs, adjectives...); (3) nouns and noun groups, often corresponding to concepts, occupy important place among the transformations.

5 Conclusion and Future Work

We proposed to work with parallel sentences differentiated by their technicity: technical and simplified contents are put in parallel. The main purpose is to describe the transformations involved during the simplification. Hence, the sentences are characterized on three dimensions: splitting and merging of sentences, semantic annotation of the transformations, and their syntactic annotation. We also propose a typology of transformations and quantify them. For instance, our work indicates that among the most frequent transformations we can find: synonymy, specification (in-

sertion of additional information), generalization (removal of information), pronominalization, substitution of adjectives by the corresponding nouns, and swich between singular and plural forms. The material prepared will be used for the creation of transformation rules joining syntactic, lexical and semantic information. These rules will be later used for the simplification of biomedical texts.

Acknowledgments

This work was funded by the French National Agency for Research (ANR) as part of the *CLEAR* project (*Communication, Literacy, Education, Accessibility, Readability*), ANR-17-CE19-0016-01.

The authors would like to thank the reviewers for their questions and comments that helped in improving the article.

References

Diana J. Arya, Elfrieda H. Hiebert, and P. David Pearson. 2011. The effects of syntactic and lexical complexity on the comprehension of elementary science texts. *Int Electronic Journal of Elementary Education*, 4(1):107–125.

Susana Bautista, Pablo Gervás, and R. Ignacio Madrid. 2009. Feasibility analysis for semi-automatic conversion of text to improve readability. In *Int Conf on Inform and Comm Technology and Accessibility (ICTA)*, pages 33–40.

B Beigman Klebanov, K Knight, and D Marcu. 2004. Text simplification for information-seeking applications. In R Meersman and Z Tari, editors, *On the Move to Meaningful Internet Systems 2004: CoopIS, DOA, and ODBASE*. Springer, LNCS vol 3290, Berlin, Heidelberg.

Catherine Blake, Julia Kampov, Andreas Orphanides, David West, and Cory Lown. 2007. Query expansion, lexical simplification, and sentence selection strategies for multi-document summarization. In *DUC*.

Laetitia Brouwers, Delphine Bernhard, Anne-Laure Ligozat, and Thomas François. 2014. Syntactic sentence simplification for French. In *PITR workshop*, pages 47–56.

Dominique Brunato, Felice Dell'Orletta, Giulia Venturi, and Simonetta Montemagni. 2014. Defining an annotation scheme with a view to automatic text simplification. In *CLICIT*, pages 87–92.

J Carroll, G Minnen, Y Canning, S Devlin, and J Tait. 1998. Practical simplification of English newspaper text to assist aphasic readers. In *AAAI-98 Workshop on Integrating Artificial Intelligence and Assistive Technology*, pages 7–10.

R Chandrasekar and B Srinivas. 1997. Automatic induction of rules for text simplification. *Knowledge Based Systems*, 10(3):183–190.

Ping Chen, John Rochford, David N. Kennedy, Soussan Djamasbi, Peter Fay, and Will Scott. 2016. Automatic text simplification for people with intellectual disabilities. In *AIST*, pages 1–9.

Jan De Belder and Marie-Francine Moens. 2010. Text simplification for children. In *Workshop on Accessible Search Systems of SIGIR*, pages 1–8.

Siobhan Devlin and John Tait. 1998. The use of psycholinguistic database in the simplification of text for aphasic readers. In *Linguistic Database*, pages 161–173.

B Drndarevic, S Stajner, and H Saggion. 2012. Reporting simply: A lexical simplification strategy for enhancing text accessibility. In *Easy to read on the web*, pages 1–6.

Ulrich Germann. 2008. Yawat: Yet another word alignment tool. In *ACL-08: HLT Demo Session*, pages 20–23, Columbus, USA.

Goran Glavas and Sanja Stajner. 2015. Simplifying lexical simplification: Do we need simplified corpora? In *ACL-COLING*, pages 63–68.

Yea-Seul Kim, Jessica Hullman, Matthew Burgess, and Eytan Adar. 2016. SimpleScience: Lexical simplification of scientific terminology. In *EMNLP*, pages 1–6.

D Laurent, S Nègre, and P Séguéla. 2009. L'analyseur syntaxique Cordial dans Passage. In *Traitement Automatique des Langues Naturelles (TALN)*.

Gondy Leroy, David Kauchak, and Obay Mouradi. 2013. A user-study measuring the effects of lexical simplification and coherence enhancement on perceived and actual text difficulty. *Int J Med Inform*, 82(8):717–730.

T Mikolov, K Chen, G Corrado, and J Dean. 2013. Efficient estimation of word representations in vector space. In *Workshop at ICLR*.

George A. Miller, Richard Beckwith, Christiane Fellbaum, Derek Gross, and Katherine Miller. 1993. Introduction to wordnet: An on-line lexical database. Technical report, WordNet.

Sergiu Nisioi, Sanja Stajner, Simone Paolo Ponzetto, and Liviu P. Dinu. 2017. Exploring neural text simplification models. In *Ann Meeting of the Assoc for Comp Linguistics*, pages 85–91.

Gustavo H. Paetzold and Lucia Specia. 2016. Benchmarking lexical simplification systems. In *LREC*, pages 3074–3080.

Jeffrey Pennington, Richard Socher, and Christopher D Manning. 2014. Glove: Global vectors for word representation. In *EMNLP 2014*, pages 1532–1543.

Rico Sennrich, Barry Haddow, and Alexandra Birch. 2016. Improving neural machine translation models with monolingual data. In *Proc of the Ann Meeting of the Assoc for Comp Linguistics*, pages 86–96, Berlin, Germany.

Matthew Shardlow. 2014. A survey of automated text simplification. *Int J Advanced Computer Science and Applications*, 1:1–13.

Ji Y. Son, Linda B. Smith, and Robert L. Goldstone. 2008. Simplicity and generalization: Shortcutting abstraction in children's object categorizations. *Cognition*, 108:626–638.

S Stymne, J Tiedemann, C Hardmeier, and J Nivre. 2013. Statistical machine translation with readability constraints. In *NODALIDA*, pages 1–12.

D Vickrey and D Koller. 2008. Sentence simplification for semantic role labeling. In *Annual Meeting of the Association for Computational Linguistics-HLT*, pages 344–352.

Marta Vila, M Antònia Mart, and Horacio Rodríguez. 2011. Paraphrase concept and typology. A linguistically based and computationally oriented approach. *Procesamiento del Lenguaje Natural*, 46:83–90.

Tong Wang, Ping Chen, Kevin Amaral, and Jipeng Qiang. 2016a. An experimental study of LSTM encoder-decoder model for text simplification. In *IJCAI*, pages 1–7.

Tong Wang, Ping Chen, John Rochford, and Jipeng Qiang. 2016b. Text simplification using neural machine translation. In *Proc of the AAAI Conference on Artificial Intelligence (AAAI-16)*, pages 4270–4271.

Chih-Hsuan Wei, Robert Leaman, and Zhiyong Lu. 2014. SimConcept: A hybrid approach for simplifying composite named entities in biomedicine. In *BCB '14*, pages 138–146.

S Wubben, A van den Bosch, and E Krahmer. 2012. Sentence simplification by monolingual machine translation. In *Annual Meeting of the Association for Computational Linguistics*, pages 1015–1024.

Wei Xu, Courtney Napoles, Ellie Pavlick, Quanze Chen, and Chris Callison-Burch. 2016. Optimizing statistical machine translation for text simplification. *Transactions of the Association for Computational Linguistics*, 4:401–415.

Qian Yu, Aurélien Max, and François Yvon. 2012. Revisiting sentence alignment algorithms for alignment visualization and evaluation. In *BUCC workshop*, pages 1–7.

Xingxing Zhang and Mirella Lapata. 2017. Sentence simplification with deep reinforcement learning. In *Proc of the Conf on Empirical Methods in Natural Language Processing*, pages 584–594, Copenhagen, Denmark.

S Zhao, H Wang, and T Liu. 2010. Leveraging multiple
MT engines for paraphrase generation. In *COLING*,
pages 1326–1334.

Z Zhu, D Bernhard, and I Gurevych. 2010. A monolingual tree-based translation model for sentence simplification. In *COLING 2010*, pages 1353–1361.

ScispaCy: Fast and Robust Models
for Biomedical Natural Language Processing

Mark Neumann, Daniel King, Iz Beltagy, Waleed Ammar
Allen Institute for Artificial Intelligence, Seattle, WA, USA
`{markn,daniel,beltagy,waleeda}@allenai.org`

Abstract

Despite recent advances in natural language processing, many statistical models for processing text perform extremely poorly under domain shift. Processing biomedical and clinical text is a critically important application area of natural language processing, for which there are few robust, practical, publicly available models. This paper describes scispaCy, a new Python library and models for practical biomedical/scientific text processing, which heavily leverages the spaCy library. We detail the performance of two packages of models released in scispaCy and demonstrate their robustness on several tasks and datasets. Models and code are available at `https://allenai.github.io/scispacy/`.

1 Introduction

The publication rate in the medical and biomedical sciences is growing at an exponential rate (Bornmann and Mutz, 2014). The information overload problem is widespread across academia, but is particularly apparent in the biomedical sciences, where individual papers may contain specific discoveries relating to a dizzying variety of genes, drugs, and proteins. In order to cope with the sheer volume of new scientific knowledge, there have been many attempts to automate the process of extracting entities, relations, protein interactions and other structured knowledge from scientific papers (Wei et al., 2016; Ammar et al., 2018; Poon et al., 2014).

Although there exists a wealth of tools for processing biomedical text, many focus primarily on named entity recognition and disambiguation. MetaMap and MetaMapLite (Aronson, 2001; Demner-Fushman et al., 2017), the two most widely used and supported tools for biomedical text processing, support entity linking with negation detection and acronym resolution. However,

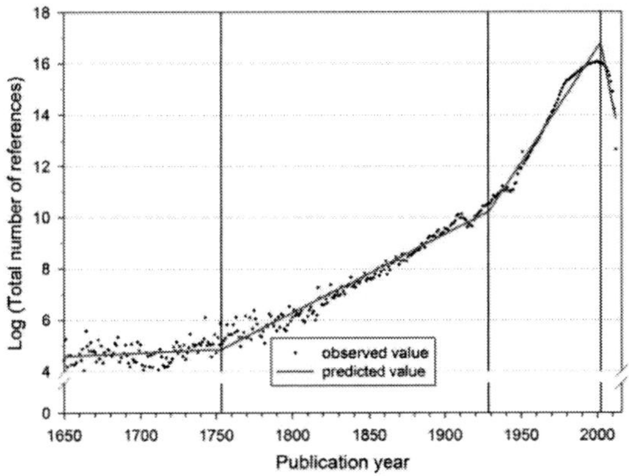

Figure 1: Growth of the annual number of cited references from 1650 to 2012 in the medical and health sciences (citing publications from 1980 to 2012). Figure from (Bornmann and Mutz, 2014).

tools which cover more classical natural language processing (NLP) tasks such as the GENIA tagger (Tsuruoka et al., 2005; Tsuruoka and Tsujii, 2005), or phrase structure parsers such as those presented in McClosky and Charniak (2008) typically do not make use of new research innovations such as word representations or neural networks.

In this paper, we introduce scispaCy, a specialized NLP library for processing biomedical texts which builds on the robust spaCy library,[1] and document its performance relative to state of the art models for part of speech (POS) tagging, dependency parsing, named entity recognition (NER) and sentence segmentation. Specifically, we:

- Release a reformatted version of the GENIA 1.0 (Kim et al., 2003) corpus converted into Universal Dependencies v1.0 and aligned with the original text from the PubMed abstracts.

---

[1] `spacy.io`

319

Proceedings of the BioNLP 2019 workshop, pages 319–327
Florence, Italy, August 1, 2019. ©2019 Association for Computational Linguistics

Model	Vocab Size	Vector Count	Min Word Freq	Min Doc Freq
en_core_sci_sm	58,338	0	50	5
en_core_sci_md	101,678	98,131	20	5

Table 1: Vocabulary statistics for the two core packages in scispaCy.

- Benchmark 9 named entity recognition models for more specific entity extraction applications demonstrating competitive performance when compared to strong baselines.

- Release and evaluate two fast and convenient pipelines for biomedical text, which include tokenization, part of speech tagging, dependency parsing and named entity recognition.

2 Overview of (sci)spaCy

In this section, we briefly describe the models used in the spaCy library and describe how we build on them in scispaCy.

spaCy. The Python-based spaCy library (Honnibal and Montani, 2017)[2] provides a variety of practical tools for text processing in multiple languages. Their models have emerged as the defacto standard for practical NLP due to their speed, robustness and close to state of the art performance. As the spaCy models are popular and the spaCy API is widely known to many potential users, we choose to build upon the spaCy library for creating a biomedical text processing pipeline.

scispaCy. Our goal is to develop scispaCy as a robust, efficient and performant NLP library to satisfy the primary text processing needs in the biomedical domain. In this release of scispaCy, we retrain spaCy[3] models for POS tagging, dependency parsing, and NER using datasets relevant to biomedical text, and enhance the tokenization module with additional rules. scispaCy contains two core released packages: **en_core_sci_sm** and **en_core_sci_md**. Models in the **en_core_sci_md** package have a larger vocabulary and include word vectors, while those in **en_core_sci_sm** have a smaller vocabulary and do not include word vectors, as shown in Table 1.

[2]Source code at https://github.com/explosion/spaCy

[3]scispaCy models are based on spaCy version 2.0.18

Software Package	Processing Times Per	
	Abstract (ms)	Sentence (ms)
NLP4J (java)	19	2
Genia Tagger (c++)	73	3
Biaffine (TF)	272	29
Biaffine (TF + 12 CPUs)	72	7
jPTDP (Dynet)	905	97
Dexter v2.1.0	208	84
MetaMapLite v3.6.2	293	89
en_core_sci_sm	32	4
en_core_sci_md	33	4

Table 2: Wall clock comparison of different publicly available biomedical NLP pipelines. All experiments run on a single machine with 12 Intel(R) Core(TM) i7-6850K CPU @ 3.60GHz and 62GB RAM. For the Biaffine Parser, a pre-compiled Tensorflow binary with support for AVX2 instructions was used in a good faith attempt to optimize the implementation. Dynet does support the Intel MKL, but requires compilation from scratch and as such, does not represent an "off the shelf" system. TF is short for Tensorflow.

Processing Speed. To emphasize the efficiency and practical utility of the end-to-end pipeline provided by scispaCy packages, we perform a speed comparison with several other publicly available processing pipelines for biomedical text using 10k randomly selected PubMed abstracts. We report results with and without segmenting the abstracts into sentences since some of the libraries (e.g., GENIA tagger) are designed to operate on sentences.

As shown in Table 2, both models released in scispaCy demonstrate competitive speed to pipelines written in C++ and Java, languages designed for production settings.

Whilst scispaCy is not as fast as pipelines designed for purely production use-cases (e.g., NLP4J), it has the benefit of straightforward integration with the large ecosystem of Python libraries for machine learning and text processing. Although the comparison in Table 2 is not an apples to apples comparison with other frameworks (different tasks, implementation languages etc), it is useful to understand scispaCy's runtime in the context of other pipeline components. Running scispaCy models *in addition to* standard Entity Linking software such as MetaMap would result in only a marginal increase in overall runtime.

In the following section, we describe the POS taggers and dependency parsers in scispaCy.

3 POS Tagging and Dependency Parsing

The joint POS tagging and dependency parsing model in spaCy is an arc-eager transition-based parser trained with a dynamic oracle, similar to Goldberg and Nivre (2012). Features are CNN representations of token features and shared across all pipeline models (Kiperwasser and Goldberg, 2016; Zhang and Weiss, 2016). Next, we describe the data we used to train it in scispaCy.

3.1 Datasets

GENIA 1.0 Dependencies. To train the dependency parser and part of speech tagger in both released models, we convert the treebank of McClosky and Charniak (2008),[4] which is based on the GENIA 1.0 corpus (Kim et al., 2003), to Universal Dependencies v1.0 using the Stanford Dependency Converter (Schuster and Manning, 2016). As this dataset has POS tags annotated, we use it to train the POS tagger jointly with the dependency parser in both released models.

As we believe the Universal Dependencies converted from the original GENIA 1.0 corpus are generally useful, we have released them as a separate contribution of this paper.[5] In this data release, we also align the converted dependency parses to their original text spans in the raw, untokenized abstracts from the original release,[6] and include the PubMed metadata for the abstracts which was discarded in the GENIA corpus released by McClosky and Charniak (2008). We hope that this raw format can emerge as a resource for practical evaluation in the biomedical domain of core NLP tasks such as tokenization, sentence segmentation and joint models of syntax.

Finally, we also retrieve from PubMed the original metadata associated with each abstract. This includes relevant named entities linked to their Medical Subject Headings (MeSH terms) as well as chemicals and drugs linked to a variety of ontologies, as well as author metadata, publication dates, citation statistics and journal metadata. We hope that the community can find interesting problems for which such natural supervision can be used.

Package/Model	GENIA
MarMoT	98.61
jPTDP-v1	98.66
NLP4J-POS	98.80
BiLSTM-CRF	98.44
BiLSTM-CRF- charcnn	98.89
BiLSTM-CRF - char lstm	98.85
en_core_sci_sm	98.38
en_core_sci_md	98.51

Table 3: Part of Speech tagging results on the GENIA Test set.

Package/Model	UAS	LAS
Stanford-NNdep	89.02	87.56
NLP4J-dep	90.25	88.87
jPTDP-v1	91.89	90.27
Stanford-Biaffine-v2	92.64	91.23
Stanford-Biaffine-v2(Gold POS)	92.84	91.92
en_core_sci_sm - SD	90.31	88.65
en_core_sci_md - SD	90.66	88.98
en_core_sci_sm	89.69	87.67
en_core_sci_md	90.60	88.79

Table 4: Dependency Parsing results on the GENIA 1.0 corpus converted to dependencies using the Stanford Universal Dependency Converter. We additionally provide evaluations using Stanford Dependencies(SD) in order for comparison relative to the results reported in (Nguyen and Verspoor, 2018).

OntoNotes 5.0. To increase the robustness of the dependency parser and POS tagger to generic text, we make use of the OntoNotes 5.0 corpus[7] when training the dependency parser and part of speech tagger (Weischedel et al., 2011; Hovy et al., 2006). The OntoNotes corpus consists of multiple genres of text, annotated with syntactic and semantic information, but we only use POS and dependency parsing annotations in this work.

3.2 Experiments

We compare our models to the recent survey study of dependency parsing and POS tagging for biomedical data (Nguyen and Verspoor, 2018) in Tables 3 and 4. POS tagging results show that both models released in scispaCy are competitive with state of the art systems, and can be considered of

---

[4] https://nlp.stanford.edu/~mcclosky/biomedical.html

[5] https://github.com/allenai/genia-dependency-trees

[6] Available at http://www.geniaproject.org/

[7] Instructions for download at http://cemantix.org/data/ontonotes.html

equivalent practical value. In the case of dependency parsing, we find that the Biaffine parser of Dozat and Manning (2016) outperforms the scispaCy models by a margin of 2-3%. However, as demonstrated in Table 2, the scispaCy models are approximately 9x faster due to the speed optimizations in spaCy. [8]

Robustness to Web Data. A core principle of the scispaCy models is that they are useful on a wide variety of types of text with a biomedical focus, such as clinical notes, academic papers, clinical trials reports and medical records. In order to make our models robust across a wider range of domains more generally, we experiment with incorporating training data from the OntoNotes 5.0 corpus when training the dependency parser and POS tagger. Figure 2 demonstrates the effectiveness of adding increasing percentages of web data, showing substantially improved performance on OntoNotes, at no reduction in performance on biomedical text. Note that mixing in web text during training has been applied to previous systems - the GENIA Tagger (Tsuruoka et al., 2005) also employs this technique.

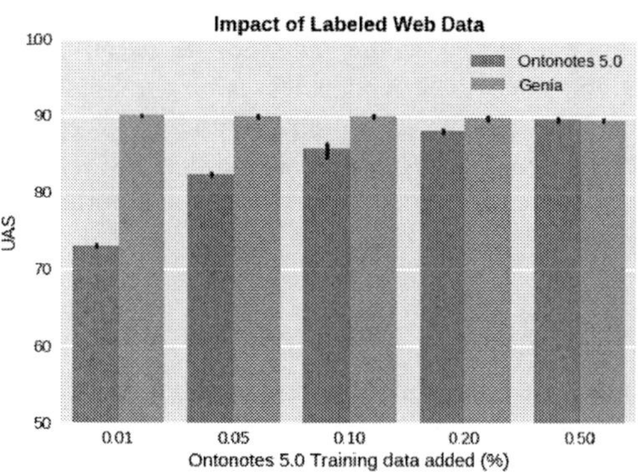

Figure 2: Unlabeled attachment score (UAS) performance for an **en_core_sci_md** model trained with increasing amounts of web data incorporated. Table shows mean of 3 random seeds.

4 Named Entity Recognition

The NER model in spaCy is a transition-based system based on the chunking model from Lample et al. (2016). Tokens are represented as hashed, embedded representations of the prefix, suffix, shape and lemmatized features of individ-

ual words. Next, we describe the data we used to train NER models in scispaCy.

4.1 Datasets

The main NER model in both released packages in scispaCy is trained on the mention spans in the MedMentions dataset (Murty et al., 2018). Since the MedMentions dataset was originally designed for entity linking, this model recognizes a wide variety of entity types, as well as non-standard syntactic phrases such as verbs and modifiers, but the model does not predict the entity type. In order to provide for users with more specific requirements around entity types, we release four additional packages **en_ner_{bc5cdr|craft|jnlpba|bionlp13cg}_md** with finer-grained NER models trained on BC5CDR (for chemicals and diseases; Li et al., 2016), CRAFT (for cell types, chemicals, proteins, genes; Bada et al., 2011), JNLPBA (for cell lines, cell types, DNAs, RNAs, proteins; Collier and Kim, 2004) and BioNLP13CG (for cancer genetics; Pyysalo et al., 2015), respectively.

4.2 Experiments

As NER is a key task for other biomedical text processing tasks, we conduct a through evaluation of the suitability of scispaCy to provide baseline performance across a wide variety of datasets. In particular, we retrain the spaCy NER model on each of the four datasets mentioned earlier (BC5CDR, CRAFT, JNLPBA, BioNLP13CG) as well as five more datasets in Crichton et al. (2017): AnatEM, BC2GM, BC4CHEMD, Linnaeus, NCBI-Disease. These datasets cover a wide variety of entity types required by different biomedical domains, including cancer genetics, disease-drug interactions, pathway analysis and trial population extraction. Additionally, they vary considerably in size and number of entities. For example, BC4CHEMD (Krallinger et al., 2015) has 84,310 annotations while Linnaeus (Gerner et al., 2009) only has 4,263. BioNLP13CG (Pyysalo et al., 2015) annotates 16 entity types while five of the datasets only annotate a single entity type.[9]

Table 5 provides a thorough comparison of the scispaCy NER models compared to a variety of models. In particular, we compare the models to

---

[8]We refer the interested reader to Nguyen and Verspoor (2018) for a comprehensive description of model architectures considered in this evaluation.

[9]For a detailed discussion of the datasets and their creation, we refer the reader to https://github.com/ cambridgeltl/MTL-Bioinformatics-2016/ blob/master/Additional%20file%201.pdf

strong baselines which do not consider the use of 1) multi-task learning across multiple datasets and 2) semi-supervised learning via large pretrained language models. Overall, we find that the scispaCy models are competitive baselines for 5 of the 9 datasets.

Additionally, in Table 6 we evaluate the recall of the pipeline mention detector available in both scispaCy models (trained on the MedMentions dataset) against all 9 specialised NER datasets. Overall, we observe a modest drop in average recall when compared directly to the MedMentions results in Table 7, but considering the diverse domains of the 9 specialised NER datasets, achieving this level of recall across datasets is already nontrivial.

Dataset	sci_sm	sci_md
BC5CDR	75.62	78.79
CRAFT	58.28	58.03
JNLPBA	67.33	70.36
BioNLP13CG	58.93	60.25
AnatEM	56.55	57.94
BC2GM	54.87	56.89
BC4CHEMD	60.60	60.75
Linnaeus	67.48	68.61
NCBI-Disease	65.76	65.65
Average	62.81	64.14

Table 6: Recall on the test sets of 9 specialist NER datasets, when the base mention detector is trained on MedMentions. The base mention detector is available in both **en_core_sci_sm** and **en_core_sci_md** models.

Model	Precision	Recall	F1
en_core_sci_sm	69.22	67.19	68.19
en_core_sci_md	70.44	67.56	68.97

Table 7: Performance of the base mention detector on the MedMentions Corpus.

5 Candidate Generation for Entity Linking

In addition to Named Entity Recognition, scispaCy contains some initial groundwork needed to build an Entity Linking model designed to link to a subset of the Unified Medical Language System (UMLS; Bodenreider, 2004). This reduced subset is comprised of sections 0, 1, 2 and 9 (SNOMED) of the UMLS 2017 AA release, which are publicly distributable. It contains 2.78M unique concepts and covers 99% of the mention concepts present in the MedMentions dataset (Murty et al., 2018).

5.1 Candidate Generation

To generate candidate entities for linking a given mention, we use an approximate nearest neighbours search over our subset of UMLS concepts and concept aliases and output the entities associated with the nearest K. Concepts and aliases are encoded using the vector of TF-IDF scores of character 3-grams which appears in 10 or more entity names or aliases (i.e., document frequency $\geq$ 10). In total, all data associated with the candidate generator including cached vectors for 2.78M concepts occupies 1.1GB of space on disk.

Aliases. Canonical concepts in UMLS have *aliases* - common names of drugs, alternative spellings, and otherwise words or phrases that are often linked to a given concept. Importantly, aliases may be shared across concepts, such as "cancer" for the canonical concepts of both "Lung Cancer" and "Breast Cancer". Since the nearest neighbor search is based on the surface forms, it returns K string values. However, because a given string may be an alias for multiple concepts, the list of K nearest neighbor strings may not translate to a list of K candidate entities. This is the correct implementation in practice, because given a possibly ambiguous alias, it is beneficial to score all plausible concepts, but it does mean that we cannot determine the exact number of candidate entities that will be generated for a given value of K. In practice, the number of retrieved candidates for a given K is much lower than K itself, with the exception of a few long tail aliases, which are aliases for a large number of concepts. For example, for K=100, we retrieve 54.26 ± 12.45 candidates, with the max number of candidates for a single mention being 164.

Abbreviations. During development of the candidate generator, we noticed that abbreviated mentions account for a substantial proportion of the failure cases where none of the generated candidates match the correct entity. To partially remedy this, we implement the unsupervised abbreviation detection algorithm of Schwartz and Hearst (2002), substituting mention candidates marked as abbreviations for their long form definitions before searching for their nearest neighbours. Figure 3 demonstrates the improved recall of gold concepts

Dataset	Baseline	SOTA	+ Resources	sci_sm	sci_md
BC5CDR (Li et al., 2016)	83.87	86.92b	89.69[bb]	78.83	83.92
CRAFT (Bada et al., 2011)	79.55	-	-	72.31	76.17
JNLPBA (Collier and Kim, 2004)	68.95	73.48[b]	75.50[bb]	71.78	73.21
BioNLP13CG (Pyysalo et al., 2015)	76.74	-	-	72.98	77.60
AnatEM (Pyysalo and Ananiadou, 2014)	88.55	91.61[**]	-	80.13	84.14
BC2GM (Smith et al., 2008)	84.41	80.51[b]	81.69[bb]	75.77	78.30
BC4CHEMD (Krallinger et al., 2015)	82.32	88.75[a]	89.37[aa]	82.24	84.55
Linnaeus (Gerner et al., 2009)	79.33	95.68[**]	-	79.20	81.74
NCBI-Disease (Dogan et al., 2014)	77.82	85.80[b]	87.34[bb]	79.50	81.65

bb: LM model from Sachan et al. (2017) **b**: LSTM model from Sachan et al. (2017)
a: Single Task model from Wang et al. (2018) **aa**: Multi-task model from Wang et al. (2018)
** Evaluations use dictionaries developed without a clear train/test split.

Table 5: Test F1 Measure on NER for the small and medium scispaCy models compared to a variety of strong baselines and state of the art models. The **Baseline** and **SOTA** (State of the Art) columns include only single models which do not use additional resources, such as language models, or additional sources of supervision, such as multi-task learning. **+ Resources** allows any type of supervision or pretraining. All scispaCy results are the mean of 5 random seeds.

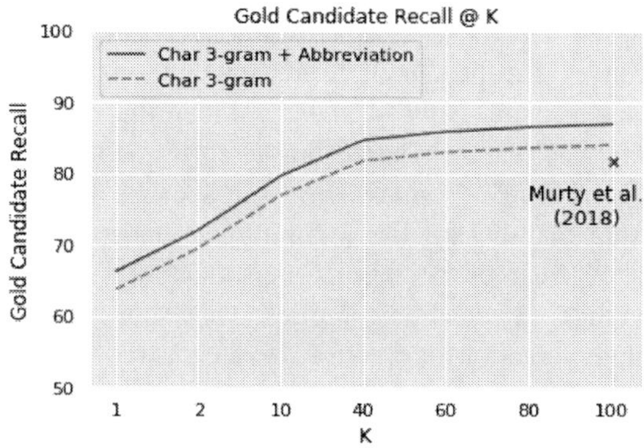

Figure 3: Gold Candidate Generation Recall for different values of K. Note that K refers to the number of nearest neighbour queries, and not the number of considered candidates. Murty et al. (2018) do not report this distinction, but for a given K the same amount of work is done (retrieving K neighbours from the index), so results are comparable. For all K, the actual number of candidates is considerably lower on average.

for various values of K nearest neighbours. Our candidate generator provides a 5% absolute improvement over Murty et al. (2018) despite generating 46% fewer candidates per mention on average.

6 Sentence Segmentation and Citation Handling

Accurate sentence segmentation is required for many practical applications of natural language processing. Biomedical data presents many dif-ficulties for standard sentence segmentation algorithms: abbreviated names and noun compounds containing punctuation are more common, whilst the wide range of citation styles can easily be misidentified as sentence boundaries.

We evaluate sentence segmentation using both sentence and full-abstract accuracy when segmenting PubMed abstracts from the raw, untokenized GENIA development set (the **Sent/Abstract** columns in Table 8).

Additionally, we examine the ability of the segmentation learned by our model to generalise to the body text of PubMed articles. Body text is typically more complex than abstract text, but in particular, it contains citations, which are considerably less frequent in abstract text. In order to examine the effectiveness of our models in this scenario, we design the following synthetic experiment. Given sentences from Cohan et al. (2019) which were originally designed for citation intent prediction, we run these sentences individually through our models. As we know that these sentences should be single sentences, we can simply count the frequency with which our models segment the individual sentences containing citations into multiple sentences (the **Citation** column in Table 8).

As demonstrated by Table 8, training the dependency parser on in-domain data (both the scispaCy models) completely obviates the need for rule-based sentence segmentation. This is a positive result - rule based sentence segmentation is a brittle, time consuming process, which we have replaced with a domain specific version of an existing pipeline component.

Both scispaCy models are released with the custom tokeniser, but without a custom sentence segmenter by default.

Model	Sent	Abstract	Citation
web-small	88.2%	67.5%	74.4%
web-small + ct	86.6%	62.1%	88.6%
web-small + cs	91.9%	77.0%	87.5%
web-small + cs + ct	92.1%	78.3%	94.7%
sci-small + ct	97.2%	81.7%	97.9%
sci-small + cs + ct	97.2%	81.7%	98.0%
sci-med + ct	97.3%	81.7%	98.0%
sci-med + cs + ct	97.4%	81.7%	98.0%

Table 8: Sentence segmentation performance for the core spaCy and scispaCy models. **cs** = custom rule based sentence segmenter and **ct** = custom rule based tokenizer, both designed explicitly to handle citations and common patterns in biomedical text.

7   Related Work

Apache cTakes (Savova et al., 2010) was designed specifically for clinical notes rather than the broader biomedical domain. MetaMap and MetaMapLite (Aronson, 2001; Demner-Fushman et al., 2017) from the National Library of Medicine focus specifically on entity linking using the Unified Medical Language System (UMLS) (Bodenreider, 2004) as a knowledge base. Buyko et al. adapt Apache OpenNLP using the GENIA corpus, but their system is not openly available and is less suitable for modern, Python-based workflows. The GENIA Tagger (Tsuruoka et al., 2005) provides the closest comparison to scispaCy due to it's multi-stage pipeline, integrated research contributions and production quality runtime. We improve on the GENIA Tagger by adding a full dependency parser rather than just noun chunking, as well as improved results for NER without compromising significantly on speed.

In more fundamental NLP research, the GENIA corpus (Kim et al., 2003) has been widely used to evaluate transfer learning and domain adaptation. McClosky et al. (2006) demonstrate the effectiveness of self-training and parse re-ranking for domain adaptation. Rimell and Clark (2008) adapt a CCG parser using only POS and lexical categories, while Joshi et al. (2018) extend a neu-

ral phrase structure parser trained on web text to the biomedical domain with a small number of partially annotated examples. These papers focus mainly of the problem of domain adaptation itself, rather than the objective of obtaining a robust, high-performance parser using existing resources.

NLP techniques, and in particular, *distant supervision* have been employed to assist the curation of large, structured biomedical resources. Poon et al. (2015) extract 1.5 million cancer pathway interactions from PubMed abstracts, leading to the development of Literome (Poon et al., 2014), a search engine for genic pathway interactions and genotype-phenotype interactions. A fundamental aspect of Valenzuela-Escarcega et al. (2018) and Poon et al. (2014) is the use of handwritten rules and triggers for events based on dependency tree paths; the connection to the application of scispaCy is quite apparent.

8   Conclusion

In this paper we presented several robust model pipelines for a variety of natural language processing tasks focused on biomedical text. The scispaCy models are fast, easy to use, scalable, and achieve close to state of the art performance. We hope that the release of these models enables new applications in biomedical information extraction whilst making it easy to leverage high quality syntactic annotation for downstream tasks. Additionally, we released a reformatted GENIA 1.0 corpus augmented with automatically produced Universal Dependency annotations and recovered and aligned original abstract metadata. Future work on scispaCy will include a more fully featured entity linker built from the current candidate generation work, as well as other pipeline components such as negation detection commonly used in the clinical and biomedical natural language processing communities.

References

Waleed Ammar, Dirk Groeneveld, Chandra Bhagavatula, Iz Beltagy, Miles Crawford, Doug Downey, Jason Dunkelberger, Ahmed Elgohary, Sergey Feldman, Vu Ha, Rodney Kinney, Sebastian Kohlmeier, Kyle Lo, Tyler Murray, Hsu-Han Ooi, Matthew E. Peters, Joanna Power, Sam Skjonsberg, Lucy Lu Wang, Chris Wilhelm, Zheng Yuan, Madeleine van Zuylen, and Oren Etzioni. 2018. Construction of the literature graph in semantic scholar. In *NAACL-HLT*.

Alan R. Aronson. 2001. Effective mapping of biomedical text to the UMLS Metathesaurus: the MetaMap program. *Proceedings. AMIA Symposium*, pages 17–21.

Michael Bada, Miriam Eckert, Donald Evans, Kristin Garcia, Krista Shipley, Dmitry Sitnikov, William A. Baumgartner, K. Bretonnel Cohen, Karin M. Verspoor, Judith A. Blake, and Lawrence Hunter. 2011. Concept annotation in the CRAFT corpus. In *BMC Bioinformatics*.

Olivier Bodenreider. 2004. The unified medical language system (UMLS): integrating biomedical terminology. *Nucleic acids research*, 32 Database issue:D267–70.

Lutz Bornmann and Rüdiger Mutz. 2014. Growth rates of modern science: A bibliometric analysis. *CoRR*, abs/1402.4578.

Ekaterina Buyko, Joachim Wermter, Michael Poprat, and Udo Hahn. Automatically adapting an NLP core engine to the biology domain. In *Proceedings of the ISMB 2006 Joint Linking Literature, Information and Knowledge for Biology and the 9th Bio-Ontologies Meeting*.

Arman Cohan, Waleed Ammar, Madeleine van Zuylen, and Field Cady. 2019. Structural scaffolds for citation intent classification in scientific publications. *CoRR*, abs/1904.01608.

Nigel Collier and Jin-Dong Kim. 2004. Introduction to the bio-entity recognition task at JNLPBA. In *NLP-BA/BioNLP*.

Gamal K. O. Crichton, Sampo Pyysalo, Billy Chiu, and Anna Korhonen. 2017. A neural network multi-task learning approach to biomedical named entity recognition. In *BMC Bioinformatics*.

Dina Demner-Fushman, Willie J. Rogers, and Alan R. Aronson. 2017. MetaMap Lite: an evaluation of a new Java implementation of MetaMap. *Journal of the American Medical Informatics Association : JAMIA*, 24 4:841–844.

Rezarta Islamaj Dogan, Robert Leaman, and Zhiyong Lu. 2014. NCBI disease corpus: A resource for disease name recognition and concept normalization. *Journal of biomedical informatics*, 47:1–10.

Timothy Dozat and Christopher D. Manning. 2016. Deep biaffine attention for neural dependency parsing. *CoRR*, abs/1611.01734.

Martin Gerner, Goran Nenadic, and Casey M. Bergman. 2009. LINNAEUS: A species name identification system for biomedical literature. In *BMC Bioinformatics*.

Yoav Goldberg and Joakim Nivre. 2012. A dynamic oracle for arc-eager dependency parsing. In *Coling 2012*.

Matthew Honnibal and Ines Montani. 2017. spaCy 2: Natural language understanding with bloom embeddings, convolutional neural networks and incremental parsing. *To appear*.

Eduard H. Hovy, Mitchell P. Marcus, Martha Palmer, Lance A. Ramshaw, and Ralph M. Weischedel. 2006. OntoNotes: The 90% solution. In *HLT-NAACL*.

Vidur Joshi, Matthew Peters, and Mark Hopkins. 2018. Extending a parser to distant domains using a few dozen partially annotated examples. In *ACL*.

Jin-Dong Kim, Tomoko Ohta, Yuka Tateisi, and Jun'ichi Tsujii. 2003. GENIA corpus - a semantically annotated corpus for bio-textmining. *Bioinformatics*, 19 Suppl 1:i180–2.

Eliyahu Kiperwasser and Yoav Goldberg. 2016. Simple and accurate dependency parsing using bidirectional lstm feature representations. *Transactions of the Association for Computational Linguistics*, 4:313–327.

Martin Krallinger, Florian Leitner, Obdulia Rabal, Miguel Vazquez, Julen Oyarzábal, and Alfonso Valencia. 2015. CHEMDNER: The drugs and chemical names extraction challenge. In *J. Cheminformatics*.

Guillaume Lample, Miguel Ballesteros, Sandeep Subramanian, Kazuya Kawakami, and Chris Dyer. 2016. Neural architectures for named entity recognition. In *HLT-NAACL*.

Jiao Li, Yueping Sun, Robin J. Johnson, Daniela Sciaky, Chih-Hsuan Wei, Robert Leaman, Allan Peter Davis, Carolyn J. Mattingly, Thomas C. Wiegers, and Zhiyong Lu. 2016. BioCreative V CDR task corpus: a resource for chemical disease relation extraction. *Database : the journal of biological databases and curation*, 2016.

David McClosky and Eugene Charniak. 2008. Self-training for biomedical parsing. In *ACL*.

David McClosky, Eugene Charniak, and Mark Johnson. 2006. Reranking and self-training for parser adaptation. In *ACL*.

Shikhar Murty, Patrick Verga, Luke Vilnis, Irena Radovanovic, and Andrew McCallum. 2018. Hierarchical losses and new resources for fine-grained entity typing and linking. In *ACL*.

Dat Quoc Nguyen and Karin Verspoor. 2018. From POS tagging to dependency parsing for biomedical event extraction. *arXiv preprint arXiv:1808.03731*.

Hoifung Poon, Chris Quirk, Charlie DeZiel, and David Heckerman. 2014. Literome: PubMed-scale genomic knowledge base in the cloud. *Bioinformatics*, 30 19:2840–2.

Hoifung Poon, Kristina Toutanova, and Chris Quirk. 2015. Distant supervision for cancer pathway extraction from text. *Pacific Symposium on Biocomputing. Pacific Symposium on Biocomputing*, pages 120–31.

Sampo Pyysalo and Sophia Ananiadou. 2014. Anatomical entity mention recognition at literature scale. In *Bioinformatics*.

Sampo Pyysalo, Tomoko Ohta, Rafal Rak, Andrew Rowley, Hong-Woo Chun, Sung-Jae Jung, Sung-Pil Choi, Jun'ichi Tsujii, and Sophia Ananiadou. 2015. Overview of the cancer genetics and pathway curation tasks of BioNLP shared task 2013. In *BMC Bioinformatics*.

Laura Rimell and Stephen Clark. 2008. Adapting a lexicalized-grammar parser to contrasting domains. In *EMNLP*.

Devendra Singh Sachan, Pengtao Xie, Mrinmaya Sachan, and Eric P. Xing. 2017. Effective use of bidirectional language modeling for transfer learning in biomedical named entity recognition. In *MLHC*.

Guergana K. Savova, James J. Masanz, Philip V. Ogren, Jiaping Zheng, Sunghwan Sohn, Karin Kipper Schuler, and Christopher G. Chute. 2010. Mayo clinical text analysis and knowledge extraction system (cTAKES): architecture, component evaluation and applications. *Journal of the American Medical Informatics Association : JAMIA*, 17 5:507–13.

Sebastian Schuster and Christopher D. Manning. 2016. Enhanced english universal dependencies: An improved representation for natural language understanding tasks. In *LREC*.

Ariel S. Schwartz and Marti A. Hearst. 2002. A simple algorithm for identifying abbreviation definitions in biomedical text. *Pacific Symposium on Biocomputing. Pacific Symposium on Biocomputing*, pages 451–62.

Larry Smith, Lorraine K. Tanabe, Rie Johnson nee Ando, Cheng-Ju Kuo, I-Fang Chung, Chun-Nan Hsu, Y Lin, Roman Klinger, Christoph M. Friedrich, Kuzman Ganchev, Manabu Torii, Hongfang Liu, Barry Haddow, Craig A. Struble, Richard J. Povinelli, Andreas Vlachos, William A. Baumgartner, Lawrence E. Hunter, Bob Carpenter, Richard Tzong-Han Tsai, Hong-Jie Dai, Feng Liu, Yifei Chen, Chengjie Sun, Sophia Katrenko, Pieter Adriaans, Christian Blaschke, Rafael Torres, Mariana Neves, Preslav Nakov, Anna Divoli, Manuel Maña-López, Jacinto Mata, and W. John Wilbur. 2008. Overview of BioCreative II gene mention recognition. *Genome Biology*, 9:S2 – S2.

Yoshimasa Tsuruoka, Yuka Tateishi, Jin-Dong Kim, Tomoko Ohta, John McNaught, Sophia Ananiadou, and Jun'ichi Tsujii. 2005. Developing a robust part-of-speech tagger for biomedical text. In *Panhellenic Conference on Informatics*.

Yoshimasa Tsuruoka and Jun'ichi Tsujii. 2005. Bidirectional inference with the easiest-first strategy for tagging sequence data. In *HLT/EMNLP*.

Marco Antonio Valenzuela-Escarcega, Ozgun Babur, Gus Hahn-Powell, Dane Bell, Thomas Hicks, Enrique Noriega-Atala, Xia Wang, Mihai Surdeanu, Emek Demir, and Clayton T. Morrison. 2018. Large-scale automated machine reading discovers new cancer-driving mechanisms. In *Database*.

Xuan Wang, Yu Zhang, Xiang Ren, Yuhao Zhang, Marinka Zitnik, Jingbo Shang, Curtis P. Langlotz, and Jiawei Han. 2018. Cross-type biomedical named entity recognition with deep multi-task learning. *Bioinformatics*.

Chih-Hsuan Wei, Yifan Peng, Robert Leaman, Allan Peter Davis, Carolyn J. Mattingly, Jiao Li, Thomas C. Wiegers, and Zhiyong Lu. 2016. Assessing the state of the art in biomedical relation extraction: overview of the BioCreative V chemical-disease relation (CDR) task. *Database : the journal of biological databases and curation*, 2016.

Ralph Weischedel, Eduard Hovy, Martha Palmer, Mitch Marcus, Robert Belvin adn Sameer Pradhan, Lance Ramshaw, and Nianwen Xue. 2011. OntoNotes: A large training corpus for enhanced processing. In Joseph Olive, Caitlin Christianson, and John McCary, editors, *Handbook of Natural Language Processing and Machine Translation*. Springer.

Yuan Zhang and David I Weiss. 2016. Stack-propagation: Improved representation learning for syntax. *CoRR*, abs/1603.06598.

Improving Chemical Named Entity Recognition in Patents with Contextualized Word Embeddings

Zenan Zhai[1], Dat Quoc Nguyen[1], Saber A. Akhondi[2], Camilo Thorne[2],
Christian Druckenbrodt[2], Trevor Cohn[1], Michelle Gregory[2], Karin Verspoor[1]
[1]The University of Melbourne, Australia; [2]Elsevier
[1]{zenan.zhai,dqnguyen,trevor.cohn,karin.verspoor}@unimelb.edu.au
[2]{s.akhondi,c.thorne.1,c.druckenbrodt,m.gregory}@elsevier.com

Abstract

Chemical patents are an important resource for chemical information. However, few chemical Named Entity Recognition (NER) systems have been evaluated on patent documents, due in part to their structural and linguistic complexity. In this paper, we explore the NER performance of a BiLSTM-CRF model utilising pre-trained word embeddings, character-level word representations and contextualized ELMo word representations for chemical patents. We compare word embeddings pre-trained on biomedical and chemical patent corpora. The effect of tokenizers optimized for the chemical domain on NER performance in chemical patents is also explored. The results on two patent corpora show that contextualized word representations generated from ELMo substantially improve chemical NER performance w.r.t. the current state-of-the-art. We also show that domain-specific resources such as word embeddings trained on chemical patents and chemical-specific tokenizers have a positive impact on NER performance.

1 Introduction

Chemical patents are an important starting point for understanding of chemical compound purpose, properties, and novelty. New chemical compounds are often initially disclosed in patent documents; however it may take 1-3 years for these chemicals to be mentioned in chemical literature (Senger et al., 2015), suggesting that patents are a valuable but underutilized resource. As the number of new chemical patent applications is drastically increasing every year (Muresan et al., 2011), it is becoming increasingly important to develop automatic natural language processing (NLP) approaches enabling information extraction from these patents (Akhondi et al., 2019). Chemical Named-Entity Recognition (NER) is a fundamental step for information extraction from chemical-related texts,

supporting relation extraction (Wei et al., 2016), reaction prediction (Schwaller et al., 2018) and retro-synthesis (Segler et al., 2018).

However, performing NER in chemical patents can be challenging (Akhondi et al., 2014). As legal documents, patents are written in a very different way compared to scientific literature. When writing scientific papers, authors strive to make their words as clear and straight-forward as possible, whereas patent authors often seek to protect their knowledge from being fully disclosed (Valentinuzzi, 2017).

In tension with this is the need to claim broad scope for intellectual property reasons, and hence patents typically contain more details and are more exhaustive than scientific papers (Lupu et al., 2011).

There are a number of characteristics of patent texts that create challenges for NLP in this context. Long sentences listing names of compounds in chemical patents are frequently used. The structure of sentences in patent claims is usually complex, and syntactic parsing in patents can be difficult (Hu et al., 2016). A quantitative analysis by Verberne et al. (2010) showed that the average sentence length in a patent corpus is much longer than in general language use. That work also showed that the lexicon used in patents usually includes domain-specific and novel terms that are difficult to understand. Some patent authorities use Optical Character Recognition (OCR) for digitizing patents, which can be problematic when applying automatic NLP approaches as the OCR errors introduces extra noise to the data (Akhondi et al., 2019).

Most NER systems for the chemical domain were developed, trained and tested on either chemical literature or only the title and abstract of chemical patents (Akhondi et al., 2019). There are substantial linguistic differences between ab-

Proceedings of the BioNLP 2019 workshop, pages 328–338
Florence, Italy, August 1, 2019. ©2019 Association for Computational Linguistics

stracts and the corresponding full text publications (Cohen et al., 2010). The performance of NER approaches on full patent documents has still not been fully explored (Krallinger et al., 2015).

Hence, this paper will focus on presenting the best NER performance achieved to date on full chemical patent corpus.

We use a combination of pre-trained word embeddings, a CNN-based character-level word representation and contextualized word representations generated from ELMo, trained on a patent corpus, as input to a BiLSTM-CRF model. The results show that contextualized word representations help improve chemical NER performance substantially. In addition, the impact of the choice of pre-trained word embeddings and tokenizers is assessed.

The results show that word embeddings that are pre-trained on chemical patents outperform embeddings pre-trained on biomedical datasets, and using tokenizers optimized for the chemical domain can improve NER performance in chemical patent corpora.

2   Related work

In this section, we summarize previous methods and empirical studies on NER in chemical patents.

Two existing Conditional Random Field (CRF)-based systems for chemical named entity recognition are tmChem (Leaman et al., 2015) and ChemSpot (Rocktäschel et al., 2012); each makes use of numerous hand-crafted features including word shape, prefix, suffix, part-of-speech and character N-grams in an algorithm based on modelling of tag sequences. A previous detailed empirical study explored the generalization performance of these systems and their ensembles (Habibi et al., 2016). The application of the tmChem model trained on chemical *literature* corpora of the BioCreative IV CHEMDNER task (Krallinger et al., 2015) and the ChemSpot model trained on a subset of the SCAI corpus (Klinger et al., 2008) resulted in a significant performance drop over chemical *patent* corpora.

Zhang et al. (2016) compared the performance of CRF- and Support Vector Machine (SVM)-based models on the CHEMDNER-patents corpus (Krallinger et al., 2015). The features constructed in that work included the binarized embedding (Guo et al., 2014), Brown clustering (Brown et al., 1992) and domain-specific features extracted by

detecting common prefixes/suffixes in chemical words. The obtained results show that the performance of CRF and SVM models can be significantly improved by incorporating unsupervised features (e.g. word embeddings, word clustering). The study also showed that the SVM model slightly outperformed the CRF model in the chemical NER task.

To perform chemical NER on the CHEMD-NER patents corpus, Akhondi et al. (2016) proposed an ensemble approach combining a gazetteer-based method and a modified version of tmChem. Here, the gazetteer-based method utilized a wide range of chemical dictionaries, while additional features such as stems, prefixes/suffixes, chemical elements were added to the original feature set of tmChem. In the ensemble approach, tokens were predicted as chemical mentions if recognized as positive by either tmChem or the gazetteer-based method. The results showed that both gazetteer-based and ensemble approaches were outperformed by the modified tmChem version in terms of overall F_1 score, although these two approaches can obtain higher recall.

Huang et al. (2015) proposed a BiLSTM-CRF based on the use of a bidirectional long-short term memory network – BiLSTM (Schuster and Paliwal, 1997) – to extract (latent) features for a CRF classifier. The BiLSTM encodes the input in both forward and backward directions and passes the concatenation of outputs from both directions as input to a linear-chain CRF sequence tagging layer. In this approach, the BiLSTM selectively encodes information and long-distance dependencies observed while processing input sentences in both directions, while the CRF layer globally optimizes the model by using information from neighbor labels.

The morphological structures within words are also important clues for identifying named entities in biological domain. Such morphological structures are widely used in systematic chemical name formats (e.g. IUPAC names) and hence particularly informative for chemical NER (Klinger et al., 2008). Character-level word representations have been developed to leverage information from these structures by encoding the character sequences within tokens. Ma and Hovy (2016) uses Convolutional Neural Networks (CNNs) to encode character sequences while Lample et al. (2016)

developed a LSTM-based approach for encoding character level information.

Habibi et al. (2017) presented an empirical study comparing three NER models on a large collection of biomedical corpora including the BioSemantics patent corpus: (1) tmChem–the CRF-based model with hand-crafted features–used as the baseline; (2) a second CRF model based on CRFSuite (Okazaki, 2007) using pre-trained word embeddings; (3) and a BiLSTM-CRF model with additional LSTM-based character-level word embeddings (Lample et al., 2016). The performance of CRFSuite- and BiLSTM-CRF-based models with different sets of pre-trained biomedical word embeddings (Pyysalo et al., 2013) were also explored. The results showed that the BiLSTM-CRF model with the combination of domain-specific pre-trained word embedding and LSTM-based character-level word embeddings outperformed the two CRF-based models on chemical NER tasks in both chemical literature and chemical patent corpora. However, this work used only a general tokenizer (i.e. OpenNLP) and word embeddings pre-trained on biomedical corpora.

Corbett and Boyle (2018) presented word-level and character-level BiLSTM networks for chemical NER in literature domain. The word-level model employed word embeddings learned by GloVe (Pennington et al., 2014) on a corpus of patent titles and abstracts. The character-level model used two different transfer learning approaches to pre-train its character-level encoder. The first approach attempts to predict neighbor characters at each time step, while the other tries to predict whether a given character sequence is an entry in the chemical database ChEBI (Degtyarenko et al., 2007). Experimental results show that the character-level model can produce better NER performance than word-level model by leveraging transfer learning. In addition, for the word-level model, using pre-trained word embeddings learned from a patent corpus helps produce better performance than using the pre-trained ones learned from a general corpus.

3 Our empirical methodology

This section presents our empirical study of NER chemical on patent datasets. We first outline the experimental datasets (Section 3.1) and the tokenizers (Section 3.2) used to pre-process these datasets, and then we introduce the BiLSTM-CRF-based models (Section 3.3) with pre-trained word embeddings (Section 3.4), character-level word embeddings (Section 3.5), contextualized word embeddings (Section 3.6) and implementation details (Section 3.7).

3.1 Dataset

We conduct experiments on 2 patent corpora: the BioSemantics patent corpus (Akhondi et al., 2014) and Reaxys gold set (Akhondi et al., 2019).

The BioSemantics patent corpus (Akhondi et al., 2014) consists of 200 full chemical patent documents with 9 different entity classes. In particular, this corpus has 170K sentences and and 360K entity annotations, which is much larger than previously used datasets, e.g. the CHEMD-NER patent abstract corpus (Krallinger et al., 2015). Therefore, this corpus can be considered as a more suitable resource for evaluating deep learning methods in which a large amount of training data is required (LeCun et al., 2015). A subset of 47 patents were annotated by multiple groups (at least 3) of annotators and evaluated through inner-annotator agreement. By harmonizing the annotations from different annotator groups, these 47 patents formed the *"harmonized"* set in the BioSemantics patent corpus. We use the harmonized set for both hyper-parameter tuning and error analysis as it has known high-quality annotations.

The Reaxys gold set (Akhondi et al., 2019) contains 131 patent snippets (parts of full chemical patent documents) from several different patent offices. The tagging scheme of this corpus includes 2 coarse-grained labels chemical class and chemical compounds, and 7 fine-grained labels of chemical compound (e.g. *mixture-part*, *prophetic*) and chemical class (e.g. *bio-molecule*, *Markush*, *mixture*, *mixture-part*). This corpus is relatively small in size, approximately 20,000 sentences in total, but very richly annotated. The relevancy score of each chemical entity and the relations between them were also annotated, which allows this corpus to be used in other tasks beyond named entity recognition.

In our experiments, each corpus is used separately. We follow Habibi et al. (2017) to use a ratio split of 60%/10%/30% for training/development/test. Note that on the BioSemantic patent corpus, our sampling of datasets may not be exactly the same as in Habibi et al. (2017).

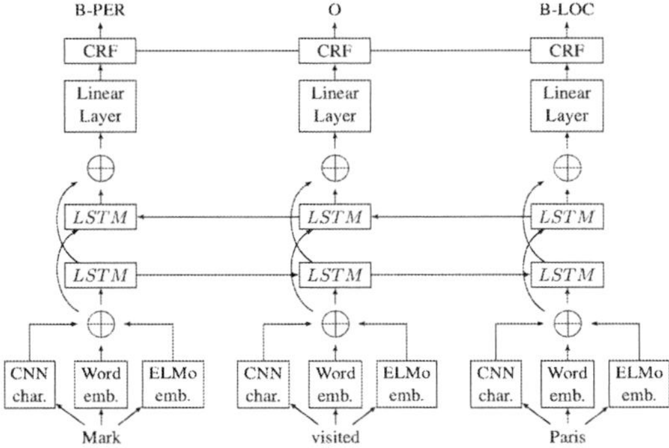

Figure 1: Architecture of EBC-CRF

3.2 Tokenizers

The morphological information captured by character-level word representations can be highly affected by tokenization quality. General-purpose tokenizers usually split tokens by spaces and punctuation. However, strict adherence to such boundaries may not be suitable for chemical texts as spaces and punctuation are commonly used in the IUPAC format for chemical names (e.g. *3-(4,5-dimethylthiazol-2-yl)-2,5-diphenyl tetrazolium bromide*) (Jessop et al., 2011). Hence, the impact of different tokenizers on NER also needs to be explored.

A pre-processing step is applied to the patent corpora including sentence detection and tokenization. Following Habibi et al. (2017), we use the OpenNLP (Morton et al., 2005) English sentence detection model. To explore the relationship between tokenization quality and final NER performance, we apply different tokenizers and train/test models with each tokenizer individually. To investigate the effect of a general domain tokenizer, following Habibi et al. (2017), we also use the OpenNLP tokenizer. To investigate whether NER performance will be affected by tokenization quality, we employ three tokenizers optimized for chemical texts including ChemTok (Akkasi et al., 2016), OSCAR4 (Jessop et al., 2011) and NBIC UMLSGeneChemTokenizer.[1]

3.3 Models

We use the BiLSTM-CNN-CRF model (Ma and Hovy, 2016) as our baseline. We extend the baseline by adding the contextualized word representations generated from ELMo (Peters et al., 2018).

For convenience, we call the extended version as EBC-CRF as illustrated in Figure 1. In particular, for EBC-CRF, we use a concatenation of pre-trained word embeddings, CNN-based character-level word embeddings and ELMo-based contextualized word embeddings as the input of a BiLSTM encoder. The BiLSTM encoder learns a latent feature vector for each word in the input. Then each latent feature vector is linearly transformed before being fed into a linear-chain CRF layer (Lafferty et al., 2001) for NER tag prediction. We assume binary potential between tags and unary potential between tags and words.

3.4 Pre-trained word embeddings

Dai et al. (2019) showed that NER performance is significantly affected by the overlap between pre-trained word embedding vocabulary and the target NER data. Therefore, we explore the effects of different sets of pre-trained word embeddings on the NER performance.

We use 200-dimensional pre-trained PubMed-PMC and Wiki-PubMed-PMC word embeddings (Pyysalo et al., 2013), which are widely used for NLP tasks in biomedical domain. Both the PubMed-PMC and Wiki-PubMed-PMC embeddings word embeddings were generated by training the Word2Vec skip-gram model (Mikolov et al., 2013) on a collection of PubMed abstracts and PubMed Central articles. Here, an additional Wikipedia dump was also used to learn the Wiki-PubMed-PMC word embeddings.

To explore whether word embeddings trained in the same domain can produce better performance in NER tasks, we learn another set of word embeddings, which we called *ChemPatent* embeddings, by applying the same model and hyperparameters from Pyysalo et al. (2013) on a collection of 84,076 full patent documents (1B tokens) across 7 patent offices (see Table 1 for details).

The pre-trained PubMed-PMC, Wiki-PubMed-PMC and ChemPatent word embeddings are fixed during training of the NER models. For a more concrete comparison, a set of 200-dimensional trainable word embeddings initialized from normal distribution is used as a baseline.

The 200-dimensional baseline word embeddings contain all words in the vocabulary of the dataset and are initialized from a normal distribution, the baseline word embeddings are learned during training process. The vocabulary of models

---

[1]NBIC UMLSGeneChemTokenizer is developed by the Netherlands Bioinformatics Center, available at `https://trac.nbic.nl/data-mining/wiki`.

Patent Office	Document	Sentence	Tokens
AU	7,743	4,662,375	156,137,670
CA	1,962	463,123	16,109,776
EP	19,274	3,478,258	117,992,191
GB	918	182,627	6,038,837
IN	1,913	261,260	9,015,238
US	41,131	19,800,123	628,256,609
WO	11,135	4,830,708	159,286,325
Total	84,076	33,687,474	1,092,836,646

Table 1: Statistics of the unannotated patent corpus used for training ChemPatent embeddings and ELMo.

Hyper-para.	Value
Optimizer	Adam
Learning rate	0.001
Mini-batch size	16
Clip Norm(L2)	1
Dropout	[0.25, 0.25]

(a) BiLSTM-CRF

Hyper-para.	Value
charEmbedSize	50
filter length	3
# of filters	30
output size	30

(b) CNN-char

Table 2: Fixed hyper-parameter configurations.

using pre-trained word embeddings is built by taking the union of words in the pre-traied word embedding file and words with frequency more than 3 in training and development sets. We do not update weights for word embeddings if pre-trained word embeddings were used.

3.5 Character-level representation

The BiLSTM-CRF model with character-level word representations (Lample et al., 2016; Ma and Hovy, 2016) has been shown to have state-of-the-art performance in NER tasks on chemical patent datasets (Habibi et al., 2017). It has been shown that the choice of using LSTM-based or CNN-based character-level word representation has little effect on final NER performance in both general and biomedical domain while the CNN-based approach has the advantage of reduced training time (Reimers and Gurevych, 2017b; Zhai et al., 2018). Hence, we use the CNN-based approach with the same hyper-parameter settings of Reimers and Gurevych (2017b) for capturing character-level information (see Table 2 for details).

3.6 ELMo

ELMo (Peters et al., 2018) and BERT (Devlin et al., 2019) can be used to generate contextualized word representations by combining internal states of different layers in neural language models. Contextualized word representation can help to improve performance in various NLP tasks by incorporating contextual information, essentially allowing for the same word to have distinct context-dependent meanings. This could be particularly powerful for chemical NER since generic chemical names (e.g. *salts, acid*) may have different meanings in other domains. We therefore explore the impact of using contextualized word representations for chemical patents.

We train ELMo on the same corpus of 84K

patents (detailed in Table 1), which we use for training the ChemPatent embeddings (described in Section 3.4). We use the ELMo implementation provided by Peters et al. (2018) with default hyper-parameters.[2] Such neural language models require a large amount of computational resources to train. In ELMo, a maximum character sequence length of tokens is set to make training feasible. However, systematic chemical names in chemical patents are often longer than the typical maximum sequence length of these neural language models. As very long tokens tend to be systematic chemical names, we reduced the max length of word from 50 to 25 and replace tokens longer than 25 characters by a special token "Long_Token".

3.7 Implementation details

Our NER model implementation is based on the AllenNLP system (Gardner et al., 2017). We learn model parameters using the training set, and we use the overall F_1 score over development set as indicator for performance improvement. All models in this paper are trained with 50 epochs in maximum, and an early stopping is applied if there are no overall F_1 score improvement observed after 10 epochs.

In Reimers and Gurevych (2017a) and Zhai et al. (2018), optimal hyper-parameters of BiLSTM-CRF models in NER tasks were explored. Hence, we fix the hyper-parameters shown in Table 2 to the suggested values in our experiments, which means that only models with 2-stacked LSTM of size 250 are evaluated.

In this study, we also consider the choice of tokenizer and word embedding source as hyper-parameters. To compare the performance of different tokenizers, we tokenize the same split of datasets with different tokenizers and evaluate the overall F_1 score over development set. After the best tokenizer for pre-processing patent corpus is determined, we use datasets tokenized by the best

---

[2] https://github.com/allenai/bilm-tf

Tokenizer	BioSemantics	Reaxys	Avg.
OpenNLP	89.36	89.43	89.40
NBIC	**+0.86**	-0.13	+0.37
ChemTok	+0.04	+1.68	+0.86
OSCAR4	+0.08	**+1.86**	**+0.97**

Table 3: Best F_1 of EBC-CRF model with different tokenizations on development sets of BioSemantics patent (harmonized set) and Reaxys Gold with ChemPatent embeddings in use. Recall that the harmonized set of 47 patents is a subset of BioSemantics, which were annotated by multiple groups (i.e. better annotation quality than remaining patents).

Embeddings	BioSemantics	Reaxys	Avg.
Baseline	88.54	90.05	89.30
PubMed-PMC	+0.61	+1.03	+0.82
Wiki-PubMed-PMC	+1.24	+0.95	+1.10
ChemPatent	**+1.68**	**+1.24**	**+1.46**

Table 4: Best F_1 of EBC-CRF model with different word embeddings on development sets of BioSemantics patent (harmonized set) and Reaxys Gold (tokenized by NBIC and OSCAR 4 tokenizer respectively)

tokenizer to train models with different pre-trained word embeddings. The best set of pre-trained word embeddings for patent corpus is determined based on the overall F_1 score over development set. We then take the best performing tokenizer and pre-trained word embeddings by comparing the marco-average F_1 score improvement on both experimental datasets.

4 Results

4.1 Main Results

Effects of different tokenizers: Table 3 shows that all 3 tokenizers optimized for the chemical domain outperform the baseline general-purpose tokenizer (i.e. OpenNLP). The best performance on BioSemantics and Reaxys Gold are achieved by using the NBIC tokenizer (+1.86 F_1 score) and the OSCAR4 tokenizer (+0.86 F_1 score), respectively. The best overall tokenizer is OSCAR4 which obtains about 1.0 absolute macro-averaged F_1 improvement in comparison to the baseline.

Effects of different sets of word embeddings: Table 4 shows results obtained by training EBC-CRF with different sets of pre-trained word embeddings. On both BioSemantics and Reaxys Gold, it is not surprising that our ChemPatent word embeddings help produce the best performance on the development set, obtaining (on average) a higher F_1 score of 1.5 as compared to the

Model	P	R	F_1
tmChem	72.56	78.37	75.35
CRFSuite	**81.93**	78.38	80.12
BiLSTM-CRF + LSTM-char	79.72	**84.42**	**82.01**
BiLSTM-CNN-CRF	83.76	85.01	84.38
EBC-CRF	**84.30**	**87.11**	**85.68**

Table 5: NER scores on full BioSemantics test set (Akhondi et al., 2014). Results in the first 3 rows were reported in Habibi et al. (2017). BiLSTM-CRF + LSTM-char denotes the BiLSTM-CRF model with additional LSTM-based character-level word embeddings (Lample et al., 2016). Recall that our models use the OSCAR4 tokenizer and pre-trained ChemPatent word embeddings.

baseline embeddings. Specifically, ChemPatent does better than the second best Wiki-PubMed-PMC with about 0.4 improvement. In the rest of the Results section, obtained results are reported with the use of the OSCAR4 tokenizer and the ChemPatent embeddings on both experimental datasets.[3]

Final results: Table 5 compared results reported in Habibi et al. (2017) and our approach on the full BioSemantics test set. It is clear that all neural models outperform conventional CRF-based models tmChem and CRFSuite. Our EBC-CRF model outperforms the BiLSTM-CRF + LSTM-char model with a 3.7 F_1 score improvement. Compared to the baseline model BiLSTM-CNN-CRF, the ELMo-based contextualized word embeddings help to produce an F_1 improvement of 1.3 points.

Table 6 details our F_1 scores for BiLSTM-CNN-CRF and EBC-CRF with respect to each entity label on both the BioSemantics patent corpus and the Reaxys Gold set. The overall results show that ELMo-based contextualized word embeddings help improve the baseline by 1.3 and 4.8 absolute F_1 score on BioSemantics and Reaxys, respectively.

In BioSemantics patent corpus, we obtain 1+ F_1 score improvements on frequent entity labels (i.e. $> 3,000$ instances) except for the entity label *Formula*, which has 0.4 absolute improvement. Higher improvements can be observed on rare entity labels (e.g. 4 points on *Mode of Actions*, 6 points on *Registry numbers* and *Trademarks*). The highest improvement at 9 points is found for the most rare entity label *CAS Number*.

---

[3]OSCAR4 helped produce the highest "macro-averaged" improvement on both datasets.

Entity label	Count†		BiLSTM-CNN-CRF			+ELMo			Δ_{F_1}
	#	%	P	R	F_1	P	R	F_1	
B (Abbreviation)	6,558	5.78	85.90	87.02	86.46	85.78	89.98	87.83	+1.37
C (CAS Number)	13	0.01	54.55	46.15	50.00	57.14	61.54	59.26	+9.26
D (Trademark)	2,290	2.01	62.58	61.79	62.18	66.44	71.40	68.83	+6.65
F (Formula)	7,935	6.99	86.05	86.81	86.42	83.07	90.91	86.82	+0.40
G (Generic)	51,313	45.20	81.45	84.56	82.98	83.84	84.44	84.14	+1.16
M (IUPAC)	39,896	35.14	88.40	87.77	88.09	87.25	91.20	89.18	+1.09
MOA (Mode of Action)	1,137	1.00	68.97	63.32	66.02	67.62	72.74	70.08	+4.06
R (Registry #)	96	0.08	55.68	51.04	53.26	65.82	54.17	59.43	+6.17
T (Target)	4,290	3.78	77.77	77.32	77.55	77.21	82.68	79.85	+2.30
Micro Avg.	113,528	100.0	83.76	85.01	84.38	84.30	87.11	85.68	+1.30

(a) BioSemantics

Entity label	Count†		BiLSTM-CNN-CRF			+ELMo			Δ_{F_1}
	#	%	P	R	F_1	P	R	F_1	
1 (chemClass)	1,476	12.36	78.35	66.46	71.92	81.96	75.75	78.73	+6.81
2 (chemClass$_{biomolecule}$)	951	7.96	71.86	70.50	71.17	76.27	78.76	77.50	+6.33
3 (chemClass$_{markush}$)	38	0.32	42.86	47.37	45.00	42.86	47.37	45.00	+0.00
4 (chemClass$_{mixture}$)	387	3.24	76.49	59.69	67.05	74.18	64.60	69.06	+2.01
5 (chemClass$_{mixture-part}$)	161	1.35	71.00	44.10	54.41	78.10	50.93	61.65	+7.24
6 (chemClass$_{polymer}$)	609	5.10	81.40	72.82	76.87	89.20	84.07	86.56	+13.74
7 (chemCompound)	6,988	58.53	89.02	92.01	90.49	91.01	94.58	92.76	+2.27
8 (chemCompound$_{mixture-part}$)	904	7.57	90.02	81.86	85.75	90.63	85.62	88.05	+2.27
9 (chemCompound$_{prophetics}$)	426	3.57	18.52	2.35	4.17	77.75	79.58	78.65	+74.48
Micro Avg.	11,940	100.0	85.12	80.36	82.67	87.41	87.53	87.47	+4.80

(b) Reaxys Gold

Table 6: F_1 score with respect to each entity label. "Count†" denotes gold-entity counts in test sets. "+ELMo" denotes scores obtained by EBC-CRF.

In the Reaxys Gold set, with ELMo we obtain 2+ F_1 score improvements on entity labels *chemCompound*, *chemCompound-mixture part* and *chemClass-mixture*. Higher improvements ($>$ 6 points) can be seen on some rare entity labels such as *chemClass*, *chemClass-biomolecule*, *chemclass-mixture-part* and *chemClass-polymer*. The improvements on entity label *chemClass-Markush* and *chemCompound-prophetics* are irregular compared to others. In particular, an absolute F_1 improvement of 74+ is achieved on entity label *chemCompound-prophetics*, while we do not find any improvement on *chemClass-Markush*.

4.2 Error Analysis

To perform error analysis on BioSemantics, we use its harmonized subset. Figure 2 (a) shows that most of the errors are confusions between non-chemical words and generic chemical names (e.g. *water, salt, acid*). For example, as illustrated in Figure 3 (a), the word "*salt*" which appears at the end of a systematic name should be identified as a part of the systematic name. However,

the same word is also widely used to describe a class of chemicals, e.g. "*pharmaceutically acceptable salt*" in Figure 3 (b). Disambiguation between chemical class and chemical compound is a challenging task even for human annotators, and is thus particularly difficult for a statistical model to learn. The confusion matrix of Reaxys Gold set in Figure 2 (b) also supports this point since most confusions are between non-chemical words, chemical classes and chemical compounds.

The Reaxys Gold set has a more complex tag set than the BioSemantics patent corpus, as it assigns separate fine-grained tags for subcategories of chemical classes (*chemClass*) and chemical compounds (*chemCompound*). As illustrated in Table 6, there is not sufficient training data for fine-grained sub-category labels. It is difficult for a high complexity neural model to learn characteristics of these sub-category labels and the key difference between the main categories and their subcategories. Figure 2 (b) shows that 50% the errors for "*chemical compound prophetics*" and 80% errors for "*chemical compound mixture part*" are

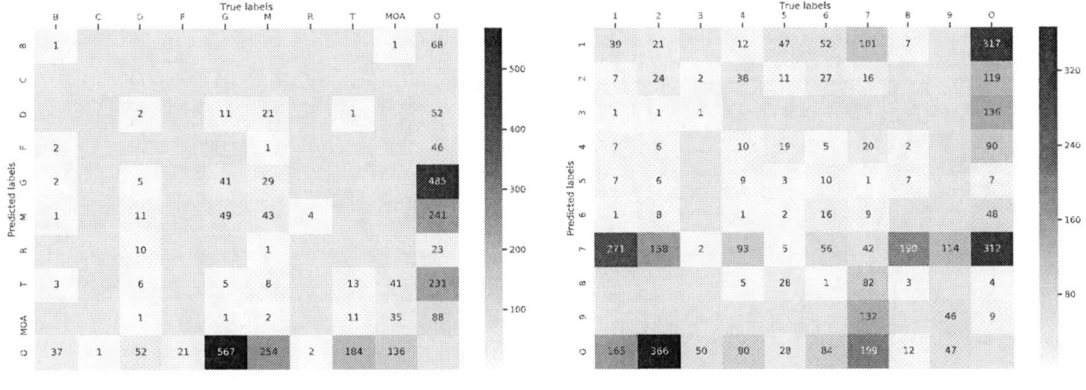

(a) BioSemantics harmonized set   (b) Reaxys Gold set

Figure 2: Confusion matrix of EBC-CRF model on BioSemantics (harmonized) and Reaxys Gold. x-axis: true labels; y-axis: predicted labels; numbers on cell where $x = y$ represent confusion between B (Begin) and I (Inside) tags. In (b) Labels 1–9 are detailed in Table 6 (b).

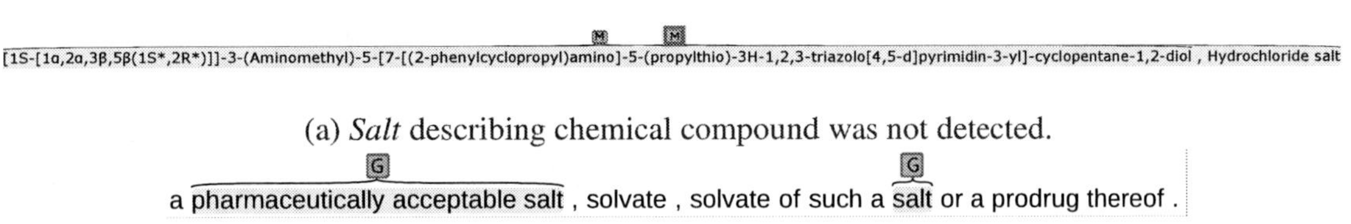

(a) *Salt* describing chemical compound was not detected.

(b) *Salt* describing chemical class being predicted as chemical compound.

Figure 3: Example of confusions caused by generic chemical names. (▣: false negatives, ▣, ▣: false positives)

due to confusion with their parent category *"chemical compound"*.

Another typical error observed frequently in BioSemantics and Reaxys is caused by participles. The most common example is word 'substituted'. In *"substituted or un-substituted alkyl"*, the token *"substituted"* refers to a specific chemical compound *"substituted alkyl"*. Whereas in *"2-pyridinyl is optionally substituted with 1-3 substituents"*, the token *"substituted"* refers to the substitution reaction.

We also observe that in both patent corpora, there are long sequences of systematic chemical names connected by comma only. Since there are no narrative words between the chemical names in such sequences, it is unlikely that the model can capture any contextual information when tagging them. This can potentially cause a "chain reaction" as shown in Figure 4, in which all chemical names fail to be recognized when the first chemical name is not tagged correctly.

4.3 Discussion

The results in Table 3 show that all chemical tokenizers outperform the OpenNLP general domain tokenizer. This is not surprising because tokenizers optimized for the chemical domain usually use either rule-based method or gazetteer-based methods to ensure that long systematic chemical names will be treated as a single token instead of being split into several tokens by symbols. This is reasonable as the character-level word representation will not be able to capture the morphological structures in a long chemical name if it is split into several tokens.

In the BioSemantics patent corpus, 80% of all entities are annotated as *Generic* or *IUPAC*. When adding ELMo-based word representations, we obtain smaller improvements in F_1 score for *Generic* and *IUPAC* than for remaining entity labels/types. This makes sense, as there are already enough training instances for these two labels in the dataset. By contrast, for rare entity labels with frequencies of less than 2 (e.g. *CAS Numbers, Trademarks, Mode of Actions, Registry numbers*), we obtain improvements of 4+ points when exploiting external information conveyed via ELMo.

The global F_1 score improvements on both experimental datasets confirm further this observation, viz., that score improvements due to ELMo

335

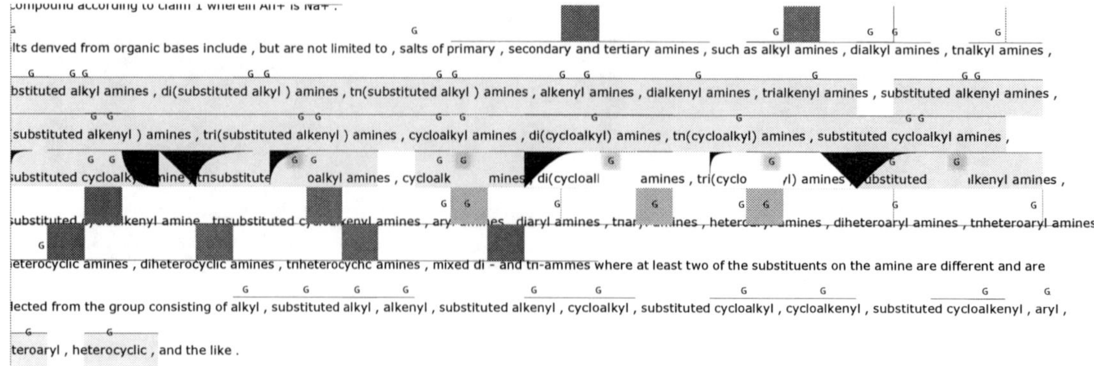

Figure 4: Example of "chain reaction" like errors. (—: false negatives, —: false positives, —: true positives)

decrease in inverse proportion to label frequency and training set size. Since the BioSemantics patent corpus contains 10 times more training instances than the Reaxys Gold set, we obtain an absolute improvement of 4.8 on Reaxys Gold set but of 1.3 points on the BioSemantics patent corpus.

Adding ELMo substantially improves the F_1 score on *chemCompound-prophetics*. This is because *chemCompound-prophetics* named entities are all long systematic chemical names which are arranged in lists. Since we replace all tokens longer than 25 characters with "Long_Token" when training ELMo, almost all sentences containing *chemCompound-prophetics* entities appear in the "Long_Token" style. This makes the ELMo-based representations of such long entities almost identical, and particularly easy to predict, thus resulting in an F_1 score improvement of 74 points for *chemCompound-prophetics*. We also observe no improvement for the *chemClass-Markush* label. The Markush structures are figures describing the structure of chemical compounds in which only a few parts/functional groups are labeled. When transforming to text, only the textual labels in the Markush structure are preserved. Thus, it is difficult for ELMo to learn any useful information from the broken Markush structures.

5 Conclusions

In this paper, we have made the following contributions towards improved chemical named entity recognition in chemical patents:

1. We improve on the current state-of-art for chemical NER in patents by +2.67 F_1 score.
2. We confirm that tokenizers optimized for chemical domain have a positive effect on NER performance by preserving informative morphological structures in systematic chemical names.
3. We demonstrate that word embeddings pre-trained on an in-domain chemical patent corpus help produce better performance than the word embeddings pre-trained on biomedical literature corpora.
4. We show that chemical NER performance can be improved by using contextualized word representations.
5. We release our ChemPatent word embeddings and an ELMo model trained from scratch on a newly collected corpus of 84K unannotated chemical patents, which can be utilized for downstream NLP tasks on chemical patents.[4]

Inspired by the patterns uncovered by our error analysis, our future work on chemical NER will focus on developing models which can be used to support disambiguation of general chemical words. In addition, it would be interesting to explore contextualized word embeddings learned by other neural models such as BERT (Devlin et al., 2019) or OpenAI GPT models (Radford et al., 2019) in future work.

Acknowledgments

This work was supported by an Australian Research Council Linkage Project grant (LP160101469) and Elsevier BV. We appreciate the contributions of the Content and Innovation team at Elsevier, including Georgios Tsatsaronis, Mark Sheehan, Marius Doornenbal, Michael Maier, and Ralph Hössel.

---

[4]`https://github.com/zenanz/ChemPatentEmbeddings`

References

Saber A Akhondi, Alexander G Klenner, Christian Tyrchan, Anil K Manchala, Kiran Boppana, Daniel Lowe, Marc Zimmermann, Sarma ARP Jagarlapudi, Roger Sayle, Jan A Kors, et al. 2014. Annotated chemical patent corpus: a gold standard for text mining. *PloS one*, 9(9):e107477.

Saber A Akhondi, Ewoud Pons, Zubair Afzal, Herman van Haagen, Benedikt FH Becker, Kristina M Hettne, Erik M van Mulligen, and Jan A Kors. 2016. Chemical entity recognition in patents by combining dictionary-based and statistical approaches. *Database*, 2016.

Saber A. Akhondi, Hinnerk Rey, Markus Schwörer, Michael Maier, John P. Toomey, Heike Nau, Gabriele Ilchmann, Mark Sheehan, Matthias Irmer, Claudia Bobach, Marius A. Doornenbal, Michelle Gregory, and Jan A. Kors. 2019. Automatic identification of relevant chemical compounds from patents. *Database*, 2019:baz001.

Abbas Akkasi, Ekrem Varoğlu, and Nazife Dimililer. 2016. Chemtok: a new rule based tokenizer for chemical named entity recognition. *BioMed research international*, 2016.

Peter F Brown, Peter V Desouza, Robert L Mercer, Vincent J Della Pietra, and Jenifer C Lai. 1992. Class-based n-gram models of natural language. *Computational linguistics*, 18(4):467–479.

K. Bretonnel Cohen, Helen L. Johnson, Karin Verspoor, Christophe Roeder, and Lawrence E. Hunter. 2010. The structural and content aspects of abstracts versus bodies of full text journal articles are different. *BMC Bioinformatics*, 11(1):492.

Peter Corbett and John Boyle. 2018. Chemlistem: chemical named entity recognition using recurrent neural networks. *Journal of cheminformatics*, 10(1):59.

Xiang Dai, Sarvnaz Karimi, Ben Hachey, and Cecile Paris. 2019. Using Similarity Measures to Select Pretraining Data for NER. In *Proceedings of NAACL-HLT 2019*, pages 1460–1470.

Kirill Degtyarenko, Paula De Matos, Marcus Ennis, Janna Hastings, Martin Zbinden, Alan McNaught, Rafael Alcántara, Michael Darsow, Mickaël Guedj, and Michael Ashburner. 2007. Chebi: a database and ontology for chemical entities of biological interest. *Nucleic acids research*, 36(suppl_1):D344–D350.

Jacob Devlin, Ming-Wei Chang, Kenton Lee, and Kristina Toutanova. 2019. BERT: Pre-training of deep bidirectional transformers for language understanding. In *Proceedings of the 2019 Conference of the North American Chapter of the Association for Computational Linguistics: Human Language Technologies*, pages 4171–4186.

Matt Gardner, Joel Grus, Mark Neumann, Oyvind Tafjord, Pradeep Dasigi, Nelson F. Liu, Matthew Peters, Michael Schmitz, and Luke S. Zettlemoyer. 2017. Allennlp: A deep semantic natural language processing platform. In *arXiv:1803.07640*.

Jiang Guo, Wanxiang Che, Haifeng Wang, and Ting Liu. 2014. Revisiting embedding features for simple semi-supervised learning. In *Proceedings of the 2014 Conference on Empirical Methods in Natural Language Processing (EMNLP)*, pages 110–120.

Maryam Habibi, Leon Weber, Mariana Neves, David Luis Wiegandt, and Ulf Leser. 2017. Deep learning with word embeddings improves biomedical named entity recognition. *Bioinformatics*, 33(14):i37–i48.

Maryam Habibi, David Luis Wiegandt, Florian Schmedding, and Ulf Leser. 2016. Recognizing chemicals in patents: a comparative analysis. *Journal of cheminformatics*, 8(1):59.

Mengke Hu, David Cinciruk, and John MacLaren Walsh. 2016. Improving automated patent claim parsing: Dataset, system, and experiments. *arXiv preprint arXiv:1605.01744*.

Zhiheng Huang, Wei Xu, and Kai Yu. 2015. Bidirectional LSTM-CRF models for sequence tagging. *arXiv preprint arXiv:1508.01991*.

David M Jessop, Sam E Adams, Egon L Willighagen, Lezan Hawizy, and Peter Murray-Rust. 2011. OSCAR4: a flexible architecture for chemical text-mining. *Journal of cheminformatics*, 3(1):41.

Roman Klinger, Corinna Kolářik, Juliane Fluck, Martin Hofmann-Apitius, and Christoph M Friedrich. 2008. Detection of iupac and iupac-like chemical names. *Bioinformatics*, 24(13):i268–i276.

Martin Krallinger, Obdulia Rabal, Analia Lourenço, Martin Perez Perez, Gael Perez Rodriguez, Miguel Vazquez, Florian Leitner, Julen Oyarzabal, and Alfonso Valencia. 2015. Overview of the CHEMDNER patents task. In *Proceedings of the Fifth BioCreative challenge evaluation workshop*, pages 63–75.

John D. Lafferty, Andrew McCallum, and Fernando C. N. Pereira. 2001. Conditional Random Fields: Probabilistic Models for Segmenting and Labeling Sequence Data. In *Proceedings of the Eighteenth International Conference on Machine Learning*, pages 282–289.

Guillaume Lample, Miguel Ballesteros, Sandeep Subramanian, Kazuya Kawakami, and Chris Dyer. 2016. Neural Architectures for Named Entity Recognition. In *Proceedings of the 2016 Conference of the North American Chapter of the Association for Computational Linguistics: Human Language Technologies*, pages 260–270.

Robert Leaman, Chih-Hsuan Wei, and Zhiyong Lu. 2015. tmChem: a high performance approach for chemical named entity recognition and normalization. *Journal of Cheminformatics*, 7(1):S3.

Yann LeCun, Yoshua Bengio, and Geoffrey Hinton. 2015. Deep learning. *nature*, 521(7553):436.

Mihai Lupu, Katja Mayer, John Tait, and Anthony J Trippe. 2011. *Current challenges in patent information retrieval*, volume 29. Springer.

Xuezhe Ma and Eduard Hovy. 2016. End-to-end Sequence Labeling via Bi-directional LSTM-CNNs-CRF. In *Proceedings of the 54th Annual Meeting of the Association for Computational Linguistics (Volume 1: Long Papers)*, pages 1064–1074.

Tomas Mikolov, Ilya Sutskever, Kai Chen, Greg S Corrado, and Jeff Dean. 2013. Distributed representations of words and phrases and their compositionality. In *Advances in neural information processing systems*, pages 3111–3119.

Thomas Morton, Joern Kottmann, Jason Baldridge, and Gann Bierner. 2005. Opennlp: A java-based nlp toolkit. In *Proceeding of the 10th Conference of the European Chapter of the Association of Computational Linguistics*.

Sorel Muresan, Plamen Petrov, Christopher Southan, Magnus J Kjellberg, Thierry Kogej, Christian Tyrchan, Peter Varkonyi, and Paul Hongxing Xie. 2011. Making every sar point count: the development of chemistry connect for the large-scale integration of structure and bioactivity data. *Drug Discovery Today*, 16(23-24):1019–1030.

Naoaki Okazaki. 2007. Crfsuite: a fast implementation of conditional random fields.

Jeffrey Pennington, Richard Socher, and Christopher Manning. 2014. Glove: Global vectors for word representation. In *Proceedings of the 2014 conference on empirical methods in natural language processing (EMNLP)*, pages 1532–1543.

Matthew E Peters, Mark Neumann, Mohit Iyyer, Matt Gardner, Christopher Clark, Kenton Lee, and Luke Zettlemoyer. 2018. Deep contextualized word representations. *arXiv preprint arXiv:1802.05365*.

Sampo Pyysalo, Filip Ginter, Hans Moen, Tapio Salakoski, and Sophia Ananiadou. 2013. Distributional semantics resources for biomedical text processing. In *Proceedings of the 5th International Symposium on Languages in Biology and Medicine, Tokyo, Japan*, pages 39–43.

Alec Radford, Jeff Wu, Rewon Child, David Luan, Dario Amodei, and Ilya Sutskever. 2019. Language models are unsupervised multitask learners.

Nils Reimers and Iryna Gurevych. 2017a. Optimal Hyperparameters for Deep LSTM-Networks for Sequence Labeling Tasks. *arXiv preprint*, arXiv:1707.06799.

Nils Reimers and Iryna Gurevych. 2017b. Reporting Score Distributions Makes a Difference: Performance Study of LSTM-networks for Sequence Tagging. In *Proceedings of the 2017 Conference on Empirical Methods in Natural Language Processing*, pages 338–348.

Tim Rocktäschel, Michael Weidlich, and Ulf Leser. 2012. Chemspot: a hybrid system for chemical named entity recognition. *Bioinformatics*, 28(12):1633–1640.

Mike Schuster and Kuldip K Paliwal. 1997. Bidirectional recurrent neural networks. *IEEE Transactions on Signal Processing*, 45(11):2673–2681.

Philippe Schwaller, Theophile Gaudin, David Lanyi, Costas Bekas, and Teodoro Laino. 2018. "found in translation": predicting outcomes of complex organic chemistry reactions using neural sequence-to-sequence models. *Chemical science*, 9(28):6091–6098.

Marwin HS Segler, Mike Preuss, and Mark P Waller. 2018. Planning chemical syntheses with deep neural networks and symbolic ai. *Nature*, 555(7698):604.

Stefan Senger, Luca Bartek, George Papadatos, and Anna Gaulton. 2015. Managing expectations: assessment of chemistry databases generated by automated extraction of chemical structures from patents. *Journal of cheminformatics*, 7(1):49.

Max E Valentinuzzi. 2017. Patents and scientific papers: Quite different concepts: The reward is found in giving, not in keeping [retrospectroscope]. *IEEE Pulse*, 8(1):49–53.

S Verberne, EKL D'hondt, NHJ Oostdijk, and CHA Koster. 2010. Quantifying the challenges in parsing patent claims. In *Proceedings of the 1st International Workshop on Advances in Patent Information Retrieval at ECIR 2010*, pages 14–21.

Chih-Hsuan Wei, Yifan Peng, Robert Leaman, Allan Peter Davis, Carolyn J Mattingly, Jiao Li, Thomas C Wiegers, and Zhiyong Lu. 2016. Assessing the state of the art in biomedical relation extraction: overview of the biocreative v chemical-disease relation (cdr) task. *Database*, 2016.

Zenan Zhai, Dat Quoc Nguyen, and Karin Verspoor. 2018. Comparing cnn and lstm character-level embeddings in bilstm-crf models for chemical and disease named entity recognition. In *Proceedings of the 9th International Workshop on Health Text Mining and Information Analysis (LOUHI 2018)*, pages 38–43.

Yaoyun Zhang, Jun Xu, Hui Chen, Jingqi Wang, Yonghui Wu, Manu Prakasam, and Hua Xu. 2016. Chemical named entity recognition in patents by domain knowledge and unsupervised feature learning. *Database*, 2016.

Improving classification of Adverse Drug Reactions through Using Sentiment Analysis and Transfer Learning

Hassan Alhuzali   **Sophia Ananiadou**
National Centre for Text Mining
School of Computer Science, The University of Manchester, United Kingdom
{hassan.alhuzali, sophia.ananiadou}@manchester.ac.uk

Abstract

The availability of large-scale and real-time data on social media has motivated research into adverse drug reactions (ADRs). ADR classification helps to identify negative effects of drugs, which can guide health professionals and pharmaceutical companies in making medications safer and advocating patients' safety. Based on the observation that in social media, negative sentiment is frequently expressed towards ADRs, this study presents a neural model that combines sentiment analysis with transfer learning techniques to improve ADR detection in social media postings. Our system is firstly trained to classify sentiment in tweets concerning current affairs, using the SemEval17-task4A corpus. We then apply transfer learning to adapt the model to the task of detecting ADRs in social media postings. We show that, in combination with rich representations of words and their contexts, transfer learning is beneficial, especially given the large degree of vocabulary overlap between the current affairs posts in the SemEval17-task4A corpus and posts about ADRs. We compare our results with previous approaches, and show that our model can outperform them by up to 3% F-score.

1   Introduction

Social media generate a huge amount of data for health and are considered to be an important source of information for pharmacovigilance (Sloane et al., 2015; Harpaz et al., 2014; Kass-Hout and Alhinnawi, 2013). ADR detection from social media has attracted a large amount of interest as a source of information regarding morbidity and mortality. In this respect, social networks are an invaluable source of information, allowing us to extract and analyse ADRs from health communication threads between thousands of users in real-time.

Several ADR systems have utilised features related to the sentiment of words to boost their system performance (Wu et al., 2018; Kiritchenko et al., 2017; Alimova and Tutubalina, 2017; Korkontzelos et al., 2016; Sarker and Gonzalez, 2015). Korkontzelos et al. (2016) analyse the impact of sentiment analysis features on extracting ADR from tweets. The authors observed that users frequently express negative sentiments when tweeting/posting about ADRs and they found the use of sentiment-aware features could improve ADR sequence labelling and classification.

It may be observed that the language used to express sentiment is often common across different domains. Consider, for example, the tweet "I hate how Vyvanse makes me over think everything and it makes me angry about things that I shouldn't even be angry about". The keywords used in this tweet to express the authors negative sentiment towards an ADR, i.e., hate and anger, are not specific to ADRs, and may be used to express sentiment towards many different kinds of topics. Based on this observation, we hypothesise that we can leverage transfer learning techniques by using sentiment analysis data to boost the detection of ADRs.

Our main research contribution is a new neural model that detects ADRs by firstly learning to classify sentiment, using a publicly available corpus of Tweets that is annotated with sentiment information and then using transfer learning to adapt this classifier to the detection of ADRs in social media postings.

Our new ADR detection model firstly trains a classifier on the SemEval17-task4A data, which consists of Tweets on the subject of current affairs. This pre-trained classifier then is adapted to the task of detecting ADRs, using datasets of social media postings that are annotated according to the presence or absence of ADRs. To our knowledge, this is the first attempt to apply transfer learning

Proceedings of the BioNLP 2019 workshop, pages 339–347
Florence, Italy, August 1, 2019. ©2019 Association for Computational Linguistics

techniques to adapt a sentiment analysis classifier to the task of detecting ADRs. In contrast to previous research, we use generalised neural methods that avoid the use of hand-crafted features, since these are time-consuming to generate, and are usually domain-dependent. We also explore different fine-tuning methods, (Howard and Ruder, 2018; Felbo et al., 2017), to determine which one performs best in our scenario.

The rest of the paper is organised as follows: Section 2 provides a review of related work. Section 3 presents the two datasets used to create our model. Section 4 describes our method and model. Section 5 reports on the analysis of results while Section 6 provides some conclusions.

2   Related Work

There is a growing body of literature concerned with the detection and classification of ADRs in social media texts (Wang et al., 2018; Huynh et al., 2016; Ebrahimi et al., 2016; Liu and Chen, 2015). Recent work has employed sentiment analysis features to improve the classification of ADRs (Wu et al., 2018; Kiritchenko et al., 2017; Alimova and Tutubalina, 2017; Korkontzelos et al., 2016; Sarker and Gonzalez, 2015).

Nikfarjam et al. (2015) exploited a set of features, including context features, ADR lexicon, part of speech (POS) and negation, to enhance the performance of ADR extraction. The authors chose Conditional Random Field as their classifier (CRF). Korkontzelos et al. (2016) followed the same research hypothesis, but focused on the evaluation of sentiment analysis features as an aid to extracting ADRs, based on the correlation between negative sentiments and ADRs. Alimova and Tutubalina (2017) built a classification system for the detection of ADRs for which they used a Support Vector Machine (SVM), instead of CRF. The authors also explored different types of features, including sentiment features and demonstrated that they improved the performance of ADR identification. Wu et al. (2018) utilised a set of hand-crafted features (i.e. sentiment features learned from lexica), similar to all of the other studies introduced above. However, the main difference is that the model is based on a neural network architecture, including word and character embeddings, Convolutional neural network (CNN), Long Short-Term Memory (LSTM) and multi-head attentions. This was the best performing system in the 2018 ADRs shared-task[1], which is part of the social media mining for health workshop (SMM4H).

In contrast to the models proposed in the above studies, it is possible to leverage sentiment analysis features automatically, without relying on any hand-crafted features. One common approach is to pre-train a classifier on a corpus annotated with sentiment information and then to adapt this pre-trained classifier to the detection of ADRs. The advantage of this approach is that the target system only needs access to the pre-trained model, but not the original sentiment corpus, which can be important for storage and data regulation issues. This method has been investigated by various researchers (Devlin et al., 2018; Howard and Ruder, 2018; Felbo et al., 2017). Felbo et al. (2017) learned a rich representation for detecting sentiment, sarcasm, and emotion using millions of emojis' dataset, acquired from Twitter. They demonstrated that this approach performs well and can achieve results that are competitive with state of the art systems. Recently, Devlin et al. (2018) built a deep bidirectional representation from transformers, which can be fine-tuned to different target tasks with an additional output layer. The model, which is called "Bert", showed significant improvements for a wide array of tasks, such as text classification, textual entailment and question answering, among others.

Compared to the above approaches, our work uses a simpler network architecture and does not require any feature engineering. Furthermore, we take advantage of transfer learning techniques acquired knowledge from sentiment analysis data. Our work is motivated by Felbo et al. (2017) who constructed a pre-trained classifier on emoji's data and then adapted to sentiment and emotion detection. The full details of our architecture are described in section 4.1.

3   Data

Several datasets have been created for ADRs. Some of these are gathered from specialised social networking forums for health (Thompson et al., 2018; Sampathkumar et al., 2014; Yates and Goharian, 2013; Yang et al., 2012), while others are collected from social media (Ginn et al., 2014; Jiang and Zheng, 2013; Bian et al., 2012).

---

[1]https://healthlanguageprocessing.org/smm4h/

In this research, we chose a widely used dataset (containing postings from Twitter and DailyStrength[2]) (Nikfarjam et al., 2015) that are annotated according to the presence or absence of ADRs in each post. The authors partitioned the data into a training (75%) and test (25%) sets. We further divided the training set into a 60% for training and 40% for validation. The validation set is used to develop our model before it is evaluated on the original test set (i.e. 25% of the complete corpus). Our model is designed to perform binary classification, to determine whether or not a given tweet or post mentions an ADR. Table 1 presents the number of tweets/posts belong to each category in the three different partitions of the data. More detailed information about the datasets can be found in Korkontzelos et al. (2016) and Nikfarjam et al. (2015).

Datasets	#ADRs	#None
Training		
DailyS.	900	417
Twitter	390	384
Validation		
DailyS.	600	278
Twitter	260	256
Test		
DailyS.	533	225
Twitter	236	192

Table 1: Data statistics (DailyS. = DailyStrength)

3.1 Sentiment Analysis corpus

We firstly train a sentiment analysis model on Twitter data from the SemEval17-task4A, which focuses on classifying the sentiment polarity of tweets on the subject of current affairs into predefined categories, e.g. positive, negative, and neutral. The dataset is partitioned into a training set of $50,000$ tweets and a test set of $12,000$ tweets (Rosenthal et al., 2017). A description of the sentiment analysis model is provided in section 4.

3.2 Preprocessing

Since Twitter data possesses specific characteristics, including informal language, misspellings, and abbreviations, we pre-process the data before applying the methods described in the next section. We use a tool that is specifically designed for the Twitter domain (Baziotis et al., 2017). The tool provides a number of different functionalities, such as tokenisation, normalisation, spelling-correction, and segmentation. We use the tool to tokenise the text, to convert words to lower-case, to correct misspellings, and to normalise user mentions, urls and repeated-characters.

4 Methods

This section discusses our model architecture, which is composed of two stages: the first stage involves building a sentiment analysis model, while the second stage adapts this model to a target task, which our case is the detection of ADRs. We describe our architectures in the following subsections.

4.1 Network Architecture

Our architecture consists of an embedding layer (Mikolov et al., 2013), a Long Short-Term Memory (LSTM) layer (Hochreiter and Schmidhuber, 1997), a self-attention mechanism (Bahdanau et al., 2014) and a classification layer. Figure 1 depicts the network architecture of our model.

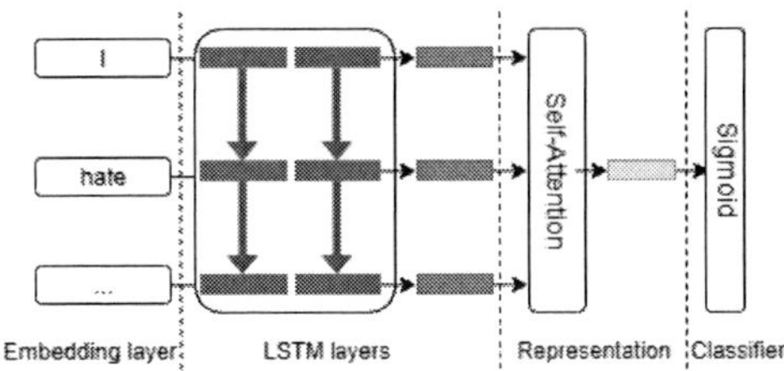

Figure 1: A description of the framework for our system.

In our different experiments, we use both an LSTM and a bi-directional LSTM (BiLSTM). Both are able to capture sequential dependencies especially in time series data, of which language can be seen as an example. The model's weights are initialized from the *word2vec* embedding with 300 dimensional size[3]. Additionally, the model consists of two LSTM/BiLSTM layers. For regularisation, we apply a dropout rate of 0.2 and 0.3 on the embedding output and after the second hidden layer, respectively, to prevent the network from over-fitting to the training set (Hinton

---

[2]DailyStrength is a specialised social networking website for health.

[3]`https://github.com/alexandra-chron/ntua-slp-semeval2018`

et al., 2012). We also choose Adam (Kingma and Ba, 2014) for optimisation and select 0.001 as the learning rate. We train the network for 10 epochs and the best performing cycle is only retained. It should be mentioned that the above set of hyper-parameters was determined using the validation set. Table 2 summarises the network architecture and hyper-parameters.

Hyper-Parameter	Value
embed-dim	300
layers	2
units	$\{200, 300, 400^*\}$
batch size	$\{32^*, 64\}$
epochs	10
sequence length	30
embed-dropout	0.2
lstm-dropout	$\{0.3, 0.4^*\}$
learning rate	0.001

Table 2: Network architecture and hyper-parameters. The asterisk (*): denotes the best performing setting

Embedding layer: T is a sequence of words $\{w_1, w_2, ..., w_n\}$ in a tweet/post and each w_i is a d dimensional word embedding for the i-th word in the sequence, where n is the number of words in the tweet. T should have the following shape n-by-d.

LSTM/Bi-LSTM layer: An LSTM layer takes as its input a sequence of word embeddings and generates word representations $\{h_1, h_2, ..., h_n\}$, where each h_i is the hidden state at time-step i, retaining all the information of the sequence up to w_i. Additionally, we experiment with a BiLSTM where the vector representation is built as a concatenation of two vectors, the first running in a forward direction $\overrightarrow{h}$ from left-to-right and the second running in a backward direction $\overleftarrow{h}$ from right-to-left $h_i = [\overrightarrow{h}; \overleftarrow{h}]$.

Self-attention: A self-attention mechanism has been shown to attend to the most informative words within a sequence by assigning a weight a_i to each hidden state h_i. The representation of the whole input is computed as follows:

$$e_i = tanh(W_h h_i + b_h) \tag{1}$$

$$a_i = softmax(e_i) \tag{2}$$

$$r = \sum_{i=1}^{T} a_i \cdot h_i \tag{3}$$

, where W_h, b_h are the attention's weights.

Classification layer: The vector r is an encoded representation of the whole input text (i.e. a tweet or post), which is eventually passed to a fully-connected layer for classification. A binary classification decision is made according to whether or not the input text mentions ADRs.

Transfer Learning: There are two common approaches to transfer learning (Peters et al., 2019). One approach is to use the last layer of a pre-trained model when fine-tuning to the target task. In this scenario, the network is used as a feature extractor. An alternative approach is to use the network for initialization, i.e., the full network is unfrozen and then fine-tuned to the target task.

In this work, After training the sentiment classification model, we exclude its output layer and replace it by an ADR output layer. Finally, the network is fine-tuned to detect the ADRs adopting the same architecture and hyper-parameters as the original model. We analyse the fine-tuning methods in section 5.2.1.

5 Results & Analysis

5.1 Results

Table 3 presents the performance of our models in terms of F-score, and compares these to the three of the best performing models from recently published research. For our own results, we report the results of three different experiments. Firstly, the baseline (LSTMA) is trained to detect ADRs using only the ADR datasets mentioned above, without the use of transfer learning. The other two models (LSTMA-TL and BiLSTMA-TL) apply transfer learning, making use of pre-training of a sentiment analysis model using the SemEval17-task4A dataset. These latter two models different in terms of whether they use a single direction or bi-directional LSTM, respectively. For experiments related to previous work, we replicated the three models following their details as described in Huynh et al. (2016), Alimova and Tutubalina (2017) and Wu et al. (2018).

5.1.1 Previous Work

Alimova and Tutubalina (2017) used an SVM model with different types of hand-crafted features (i.e. sentiment and corpus-based features). Their model performed to a high degree of accuracy, which is not surprising, due to the power of the SVM model when applied to small data. Similarly,

Huynh et al. (2016) exploited different neural networks, i.e CNN and a combination of both CNN and Gated Recurrent Units (GRU). They found that CNN obtained the best performance. For this reason, the results reported in Table 3 are those obtained for the CNN model. On the Twitter dataset, the performance of the CNN is even lower than the performance of our baseline model on this dataset. However, the performance on the DailyStrength dataset is considerably higher. The model developed by Wu et al. (2018) obtained the best results among the three compared systems; indeed, the results reach the same level as our baseline system. However, it is important to note that in contrast to our model architecture, that of Wu et al. (2018) is more complex and it relies on hand-crafted features as well as deep neural architectures.

5.1.2 Contextualised Word Embedding

In this work, we also compared our model to contextualized embedding (i.e. Bert) since it has been shown to achieve high results for various NLP tasks, including text classification (Devlin et al., 2018). We use the open-source PyTorch implementations [4] and only consider the "bert-base-uncase" model. The model is trained on the default hyper-parameters except that the number of batch-size and sequence length are chosen as follows 32 and 30, respectively, to match our model hyper-parameters for these two values. As shown in Table 3, Bert model achieves the same performance as our best model "LSTMA-TL" when applied to the Twitter data, although its performance is 3% lower than our best performing model when applied to the DailyStrength dataset. Even though transfer learning is beneficial, it can achieve better performance when learned from a related domain to the problem under investigation.

5.1.3 This Work

As Table 3 demonstrates, our proposed model is able to outperform all compared systems on the DailyStrength dataset, and all systems apart from Bert when applied to the Twitter Dataset. More specifically, the "LSTMA-TL" obtained the best results, thus demonstrating the utility and advantages of transfer learning techniques. The "BiLSTMA-TL" also demonstrates competitive results for the DailyStrength dataset, but it is 1% less than the "LSTMA-TL" for the Twitter dataset.

<hr>

[4]https://github.com/huggingface/
pytorch-pretrained-BERT

This may be due to the size of data and the architecture used in this work. Although the sentiment analysis model is trained on Twitter data, our ADR detection system still demonstrated substantial improvement on the DailyStrength dataset. Specifically, we obtained 3% and 2% improvement over our baseline model (i.e. LSTMA) on the Twitter and Dailystrength datasets, respectively.

Even though our experiments are based on a small dataset, the model demonstrated strong performance for ADR classification. Recent research claims that transfer learning techniques (i.e. fine-tuning) are beneficial for downstream tasks even if the target data size is small (Howard and Ruder, 2018; Alhuzali et al., 2018).

Datasets	DailyS.	Twitter
Models	F1	F1
Previous Work		
Huynh et al. (2016)	0.89	0.75
Alimova (2017)	0.89	0.78
Wu et al. (2018)	0.90	0.79
Contextualized W.E.		
Devlin et al. (2018)	0.89	**0.82**
This Work		
LSTMA (baseline)	0.90	0.79
LSTMA-TL	**0.92**	**0.82**
BiLSTMA-TL	**0.92**	0.81

Table 3: Comparison of our models to those reported in previous work. **LSTMA**: refers to LSTM with self−attention mechanism, while **LSTMA-TL**: means the same thing except the addition of transfer learning model. **BiLSTM-TF**: uses a BiLSTM with transfer learning model. Alimova (2017): Alimova and Tutubalina (2017). Best: bold.

5.2 Analysis

5.2.1 Impact of fine-tuning

We evaluate different methods to fine-tune our model, i.e. Last, Chain-thaw, Full and Simple Gradual unfreezing (GU). The first three techniques are adopted from Felbo et al. (2017) while the fourth one is described by Chronopoulou et al. (2019). "Last" refers to the process of only fine-tune the last layer (i.e. output layer), while the other layers are kept frozen. "Chain-thaw" method aims to firstly fine-tuned each layer independently and then fine-tuned the whole network simultaneously. "GU" is similar to the Chain-thaw method except that the fine-tuning is performed at different epochs. In this work, we experimented with these methods and selected the one that achieved the highest results for both datasets (i.e. Twitter and DailyStrength). The results of these four methods are reported in Figure 2.

"Last", which is the standard technique in fine-tuning, achieved the lowest performance; this is not surprising, because it contains the least general knowledge. In contrast, "Chain-thaw" achieved better results than "Last". The "Full" and "GU" obtained the best results for ADR classification. When we fine-tuned the whole network, we modified the "Full" method such that the embedding layer is frozen and we called it "Full-no-Emb", instead. The intuition behind this is that the embedding layer computes a word-based representation, which does not take into account the context of a word. This method obtains the best performance for both Twitter and DailyStrength datasets.

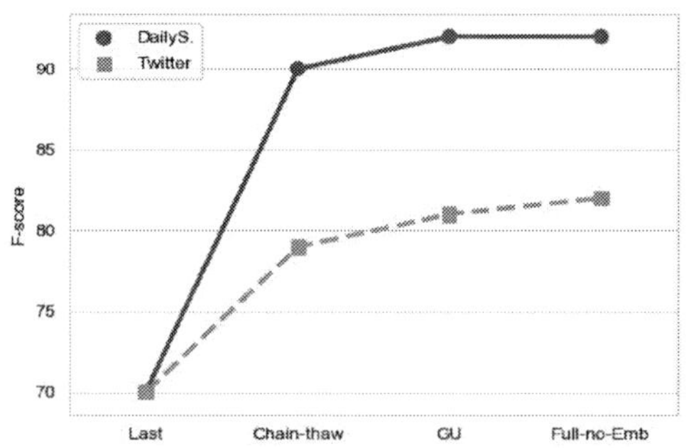

Figure 2: F-score for our model with a different set of fine-tuning methods.

5.2.2 Word Coverage

We observed that the vocabularies used in the sentiment analysis dataset and the ADR datasets share a large proportion of common words. To further investigate this, we measured the degree of common word coverage between the training and test parts of each dataset (i.e. Twitter and DailyStrength). The SemEval17-task4A training set is also included in this comparison. It should be noted that we compute the word coverage after pre-processing the data. Table 4 shows percentage of shared-vocabulary between the datasets. As shown in Table 4, the percentage of shared words between the training and test set of ADR Twitter data is 56.50%, while it is 74.22% between the SemEval17-task4A training set and the ADR Twitter test set. A similar pattern is also observed for the DailyStrength dataset, although there is a greater proportion of shared vocabulary between the training and test sets of DailyStrength. The vocabulary of the SemEval17-task4A dataset exhibits a large degree of overlap with the test sets of both Twitter and DailyStrength.

We hypothesise a number of reasons could account for this finding. Intuitively, users often use none-technical keywords when they post or tweet about ADRs. In other words, they do not employ terms found in medical lexicons. This allows users to express their opinion towards ADRs using terms which may be used to express sentiment towards other different topics. Additionally, several datasets have been collected for ADRs. However, most of them have not been made available for the research community. In contrast, there are dozens of sentiment analysis datasets available online, including SemEval17-task4A[5], Yelp reviews [6], Amazon reviews[7] and Stanford[8], among others. Thus, this confirms our initial observations and helps to reinforce that ADR system can benefit from the proliferation of sentiment analysis data available online, which is the primary motivation of this work.

Dataset	Train	SEl17-4A	Δ %
Twitter test	56.50%	74.22%	17.72%
DailyS. test	68.03%	78.22%	10.19%

Table 4: Word coverage. "SEl17-4A": corresponds to the training set of the SemEval17-task4A. Δ%: represents the difference between the two percentages for each dataset in a row.

5.2.3 Error Analysis

We experiment with small data in this work and this may limit our interpretation and analysis in this section. Nevertheless, performing error analysis can reveal some strengths and weaknesses of the proposed models and identify room for future work.

For error analysis, we selected examples which are incorrectly classified by the proposed model in this paper (i.e. LSTMA-TL) and previous work (i.e. (Huynh et al., 2016; Alimova and Tutubalina,

---

[5] http://alt.qcri.org/semeval2017/task4/index.php?id=data-and-tools

[6] https://www.yelp.com/dataset

[7] https://s3.amazonaws.com/amazon-reviews-pds/readme.html

[8] https://nlp.stanford.edu/sentiment/index.html

2017). Figure 3 and 4 present the number of false positive and false negative classifications for each model. As can be seen in Figure 3 that the number of miss-classified examples as false negative is higher than false positive for the DailyStrength dataset, while the opposite pattern is observed for the Twitter dataset as shown in Figure 4. Our model also demonstrated balanced error classifications for both false positive and false negative. In contrast, the other two models, proposed by previous research, obtained unbalanced error classifications except Alimova and Tutubalina (2017)'s model achieved quite balanced errors for the Twitter dataset. For future work, it might be useful to investigate different ensemble methods that can help to reduce the false positive and false negative classifications and improve the classification of ADR.

In addition, we analysed examples only classified correctly by our model. We observed that our model is able to classify examples carrying none-specific keywords to ADRs, but to sentiments in general. This shows the importance of sentiment features to ADRs. Examples 1-3 below illustrate the instances that are correctly predicted by our proposed model. The first two examples are part of the Twitter test set, while the third example is part of the DailyStrength test set.

- Example 1: is it hot in here or is [durg_name] just kicking in?.

- Example 2: anyone ever taken [durg_name]? i've been on it for a week, not too sure how i feel about it yet. anyone want to share their experience?.

- Example 3: loved it , except for not being able to be woken up at night . . yeah that blew.

On the other hand, we inspected examples that our model failed to correctly classify. For instance, example (4) below was extracted from the Twitter test set and it was predicted as negative for the presence of ADR, whereas the true label is positive for the presence of ADR. Examples (5) also illustrates the same observation, but is part of the DailyStrength test set. We anticipate that our model failed to classify example (4) and (5) due to the lack of context and unambiguous keywords. Example (4) can also be interpreted as either positive or negative for the presence of ADRs. This may explain that the true label can be sometimes misleading and requires further examination.

- Example 4: moved on to something else when it quit working.

- Example 5: i'm with you. even though the [durg_name] works, i still don't feel fully human.

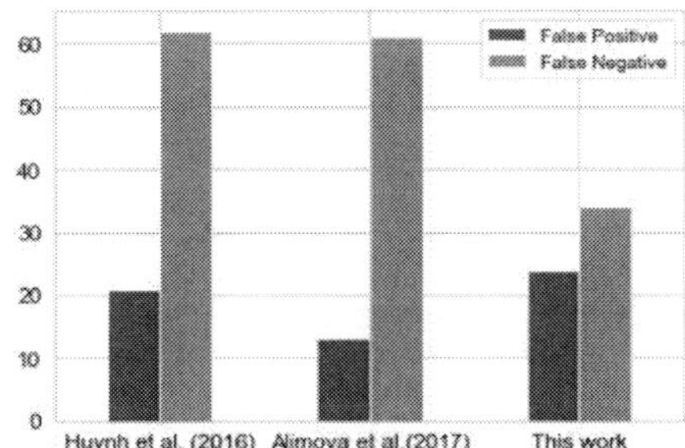

Figure 3: The number of miss-classified examples by the proposed models of this work and previous research for the DailyStrength dataset. This work: refers to the proposed model in this paper (i.e. LSTMA-TL).

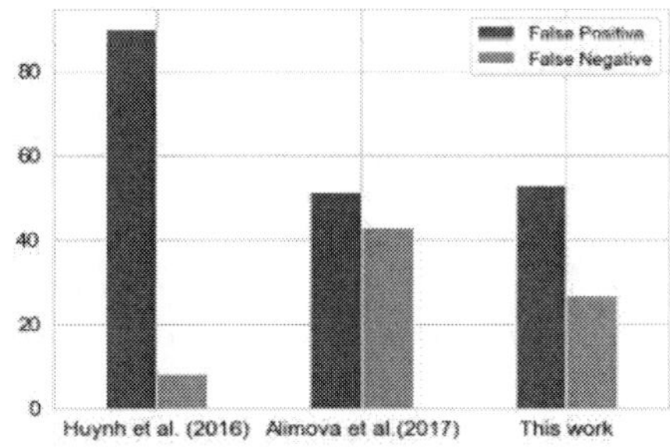

Figure 4: The number of miss-classified examples by the proposed models of this work and previous research for the Twitter dataset. This work: refers to the proposed model in this paper (i.e. LSTMA-TL).

6 Conclusion

In this work, we proposed a novel neural network architecture for ADR identification. Our approach exploits the fact that in social media, ADRs are frequently expressed with negative sentiment. Taking advantage of the readily available sentiment analysis datasets that are available online, our architecture firstly trains a sentiment analysis classifier on Tweets concerned with current affairs, and then adapts this to detect ADRs in social media. Our empirical results have demonstrated that the application of the fine-tuned model to ADR datasets obtains a substantial improvement over previously published models. It also achieved higher results than Bert on DailyStrength dataset. Additionally, the word coverage analyses revealed that sentiment analysis dataset shares a significant amount of vocabulary with ADR dataset, which is even higher than the correlation between the words in training and test sets of the same ADR dataset. This paper has empirically discussed the advantages and utility of both sentiment analysis datasets and transfer learning techniques for improving the performance of ADR detection in social media and specialised health-related forums. Finally, we provided some error analyses and potential future work.

7 Acknowledgement

We thank Prof. Graciela Gonzalez-Hernandez, University of Pennsylvania, for sharing the Twitter and DailyStrength datasets with us. We would like also to thank Paul Thompson for his valuable comments and suggestions. The first author is supported by the Ministry of Higher Education of the Kingdom of Saudi Arabia.

References

Hassan Alhuzali, Mohamed Elaraby, and Muhammad Abdul-Mageed. 2018. Ubc-nlp at iest 2018: Learning implicit emotion with an ensemble of language models. In *Proceedings of the 9th Workshop on Computational Approaches to Subjectivity, Sentiment and Social Media Analysis*, pages 342–347.

Ilseyar Alimova and Elena Tutubalina. 2017. Automated detection of adverse drug reactions from social media posts with machine learning. In *International Conference on Analysis of Images, Social Networks and Texts*, pages 3–15. Springer.

Dzmitry Bahdanau, Kyunghyun Cho, and Yoshua Bengio. 2014. Neural machine translation by jointly learning to align and translate. *arXiv preprint arXiv:1409.0473*.

Christos Baziotis, Nikos Pelekis, and Christos Doulkeridis. 2017. Datastories at semeval-2017 task 4: Deep lstm with attention for message-level and topic-based sentiment analysis. In *Proceedings of the 11th International Workshop on Semantic Evaluation (SemEval-2017)*, pages 747–754.

Jiang Bian, Umit Topaloglu, and Fan Yu. 2012. Towards large-scale twitter mining for drug-related adverse events. In *Proceedings of the 2012 international workshop on Smart health and wellbeing*, pages 25–32. ACM.

Alexandra Chronopoulou, Christos Baziotis, and Alexandros Potamianos. 2019. An embarrassingly simple approach for transfer learning from pretrained language models. *arXiv preprint arXiv:1902.10547*.

Jacob Devlin, Ming-Wei Chang, Kenton Lee, and Kristina Toutanova. 2018. Bert: Pre-training of deep bidirectional transformers for language understanding. *arXiv preprint arXiv:1810.04805*.

Monireh Ebrahimi, Amir Hossein Yazdavar, Naomie Salim, and Safaa Eltyeb. 2016. Recognition of side effects as implicit-opinion words in drug reviews. *Online Information Review*, 40(7):1018–1032.

Bjarke Felbo, Alan Mislove, Anders Søgaard, Iyad Rahwan, and Sune Lehmann. 2017. Using millions of emoji occurrences to learn any-domain representations for detecting sentiment, emotion and sarcasm. In *Proceedings of the 2017 Conference on Empirical Methods in Natural Language Processing*, pages 1615–1625.

Rachel Ginn, Pranoti Pimpalkhute, Azadeh Nikfarjam, Apurv Patki, Karen OConnor, Abeed Sarker, Karen Smith, and Graciela Gonzalez. 2014. Mining twitter for adverse drug reaction mentions: a corpus and classification benchmark. In *Proceedings of the fourth workshop on building and evaluating resources for health and biomedical text processing*, pages 1–8. Citeseer.

Rave Harpaz, Alison Callahan, Suzanne Tamang, Yen Low, David Odgers, Sam Finlayson, Kenneth Jung, Paea LePendu, and Nigam H. Shah. 2014. Text mining for adverse drug events: the promise, challenges, and state of the art. *Drug Safety*, 37(10):777–790.

Geoffrey E Hinton, Nitish Srivastava, Alex Krizhevsky, Ilya Sutskever, and Ruslan R Salakhutdinov. 2012. Improving neural networks by preventing coadaptation of feature detectors. *arXiv preprint arXiv:1207.0580*.

Sepp Hochreiter and Jürgen Schmidhuber. 1997. Long short-term memory. *Neural computation*, 9(8):1735–1780.

Jeremy Howard and Sebastian Ruder. 2018. Universal language model fine-tuning for text classification. In *Proceedings of the 56th Annual Meeting of the Association for Computational Linguistics (Volume 1: Long Papers)*, volume 1, pages 328–339.

Trung Huynh, Yulan He, Alistair Willis, and Stefan Rüger. 2016. Adverse drug reaction classification with deep neural networks. Coling.

Keyuan Jiang and Yujing Zheng. 2013. Mining twitter data for potential drug effects. In *International conference on advanced data mining and applications*, pages 434–443. Springer.

Taha A Kass-Hout and Hend Alhinnawi. 2013. Social media in public health. *Br Med Bull*, 108(1):5–24.

Diederik P Kingma and Jimmy Ba. 2014. Adam: A method for stochastic optimization. *arXiv preprint arXiv:1412.6980*.

Svetlana Kiritchenko, Saif M Mohammad, Jason Morin, and Berry de Bruijn. 2017. Nrc-canada at smm4h shared task: classifying tweets mentioning adverse drug reactions and medication intake. *arXiv preprint arXiv:1805.04558*.

Ioannis Korkontzelos, Azadeh Nikfarjam, Matthew Shardlow, Abeed Sarker, Sophia Ananiadou, and Graciela H Gonzalez. 2016. Analysis of the effect of sentiment analysis on extracting adverse drug reactions from tweets and forum posts. *Journal of biomedical informatics*, 62:148–158.

Xiao Liu and Hsinchun Chen. 2015. A research framework for pharmacovigilance in health social media: identification and evaluation of patient adverse drug event reports. *Journal of biomedical informatics*, 58:268–279.

Tomas Mikolov, Ilya Sutskever, Kai Chen, Greg S Corrado, and Jeff Dean. 2013. Distributed representations of words and phrases and their compositionality. In *Advances in neural information processing systems*, pages 3111–3119.

Azadeh Nikfarjam, Abeed Sarker, Karen Oconnor, Rachel Ginn, and Graciela Gonzalez. 2015. Pharmacovigilance from social media: mining adverse drug reaction mentions using sequence labeling with word embedding cluster features. *Journal of the American Medical Informatics Association*, 22(3):671–681.

Matthew Peters, Sebastian Ruder, and Noah A. Smith. 2019. To tune or not to tune? adapting pre-trained representations to diverse tasks. *CoRR*, abs/1903.05987.

Sara Rosenthal, Noura Farra, and Preslav Nakov. 2017. Semeval-2017 task 4: Sentiment analysis in twitter. In *Proceedings of the 11th International Workshop on Semantic Evaluation (SemEval-2017)*, pages 502–518.

Hariprasad Sampathkumar, Xue-wen Chen, and Bo Luo. 2014. Mining adverse drug reactions from online healthcare forums using hidden markov model. *BMC medical informatics and decision making*, 14(1):91.

Abeed Sarker and Graciela Gonzalez. 2015. Portable automatic text classification for adverse drug reaction detection via multi-corpus training. *Journal of biomedical informatics*, 53:196–207.

Richard Sloane, Orod Osanlou, David Lewis, Danushka Bollegala, Simon Maskell, and Munir Pirmohamed. 2015. Social media and pharmacovigilance: a review of the opportunities and challenges. *British journal of clinical pharmacology*, 80(4):910–920.

Paul Thompson, Sophia Daikou, Kenju Ueno, Riza Batista-Navarro, Junichi Tsujii, and Sophia Ananiadou. 2018. Annotation and detection of drug effects in text for pharmacovigilance. *Journal of cheminformatics*, 10(1):37.

C. Wang, H. Dai, F. Wang, and E. C. Su. 2018. Adverse drug reaction post classification with imbalanced classification techniques. In *2018 Conference on Technologies and Applications of Artificial Intelligence (TAAI)*, pages 5–9.

Chuhan Wu, Fangzhao Wu, Junxin Liu, Sixing Wu, Yongfeng Huang, and Xing Xie. 2018. Detecting tweets mentioning drug name and adverse drug reaction with hierarchical tweet representation and multi-head self-attention. In *Proceedings of the 2018 EMNLP Workshop SMM4H: The 3rd Social Media Mining for Health Applications Workshop and Shared Task*, pages 34–37.

Christopher C Yang, Haodong Yang, Ling Jiang, and Mi Zhang. 2012. Social media mining for drug safety signal detection. In *Proceedings of the 2012 international workshop on Smart health and wellbeing*, pages 33–40. ACM.

Andrew Yates and Nazli Goharian. 2013. Adrtrace: detecting expected and unexpected adverse drug reactions from user reviews on social media sites. In *European Conference on Information Retrieval*, pages 816–819. Springer.

Exploring Diachronic Changes of Biomedical Knowledge using Distributed Concept Representations

Gaurav Vashisth[1,2], **Jan-Niklas Voigt-Antons**[1,2], **Michael Mikhailov**[1] and **Roland Roller**[1]
[1]German Research Center for Artificial Intelligence (DFKI), Berlin, Germany
[2]Technische Universität Berlin
`firstname.lastname@dfki.de`

Abstract

In research best practices can change over time as new discoveries are made and novel methods are implemented. Scientific publications reporting about the latest facts and current state-of-the-art can be possibly outdated after some years or even proved to be false. A publication usually sheds light only on the knowledge of the period it has been published. Thus, the aspect of time can play an essential role in the reliability of the presented information. In Natural Language Processing many methods focus on information extraction from text, such as detecting entities and their relationship to each other. Those methods mostly focus on the facts presented in the text itself and not on the aspects of knowledge which changes over time.

This work instead examines the evolution in biomedical knowledge over time using scientific literature in terms of diachronic change. Mainly the usage of temporal and distributional concept representations are explored and evaluated by a proof-of-concept.

1 Introduction

Scientific literature presents knowledge for a particular time period it has been published. Various studies have been performed to explore such knowledge from scientific literature, where work by Swanson (1986) led to the discovery of a new drug to treat Raynaud's disease. Similarly, a study by Zhu et al. (2013) has concluded that drug discovery using scientific literature plays a pivotal role in the treatment of cancer, which can improve the quality of life of patients (Cummings et al., 2011). Although scientific literature is an excellent source of information, there has been an explosion in the number of publications each year. This poses a challenge for biomedical researchers and practitioners to keep themselves informed of recent developments. The increasing number at the same time provides an opportunity to automatically explore the data on how a change in knowledge has evolved. Some studies have tried to explore such changed knowledge by investigating temporal information (Zhou and Hripcsak, 2007; He and Chen, 2018), studying the diachronic change in the meaning of the word. A diachronic semantic change in language is associated with progression in the meaning of the word which is estimated by exploring its usage over time.

This work aims to automatically explore the advances in medical knowledge extracted from the abstracts of scientific research by using word/concept embeddings. Especially, we examine how treatments of pathological conditions have changed over time. For this reason we focus on concepts rather than words, as biomedical concepts can be mentioned in text in different ways (e.g. 'headache', 'cephalgia' or 'pain in the head'). Moreover, biomedical concepts help to encapsulate noun phrases represented by more than one word, for example, 'eye lens' or 'lung cancer'. An analysis on word level instead would take all situations the single words occur into account, and therefore would be more general. To quantify such changes we measure how the usage of a biomedical concept has (semantically) changed over time by comparing different embedding periods.

The rest of the work is structured as follows: The next section presents related work in the context of diachronic changes in and outside the biomedical domain. Then, in Section 3 we present how the biomedical concept embeddings are generated and how the time aspect is taken into account. Section 4 shows the usage of our embeddings to explore diachronic changes as a proof-of-concept. Then we apply the temporal embeddings to explore some exemplary relational data of UMLS, followed by a conclusion.

Proceedings of the BioNLP 2019 workshop, pages 348–358
Florence, Italy, August 1, 2019. ©2019 Association for Computational Linguistics

2   Related Work

Human language is a complex system which has been evolving from the point of its origin whether it is because of social or cultural (Hamilton et al., 2016a,b) or technological (Phillips et al., 2017) reasons. Some words acquire new meaning much faster than other words (Blank, 1999) for example words like *broadcast*, *gay*, and *awful* have been used in a different context in the present time as compared to an earlier time.

To study the semantic change for words, initially, co-occurrence matrices (Sagi et al., 2009; Wijaya and Yeniterzi, 2011; Jatowt and Duh, 2014), K-means clustering (Wijaya and Yeniterzi, 2011), Frequency-based methods (Kulkarni et al., 2014) were used.   Representations using co-occurrence matrices are based on the notion of word co-occurring in the same context. The co-occurrence matrix assumes that words occurring in same context tend to have the same meaning (Firth, 1957) and are represented by methods such pointwise mutual information (Turney and Pantel, 2010), Singular Value Decomposition matrices, and Latent Semantic Analysis.

Another popular method to represent words are distributed representations. Words are represented in a dense and continuous form, that enables us to capture the meaning in a condensed form. There are various methods such Word2Vec (Mikolov et al., 2013b,a) and Global Vectors for Word Representation (Glove) (Pennington et al., 2014) which create a distributed representation of words. Distributed methods consume less memory compared to co-occurrence matrices because of their compact size and ranges between 100 dimensions to 1000. Moreover, the distributed methods are robust baseline methods with their proven success in capturing linguistic meaning (Mikolov et al., 2013b).

Kim et al. (2014) explored the temporal changes in the meaning of word using Skip-gram negative sampling (SGNS) method. To generate word embedding for each time frame the embeddings from previous time frame was used to initialize the embedding for the next successive time frame. Hamilton et al. (2016b) try to answer two questions, first whether the frequency of a word affects the change in meaning, which has been long studied (Bybee et al., 2007; Pagel et al., 2007; Lieberman et al., 2007).   Second, whether there is a relationship between a polysemous and semantic change of a word.

Also in the biomedical domain semantic changes in scientific abstracts have been explored (Yan and Zhu, 2018).   In the study, the authors explored semantic changes for a set of words using their occurrence frequency and their distribution across different topics. Scientific literature has also motivated studies using biomedical concepts instead of free text; however, they only measure the similarity and relatedness between different concepts using different embedding methods (De Vine et al., 2014; Choi et al., 2016; Liu et al., 2018; Beam et al., 2018).

Our study draws motivation from previous studies. However, different to other work we try to explore diachronic change using biomedical concepts. Particularly we would like to use diachronic change to assist the exploration of knowledge changes in the biomedical domain.

3   Temporal Concept Embeddings

In the following the generation of the biomedical temporal concept embeddings used to identify semantic changes is introduced.

3.1   Data Resources

The MEDLINE repository[1] is a bibliographic database from life sciences containing around 26 millions articles dating back to 1809. MEDLINE is quickly growing as the number of publications added to the repository each year are increasing (see Figure 1).   Title and abstracts within the MEDLINE repository define the source to generate the embeddings in this work.

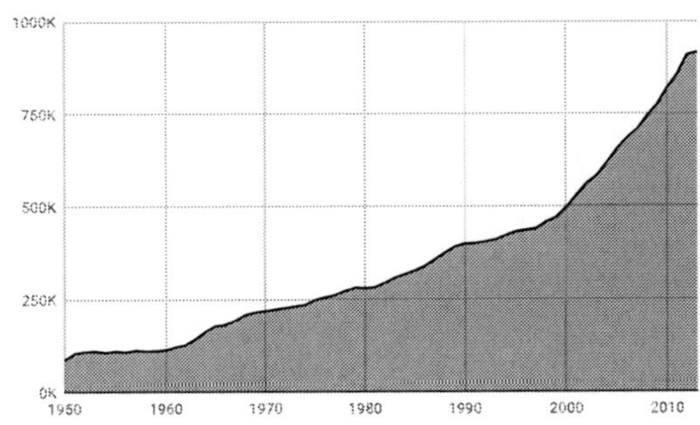

Figure 1: Number of MEDLINE abstracts published each year on PUBMED between 1950 and 2014.

Another relevant resource is the Unified Medical Language System (UMLS) (Bodenreider,

---

[1]https://mbr.nlm.nih.gov/

349

2004), a biomedical knowledge base which defines a large number of biomedical concepts and their relations to each other. Each concept is represented by a unique concept identifier known as CUI and includes word variations and synonyms. As we focus on the generation of concept embeddings, we normalize text mentions from MEDLINE abstracts to UMLS.

The concept normalization is carried out using MetaMap (Aronson, 2001), a popular named entity recognition system for biomedical text. However, to avoid processing millions of sentences with MetaMap, we use the MetaMapped 2015 MEDLINE Baseline Results, a MEDLINE subset already enriched by MetaMap Machine Output (MMO). In addition to that, we also use annual baseline files from the MEDLINE/PUBMED Baseline Repository (MBR) which contain meta-information about each publication such as publication ID, publication year and author name(s).

3.2 Data Preprocessing

First, publications from MMO are enriched with publication year (PubYear) by using the publication ID and the information from the MBR files. Then, the text occurrence of each medical abstract and its title are mapped and replaced with their concept ID, using the offset information provided in MMO (Figure 2). In this way, we create a text to train our embeddings. Since we do not consider character embeddings, we can treat concept IDs as words without any disadvantage.

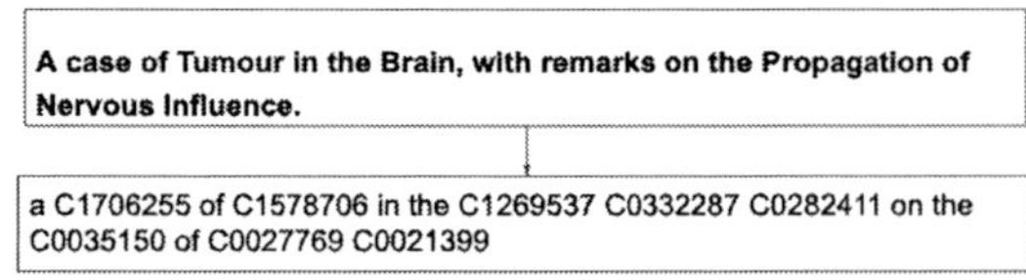

Figure 2: Shows mapping of medical text to their corresponding concept ID for Publication ID:20895112.

To create temporal embeddings, the preprocessed MEDLINE abstracts are split into different time depended subsets using PubYear. Embeddings are then trained on those splits. Ideally, we would like to train models using equally sized time ranges, such as embeddings per year or decade. However, this is not easily possible for various reasons: Firstly, as seen in Figure 1 the number of publications is constantly increasing. A consistent split into equal time frames would result in highly unbalanced splits regarding the

number of included abstracts. In addition to that PUBMED includes mainly titles and no abstracts before 1975, which further reduces the number of text for the lower represented period.

Period	# Publications
1809-1970	3,374,099
1971-1975	1,162,030
1976-1980	1,346,833
1981-1985	1,528,475
1986-1990	1,863,659
1991-1995	2,065,386
1996-2000	2,297,006
2001-2005	2,938,855
2006-2010	3,721,166
2011-2012	1,762,603
2013-2015	1,283,218

Table 1: Distribution of publications in each period

Conversely, the generation of equally sized splits (according to the number of abstracts/sentences) has the disadvantage that it will be more challenging to differentiate between particular years. Rounding up or down the number of included publications might also be not a satisfying solution, as the time ranges might differ too much. For this reason, we mainly focus on time range splits including 5 years of MEDLINE abstracts. As the number of publications is lower at the beginning of the 20th century and publications often do not contain any abstract, we combine the 'early' MEDLINE data into one big split (1809-1970). Moreover, as the number of publications steadily increase we create smaller splits from 2011. The final split into periods is presented in Table 1, including their corresponding number of abstracts.

3.3 Temporal Embeddings

To generate temporal embeddings, we use FastText (Bojanowski et al., 2016) in Skip-gram negative sampling (SGNS) mode, which predicts context words corresponding to a given target word occurring in its neighborhood. The values of the hyperparameters base on the recommendation of Levy et al. (2015). The authors did an extensive set of experiments using different representation methods and analyzed the effect of hyperparameters on the embeddings generated by them. We chose negative samples as 10, the minimum occurrence of concepts is 5, learning rate as .05, sampling threshold as .0001, dimension to 300 and context window to 10.

The different temporal embeddings were trained sequentially, starting from the first period (1809-1970) and ending with (2013-2015). We started the training of the first period with random initialization of the embeddings. All other embeddings were then initialized by the values of the former time embedding. This incremental training process has been applied as the training of a particular time period can build on the knowledge seen in earlier periods. Incremental training can be seen as an analogy of how human knowledge evolves over time. The temporal concept embeddings used in this work can be downloaded here[2].

3.4 Measuring Semantic Changes

To measure the semantic change between a concept pair we use cosine distance (similarity) at different periods as also described in Hamilton et al. (2016b). A cosine distance closer to 1 shows a stronger similarity/relation between the two concepts than a distance closer to 0. In this work, however, we are particularly interested in examining whether the semantic shift can be used to explore how treatments (of particular diseases) evolved. Therefore, we selected particular concept pairs and explore how their similarity score evolves.

In addition to cosine similarity we use Positive Pointwise Mutual Information (PPMI) matrix (Levy and Goldberg, 2014), as reference measure. A PPMI matrix is a variant of Pointwise Mutual Information (PMI) and provides an association between two words occurring together in a corpus and how strongly they are related to each other (Church and Hanks, 1990). When a specific word pair co-occurs more frequently they have a higher PPMI score and vice-versa. PPMI is still widely used co-occurrence matrix method and in this work we have used a normalized PPMI score which ranges between 0 to 1, whereas 1 indicates more frequent pairs.

4 Exploring Biomedical Knowledge Changes

In this section, we examine the usage of temporal concept embeddings to detect diachronic changes

in the context of altering knowledge in biomedical literature. Particularly, we explore whether the embeddings reveal known changes in treatments in biomedical history, as a proof-of-concept. For instance, we would like to know whether it is possible to see a relative change in terms of cosine similarity, i.e., if a preferred treatment for some *Disease X* changes at time t from one medication to a new one (see example in Figure 3). Our assumption is that the usage of temporal concept embeddings reveal a similar pattern. Before time t we assume, that the old treatment has got a higher cosine similarity compared to the new treatment. And then after some decrease the new medication outperforms the other one. In the following, we will explore this phenomenon based on various examples.

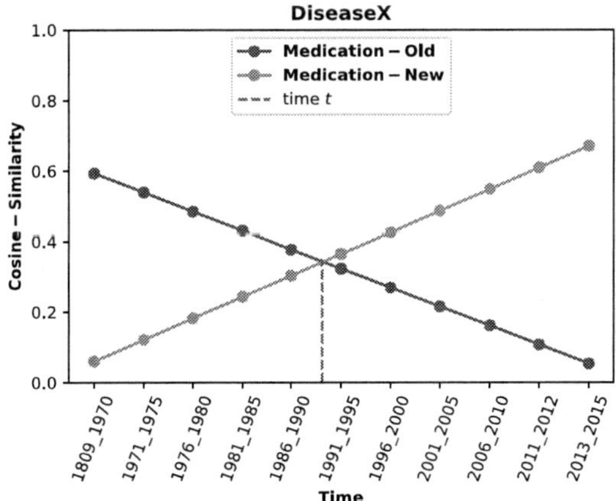

Figure 3: Shows a treatment change of some Disease X from *Medication-old* to *Medication-new* at time t.

4.1 Proof-of-Concept

In this section, different examples are presented to explore the usage of temporal concept embeddings to detect knowledge changes. We use those examples as a proof-of-concept. Each example includes a high-level introduction, followed by an investigation of the similarity scores over time and an explanation of the presented results.

In order to provide reliable insights, presented results are supported through a significance test (Welch's T-test) using a confidence interval of 99% (p_value < 0.01). The significance test relies on 15 different complete sets of temporal embeddings (all periods) which were trained from scratch.

---

[2]http://biomedical.dfki.de/

4.1.1 Minoxidil

Minoxidil (CUI=C0026196) is a medication, initially used for treating high blood pressure (Hypertension) (Stoehr et al., 2019) . Nowadays Minoxidil is still used as a drug of last resort for treatment of resistant hypertension (remains above a target level, in spite of being prescribed three or more anti-hypertensive drugs simultaneously with different mechanisms of action). However, in 1988 FDA approved the medication also for treating hair loss problems. Presently, Minoxidil is used mainly to treat early baldness pattern such as *Androgenic Alopecia* (C0162311) and *Scalp Hair Loss* (C0574769).

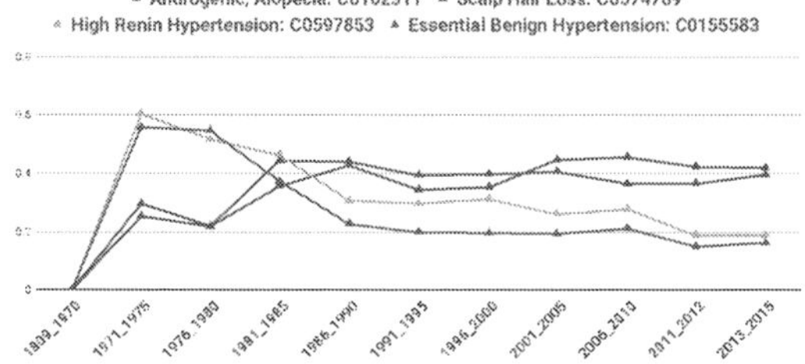

Figure 4: The similarity score of minoxidil with medical conditions from 1809 to 2015. Change in usage ocurs in 1986. Where **Old Usage** was *High Blood Pressure* and **New Usage** is *hair fall*

The exploration of the cosine similarity for minoxidil and its change in treatment is presented in Figure 4. The figure depicts a high similarity to hypertension in the '70s, which is significantly higher than the high blood pressure. However, after 1980 the similarity slowly decreases in the next following years. Around 1985 we can see a big drop. At the same time the similarity of *alopecia* and *scalp hair loss* strongly increase around 1985. From the following period, the similarity score of both concepts outperforms hypertension and are significantly higher than hypertension.

4.1.2 Microprolactinoma

Microprolactinoma (Prolactinoma)[3] is a type of benign tumor that occurs in the pituitary gland of the brain (Casanueva et al., 2006; Glezer and Bronstein, 2015). Its treatment has changed notably over time. Until the 1970's this tumor was removed by a surgical method known as *Transethmoidal Hypophysectomy* (C0405509) (Richards et al., 1974). Beginning from the late 1970's

a new class of medical therapy with *Dopamine Agonists* was introduced to treat Microprolactinoma (C0344452) without having to undergo a surgery. *Dopamine Agonists* is a class of drugs that activate dopamine receptors. The treatment using *Dopamine Agonists* has a cure rate of more than 80%. The most effective *Dopamine Agonists* used as a main treatment drugs are *Cabergoline* (C0107994) and *bromocriptine* (C0006230) (Tirosh and Shimon, 2016; Glezer and Bronstein, 2015) which are D2 dopamine agonists that inhibit prolactin secretion. Only if patients do not respond to medications, a surgical method called *Transsphenoidal surgery*[4] (C2985562) is used (Tirosh and Shimon, 2016).

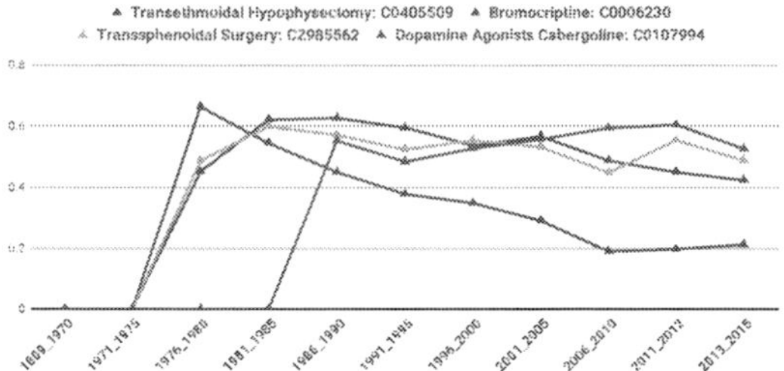

Figure 5: Similarity Score of Microprolactinoma with different treatment methods from 1809 until 2015. Change in the medication occurs after 1976. Where **Old Method** was *Transethmoidal hypophysectomy* and **New Methods** are *Transsphenoidal surgery, Bromocriptine, dopamine agonists cabergoline*.

Figure 5 presents the semantic shift in the use of different treatment methods for Microprolactinoma. The first embedding point is seen from the period 1976-1980. Before that period none of the concepts occurred frequently enough to be considered in the embedding. Within the first period of occurrence (1976-1980) we have a significantly higher similarity score of Microprolactinoma with *Transethmoidal Hypophysectomy* in comparison to *Bromocriptine* concepts and *Transsphenoidal surgery* which starts decreasing in the next following years. After 1980, we see an increase in the similarity for all the *bromocriptine* concepts along with *Transsphenoidal surgery*, which shows a change in the treatment method for Microprolactinoma. The similarity score of both the *Bromocriptine* and *Transsphenoidal surgery* concepts have significantly higher similarity score than *Transethmoidal Hypophysectomy* from 1981. Whereas

---

[3] Microadenoma of a pituitary gland

[4] A surgical method used to remove tumors of pituitary glands.

from 1986, after the induction of *cabergoline*, both of the *dopamine agonists* and *Transsphenoidal surgery* have a higher similarity score than *Transethmoidal Hypophysectomy*. Also *Cabergoline* is getting more popular after 2006 and is then significantly higher than other treatments.

4.1.3 White Blood Cell Cancer

A subtype of cancer of white blood cells known as chronic myeloid leukemia (CML) or *Chronic Myelosis* (C0023473) is a medical condition. In this condition there is an abnormal increase in the number of white blood cells (WBC) compared to red blood cells (RBC). WBC are responsible for protecting the body against infections, but when produced in large numbers, they start accumulating in blood and bone marrow. This prohibits the growth of RBC and causes weight loss, spleen enlargement and bone pain (Radich et al., 2018).

Before 2001, *Chronic Myelosis* was treated predominantly by chemotherapy using alkylating antineoplastic agents, such as *Mitobronitol* (C0026236) and *Myelobromol* (C0700014). The introduction of targeted therapy led to the improved survival rate of patients compared to the earlier generation of medication. The new targeted therapy method includes a class of drugs called *Tyrosine Kinase Inhibitors* (TKI) (C1268567), whereas *Imatinib* (C0935989) is one of the most important representatives of this class. *Tyrosine Kinase Inhibitors* were first synthesized in 1998 (Yaish et al., 1988), and *Imatinib* was first approved in 2001 to treat this type of blood cancer.

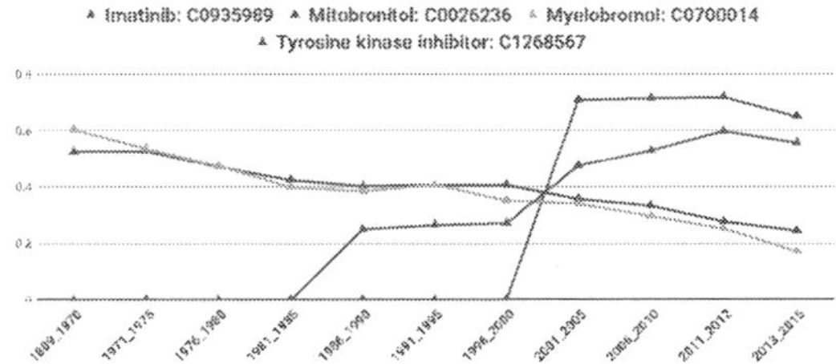

Figure 6: Similarity Score of White blood cell cancer with different treatment methods from 1809 until 2015. Change in the medication occurs in 2001. **Old Methods** were *Mitobronitol, Myelobromol* and **New Methods** are *Tyrosine Kinase Inhibitor* and *Imatinib*.

Figure 6 depicts different treatments used for white blood cell cancer. The similarity score for both *Mitobronitol* and *Myelobromol* is high in

'70s. However, after '70s their score starts decreasing but are still significantly higher than *Tyrosine Kinase Inhibitor* from 1990's to 2000. From 2001 there is a significantly higher similarity for both *Tyrosine Kinase Inhibitor* and *Imatinib* as compared to both *Mitobronitol* and *Myelobromol*.

4.1.4 Hepatitis-C

Hepatitis-C (C0220847) is an infectious blood-borne disease which is caused by the hepatitis C virus (HCV). Hepatitis-C mainly affects the liver which can cause liver diseases and eventually lead to liver failure. HCV spreads mostly through infected blood transfusions or poorly sterilized injection needles, also during intravenous injection of drugs. (Maheshwari and Thuluvath, 2010).

Presently there is no vaccine to prevent HCV virus, however chronic infections are treated by antiviral medications (Webster and Klenerman, 2015). Until 2011, *Polyethylene Interferon Alpha-2a* (C0391001), *Polyethylene Interferon Alpha-2b* (C0796545) in combination with *Ribavirin*[5] (C0035525) were used to treat hepatitis-C and had a cure rate of less than 50%. From 2011, the second generation of antiviral medication known as Direct Antiviral Agents (DAA) was approved by the FDA. DAA directly interfere with the machinery of Hepatitis-C virus, thus inhibiting its growth and transmission. There are several classes of DAA that are used at different stages in the treatment of Hepatitis-C such as *Telaprevir, Boceprevir, Daclatasvir*. However, for current work we just show *Telaprevir* (C1876229). This DAA is used in combination with *Ribavirin* which have a cure rate of more than 90% (Rivett and Alexander, 2019).

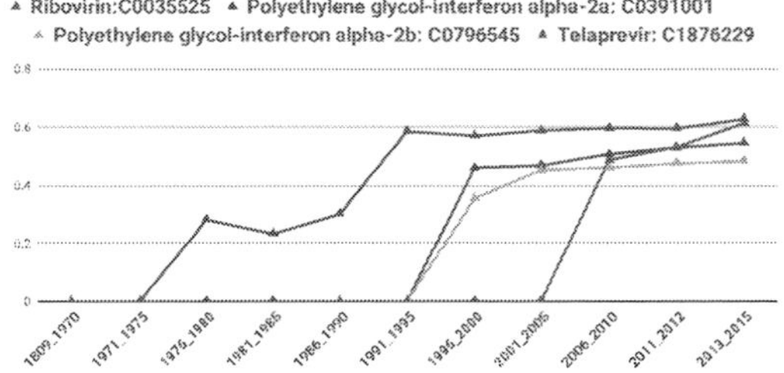

Figure 7: Similarity Score of Hepatitis-C with different treatment methods from 1809 to 2015, change in the medication occurs in 2011. **Old Methods** were *Polyethylene interferon alpha-(2a and 2b)* and **New Methods** is *Telaprevir*.

---

[5]First generation of antivirals.

Figure 7 shows a rise in the similarity of second generation of antivirals (*Telaprevir*) from 2011 as compared with first generations (*Polyethylene Interferon Alpha-(2a,2b)*) where there is a decrease in the similarity. From 2011 the similarity score of *Telaprevir* is significantly higher than the both *Polyethylene interferon alpha-(2a,2b)*, respectively. We can also notice, that the similarity score *Ribavirin* is high this is because it is still used in combination with the new generation of antiviral medications as well. Before 1976 the occurrence of any antivirals medication concepts that appears close to Hepatitis-C is not high enough as such concepts are not present.

4.2 Concept Embeddings vs. Co-occurrence

As seen in the examples above, temporal concept embeddings can be used to identify diachronic changes. In comparison to that, those changes can be also identified using a simple co-occurence metrics, such as PPMI. Figure 8 shows an example for White Blood Cell Cancer. However, in comparison to the example in Section 4.1.3, changes can be much stronger and values can quickly decrease to zero, if the co-occurrence of two concepts suddenly decreases.

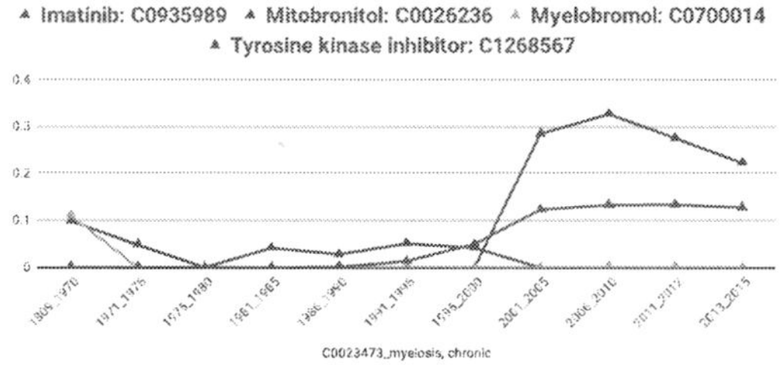

Figure 8: Similarity Score using PPMI matrix for WBC with different treatment methods from 1809 to 2015.

The score of *Mitobronitol* for instance suddenly drops to 0 in 1976-1980 and then increases in 1981. Conversely, the concept embeddings show a slow decrease in similarity for same pair at 1976-1980. This is can have several reasons: Firstly, even if concepts do not occur together within the same context window, they might occur within the same context which is considered by concept embedding. Moreover, the initialization of embeddings for 1976-1980 build on top of the previous period (1970-1975). The incremental learning mechanism helps concept embeddings to overcome the drawback of sudden drop in the similarity of a concept-pair if they do not co-occur in a specific period.

4.3 Discussion

The previous examples showed that we can use diachronic semantic changes to identify medical knowledge change. To measure the change in treatment of some disease from an old medication to a new medication was not as simple as our initial assumption was. Originally, examples were provided by a medical student on a rather high level. Given these examples the corresponding concepts and concept IDs had to be identified within UMLS. In various cases those concepts were ambiguous and the most appropriate concept had to be selected, e.g. Hepatitis-C in UMLS is represented as Hepatitis C virus (C0220847) as well as Hepatitis C (C0019196).

It also happened that a concept mention did not show the effect we were interested in (no occurrence, low similarity scores, no increase/decrease). This caused a more detailed manual analysis to find out why. In some cases, if a concept did not show the effect we were searching for, it turned out that a more specific concept instead showed the expected effect. For instance we found that particular derivatives of *dopamine agonists* such *Cabergoline* and *Bromocriptine* were more talked about in the context of Microprolactinoma than *dopamine agonists*. This is an interesting aspect of how information are connected and which are actually mentioned in the scientific text. Unfortunately this is difficult to solve given our high level examples and a method solely based on general literature.

However, even though the examples above were manually selected with a lot of domain knowledge, we can clearly show that knowledge changes are present in our temporal concept embeddings. In order to address possible concerns, the next section explores knowledge changes of known UMLS pairs.

5 Exploring Existing Medical Knowledge

In the previous section, we showed that changes in biomedical knowledge and particularly changes of treatments could be reflected within temporal concept embeddings. However, those examples were manually selected by a medical expert. In this section instead we apply the technique to explore known drug-disease pair relations of the UMLS

	Concept Embeddings					Co-occurrence			
Period	#	POS (MAX)	#	NEG (MAX)	#	POS (MAX)	#	NEG (MAX)
1809-1970	7	0.330 (0.754)	56	0.208 (0.588)	225	0.026 (0.286)	573	0.003 (0.171)
1971-1975	2	0.318 (0.721)	52	0.197 (0.596)	644	0.022 (0.354)	984	0.003 (0.086)
1976-1980	11	0.307 (0.742)	106	0.168 (0.564)	360	0.026 (0.325)	680	0.003 (0.134)
1981-1985	10	0.310 (0.711)	137	0.157 (0.530)	355	0.029 (0.330)	663	0.003 (0.150)
1986-1990	12	0.304 (0.681)	135	0.155 (0.553)	388	0.028 (0.310)	729	0.002 (0.139)
1991-1995	16	0.301 (0.672)	150	0.149 (0.505)	527	0.026 (0.266)	761	0.002 (0.073)
1996-2000	12	0.297 (0.680)	157	0.149 (0.510)	566	0.025 (0.337)	780	0.002 (0.149)
2001-2005	7	0.287 (0.689)	147	0.146 (0.499)	536	0.024 (0.309)	767	0.002 (0.121)
2006-2010	10	0.271 (0.695)	177	0.144 (0.476)	655	0.021 (0.300)	832	0.002 (0.077)
2011-2012	13	0.272 (0.730)	146	0.153 (0.467)	957	0.017 (0.355)	1178	0.002 (0.088)
2013-2015	15	0.265 (0.696)	136	0.152 (0.425)	1158	0.015 (0.246)	1264	0.002 (0.077)

Table 2: Exploration of known (positive) and unknown (negative) drug-disease concept pairs of UMLS across different time periods. The table shows the mean and its maximum scores below POS and NEG in terms of cosine similarity and PPMI. In addition to that, that table shows the number of concept pairs (#) which do not occur together within the set of 3,000 drug-disease pairs.

Metathesaurus. First we explore known concept pairs with cosine similarity for concept embeddings in comparison to PPMI. After that we examine selected relations of UMLS and track their similarity across different periods.

5.1 Exploring known Drug-Disease Pairs: Concept Embeddings vs. Co-occurrence

In the following we examine concept embeddings using cosine similarity in comparison to the co-occurrence metric PPMI on known drug-disease relations of UMLS. To do so, we use *may-treat* and *may-prevent* relations of UMLS and selected randomly for each time period a set of 3,000 concept pairs. We made sure, that both concepts occurred within that time slice. Then we randomly generated a set of negative concept pairs (unknown according to UMLS) with the same size. Next we use both sets (positive and negative) to calculate cosine similarity using concept embedding and PPMI matrix .

The results are presented in Table 2 and show, that the average score is higher for the known relations pairs (positive) in comparison to the randomly generated negative pairs. This is valid for cosine similarity and PPMI. Moreover we can see, that the average cosine score for concept embedding is above the PPMI, as well as for the corresponding MAX scores. However, both scores can not be directly compared.

Interestingly, the table shows a varying number of concept pairs which are not covered by a metric (lower than .05 for concept embedding and zero for PPMI). Particularly the co-occurrence metric PPMI has fewer information about various con-cept pairs in comparison to concept embedding. For instance, in period 2013-2015 while the cosine similarity for concept embedding score for only 15 positive concept pairs is below .05, 1158 concept pairs are not considered by co-occurrence, as concepts do not occur together frequent enough. Note, the low PPMI scores might be related to the sparseness of the PPMI matrix.

Overall, the results show, that the incremental temporal concept embeddings have got an advantage over the co-occurrence metric PPMI. As the concept embedding uses knowledge from previous time slices and considers contextual information it is able to better cope with the situation if concept pairs do not frequently together.

5.2 Exploring Drug-Disease Pairs across different Time Periods

In the following we use temporal concept embeddings to explore changes in biomedical knowledge. We apply the technique to explore known drug-disease pair relations *may_treat* and *may_prevent* of the UMLS Metathesaurus. An increase over time might indicate[6] a higher use of drug against the corresponding disease in present time as compared to previous periods; whereas a decrease can indicate new treatment therapy for the disease from disease-drug pair. This might be interesting as often various treatments exist for a disease. In this way, it might be possible to identify a more popular treatment (according to similarity score) which is at the same time also encoded within the embeddings.

---

[6] Of course it could also mean something different.

Drug	Disease	1809-1970	1971-1975	1991-1995	1996-2000	2011-2012	2013-2015
Oxymetholone	Anemias	0.36	0.43	0.23	0.18	0.23	0.23
Epoetin Alfa Recombinant		0.00	0.00	0.43	0.37	0.35	0.42
Sodium Cromoglycate	Bronchitic Asthma	0.58	0.60	0.51	0.45	0.45	0.29
Aalmeterol		0.00	0.00	0.56	0.55	0.50	0.56
Tolazamide	Type 2 Diabetes Mellitus	0.63	0.46	0.48	0.45	0.28	0.27
Sitagliptin		0.00	0.00	0.00	0.00	0.54	0.62
Pramipexole	Syndrome Parkinson's	0.00	0.00	0.39	0.48	0.46	0.46
Amantadine Hydrochloride		0.41	0.51	0.41	0.36	0.17	0.20
Risperidone	Type Schizophrenia	0.00	0.00	0.53	0.60	0.58	0.59
Acetophenazine Maleate		0.46	0.41	0.35	0.32	0.20	0.20
Tamoxifen	Tumor of Breast	0.00	0.28	0.51	0.48	0.49	0.49
Testolactone		0.43	0.38	0.25	0.22	0.12	0.10

Table 3: Decrease (upper part) and increase (lower part) in similarity for *may-treat* and *may-prevent* drug-disease pairs across different time periods

Table 3 presents results for particular diseases in terms of increasing and decreasing similarity scores for known *may-treat* and *may-prevent* drug-disease pairs. The similarity scores shown here are for the first two periods (1809-1970, 1971-1975), two periods from the middle (1991-1995, 1996-2000) and the last two ones (2011-2012, 2013-2015). Each row contains a two different known drugs related to a disease. The upper part presents a scenario with a decreasing similarity score (relative to the disease) and the lower part an increasing score. For example, the table shows that the similarity between *Tolazamide* and *Type 2 Diabetes Mellitus* is .63 in 1809-1970. With each succeeding period the value decreases and eventually reaches .27 in 2013-2015. On the other hand, the similarity between the *Sitagliptin* with *Type 2 Diabetes Mellitus* is 0 until 1996-2000 due to its absence in this period. However, from 2011 we see a sudden and strong increase.

The table shows that we can detect changes of known relational facts. The results are also in line with our original hypothesis that scientific journals reflect the change in medical knowledge since each journal provide current medical facts. As scientific research around these fact evolves, we witness a change in medical knowledge which is present in the scientific journals.

6   Conclusion

In the present work, we have successfully shown that it is possible to explore the diachronic semantic change on a biomedical concept level. The automatic exploration of knowledge changes might be particularly useful to extend structured knowledge, such as UMLS potentially. For instance, UMLS often includes an extensive range of different treatments or preventions for a disease. However, all relations have the same importance and the same weighting. Thus it is not necessarily obvious which one is the treatment of choice (also depending on time, but also co-morbidities or other symptoms). Our proposed method could be a first (and simplistic) step to highlight particular concept pairs. For instance, temporal concept embeddings could be used to support (distantly supervised) relation extraction (Roller and Stevenson, 2014) or to spot particular trends automatically (Chen et al., 2007).

However, our current approach has got some limitations as it is unable to detect the negative polarity between the pairs. In terms of this we assume that a higher similarity is correlated with a stronger use, which is not necessarily correct. Future work could take this into account.

Finally, as mentioned in Section 4.3, it would be interesting to address the problem that sometimes only particular child concepts show an effect we are interested in. It might be possible to overcome this by including graph embeddings in addition to the text based temporal ones.

Acknowledgments

This project has received funding from the European Unions Horizon 2020 research and innovation programme under grant agreement No 780495 (BigMedilytics). In addition to that we would like to thank our colleagues for their feedback and suggestions.

References

Alan R Aronson. 2001. Effective Mapping of Biomedical Text to the UMLS Metathesaurus: The MetaMap Program. In *Proceedings of the AMIA Symposium*, page 17. American Medical Informatics Association.

Andrew L. Beam, Benjamin Kompa, Inbar Fried, Nathan P. Palmer, Xu Shi, Tianxi Cai, and Isaac S. Kohane. 2018. Clinical Concept Embeddings Learned from Massive Sources of Medical Data. *CoRR*, abs/1804.01486.

Andreas Blank. 1999. Why do new meanings occur? A cognitive typology of the motivations for lexical semantic change. *Historical semantics and cognition*, 13:6.

Olivier Bodenreider. 2004. The Unified Medical Language System (UMLS): Integrating Biomedical Terminology. *Nucleic acids research*, 32(Database issue):D267–70.

Piotr Bojanowski, Edouard Grave, Armand Joulin, and Tomas Mikolov. 2016. Enriching Word Vectors with Subword Information. *arXiv preprint arXiv:1607.04606*.

Joan Bybee et al. 2007. *Frequency of Use and the Organization of Language*. Oxford University Press on Demand.

Felipe F Casanueva, Mark E Molitch, Janet A Schlechte, Roger Abs, Vivien Bonert, Marcello D Bronstein, Thierry Brue, Paolo Cappabianca, Annamaria Colao, Rudolf Fahlbusch, et al. 2006. Guidelines of the Pituitary Society for the diagnosis and management of prolactinomas. *Clinical endocrinology*, 65(2):265–273.

Elizabeth S Chen, Peter D Stetson, Yves A Lussier, Marianthi Markatou, George Hripcsak, and Carol Friedman. 2007. Detection of practice pattern trends through natural language processing of clinical narratives and biomedical literature. In *AMIA Annual Symposium Proceedings*, volume 2007, page 120. American Medical Informatics Association.

Youngduck Choi, Chill Yi-I Chiu, and David Sontag. 2016. Learning Low-Dimensional Representations of Medical Concepts. *AMIA Summits on Translational Science Proceedings*, 2016:41.

Kenneth Ward Church and Patrick Hanks. 1990. Word Association Norms, Mutual Information, and Lexicography. *Computational linguistics*, 16(1):22–29.

Greta G. Cummings, Susan Armijo Olivo, Patricia D. Biondo, Carla R. Stiles, Ozden Yurtseven, Robin L. Fainsinger, and Neil A. Hagen. 2011. Effectiveness of Knowledge Translation Interventions to Improve Cancer Pain Management. *J. Pain Symptom Manage.*, 41(5):915–939.

Lance De Vine, Guido Zuccon, Bevan Koopman, Laurianne Sitbon, and Peter Bruza. 2014. Medical Semantic Similarity with a Neural Language Model. In *Proceedings of the 23rd ACM International Conference on Conference on Information and Knowledge Management*, pages 1819–1822. ACM.

John R Firth. 1957. A Synopsis of Linguistic Theory, 1930-1955. *Studies in linguistic analysis*.

Andrea Glezer and Marcello D. Bronstein. 2015. Prolactinomas. *Endocrinol. Metab. Clin. North Am.*, 44(1):71–78.

William L. Hamilton, Jure Leskovec, and Dan Jurafsky. 2016a. Cultural Shift or Linguistic Drift? Comparing Two Computational Measures of Semantic Change. *Proceedings of the 2016 Conference on Empirical Methods in Natural Language Processing*.

William L. Hamilton, Jure Leskovec, and Dan Jurafsky. 2016b. Diachronic Word Embeddings Reveal Statistical Laws of Semantic Change. In *Proc. Assoc. Comput. Ling. (ACL)*.

Jiangen He and Chaomei Chen. 2018. Predictive Effects of Novelty Measured by Temporal Embeddings on the Growth of Scientific Literature. *Frontiers in Research Metrics and Analytics*, 3:9.

Adam Jatowt and Kevin Duh. 2014. A Framework for Analyzing Semantic Change of Words across Time. In *Proceedings of the 14th ACM/IEEE-CS Joint Conference on Digital Libraries*, pages 229–238. IEEE Press.

Yoon Kim, Yi-I Chiu, Kentaro Hanaki, Darshan Hegde, and Slav Petrov. 2014. Temporal Analysis of Language through Neural Language Models. In *Proceedings of the Workshop on Language Technologies and Computational Social Science@ACL 2014, Baltimore, MD, USA, June 26, 2014*, pages 61–65. Association for Computational Linguistics.

Vivek Kulkarni, Rami Al-Rfou, Bryan Perozzi, and Steven Skiena. 2014. Statistically Significant Detection of Linguistic Change. *CoRR*, abs/1411.3315.

Omer Levy and Yoav Goldberg. 2014. Neural Word Embedding as Implicit Matrix Factorization. In *Advances in Neural Information Processing Systems*, pages 2177–2185.

Omer Levy, Yoav Goldberg, and Ido Dagan. 2015. Improving Distributional Similarity with Lessons Learned from Word Embeddings. *Transactions of the Association for Computational Linguistics*, 3:211–225.

Erez Lieberman, Jean-Baptiste Michel, Joe Jackson, Tina Tang, and Martin A Nowak. 2007. Quantifying the Evolutionary Dynamics of Language. *Nature*, 449(7163):713.

Yue Liu, Tao Ge, Kusum S Mathews, Heng Ji, and Deborah L McGuinness. 2018. Exploiting Task-Oriented Resources to Learn Word Embeddings for Clinical Abbreviation Expansion. *arXiv preprint arXiv:1804.04225*.

Anurag Maheshwari and Paul J. Thuluvath. 2010. Management of acute hepatitis C. *Clin. Liver Dis.*, 14(1):169–176.

Tomas Mikolov, Kai Chen, Greg Corrado, and Jeffrey Dean. 2013a. Efficient Estimation of Word Representations in Vector Space. *arXiv preprint arXiv:1301.3781*.

Tomas Mikolov, Quoc V Le, and Ilya Sutskever. 2013b. Exploiting similarities among languages for machine translation. *arXiv preprint arXiv:1309.4168*.

Mark Pagel, Quentin D Atkinson, and Andrew Meade. 2007. Frequency of Word-Use Predicts Rates of Lexical Evolution throughout Indo-European History. *Nature*, 449(7163):717.

Jeffrey Pennington, Richard Socher, and Christopher D. Manning. 2014. Glove: Global Vectors for Word Representation. In *Proceedings of the 2014 Conference on Empirical Methods in Natural Language Processing, EMNLP 2014, October 25-29, 2014, Doha, Qatar, A Meeting of SIGDAT, a Special Interest Group of the ACL*, pages 1532–1543. ACL.

Lawrence Phillips, Kyle Shaffer, Dustin Arendt, Nathan Hodas, and Svitlana Volkova. 2017. Intrinsic and extrinsic evaluation of spatiotemporal text representations in Twitter streams. In *Proceedings of the 2nd Workshop on Representation Learning for NLP*, pages 201–210.

Jerald P Radich, Michael Deininger, Camille N Abboud, Jessica K Altman, Ellin Berman, Ravi Bhatia, Bhavana Bhatnagar, Peter Curtin, Daniel J DeAngelo, Jason Gotlib, et al. 2018. Chronic myeloid leukemia, version 1.2019, nccn clinical practice guidelines in oncology. *Journal of the National Comprehensive Cancer Network*, 16(9):1108–1135.

S H Richards, J P Thomas, and D Kilby. 1974. Transethmoidal hypophysectomy for pituitary tumours. *Proceedings of the Royal Society of Medicine*, 67(9):889–892.

Lucy Rivett and Graeme Alexander. 2019. Is the conquest of hepatitis c imminent? *Expert reviews in molecular medicine*, 21.

Roland Roller and Mark Stevenson. 2014. Self-supervised Relation Extraction Using UMLS. In *Information Access Evaluation. Multilinguality, Multimodality, and Interaction*, pages 116–127, Cham. Springer International Publishing.

Eyal Sagi, Stefan Kaufmann, and Brady Clark. 2009. Semantic density analysis: Comparing word meaning across time and phonetic space. In *Proceedings of the Workshop on Geometrical Models of Natural Language Semantics*, pages 104–111. Association for Computational Linguistics.

Jenna R Stoehr, Jennifer N Choi, Maria Colavincenzo, and Stefan Vanderweil. 2019. Off-label use of topical minoxidil in alopecia: A review. *American journal of clinical dermatology*, pages 1–14.

Don R Swanson. 1986. Fish oil, raynaud's syndrome, and undiscovered public knowledge. *Perspectives in biology and medicine*, 30(1):7–18.

Amit Tirosh and Ilan Shimon. 2016. Current approach to treatments for prolactinomas. *Minerva Endocrinol.*, 41(3):316–323.

Peter D Turney and Patrick Pantel. 2010. From Frequency to Meaning: Vector Space Models of Semantics. *Journal of artificial intelligence research*, 37:141–188.

Daniel P Webster and Paul Klenerman. 2015. Hepatitis c. *Hepatitis C. Lancet*, 385(9973):1124–1135.

Derry Tanti Wijaya and Reyyan Yeniterzi. 2011. Understanding Semantic Change of Words over Centuries. In *Proceedings of the 2011 International Workshop on DETecting and Exploiting Cultural diversiTy on the Social Web*, pages 35–40. ACM.

P. Yaish, A. Gazit, C. Gilon, and A. Levitzki. 1988. Blocking of EGF-dependent cell proliferation by EGF receptor kinase inhibitors. *Science*, 242(4880):933–935.

Erjia Yan and Yongjun Zhu. 2018. Tracking Word Semantic Change in Biomedical Literature. *International journal of medical informatics*, 109:76–86.

Li Zhou and George Hripcsak. 2007. Temporal Reasoning with Medical Data—a Review with Emphasis on Medical Natural Language Processing. *Journal of biomedical informatics*, 40(2):183–202.

Fei Zhu, Preecha Patumcharoenpol, Cheng Zhang, Yang Yang, Jonathan Chan, Asawin Meechai, Wanwipa Vongsangnak, and Bairong Shen. 2013. Biomedical Text Mining and Its Applications in Cancer Research. *Journal of biomedical informatics*, 46(2):200–211.

Extracting relations between outcomes and significance levels in Randomized Controlled Trials (RCTs) publications

Anna Koroleva

LIMSI, CNRS, Université Paris-Saclay,

F-91405 Orsay, France

Academic Medical Center, University of Amsterdam,

Amsterdam, the Netherlands

koroleva@limsi.fr

Patrick Paroubek

LIMSI, CNRS,

Université Paris-Saclay,

F-91405 Orsay, France

pap@limsi.fr

Abstract

Randomized controlled trials assess the effects of an experimental intervention by comparing it to a control intervention with regard to some variables - trial outcomes. Statistical hypothesis testing is used to test if the experimental intervention is superior to the control. Statistical significance is typically reported for the measured outcomes and is an important characteristic of the results. We propose a machine learning approach to automatically extract reported outcomes, significance levels and the relation between them. We annotated a corpus of 663 sentences with 2,552 outcome - significance level relations (1,372 positive and 1,180 negative relations). We compared several classifiers, using a manually crafted feature set, and a number of deep learning models. The best performance (F-measure of 94%) was shown by the BioBERT fine-tuned model.

1 Introduction

In clinical trials, outcomes are the dependent variables that are monitored to assess how they are influenced by other, independent, variables (treatment used, dosage, patient characteristics). Outcomes are a central notion for clinical trials.

To assess the impact of different variables on the outcomes, statistical hypothesis testing is commonly used, giving an estimation of statistical significance – the likelihood that a relationship between two or more variables is caused by something other than a chance (Schindler, 2015). Statistical significance levels are typically reported along with the trial outcomes as p-values, with a certain set threshold, where a p-value below the threshold means that the results are statistically significant, while a p-value above the threshold presents non-significant results. Hypothesis testing in clinical trials is used in two main cases:

1. In a trial comparing several treatments given to different groups of patients, a difference in value of an outcome observed between the groups at the end of the trial is evaluated by hypothesis testing to determine if the difference is due to the difference in medication. If the difference is statistically significant, the null hypothesis (the difference between treatments is due to a chance) is rejected, i.e. the superiority of one treatment over the other is considered to be proved.

2. When an improvement of an outcome is observed within a group of patients taking a treatment, hypothesis testing is used to determine if the difference in the outcome at different time points within the group is due to the treatment. If the results are statistically significant, it is considered to be proven that the treatment has a positive effect on the outcome in the given group of patients.

Although p-values are often misused and misinterpreted (Head et al., 2015), extracting significance levels for trial outcomes is still vital for a number of tasks, such as systematic reviews, detection of bias and spin. In particular, our application of interest is automatic detection of spin, or distorted reporting of research results, that consists in presenting an intervention studied in a trial as having higher beneficial effects than the research has proved. Spin is an alarming problem in health care as it causes overestimation of the intervention by clinicians (Boutron et al., 2014) and unjustified positive claims regarding the intervention is health news and press releases (Haneef et al., 2015; Yavchitz et al., 2012).

Spin is often related to a focus on significant outcomes, and occurs when the primary outcome (the main variable monitored during a trial) is not significant. Thus, to detect spin, it is important to identify the significance of outcomes, and espe-

Proceedings of the BioNLP 2019 workshop, pages 359–369

Florence, Italy, August 1, 2019. ©2019 Association for Computational Linguistics

cially of the primary outcome. To our best knowledge, no previous work addressed the extraction of the relation between outcomes and significance levels. In this paper, we present our approach towards extracting outcomes, significance levels and relations between them, that can be incorporated into a spin detection pipeline.

2   State of the art

Extraction of outcome - significance level relations consists of two parts: entity extraction (reported outcomes and significance levels) and extraction of the relationship between the entities. In this section, we present the previous works on these or similar tasks.

2.1   Entity extraction

The number of works addressing automatic extraction of significance levels is limited.

(Hsu et al., 2012) used regular expressions to extract statistical interpretation, p-values, confidence intervals, and comparison groups from sentences categorized as "outcomes and estimation". The authors report precision of 93%, recall of 88% and F-measure of 90% for this type of information.

(Chavalarias et al., 2016) applied text mining to evaluate the p-values reported in the abstracts and full texts of biomedical articles published in 1990 – 2015. The authors also assessed how frequently statistical information is presented in ways other than p-values. P-values were extracted using a regular expression; the system was evaluated on a manually annotated dataset. The reported sensitivity (true positive rate) is 96.3% and specificity (true negative rate) is 99.8%. P-values and qualitative statements about significance were more common ways of reporting significance than confidence intervals, Bayes factors, or effect sizes.

A few works focused on extracting outcome-related information, addressing it either as a sentence classification, or as entity extraction task.

(Demner-Fushman et al., 2006) defined an outcome as *"The sentence(s) that best summarizes the consequences of an intervention"* and thus adopted a sentence classification approach to extract outcome-related information from medical articles, using a corpus of 633 MEDLINE citations. The authors tested Naive Bayes, linear SVM and decision-tree classifiers. Naive Bayes showed the best performance. The reported classification accuracy ranged from 88% to 93%.

One of the notable recent works addressing outcome identification as an entity extraction task, rather than sentence classification, is (Blake and Lucic, 2015). The authors addressed a particular type of syntactic constructions – comparative sentences – to extract three items: the compared entities, referred to as the agent and the object, and the ground for comparison, referred to as the endpoint (synonymous to outcome). The aim of this work was to extract corresponding noun phrases. The dataset was based on full-text medical articles and included only the sentences that contain all the three entities (agent, object and endpoint). The training set comprised 100 sentences that contain 656 noun phrases. The algorithm proceeds in two steps: first, comparative sentences are detected with the help of a set of adjectives and lexico-syntactic patterns. Second, the noun phrases are classified according to their role (agent, object, endpoint) using SVM and generalized linear model (GLM). On the training set, SVM showed better performance than GLM, with an F-measure of 78% for the endpoint. However, on the test set the performance was significantly lower: SVM showed an F-measure of only 51% for the endpoint. The performance was higher on shorter sentences (up to 30 words) than on the longer ones.

A following work (Lucic and Blake, 2016) aimed at improving the recognition of the first entity and of the endpoint. The authors propose to use in the classification the information on whether the head noun of the candidate noun phrase denotes an amount or a measure. The annotation of the corpus was enriched by the corresponding information. As a result, precision of the endpoint detection improved to 56% on longer sentences and 58% on shorter ones; recall improved to 71% on longer sentences and 74% on shorter ones.

2.2   Relation extraction

To our knowledge, extraction of the relation between outcomes and significance levels has not been addressed yet. In this section, we overview some frameworks for relation extraction and outline some common features of different approaches in the biomedical relation extraction.

A substantial number of works addressed extracting binary relations, such as protein-protein interactions or gene-phenotype relation, or complex relations, such as biomolecular events. A common feature of the works in this domain, noted by (Zhou et al., 2014; Lever and Jones, 2017) and still relevant for recent works e.g. (Peng and Lu, 2017; Asada et al., 2017), consists in assuming that entities of interest are already extracted and provided to the relation extraction system as input. Thus, the relation extraction is assessed separately, without taking into account the performance of entity extraction. We adopt this approach for relation extraction evaluation in our work, but we provide separate assessment for our algorithms of entity extraction.

One of the general frameworks for relation extraction in the biomedical domain is proposed by (Zhou et al., 2014). The authors suggest using trigger words to determine the type of a relation, noting that for some relation types trigger words can be extracted simply with a dictionary, while for other types, rule-based or machine-learning approaches may be required. For relation extraction, rule-based methods can be applied, often employing regular expressions using words or POS tags. Rules can be crafted manually or learned automatically. The machine learning approaches to binary relation extraction, as the authors note, usually treat the task as a classification problem. Features for classification often use output of textual analysis algorithms such as POS-tagging and syntactic parsing. Machine learning approaches can be divided into feature-based approaches (using syntactic and semantic features) and kernel approaches (calculating similarity between input sequences based on string or syntactic representation of the input). Supervised machine learning is a highly successful approach for binary relation extraction, but its main drawback consists in the need of large amount of annotated data.

A framework for pattern-based relation extraction is introduced by (Peng et al., 2014). The approach aims at reducing the need for manual annotation. The approach is based on a user-provided list of trigger words and specifications (the definition of arguments for each trigger). Variations of lexico-syntactic patterns are derived using this information and are matched with the input text, detecting the target relations. Some interesting features of the framework include the following: the use of text simplification to avoid writing rules for all existing constructions; the use of referential relations to find the best phrase referring to an entity.

The authors state that their system is characterized by good generalizability due to the use of language properties and not of task-specific knowledge.

A recent work (Björne and Salakoski, 2018) reports on the development of convolutional neural networks (CNNs) for event and relation extraction, using Keras (Chollet et al., 2015) with Tensorflow backend (Abadi et al., 2016). Parallel convolutional layers process the input, using sequence windows centered around the candidate entity, relation or event. Vector space embeddings are built for input tokens, including features such as word vectors, POS, entity features, relative position, etc. The system was tested on several tasks and showed improved performance and good generalizability.

3 Our dataset

3.1 Corpus creation and annotation

In our previous work on outcome extraction, we manually annotated a corpus for reported outcomes comprising 1,940 sentences from the Results and Conclusions sections of PMC article abstracts. We used this corpus as a basis for a corpus with annotations for outcome significance level relations.

Our corpus contains 2,551 annotated outcomes. Out of the sentences with outcomes, we selected those where statistical significance levels are supposedly reported (using regular expressions) and manually annotated relations between outcomes and significance levels. The annotation was done by one annotator (AK), in consultation with a number of domain experts, due to infeasibility of recruiting several annotators with sufficient level of expertise within a reasonable time frame.

The final corpus contains 663 sentences with 2,552 annotated relations, out of which 1,372 relations are positive (the significance level is related to the outcome) and 1,180 relations are negative (the significance level is not related to the outcome). The corpus is publicly available (Anna, 2019).

3.2 Data description

There are three types of data relevant for this work: outcomes, significance levels, and relationship between them. In this section, we describe these types of data and the observed variability in the ways of presenting them.

1. Outcomes

A trial outcome is, in broad sense, a measure or variable monitored during a trial. It can be binary (presence of a symptom or state), numerical ("*temperature*") or qualitative ("*burden of disease*"). Apart from the general term denoting the outcome, there are several aspects that define it: a measurement tool (questionnaire, score, etc.) used to measure the outcome; time points at which the outcome is measured; patient-level analysis metrics (change from baseline, time to event); population-level aggregation method (mean, median, proportion of patients with some characteristic).

Generally, there are two main contexts in which outcomes of a clinical trial can be mentioned: a definition of what the outcomes of a trial were ("***Quality of life*** *was selected as the primary outcome.*"), and reporting results for an outcome ("***Quality of life*** *was higher in the experimental group than in the control group.*"). In both cases, a mention of an outcome may contain the aspects listed above, but does not necessarily include all of them. In this work, we are interested in the second type of context.

The ways of reporting outcomes are highly diverse. Results for an outcome may be reported as a value of the outcome measure: for binary outcomes, it refers to presence/absence of an event or state; for numerical outcome, it is a numerical value; for qualitative outcome, it is often a value obtained on the associated measurement tool. As the primary goal of RCTs is to compare two or more interventions, results for an outcome can be reported as a comparison between the interventions/patient groups, with or without actual values of the outcome measure. Syntactically, an outcome may be represented by a noun phrase, a verb phrase, an adjective or a clause. We provide here some examples of outcome reporting, to give an idea of variability of expressions.

The outcome is reported as a numerical value:

a) *The median* ***progression-free survival*** *was 32 days.*

The outcome is reported as a comparison between groups, without the values for groups:

b) *MMS resulted in more* ***stunting*** *than standard Fe60F (p = 0.02).*

The outcome is reported as a numerical value with comparison between groups:

c) *The average* ***birth weight*** *was 2694 g and* ***birth length*** *was 47.7 cm, with no difference among intervention groups.*

d) *The crude incidence of* ***late rectal toxicity*** $\geq$ ***G2*** *was 14.0% and 12.3% for the arm A and B, respectively.*

e) *More than 96% of patients who received DPT were* ***apyrexial*** *48 hours after treatment compared to 83.5% in the AL group (p < 0.001).*

f) *The proportion of patients who* ***remained relapse-free at Week 26*** *did not differ significantly between the placebo group (5/16, 31%) and the IFN beta-1a 44 mcg biw (6/17, 35%; p = 0.497), 44 mcg tw (7/16, 44%; p = 0.280) or 66 mcg tw (2/18, 11%; p = 0.333) groups.*

In the latter case, the variation is especially high, and the same outcome may be reported in several different ways (cf. the examples **d**, **e** and **f** that all talk about a percentage of patients in which a certain event occurred, but the structure of the phrases differs).

Identifying the textual boundaries of an outcome presents a challenge: for the example **d**, it can be "*the crude incidence of late rectal toxicity* $\geq$ *G2*" or "*late rectal toxicity* $\geq$ *G2*"; for the example **f**, it can be "*the proportion of patents who remained relapse-free at Week 26*", or "*remained relapse-free at Week 26*", or simply "*relapse-free*". This variability poses difficulties for both annotation and extraction of reported outcomes. In our annotation, we aimed at annotating the minimal possible text span describing an outcome, not including time points, aggregation and analysis metrics.

2. Significance levels

The ways of presenting significance levels are less diverse than the ways of reporting outcomes. Typically, significance levels are reported via p-values. Another way of determining significance of the results is the confidence interval (CI), where a CI comprising zero denotes non-significant results. In this work, we do not address CIs as they are less frequently reported (Chavalarias et al., 2016).

Statistical significance can be reported as an exact value of P (*"p=0.02"*), as P-value relative to a pre-set threshold (*"p<0.05"*), or in qualitative form (*"significant"/"non-significant"*). We address all these forms of reporting significance.

Although in general the ways of presenting statistical significance are rather uniform, there are a few cases to be noted:

- Coordinated p-values:
  For the non-HPD stratum, the intent-to-treat relative risks of spontaneous premature birth at < 34 and < 37 weeks' gestation were 0.33 (0.03, 3.16) and 0.49 (0.17, 1.44), respectively, and they were non-significant (ns) with **p = 0.31 and 0.14**.
- Significance level in score of a negation:
  The respiratory rate, chest indrawing, cyanosis, stridor, nasal flaring, wheeze and fever in both groups recorded at enrollment and parameters **did not differ significantly** *between the two groups.*
  A particular difficulty is presented by the cases in which a negation marker occurs in the main clause and a significance level in the dependent clause, thus the significance level is within the scope of the negation, but there is a big linear distance between them:
  Results There was **no evidence** *that an incentive (52% versus 43%, Risk Difference (RD) -8.8 (95%CI 22.5, 4.8); or abridged questionnaire (46% versus 43%, RD 2.9 (95%CI 16.5, 10.7);* **statistically significantly** *improved dentist response rates compared to a full length questionnaire in RCT A.*

3. Relationship between outcomes and significance levels

The correspondence between outcomes and significance levels in a sentence is often not one-to-one: multiple outcomes can be linked to the same significance level, and vice versa. Several outcomes are linked to one significance level when outcomes are coordinated:

No significant *improvements in* ***lung function***, ***symptoms***, *or* ***quality of life*** *were seen.*

Several significance levels can be associated to one outcome in a number of cases:

- one outcome is linked to two significance levels when a significance level is presented in both qualitative and numerical form:
  Results **The response rates were not significantly** *different Odds Ratio 0.88 (95% confidence intervals 0.48 to 1.63)* **p = 0.69**.
- in the case of comparison between patient groups taking different medications, when there are more than 2 groups, significance can be reported for all pairs of groups;
- significance level for difference observed within groups of patients receiving a particular medication:
  *[**Na**] increased* **significantly** *in the 0.9% group (+0.20 mmol/L/h [IQR +0.03, +0.4];* **P = 0.02**) *and increased, but* **not significantly**, *in the 0.45% group (+0.08 mmol/L/h [IQR -0.15, +0.16];* **P = 0.07**).
- significance reported for both between- and within-group comparison:
  PTEF *increased* **significantly** *both after albuterol and saline treatments but the difference between the two treatments was* **not significant (P = 0.6)**.
- significance for differences within subgroups of patients (e.g. gender or age subgroups) receiving a medication;
- significance for different types of analysis: intention-to-treat / per protocol:
  Results For **BMD**, **no** *intent-to-treat analyses were* **statistically significant**; *however, per protocol analyses (ie, only including TC participants who completed ≥ 75% training requirements) of* **femoral neck BMD** *changes were* **significantly** *different between TC and UC (+0.04 vs -0.98%; P = 0.05).*
- significance for several time points:
  Results A **significant** *main effect of time (p < 0.001) was found for* **step-counts** *attributable to significant increases in steps/day between: pre-intervention (M*

= 6941, SD = 3047) and 12 weeks (M
= 9327, SD = 4136), t (78) = - 6.52, **p
< 0.001**, d = 0.66; pre-intervention and
24 weeks (M = 8804, SD = 4145), t (78)
= - 4.82, **p < 0.001**, d = 0.52; and pre-
intervention and 48 weeks (M = 8450,
SD = 3855), t (78) = - 4.15, **p < 0.001**,
d = 0.44.*

- significance level for comparison of various analysis metrics (mean, AUC, etc.)

4 Methods

To extract the relation between an outcome and its
significance level, we propose a 3-step algorithm:
1) extracting reported outcomes; 2) extracting significance levels; 3) classification of pairs of outcomes and significance levels to detect those related to each other.

As significance levels are not characterized by
high variability, we follow the previous research
in using rules (regular expressions and sequential
rules using information from pos-tagging) to extract significance levels.

We present our methods and results for outcome
extraction in detail elsewhere, here we provide
a brief summary. We tested several approaches:
a baseline approach using sequential rules using
information from pos-tagging; an approach using rules based on syntactic structure provided by
spaCy dependency parser (Honnibal and Johnson,
2015); a combination of bi-LSTM, CNN and CRF
using GloVe(Pennington et al., 2014) word embeddings and character-level representations (Ma
and Hovy, 2016); and a fine-tuned bi-LSTM using
BERT (Devlin et al., 2018) vector word representations.

BERT (Bidirectional Encoder Representations
from Transformers) is a recently introduced approach to pre-training language representations,
using a masked language model (MLM) which
randomly masks some input tokens, allowing to
pre-train a deep bidirectional Transformer using
both left and right context. The pre-trained BERT
models can be fine-tuned for supervised downstream tasks by adding one output layer.

BERT was trained on a dataset of 3.3B words
combining English Wikipedia and BooksCorpus.
Two domain-specific versions of BERT are available, pre-trained on a combination of the initial BERT corpus and additional domain-specific
datasets: BioBERT (Lee et al., 2019), adding a

large biomedical corpus of PubMed abstracts and
PMC full-text articles comprising 18B tokens; and
SciBERT (Beltagy et al., 2019), adding a corpus
of 1.14M full-text papers from Semantic Scholar
with the total of 3.1B tokens. Both BioBERT and
SciBERT outperform BERT on biomedical tasks.

BERT provides several models: uncased
(trained on lower-cased data) and cased (trained
on unchanged data); base and large (differing in
model sizes). BioBERT is based on the BERT-
base cased model and provides three versions
of models: pre-trained on PubMed abstracts, on
PMC full-text articles, or on combination of both.
SciBERT has both cased and uncased models and
provides two versions of vocabulary: BaseVocab
(the initial BERT vocabulary) and SciVocab (the
vocabulary from the SciBERT corpus). We fine-
tuned and tested the BioBERT model trained on
the whole corpus, and both cased and uncased base
models for BERT and SciBERT (using SciVocab).
We did not perform experiments with BERT-Large
as we do not have enough resources. We used the
code provided by BioBERT for the entity extraction task[1].

The relation extraction assumes that the entities
have already been extracted and are given as an
input to the algorithm, with the sentence in which
they occur. To predict the tag for outcome - significance level pair, we use machine learning.

As the first approach, we compared several clas-
sifiers available in the Python scikit-learn library
(Pedregosa et al., 2011): Support Vector Machine (SVM) (Cortes and Vapnik, 1995); Deci-
sionTreeClassifier (Rokach and Maimon, 2008);
MLPClassifier(von der Malsburg, 1986); Kneigh-
borsClassifier (Altman, 1992); GaussianProcess-
Classifier (Rasmussen and Williams, 2005); RandomForestClassifier (Breiman, 2001); AdaBoost-
Classifier (Freund and Schapire, 1997); Extra-
TreesClassifier (Geurts et al., 2006); Gradient-
BoostingClassifier (Friedman, 2002). Feature engineering was performed manually and was based
on our observations on the corpus.

Evaluation was performed using 10-fold cross-
validation. To account for different random states,
the experiments were run 10 times, we report the
average results of the 10 runs. We performed hyperparameters tuning via exhaustive grid search
(with the help of the scikit-learn GridSearchCV

---

[1]https://github.com/dmis-
lab/biobert/blob/master/run_ner.py

function).

As the second approach, we employed a deep learning approach to relation extraction, fine-tuning BERT-based models on this task. We tested the same models as for the outcome extraction. We used the code provided by BioBERT for relation extraction task[2]. The algorithm takes as input sentences with the two target entities replaced by masks ("@outcome$" and "@significance$") and positive/negative relation labels assigned to the sentence.

Hyperparameters for entity and relation extraction with BERT-based algorithms are shown in the Table 1. We tested both possible values (True/False) of the hyperparameter "do_lower_case" (lower-casing the input) for all the models.

Hyperparameter	Entity extraction	Relation extraction
max_seq_length	128	
train_batch_size	32	
eval_batch_size	8	
predict_batch_size	8	
use_tpu	False	
learning_rate	5e-5	2e-5
num_train_epochs	10.0	3.0
warmup_proportion	0.1	
save_checkpoints_steps	1000	
iterations_per_loop	1000	
tf.master	None	

Table 1: BERT/BioBERT/SciBERT hyperparameters

5 Features

Features are calculated for each pair of outcome and significance level. They are based both on the information about these entities (their position, text, etc.) and on the contextual information (presence of other entities in the sentence, etc.). We used the following binary (True/False) features:

1. only_out: whether the outcome is the only outcome present in the sentence. If yes, it is the only candidate that can be related to the present statistical significance values.

2. only_signif: whether the significance level is the only significance level in the sentence. If yes, it is the only candidate that can be related to the present outcomes.

3. signif_type_num: whether the significance level is expressed in the numerical form;

[2]https://github.com/dmis-lab/biobert/blob/master/run_re.py

Algorithm	do_lower _case	Precision	Recall	F1
SciBERT uncased	True	81.17	78.09	79.42
BioBERT	True	80.38	77.85	78.92
BioBERT	False	79.61	77.98	78.6
SciBERT cased	False	79.6	77.65	78.38
SciBERT cased	True	79.24	76.61	77.64
SciBERT uncased	False	79.51	75.5	77.26
BERT uncased	True	78.98	74.96	76.7
BERT cased	False	76.63	74.25	75.18
BERT cased	True	76.7	73.97	75.1
BERT uncased	False	77.28	72.25	74.46
Bi-LSTM-CNN-CRF		51.12	44.6	47.52
Rule-based		26.69	55.73	36.09

Table 2: Reported outcome extraction results

4. signif_type_word: whether the significance level is expressed in the qualitative form;

5. signif_exact: whether the exact value of significance level is given ($P = 0.049$), or it is presented only as comparison to a threshold ($P < 0.05$). Significance levels expressed in the word form always have "False" value for this feature. We assumed that significance levels with exact numerical value are less likely to be related to several outcomes that significance levels with inexact value: obtaining exactly same significance level for several outcomes seems unlikely.

6. signif_precedes: whether the significance level precedes the outcome. It is especially pertinent for numerical significance values as they most often follow the related outcome.

7. out_between: whether there is another outcome between the outcome and significance level in the given pair. The outcome that is closer to a significance level is a more likely candidate to be related to it.

8. signif_between: whether there is another significance level between the outcome and the significance level in a given pair. The significance level that is closer to an outcome is a more likely candidate to be related to it.

9. concessive_between: whether there are words

Classifier	Hyperparameters	Precision	Recall	F1
RandomForestClassifier	max_depth = 15, min_samples_split = 10, n_estimators = 300	90.16	92.6	91.33
ExtraTreesClassifier	default	89.74	88.53	89.08
GradientBoostingClassifier	learning_rate = 0.25, max_depth = 23.0, max_features = 7, min_samples_leaf = 0.1, min_samples_split = 0.2, n_estimators = 200	88.44	89.8	89.07
RandomForestClassifier	default	89.54	88.64	89.03
GaussianProcessClassifier	1.0 * RBF(1.0)	86.99	90.38	88.64
GradientBoostingClassifier	default	87.75	89.14	88.4
SVC	C = 1000, gamma = 0.0001, kernel = 'rbf'	86.14	89.65	87.79
DecisionTreeClassifier	default	87.85	86.83	87.27
MLPClassifier	activation = 'tanh', alpha = 0.0001, hidden_layer_sizes = (50, 100, 50), learning_rate = 'constant', solver = 'adam'	84.06	85.15	84.44
MLPClassifier	default	84.4	83.34	83.47
KNeighborsClassifier	n_neighbors = 7, p = 1	83.37	81.27	82.21
AdaBoostClassifier	learning_rate = 0.1, n_estimators = 500	81.34	83.09	82.16
AdaBoostClassifier	default	80.85	82.36	81.53
KNeighborsClassifier	default	81.39	79.88	80.55
GaussianProcessClassifier	default	79.41	78.86	79.1
SVC	default	87.24	64.06	73.77
baseline (majority class)		53.76	100	69.92

Table 3: Results of classifiers

Feature	Weight
only_signif	0.21663222
signif_type_num	0.21341347
signif_exact	0.15207938
signif_type_word	0.10103105
dist_min_out_preceding	0.0919397
out_between	0.05683003
dist_min_out_following	0.04683059
concessive_between	0.04260114
only_out	0.02336161
dist	0.02043495
dist_min_graph	0.01794923
signif_precedes	0.01631646
signif_between	0.00058017

Table 4: Feature ranking

Algorithm	do_lower_case	Precision	Recall	F1
BioBERT	True	94.3	94	94
SciBERT cased	True	93.9	93.6	93.8
SciBERT cased	False	93.5	93.1	93.3
SciBERT uncased	False	94.2	92.3	93.3
SciBERT uncased	True	94	92.8	93.2
BioBERT	False	92.8	89.7	91.1
BERT cased	False	91.6	90.2	90.9
BERT uncased	True	90.9	90.9	90.8
BERT uncased	False	90.4	89.8	90
BERT cased	True	89.6	90.5	89.8

Table 5: Results of relation extraction with BERT/BioBERT/SciBERT

(conjunctions) with consessive semantics (*but*, *however*, *although*, etc.) between the outcome and the significance level in the pair.

We used the following numerical features:

1. dist: the distance in characters between the outcome and the significance level in the pair;

2. dist_min_graph: the minimal syntactic distance between the words in the outcome and the words in the significance level;

3. dist_min_out_preceding: the distance from the outcome of the pair to the nearest preceding outcome.

4. dist_min_out_following: the distance from the outcome of the pair to the nearest following outcome. The two last features are designed to reflect the information about coordination of outcomes (the distances between

coordinated entities is typically small), as co-ordinated outcomes are likely to be related to the same significance level.

We assessed the importance of the features with the attribute "feature_importances_" of the RandomForestClassifier classifier. The results are presented in the Table 4.

6 Evaluation

6.1 Entity extraction

The rule-based extraction of significance levels shows the following per-token performance: precision of 99.18%, recall of 96.58% and F-measure of 97.86%.

The results of all the tested approaches to the extraction of reported outcomes are reported in the Table 2. The best performance was achieved by the fine-tuned SciBERT uncased model: precision was 81.17%, recall was 78.09% and F-measure was 79.42%.

6.2 Relation extraction

The baseline value is based on assigning the majority (positive) class to all the entity pairs. Baseline precision is 53.76%, recall is 100% and F-measure is 69.95%.

The results of the classifiers are presented in the Table 3. We present the performance of the default classifiers and of the classifiers with tuned hyperparameters. All the classifiers outperformed the baseline. Random Forest Classifier with tuned hyperparameters (max_depth = 15, min_samples_split = 10, n_estimators = 300) showed the best results, with F-measure of 91.33%, which is by 21.41% higher than the baseline.

It is interesting to compare the deep learning approach using BERT-based fine-tuned models (Table 5) to the feature-based classifiers: none of the Google BERT models outperformed the Random Forest Classifier, neither did BioBERT with unchanged input data. However, all the SciBERT fine-tuned models and the BioBERT model with lower-cased input outperformed the Random Forest Classifier. Interestingly, BioBERT, which only has a cased model pre-trained on unchanged data and is thus meant to work with unchanged input, showed the best performance on lower-cased input for the relation extraction task, achieving the F-measure of 94%.

7 Conclusion and future work

In this paper, we presented a first approach towards the extraction of the relation between outcomes of clinical trials and their reported significance levels. We presented our annotated corpus for this task and described the ways of reporting outcomes, significance levels and their relation in a text. We pointed out the difficulties posed by the high diversity of the data.

We crafted a feature set for relation extraction and trained and tested a number of classifiers for this task. The best performance was shown by the Random Forest classifier, with the F-measure of 91.33%. Further, we fine-tuned and evaluated a few deep learning models (BERT, SciBERT, BioBERT). The best performance was achieved by the BioBERT model fine-tuned on lower-cased data, with F-measure of 94%.

Our relation extraction algorithm assumes that the entities have been previously extracted and provided as input. An interesting direction for future experiments is building an end-to-end system extracting both entities and relations, as proposed by (Miwa and Bansal, 2016) or (Pawar et al., 2017).

As in our algorithm the extraction of the relevant entities (reported outcomes and significance levels) is essential for extracting the relations, we reported the results of our experiments for extracting this task. Extraction of significance levels reaches the F-measure of 97.86%, while the extraction of reported outcomes shows the F-measure of only 79.42%. Thus, improving the outcome extraction is the main direction of the future work.

Besides, a very important task for clinical trial data analysis consists in determining the significance level for the primary outcome. This task requires two additional steps: 1) identifying the primary outcome, and 2) establishing the correspondence between the primary outcome and a reported outcome. We will present our algorithms for these tasks in a separate paper.

8 Acknowledgements

We thank Sanjay Kamath for his help in conducting experiments with BERT.

This project has received funding from the European Unions Horizon 2020 research and innovation programme under the Marie Sklodowska-Curie grant agreement No 676207.

References

M. Abadi, P. Barham, J. Chen, Z. Chen, A. Davis, J. Dean, M. Devin, S. Ghemawat, G. Irving, M. Isard, M. Kudlur, J. Levenberg, R. Monga, S. Moore, D. G. Murray, B. Steiner, P. Tucker, V. Vasudevan, P. Warden, M. Wicke, Y. Yu, and X. Zheng. 2016. Tensorflow: A system for large-scale machine learning.

N.S. Altman. 1992. An introduction to kernel and nearest-neighbor nonparametric regression. *American Statistician - AMER STATIST*, 46:175–185.

Koroleva Anna. 2019. Annotated corpus for the relation between reported outcomes and their significance levels.

Masaki Asada, Makoto Miwa, and Yutaka Sasaki. 2017. Extracting drug-drug interactions with attention CNNs. In *BioNLP 2017*, pages 9–18, Vancouver, Canada,. Association for Computational Linguistics.

Iz Beltagy, Arman Cohan, and Kyle Lo. 2019. Scibert: Pretrained contextualized embeddings for scientific text.

Jari Björne and Tapio Salakoski. 2018. Biomedical event extraction using convolutional neural networks and dependency parsing. In *BioNLP 2018 workshop*, pages 98–108, Melbourne, Australia. ACL.

C. Blake and A. Lucic. 2015. Automatic endpoint detection to support the systematic review process. *J. Biomed. Inform.*

Isabelle Boutron, Douglas Altman, Sally Hopewell, Francisco Vera-Badillo, Ian Tannock, and Philippe Ravaud. 2014. Impact of spin in the abstracts of articles reporting results of randomized controlled trials in the field of cancer: the spiin randomized controlled trial. *Journal of Clinical Oncology.*

Leo Breiman. 2001. Random forests. *Mach. Learn.*, 45(1):5–32.

David Chavalarias, Joshua D Wallach, Alvin Ho Ting Li, and John P. A. Ioannidis. 2016. Evolution of reporting p values in the biomedical literature, 1990-2015. *JAMA*, 315 11:1141–8.

François Chollet et al. 2015. Keras.

Corinna Cortes and Vladimir Vapnik. 1995. Support-vector networks. *Machine Learning*, 20(3):273–297.

D. Demner-Fushman, B. Few, S.E. Hauser, and G. Thoma. 2006. Automatically identifying health outcome information in medline records. *Journal of the American Medical Informatics Association.*

Jacob Devlin, Ming-Wei Chang, Kenton Lee, and Kristina Toutanova. 2018. BERT: pre-training of deep bidirectional transformers for language understanding. *CoRR*, abs/1810.04805.

Yoav Freund and Robert E Schapire. 1997. A decision-theoretic generalization of on-line learning and an application to boosting. *Journal of Computer and System Sciences*, 55(1):119 – 139.

Jerome H. Friedman. 2002. Stochastic gradient boosting. *Comput. Stat. Data Anal.*, 38(4):367–378.

Pierre Geurts, Damien Ernst, and Louis Wehenkel. 2006. Extremely randomized trees. *Mach. Learn.*, 63(1):3–42.

R. Haneef, C. Lazarus, P. Ravaud, A. Yavchitz, and I. Boutron. 2015. Interpretation of results of studies evaluating an intervention highlighted in google health news: a cross-sectional study of news. *PLoS ONE.*

Megan L Head, Luke Holman, Rob Lanfear, Andrew T. Kahn, and Michael D. Jennions. 2015. The extent and consequences of p-hacking in science. In *PLoS biology.*

Matthew Honnibal and Mark Johnson. 2015. An improved non-monotonic transition system for dependency parsing. In *Proc. of EMNLP 2015*, pages 1373–1378, Lisbon, Portugal. Association for Computational Linguistics.

William Hsu, William Speier, and Ricky K. Taira. 2012. Automated extraction of reported statistical analyses: Towards a logical representation of clinical trial literature. *AMIA Annual Symposium*, 2012:350–359.

Jinhyuk Lee, Wonjin Yoon, Sungdong Kim, Donghyeon Kim, Sunkyu Kim, Chan Ho So, and Jaewoo Kang. 2019. Biobert: a pre-trained biomedical language representation model for biomedical text mining. *arXiv preprint arXiv:1901.08746.*

Jake Lever and Steven Jones. 2017. Painless relation extraction with kindred. In *BioNLP 2017*, pages 176–183, Vancouver, Canada,. Association for Computational Linguistics.

A. Lucic and C. Blake. 2016. Improving endpoint detection to support automated systematic reviews. In *AMIA Annu Symp Proc.*

Xuezhe Ma and Eduard Hovy. 2016. End-to-end sequence labeling via bi-directional lstm-cnns-crf. In *Proceedings of the 54th Annual Meeting of the ACL (Volume 1: Long Papers)*, pages 1064–1074, Berlin, Germany. Association for Computational Linguistics.

Christoph von der Malsburg. 1986. Frank rosenblatt: Principles of neurodynamics: Perceptrons and the theory of brain mechanisms. *Brain Theory*, pages 245–248.

Makoto Miwa and Mohit Bansal. 2016. End-to-end relation extraction using LSTMs on sequences and tree structures. In *Proceedings of the 54th Annual Meeting of the Association for Computational Linguistics*

(Volume 1: Long Papers), pages 1105–1116, Berlin, Germany. Association for Computational Linguistics.

Sachin Pawar, Pushpak Bhattacharyya, and Girish Palshikar. 2017. End-to-end relation extraction using neural networks and Markov logic networks. In *Proceedings of the 15th Conference of the European Chapter of the Association for Computational Linguistics: Volume 1, Long Papers*, pages 818–827, Valencia, Spain. Association for Computational Linguistics.

F. Pedregosa, G. Varoquaux, A. Gramfort, V. Michel, B. Thirion, O. Grisel, M. Blondel, P. Prettenhofer, R. Weiss, V. Dubourg, J. Vanderplas, A. Passos, D. Cournapeau, M. Brucher, M. Perrot, and E. Duchesnay. 2011. Scikit-learn: Machine learning in Python. *Journal of Machine Learning Research*, 12:2825–2830.

Yifan Peng and Zhiyong Lu. 2017. Deep learning for extracting protein-protein interactions from biomedical literature. In *BioNLP 2017*, pages 29–38, Vancouver, Canada,. Association for Computational Linguistics.

Yifan Peng, Manabu Torii, Cathy Wu, and K Vijay-Shanker. 2014. A generalizable nlp framework for fast development of pattern-based biomedical relation extraction systems. *BMC bioinformatics*, 15:285.

Jeffrey Pennington, Richard Socher, and Christopher Manning. 2014. Glove: Global vectors for word representation. In *Proc. of EMNLP 2014*, pages 1532–1543, Doha, Qatar. Association for Computational Linguistics.

Carl Edward Rasmussen and Christopher K. I. Williams. 2005. *Gaussian Processes for Machine Learning (Adaptive Computation and Machine Learning)*. The MIT Press.

Lior Rokach and Oded Maimon. 2008. *Data Mining with Decision Trees: Theroy and Applications*. World Scientific Publishing Co., Inc., River Edge, NJ, USA.

Thomas M. Schindler. 2015. Hypothesis testing in clinical trials. *AMWA Journal*, 30(2).

A. Yavchitz, I. Boutron, A. Bafeta, I. Marroun, P. Charles, J. Mantz, and P. Ravaud. 2012. Misrepresentation of randomized controlled trials in press releases and news coverage: a cohort study. *PLoS Med*.

Deyu Zhou, Dayou Zhong, and Yulan He. 2014. Biomedical relation extraction: From binary to complex. In *Comp. Math. Methods in Medicine*.

Overview of the MEDIQA 2019 Shared Task on Textual Inference, Question Entailment and Question Answering

Asma Ben Abacha[1]
asma.benabacha@nih.gov

Chaitanya Shivade[2]
cshivade@us.ibm.com

Dina Demner-Fushman[1]
ddemner@mail.nih.gov

[1]LHC, NLM, Bethesda, MD   [2]IBM, Almaden Research Center, San Jose, CA

Abstract

This paper presents the MEDIQA 2019 shared task organized at the ACL-BioNLP workshop. The shared task is motivated by a need to develop relevant methods, techniques and gold standards for inference and entailment in the medical domain, and their application to improve domain specific information retrieval and question answering systems. MEDIQA 2019 includes three tasks: Natural Language Inference (NLI), Recognizing Question Entailment (RQE), and Question Answering (QA) in the medical domain. 72 teams participated in the challenge, achieving an accuracy of 98% in the NLI task, 74.9% in the RQE task, and 78.3% in the QA task. In this paper, we describe the tasks, the datasets, and the participants' approaches and results. We hope that this shared task will attract further research efforts in textual inference, question entailment, and question answering in the medical domain.

1 Introduction

The first open-domain challenge in Recognizing Textual Entailment (RTE) was launched in 2005 (Dagan et al., 2005) and has prompted the development of a wide range of approaches (Bar-Haim et al., 2014). Recently, large-scale datasets such as SNLI (Bowman et al., 2015) and MultiNLI (Williams et al., 2018) were introduced for the task of Natural Language Inference (NLI) targeting three relations between sentences: Entailment, Neutral, and Contradiction. Few efforts have studied the benefits of RTE and NLI in other NLP tasks such as text exploration (Adler et al., 2012), identifying evidence for eligibility criteria satisfaction in clinical trials (Shivade et al., 2015), and the summarization of PMC articles (Chachra et al., 2016).

NLI can also be beneficial for Question Answering (QA). Harabagiu and Hickl (2006) presented entailment-based methods to filter and rank answers and showed that RTE can enhance the performance of open-domain QA systems and provide the inferential information needed to validate the answers. Çelikyilmaz et al. (2009) presented a graph-based semi-supervised method for QA exploiting entailment relations between questions and candidate answers and demonstrated that the use of unlabeled entailment data can improve answer ranking. Ben Abacha and Demner-Fushman (2016) noted that the requirements of question entailment in QA are different from general question similarity, and introduced the task of Recognizing Question Entailment (RQE) in order to answer new questions by retrieving entailed questions with pre-existing answers. Ben Abacha and Demner-Fushman (2019) proposed a novel QA approach based on RQE, with the introduction of the MedQuAD medical question-answer collection, and showed empirical evidence supporting question entailment for QA.

Although the idea of using entailment in QA has been introduced, research investigating methods to incorporate textual inference and question entailment into QA systems is still limited in the literature. Moreover, despite a few recent efforts to design RTE methods and datasets from MEDLINE abstracts (Ben Abacha et al., 2015) and to create the MedNLI dataset from clinical data (Romanov and Shivade, 2018), the entailment and inference tasks remain less studied in the medical domain.

MEDIQA 2019[1] aims to highlight further the NLI and RQE tasks in the medical domain, and their applications in QA and NLP. Figure 2 presents the MEDIQA tasks in the AIcrowd platform[2]. For the QA task, participants were tasked to filter and re-rank the provided answers. Reuse of the systems developed in the first and second tasks was highly encouraged.

---

[1]https://sites.google.com/view/mediqa2019

[2]https://www.aicrowd.com/organizers/mediqa-acl-bionlp

370

Proceedings of the BioNLP 2019 workshop, pages 370–379
Florence, Italy, August 1, 2019. ©2019 Association for Computational Linguistics

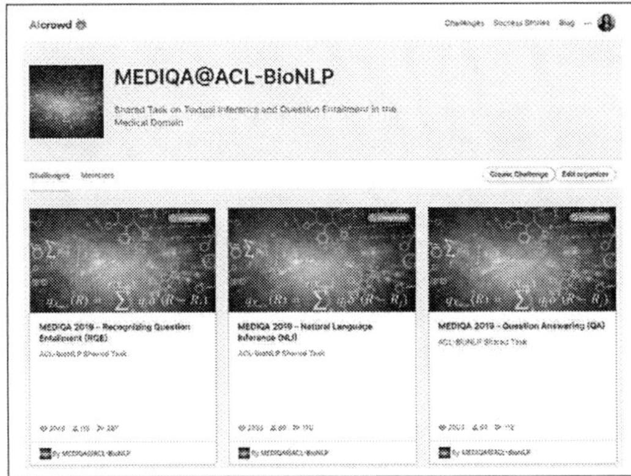

Figure 1: MEDIQA tasks on AIcrowd

2 Tasks

2.1 Natural Language Inference (NLI)

The first task focuses on Natural Language Inference (NLI) in the medical domain. We use three labels for the relation between two sentences: Entailment, Neutral and Contradiction.

2.2 Recognizing Question entailment (RQE)

The second task tackles Recognizing Question entailment (RQE) in the medical domain. We use the following definition tailored to QA: "a question A entails a question B if every answer to B is also a complete or partial answer to A" (Ben Abacha and Demner-Fushman, 2016).

2.3 Question Answering (QA)

The objective of this task is to filter and improve the ranking of automatically retrieved answers. The input ranks are generated by the medical QA system CHiQA[3]. We highly recommended the reuse of the RQE and NLI systems (first tasks). For instance (i) the RQE system could be used to retrieve answered questions (e.g. from the MedQuAD dataset[4]) that are entailed from the original questions and use their answers to validate the system's answers and re-rank them; and (ii) the NLI system could be used to identify the relations (i.e. entailment, contradiction, neutral) between the answers of the same question, as well as the answers of the questions related by the entailment relation. We encouraged all other ideas and approaches for using textual inference and question entailment to filter and re-rank the retrieved answers.

3 Data Description

3.1 NLI Datasets

The MEDIQA-NLI test set consists of 405 text-hypothesis pairs. The training set is the MedNLI dataset, which includes 14,049 clinical sentence pairs derived from the MIMIC-III database (Romanov and Shivade, 2018). Both datasets are publicly available[5].

3.2 RQE Datasets

The MEDIQA-RQE test set consists of 230 pairs of Consumer Health Questions (CHQs) received by the U.S. National Library of Medicine (NLM) and Frequently Asked Questions (FAQs) from NIH institutes. The collection was created automatically and double validated manually by medical experts. Table 1 presents positive and negative examples from the test set. The RQE training and validation sets contain respectively 8,890 and 302 medical question pairs created by (Ben Abacha and Demner-Fushman, 2016) using a collection of clinical questions (Ely et al., 2000) for the training set and pairs of CHQs and FAQs pairs for the validation set. All the RQE training, validation and test sets are publicly available[6].

3.3 QA Datasets

The MEDIQA-QA training, validation and test sets were created by submitting medical questions to the consumer health QA system CHiQA (Demner-Fushman et al., 2019), and then rating and re-ranking the retrieved answers manually by medical experts to provide reference ranks (1 to 11) and scores (4: Excellent Answer, 3: Correct but Incomplete, 2: Related, 1: Incorrect).

We provided two training sets for the QA task:

- 104 consumer health questions from the TREC-2017-LiveQA medical data (Ben Abacha et al., 2017) covering different topics such as diseases and drugs, and 839 associated answers retrieved by CHiQA and manually rated and re-ranked.

- 104 simple questions about the most frequent diseases (dataset named Alexa), and 862 associated answers.

ID (Label)	Type	Question
Pair#1 (True)	Premise	I have a list of questions about Tay sachs disease and clubfoot 1. what is TSD/Clubfoot, and how does it effect a baby 2. what causes both? can it be prevented, treated, or cured 3. How common is TSD? how common is Clubfoot 4. How can your agency help a women/couple who are concerned about this congenital condition, and is there a cost? If you can answer these few questions I would be thankful, please get back as soon as you can.
	Hypothesis	How does congenital talipes equinovarus affect a child?
Pair#2 (True)	Premise	When and how do you know when you have congenital night blindness?
	Hypothesis	What are the symptoms of X-linked congenital stationary night blindness ?
Pair#3 (True)	Premise	Polycystic ovarian syndrome Is it possible for parents to pass this on in the genes to their children - is there any other way this can be acquired?
	Hypothesis	Can polycystic ovary syndrome be inherited ?
Pair#4 (True)	Premise	polymicrogyria. My 16 month old son has this. Does not sit up our crawl yet but still trying and is improving in grabbing things etc etc. Have read about other cases that seem 10000 time worse. It's it possible for this post of his brain to grown to normal and he grow out of it?
	Hypothesis	What is the outlook for Polymicrogyria ?
Pair#5 (False)	Premise	spina bifida; vertbral fusion;syrinx tethered cord. can u help for treatment of these problem
	Hypothesis	Does Spina Bifida cause vertebral fusion?
Pair#6 (False)	Premise	varicella shingles How can I determine whether or not I've had chicken pox. If there is a test for it, what are the results of the tests I need to know that will tell me whether or not I have had chicken pox? I want to know this to determine if I should have shingles vaccine (Zostavax) Thank you.
	Hypothesis	Who can catch shingles ?
Pair#7 (False)	Premise	Would appreciate any good info on Lewy Body Dementia, we need to get people aware of this dreadful disease, all they talk about is alzheimers. Thank you
	Hypothesis	What is alzheimer's ?
Pair#8 (False)	Premise	Can you please send me as much information as possible on hypothyroidism. I was recently diagnosed with the disease and I am struggling to figure out what it is and how I got it (...)
	Hypothesis	How is Hypothyroidism diagnosed?

Table 1: Positive and negative examples from the MEDIQA-RQE test set.

The MEDIQA-QA validation set consists of 25 consumer health questions and 234 associated answers returned by CHiQA and judged manually.

The MEDIQA-QA test set consists of 150 consumer health questions and 1,107 associated answers.

All the QA training, validation and test sets are publicly available[7].

In addition, the MedQuAD dataset of 47K medical question-answer pairs (Ben Abacha and Demner-Fushman, 2019) can be used to retrieve answered questions that are entailed from the original questions.

The validation sets of the RQE and QA tasks were used for the first (validation) round on AIcrowd. The test sets were used for the official and final challenge evaluation.

4 Evaluation

4.1 Evaluation Metrics

The evaluation of the NLI and RQE tasks was based on accuracy. In the QA task, participants were tasked to filter and re-rank the provided answers. The QA evaluation was based on accuracy, Mean Reciprocal Rank (MRR), Precision, and Spearmans Rank Correlation Coefficient (Spearman's rho).

4.2 Baseline Systems

- The NLI baseline is the InferSent system (Conneau et al., 2017) based on fasttext (Bojanowski et al., 2017) word embeddings trained on the MIMIC-III data Romanov and Shivade (2018).

- The RQE baseline is a feature-based SVM classifier relying on similarity measures and semantic features (Ben Abacha and Demner-Fushman, 2016).

- The QA baseline is the CHiQA question-answering system (Demner-Fushman et al., 2019). The system was used to provide the answers for the QA task.

---

[7]https://github.com/abachaa/
MEDIQA2019/tree/master/MEDIQA_Task3_QA

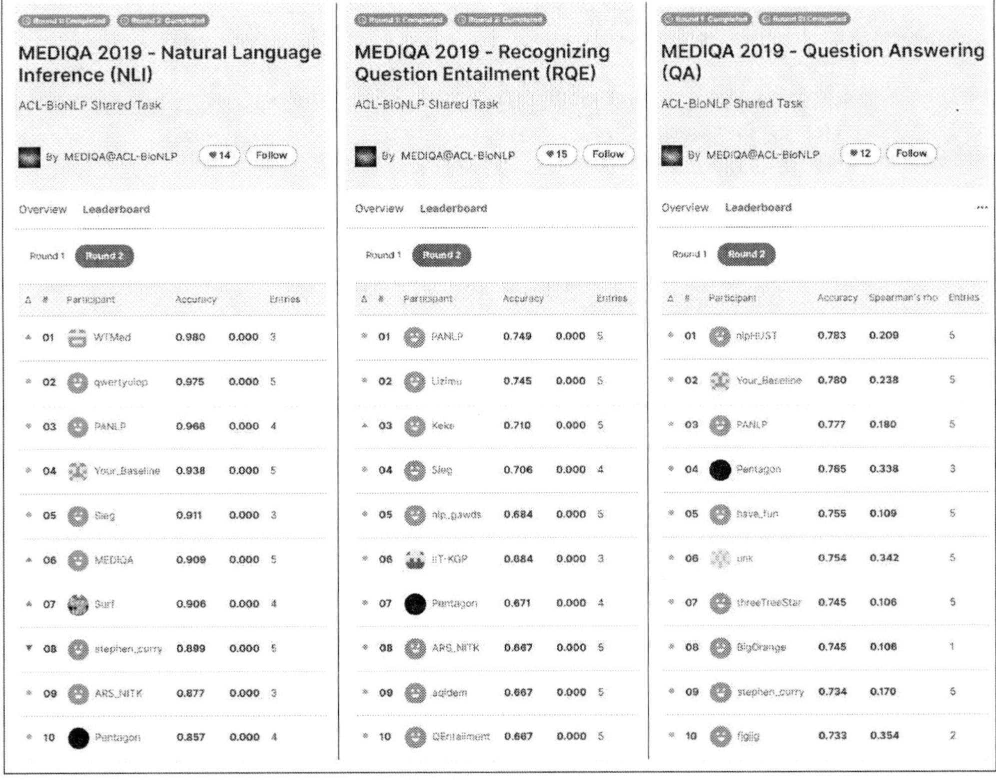

Figure 2: Top-10 results of the three tasks in MEDIQA 2019 among 72 participating teams on AIcrowd

5 Official Results

Seventy two teams participated in the challenge on the AIcrowd platform. Figure 2 presents the original top-10 scores for each task.

The official scores include only the teams who sent a working notes paper describing their approach. The accepted teams are presented in table 2. The official scores for the MEDIQA NLI, RQE, and QA tasks are presented respectively in tables 3, 4, and 5.

5.1 NLI Approaches & Results

Seventeen official teams submitted runs along with a paper describing their approaches among 43 participating teams on NLI@AIcrowd[8]. Most systems build up on the BERT model (Devlin et al., 2019). This model is pretrained on a large open-domain corpus. However, since MedNLI is from the clinical domain following variations of BERT were used.

SciBERT (Beltagy et al., 2019) is a set of variants of the original BERT trained with full text scientific articles, primarily from PubMed. Variants of the model either use the vocabulary of the original BERT model or a new vocabulary learnt specifically for this corpus.

BioBERT (Lee et al., 2019a) is initialized with the original BERT model and then pretrained on biomedical articles from PMC full text articles and PubMed abstracts. BioBERT can be fine-tuned for specific tasks like named entity recognition, relation extraction, and question answering. The data used for pretraining BioBERT is much larger (4.5B words from abstracts and 13.5B words from full text articles) than that used for SciBERT (3.1B words).

ClinicalBERT (Huang et al., 2019) is initialized with the original BERT model and then pretrained on clinical notes from the MIMIC-III dataset. Alsentzer et al. (2019) also released another resource with the same name. These are BERT and BioBERT models further pretrained on the full set of MIMIC-III notes and a subset of discharge summaries.

[8]www.aicrowd.com/challenges/mediqa-2019-natural-language-inference-nli/leaderboards

Table 2: Official teams in MEDIQA 2019 among 72 participating teams on AIcrowd

Team	Task(s)
ANU-CSIRO (Nguyen et al., 2019)	NLI, RQE, QA
ARS_NITK (Agrawal et al., 2019)	NLI, RQE, QA
DoubleTransfer (Xu et al., 2019)	NLI, RQE, QA
Dr.Quad (Bannihatti Kumar et al., 2019)	NLI, RQE, QA
DUT-BIM (Zhou et al., 2019a)	QA
DUT-NLP (Zhou et al., 2019b)	RQE, QA
IITP (Bandyopadhyay et al., 2019)	NLI, RQE, QA
IIT-KGP (Sharma and Roychowdhury, 2019)	RQE
KU_ai (Cengiz et al., 2019)	NLI
lasigeBioTM (Lamurias and Couto, 2019)	NLI, RQE, QA
MSIT_SRIB (Chopra et al., 2019)	NLI
NCUEE (Lee et al., 2019b)	NLI
PANLP (Zhu et al., 2019)	NLI, RQE, QA
Pentagon (Pugaliya et al., 2019)	NLI, RQE, QA
Saama Research (Kanakarajan, 2019)	NLI
Sieg (Bhaskar et al., 2019)	NLI, RQE
Surf (Nam et al., 2019)	NLI
UU_TAILS (Tawfik and Spruit, 2019)	NLI, RQE
UW-BHI (Kearns et al., 2019)	NLI
WTMED (Wu et al., 2019)	NLI

Table 3: Official Results of the MEDIQA-NLI Task

Rank	Team	Accuracy
1	WTMED	0.980
2	PANLP	0.966
3	DoubleTransfer	0.938
4	Sieg	0.911
5	Surf	0.906
6	ARS_NITK	0.877
7	Pentagon	0.857
8	Dr.Quad	0.855
9	UU_TAILS	0.852
10	KU_ai	0.847
11	NCUEE	0.840
12	IITP	0.818
13	MSIT_SRIB	0.813
14	uw-bhi	0.813
15	ANU-CSIRO	0.800
16	Saama Research	0.783
17	lasigeBioTM	0.724
-	*NLI-Baseline*	*0.714*

Table 4: Official Results of the MEDIQA-RQE Task

Rank	Team	Accuracy
1	PANLP	0.749
2	Sieg	0.706
3	IIT-KGP	0.684
4	Pentagon	0.671
5	ARS_NITK	0.667
5	Dr.Quad	0.667
7	DoubleTransfer	0.662
8	DUT-NLP	0.636
9	UU_TAILS	0.584
10	IITP	0.532
11	ANU-CSIRO	0.489
12	lasigeBioTM	0.485
-	*RQE-Baseline*	*0.541*

Table 5: Official Results of the MEDIQA-QA Task

Rank	Team	Accuracy	Precision	MRR	Spearman's rho
1	DoubleTransfer	0.780	0.8191	0.9367	0.238
2	PANLP	0.777	0.7806	0.9378	0.180
3	Pentagon	0.765	0.7766	0.9622	0.338
4	DUT-BIM	0.745	0.7466	0.9061	0.106
4	DUT-NLP	0.745	0.7466	0.9061	0.106
6	IITP	0.717	0.7936	0.8611	0.024
7	lasigeBioTM	0.637	0.5975	0.91	0.211
8	ANU-CSIRO	0.584	0.5568	0.7843	0.122
9	Dr.Quad	0.565	0.6679	0.6069	0.009
10	ARS_NITK	0.536	0.5596	0.6293	0.196
-	*Provided Answers*	*0.517*	*0.5167*	*0.895*	*0.315*

Another common model used by participating systems was the Multi-Task Deep Neural Network MT-DNN (Liu et al., 2019) which builds up on BERT to perform multi-task learning and is evaluated on the GLUE benchmark (Wang et al., 2018). A common theme across all the papers was training of multiple models and then using an ensemble as the final system which performed better than the individual models. Tawfik and Spruit (2019) trained 30 different models as candidates to the ensemble and experimented with various aggregation techniques. Some teams also leveraged dataset-specific properties to enhance the performance. The WTMED team (Wu et al., 2019) modeled parameters specific to the index of the text-hypothesis pair in the dataset which shows a significant boost in performance.

5.2 RQE Approaches & Results

Twelve official teams participated in MEDIQA-RQE among 53 participating teams in the second round on RQE@AIcrowd[9]. The results of the RQE task were surprisingly good knowing the challenges of the test set. For instance, positive question pairs can use different synonyms of the same medical entities (e.g. Pair#1 in table 1) and/or express differently the same information needs (e.g. Pair#4), while negative pairs can use similar language (e.g. Pair#8). Also, the test set is a realistic dataset consisting of actual consumer health questions including one or multiple sub-questions, when the training set consisted of automatically generated question pairs created from doctors' questions. This highlights the fact that

several of the proposed deep networks reached relevant generalizations and abstractions of the questions.

The best results on the RQE task were obtained by the PANLP team (Zhu et al., 2019) with an approach based on multi-task learning. More specifically, their approach relied on a language model learned by the recent MT-DNN (Liu et al., 2019). In a post-processing step, they applied re-ranking heuristics based on grouping observations from the NLI and RQE datasets. E.g., for NLI the text pairs came in groups of three, where a given premise text had three counter-parts for the three relation types: entailment, neutral, and contradiction. Their heuristic re-ranking approach eliminated potential conflicts in the resuls accorcing to the group observation, and led to an increase of 5.1% in accuracy.

More generally, approaches combining ensemble methods and transfer learning of multi-task language models were the clear winners of the competition for RQE with the first and second scores (Zhu et al., 2019; Bhaskar et al., 2019). Approaches that used ensemble methods without multi-task language models (Sharma and Roychowdhury, 2019) or multi-task learning without ensemble methods (Pugaliya et al., 2019) performed worse than the first category but made it to the top 4.

Domain knowledge was also used in several participating approaches with a clear positive impact. For instance, several systems used the UMLS (Bodenreider, 2004) to expand acronyms or to replace mentions of medical entities (Bhaskar et al., 2019; Bannihatti Kumar et al., 2019). Data augmentation also played a key role for several

[9]www.aicrowd.com/challenges/mediqa-2019-recognizing-question-entailment-rqe/leaderboards

systems that used external data to extend batches of in-domain data (Xu et al., 2019), created synthetic data (Bannihatti Kumar et al., 2019), or used models trained on external datasets (e.g. MultiNLI) in ensemble methods (Bhaskar et al., 2019; Sharma and Roychowdhury, 2019).

5.3 QA Approaches & Results

Ten official teams participated in the QA task among 23 participating teams in the second round on QA@AIcrowd[10]. The relevant answer classification problem was relatively challenging with a best accuracy of 78%, however most systems did well on the first answer ranking with a best MRR of 96.22%. Precision also ranged from 79.3% to 81.9% for the six first systems. Many teams used their RQE and/or NLI models in the QA task (Bannihatti Kumar et al., 2019; Pugaliya et al., 2019; Zhu et al., 2019; Nguyen et al., 2019). The DUT-NLP team (Zhou et al., 2019b) used an adversarial multi-task network to jointly model RQE and QA.

The approach that had the best accuracy and precision in the QA task (Xu et al., 2019) relied on multi-task language models (MT-DNN) and ensemble methods. To avoid overfitting, the Double-Transfer team proposed a method, called Multi-Source, that enriches the data batches during training from external datasets by a 50% ratio and random selection. The final ensemble method further combines the Multi-Source method with pre-trained MT-DNN and SciBERT models by taking the majority vote from their predictions and resolving ties by summing the prediction probabilities for each label. The PANLP team's best run (Zhu et al., 2019) ranked second in the QA task despite the fact that the QA data do not have a group structure that could be used in re-ranking heuristics. This shows that their core model is a strong approach, and highlights further the outstanding performance of ensemble methods and multi-task language models for transfer learning for natural language understanding tasks.

Interestingly, the runs that did best on accuracy and precision did not have the best performance in terms of MRR and Spearman's rank correlation coefficient. The best team on these two metrics, Pentagon (Pugaliya et al., 2019), used the MedQuAD and the iCliniq datasets to retrieve entailed answers and used them to build more general embeddings of the considered answer. They also integrated the top-3 RQE candidates from these datasets for the considered question to build joint embeddings. The final answer embeddings were enriched with metadata such as the candidate answer source, answer length, and the original system rank. The same joint embeddings are then used in a filtering classifier for answer relevance and in a binary answer-to-answer classifier that decides if an answer is better than another. These generalized joint answer embeddings and the focus on the answer-to-answer relationship are likely to be the key elements that led to the best performance in MRR and Spearman's rho, despite the fact that the approach did not rely on the state-of-the-art ensemble models from the NLI and RQE tasks.

5.4 Multi-Tasking & External Resources

One of the aims of the MEDIQA 2019 shared task was to investigate ideas that can be reused across the three tasks. Of the twenty working notes papers, ten papers describe systems attempting more than one task. Eight papers describe systems attempting all three tasks. The multi-task nature of MEDIQA 2019 was leveraged by teams to train models such as MT-DNN (e.g. (Bannihatti Kumar et al., 2019; Xu et al., 2019; Zhu et al., 2019)). The Sieg team (Bhaskar et al., 2019) trained a model with shared layers being trained for the NLI and RQE tasks. Some teams also reused models across the three tasks. Pugaliya et al. (2019) used models developed for NLI and RQE as feature extractors in the QA task, which led to the best performance in MRR and Spearman's rho.

The shared task also encouraged the use of external resources other than the training data provided for the three tasks. Below is a non-exhaustive list resources used by various teams.

- **Abbreviation expansion** Many teams pre-processed the training data with UMLS for abbreviation expansion. While Nguyen et al. (2019) used the ADAM database (Zhou et al., 2006) for this task, Bannihatti Kumar et al. (2019) used a CAMC[11] gazzeteer.

- **External datasets** Bannihatti Kumar et al. (2019) used the Quora question pairs dataset (Shankar Iyer and Csernai, 2017) to boost the training for the RQE task, applied

---

[10] www.aicrowd.com/challenges/mediqa-2019-question-answering-qa/leaderboards

[11] http://www.camc.org

MetaMap[12] to recognize medical entities, and synthetically created new questions and paraphrases. Bhaskar et al. (2019) and Pugaliya et al. (2019) used the online iCliniq forum to augment training data for the RQE task. Pugaliya et al. (2019), Xu et al. (2019), Lamurias and Couto (2019), and Nguyen et al. (2019) used the MedQuAD[13] dataset of medical questions and answers (Ben Abacha and Demner-Fushman, 2019).

- **Word Embeddings** While many teams used BERT (Lamurias and Couto, 2019; Zhou et al., 2019a; Bandyopadhyay et al., 2019; Nguyen et al., 2019; Sharma and Roychowdhury, 2019)[14], some teams also used word embeddings as the input to their models. Bhaskar et al. (2019) used biomedical word embeddings from Chen et al. (2018) while Kearns et al. (2019) used cui2vec (Beam et al., 2018).

6   Conclusions

We presented the MEDIQA 2019 shared task on Natural Language Inference (NLI), Recognizing Question Entailment (RQE), and Question answering (QA) in the medical domain. The runs submitted to the challenge by 20 official teams among 72 participating teams achieved promising results and highlighted the strength of multi-task language models, transfer learning, and ensemble methods. Integrating domain knowledge and targeted data augmentation were also key factors for best performing systems. We hope that further research works and insights will be developed in the future from the MEDIQA tasks and their publicly available datasets.

Acknowledgments

This work was supported by the intramural research program at the U.S. National Library of Medicine, National Institutes of Health.
We would like to thank Sharada Mohanty, CEO and co-founder of AIcrowd, and Yassine Mrabet from the NLM for his support with the CHiQA system. We are also thankful to Vandana Mukherjee from IBM Research for supporting the project.

---

[12]metamap.nlm.nih.gov
[13]github.com/abachaa/MedQuAD
[14]github.com/Team-IIT-KGP/Qspider

References

Meni Adler, Jonathan Berant, and Ido Dagan. 2012. Entailment-based text exploration with application to the health-care domain. In *The 50th Annual Meeting of the Association for Computational Linguistics, System Demonstrations, 2012, Korea.*

Anumeha Agrawal, Rosa Anil George, Selvan Sunitha Ravi, Sowmya Kamath, and Anand Kumar. 2019. Ars_nitk at mediqa 2019:analysing various methods for natural language inference, recognising question entailment and medical question answering system. In *Proceedings of the BioNLP 2019 workshop, Florence, Italy, August 1, 2019*. Association for Computational Linguistics.

Emily Alsentzer, John R Murphy, Willie Boag, Wei-Hung Weng, Di Jin, Tristan Naumann, and Matthew McDermott. 2019. Publicly available clinical bert embeddings. *arXiv preprint arXiv:1904.03323.*

Dibyanayan Bandyopadhyay, Baban Gain, Tanik Saikh, and Asif Ekbal. 2019. Iitp at mediqa 2019: Systems report for natural language inference, question entailment and question answering. In *Proceedings of the BioNLP 2019 workshop, Florence, Italy, August 1, 2019*. Association for Computational Linguistics.

Vinayshekhar Bannihatti Kumar, Ashwin Srinivasan, Aditi Chaudhary, James Route, Teruko Mitamura, and Eric Nyberg. 2019. Dr.quad at mediqa 2019: Towards textual inference and question entailment using contextualized representations. In *Proceedings of the BioNLP 2019 workshop, Florence, Italy, August 1, 2019*. Association for Computational Linguistics.

Roy Bar-Haim, Ido Dagan, and Idan Szpektor. 2014. Benchmarking applied semantic inference: The PASCAL recognising textual entailment challenges. In *Language, Culture, Computation. Computing - Theory and Technology - Essays Dedicated to Yaacov Choueka on the Occasion of His 75th Birthday.*

Andrew L Beam, Benjamin Kompa, Inbar Fried, Nathan P Palmer, Xu Shi, Tianxi Cai, and Isaac S Kohane. 2018. Clinical concept embeddings learned from massive sources of multimodal medical data. *arXiv preprint arXiv:1804.01486.*

Iz Beltagy, Arman Cohan, and Kyle Lo. 2019. Scibert: Pretrained contextualized embeddings for scientific text. *arXiv preprint arXiv:1903.10676.*

Asma Ben Abacha, Eugene Agichtein, Yuval Pinter, and Dina Demner-Fushman. 2017. Overview of the medical question answering task at TREC 2017 LiveQA. In *Proceedings of The Twenty-Sixth Text REtrieval Conference, TREC 2017, Gaithersburg, Maryland, USA, November 15-17, 2017.*

Asma Ben Abacha and Dina Demner-Fushman. 2016. Recognizing question entailment for medical question answering. In *AMIA 2016, American Med-*

ical Informatics Association Annual Symposium, Chicago, IL, USA, November 12-16, 2016.

Asma Ben Abacha and Dina Demner-Fushman. 2019. A question-entailment approach to question answering. *CoRR*, abs/1901.08079.

Asma Ben Abacha, Duy Dinh, and Yassine Mrabet. 2015. Semantic analysis and automatic corpus construction for entailment recognition in medical texts. In *Artificial Intelligence in Medicine - 15th Conference on Artificial Intelligence in Medicine, AIME 2015, Pavia, Italy, June 17-20, 2015.*

Sai Abishek Bhaskar, Rashi Rungta, James Route, Eric Nyberg, and Teruko Mitamura. 2019. Sieg at mediqa 2019: Multi-task neural ensemble for biomedical inference and entailment. In *Proceedings of the BioNLP 2019 workshop, Florence, Italy, August 1, 2019.* Association for Computational Linguistics.

Olivier Bodenreider. 2004. The unified medical language system (UMLS): integrating biomedical terminology. *Nucleic Acids Research*, 32(Database-Issue):267–270.

Piotr Bojanowski, Edouard Grave, Armand Joulin, and Tomas Mikolov. 2017. Enriching word vectors with subword information. *Transactions of the Association for Computational Linguistics*, 5:135–146.

Samuel R. Bowman, Gabor Angeli, Christopher Potts, and Christopher D. Manning. 2015. A large annotated corpus for learning natural language inference. In *Proceedings of the 2015 Conference on Empirical Methods in Natural Language Processing, EMNLP 2015, Lisbon, Portugal.*

Asli Çelikyilmaz, Marcus Thint, and Zhiheng Huang. 2009. A graph-based semi-supervised learning for question-answering. In *ACL 2009, Proceedings of the 47th Annual Meeting of the Association for Computational Linguistics and the 4th International Joint Conference on Natural Language Processing of the AFNLP, 2-7 August 2009, Singapore.*

Cemil Cengiz, Ula Sert, and Deniz Yuret. 2019. Ku_ai at mediqa 2019: Domain-specific pre-training and transfer learning for medical nli. In *Proceedings of the BioNLP 2019 workshop, Florence, Italy, August 1, 2019.* Association for Computational Linguistics.

Suchet K. Chachra, Asma Ben Abacha, Sonya E. Shooshan, Laritza Rodriguez, and Dina Demner-Fushman. 2016. A hybrid approach to generation of missing abstracts in biomedical literature. In *COLING 2016, 26th International Conference on Computational Linguistics, 2016, Osaka, Japan.*

Qingyu Chen, Yifan Peng, and Zhiyong Lu. 2018. Biosentvec: creating sentence embeddings for biomedical texts. *arXiv preprint arXiv:1810.09302.*

Sahil Chopra, Ankita Gupta, and Anupama Kaushik. 2019. Msit_srib at mediqa 2019: Knowledge directed multi-task framework for natural language inference in clinical domain. In *Proceedings of the BioNLP 2019 workshop, Florence, Italy, August 1, 2019.* Association for Computational Linguistics.

Alexis Conneau, Douwe Kiela, Holger Schwenk, Loïc Barrault, and Antoine Bordes. 2017. Supervised learning of universal sentence representations from natural language inference data. In *Proceedings of the 2017 Conference on Empirical Methods in Natural Language Processing*, pages 670–680, Copenhagen, Denmark. Association for Computational Linguistics.

Ido Dagan, Oren Glickman, and Bernardo Magnini. 2005. The PASCAL recognising textual entailment challenge. In *Machine Learning Challenges, Evaluating Predictive Uncertainty, Visual Object Classification and Recognizing Textual Entailment, First PASCAL Machine Learning Challenges Workshop, MLCW 2005, Southampton, UK.*

Dina Demner-Fushman, Asma Ben Abacha, and Yassine Mrabet. 2019. Consumer health information and question answering: Helping consumers find answers to their health-related information needs. *Submitted to the Journal of the American Medical Informatics Association (JAMIA).*

Jacob Devlin, Ming-Wei Chang, Kenton Lee, and Kristina Toutanova. 2019. Bert: Pre-training of deep bidirectional transformers for language understanding. In *Proceedings of NAACL.*

John W. Ely, Jerome A. Osheroff, Paul N. Gorman, Mark H. Ebell, M. Lee Chambliss, Eric A. Pifer, and P. Zoe Stavri. 2000. A taxonomy of generic clinical questions: classification study. *British Medical Journal*, 321:429–432.

Sanda M. Harabagiu and Andrew Hickl. 2006. Methods for using textual entailment in open-domain question answering. In *ACL 2006, 21st International Conference on Computational Linguistics and 44th Annual Meeting of the Association for Computational Linguistics, Sydney, Australia.*

Kexin Huang, Jaan Altosaar, and Rajesh Ranganath. 2019. Clinicalbert: Modeling clinical notes and predicting hospital readmission. *arXiv:1904.05342.*

Kamal raj Kanakarajan. 2019. Saama research at mediqa 2019: Pre-trained biobert with attention visualisation for medical natural language inference. In *Proceedings of the BioNLP 2019 workshop, Florence, Italy, August 1, 2019.* Association for Computational Linguistics.

William Kearns, Wilson Lau, and Jason Thomas. 2019. Uw-bhi at mediqa 2019: An analysis of representation methods for medical natural language inference. In *Proceedings of the BioNLP 2019 workshop, Florence, Italy, August 1, 2019.* Association for Computational Linguistics.

Andre Lamurias and Francisco Couto. 2019. Lasige-biotm at mediqa 2019: Biomedical question answering using bidirectional transformers and named entity recognition. In *Proceedings of the BioNLP 2019 workshop, Florence, Italy, August 1, 2019*. Association for Computational Linguistics.

Jinhyuk Lee, Wonjin Yoon, Sungdong Kim, Donghyeon Kim, Sunkyu Kim, Chan Ho So, and Jaewoo Kang. 2019a. Biobert: pre-trained biomedical language representation model for biomedical text mining. *arXiv preprint arXiv:1901.08746*.

Lung-Hao Lee, Yi Lu, Po-Han Chen, Po-Lei Lee, and Kuo-Kai Shyu. 2019b. Ncuee at mediqa 2019: Medical text inference using ensemble bert-bilstm-attention model. In *Proceedings of the BioNLP 2019 workshop, Florence, Italy, August 1, 2019*. Association for Computational Linguistics.

Xiaodong Liu, Pengcheng He, Weizhu Chen, and Jianfeng Gao. 2019. Multi-task deep neural networks for natural language understanding. *arXiv preprint arXiv:1901.11504*.

Jiin Nam, Seunghyun Yoon, and Kyomin Jung. 2019. Surf at mediqa 2019: Improving performance of natural language inference in the clinical domain by adopting pre-trained language model. In *Proceedings of the BioNLP 2019 workshop, Florence, Italy, August 1, 2019*. Association for Computational Linguistics.

Vincent Nguyen, Sarvnaz Karimi, and Zhenchang Xing. 2019. Anu-csiro at mediqa 2019: Question answering using deep contextual knowledge. In *Proceedings of the BioNLP 2019 workshop, Florence, Italy, August 1, 2019*. Association for Computational Linguistics.

Hemant Pugaliya, Karan Saxena, Shefali Garg, Sheetal Shalini, Prashant Gupta, Eric Nyberg, and Teruko Mitamura. 2019. Pentagon at mediqa 2019: Multi-task learning for filtering and re-ranking answers using language inference and question entailment. In *Proceedings of the BioNLP 2019 workshop, Florence, Italy, August 1, 2019*. Association for Computational Linguistics.

Alexey Romanov and Chaitanya Shivade. 2018. Lessons from natural language inference in the clinical domain. In *Proceedings of the 2018 Conference on Empirical Methods in Natural Language Processing, Brussels, Belgium, 2018*.

Nikhil Dandekar Shankar Iyer and Kornl Csernai. 2017. First quora dataset release: Question pairs.

Prakhar Sharma and Sumegh Roychowdhury. 2019. Iit-kgp at mediqa 2019: Recognizing question entailment using sci-bert stacked with a gradient boosting classifier. In *Proceedings of the BioNLP 2019 workshop, Florence, Italy, August 1, 2019*. Association for Computational Linguistics.

Chaitanya Shivade, Courtney Hebert, Marcelo A. Lopetegui, Marie-Catherine de Marneffe, Eric Fosler-Lussier, and Albert M. Lai. 2015. Textual inference for eligibility criteria resolution in clinical trials. *Journal of Biomedical Informatics*, 58.

Noha Tawfik and Marco Spruit. 2019. Uu_tails at mediqa 2019: Learning textual entailment in the medical domain. In *Proceedings of the BioNLP 2019 workshop, Florence, Italy, August 1, 2019*. Association for Computational Linguistics.

Alex Wang, Amapreet Singh, Julian Michael, Felix Hill, Omer Levy, and Samuel R Bowman. 2018. GLUE: A multi-task benchmark and analysis platform for natural language understanding. *arXiv preprint arXiv:1804.07461*.

Adina Williams, Nikita Nangia, and Samuel R. Bowman. 2018. A broad-coverage challenge corpus for sentence understanding through inference. In *Proceedings of the 2018 Conference of the North American Chapter of the Association for Computational Linguistics: Human Language Technologies, NAACL-HLT 2018, New Orleans, Louisiana, USA*.

Zhaofeng Wu, Yan Song, Sicong Huang, Yuanhe Tian, and Fei Xia. 2019. Wtmed at mediqa 2019: A hybrid approach to biomedical natural language inference. In *Proceedings of the BioNLP 2019 workshop, Florence, Italy, August 1, 2019*. Association for Computational Linguistics.

Yichong Xu, Xiaodong Liu, Chunyuan Li, Hoifung Poon, and Jianfeng Gao. 2019. Doubletransfer at mediqa 2019: Multi-source transfer learning for natural language understanding in the medical domain. In *Proceedings of the BioNLP 2019 workshop, Florence, Italy, August 1, 2019*. Association for Computational Linguistics.

Huiwei Zhou, Bizun Lei, Zhe Liu, and Zhuang Liu. 2019a. Dut-bim at mediqa 2019: Utilizing transformer network and medical domain-specific contextualized representations for question answering. In *Proceedings of the BioNLP 2019 workshop, Florence, Italy, August 1, 2019*. Association for Computational Linguistics.

Huiwei Zhou, Weihong Yao Xuefei Li, Chengkun Lang, and Shixian Ning. 2019b. Dut-nlp at mediqa 2019:an adversarial multi-task network to jointly model recognizing question entailment and question answering. In *Proceedings of the BioNLP 2019 workshop, Florence, Italy, August 1, 2019*. Association for Computational Linguistics.

Wei Zhou, Vetle I Torvik, and Neil R Smalheiser. 2006. Adam: another database of abbreviations in medline. *Bioinformatics*, 22(22):2813–2818.

Wei Zhu, Xiaofeng Zhou, Keqiang Wang, Xun Luo, Xiepeng Li, Yuan Ni, and Guotong Xie. 2019. Panlp at mediqa 2019: Pre-trained language models, transfer learning and knowledge distillation. In *Proceedings of the BioNLP 2019 workshop, Florence, Italy, August 1, 2019*.

PANLP at MEDIQA 2019: Pre-trained Language Models, Transfer Learning and Knowledge Distillation

Wei Zhu, Xiaofeng Zhou, Keqiang Wang, Xun Luo, Xiepeng Li, Yuan Ni, Guotong Xie

Pingan Health Tech, Shanghai, China

{zhuwei972, zhouxiaofeng824, wangkeqiang265, luoxun492, lixiepeng538,
niyuan442, xieguotong}@pingan.com.cn

Abstract

This paper describes the models designated for the MEDIQA 2019 shared tasks by the team PANLP. We take advantages of the recent advances in pre-trained bidirectional transformer language models such as BERT (Devlin et al., 2018) and MT-DNN (Liu et al., 2019b). We find that pre-trained language models can significantly outperform traditional deep learning models. Transfer learning from the NLI task to the RQE task is also experimented, which proves to be useful in improving the results of fine-tuning MT-DNN large. A knowledge distillation process is implemented, to distill the knowledge contained in a set of models and transfer it into an single model, whose performance turns out to be comparable with that obtained by the ensemble of that set of models. Finally, for test submissions, model ensemble and a re-ranking process are implemented to boost the performances. Our models participated in all three tasks and ranked the 1st place for the RQE task, and the 2nd place for the NLI task, and also the 2nd place for the QA task.

1 Introduction

There are three tasks in the MEDIQA 2019 shared tasks (see Ben Abacha et al. (2019) for details of the tasks). The first one, NLI, consists in identifying three inference relations between two sentences: *Entailment*, *Neutral* and *Contradiction*. The second one, RQE, requires one to identify whether one question entails the other, where the definition of *entailment* is that *a question A entails a question B if every answer to B is also a complete or partial answer to A*. The third task, QA, considers not only the identification of entailment for the asked question among a set of retrieved questions, but also the ranks of retrieved answers.

In this work, we demonstrate that we can achieve significant performance gains over traditional deep learning models like ESIM (Chen et al., 2016), by adapting pre-trained language models into the medical domain. Language model pre-training has shown to be effective for learning universal language representations by leveraging large amounts of unlabeled data. Some of the most famous examples are GPT-V2 (see Radford et al., 2019) and BERT (by Devlin et al., 2018). These are neural network language models trained on text data using unsupervised objectives. For example, BERT is based on a multi-layer bidirectional Transformer, and is trained on plain text for masked word prediction and next sentence prediction tasks. To apply a pre-trained model to specific NLU tasks such as tasks for MEDIQA 2019 shared tasks, we often need to fine-tune the model with additional task-specific layers using task-specific training data. For example, Devlin et al. (2018) show that BERT can be fine-tuned this way to create state-of-the-art models for a range of NLU tasks, such as question answering and natural language inference.

We also tryout a transfer learning procedure, where an intermediate model obtained on the NLI task is used to be fine-tuned on the RQE task. Although this procedure cannot consistently improve the dev set performance for all the models, it is proven to be beneficial on the test set by adding variety to the model pool.

To further improve the performance of single models, we implement a knowledge distillation procedure on the RQE task and the NLI task. Knowledge distillation distills or transfers the knowledge from a (set of) large, cumbersome model(s) to a lighter, easier-to-deploy single model, without significant loss in performance (Liu et al., 2019a; Tan et al., 2019). Knowledge distillation recently has attracted a lot of attentions. We believe it is interesting and of great importance to explore this method on the applications of the medical domain.

Proceedings of the BioNLP 2019 workshop, pages 380–388
Florence, Italy, August 1, 2019. ©2019 Association for Computational Linguistics

For test submissions, model ensemble is used to obtain more stable and unbiased predictions. We only adopt a simple ensemble model, that is, averaging the class probabilities of different models. After obtaining test predictions, for the NLI and RQE task, simple re-ranking operations among pairs with the same premise are used to boost the performance metrics.

The rest of the paper is organized as follows. In Section 2 , we demonstrate our experiments on the three tasks. In Section 3, transfer learning from NLI to RQE is presented. Section 4 elaborates on the knowledge distillation and the corresponding experimental results. Section 5 and Section 6 present the model ensemble technique and the re-ranking strategies. Section 7 explains our submission records in detail. Section 8 concludes and discusses future work.

2 Pairwise Text Modeling

This section elaborates on the fundamental methods we used for the three tasks.

2.1 RQE

The RQE task, as a pairwise text classification task, defined here involves a premise $P = (p_1, p_2, ..., p_m)$ of m words, which is a medical question posted online, and a hypothesis $H = (h_1, h_2, ..., h_n)$ of n words, which is a standard frequently asked question that is collected to build a QA system, and aims to find a logical relationship R between P and H. For the RQE task, relationship R is either *true* or *false*, indicating whether the premise entails the hypothesis or not. We mainly experiment on two groups of models, one using fixed pre-trained embedding[1], the other employing pre-trained language models.

Traditional deep learning models typically use a fixed pre-trained word embedding to map words into low-dimensional vector space, and then use some kind of encoders to encode and pool the contexts of the premise to vector r_1 and hypothesis H to r_2. And the features provided to the classification layer is $concat(r_1, r_2, ||r_1 - r_2||, r_1 * r_2)$. (see Bowman et al., 2015) Then the classification output layer is usually a dense layer with soft-max output. We experiment with the following 4 traditional deep learning models. The first model, which will be called Weighted-

Transformer-NLI model, encodes the sentences via a shared Weighted Transformer module (see Ahmed et al., 2017 for details). The second model, called RCNN-NLI, encodes the premise and hypothesis via the RCNN model (see Lai et al., 2015). The third model we consider, is the decomposable attention model by Parikh et al. (2016). The fourth model is the ESIM model by Chen et al. (2016), which is one of the most popular models in the natural language inference task. We will not elaborate on the specific architecture of the last two models since readers can refer to the original papers for details.

For the RQE task, the pre-trained language models we considered are as follows: (a) the original BERT models (both base and large models); (b) the Bio-BERT model by Lee et al. (2019) which is pre-trained on scientific literature in biomedical domain; (c) the Sci-BERT model by Beltagy et al. (2019) which is trained on academic papers from the corpus of *semanticscholar.org*; (d) MT-DNN models (see Liu et al., 2019b), which are based on BERT and go through a multi-task learning procedure on the GLUE benchmark. On top of the transformer encoders from the pre-trained language model, we implement two kinds of output modules: (a) linear projection, which will be referred to as LP0, which is to take the hidden state corresponding to the first token $[CLS]$ of the sentence pair; (b) a more sophisticated classification module called stochastic answer network (henceforth SAN) proposed by Liu et al. (2017). Rather than directly predicting the entailment given the input, SAN maintains a state and iteratively refines its predictions.

When implementing the traditional deep learning models, the Glove embedding (Pennington et al., 2014) is used. Before training, we use the Unified Medical System (UMLS) provided by provided by the National Library of Medicine [2] to replace all the abbreviations (e.g., *IBS*) of a medical concept or entity to its full name, or to the same name that appears in the same pair. We tune the hyper-parameters on the dev set, and report the best performance obtained by each model in Table 1.

Among the four traditional models, RCNN-NLI performs the worst. Although a powerful model as shown in Ahmed et al. (2017),

---

[1] We will refer to this type of models as traditional deep learning models

[2] https://www.nlm.nih.gov/research/umls/

Model	valid acc
Weighted-Transformer-NLI	0.6821
RCNN-NLI	0.5530
Decomposable attention	0.6854
ESIM	0.7218
BERT base + linear projection	0.7815
BERT base + SAN	0.7119
BERT large + linear projection	0.7782
BERT large + SAN	0.7682
Bio-BERT + linear projection	0.4338
Bio-BERT + SAN	0.4305
Sci-BERT + linear projection	0.7547
Sci-BERT + SAN	0.5993
MT-DNN base + linear projection	**0.8378**
MT-DNN base + SAN	0.7715
MT-DNN large + linear projection	0.7881
MT-DNN large + SAN	0.7815

Table 1: performances of different models on the valid set of the RQE task.

Weighted-Transformer-NLI cannot perform very well on this dataset. The ESIM model performs the best among the four. However the traditional deep learning models cannot perform well enough when compared with the results on the Round 1 leader board. We believe the reasons are as follows. First, the dataset is relatively small, thus models like Weighted-Transformer-NLI will immediately over-fit. [3] Second, the distribution of training data for RQE task is different from the distributions of the dev and test data. We see most of the pairs in train set have approximately equal length, and there are 1, 445 pairs in which the premise and hypothesis are exactly the same. Meanwhile, in dev and test sets, the premise is usually much longer than the hypothesis.

When compared with traditional deep learning models, the pre-trained language models perform significantly better on the dev set. In addition, one can see that adding a sophisticated output module like SAN on top of the pre-trained language model tends to worsen the dev performance. Among all the BERT model family, the MT-DNN model (base model) performs best, and the original BERT base model performs slightly worse. Since the MT-DNN family are BERT models fine-tuned on GLUE benchmark via a multi-task learning mechanism, and in GLUE eight out of nine

layers to freeze	valid acc
0	0.7782
1	0.8013
3	0.7914
6	0.7881
9	0.8179
10	0.8344
11	**0.8378**

Table 2: performances of the MT-DNN base model with linear projection, when different number of layers are freezed during fine-tuning on the RQE dataset

tasks are pairwise text modeling tasks, MT-DNN are more equipped to model pairwise text classification tasks on different domains than the original BERT model. And we can see that MT-DNN base performs better than MT-DNN large, which is in contradiction to the results on the GLUE benchmark reported in Liu et al. (2019b). Sci-BERT and Bio-BERT model does not perform well. We believe the reasons are that the Sci-BERT and Bio-BERT models share the same feature that they are trained on scientific literature, in which the language is more formal and rigid. However, texts in RQE is drawn from online questions from medical forums, thus Sci-BERT and Bio-BERT are not suitable for this task.

We also notice that freezing the lower bi-directional transformer layers of MT-DNN significantly improves the dev set accuracy. In Table 2,

[3] Readers can refer to Guo et al. (2019) for more detailed discussions on how transformer models performs unsatisfyingly on medium or small datasets, when directly trained from scratch.

Model	valid acc
ESIM (by Romanov and Shivade, 2018)	0.7440
InferSent (by Romanov and Shivade, 2018)	0.7600
BERT base + linear projection	0.8186
BERT base + SAN	0.8143
BERT large + linear projection	0.8229
BERT large + SAN	0.8280
Bio-BERT + linear projection	0.6824
Bio-BERT + SAN	0.6882
Sci-BERT + linear projection	**0.8466**
Sci-BERT + SAN	0.8251
MT-DNN base + linear projection	0.8265
MT-DNN base + SAN	0.8287
MT-DNN large + linear projection	0.8420
MT-DNN large + SAN	0.8327

Table 3: performances of different models on the valid set of the NLI task.

we can see that freezing 11 lower layers of the MT-DNN base performs best. During training of different models, even traditional deep learning models, we notice that a model can easily over-fit on the training set of RQE, fine-tuning the whole language model will introduce much bias into the model. Meanwhile freezing the lower layers can alleviate over-fitting and maintain the generalization ability of the pre-trained models.

2.2 NLI

For the NLI task, we are tasked to identify the relationship R between the premise and the hypothesis, which is among the following three: *entailment*, *neutral* or *contradiction*. Romanov and Shivade (2018) has done a thorough investigation on how traditional deep learning models like ESIM and InferSent perform on the original NLI datasets. Thus to save time, we only implement with pre-trained language models for this task.

The BERT based models we tried are the same as we investigate on the RQE datasets, whose results are reported in Table 3. It turns out, the BERT-based model significantly outperforms the traditional models. MT-DNN models still perform quite well, but the Sci-BERT with linear projection achieves the highest accuracy on the dev set. The Bio-BERT model still cannot achieve satisfying results. We find that models behave quite differently on NLI compared with the RQE datasets. First, on the NLI dataset, BERT large and the MT-DNN large, which is derived from BERT large, perform better than their base counterparts, BERT

base and MT-DNN base. Second, during tuning the hyper-parameters, we find that freezing layers leads to performance loss. Third, the SAN output module does not lead to significant performance change except for Sci-BERT, whereas on the RQE dataset adding SAN module usually leads to significant performance loss.

2.3 QA

On the basis of the results obtained on RQE and NLI task, we found that the MT-DNN models outperform other pre-trained language models. Thus, with limited time, in the QA task we chose to directly look into the MT-DNN models on the QA datasets.

The QA task requires us not only give a binary label to an answer, but also rank the answers of the same questions. There are two perspectives of treating such a task: classification and regression. The classification model just distinguishes whether the question and the answer match, and the output of Softmax layer can be used to rank the answers. However, the regression model is able to predict the matching degree between questions and answers, and rank the answers according to the matching degree. The final result achieved is a combination of two models.

From the perspective of the classification model, answers with *ReferenceScore* less than 3 are given a *not entailment* label, and the rest are labeled *entailment*. The dataset obtained with this treatment is called the QA-C dataset. Table 4 reports the performance on the dev set. To align

Model	acc	Spearman's Rank Corr
MT-DNN base on QA-R	0.8248	0.1478
MT-DNN large on QA-R	0.8333	0.2054
MT-DNN base + linear projection on QA-C	0.7479	0.0557
MT-DNN base + SAN on QA-C	0.7607	-0.0108
MT-DNN large + linear projection on QA-C	0.8333	0.0803
MT-DNN large + SAN on QA-C	0.8120	0.2146

Table 4: performances of different models on the valid set of the QA task. Here accuracy is calculated on the whole dev set.

Model	dev acc
MT-DNN base	0.8378
MT-DNN base + transfer learning on NLI	0.8220
MT-DNN large	0.7881
MT-DNN large + transfer learning on NLI	0.7957

Table 5: The performance on the RQE dev set, when we apply transfer learning, compared with the performances obtained by directly fine-tuning the MT-DNN models on the RQE dataset.

with the leader board, we calculated accuracy and Spearman's Rank Correlation Coefficient (henceforth SRCC). As is shown in Table 4, BERT base can achieve accuracy of 0.7478 after fine-tuning. However, SRCC is 0.057, which is quite poor. The results demonstrate that a binary classification model helps us to get a fair accuracy score, but it omits all the ranking information like *ReferenceRank* and *ReferenceScore* from the original data. Thus the resulting model can not tell whether an answer is better than another. Bearing that in mind, we decided to introduce a related but different model to specialize in providing ranking information, while leave the accuracy metric to the classification model.

The new model we are introducing treats the task at hand as a regression task. For a sample data, the input is a pair composed of a query and an answer. The target value is the relevance score between the query and the answer, which is defined as follows:

$$score = ReferenceScore + 1/ReferenceRank. \tag{1}$$

The reciprocal of the *ReferenceRank* is used to enlarge the gaps of relevance scores among different answers. The dataset obtained with the above modification is called the QA-R dataset. The regression model is also built on the pre-trained language models by replacing the classification output module with a regression task header (see equation (2) of Liu et al., 2019b). Table 4 shows that we can obtain a huge bump on SRCC with

the regression model. The best dev SRCC we can obtain is 0.148, which is the result of fine-tuning the MT-DNN large model. With a threshold for the relevance score, we can also get the classification label from the regression label. After adjusting the threshold, we can also get accuracy of 0.8247. Thus, we can conclude that the regression model works better in capturing the ranking information without reducing the accuracy of the model.

By observing the SRCC obtained at each epoch during training, we can see the following phenomenon: SRCC can improve from 0.125 to 0.273 after a single epoch, and suddenly drop to 0.023 on the next one. SRCC seems to be quite unstable, which will be problematical when making predictions for the unknown test set. This is a problem that we fail to solve at the end of competition and requires further investigations.

3 Transfer learning

We also experimented with transfer learning for the RQE task. The procedure is to first fine-tune a MT-DNN model on the NLI dataset for a certain number of epochs, then the obtained model will further be fine-tuned on the RQE dataset. Our motivation is that first fine-tuning on the NLI task can help the pre-trained language model to adapt to the medical domain, thus making the training on RQE more stable. Table 5 reports that after the transfer learning procedure, MT-DNN base model performs worse, but it makes the MT-DNN large model perform slightly better.

4 knowledge distillation

In this section, we experiment on the idea of knowledge distillation (Hinton et al., 2015), to further boost the performance of single models. We implement knowledge distillation on each task separately.[4] The procedure is as follows:

- train a set of models on each tasks. Following Liu et al. (2019a), the set of models are: MT-DNN base and MT-DNN large, with different dropout rates ranged in $0.1, 0.3, 0.5$ for the task specific output layers, while keeping the hyper-parameters of lower BERT encoders the same with those in the previous section.

- ensemble the above models to get a label model (Ratner et al., 2018)[5]. This so-called label model is constructed by modeling a generative model over all the label functions, i.e., the single models, to maximize the log likelihood, give the label matrix (Ratner et al., 2017). The label model is a generalization of the so-called teacher model in (Liu et al., 2019a), where the teacher model is simply the average of class probabilities.

- The end model (or called the student model by Liu et al., 2019a) is trained on the soft targets given out by the label model. Here, training on the soft targets means the cross-entropy loss is averaged with the class probabilities as weights.

- Inference is the same for end model with other normal models.

In Table 6, we can see that knowledge distillation can significantly improve the performance on the NLI task, and can even achieve better results than model ensemble. However, on the RQE task, knowledge distillation cannot perform better than model ensemble, but still outperforms the best single model.

5 Ensemble

Since the test set is small, one single model is too biased to achieve great results on the test dataset. Ensemble learning is an effective approach to improve model generalization, and has been used to achieve new state-of-the-art results in a wide range of natural language understanding (NLU) tasks (Devlin et al., 2018, Liu et al., 2017).

For the MEDIQA 2019 shared task, we only adopt a simple ensemble approach, that is, averaging the softmax outputs from different models, or different runs or epochs of the same model, and makes prediction based on these averaged class probabilities. All our submissions follow this ensemble strategy. [6]

6 Re-ranking strategies for the NLI and RQE tasks

The previous sections demonstrate how deep learning models perform on the task datasets. However, in order to obtain more competitive results, one could adopt some simple heuristics.

For the NLI task, after observing the task datasets, we can see that one premise is grouped with three different hypothesis, and the latter are labeled with *entailment*, *neutral* and *contradiction* respectively. We call the three pairs with the same premise a group. Our sentence pair model does not know the idea of groups, thus the labels corresponding to the maximum class probabilities obtained by soft-max layer can conflict with one another. For example, two pairs in the same group may both be labeled as *entailment*. To eliminate the above conflicts, we adopt the following heuristic post-processing procedure:

- obtain the label predictions directly from the softmax output. If there is no conflict in a group, accept the predictions. Otherwise, in this group:

- Give the *contradiction* label to the pair with the highest score for this label

- Between the remaining two pairs, decide which one should get the *neutral* label via the scores for this label

---

[4]Liu et al. (2019a) extends the knowledge distillation to multi-task learning setting, which is a direction we need to explore in future work.

[5]There are alternative terminologies for knowledge distillation. We mainly follow Ratner et al. (2018).

[6]We definitely can try some more sophisticated ensemble methods, but we believe experimenting different learning strategies like MTL and knowledge distillation is more meaningful for research purpose, and is in alignment with the objective of the MEDIQA 2019 share tasks.

Model	NLI	RQE
best single model	0.8466	0.8378
model ensemble	0.8638	0.8477
knowledge distillation	0.8667	0.8411

Table 6: Comparison of performances on the dev sets, among the best single model, ensemble model and the model obtained by knowledge distillation.

- the remaining pair get the *entailment* label

For the RQE task, since the label is binary, and the number of pairs in a group in this task varies, the re-ranking heuristic is a little different, which is elaborated as follows.

- obtain the score of the *entailment* label from the model

- for each group, rank the pairs by their scores.

- denote the number of pairs in a group as n, then we directly label the last $max(1, [n/2] - 1)$ as negative pairs. and the top pair as positive pair

- For the rest of pairs, we choose a threshold t, if the score of a pair is higher than t, it is labeled *entailment*, otherwise it is labeled as *not entailment*. We choose the threshold to obtain the highest accuracy on the dev set

7 Submission results

This section discusses the submission results on the leader boards.

First, let us look at the submission history on the RQE task (presented here in Table 7). The first submission is a single MT-DNN base model trained only on the training data, with re-ranking. On the second submission, we add the available dev set in, and re-train all the models. The ensemble of a MT-DNN base and a MT-DNN large after re-ranking push the test accuracy to 0.736. Then we tryout transfer learning on the third run, two runs of MT-DNN large, which go through the transfer learning process described in Section 3, achieves 0.745 after re-ranking. Adding the end model after knowledge distillation to the combination in the third run makes the performance drops slightly to 0.740. For the final submission, we just ensemble all the models available, and achieve 0.749 on the test set, which ranks the first on the RQE task.

Table 8 presents the submission records on the NLI task. On the first submission, we experiment the model obtained by knowledge distillation, which obtains 0.865 on accuracy. The second submission, we use a single MT-DNN large fine-tuned on the train set and post-processed for re-ranking. The accuracy is 0.916 for this submission. Then the ensemble of four models, the 8-th epoch of 2 different runs of MT-DNN large, the 10-th epoch of 2 different runs of Sci-BERT, achieves an accuracy of 0.946 after re-ranking. The final submission combines MT-DNN large, Sci-BERT, MT-DNN large after knowledge distillation, obtains 0.966 after re-ranking, which ranks the third on the leader board.

For the QA task, the first two submissions are based on a single MT-DNN large model fine-tuned on QA-R data set, chosen from two different training epochs. The first submission with accuracy of 0.73 is chosen because in this epoch of training, we achieved the best Spearman's rho result on the dev dataset; Similarly, the second submission with accuracy of 0.733 is chosen at the epoch where we achieved best ACC result on the dev dataset. From the third round, we started applying ensemble strategy by considering some well performing epochs at different runs together. The two submissions with accuracy of 0.774 and 0.777 are the results of different processing strategies: max score and mean score. According to the results obtained, we find that "max score" strategy performs slightly better on SRCC, while "mean score" works better on ACC.

8 Conclusion and discussions

To conclude, we have shown that domain adaptation with the pre-trained language models achieves significant improvement over traditional deep learning models on the MEDIQA 2019 shared tasks. We also experimented transfer learning from the NLI task to the RQE task. Knowledge distillation obtains a single model which significantly outperforms the single models trained

Submission No.	test acc	details
1	0.675	1 * MT-DNN base (trained on train set) + re-rank
2	0.736	1 * MT-DNN base + 1 * MT-DNN large + re-rank
3	0.745	2 * MT-DNN large (TL) + re-rank
4	0.740	1 * MT-DNN large (KD) + 2 * MT-DNN large (TL) + re-rank
5	0.749	2 * MT-DNN base + 2 * MT-DNN large (TL) + 1 * MT-DNN large (KD) + 1 * MT-DNN large + re-rank

Table 7: The submission results on the RQE task. Multiplication symbol "*" here means multiple runs or epochs of the same model (with different random seed). "TL" means the model go through transfer learning on the NLI task. "KD" means the model is obtained via knowledge distillation. Without declaration, all the models here are trained on the train and dev set.

Submission No.	test acc	details
1	0.865	1 * MT-DNN large (KD)
2	0.916	1 * MT-DNN large (on train set) + re-rank
3	0.946	2 * MT-DNN large + 2 * Sci-BERT + re-rank
4	0.966	4 * MT-DNN large + 4 * Sci-BERT + 2 * MT-DNN large (KD) + re-rank

Table 8: The submission records on the NLI task. Multiplication symbol "*" here means multiple runs or epochs of the same model (with different random seed). "KD" means the model is obtained via knowledge distillation. Without declaration, all the models here are trained on the train and dev set.

Submission No.	test acc	test Spearman's rho	details
1	0.730	0.236	MT-DNN large (epoch with best training SRCC)
2	0.736	0.204	MT-DNN large (epoch with best training ACC)
3	0.774	0.22	MT-DNN large ensemble(rank by max socre)
4	0.777	0.18	MT-DNN large ensemble(rank by mean socre)
5	0.772	0.204	MT-DNN large ensemble(rank by mean socre)

Table 9: The submission results on the QA task.

in the usual way. Our submission results, although including model ensemble and re-ranking, are strong demonstration of the power of language model pre-training, transfer learning and knowledge distillation.

However, due to the limited time and the fact that we participate all three tasks at once, we haven't exhaustively explore all the possible ways to boost the performance on the leader board, e.g., utilizing external sources such as medical knowledge bases to rule out false positive answers. Multi-task learning is also a direction that we need to pay more attention to.

In addition, the heuristics adopted in the re-ranking strategies resemble the relevance ranking task (Huang et al., 2013), where one compares different pairs in a group to obtain the final decisions. Due to time constraint, we didn't implement a pairwise relevance ranking model on top of the MT-DNN model, but this research direction will be investigated by us in future work.

References

Karim Ahmed, Nitish Shirish Keskar, and Richard Socher. 2017. Weighted Transformer Network for Machine Translation. *arXiv e-prints*, page arXiv:1711.02132.

Iz Beltagy, Arman Cohan, and Kyle Lo. 2019. SciBERT: Pretrained Contextualized Embeddings for Scientific Text. *arXiv e-prints*, page arXiv:1903.10676.

Asma Ben Abacha, Chaitanya Shivade, and Dina Demner-Fushman. 2019. Overview of the mediqa 2019 shared task on textual inference, question entailment and question answering. In *Proceedings of the BioNLP 2019 workshop, Florence, Italy, August 1, 2019*. Association for Computational Linguistics.

Samuel R. Bowman, Gabor Angeli, Christopher Potts,

and Christopher D. Manning. 2015. A large annotated corpus for learning natural language inference. In *Proceedings of the 2015 Conference on Empirical Methods in Natural Language Processing*, pages 632–642, Lisbon, Portugal. Association for Computational Linguistics.

Qian Chen, Xiaodan Zhu, Zhenhua Ling, Si Wei, Hui Jiang, and Diana Inkpen. 2016. Enhanced LSTM for Natural Language Inference. *arXiv e-prints*, page arXiv:1609.06038.

Jacob Devlin, Ming-Wei Chang, Kenton Lee, and Kristina Toutanova. 2018. BERT: Pre-training of Deep Bidirectional Transformers for Language Understanding. *arXiv e-prints*, page arXiv:1810.04805.

Qipeng Guo, Xipeng Qiu, Pengfei Liu, Yunfan Shao, Xiangyang Xue, and Zheng Zhang. 2019. Star-Transformer. *arXiv e-prints*, page arXiv:1902.09113.

Geoffrey Hinton, Oriol Vinyals, and Jeff Dean. 2015. Distilling the Knowledge in a Neural Network. *arXiv e-prints*, page arXiv:1503.02531.

Po-Sen Huang, Xiaodong He, Jianfeng Gao, Li Deng, Alex Acero, and Larry Heck. 2013. Learning deep structured semantic models for web search using clickthrough data. In *Proceedings of the 22nd ACM international conference on Information & Knowledge Management*, pages 2333–2338. ACM.

Siwei Lai, Liheng Xu, Kang Liu, and Jun Zhao. 2015. Recurrent convolutional neural networks for text classification.

Jinhyuk Lee, Wonjin Yoon, Sungdong Kim, Donghyeon Kim, Sunkyu Kim, Chan Ho So, and Jaewoo Kang. 2019. BioBERT: a pre-trained biomedical language representation model for biomedical text mining. *arXiv e-prints*, page arXiv:1901.08746.

Xiaodong Liu, Pengcheng He, Weizhu Chen, and Jianfeng Gao. 2019a. Improving Multi-Task Deep Neural Networks via Knowledge Distillation for Natural Language Understanding. *arXiv e-prints*, page arXiv:1904.09482.

Xiaodong Liu, Pengcheng He, Weizhu Chen, and Jianfeng Gao. 2019b. Multi-Task Deep Neural Networks for Natural Language Understanding. *arXiv e-prints*, page arXiv:1901.11504.

Xiaodong Liu, Yelong Shen, Kevin Duh, and Jianfeng Gao. 2017. Stochastic Answer Networks for Machine Reading Comprehension. *arXiv e-prints*, page arXiv:1712.03556.

Ankur P. Parikh, Oscar Täckström, Dipanjan Das, and Jakob Uszkoreit. 2016. A Decomposable Attention Model for Natural Language Inference. *arXiv e-prints*, page arXiv:1606.01933.

Jeffrey Pennington, Richard Socher, and Christopher D. Manning. 2014. Glove: Global vectors for word representation. In *Empirical Methods in Natural Language Processing (EMNLP)*, pages 1532–1543.

Alec Radford, Jeffrey Wu, Rewon Child, David Luan, Dario Amodei, and Ilya Sutskever. 2019. Language models are unsupervised multitask learners. *OpenAI Blog*, 1:8.

Alexander Ratner, Stephen H. Bach, Henry Ehrenberg, Jason Fries, Sen Wu, and Christopher Ré. 2017. Snorkel: Rapid Training Data Creation with Weak Supervision. *arXiv e-prints*, page arXiv:1711.10160.

Alexander Ratner, Braden Hancock, Jared Dunnmon, Frederic Sala, Shreyash Pandey, and Christopher Ré. 2018. Training Complex Models with Multi-Task Weak Supervision. *arXiv e-prints*, page arXiv:1810.02840.

Alexey Romanov and Chaitanya Shivade. 2018. Lessons from Natural Language Inference in the Clinical Domain. *arXiv e-prints*, page arXiv:1808.06752.

Xu Tan, Yi Ren, Di He, Tao Qin, Zhou Zhao, and Tie-Yan Liu. 2019. Multilingual Neural Machine Translation with Knowledge Distillation. *arXiv e-prints*, page arXiv:1902.10461.

Pentagon at MEDIQA 2019: Multi-task Learning for Filtering and Re-ranking Answers using Language Inference and Question Entailment

Hemant Pugaliya,[*] **Karan Saxena,**[*] **Shefali Garg,**[*] **Sheetal Shalini,**[*]
Prashant Gupta,[*] **Eric Nyberg, Teruko Mitamura**

{hpugaliy, karansax, shefalig, sshalini, prashang, ehn, teruko}@cs.cmu.edu

Language Technologies Institute (LTI),
Carnegie Mellon University (CMU)

Abstract

Parallel deep learning architectures like fine-tuned BERT and MT-DNN, have quickly become the state of the art, bypassing previous deep and shallow learning methods by a large margin. More recently, pre-trained models from large related datasets have been able to perform well on many downstream tasks by just fine-tuning on domain-specific datasets (similar to transfer learning).

However, using powerful models on non-trivial tasks, such as ranking and large document classification, still remains a challenge due to input size limitations[1] of parallel architecture and extremely small datasets (insufficient for fine-tuning).

In this work, we introduce an end-to-end system, trained in a multi-task setting, to filter and re-rank answers in medical domain. We use task-specific pre-trained models as deep feature extractors. Our model achieves the highest Spearman's Rho and Mean Reciprocal Rank of 0.338 and 0.9622 respectively, on the ACL-BioNLP workshop MediQA Question Answering shared-task.

1 Introduction

In this work, we study the problem of re-ranking and filtering in medical domain Information Retrieval (IR) systems. Historically, re-ranking is generally treated as a 'Learning to Rank' problem while filtering is posed as a 'Binary Classification' problem. Traditional methods have used handcrafted features to train such systems. However, recently deep learning methods have gained popularity in the Information retrieval (IR) domain (Mitra and Craswell, 2017).

The ACL-BioNLP workshop MediQA shared task (Ben Abacha et al., 2019) aims to develop relevant techniques for inference and entailment in medical domain to improve domain specific IR and QA systems. The challenge consists of three tasks which are evaluated separately.

The first task is the Natural Language Inference (NLI) task which focuses on determining whether a natural language hypothesis can be inferred from a natural language premise. The second task is to recognize question entailment (RQE) between a pair of questions. The third task is to filter and improve the ranking of automatically retrieved answers.

For the NLI and RQE tasks, we use transfer learning on prevalent pre-trained models like BERT (Devlin et al., 2018) and MT-DNN (Liu et al., 2019). These models play a pivotal role to gain deeper semantic understanding of the content for the final task (filtering and re-ranking) of the challenge (Demszky et al., 2018). Besides using usual techniques for candidate answer selection and re-ranking, we use features obtained from NLI and RQE models. We majorly concentrate on the novel multi-task approach in this paper. We also succinctly describe our NLI and RQE models and their performance on the final leaderboard.

2 Related Work

Past research demonstrates a simple architecture for filtering and re-ranking, where the system returns the best answer based on the Information Retrieval and Question Entailment Scores [from a corpus of FAQs scraped from medical websites MediQUAD (Ben Abacha and Demner-Fushman, 2019)]. This system outperformed all the systems participating in the TREC Medical LiveQA17

---

* Equal contribution, randomly sorted. Karan and Shefali took ownership of the NLI module while Sheetal and Prashant worked on the RQE module. Hemant researched and implemented the Question-Answering system including baseline and multi-task learning. Sheetal and Hemant worked on scraping data from icliniq. Karan and Prashant helped with integration of NLI and RQE module respectively into the multi-task system.

[1] https://github.com/google-research/bert/issues/27

Proceedings of the BioNLP 2019 workshop, pages 389–398
Florence, Italy, August 1, 2019. ©2019 Association for Computational Linguistics

challenges. (Harabagiu and Hickl, 2006) successfully shows the use of Natural Language Inference (NLI) in passage retrieval, answer selection and answer re-ranking to advance open-domain question answering. (Tari et al., 2007) shows effective use of UMLS (Bodenreider, 2004), a Unified Medical Language System to asses passage relevancy through semantic relatedness. All these methods work well independently, but to the best of our knowledge, there hasn't been much work in using NLI and RQE systems in tandem for the tasks of filtering and re-ranking.

As noted in (Romanov and Shivade, 2018), the task of Natural Language Inference is not domain agnostic, and thus is not able to transfer well to other domains. The authors use a gradient boosting classifier (Mason et al., 2000) with a variety of hand crafted features for baselines. They then use Infersent (Conneau et al., 2017) as a sentence encoder. The paper also reports results on the ESIM Model (Chen et al., 2017) but with no visible improvements. They also discuss transfer learning and external knowledge based methods.

Most traditional approaches to Question Entailment use bag-of-word pair classifiers (Tung and Xu, 2017) using only lexical similarity. However, in the recent past, neural models (Mishra and Bhattacharyya, 2018) have been employed to determine entailment between questions incorporating their semantic similarity as well. These techniques work by generating word-level representations for both the questions, which are then combined into independent question representations by passing it through a recurrent cell like Bi-LSTM (Liu et al., 2016). However, current state-of-the-art methods like BERT (Devlin et al., 2018) and MT-DNN (Huang et al., 2013) learn a joint embedding of the two questions, which is then used for classification.

(Abacha and Demner-Fushman, 2016) implemented the SVM, Logistic Regression, Naive Bayes and J48 models as baselines for Question Entailment task. They use a set of handcrafted lexical features, like word overlap and bigram similarity, and semantic features like number of medical entities (problems, treatments, tests) using a CRF classifier trained on i2b2 (Uzuner et al., 2011) and NCBI corpus (Doğan et al., 2014).

3 Dataset & Evaluation

The dataset for re-ranking and filtering has been provided by the MediQA Shared task (Ben Abacha et al., 2019) in ACL-BioNLP 2019 workshop. It consists of medical questions and their associated answers retrieved by CHiQA [2]. The training dataset consists of 208 questions while the validation and test datasets have 25 and 150 questions respectivley. Each question has upto 10 candidate answers, with each answer having the following attributes :

1. SystemRank: It corresponds to CHiQA's rank.

2. ReferenceRank: It corresponds to the correct rank.

3. ReferenceScore: This is an additional score that is provided only in the training and validation sets, which corresponds to the manual judgment/rating of the answer [4: Excellent, 3: Correct but Incomplete, 2: Related, 1: Incorrect].

For the answer classification task, answers with scores 1 and 2 are considered as incorrect (label 0), and answers with scores 3 and 4 are considered as correct (label 1). The evaluation metrics for filtering task is Accuracy and Precision while metrics for re-ranking task is Mean Reciprocal Rank (MRR) and Spearman's Rank Correlation Coefficient.

To train the Natural Language Inference and Question Entailment module of our system we again use the data from MediQA shared task (Ben Abacha et al., 2019).

For Natural Language Inference (NLI), we use MedNLI (Romanov and Shivade, 2018) dataset. It is a dataset for natural language inference in clinical domain that is analogous to SNLI. It includes 15,473 annotated clinical sentence pairs. For our model, we create a training set of 14,050 pairs and a held out validation set of 1,423 pairs. The evaluation metric for NLI is accuracy.

The dataset used for Question Entailment (RQE) consists of paired customer health questions (CHQ) and Frequently Asked Questions (FAQ) (Ben Abacha and Demner-Fushman, 2016). We are provided labels for whether FAQ entails CHQ or not. The RQE training dataset consists

---

[2] https://chiqa.nlm.nih.gov/

of 8,588 medical question pairs. The validation set comprises of 302 pairs. The evaluation metric used for RQE is accuracy.

We also augment the data from a popular medical expert answering website called [3] icliniq. It is a forum where users can delineate their medical issues, which are then paraphrased as short queries by medical experts. The user queries are treated as CHQs whereas the paraphrased queries are treated as FAQs. We extract 9,958 positive examples and generate an equal number of negative examples by random sampling. The average CHQ length is 180 tokens whereas the average FAQ length is 11 tokens. In addition, the expert answers are used to augment the MediQUAD corpus (Ben Abacha and Demner-Fushman, 2019).

4 Approach/System Overview

We use pretrained RQE and NLI modules as feature extractors to compute best entailed questions and best candidate answers in our proposed pipeline.

4.1 Pretraining NLI and RQE modules

Both the NLI and RQE modules use MediQA shared task (Ben Abacha et al., 2019) for training (fine-tuning) and computing the inference and entailment scores. For both the tasks, we use the following approaches to preprocess the datasets:

1. Replacing medical terms with their preferred UMLS name. We augment the terms like *Heart attack* in the sentence with *Myocardial infarction* extracted from UMLS.

2. Expanding abbreviations for medical terms in order to normalize the data. The list of medical abbreviations is scraped from Wikipedia. Since this list of abbreviations also contains full forms of stop words like "IS", "BE", we manually curate the list to contain only the relevant acronyms.

For fine-tuning the NLI and RQE modules, we use the dataset for NLI and RQE tasks of MediQA shared task (Ben Abacha et al., 2019) respectivley. We also augment the RQE dataset with data from icliniq during fine-tuning.

4.2 Preprocessing

A lot of answers have spurious trailing lines about FAQs being updated. Any trailing sentences in the

answers having "Updated by:" are removed. A co-reference resolution is run on each answer using Stanford CoreNLP (Manning et al., 2014) and all the entity-mentions are replaced with their corresponding names.

4.3 Using RQE module

For each question in the training set we get upto N entailing questions (along with their scores and embeddings) and answers with a threshold T for confidence using RQE module. We use this system both in the baseline and the multi-task learning system. The complete process is highlighted in Figure 1.

4.4 Baseline: Feature-Engineered System

We develop a feature-engineered system as a baseline. This system uses the following features:

1. Answer Source (One-hot)

2. Answer Length In Sentences

3. ChiQA Rank

4. Bag of Words(BoW) TF-IDF scores of Candidate Answer (trained on MediQUAD)

5. Bag of Words (BoW) TF-IDF scores of 1-best Entailed answer (trained on Medi-QUAD)

6. N-best RQE Scores

7. N-best RQE embeddings

8. N-best Average NLI Scores

Average NLI score between the candidate answer 's' containing 'S' sentences and entailed answer 'p' containing 'P' sentences is defined as:

$$ANLI(s,p) = \frac{\sum_S(\max_P(NLI(S,P)))}{|S|}$$

where $|S|$ symbolizes the total number of sentences in candidate answer.
For a given confidence threshold T, if N candidates are not obtained from RQE model we set the corresponding features to 0.

We train the system using the above features with Logistic regression for filtering and use the scores to rank the answers. We also train a system with same features using SVM-rank (Joachims, 2006) to improve our ranking metrics. All the results have been discussed in Section 5.

---

[3]https://www.icliniq.com/qa/medical-conditions

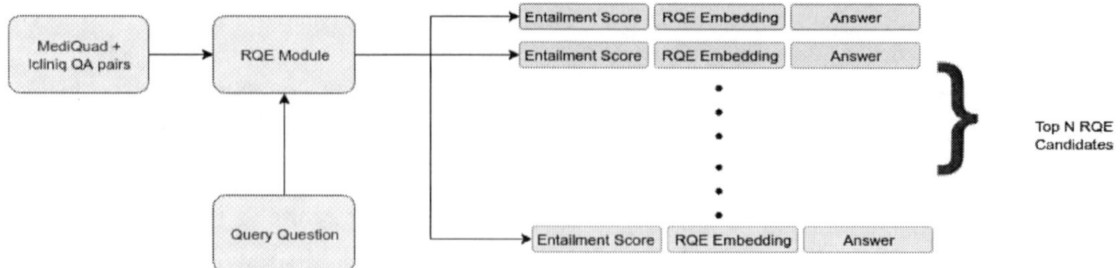

Figure 1: Finding entailed questions from MediQuad and Icliniq QA pairs, for a particular query. We get an entailment score, RQE entailment embedding, and answer for each QA pair in MediQuad and Iclinq data. We then pick up the top N entailed questions.

4.5 Jointly Learning to Filter and Re-rank

Multitask learning is defined as, "A learning paradigm in machine learning whose aim is to leverage useful information contained in multiple related tasks to help improve the generalization performance of all the tasks". (Zhang and Yang, 2017) As the tasks of both filtering and re-ranking are highly related and can benefit from shared feature space, we propose a multi-task learning based system to both rank and filter our candidate answers.

In this system we use the MT-DNN (Liu et al., 2019) based models developed for NLI and RQE (described in Sectition 4.1) as feature extractors. The embedding generated for classification from both the models is used as features. In addition we also use the scores from RQE models to get RQE candidates from MediQUAD (Ben Abacha and Demner-Fushman, 2019) corpus. Going forward we refer to these features as embeddings.

Our initial step is the same as our baseline system and is summarized in Section 4.3. For each candidate answer in training set and the retrieved entailed answer we obtain the following embeddings:

1. **NLI Embedding**: If an entailing question's answer A has a sentences and candidate answers C have c sentences, then a tensor of

$$a * c * 768$$

is extracted to make an embedding matrix using the NLI module. Each sentence in entailing Answer A is combined with every sentence of candidate answer C and passed to the MT-DNN NLI model to build this tensor. We then run a convolution encoder on this matrix to obtain an NLI embedding. The final layer is an average pooling layer which averages each of the four quadrants of 256 channel feature map and concatenates them to obtain an NLI embedding of size 1024. This step is necessary to convert varied size (due to varying a and c above) feature maps to a single embedding of size 1024.

2. **RQE Embedding**: This is the embedding obtained from the RQE model while searching for the entailed questions.

3. **Metadata Embedding**: This embedding encodes metadata features for the pair. We encode the candidate answer source (one-hot), the entailed answer source (one-hot), candidate answer length, entailed answer length, candidate answer system rank, and TF-IDF scores of 2000 words (trained on Medi-QUAD) for the candidate answer.

We concatenate the above embeddings for each candidate answer (referred to as joint embedding going forward). For a given entailed answer, one joint embedding is obtained for each candidate answer. The entire process of converting a single entailing answer to a set of joint embeddings for candidates is summarized in Figure 2.

Using the joint embeddings obtained above we train two binary classifiers, which are fully connected neural networks, as follows:

1. **Filtering classifier**: This classifier takes in the joint embedding for a single candidate answer and classifies it as relevant or irrelevant.

2. **Pairwise ranking classifier**: This classifier takes in the joint embedding of two candidate answers and classifies if the first candidate ranks higher or not.

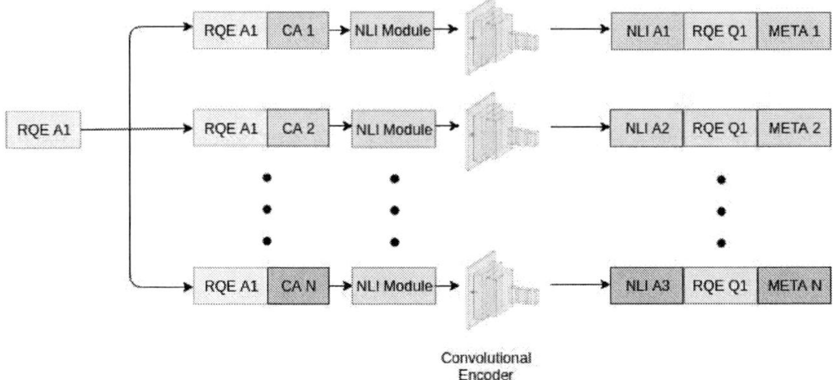

Figure 2: Creating NLI embedding for each RQE Answer A1 by concatenating it with every candidate answer (CA) and passing it through the NLI module and convolutional encoder. This NLI embedding is then concatenated with the corresponding RQE Question (Q1) embedding and Metadata embedding to obtain the joint embedding.

Architecture details are provided in Appendix A.

4.6 Training and Inference details

For a given confidence threshold T, if less than N questions are obtained from the RQE model, we only use the questions which satisfy the threshold. In case no entailed question is returned, we use the top entailed question despite confidence being below the threshold. Unlike the baseline, here joint embedding is extracted separately for each entailed answer. Hence this allows for having different number of entailed answers for each question.

For training we consider each question and candidate answers as a batch. We define the final training loss as follows:

$$L_{total} = \sum_N (\sum_c L_{filter}(c)$$
$$+ \alpha \sum_{p \in c^2 - 1 pairs} L_{pair}(p)) \quad (1)$$

where N is the number of RQE candidates we have for this question, c is the number of candidate answers, L_{filter} is the loss obtained from filtering classifiers, L_{pair} is the loss obtained from pairwise classifier. We use Cross-Entropy Loss for both L_{filter} and L_{pair}. Here we use $\alpha = 2$ to focus more on the re-ranking task as it is considered tougher than filtering. To augment the training data we use higher-ranked candidate answers as entailed answers to create training instances with lower-ranked candidate answers. While inference we use the ensemble from different RQE candi-dates to decide upon filtering and pairwise ranking, by summing the scores from candidates.

5 Experiments and Results

For this task we perform multiple experiments on feature-engineered system in Section 4.4 to asses the usefulness of the designed features. These experiments later help us incorporate these features into Metadata Embedding defined in Section 4.5.

Firstly we run the experiments on Metedata features, BoW, Coarse-grained RQE and NLI scores. The results are shown in Table 1. We later incorporate the RQE embeddings from RQE system and the results are shown in Table 2. Here we evaluate the system with different number of RQE candidates at different threshold settings. Previous experiments were conducted on the filtering task only . For ranking task we train SVM-Rank (Joachims, 2006) based systems to learn pair-wise ranking, using the same features as the filtering task. Experiments with SVM-Rank (Joachims, 2006) were performed with N=3 RQE candidates and the results are shown in Table 4.

Moving to jointly learning system introduced in Section 4.5, we train it with different parameter settings. Due to lack of resources, we could evaluate only a few hyperparameter settings where N is the most number of RQE candidates considered while training and T is the the threshold for retrieving the candidates. In addition we also evaluate the results with augmented datasets from icliniq. We share the results on validation data in Table 6 and results on test set in Table 5.

Metrics	Accuracy	Spearmans Rho
Metadata	50.12	0.091
Metadata + BoW + RQE Scores	61.23	0.125
Metadata + BoW + RQE scores + Avg NLI	62.17	0.127

Table 1: Results with features except RQE embeddings using Logistic Regression

	No. of RQE Candidates		
RQE Threshold	N=1	N=3	N=5
No Threshold	63.12	67.17	64.92
T=0.9	63.21	65.12	63.9
T=0.7	64.91	**69.67**	**66.123**
T=0.5	**65.18**	68.96	65.031

Table 2: Accuracy obtained on including RQE embeddings.

RQE Threshold	Coverage
T=0.5	186/208
T=0.7	175/208
T=0.9	150/208

Table 3: Coverage of Validation set based on RQE threshold for Task 3.

System	Spearman's Rho
LR based filtering	0.2327
Rank-SVM(T=0.9)	0.2627
Rank-SVM(T=0.7)	**0.2972**
Rank-SVM(T=0.5)	0.2812

Table 4: Rank-SVM results with Fine-grained features for N=3 candidates with different threshold levels for Task 3.

6 Discussion

We design the experiments to see if the answering and re-ranking tasks can be improved upon using RQE and NLI tasks. This hypothesis was proved by seeing the improved performance on including RQE and NLI features in Section 4.4 as shown in Table 1. Moreover we see that on including RQE embeddings we get the performance boost as seen in Table 2.

Another question which we can ask ourselves is how many entailed answers are good enough for performing filtering and re-ranking and how confident do we need to be about the entailment to consider a candidate. Experimental results shown in Table 2 show that we can't take too high number of candidates as well as the threshold can't be too high. We see in Table 3 that if we take too high threshold for entailment, we might not find an entailing answer altogether. Hence going forward for all experiments we have taken threshold as 0.7 and number of candidates as 3.

While the feature sets discussed in the above experiments perform well in filtering tasks, they do not do well when their re-ranking is done based on their filtering scores. In further experiments we train a specialized ranking system using SVM-Rank (Joachims, 2006) and the results are shared in Table 4. We see that the same exact feature set could learn well to re-rank when trained with specialized algorithm. Improved results in Table 2 and Table 4 by learning on the same feature set but using different algorithms motivated us to design our approach in Section 4.5 which would learn a joint high-dimensional feature space for both the tasks.

Experiments on multi-task learning clearly show that this technique is superior to feature-engineered approach in both re-ranking and filtering. We attribute this increase in performance to mainly two factors: Firstly the multi-task setting allows it to learn more generalized features. Secondly, inclusion of high-dimensional NLI features in the architecture which was previously not possible with feature-engineered approach. However the computationally expensive nature of this approach did not let us experiment with many hyper-parameter settings. The results on Validation data and Test data are shown in Table 6 and Table 5 respectively.

From the results in Table 6 and Table 5 we see that it reinforces our analysis done about the can-

Hyperparameters	Accuracy	Spearman's Rho	MRR	Precision
N=3, T=0.7, Corpus = Mediquad + Icliniq	**0.765**	0.338	**0.962**	**0.776**
N=3 , T=0.7 , Corpus = Mediquad	0.733	**0.354**	0.955	0.741
N=5 , T=0.7, Corpus = Mediquad	0.7	0.317	0.97	0.709

Table 5: Multi-Task learning results with different parameter settings on Test data for Task 3.

Hyperparameters	Accuracy	Spearman's Rho
N=3, T=0.7, Corpus = Mediquad + Icliniq	**78.12**	0.351
N=3 , T=0.7 , Corpus = Mediquad	76.1	**0.372**
N=5 , T=0.7, Corpus = Mediquad	71.1	0.331

Table 6: Multi-Task learning results with different parameter settings on Validation data for Task 3.

didate and threshold settings based on Table 3. We also see that adding additional data from Icliniq improves the accuracy but decreases the Spearman's Rho. This can be attributed to the language style difference between ICliniq and MediQUAD (Ben Abacha and Demner-Fushman, 2019). As re-ranking is a tougher task, it's performance takes a hit while the accuracy does improve owing to better RQE coverage.

7 Shared Task Performance

To evaluate our performance on the test sets, we submitted our NLI, RQE and Re-ranking & Filtering model independently on the shared task leaderboard. For Task 1, i.e. the NLI task, we achieved an accuracy of 85.7 on the test set. For Task 2, i.e. the RQE task, we observed that the test set varied greatly as compared to the training set, leading to poor results on test dataset. To account for this difference, we discarded the training data and trained our model only on the validation and augmented data. This model gave us an accuracy of 67.1 on the test set. The best model for both the tasks is the ensemble of Infersent (Conneau et al., 2017), BERT fine-tuned (last 4 layers) (Devlin et al., 2018) and MT-DNN (Huang et al., 2013). For both For Task 3, i.e. the re-ranking and filtering task, the results are shown in Table 5.

In the NLI task , our system ranked 7th (out of 17), showing an improvement of 20% over the task baseline. In the RQE task, our system ranked 4th (out of 12), showing an improvement of 24% over the task baseline. In the Question Answering Task, our system ranked 3rd (out of 10) in filtering metrics (both Accuracy and Precision) while it ranked 1st (out of 10) in the ranking metrics (both Mean Reciprocal Rank and Spearmans̈ Rho). The system performs significantly better than others in ranking metrics, showing an improvement of 2.6% and 42% in Mean Reciprocal Rank and Spearman's Rho respectively over the next best scores from the participating teams. Interestingly, our system is the only participating system which outperforms the baseline (ChiQA provided answers) on Spearman's Rho. However, this is not surprising as we take ChiQA rank as one of our input features.

8 Error Analysis

In case of NLI, we observe that the model generally fails in two major settings explained below. Since most of our training data has negation words like 'do not', 'not' etc for the contradicting hypothesis, the model assigns the label as contradiction whenever it sees a confusing example with negation term as shown in Figure 3.

Premise: on return to the floor , pt was given narcan without any significant change in mental status .

Hypothesis: the patient did not have an opioid overdose .

Predicted: Contradiction

Actual: Neutral

Figure 3: NLI model Incorrectly Predicting Contradiction on Test Set

The model also fails while trying to differentiate between statements that are neutral versus those that entail each other. The model generally relies on lexical overlap between the hypothesis and the premise, and in cases, when it is unable to find one, falls back to assigning the label as neutral as shown in Figure 4.

For the RQE task, we observe that our model la-

Premise: while on the floor he was started on azithromycin, solumedrol 125 mg iv tid , albuterol and ipratropium nebulizers .

Hypothesis: the patient is being treated for a chf exacerbation .

Predicted: Neutral

Acutal: Entailment

Figure 4: NLI model Incorrectly Predicting Neutral on Test Set

bels the CHQ-FAQ pairs as entailment when they have a high lexical overlap of the medical entities and not entailment otherwise. We confirm this with some examples from the RQE test set.

CHQ: know more about My Daughter have Distal renal tubular acidosis. we are from Mexico, and we ae woundering if can send to us more infomation. maybe you can reccommend to us a association???? i don?t know

FAQ: What is the best treatment of distal renal tubular acidosis in Mexico ?

Predicted: Not entailment

Actual: Entailment

Figure 5: RQE model incorrectly predicting True on test set

CHQ: how can someone have neurosyphilis and there partner not get syphilis

FAQ: How is syphilis prevented?

Predicted: Entailment

Actual: Not entailment

Figure 6: RQE model incorrectly predicting False on test set

The example shown in Figure 5 has a unigram overlap of 6 and bigram overlap of 3. So our model predicts the label as True, whereas the ground truth label is False because even though the same disease is being referred to in both the CHQ and FAQ, the questions being asked about it are different.

The example shown in Figure 6 has a unigram overlap of 2 and bigram overlap of 0. So our model predicts the label as False, whereas the ground truth label is True because the FAQ is sort of like an abstractive summary of the CHQ with less lexical overlap.

Above analysis suggests that RQE or NLI models are baised to the lexical overlap of medical entities. To overcome this, we could extract medical entities using Metamap (Aronson, 2006) and mask

them randomly during training so that the model learns the semantic representation even without the medical entities. Masking entities has been shown to generalize better in ERNIE(Zhang et al., 2019) in comparison to BERT(Devlin et al., 2018).

For the re-ranking and filtering tasks we look into the macro-trends and investigate what qualifies as tougher problems for both the tasks. From Figure 7, it is clear that lower ranked valid answers are generally harder answers for filtering. Observing the valid answers with low ranks, we see that they generally have only 1-2 relevant sentences each, which might be hard for the model especially in cases where the answers have a lot of sentences. Similar analysis for the filtering tasks based on the

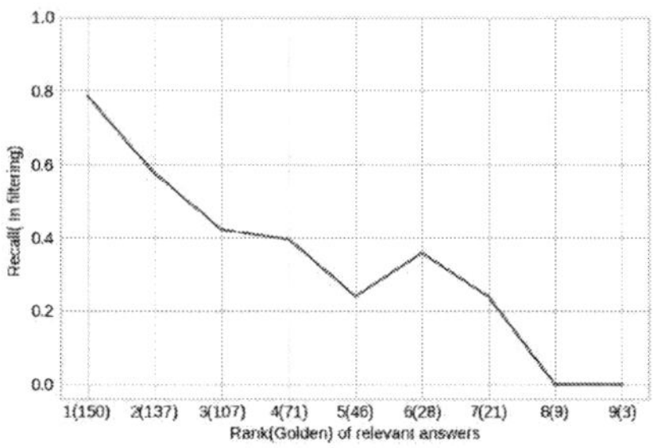

Figure 7: Relationship between the rank of the valid answer and it's filtering recall. The number in parenthesis denotes number of such examples seen in the test dataset.

number of sentences in the answers show some interesting trends, as shown in Figure 8. Interestingly, the model performs really well for filtering longer answers with more than 80 sentences. On further analysis, it is seen that generally the entailed answers can be entirely found in these large candidate answers for the valid answers.

We also observe that the spearman's rho is sensitive to the number of valid candidates for each question. Especially when the number of valid candidates are less, the metric can vary considerably even with a small error. When analyzing the spearman's rho on per question basis, it is seen that the questions with just two valid answers get a score of -1 on getting the order wrong, while the score is 1 if the order is right. This variability is captured in Figure 9. The accuracy however, varies only slightly based on the number of valid answers.

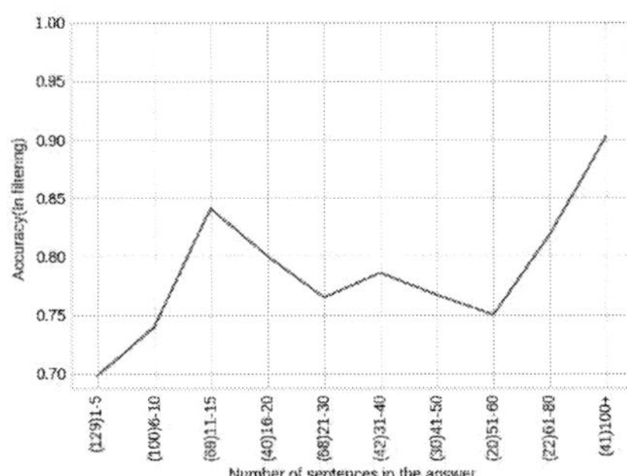

Figure 8: Relationship between the number of sentences in an answer and the filtering accuracy. The number in parenthesis denotes number of such examples seen in the test dataset.

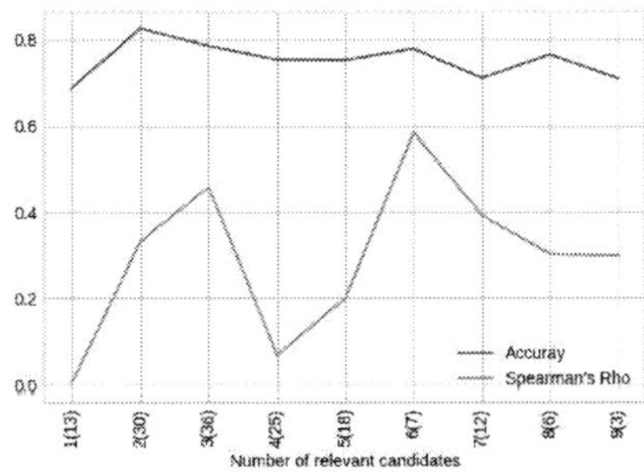

Figure 9: Trends in accuracy and spearman's rho based on the number of valid answers for a question. In case of 1 valid answer, spearman's rho is always taken as 0. The number in parenthesis denotes number of such examples seen in the test dataset.

9   Conclusion and Future work

Our results show that learning to re-rank and filter answers in a multi-task setting help learn a joint feature space which improves performance on both the tasks. In addition, we show that we can harness the power of pre-trained models by fine-tuning them for a specific task and using them as feature extractors to assist in non-trivial tasks such as re-ranking and large document classification. We see that an increase in the size of the corpus with augmented data leads to improved results, hence some more work can be done to build upon the work of (Ben Abacha and Demner-Fushman, 2019). Additionally, we could improve the NLI and RQE systems by tackling the bias created due to the lexical overlap of medical entities among the two sentences/questions, as these were the predominant errors made by our models. This would indirectly translate to an improved performance of the filtering and re-ranking system.

References

Asma Ben Abacha and Dina Demner-Fushman. 2016. Recognizing question entailment for medical question answering. In *AMIA Annual Symposium Proceedings*, volume 2016, page 310. American Medical Informatics Association.

Alan R Aronson. 2006. Metamap: Mapping text to the umls metathesaurus. *Bethesda, MD: NLM, NIH, DHHS*, pages 1–26.

Asma Ben Abacha and Dina Demner-Fushman. 2016. Recognizing question entailment for medical question answering. In *AMIA 2016, American Medical Informatics Association Annual Symposium, Chicago, IL, USA, November 12-16, 2016*.

Asma Ben Abacha and Dina Demner-Fushman. 2019. A question-entailment approach to question answering. *arXiv e-prints*.

Asma Ben Abacha, Chaitanya Shivade, and Dina Demner-Fushman. 2019. Overview of the mediqa 2019 shared task on textual inference, question entailment and question answering. In *ACL-BioNLP 2019*.

Olivier Bodenreider. 2004. The unified medical language system (umls): Integrating biomedical terminology. *Nucleic acids research*, 32:D267–70.

Qian Chen, Xiaodan Zhu, Zhen-Hua Ling, Si Wei, Hui Jiang, and Diana Inkpen. 2017. Enhanced lstm for natural language inference. *Proceedings of the 55th Annual Meeting of the Association for Computational Linguistics (Volume 1: Long Papers)*.

Alexis Conneau, Douwe Kiela, Holger Schwenk, Loic Barrault, and Antoine Bordes. 2017. Supervised learning of universal sentence representations from natural language inference data. *arXiv preprint arXiv:1705.02364*.

Dorottya Demszky, Kelvin Guu, and Percy Liang. 2018. Transforming Question Answering Datasets Into Natural Language Inference Datasets. *arXiv e-prints*, page arXiv:1809.02922.

Jacob Devlin, Ming-Wei Chang, Kenton Lee, and Kristina Toutanova. 2018. Bert: Pre-training of deep bidirectional transformers for language understanding. *arXiv preprint arXiv:1810.04805*.

Rezarta Islamaj Doğan, Robert Leaman, and Zhiyong Lu. 2014. Ncbi disease corpus: a resource for disease name recognition and concept normalization. *Journal of biomedical informatics*, 47:1–10.

Sanda Harabagiu and Andrew Hickl. 2006. Methods for using textual entailment in open-domain question answering. In *Proceedings of the 21st International Conference on Computational Linguistics and 44th Annual Meeting of the Association for Computational Linguistics*, pages 905–912, Sydney, Australia. Association for Computational Linguistics.

Yan Huang, Wei Wang, Liang Wang, and Tieniu Tan. 2013. Multi-task deep neural network for multi-label learning. In *2013 IEEE International Conference on Image Processing*, pages 2897–2900. IEEE.

Thorsten Joachims. 2006. Training linear svms in linear time. In *Proceedings of the 12th ACM SIGKDD International Conference on Knowledge Discovery and Data Mining*, KDD '06, pages 217–226, New York, NY, USA. ACM.

Xiaodong Liu, Pengcheng He, Weizhu Chen, and Jianfeng Gao. 2019. Multi-task deep neural networks for natural language understanding. *arXiv preprint arXiv:1901.11504*.

Yang Liu, Chengjie Sun, Lei Lin, and Xiaolong Wang. 2016. Learning natural language inference using bidirectional lstm model and inner-attention. *arXiv preprint arXiv:1605.09090*.

Christopher D. Manning, Mihai Surdeanu, John Bauer, Jenny Finkel, Steven J. Bethard, and David McClosky. 2014. The Stanford CoreNLP natural language processing toolkit. In *Association for Computational Linguistics (ACL) System Demonstrations*, pages 55–60.

Llew Mason, Jonathan Baxter, Peter L Bartlett, and Marcus R Frean. 2000. Boosting algorithms as gradient descent. In *Advances in neural information processing systems*, pages 512–518.

Anish Mishra and Pushpak Bhattacharyya. 2018. Deep learning techniques in textual entailment.

Bhaskar Mitra and Nick Craswell. 2017. Neural models for information retrieval. *arXiv preprint arXiv:1705.01509*.

Alexey Romanov and Chaitanya Shivade. 2018. Lessons from natural language inference in the clinical domain.

Luis Tari, Phan Tu, Barry Lumpkin, Robert Leaman, Graciela Gonzalez, and Chitta Baral. 2007. Passage relevancy through semantic relatedness.

Albert Tung and Eric Xu. 2017. Determining entailment of questions in the quora dataset.

Özlem Uzuner, Brett R South, Shuying Shen, and Scott L DuVall. 2011. 2010 i2b2/va challenge on concepts, assertions, and relations in clinical text. *Journal of the American Medical Informatics Association*, 18(5):552–556.

Yu Zhang and Qiang Yang. 2017. A Survey on Multi-Task Learning. *arXiv e-prints*, page arXiv:1707.08114.

Zhengyan Zhang, Xu Han, Zhiyuan Liu, Xin Jiang, Maosong Sun, and Qun Liu. 2019. ERNIE: Enhanced Language Representation with Informative Entities. *arXiv e-prints*, page arXiv:1905.07129.

A Appendix

Classifier	Architecture
Filtering	3824-2048:bn:a 2048-1024:bn:a 1024-512:bn:a 512-512:bn:a 512-256:bn:a 256-64:bn:a 64-1:a
Pairwise Ranking	7648-3824:bn:a 3824-2048:bn:a 2048-1024:bn:a 1024-512:bn:a 512-512:bn:a 512-256:bn:a 256-64:bn:a 64-1:a

Table 7: Classifier Specifications: 'X1-X2' - denotes a linear layer with X1 input features and X2 output features. 'bn - with batch normalization, 'a': denotes activation, ':' - separates two layers. Activation used everywhere is ReLU except for the output layer where sigmoid is used.

Convolution Encoder Layers
Input : c :768
c:768, k:(1,1), s:(1,1),p:(1,1), bn
c:512, k:(3,3), s:(1,1), p:(2,2), bn
c:512, k:(3,3), s:(2,2), p:(1,1)
c:256, k:(2,2), s:(1,1), p:(1,1), bn
c:,256 k:(3,3), s:(1,1), p:(2,2),
Quadrant Pooling

Table 8: Convolution Encoder Specification. 'c': number of filters, 'k': kernel size, 's': stride size, 'p': padding size, 'bn': with batch normalization. The sizes are in order (height, width). ReLU activation function is used after each layer except for the input and output layer. Quadrant Pooling is described in Section 4.

DoubleTransfer at MEDIQA 2019:
Multi-Source Transfer Learning for Natural Language Understanding in the Medical Domain

Yichong Xu[1], Xiaodong Liu[2], Chunyuan Li[2], Hoifung Poon[2] and Jianfeng Gao[2]

[1] Carnegie Mellon University

[2] Microsoft Research

`yichongx@cs.cmu.edu`

`{xiaodl, Chunyuan.Li, hoifung, jfgao}@microsoft.com`

Abstract

This paper describes our competing system to enter the MEDIQA-2019 competition. We use a multi-source transfer learning approach to transfer the knowledge from MT-DNN (Liu et al., 2019b) and SciBERT (Beltagy et al., 2019) to natural language understanding tasks in the medical domain. For transfer learning fine-tuning, we use multi-task learning on NLI, RQE and QA tasks on general and medical domains to improve performance. The proposed methods are proved effective for natural language understanding in the medical domain, and we rank the first place on the QA task.

1 Background

The MEDIQA 2019 shared tasks (Ben Abacha et al., 2019) aim to improve the current state-of-the-art systems for textual inference, question entailment and question answering in the medical domain. This ACL-BioNLP 2019 shared task is motivated by a need to develop relevant methods, techniques and gold standards for inference and entailment in the medical domain and their application to improve domain-specific information retrieval and question answering systems. The shared task consists of three parts: i) natural language inference (NLI) on MedNLI, ii) Recognizing Question Entailment (RQE), and iii) Question Answering (QA).

Recent advancement in NLP such as BERT (Devlin et al., 2018) has facilitated great improvements in many Natural Language Understanding (NLU) tasks (Liu et al., 2019b). BERT first trains a language model on an unsupervised large-scale corpus, and then the pretrained model is fine-tuned to adapt to downstream NLU tasks. This fine-tuning process can be seen as a form of transfer learning, where BERT learns knowledge from the large-scale corpus and transfer it to downstream tasks.

We investigate NLU in the medical (scientific) domain. From BERT, we need to adapt to i) The change from general domain corpus to scientific language; ii) The change from low-level language model tasks to complex NLU tasks. Although there is limited training data in NLU in the medical domain, we fortunately have pre-trained models from two intermediate steps:

- General NLU embeddings: We use MT-DNN (Liu et al., 2019b) trained on GLUE benchmark(Wang et al., 2019). MT-DNN is trained on 10 tasks including NLI, question equivalence, and machine comprehension. These tasks correspond well to the target MEDIQA tasks but in different domains.

- Scientific embeddings: We use SciBERT (Beltagy et al., 2019), which is a BERT model, but trained on SemanticScholar scientific papers. Although SciBERT obtained state-of-the-art results on several single-sentence tasks, it lacks knowledge from other NLU tasks such as GLUE.

In this paper, we investigate different methods to combine and transfer the knowledge from the two different sources and illustrate our results on the MEDIQA shared task. We name our method as DoubleTransfer, since it transfers knowledge from two different sources. Our method is based on fine-tuning both MT-DNN and SciBERT using multi-task learning, which has demonstrated the efficiency of knowledge transformation (Caruana, 1997; Liu et al., 2015; Xu et al., 2018; Liu et al., 2019b), and integrating models from both domains with ensembles.

Related Works. Transfer learning has been widely used in training models in the medical do-

Proceedings of the BioNLP 2019 workshop, pages 399–405
Florence, Italy, August 1, 2019. ©2019 Association for Computational Linguistics

Algorithm 1 Multi-task Fine-tuning with External Datasets

Require: In-domain datasets $\mathcal{D}_1, ..., \mathcal{D}_{K_1}$, External domain datasets $\mathcal{D}_{K_1+1}, ..., \mathcal{D}_{K_2}$, max_epoch, mixture ratio α

1: Initialize the model $\mathcal{M}$
2: **for** epoch$= 1, 2, ...,$ max_epoch **do**
3:     Divide each dataset $\mathcal{D}_k$ into N_k mini-batches $\mathcal{D}_k = \{b_1^k, ..., b_{N_k}^k\}, 1 \leq k \leq K_2$
4:     $S \leftarrow \mathcal{D}_1 \cup \mathcal{D}_2 \cup \cdots \cup \mathcal{D}_{K_1}$
5:     $N \leftarrow N_1 + N_2 + \cdots + N_{K_1}$
6:     Randomly pick $\lfloor \alpha N \rfloor$ mini-batches from $\bigcup_{k=K_1}^{K_2} \mathcal{D}_k$ and add to S
7:     Assign mini-batches in S in a random order to obtain a sequence $B = (b_1, ..., b_L)$, where $L = N + \lfloor \alpha N \rfloor$
8:     **for** each mini-batch $b \in B$ **do**
9:         Perform gradient update on $\mathcal{M}$ with loss $l(b) = \sum_{(s_1, s_2) \in b} l(s_1, s_2)$
10:     **end for**
11:     Evaluate development set performance on $\mathcal{D}_1, ..., \mathcal{D}_{K_1}$
12: **end for**
Ensure: Model with best evaluation performance

---

main. For example, Romanov and Shivade (2018) leveraged the knowledge learned from SNLI to MedNLI; a transfer from general domain NLI to medical domain NLI. They also employed word embeddings trained on MIMIC-III medical notes, which can be seen as a language model in the scientific domain. SciBERT (Beltagy et al., 2019) studies transferring knowledge from SciBERT pretrained model to single-sentence classification tasks. Our problem is unique because of the prohibitive cost to train BERT: Either BERT or SciBERT requires a very long time to train, so we only explore how to combine the existing embeddings from SciBERT or MT-DNN. Transfer learning is also widely used in other tasks of NLP, such as machine translation (Bahdanau et al., 2014) and machine reading comprehension (Xu et al., 2018).

2 Methods

We propose a multi-task learning method for the medical domain data. It employs datasets/tasks from both medical domain and external domains, and leverage the pre-trained model such as MT-DNN and SciBERT for fine-tuning. An overview of the proposed method is illustrated in Figure 1. To further improve the performance, we propose to ensemble models trained from different initialization in the evaluation stage. Below we detail our methods for fine-tuning and ensembles.

2.1 Fine-tuning details

Algorithm. We fine-tune the two types of pretrained models on all the three tasks using multi-task learning. As suggested by MEDIQA paper, we also fine-tune our model on MedQuAD (Abacha and Demner-Fushman, 2019), a medical QA dataset. We will provide details for fine-tuning on these datasets in Section 2.3. We additionally regularize the model by also training on MNLI (Williams et al., 2018). To prevent the negative transfer from MNLI, we put a larger weight on MEDIQA data by sampling MNLI data with less probability. Our algorithm is presented in Algorithm 1 and illustrated as Figure 1, which is a mixture ratio method for multi-task learning inspired by Xu et al. (2018). We start with in-domain datasets $\mathcal{D}_1, ...\mathcal{D}_{K_1}$ (i.e., the MEDIQA tasks, $K_1 = 3$) and external datasets $\mathcal{D}_{K_1+1}, ..., \mathcal{D}_{K_2}$ (in this case MNLI). We cast all the training samples as sentence pairs $(s_1, s_2) \in \mathcal{D}_k, k = 1, 2, ..., K_2$. In each epoch of training, we use all mini-batches from in-domain data, while only a small proportion (controlled by α) of mini-batches from external datasets are used to train the model. In our experiments, the mixture ratio α is set to 0.5. We use MedNLI, RQE, QA, and MedQuAD in medical domain as in-domain data and MNLI as external data. For MedNLI, we additionally find that using MedNLI as in-domain data and RQE, QA, MedQuAD as external data can also help boost performance. We use models trained using both setups of external data for en-

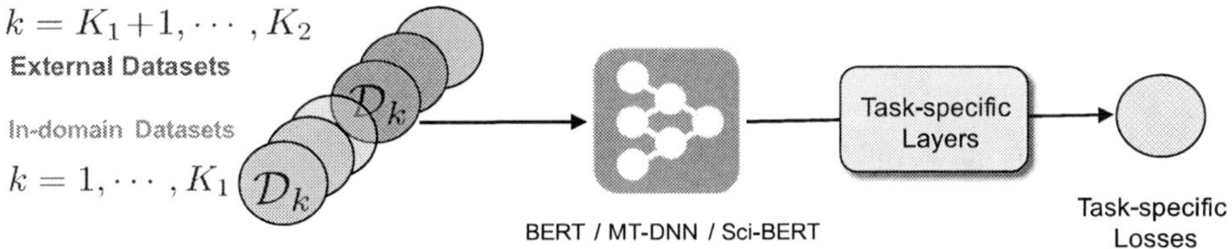

Figure 1: Illustration of the proposed multi-source multi-task learning method.

sembling.

Pre-trained Models. We use three different types of initialization as the starting point for fine-tuning: i) the uncased MT-DNN large model from Liu et al. (2019b), ii) the cased knowledge-distilled MT-DNN model from Liu et al. (2019a), and iii) the uncased SciBERT model (Beltagy et al., 2019). We add a simple softmax layer (or linear layer for QA and MedQuAD tasks) atop BERT as the answer module for fine-tuning. For initialization in step 1 in Algorithm 1, we initialize all BERT weights with the pretrained weights, and randomly initialize the answer layers. After multi-task fine-tuning, the joint model is further fine-tuned on each specific task to get better performance. We detail the training loss and fine-tuning process for each task in Section 2.3.

Objectives. MedNLI and RQE are binary classification tasks, and we use a cross-entropy loss. Specifically, for a sentence pair X we compute the loss

$$\mathcal{L}(X) = -\sum_c \mathbb{1}(X, c) \log(P_r(c|X)),$$

where c iterates over all possible classes, $\mathbb{1}(X, c)$ is the binary indicator (0 or 1) if class label c is the correct classification for X, and $P_r(c|X)$ is the model prediction for probability of class c for sample X.

We formulate QA and MedQuAD as regression tasks, and thus a MSE loss is used. Specifically, for a question-answer pair (Q, A) we compute the MSE loss as

$$\mathcal{L}(Q, A) = (y - \texttt{score}(Q, A))^2,$$

where y is the target relevance score for pair (Q, A), and $\texttt{score}(Q, A)$ is the model prediction for the same pair.

2.2 Model Ensembles

After fine-tuning, we ensemble models trained from MT-DNN and SciBERT, and using different setups of in-domain and external datasets. The traditional methods typically fuse models by averaging the prediction probability of different models. For our setting, the in-domain data is very limited and it tends to overfit; this means the predictions can be arbitrarily close to 1, favoring to more over-fitting models. To prevent over-fitting, we ensemble the models by using a majority vote on their predictions, and resolving ties using sum of prediction probabilities. Suppose we have M models, and the m-th model predicts the answer $\hat{p}_m$ for a specific question. For the classification task (MedNLI and RQE), we have $\hat{p}_m \in \mathbb{R}^C$, where C is the number of categories. Let $\hat{y}_m = \arg\max_i \hat{p}_m^{(i)}$ be the prediction of model m, where $\hat{p}_m^{(i)}$ is the i-th dimension of $\hat{p}_m$. The final prediction is chosen as

$$\hat{y}_{\text{ensemble}} = \arg\max_{y \in \text{maj}(\{\hat{y}_m\}_{m=1}^M)} \sum_{m=1}^M \hat{p}_m^{(y)}.$$

In other words, we first obtain the majority of predictions by computing the majority $\text{maj}(\{\hat{y}_m\}_{m=1}^M)$, and resolve the ties by computing the sum of prediction probabilities $\sum_{m=1}^M \hat{p}_m^{(y)}$. For QA tasks (QA and MedQuAD), the task is cast as a regression problem, where a positive number means correct answer, and negative otherwise. We have $\hat{p}_m \in \mathbb{R}$. We first compute the average score $\hat{p}_{\text{ensem}} = \frac{1}{M} \sum_{m=1}^M \hat{p}_m$. We also compute the prediction as $\hat{y}_m = I(\hat{p}_m \geq 0)$, where I is the indicator function. We compute the ensemble prediction through a similar majority vote as the classification case:

$$\hat{y}_{\text{ensem}} = \begin{cases} 1, & \text{if } \sum_{m=1}^M \hat{y}_m > M/2 \\ 0, & \text{if } \sum_{m=1}^M \hat{y}_m < M/2 \\ I(\hat{p}_{\text{ensem}} > 0), & \text{otherwise.} \end{cases}$$

To be precise, we predict the majority if a tie does not exist, or the sign of $\hat{p}_{\text{ensem}}$ otherwise. The final ranking of answers is carried out by first rank the (predicted) positive answers, and then the (predicted) negative answers.

2.3 Dataset-Specific Details

MedNLI: Since the MEDIQA shared task uses a different test set than the original MedNLI dataset, we merge the original MedNLI development set into the training set and use evaluation performance on the original MedNLI test set. Furthermore, MedNLI and MNLI are the same NLI tasks, thus, we shared final-layer classifiers for these two tasks. For MedNLI, we find that each consecutive 3 samples in all the training set contain the same premise with different hypothesizes, and contains exactly 1 entail, 1 neutral and 1 contradiction. To the end, in our prediction, we constrain the three predictions to be one of each kind, and use the most likely prediction from the model prediction probabilities.

RQE: We use the clinical question as the premise and question from FAQ as the hypothesis. We find that the test data distribution is quite different from the train data distribution. To mitigate this effect, we randomly shuffle half of the evaluation data into the training set and evaluate on the remaining half.

QA: We use the answer as the premise and the question as the hypothesis. The QA task is cast as both a ranking task and a classification task. Each question is associated with a relevance score in $\{1, 2, 3, 4\}$, and an additional rank over all the answers for a specific question is given. We use a modified score to incorporate both information: suppose there are m questions with relevance score $s \in \{1, 2, 3, 4\}$. Then the i-th most relevant answer in these m questions get modified score $s - \frac{i-1}{m}$. In this way the scores are uniformly distributed in $(s - 1, s]$. We shift all scores by -2 so that a positive score leads to a correct answer and vice versa. We also tried pairwise losses to incorporate the ranking but did not find it to boost the performance very much.

We find that the development set distribution is inconsistent with test data - the training and test set consist of both LiveQAMed and Alexa questions, whereas the development set seems to only contain LiveQAMed questions. We shuffle the training and development set to make them similar: We use the last 25 questions in original development set (LiveQAMed questions) and the last 25 Alexa questions (from the original training set) as our development set, and use the remaining questions as our training set. This results in 1,504 training pairs and 431 validation pairs. Due to the limited size of the QA dataset, we use cross-validation that divides all pairs into 5 slices and train 5 models by using each slice as a validation set. We train MT-DNN and SciBERT on both these 5 setups and obtain 10 models, and ensemble all the 10 models obtained.

MedQuAD: We use 10,109 questions from MedQuAD because the remaining questions are not available due to copyright issues. The original MedQuAD dataset only contains positive question pairs. We add negative samples to the dataset by randomly sampling an answer from the same web page. For each positive QA pair, we add two negative samples. The resulting 30,327 pairs are randomly divided into 27,391 training pairs and 2,936 evaluation pairs. Then we use the same method as QA to train MedQuAD; we also share the same answer module between QA and MedQuAD.

2.4 Implementation and Hyperparameters

We implement our method using PyTorch[1] and Pytorch-pretrained-BERT[2], as an extension to MT-DNN[3]. We also use the pytorch-compatible SciBERT pretrained model provided by AllenNLP[4]. Each training example is pruned to at most 384 tokens for MT-DNN models and 512 tokens for SciBERT models. We use a batch size of 16 for MT-DNN, and 40 for SciBERT. For fine-tuning, we train the models for 20 epochs using a learning rate of 5×10^{-5}. After that, we further fine-tune the model from the best multi-task model for 6 epochs for each dataset, using a learning rate of 5×10^{-6}. We ensemble all models with an accuracy larger than 87.7 for MedNLI, 83.5 for shuffled RQE, and 83.0 for QA. We ensemble 4 models for MedNLI, 14 models for RQE. For QA, we ensemble 10 models from cross-validation and 7 models using the normal training-validation approach.

3 Results

In this section, we provide the leaderboard performance and conduct an analysis of the effect of ensemble models from different sources.

---

[1] https://pytorch.org/
[2] https://github.com/huggingface/
pytorch-pretrained-BERT
[3] https://github.com/namisan/mt-dnn
[4] https://github.com/allenai/scibert

Model	Dev Set	Test Set
WTMed	-	**98.0**
PANLP	-	96.6
Ours	**91.7**	**93.8**
Sieg	-	91.1
SOTA	76.6	-

Table 1: The leaderboard for MedNLI task (link). Scores are accuracy(%). Our method ranked the 3rd on the leaderboard. Previous SOTA method was from (Romanov and Shivade, 2018), on the original MedNLI test set (used as dev set here).

Model	Dev Set	Test Set
PANLP	-	**74.9**
Sieg	-	70.6
IIT-KGP	-	68.4
Ours	**91.7**	**66.2**

Table 2: The leaderboard for RQE task (link). Scores are accuracy(%). Our method ranked the 7th on the leaderboard.

3.1 Test Set Performance and LeaderBoards

The results for MedNLI dataset is summarized in Table 1. Our method ends up the 3rd place on the leaderboard and substantially improving upon previous state-of-the-art (SOTA) methods.

The results for RQE dataset is summarized in Table 2. Our method ends up the 7th place on the leaderboard. Our method has a very large discrepancy between the dev set performance and test set performance. We think this is because the test set is quite different from dev set, and that the dev set is very small and easy to overfit to.

The results for QA dataset is summarized in Table 3. Our method reaches the first place on the leaderboard based on accuracy and precision score and 3rd-highest MRR. We note that the Spearman score is not consistent with other scores in the leaderboard; actually, the Spearman score is computed just based on the predicted positive answers, and a method can get very high Spearman score by never predict positive labels.

3.2 Ensembles from Different Sources

We compare the effect of ensembling from different sources in Table 4. We train 6 different models with different randomizations, with initializations from MT-DNN (#1,#2,#3) and SciBERT (#4, #5,#6) respectively. If we ensemble

Model	Acc	Spearman	Precision	MRR
Ours	**78.0**	0.238	**81.91**	0.937
PANLP	77.7	0.180	78.1	0.938
Pentagon	76.5	0.338	77.7	**0.962**
DUT-BIM	74.5	0.106	74.7	0.906

Table 3: The leaderboard for QA task (link). Our method ranked #1 on the leaderboard in terms of Acc (accuracy). The Spearman score is not consistent with other scores in the leaderboard.

models with the same MT-DNN architecture, the resulting model only has around 1.5% improvement in accuracy, compared to the numerical average of the ensemble model accuracies (#1+#2+#3 and #4+#5+#6 in Table 4). On the other hand, if we ensemble three models from different sources (#1+#2+#5 and #1+#5+#6 in Table 4), the resulting model gains more than 3% in accuracy compared to the numerical average. This shows that ensembling from different sources has a great advantage than ensembling from single-source models.

Model	Avg. Acc	Esm. Acc
Single Model		
#1, MT-DNN	-	88.61
#2, MT-DNN	-	88.33
#3, MT-DNN	-	87.84
#4, SciBERT	-	88.19
#5, SciBERT	-	87.70
#6, SciBERT	-	87.21
Ensemble Model		
#1+#2+#3, MT-DNN	88.26	89.7
#4+#5+#6, SciBERT	87.70	89.2
#1+#2+#5, MultiSource	88.21	91.6
#1+#5+#6, MultiSource	87.84	90.4
#1-6, MultiSource	87.98	91.3

Table 4: Comparison of ensembles from different sources. Avg.Acc stands for average accuracy, the numerical average of each individual model's accuracy. Esm.Acc stands for ensemble accuracy, the accuracy of the resulting ensemble model. For ensembles, MT-DNN means all the three models are from MT-DNN, and similarly for SciBERT; MultiSource denotes the ensemble models come from two different sources.

3.3 Single-Model Performance

For completeness, we report the single-model performance on the MedNLI development set under various multi-task learning setups and initializations in Table 5. (1) The *Naïve* approach denotes only MedNLI, RQE, QA, MedQuAD is considered as in-domain data in Algorithm 1 without any external data; (2) The *Ratio* approach denotes that we consider MedNLI as in-domain data, and RQE, QA, MedQuAD as external data in Algorithm 1; (3) The *Ratio+MNLI* approach denotes that we consider MedNLI, RQE, QA, MedQuAD as in-domain data and MNLI as external data in Algorithm 1. Note that MNLI is much larger than the medical datasets, so if we use RQE, QA, MedQuAD, MNLI as external data, the performance is very similar to the third setting. We did not conduct experiments on single-dataset settings, as previous works have suggested that multi-task learning can obtain much better results than single-task models (Liu et al., 2019b; Xu et al., 2018).

Overall, the best results are achieved via using SciBERT as the pre-trained model, and multi-task learning with MNLI. The models trained by mixing in-domain data (the second setup) is also competitive. We therefore use models from both setups for ensemble.

Init Model	Naïve	Ratio	Ratio+MNLI
MT-DNN	86.9	86.2	87.8
MT-DNN-KD	87.5	88.2	**88.8**
SciBERT	87.1	87.0	**89.4**

Table 5: Single model performance on MedNLI developlment data. *Naïve* means simply integrating all medical-domain data; *Ratio* means using MedNLI as in-domain data and other medical domain data as external data; *Ratio+MNLI* means using medical domain data as in-domain and MNLI as external.

4   Conclusion

We present new methods for multi-source transfer learning for the medical domain. Our results show that ensembles from different sources can improve model performance much more greatly than ensembles from a single source. Our methods are proved effective in the MEDIQA2019 shared task.

References

Asma Ben Abacha and Dina Demner-Fushman. 2019. A question-entailment approach to question answering. *arXiv preprint arXiv:1901.08079.*

Dzmitry Bahdanau, Kyunghyun Cho, and Yoshua Bengio. 2014. Neural machine translation by jointly learning to align and translate. *arXiv preprint arXiv:1409.0473.*

Iz Beltagy, Arman Cohan, and Kyle Lo. 2019. Scibert: Pretrained contextualized embeddings for scientific text. *arXiv preprint arXiv:1903.10676.*

Asma Ben Abacha, Chaitanya Shivade, and Dina Demner-Fushman. 2019. Overview of the mediqa 2019 shared task on textual inference, question entailment and question answering. In *Proceedings of the BioNLP 2019 workshop, Florence, Italy, August 1, 2019.* Association for Computational Linguistics.

Rich Caruana. 1997. Multitask learning. *Machine learning*, 28(1):41–75.

Jacob Devlin, Ming-Wei Chang, Kenton Lee, and Kristina Toutanova. 2018. Bert: Pre-training of deep bidirectional transformers for language understanding. *arXiv preprint arXiv:1810.04805.*

Xiaodong Liu, Jianfeng Gao, Xiaodong He, Li Deng, Kevin Duh, and Ye-Yi Wang. 2015. Representation learning using multi-task deep neural networks for semantic classification and information retrieval. In *Proceedings of the 2015 Conference of the North American Chapter of the Association for Computational Linguistics: Human Language Technologies*, pages 912–921.

Xiaodong Liu, Pengcheng He, Weizhu Chen, and Jianfeng Gao. 2019a. Improving multi-task deep neural networks via knowledge distillation for natural language understanding.

Xiaodong Liu, Pengcheng He, Weizhu Chen, and Jianfeng Gao. 2019b. Multi-task deep neural networks for natural language understanding. *arXiv preprint arXiv:1901.11504.*

Alexey Romanov and Chaitanya Shivade. 2018. Lessons from natural language inference in the clinical domain. In *Proceedings of the 2018 Conference on Empirical Methods in Natural Language Processing*, pages 1586–1596.

Alex Wang, Amanpreet Singh, Julian Michael, Felix Hill, Omer Levy, and Samuel R. Bowman. 2019. Glue: A multi-task benchmark and analysis platform for natural language understanding. In the Proceedings of ICLR.

Adina Williams, Nikita Nangia, and Samuel Bowman. 2018. A broad-coverage challenge corpus for sentence understanding through inference. In *Proceedings of the 2018 Conference of the North American Chapter of the Association for Computational Linguistics: Human Language Technologies, Volume 1 (Long Papers)*, pages 1112–1122. Association for Computational Linguistics.

Yichong Xu, Xiaodong Liu, Yelong Shen, Jingjing Liu, and Jianfeng Gao. 2018. Multi-task learning for machine reading comprehension. *arXiv preprint arXiv:1809.06963*.

Surf at MEDIQA 2019: Improving Performance of Natural Language Inference in the Clinical Domain by Adopting Pre-trained Language Model

Jiin Nam
AI Core Team
Samsung Research
Seoul, Korea
jiin.nam@samsung.com

Seunghyun Yoon
Dept. ECE
Seoul National University
Seoul, Korea
mysmilesh@snu.ac.kr

Kyomin Jung
Dept. ECE
Seoul National University
Seoul, Korea
kjung@snu.ac.kr

Abstract

While deep learning techniques have shown promising results in many natural language processing (NLP) tasks, it has not been widely applied to the clinical domain. The lack of large datasets and the pervasive use of domain-specific language (i.e. abbreviations and acronyms) in the clinical domain causes slower progress in NLP tasks than that of the general NLP tasks. To fill this gap, we employ word/subword-level based models that adopt large-scale data-driven methods such as pre-trained language models and transfer learning in analyzing text for the clinical domain. Empirical results demonstrate the superiority of the proposed methods by achieving 90.6% accuracy in medical domain natural language inference task. Furthermore, we inspect the independent strengths of the proposed approaches in quantitative and qualitative manners. This analysis will help researchers to select necessary components in building models for the medical domain.

1 Introduction

Natural language processing (NLP) has broadened its applications rapidly in recent years such as question answering, neural machine translation, natural language inference, and other language-related tasks. Unlike other tasks in NLP area, the lack of large labeled datasets and restricted access in the clinical domain have discouraged active participation of NLP researchers for this domain (Romanov and Shivade, 2018). Furthermore, the pervasive use of abbreviations and acronyms in the clinical domain causes the difficulty of text normalization and makes the related tasks more difficult (Pakhomov, 2002).

In building NLP models, a word embedding layer that transforms a sequence of tokens in text into a vector representation is considered as one of the fundamental components. In recent studies, it has been shown that the pre-trained language models by using a huge diversity of corpus (i.e. BERT (Devlin et al., 2018) and ELMo (Peters et al., 2018)) generate deep contextualized word representations. These methods have shown to be very effective for improving the performance of a wide range of NLP tasks by enabling better text understanding and have become a crucial part of the tasks since they have published.

To stimulate the research in the clinical domain, researchers have further investigated to transform the pre-trained language models from general purpose version into the medical domain-specific version. Lee et al. (2019) propose BioBERT that utilizes large-scale bio-medical corpora, PubMed abstracts (PubMed) and PubMed Central full-text articles (PMC), to obtain a medical domain specific language representation through fine-tuning the BERT. Similarity, a PubMed-ELMo[1], trained with medical domain corpus, is released as one of the contributed ELMo models for medical domain researchers. However, these models are not yet fully explored in medical domain tasks.

Besides these general efforts in building better word representations, Romanov and Shivade (2018) introduce a large and publicly available natural language inference (NLI) dataset, called MedNLI, for the medical domain (see table 1). Considering the expensive annotation cost of medical text due to the sparsity of the clinical-domain experts, the medical NLI task plays an import role in boosting existing datasets for medical question answering systems by retrieving similar questions that are already answered by human experts. Along with this effort, ACL-BioNLP 2019 committee announced a shared task, NLI for the medical domain, motivated by a need to develop

---

[1]https://allennlp.org/elmo

Proceedings of the BioNLP 2019 workshop, pages 406–414
Florence, Italy, August 1, 2019. ©2019 Association for Computational Linguistics

#	Premise	Hypothesis	Label
1	She was treated with Magnesium Sulfate, Labetalol, Hydralazine and bedrest as well as betamethasone.	The patient is pregnant.	entailment
2	Denied headache, sinus tenderness, rhinorrhea or congestion.	Patient has history of dysphagia	contradiction
3	Type II Diabetes Mellitus 3.	The patient does not require insulin.	neutral
4	Ruled in for NSTEMI with troponin 0.11.	The patient has myocardial ischemia.	entailment
5	Her CXR was clear and it did not appear she had an infection.	Chest x-ray showed infiltrates	contradiction
6	CHF, EF 55% 6.	complains of shortness of breath	neutral

Table 1: Examples from the development set of MedNLI.

relevant methods, techniques and gold standards for inference and entailment (Ben Abacha et al., 2019). The newly released dataset is larger in size than that of any other previous medical domain NLI dataset, however, it is still not enough to train complicated neural network based models.

To fill this gap, we propose a combination approach of NLP models and machine learning methods to tackle the medical domain NLI task. Our contributions are summarized as follows:

- We adopt the pre-trained language models (BioBERT, PubMed-ELMo) to overcome the shortage of training data which is a common problem in the clinical domain.

- We apply the transfer learning method with two general domain NLI datasets and show that a source task in a domain can benefit learning a target task in a different domain.

- We show the independent strengths of the proposed approaches in quantitative and qualitative manners. This analysis will help researchers to select necessary components in building models for the clinical domain.

2 Related Work

Researchers have investigated NLI tasks. Most of the works employed a recurrent neural network to encode each pair of sentences and to compute the similarity between them (Conneau et al., 2017; Subramanian et al., 2018). Recently, Liu et al. (2019) proposed multi-task learning for natural language tasks and achieved the best results on NLI tasks. In the medical domain, Romanov and Shivade (2018) adopted the ESIM (Chen et al., 2017) model to the MedNLI task. The ESIM model employs two bidirectional LSTM to encode each sentence independently and to calculate a matching score between the sentences by using alignment and pooling methods. They also applied transfer learning with SNLI (Bowman et al., 2015) and MNLI (Williams et al., 2018) datasets to improve model performance in the MedNLI task.

Recently, pre-trained language models were proposed (Peters et al., 2018; Devlin et al., 2018). The multi-task benchmark for natural language understanding (Wang et al., 2018) has shown that these pre-trained language models brought additional performance gain by providing deep contextualized word representations. Upon this success, researchers further extended previous pre-trained language models to medical domain-specific versions such as BioBERT (Lee et al., 2019) and PubMed-ELMo (Peters, 2018).

However, none of these researches directly applied the pre-trained language models of the medical domain to the MedNLI task.

3 Dataset and Problem

MedNLI (Romanov and Shivade, 2018), a large publicly available and expert annotated dataset, has been recently published for the MEDIQA 2019 shared task. This dataset comprises of tuples $<P, H, Y>$ where: P and H are a clinical sentence pair, (premise and hypothesis, respectively); Y indicates whether a given hypothesis can be inferred from a given premise. In particular, Y is categorized as one of three classes: *"entailment"*, *"contradiction"*, and *"neutral"*. Table 1 shows examples of the MedNLI dataset. A total of 14,049 pairs, (11,232, 1,395, 1,422 for training, development, and test, respectively), are created based on the past medical history section of MIMIC-

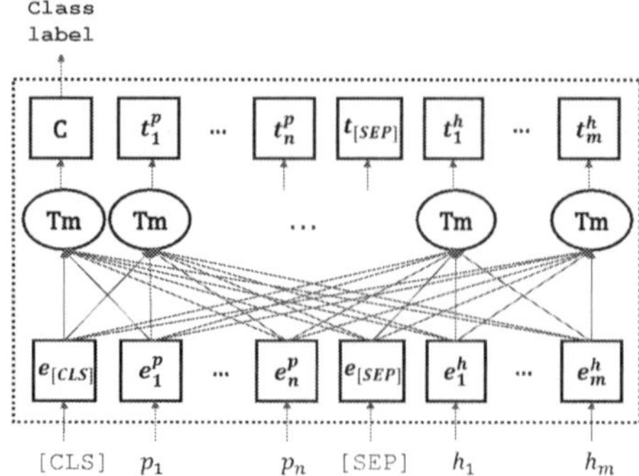

Figure 1: Overview of the BERT model.

III (Johnson et al., 2016).

In this research, we are interested in building a model that classifies the given sentence pair into the corresponding category. First, we consider a **point-wise** approach that classifies each pair of data independently into one of the three classes. Next, we re-organize the dataset into the set of a list that contains one of each class sentence pair. Then we apply **list-wise** classification that classifies three sentence pair into each "*entailment*", "*contradiction*", and "*neutral*" class exclusively.

4   Methods

As the size of the MedNLI dataset is limited to train the whole weight parameters in complicated neural network based models, we first choose a BERT (Devlin et al., 2018) based model that provides pre-trained model parameters from a large corpus. To further explore the performance of modern neural network based models, we extend the compare aggregate model (Wang and Jiang, 2016) with another type of pre-trained word-level embedding, ELMo (Peters et al., 2018). Additionally, we apply transfer learning from similar NLI tasks (Bowman et al., 2015; Williams et al., 2018), and we try to expand medical abbreviations to deal with the general problem in the medical domain.

4.1   BioBERT

As a baseline model, we choose BioBERT (Lee et al., 2019) since MedNLI is a bio-domain specific NLI task. It shows strength in understanding medical domain text as it is fine-tuned with bio datasets such as PubMed and PMC. The BioBERT adopts the same architecture as BERT, as shown in figure 1, that takes WordPiece embeddings from textual input and generates a language representation using a transformer model (Vaswani et al.,

2017).

WordPiece embedding: BioBERT utilizes the WordPiece dictionary of BERT generated from general domain corpus. Each premise **P** and hypothesis **H** turn into sub-word embeddings, $\mathbf{E}^P \in \mathbb{R}^{n \times d_e}$ and $\mathbf{E}^H \in \mathbb{R}^{m \times d_e}$, using the dictionary where d_e is a dimension of sub-word embedding vectors and n and m are the length of the sequences of **P** and **H**, respectively.

$$\mathbf{E}^P = \text{WordPiece\_embedding}(\mathbf{P}),$$
$$\mathbf{E}^H = \text{WordPiece\_embedding}(\mathbf{H}). \tag{1}$$

BioBERT adds the special classification embedding "[CLS]" as the first token of every sentence and separates $\mathbf{E}^P$ and $\mathbf{E}^H$ with a special token "[SEP]". The final input representation fed to transformer blocks is the sum of the token embeddings ($\mathbf{E}^T$), position embeddings ($\mathbf{E}^{Po}$), and segmentation embeddings ($\mathbf{E}^S$) as follow.

$$\mathbf{E} = \mathbf{E}^T + \mathbf{E}^{Po} + \mathbf{E}^S,$$
$$\mathbf{E}^T = [\mathbf{E}_{[CLS]}, \ \mathbf{E}^P, \ \mathbf{E}_{[SEP]}, \ \mathbf{E}^H]. \tag{2}$$

Transformer encoder: The transformer encoder consists of multiple transformer blocks. Each block uses Multi-Head Attention (MHA) generating h different attentions. All the attention heads calculated with different weights are concatenated. A linear layer with a weight matrix $\mathbf{W}^H \in \mathbb{R}^{(h \times d_v) \times d_e}$ computes the MHA ($\mathbb{R}^{\text{input\_length} \times d_e}$) with the concatenated attention heads as follows:

$$\text{MHA}(\mathbf{Q}, \mathbf{K}, \mathbf{V}) = (\text{concat}\{hd_1, .., hd_n\})\mathbf{W}^H,$$
$$hd_i = \text{attn}(\mathbf{Q}_i, \mathbf{K}_i, \mathbf{V}_i),$$
$$\text{attn}(\mathbf{Q}_i, \mathbf{K}_i, \mathbf{V}_i) = \text{softmax}(\frac{\mathbf{Q}_i \mathbf{K}_i^T}{\sqrt{d_k}})\mathbf{V}_i, \tag{3}$$

where $Q = [Q_1, ..., Q_h], Q_i \in \mathbb{R}^{n \times \frac{d_e}{h}}$,
$K = [K_1, ..., K_h], K_i \in \mathbb{R}^{n \times \frac{d_e}{h}}$,
$V = [V_1, ..., V_h], V_i \in \mathbb{R}^{n \times \frac{d_e}{h}}$.

4.2   Compare Aggregate (CompAggr)

As we focus on the task that classifies the relationship between two sentences **P** and **H** (premise and hypothesis) into one of three classes (entailment, contradiction, or neutral), we adopt the compare aggregate (CompAggr) model that is widely used for a text sequence matching task (Wang

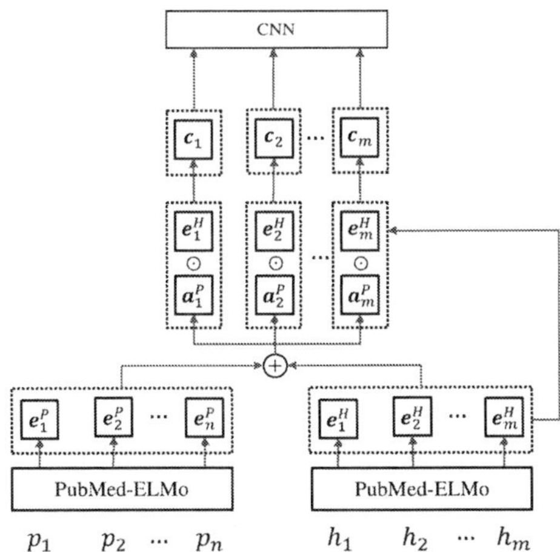

Figure 2: Overview of the CompAggr model.

and Jiang, 2016). In addition to the CompAggr model, we adopt PubMed-ELMo, that is trained with medical domain corpus and released as one of contributed ELMo models (Peters et al., 2018; Peters, 2018), to alleviate the lack of training corpus for the shared task. The final model consists of four parts which are shown in figure 2.

Word representation: Premise $\mathbf{P} \in \mathbb{R}^{d \times n}$ and hypothesis $\mathbf{H} \in \mathbb{R}^{d \times m}$, (where d is a dimensionality of word embedding and n, m are length of the sequences in $\mathbf{P}$ and $\mathbf{H}$, receptively), are processed to capture contextual information within the sentence by using pretrained PubMed ELMo (Peters et al., 2018) as follows:

$$\mathbf{E}^P = \text{PubMed-ELMo}(\mathbf{P}),$$
$$\mathbf{E}^H = \text{PubMed-ELMo}(\mathbf{H}). \tag{4}$$

Attention: The soft aliment of the $\mathbf{E}^P$ and $\mathbf{E}^H$ are computed by applying an attention mechanism over the column vector in $\mathbf{E}^P$ for each column vector in $\mathbf{E}^H$. Using an attention weight α_i for each column vector in $\mathbf{E}^P$, we obtain a corresponding vector $\mathbf{A}^P \in \mathbb{R}^{d \times m}$ from weighted sum of the column vectors of $\mathbf{E}^P$.

$$\mathbf{A}^P = \mathbf{E}^P \cdot \text{softmax}((\mathbf{WE}^P)^\mathsf{T} \mathbf{E}^H), \tag{5}$$

where $\mathbf{W}$ is a learned model parameter matrix.

Comparison: We use an element-wise multiplication as a comparison function to combine each pair of $\mathbf{A}^P$ and $\mathbf{E}^H$ into a vector $\mathbf{C} \in \mathbb{R}^{d \times m}$.

Aggregation: Finally Kim (2014)'s CNN with n-types of filters is applied to aggregate all the information followed by another fully connected layer

to classify the $\mathbf{P}$ and $\mathbf{H}$ pair as follow:

$$\mathbf{R} = \text{CNN}(\mathbf{C}), \quad (R \in \mathbb{R}^{nd})$$
$$\hat{y}_c = \text{softmax}((\mathbf{R})^\mathsf{T} \mathbf{W} + \mathbf{b}), \tag{6}$$

where $\hat{y}_c$ is the predicted probability distribution for the target classes and the $\mathbf{W} \in \mathbb{R}^{nd \times 3}$ and bias $\mathbf{b}$ are learned model parameters.

Our loss function is cross-entropy between predicted labels and true-labels as follow:

$$\mathcal{L} = -\log \prod_{i=1}^{N} \sum_{c=1}^{C} y_{i,c} \log(\hat{y}_{i,c}), \tag{7}$$

where $y_{i,c}$ is the true label vector, and $\hat{y}_{i,c}$ is the predicted probability from the softmax layer. C is the total number of classes (entailment, contradiction, and neutral for this task), and N is the total number of samples used in training.

4.3 Transfer learning

Pan and Yang (2010) provide definitions of transfer learning as follows:

Definition 1 (*Transfer Learning*) Given a source domain $\mathcal{D}_S$ and learning task $\mathcal{T}_S$, a target domain $\mathcal{D}_T$ and learning task $\mathcal{T}_T$, *transfer learning* aims to help improve the learning of the target predictive function $f_T(\cdot)$ in $\mathcal{D}_T$ using the knowledge in $\mathcal{D}_S$ and $\mathcal{T}_S$, where $\mathcal{D}_s \neq \mathcal{D}_T$, or $\mathcal{T}_S \neq \mathcal{T}_T$.

While MedNLI has a relatively large amount of training data in the clinical domain, NLI tasks in general domain such as SNLI (Bowman et al., 2015) and MNLI (Williams et al., 2018) have way larger training data than MedNLI has. Since a source and a target task in different domains can improve a model performance if they are related to each other we decide to use the two general domain NLI tasks to train BERT and BioBERT to transfer their knowledge for MedNLI. Our case is $\mathcal{D}_S \neq \mathcal{D}_T$ where the feature spaces between the domains are different or the marginal probability distributions between domain datasets are different ($P(X_S) \neq P(X_T)$).

4.4 Abbreviation expansion

Not unlike other medical text, abbreviations and acronyms are easily found throughout the text in MedNLI as table 1 shows from # 4 to 6. In order to understand the effect of expanded forms for clinical abbreviations, we replace the abbreviations with corresponding expanded forms. As Liu et al. (2015) mentions that no universal rules

Dataset	Accuracy	
	dev	test
+PMC	80.50	78.97
+PubMedd	81.14	78.83
+PubMed+PMC	**82.15**	**79.04**

Table 2: The BioBERT performance on the MedNLI task. Each model is trained on three different combinations of PMC and PubMed datasets (top score marked as bold).

Model	Accuracy	
	dev	test
BioBERT	82.15	79.04
CompAggr	80.40	75.80
BioBERT (transferred)	83.51	**82.63**
BioBERT (expanded)	**83.87**	79.95

Table 3: The model performance of four different methods (top score marked as bold). BioBERT (transferred) and BioBERT (expanded) refer to the best results of transfer learning experiments and the result of MedNLI with abbreviation expansion on BioBERT respectively.

or dictionary for clinical abbreviations is available we gather and exploit the public medical abbreviations from Taber's Online[2].

5 Experiments

We explore three kinds of BioBERT that are fine-tuned from the original BERT with PMC, PubMed, and PMC+PubMed datasets. As shown in table 2, BioBERT trained on PubMed+PMC performs the best. Thus we select it as a base BioBERT model for the rest of the experiments. Depends on a need for comparison or better understanding, we also include original BERT in the experiments and report the results. The overall results of MedNLI are shown in table 3.

5.1 Experimental Setup

All experiments based on BioBERT and BERT have a fixed learning rate 2e-5. We add early stopping to stop the models from learning if evaluation loss has not decreased for 4 steps where 1 step is defined 20% of the whole training data. Other than the learning rate and early stopping, all settings are the same as they are in BioBERT and BERT.

For the CompAggr model, we use a context projection weight matrix with 100 dimensions. In the aggregation part, we use 1-D CNN with a total of 500 filters, which involved five types of filters $K \in \mathbb{R}^{\{1,2,3,4,5\} \times 100}$, 100 per type. The weight matrices for the filters were initialized using the Xavier method (Glorot and Bengio, 2010). We use the Adam optimizer (Kingma and Ba, 2014) including gradient clipping by norm at a threshold of 5. For the purpose of regularization, we applied dropout (Srivastava et al., 2014) with a ratio of 0.7.

5.2 Performance evaluation

Transfer learning: We conduct transfer learning on four different combinations of MedNLI, SNLI, and MNLI as it shown in the table 4 (line 4 to 7) and also add the results of general domain tasks (MNLI, SNLI) for comparison. As expected, BERT performs better on tasks in the general domain while BioBERT performs better on MedNLI which is in the clinical domain.

In overall, positive transfer occurs on MedNLI. There are three things we can observe from the results. First of all, even though BioBERT is fine-tuned on general domain tasks before MedNLI, transfer learning shows better results than that fine-tuned on MedNLI directly. It implies that the same tasks in different domains have overlapping knowledge and transfer learning between the tasks effects positively on each other as the definition of transfer learning mentions in section 4. Second, the domain specific language representations from BioBERT are maintained while fine-tuning on general domain tasks by showing that the transfer learning results of MedNLI on BioBERT have better performance than the results on BERT (line 4 to 7). Lastly, the accuracy of MNLI and SNLI on BioBERT is lower than the accuracy on BERT. The lower accuracy indicates that BioBERT captures different features such as medical terms and generate different representations than what BERT does which are helpful for the clinical domain task, MedNLI, but not for the other two tasks.

The best combination is SNLI $\rightarrow$ MNLI $\rightarrow$ MedNLI on BioBERT. We refer to the best result of transfer learning as BioBERT (transferred).

Results analysis for different models: There are fundamental differences between the two models we apply. BioBERT tokenizes an input sentence

---

[2]https://www.tabers.com/tabersonline/view/Tabers-Dictionary/767492/all/Medical Abbreviations

Dataset	BERT		BioBERT	
	dev	test	dev	test
MedNLI	79.56	77.49	82.15	79.04
MNLI (M)	83.52	-	81.23	-
SNLI (S)	90.39	-	89.10	-
M → MedNLI	80.14	**78.62**	82.72	80.80
S → MedNLI	80.28	78.19	83.29	81.29
M → S → MedNLI	80.43	78.12	83.29	80.30
S → M → MedNLI	**81.72**	77.98	**83.51**	**82.63**
MedNLI (expanded)	79.13	77.07	**83.87**	79.95
S → M → MedNLI (expanded)	**82.15**	**79.95**	83.08	**81.85**

Table 4: All experiment results of transfer learning and abbreviation expansion (top-2 scores marked as bold). MedNLI (expanded) denotes MedNLI with abbreviation expansion.

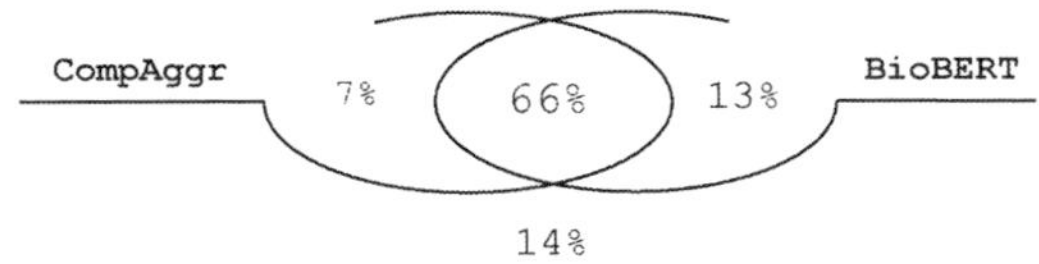

Figure 3: Venn diagram for the test results of Compare Aggregate model and BioBERT.

to sub-word level and uses the transformer model while CompAggr uses word-level embeddings and Compare&Aggregate model. In light of the dissimilar nature, we expect each model captures different features and generates different language representations.

Figure 3 shows the percentage for each area takes of the test set. CompAggr correctly classifies 97 examples (7% of the test set) which BioBERT classifies them incorrectly while BioBERT classifies 188 examples correctly (13% of the test set) which CompAggr does not. It demonstrates that both models have different strength on the MedNLI task.

We manually examine all promise and hypothesis pairs of each portion of 7% and 13% of the test set with high confidence and "element" label. For CompAggr, we pick pairs with the probability higher than 0.80 which are 6 pairs. For BioBERT, we select pairs with top 10 probabilities. Interestingly, each pair from CompAggr does not have overlapping words between premise and hypothesis. It appears that CompAggr's strength is in it's ability to capture the relationship between two sentences even though there is no word overlap while BioBERT labels them *"neutral"* except one

pair as you can see in table 5. In contrast, the majority of the pairs, 7 out of 10, from BioBERT have overlapping words between them. Biobert shows strong confidence when premise and hypothesis have overlapping words as below.

- (*Premise*) En route to the Emergency Department, she developed worsening substernal chest pain without any radiation.

- (*Hypothesis*) patient has chest pain

Lastly, we compute the average conditional probability of the correct results to check the confidence of each model. The results are 0.87 and 0.82 for BioBERT and CompAggr showing that BioBERT predicts labels with higher confidence.

Abbreviation expansion: We refer to the dataset of MedNLI with abbreviation expansion as MedNLI (expanded). The inconsistency of the experiment results on MedNLI (expanded) makes it difficult to observe their effects. MedNLI (expanded) shows better performance than MedNLI on BioBERT while MedNLI works better on BERT (see table 4). Furthermore, the performance of MedNLI (expanded) with transfer learning is higher on BERT and lower on BioBERT than the performance of MedNLI with transfer learning.

We examine the test results to figure out the inconsistency and observe an interesting phenomenon that the abbreviation expansion changes the conditional probability distribution $P(Y|X)$, where X and Y represent input texts and their expected labels, respectively. The same input texts with no expansion are classified into different

Premise	Hypothesis	CompAggr	BioBERT
He denies any fever, diarrhea, chest pain, cough, URI symptoms, or dysuria.	He denies any fever, diarrhea, chest pain, cough, URI symptoms, or dysuria.	entailment	neutral
This quickly became ventricular fibrillation and he was successfully shocked X 1 360J with return of rhythm and circulation.	Patient has NSR post-cardioversion	entailment	contradiction
PAST MEDICAL HISTORY: Coronary artery disease status post MI [**09**] years ago, status post angioplasty.	History of heart attack	entailment	neutral
A MRA prior to discharge showed increased ... of single and rector spinal muscles at T3-4 adjacent to facets and anterior within the right psoas.	the patient has degenerative changes of the spine	entailment	neutral
The patient now presents with metastatic recurrence of squamous cell carcinoma of the right mandible with extensive lymph node involvement.	The patient has oropharyngeal carcinoma.	entailment	neutral
The transbronchial biopsy was nondiagnostic.	Patient has a mediastinal mass	entailment	neutral

Table 5: Examples with the highest probabilities showing the strength of CompAggr.

Rank	Team	Accuracy
1	WTMED	98.0
2	PANLP	96.6
3	Double Transfer	93.8
4	Sieg	91.1
5	**Surf (ours)**	**90.6**
6	ARS_NITK	87.7
7	Pentagon	85.7
8	Dr.Quad	85.5
9	UU_TAILS	85.2
10	KU_ai	84.7

Table 6: Performance comparison among the top-10 participants (official) of the NLI shared task. Teams [1-4, 6-10] are from (Wu et al., 2019; Zhu et al., 2019; Xu et al., 2019; Bhaskar et al., 2019; Agrawal et al., 2019; Pugaliya et al., 2019; Bannihatti Kumar et al., 2019; Tawfik and Spruit, 2019; Cengiz et al., 2019), respectively.

classes. For instance, a pair of *Premise* and *Hypothesis* like below is not changed after abbreviation expansion since it does not contain any abbreviations or acronyms.

- (*Premise*) He denied headache or nausea or vomiting.

- (*Hypothesis*) He is afebrile.

However, the results are different. It is originally classified into *"neutral"* which is the right label for the pair but it is classified into *"entailment"* when we use MedNLI (expanded).

5.3 MEDIQA-NLI shared task

We are participating in a shared task MEDIQA-NLI of the bioNLP workshop at ACL 2019. In order to solve the task, we try four different **point-wise** approaches, CompAggr, BioBERT, transfer learning, and abbreviation expansion. We run each model several times to obtain the best result out of each. Our best result, which is ranked 5th on the leaderboard of the task, is obtained by applying **list-wise** approach (in section 3) with the best result (BioBERT (transferred)). Table 6 shows the model performance of each participant in the leaderboard.

6 Conclusion

In this paper, we study natural language inference in the clinical domain where training corpora is insufficient due to its domain nature. To tackle the problem, we propose approaches that adopts pre-trained language models, transfer learning method and data-augmentation to boost the train instances. To this end, we observe that the BioBERT pre-trained on bio-medical corpus shows better performance than that of the BERT on the general domain corpus. The CompAggr with bio-ELMO and the BioBERT behave differently in classifying the MedNLI dataset due to the difference in their own architecture. Transfer learning with NLI tasks in general domain, (MNLI, SNLI), does not hurt the ability of the BioBERT capturing language representations of the clinical domain. In addition, we observe that it transfers positive knowledge from general NLI tasks to the MedNLI task. In contrast, a abbreviation expansion method needs particular care when adopting since it may hurt the model to predict the conditional probability distribution of the task.

Acknowledgments

We sincerely thank the reviewers for their in depth feedback that helped improve the paper. K. Jung is with Automation and Systems Research Institute (ASRI), Seoul National University, Seoul, Korea, and was supported by the Ministry of Trade, Industry & Energy (MOTIE, Korea) under Industrial Technology Innovation Program (No.10073144).

References

Anumeha Agrawal, Rosa Anil George, Selvan Sunitha Ravi, Sowmya Kamath, and Anand Kumar. 2019. Ars_nitk at mediqa 2019:analysing various methods for natural language inference, recognising question entailment and medical question answering system. In *Proceedings of the BioNLP 2019 workshop, Florence, Italy, August 1, 2019*. Association for Computational Linguistics.

Vinayshekhar Bannihatti Kumar, Ashwin Srinivasan, Aditi Chaudhary, James Route, Teruko Mitamura, and Eric Nyberg. 2019. Dr.quad at mediqa 2019: Towards textual inference and question entailment using contextualized representations. In *Proceedings of the BioNLP 2019 workshop, Florence, Italy, August 1, 2019*. Association for Computational Linguistics.

Asma Ben Abacha, Chaitanya Shivade, and Dina Demner-Fushman. 2019. Overview of the mediqa 2019 shared task on textual inference, question entailment and question answering. In *Proceedings of the BioNLP 2019 workshop, Florence, Italy, August 1, 2019*. Association for Computational Linguistics.

Sai Abishek Bhaskar, Rashi Rungta, James Route, Eric Nyberg, and Teruko Mitamura. 2019. Sieg at mediqa 2019: Multi-task neural ensemble for biomedical inference and entailment. In *Proceedings of the BioNLP 2019 workshop, Florence, Italy, August 1, 2019*. Association for Computational Linguistics.

Samuel R Bowman, Gabor Angeli, Christopher Potts, and Christopher D Manning. 2015. A large annotated corpus for learning natural language inference. In *Proceedings of the 2015 Conference on Empirical Methods in Natural Language Processing*, pages 632–642.

Cemil Cengiz, Ulaş Sert, and Deniz Yuret. 2019. Ku_ai at mediqa 2019: Domain-specific pre-training and transfer learning for medical nli. In *Proceedings of the BioNLP 2019 workshop, Florence, Italy, August 1, 2019*. Association for Computational Linguistics.

Qian Chen, Xiaodan Zhu, Zhen-Hua Ling, Si Wei, Hui Jiang, and Diana Inkpen. 2017. Enhanced lstm for natural language inference. In *Proceedings of the 55th Annual Meeting of the Association for Computational Linguistics (Volume 1: Long Papers)*, pages 1657–1668.

Alexis Conneau, Douwe Kiela, Holger Schwenk, Loïc Barrault, and Antoine Bordes. 2017. Supervised learning of universal sentence representations from natural language inference data. In *Proceedings of the 2017 Conference on Empirical Methods in Natural Language Processing*, pages 670–680.

Jacob Devlin, Ming-Wei Chang, Kenton Lee, and Kristina Toutanova. 2018. Bert: Pre-training of deep bidirectional transformers for language understanding. *arXiv preprint arXiv:1810.04805*.

Xavier Glorot and Yoshua Bengio. 2010. Understanding the difficulty of training deep feedforward neural networks. In *Proceedings of the AISTATS*.

Alistair EW Johnson, Tom J Pollard, Lu Shen, H Lehman Li-wei, Mengling Feng, Mohammad Ghassemi, Benjamin Moody, Peter Szolovits, Leo Anthony Celi, and Roger G Mark. 2016. Mimic-iii, a freely accessible critical care database. *Scientific data*, 3:160035.

Yoon Kim. 2014. Convolutional neural networks for sentence classification. In *Proceedings of the 2014 Conference on Empirical Methods in Natural Language Processing (EMNLP)*, pages 1746–1751.

Diederik Kingma and Jimmy Ba. 2014. Adam: A method for stochastic optimization. *arXiv preprint arXiv:1412.6980*.

Jinhyuk Lee, Wonjin Yoon, Sungdong Kim, Donghyeon Kim, Sunkyu Kim, Chan Ho So, and Jaewoo Kang. 2019. Biobert: pre-trained biomedical language representation model for biomedical text mining. *arXiv preprint arXiv:1901.08746*.

Xiaodong Liu, Pengcheng He, Weizhu Chen, and Jianfeng Gao. 2019. Multi-task deep neural networks for natural language understanding. *arXiv preprint arXiv:1901.11504*.

Yue Liu, Tao Ge, Kusum Mathews, Heng Ji, and Deborah McGuinness. 2015. Exploiting task-oriented resources to learn word embeddings for clinical abbreviation expansion. In *Proceedings of BioNLP 15*, pages 92–97.

Serguei Pakhomov. 2002. Semi-supervised maximum entropy based approach to acronym and abbreviation normalization in medical texts. In *Proceedings of the 40th annual meeting on association for computational linguistics*, pages 160–167. Association for Computational Linguistics.

Sinno Jialin Pan and Qiang Yang. 2010. A survey on transfer learning. *IEEE Transactions on knowledge and data engineering*, 22(10):1345–1359.

Matthew Peters. 2018. PubMed ELMo. `https://allennlp.org/elmo`. [Online; accessed 09-May-2019].

Matthew Peters, Mark Neumann, Mohit Iyyer, Matt Gardner, Christopher Clark, Kenton Lee, and Luke Zettlemoyer. 2018. Deep contextualized word representations. In *Proceedings of the 2018 Conference of the North American Chapter of the Association for Computational Linguistics: Human Language Technologies, Volume 1 (Long Papers)*, pages 2227–2237.

Hemant Pugaliya, Karan Saxena, Shefali Garg, Sheetal Shalini, Prashant Gupta, Eric Nyberg, and Teruko Mitamura. 2019. Pentagon at mediqa 2019: Multi-task learning for filtering and re-ranking answers using language inference and question entailment. In *Proceedings of the BioNLP 2019 workshop, Florence, Italy, August 1, 2019*. Association for Computational Linguistics.

Alexey Romanov and Chaitanya Shivade. 2018. Lessons from natural language inference in the clinical domain. In *Proceedings of the 2018 Conference on Empirical Methods in Natural Language Processing*, pages 1586–1596.

Nitish Srivastava, Geoffrey E Hinton, Alex Krizhevsky, Ilya Sutskever, and Ruslan Salakhutdinov. 2014. Dropout: a simple way to prevent neural networks from overfitting. *Journal of machine learning research*, 15(1):1929–1958.

Sandeep Subramanian, Adam Trischler, Yoshua Bengio, and Christopher J Pal. 2018. Learning general purpose distributed sentence representations via large scale multi-task learning. In *ICLR*.

Noha Tawfik and Marco Spruit. 2019. Uu_tails at mediqa 2019: Learning textual entailment in the medical domain. In *Proceedings of the BioNLP 2019 workshop, Florence, Italy, August 1, 2019*. Association for Computational Linguistics.

Ashish Vaswani, Noam Shazeer, Niki Parmar, Jakob Uszkoreit, Llion Jones, Aidan N Gomez, Łukasz Kaiser, and Illia Polosukhin. 2017. Attention is all you need. In *Advances in neural information processing systems*, pages 5998–6008.

Alex Wang, Amapreet Singh, Julian Michael, Felix Hill, Omer Levy, and Samuel R Bowman. 2018. Glue: A multi-task benchmark and analysis platform for natural language understanding. *arXiv preprint arXiv:1804.07461*.

Shuohang Wang and Jing Jiang. 2016. A compare-aggregate model for matching text sequences. *arXiv preprint arXiv:1611.01747*.

Adina Williams, Nikita Nangia, and Samuel Bowman. 2018. A broad-coverage challenge corpus for sentence understanding through inference. In *Proceedings of the 2018 Conference of the North American Chapter of the Association for Computational Linguistics: Human Language Technologies, Volume 1 (Long Papers)*, pages 1112–1122.

Zhaofeng Wu, Yan Song, Sicong Huang, Yuanhe Tian, and Fei Xia. 2019. A hybrid approach to biomedical natural language inference. In *Proceedings of the BioNLP 2019 workshop, Florence, Italy, August 1, 2019*. Association for Computational Linguistics.

Yichong Xu, Xiaodong Liu, Chunyuan Li, Hoifung Poon, and Jianfeng Gao. 2019. Doubletransfer at mediqa 2019: Multi-source transfer learning for natural language understanding in the medical domain. In *Proceedings of the BioNLP 2019 workshop, Florence, Italy, August 1, 2019*. Association for Computational Linguistics.

Wei Zhu, Xiaofeng Zhou, Keqiang Wang, Xun Luo, Xiepeng Li, Yuan Ni, and Guotong Xie. 2019. Panlp at mediqa 2019: Pre-trained language models, transfer learning and knowledge distillation. In *Proceedings of the BioNLP 2019 workshop, Florence, Italy, August 1, 2019*. Association for Computational Linguistics.

WTMED at MEDIQA 2019: A Hybrid Approach to Biomedical Natural Language Inference

Zhaofeng Wu
Paul G. Allen School of CSE
University of Washington
zfw7@cs.washington.edu

Yan Song
Tencent AI Lab
clksong@gmail.com

Sicong Huang
Department of ECE
University of Washington
huangs33@uw.edu

Yuanhe Tian
Department of Linguistics
University of Washington
yhtian@uw.edu

Fei Xia
Department of Linguistics
University of Washington
fxia@uw.edu

Abstract

Natural language inference (NLI) is challenging, especially when it is applied to technical domains such as biomedical settings. In this paper, we propose a hybrid approach to biomedical NLI where different types of information are exploited for this task. Our base model includes a pre-trained text encoder as the core component, and a syntax encoder and a feature encoder to capture syntactic and domain-specific information. Then we combine the output of different base models to form more powerful ensemble models. Finally, we design two conflict resolution strategies when the test data contain multiple (premise, hypothesis) pairs with the same premise. We train our models on the MedNLI dataset, yielding the best performance on the test set of the MEDIQA 2019 Task 1.

1 Introduction

Natural language inference (NLI) (MacCartney and Manning, 2009), also known as textual entailment, is an important natural language processing (NLP) task that has long been studied (Bowman et al., 2015; Parikh et al., 2016; Chen et al., 2016; Conneau et al., 2017; Tay et al., 2018). It aims to capture the relationship between two sentences, identifying whether a given *premise* entails, contradicts, or is neutral to a given *hypothesis*. Success in NLI is crucial for achieving semantic comprehension of human language, which in turn is a prerequisite to accomplish natural language understanding (NLU). In general, accurate NLI systems facilitate many downstream tasks, such as commonsense reasoning (Zellers et al., 2018) and question answering (Abacha and Demner-Fushman, 2016, 2017).

Most of existing NLI studies are conducted in the general domain (Marelli et al., 2014; Bowman et al., 2015; Williams et al., 2018), with limited attention paid to domain-specific scenarios. Nevertheless, there has been increasing demand for information processing in the biomedical domain such as biomedical question answering (Abacha and Demner-Fushman, 2019) and cohort selection (Glicksberg et al., 2018). Many biomedical NLP applications require automatic understanding of symptom descriptions and examination reports (Abacha and Demner-Fushman, 2016, 2017) and therefore can greatly benefit from accurate biomedical NLI systems.

In this study, we propose a hybrid approach to biomedical NLI, which includes three main components, as illustrated in Figure 1. The main component is the base model (the largest box in the figure), which includes three encoders: an MT-DNN (Liu et al., 2019c) based text encoder, a syntax encoder that captures structural information, and a feature encoder which injects some degree of domain knowledge into the model (see §3). We conduct unsupervised pre-training for the text encoder on biomedical corpora to compensate for the lack of domain-specific supervision (Lee et al., 2019). To enhance our model, we also use model ensemble and conflict resolution strategies, corresponding to the two top dashed boxes in Figure 1 and are explained in §4. The datasets and implementation detail are described in §5. The experimental results on the MedNLI dataset (Romanov and Shivade, 2018) and the MEDIQA 2019 shared task 1 (Ben Abacha et al., 2019) are reported in §6.[1]

2 Related Work

A common neural network approach to address the NLI task is sentence pair modeling (Lan and

---

[1] Our code is publicly available at https://github.com/ZhaofengWu/MEDIQA_WTMED

Proceedings of the BioNLP 2019 workshop, pages 415–426
Florence, Italy, August 1, 2019. ©2019 Association for Computational Linguistics

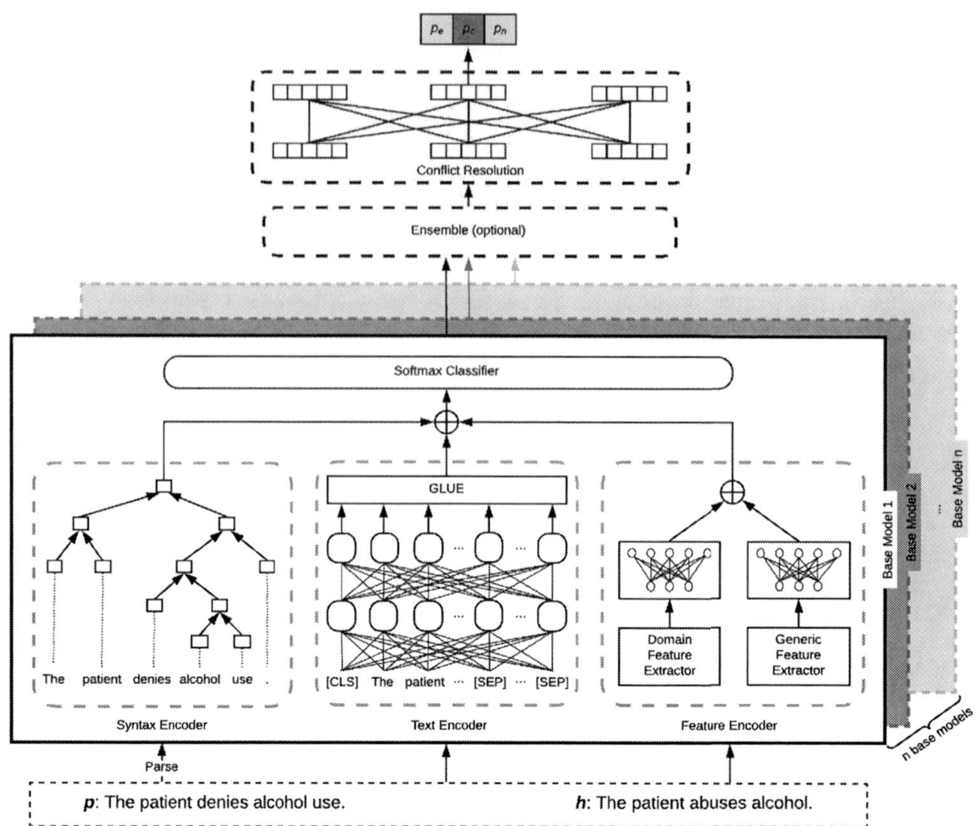

Figure 1: Our overall system. Our base model consists of three encoders and a softmax classifier: a syntax encoder that encodes the constituency parses provided by the dataset to a vector representation via Tree-LSTM; an MT-DNN based text encoder; a feature encoder that encodes domain and generic string-based features through fully-connected layers; and a softmax classifier that takes in the concatenation ($\oplus$) of the three encoders' output and generates a prediction. The output of base models is sent to the ensemble and conflict resolution modules (the multimodal attention method is depicted here as an example) to make a final prediction.

Xu, 2018). The premise and hypothesis are separately embedded (e.g. via GloVe (Pennington et al., 2014) or ELMo (Peters et al., 2018)) and encoded (e.g. via CNN or LSTM). Typically an interaction layer is employed to add information alignment between the premise and the hypothesis. For example, between the two baseline models used in the MedNLI dataset, InferSent (Conneau et al., 2017) computes the interaction vector via $[p; h; |p - h|; p * h]$ and ESIM (Chen et al., 2016) uses an attention matrix to softly align the two representations. ESIM also appends an inference composition layer to propagate the local attended information. A softmax layer is used to classify the final representation.

The recent Transformer-based models have been demonstrated to be a better encoder at NLI than CNN and LSTM by fully attending over the two sentences (Radford, 2018; Devlin et al.,

2018). BERT (Devlin et al., 2018) pre-trains the model with large unlabeled corpora which allows better text representations. MT-DNN (Liu et al., 2019c) leverages multi-task learning (Liu et al., 2015) to fine-tune the BERT weights using the GLUE datasets (Wang et al., 2018). The authors showed that resulting representations outperform BERT on many NLU tasks.

On top of this sentence pair modeling scheme, previous studies have independently leveraged syntax (Chen et al., 2016), external knowledge (Chen et al., 2018; Lu et al., 2019), ensemble methods (Ghaeini et al., 2018b), and language model fine-tuning (Alsentzer et al., 2019) to improve the performance of NLI systems. Nonetheless, to our knowledge, there have been no empirical results on the effect of combining these additions simultaneously. Additionally, as recent studies have pointed out that pre-trained contextualized representations contain rich linguistic signals (Hewitt and Manning, 2019; Liu et al., 2019b), it is reasonable to ask whether explicitly integrating knowledge will continue to augment such representations. Our work can be seen as an empirical study to examine the efficacy of applying multiple additions on top of Transformer-based models.

3 Base Model

NLI is generally treated as a three-way classification task that models whether a given premise p entails, contradicts, or is neutral to a hypothesis h. A classifier f is learned taking p and h as input to predict the class probabilities

$$f(p, h) = \begin{bmatrix} P_e & P_c & P_n \end{bmatrix}^\top \quad (1)$$

with $P_r; r \in \{e, c, n\}$ representing the probability for *entailment*, *contradiction*, and *neutral*. The final result is the class with the highest probability

$$y_{p,h} = \underset{r \in \{e,c,n\}}{\arg\max} P_r \quad (2)$$

As illustrated in Figure 1, our base model contains three modules. The widely used pre-trained Transformer model (Devlin et al., 2018; Liu et al., 2019c) serves as the basic text encoder to represent p and h. A syntax encoder and a feature encoder are also utilized to augment the basic representation by extracting and encoding more information from the input. The details of these encoders and how they are combined for f are discussed in the following subsections.

3.1 Text Encoder

Text representation is crucial to facilitate downstream tasks (Song et al., 2017, 2018). As a part of recent advancements in NLP, pre-trained models provide strong baselines for sentence representations and allow great generalizability for the represented text. Therefore, to represent p and h, we adopt a pre-trained Transformer model, MT-DNN (Liu et al., 2019c), as the text encoder in our base model. MT-DNN is based on BERT (Devlin et al., 2018) and additionally fine-tuned on GLUE (Wang et al., 2018), a set of NLU datasets including NLI subsets. Through its multi-task learning objective, MT-DNN allows a more general and powerful representation for natural language understanding than BERT (Liu et al., 2019c). Formally, one can briefly describe the encoder as

$$\mathcal{V}_{TE}(p, h) = \text{MT-DNN}(p, h) \quad (3)$$

with $\mathcal{V}_{TE}(p, h)$ referring to the output of the text encoder, a vector representing p and h.

Pre-training on large unlabeled corpora with a language modeling objective has facilitated many recent state-of-the-art advancements (Peters et al., 2018; Radford, 2018; Devlin et al., 2018; Lee et al., 2019; Radford et al., 2019). Inspired by these results, we enhance the MT-DNN representation by further fine-tuning on unlabeled biomedical data to mitigate the lack of in-domain supervision.

3.2 Syntax Encoder

Linguistic understandings, for example coreference relations (Zhang et al., 2019a,b), could aid the interpretation of a sentence. Syntactic structures are often useful for deciding the entailment of a sentence pair (Chen et al., 2016). There exist numerous NLI examples where a hypothesis is merely the premise with adjunct phrases removed. The syntax encoder also mitigates the out-of-vocabulary issue which is common in specific domains (Liu et al., 2019a) by capturing the structural information. Therefore, we include a syntax encoder in our base model.

We use Tree-LSTM (Tai et al., 2015) to model constituency parse trees of p and h. For each sentence, we encode it according to its tree structure and take the final state of the root node to represent the entire sentence. Formally, taking p as an example, the syntax encoder can be formulated as

$$\mathcal{V}_{SE}(p) = \text{Tree-LSTM}(\text{Parse}(p)) \quad (4)$$

where $\mathcal{V}_{SE}(p)$ is the output vector. Once p and h are encoded, the final output of this encoder is the concatenation of the two output vectors

$$\mathcal{V}_{SE}(p, h) = \mathcal{V}_{SE}(p) \oplus \mathcal{V}_{SE}(h) \quad (5)$$

3.3 Feature Encoder

The explicit integration of entity-level external knowledge has been used to improve many NLP models' performance (Das et al., 2017; Sun et al., 2018). Domain knowledge has also been demonstrated to be useful for in-domain tasks (Romanov and Shivade, 2018; Lu et al., 2019). Therefore, in addition to generic encoders such as MT-DNN and Tree-LSTM, we further enhance the model with domain-specific knowledge through indirectly leveraging labeled biomedical data for other tasks. To do that, we propose a domain feature encoder that identifies and vectorizes biomedical named entities using pre-trained medical taggers and counts (1) the number of each entity type in p and h; and (2) the number of shared entities and shared entity types in a (p, h) pair.

In addition to domain knowledge, inspired by Bowman et al. (2015) and Abacha and Demner-Fushman (2016), we also extract generic string features and use them to capture the similarity between p and h and then convert the results into vectors. Such similarity information includes n-gram overlap, Levenshtein distance (Levenshtein, 1966), Jaccard similarity (Jaccard, 1901), ROUGE (Lin, 2004) and BLEU (Papineni et al., 2001) scores, and absolute length difference.[2]

To encode the aforementioned features into vectors, each extracted feature is represented by a single scalar and then grouped with others into an array, denoted by $\mathbf{v}^{(d)}$ and $\mathbf{v}^{(g)}$ for domain and generic features, respectively. Later, they are converted into dense representations by linear transformations and a ReLU nonlinearty. For domain features, this process can be formulated by

$$\mathcal{V}_{FE}^{(d)}(p, h) = \text{ReLU}(\mathbf{W}^{(\mathbf{d})}\mathbf{v}^{(d)} + \mathbf{b}^{(d)}) \quad (6)$$

and $\mathcal{V}_{FE}^{(g)}(p, h)$ is obtained for generic features in a similar way. As a result, the final output of the feature encoder is the concatenation of vectors with domain and generic knowledge

$$\mathcal{V}_{FE}(p, h) = \mathcal{V}_{FE}^{(d)}(p, h) \oplus \mathcal{V}_{FE}^{(g)}(p, h) \quad (7)$$

3.4 Softmax Classifier

Once the outputs from the aforementioned encoders are obtained, a final representation of p and h is concatenated using the encoded vectors

$$\mathcal{V}(p, h) = \begin{bmatrix} \mathcal{V}_{TE}(p, h) \\ \mathcal{V}_{SE}(p, h) \\ \mathcal{V}_{FE}(p, h) \end{bmatrix} \quad (8)$$

Then, a softmax classifier is used to compute the class-wise probability distribution from $\mathcal{V}(p, h)$

$$f(p, h) = \text{softmax}(\mathbf{W}\mathcal{V}(p, h) + \mathbf{b}) \quad (9)$$

Among the three encoders, our base model always includes the text encoder. The other two encoders are optional, leading to different base models, whose performance will be compared in §6.1.

4　Model Enhancement

We enhance the base models discussed above with two techniques, namely model ensemble and conflict resolution: ensemble models combine predictions made by different base models, and conflict resolution takes advantage of NLI datasets where multiple (p, h) pairs share the same premise p.

4.1　Model Ensemble

Model ensemble is a common technique to combine predictions of multiple classifiers for better results (Maclin and Opitz, 1999). In NLI, model ensemble has also been proven helpful (Ghaeini et al., 2018a). In our work, when multiple base models are trained, we follow the strategy in Lee et al. (2015) and Lakshminarayanan et al. (2017) and average the models' predictions by

$$f^{(ME)}(p, h) = \frac{1}{n}\sum_{i=1}^{n} f_i(p, h) \quad (10)$$

with n denoting the number of ensembled base models and $f_i(p, h)$ being the probability distribution produced by the ith base model.

4.2　Conflict Resolution

Due to the special data collection strategy of MedNLI (see Romanov and Shivade (2018)), each premise is always paired with three hypotheses, each forming an entailment, a neutral, and a contradiction pair with the premise. For example, the premise "Labs were notable for Cr 1.7 (baseline 0.5 per old records) and lactate 2.4." appears three times in the dataset, each pairing with a different hypothesis: (1) "Patient has elevated Cr" (2) "Patient has normal Cr" and (3) "Patient has elevated BUN". The three hypotheses each forms a distinct relationship with the premise. We say the three (p, h) pairs with the same premise form a *group*.

For every group, there are six possible non-conflicting combinations of predictions: $\mathcal{C} = \{\langle e,c,n\rangle, \langle e,n,c\rangle, \langle n,e,c\rangle, \langle n,c,e\rangle, \langle c,n,e\rangle, \langle c,e,n\rangle\}$. Ideally, a model should yield non-conflicting group predictions; that is, $\langle y_{p,h_1}, y_{p,h_2}, y_{p,h_3}\rangle \in \mathcal{C}$ where h_1, h_2, h_3 are the three hypotheses in a group. However, our model determines the label of each pair independently from other pairs in the same group, and thus the three labels could be in conflict. To resolve this conflict, we propose two methods: heuristic processing and multimodal attention. Note that when resolving the conflict, both methods could potentially change the predictions for all three pairs in a group even when only two pairs have conflicting labels.

Heuristic Processing (HP): We first use our base or ensemble model to compute the class-wise probability distribution for each (p, h_i) pair

$$f(p, h_i) = \begin{bmatrix} P_e^{(i)} & P_c^{(i)} & P_n^{(i)} \end{bmatrix}^\top \quad (11)$$

where $i \in \{1, 2, 3\}$, and $P_r^{(i)}; r \in \{e, c, n\}$ is the probability of the i-th pair having relationship r. Then we compute the probability of each non-conflicting combination under this model by

$$P_{\langle r_1, r_2, r_3 \rangle} = \frac{1}{|\mathcal{C}|}(P_{r_1}^{(1)} + P_{r_2}^{(2)} + P_{r_3}^{(3)}) \quad (12)$$

where $\langle r_1, r_2, r_3 \rangle \in \mathcal{C}$.

Finally, we adjust the group predictions taking

$$\langle y_{p,h_1}^{(HP)}, y_{p,h_2}^{(HP)}, y_{p,h_3}^{(HP)} \rangle = \underset{\langle r_1, r_2, r_3 \rangle \in \mathcal{C}}{\arg\max} \, P_{\langle r_1, r_2, r_3 \rangle} \quad (13)$$

Intuitively, for each non-conflicting combination, we add up the prediction probabilities using the model output to derive a combination probability. We take the highest one as the final prediction.

Multimodal Attention (MA): We also trained an attention-based neural network to be responsible for conflict resolution so that it can be more expressive at intra-group interactions. It takes the probability distribution from the previous model as well as a positional encoding for input. We added the positional encoding aiming to capture patterns present in the dataset. For each pair, the input of our MA method is

$$\mathbf{p}_i = \begin{bmatrix} P_e^{(i)} & P_c^{(i)} & P_n^{(i)} & i \end{bmatrix}^\top \quad (14)$$

where $i \in \{1, 2, 3\}$ is the index of the pair. We first map it to a hidden space

$$\mathbf{h_i} = \mathbf{W}^{(h)}\mathbf{p}_i + \mathbf{b}^{(h)} \quad (15)$$

We compute intra-group attention by dot-product

$$a_{ij} = \mathbf{h_i} \cdot \mathbf{h_j} \quad (16)$$

Then, we compute attended hidden states by

$$\mathbf{h'_i} = \sum_{j=1}^{3} \frac{\exp(a_{ij})}{\sum_{k=1}^{3} \exp(a_{ik})} \mathbf{h_j} \quad (17)$$

The output probability distribution of i-th pair is

$$f^{(MA)}(p, h_i) = \mathrm{softmax}(\mathbf{W}^{(o)}\mathbf{h'_i} + \mathbf{b}^{(o)}) \quad (18)$$

Finally, the prediction is computed by Eq. (2).

	Train	**Dev**	**Test**
# of pairs	11,232	1,395	1,422
# of tokens in p	215k	29k	26k
# of tokens in h	66k	8k	8k
Max. p length	176	110	87
Max. h length	18	15	16
Avg. p length	19.2	20.4	18.6
Avg. h length	5.8	5.7	5.7

Table 1: Key statistics of the MedNLI dataset. We tokenize the sentences with NLTK (Loper and Bird, 2002).

5 Experiment Settings

5.1 Data

We use MedNLI as our main training dataset, for it is the official training set of MEDIQA. We also pre-train the text encoder on MIMIC-III discharge summaries (Johnson et al., 2016) using BERT's language modeling objectives (see §3.1).

MedNLI: The MedNLI dataset (Romanov and Shivade, 2018) presents unique challenges that require reasoning over biomedical domain knowledge. We use it to train out models and show its statistics in Table 1.

MIMIC-III: MIMIC-III (Medical Information Mart for Intensive Care) (Johnson et al., 2016) is a large database with information about patient admission to critical care units. We pre-train on its discharge summaries portion to obtain a better biomedical text representation. After some basic text cleaning, we obtain a corpus with around 7M sentences, 83M words, and 546M characters.

5.2 Data Pre-Processing

For pre-processing, we lowercase all our data and use the uncased pre-trained models unless otherwise specified. We replace masked patient health information (PHI) in the form of "[** *text* **]" with pseudo-value generated from gazetteers according to the PHI type[3]. For example, "[** Last Name **]" is replaced with a random last name such as "Smith".

5.3 Implementation

For MT-DNN, we use its own hyperparameters without modification. By default, we use 300-dimensional GloVe embeddings trained on Wikipedia and Gigawords (Pennington et al.,

---

[3]With the tool https://github.com/jtourille/mimic-tools

2014) to initialize the Tree-LSTM, which reduces each parse tree into a 100-dimensional vector. In the feature encoder, we use scispaCy (Neumann et al., 2019) to extract 38 domain features[4]. We also extract 27 linguistic features from the 6 categories specified in §3.3. We project the 38 domain features into $38 \times 20 = 760$ dimensions and the 27 linguistic features into $27 \times 20 = 540$ dimensions with fully-connected layers (See Equation (6)).

We fine-tune the text encoder with MIMIC-III discharge summaries using the same objectives as BERT, i.e. masked language model and next sentence prediction, for 8 epochs.

For training, we use the AdaMax optimizer (Kingma and Ba, 2014) with learning rate 5×10^{-5}. We use a batch size of 16 and train each model for 15 epochs. All other training hyperparameters are the same as the MT-DNN work.

6 Experimental Results

For our experiments, we first find the best configuration for a single base model, and then apply ensemble and conflict resolution on top of it. We run all these experiments with *MT-DNN base* for faster iterations. In order to maximally leverage the MedNLI dataset, unless otherwise specified, all experiments use the MedNLI training and development sets as the training data, and evaluate the performance directly on the MedNLI test set.

After obtaining the best configuration according to the development set performance, we retrain the whole system with that configuration on *MT-DNN large* using the whole MedNLI dataset (i.e. training+development+test). We run it on the MEDIQA Task 1 test set for the final submission (§6.4).

6.1 Base Model Results

The base model has many configurations depending on choices of the three encoders, whether to perform language model fine-tuning, and the embedding to use for Tree-LSTM initialization. To find a good, albeit not necessarily optimal, model configuration, we experiment with each modeling decision individually, and greedily use the best option found in the preceding experiments for the ones that follow. We then report ablation results to show the resulting configuration to be a local optimum.

Text Encoder	SE	FE	Acc.
BERT			79.68
	✓		79.89
		✓	79.54
	✓	✓	79.96
BioBERT			80.87
	✓		81.01
		✓	81.01
	✓	✓	81.29
MT-DNN			81.22
	✓		81.43
		✓	81.58
	✓	✓	**81.72**

Table 2: Performance of the base model with different configurations of the three encoders: text encoder (*TE*), syntax encoder (*SE*), and feature encoder (*FE*). We use GloVe (Embedding I) for Tree-LSTM initialization, and the experiments do not include language model fine-tuning and conflict resolution.

Pre-Training	Acc.
w/o LM fine tuning	81.72
with LM fine tuning	**83.26**

Table 3: The effect of pre-training. The first row is the best configuration from Table 2 (MT-DNN + SE + FE + Embedding I). The second row is the same system but pre-trained on MIMIC-III discharge summaries.

Encoders: Among the three encoders, the text encoder is the most important, so we will always include it in the base model. We compare three text encoders, including BERT, BioBERT[5] (Lee et al., 2019), and MT-DNN. As for syntax and feature encoders, we compare base models with or without them. The performance of all the combinations are in Table 2, which shows that MT-DNN outperforms BERT and BioBERT, and adding syntax and feature encoders to MT-DNN provides a small improvement[6]. The best result (81.72%) is in the last row and we will refer its configuration as MT-DNN + SE + FE from now on.

Language Model Fine-Tuning (LMFT): Using language modeling objective, we fine-tune the text encoder with MIMIC-III discharge summaries. The result is in Table 3, and it demonstrates that the language model fine-tuning scheme

---

[4] We use scispaCy to identify 18 types of biomedical named entities and turn them into features as mentioned in §3.3. Thus, there are totally $18 \times 2 + 2 = 38$ features.

[5] Because the BioBERT authors only released cased models, we maintain our data casing in relevant experiments.

[6] We also experimented with initializing text encoder word embedding weights with pre-trained static embeddings but it degraded the performance significantly.

Embedding for Tree-LSTM	Acc.
Embedding I	**83.26**
Embedding II	82.91
Embedding III	82.84

Table 4: Effect of different embeddings for Tree-LSTM initialization in the syntax encoder. The first row is the best result from Table 3. The last two rows are the same system but with different embeddings.

Base Model Configuration	Acc.
MT-DNN + SE + FE + LMFT + Emb I	**83.26**
MT-DNN → BERT	82.14
MT-DNN → BioBERT	82.84
– SE	82.28
– FE	82.49
– LMFT	81.72
Emb I → Emb II	82.91
Emb I → Emb III	82.84

Table 5: The ablation results on top of the best base model. LMFT denotes language model fine tuning.

brings a significant performance increase. This finding aligns with previous studies (Radford et al., 2019; Devlin et al., 2018; Lee et al., 2019; Alsentzer et al., 2019).

Syntax Encoder Embeddings: We used 300-dimensional GloVe embeddings to initialize the Tree-LSTM for Table 2 and 3, and we call it *Embedding I*. Romanov and Shivade (2018) used embeddings trained on biomedical corpora and observed non-trivial accuracy gain over general-domain embeddings. Thus, we also experimented with two domain-specific word embeddings that they used and released to initialize the Tree-LSTM, and we will call them *Embedding II* and *III*. Here is a quick summary of the embeddings:

I. GloVe embedding trained on Wikipedia 2014 + Gigaword 5;

II. Embedding initialized with common crawl[7] GloVe and fine-tuned on BioASQ and then MIMIC-III;

III. Embedding initialized with fastText (Bojanowski et al., 2017) trained on Wikipedia and fine-tuned on MIMIC-III.

Table 4 shows the effect of these embeddings. The first row is the best result from Table 3, which uses Embedding I, and the next two rows are the results when the embedding is changed. The table shows that using specific in-domain embeddings (the second and the third rows in Table 4) does not improve the performance. This is somewhat surprising, but also understandable since these in-domain embeddings are used only in the syntax encoder, instead of being used to initialize the main encoder as in Romanov and Shivade (2018).

Single Model Ablation: Table 2-4 show that the best configuration for the base model is MT-DNN + SE + FE + LMFT + Embedding I; that is, it uses

all three encoders, is fine-tuned with MIMIC-III discharge summaries, and uses regular GloVe embeddings to initialize the Tree-LSTM.

Because we followed a greedy process for various modeling decisions, there is no guarantee that this configuration is globally or even locally optimal. To test the optimality of the resulting model, we conducted ablations by individually changing each modeling decision on top of the best base model and compare the performance. The results are in Table 5, which show that the greedily found configuration is still the best-performing one among the ablations. In other words, while this configuration is still not guaranteed to be globally optimal, it is at least a locally optimal one.

6.2 Model Enhancement Results

We want a diverse set of member models to achieve better ensemble performance. We present ones that lead to better ensemble performance in Table 6. We also report the ensemble models and conflict resolution results in Table 6.

Ensemble: With the large number of possible configurations for the base model, it is infeasible to test out all ensemble combinations. On the other hand, the performance of different ensembles does not vary much. We ran all $2^9 - 1 = 511$ ensembles corresponding to all the non-empty subset of the 9 base models A-I, and found that on average on the development set (i.e. original MedNLI test set), ensemble models improve over their best-performing member by $0.86\% \pm 0.51\%$, and over the member average by $1.47\% \pm 0.60\%$. These results demonstrate the general usefulness of the ensemble stage. In Table 6, we show some of the ensemble models, most of which outperform their member models.

---

[7]https://commoncrawl.org, a corpus that contains 840 billion tokens of web data.

Model ID							Dev (i.e. MedNLI Test)			MEDIQA Test		
(R & S, 2018)	TE	SE	FE	LMFT	Emb	Prepro	Raw	HP	MA	Raw	HP	MA
	InferSent				*		73.5	-	-	-	-	-
	InferSent				III		76.6	-	-	-	-	-
Base Model	TE	SE	FE	LMFT	Emb	Prepro	Raw	HP	MA	Raw	HP	MA
A	MT-DNN	✓			I		81.36	85.16	96.20	80.49	87.16	97.28
B	MT-DNN	✓			II		81.50	85.94	96.62	78.77	87.41	97.53
C	MT-DNN	✓		✓	I		82.28	87.90	97.61	**82.47**	**90.86**	98.02
D	MT-DNN	✓		✓	I	✓	82.49	86.36	97.75	**82.47**	88.64	97.53
E	MT-DNN	✓	✓	✓	I		82.35	86.57	97.47	80.99	88.40	**99.51**
F	MT-DNN	✓	✓	✓	I	✓	**83.26**	87.62	98.17	81.23	86.91	97.53
G	MT-DNN	✓	✓	✓	II	✓	82.91	86.57	97.61	81.48	89.88	98.52
H	MT-DNN	✓	✓	✓	III	✓	82.84	86.50	97.61	80.49	89.38	98.02
I	BioBERT	✓	✓	✓	I	✓	82.84	**88.96**	**98.31**	78.03	83.46	99.01
Ensemble	Members						Raw	HP	MA	Raw	HP	MA
J	A + C + E						83.68	88.19	**98.17**	**83.95**	93.33	**99.01**
K	A + B + C + E						83.47	88.82	97.68	83.46	**93.33**	98.02
L	A + C + D						83.76	88.40	97.89	82.96	92.84	98.52
M	F + G + H						83.54	87.62	98.03	80.99	88.89	98.02
N	F + I						**83.97**	**89.94**	**98.17**	82.22	88.64	**99.01**
Avg Gain	-						-	4.59	14.79	-	7.80	16.82

Table 6: The performance of different ensemble combinations and conflict resolution strategies on our development set (i.e., the original MedNLI test set) and on the MEDIQA shared task test set. All our models in this table (i.e. the *Base Model* and *Ensemble* sections) use MedNLI training and development sets as the training set, while (R & S, 2018) models (Romanov and Shivade, 2018) use only the MedNLI training set for training and MedNLI development set for tuning. The *Prepro* column refers to whether data pre-processing is used (see §5.2). The *Raw*, *HP*, and *MA* columns refer to model performance without and with the two conflict resolution strategies. The results on the MEDIQA test set are computed after the shared task ended and its gold-standard labels were distributed. We report the baseline result and the best extension from Romanov and Shivade (2018) in the first two rows of the table. Their baseline uses the Common Crawl Glove embedding (denoted as *). Note that their results are not directly comparable with ours because they used the MedNLI Test as their test set whereas we use it as our development set. Finally, the last row, *Avg Gain*, is the average gain of HP and MA over Raw when averaged over all the base models and ensembles.

Conflict Resolution: We apply heuristic processing (HP) and multimodal attention (MA) to the base models or the ensembles. Both methods improve the performance by large margins.

To our surprise, multimodal attention works much better than heuristic processing, with around 10% absolute difference in accuracy. After a close examination of the training data and the model output, we realize that the MedNLI dataset has a clear label pattern[8] for pairs in the same group (the label sequence being *entailment*, *contradiction*, and *neutral*). Such a pattern is captured by the MA model, but not by the HP one. This finding not only explains the different performance of the two methods, but also reminds us that the high performance of the MA method is largely due to the pattern (or the bias) of this particular dataset.

Taking the best ensemble model N as an exam-

ple, we study exactly how the two conflict resolution strategies help on the development set. We show relevant statistics in Table 7. As expected, the less conflict there is in a group, the higher the raw accuracy is. We also see that the majority of HP changes are correct for groups with 2 conflicting predictions, but HP does not help groups where all raw predictions are the same. In contrast, because MA takes advantage of the inherent bias of the dataset, all its produced labels are correct. Nevertheless, MA accuracy is still below 100%, because it does not process groups with no conflicts, and raw accuracy on such groups is not at 100%.

6.3 Error Analysis

In real use cases, the input to an NLI system is more likely to be standalone (p, h) pairs instead of groups of three (p, h) pairs. Therefore, we conduct error analysis on the output of ensemble systems without conflict resolution. Figure 2 shows

---

[8] We checked the percentage of groups observing this pattern after the gold standard for test set is released, and it turns out 100% of the groups follow this pattern.

Conflict Type	# of Groups	# of Pairs	Raw Acc.	HP			MA		
				✗→✓	✓→✗	✗→✗	✗→✓	✓→✗	✗→✗
0	295	885	97.06%	0	0	0	0	0	0
2	172	516	63.57%	124	43	8	188	0	0
3	7	21	33.33%	7	3	4	14	0	0

Table 7: Conflict resolution results on model N on our development set (i.e., the MedNLI test set). Groups are categorized by **Conflict Type** (i.e., the number of sentence pairs with the same label), which could be 0, 2, or 3. Each group always has three sentence pairs. "Raw Acc." refers to the accuracy without post-processing. For each conflict resolution strategy, we find the (p, h) pairs whose labels are modified by HP or MA, categorize them based on how the updated predictions differ from the raw predictions, and report the number of (p, h) pairs in each category.

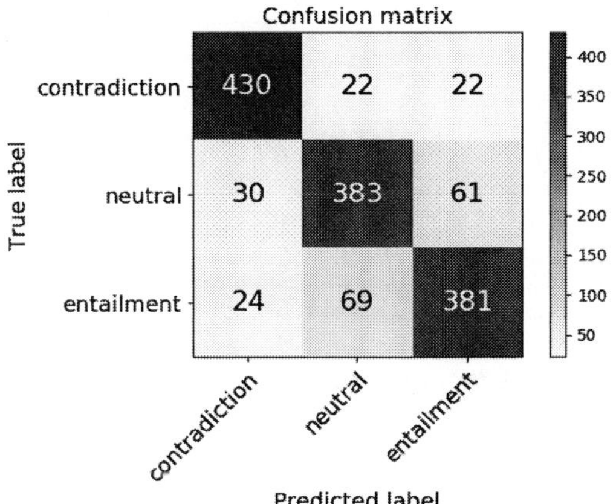

Figure 2: The confusion matrix of Model N before applying conflict resolution strategies.

the confusion matrix for Model N, the best performing ensemble model in Table 6, before conflict resolution. The confusion matrix shows that the model tends to confuse between entailment and neutral. Below are two examples where the model misidentifies entailment pairs to be neutral:

1. *p*: The patient now presents with metastatic recurrence of squamous cell carcinoma of the right mandible with extensive lymph node involvement.

   h: The patient has oropharyngeal

2. *p*: In the ED, initial VS revealed T 98.9, HR 73, BP 121/90, RR 15, O2 sat 98% on RA.

   h: The patient is hemodynamically stable.

Both examples contain many medical terms and determining the relationship for the (p, h) pairs is challenging for anyone without medical expertise. Many errors made by the model fall into this category, and fixing them would require the model to be enhanced with deeper domain knowledge.

Model ID	Conflict Resolution	Acc.
J	None	87.2
K	HP	94.8
L	MA	**98.0**

Table 8: The results of three models we submitted to MEDIQA Task 1. *Model ID* refers to the model ID in Table 6. The 2nd column denotes different conflict resolution strategies. The *Acc* column is the accuracy on MEDIQA Task 1 test set, which was calculated automatically by the shared task submission site.

6.4 Results on MEDIQA Task 1 Test Set

At the time of the shared task submission, we had not completed the systematic experiments as laid out in this paper. We used our then-best ensemble models, re-trained them on *MT-DNN large* using the whole MedNLI set (i.e. training+development+test), and ran them on the MEDIQA Task 1 test set. The results are shown in Table 8. Our best model achieves 98.0% accuracy on the MEDIQA Task 1 test set, the best among all participants.

7 Conclusion

We have presented a hybrid architecture for in-domain NLI. Our approach extends current efforts in biomedical NLP (Romanov and Shivade, 2018; Lee et al., 2019) through incorporating auxiliary encoders, domain-specific language model fine-tuning, ensembling, and conflict resolution. We dissected the usefulness of these modeling decisions and provided detailed and systematic ablations. These components work together to form the best performing model on MEDIQA Task 1.

The current system tends to make wrong predictions when in-depth domain-specific knowledge or reasoning is required. For future work, we plan to extend the system to incorporate deeper domain knowledge.

References

Asma Ben Abacha and Dina Demner-Fushman. 2016. Recognizing question entailment for medical question answering. In *AMIA Annual Symposium Proceedings*, volume 2016, page 310. American Medical Informatics Association.

Asma Ben Abacha and Dina Demner-Fushman. 2017. Nlm_nih at semeval-2017 task 3: from question entailment to question similarity for community question answering. In *Proceedings of the 11th International Workshop on Semantic Evaluation (SemEval-2017)*, pages 349–352.

Asma Ben Abacha and Dina Demner-Fushman. 2019. A question-entailment approach to question answering. *CoRR*, abs/1901.08079.

Emily Alsentzer, John R. Murphy, Willie Boag, Wei-Hung Weng, Di Jin, Tristan Naumann, and Matthew B. A. McDermott. 2019. Publicly available clinical bert embeddings. *CoRR*, abs/1904.03323v2.

Asma Ben Abacha, Chaitanya Shivade, and Dina Demner-Fushman. 2019. Overview of the mediqa 2019 shared task on textual inference, question entailment and question answering. In *Proceedings of the BioNLP 2019 workshop, Florence, Italy, August 1, 2019*. Association for Computational Linguistics.

Piotr Bojanowski, Edouard Grave, Armand Joulin, and Tomas Mikolov. 2017. Enriching word vectors with subword information. *Transactions of the Association for Computational Linguistics*, 5:135–146.

Samuel R. Bowman, Gabor Angeli, Christopher Potts, and Christopher D. Manning. 2015. A large annotated corpus for learning natural language inference. In *EMNLP*.

Qian Chen, Xiaodan Zhu, Zhen-Hua Ling, Diana Inkpen, and Si Wei. 2018. Neural natural language inference models enhanced with external knowledge. In *Proceedings of the 56th Annual Meeting of the Association for Computational Linguistics (Volume 1: Long Papers)*, pages 2406–2417, Melbourne, Australia. Association for Computational Linguistics.

Qian Chen, Xiaodan Zhu, Zhenhua Ling, Si Wei, Hui Jiang, and Diana Inkpen. 2016. Enhanced lstm for natural language inference. *arXiv preprint arXiv:1609.06038*.

Alexis Conneau, Douwe Kiela, Holger Schwenk, Loic Barrault, and Antoine Bordes. 2017. Supervised learning of universal sentence representations from natural language inference data. *arXiv preprint arXiv:1705.02364*.

Rajarshi Das, Manzil Zaheer, Siva Reddy, and Andrew McCallum. 2017. Question answering on knowledge bases and text using universal schema and memory networks. *arXiv preprint arXiv:1704.08384*.

Jacob Devlin, Ming-Wei Chang, Kenton Lee, and Kristina Toutanova. 2018. Bert: Pre-training of deep bidirectional transformers for language understanding. *CoRR*, abs/1810.04805.

Reza Ghaeini, Sadid A Hasan, Vivek Datla, Joey Liu, Kathy Lee, Ashequl Qadir, Yuan Ling, Aaditya Prakash, Xiaoli Z Fern, and Oladimeji Farri. 2018a. Dr-bilstm: Dependent reading bidirectional lstm for natural language inference. *arXiv preprint arXiv:1802.05577*.

Reza Ghaeini, Sheik Arick Hasan, Vivek Datla, Joey Liu, Kathy Y. S. Lee, Ashequl Qadir, Yuan Ling, Aaditya Prakash, Xiaoli Z. Fern, and Oladimeji Farri. 2018b. Dr-bilstm: Dependent reading bidirectional lstm for natural language inference. In *NAACL-HLT*.

Benjamin Glicksberg, Riccardo Miotto, Kipp Johnson, Shameer Khader, li li, Rong Chen, and Joel T Dudley. 2018. Automated disease cohort selection using word embeddings from electronic health records. *Pacific Symposium on Biocomputing. Pacific Symposium on Biocomputing*, 23:145–156.

John Hewitt and Christopher D Manning. 2019. A structural probe for finding syntax in word representations. In *Proceedings of the 2019 Conference of the North American Chapter of the Association for Computational Linguistics: Human Language Technologies, NAACL-HLT, Minneapolis, Minnesota, USA*, volume 2.

Paul Jaccard. 1901. Etude comparative de la distribution florale dans une portion des alpes et du jura.

Alistair Edward William Johnson, Tom J. Pollard, Lu Shen, Li wei H. Lehman, Mengling Feng, Mohammad M. Ghassemi, Benjamin Moody, Peter Szolovits, Leo Anthony Celi, and Roger G. Mark. 2016. Mimic-iii, a freely accessible critical care database. In *Scientific data*.

Diederik P. Kingma and Jimmy Ba. 2014. Adam: A method for stochastic optimization. *CoRR*, abs/1412.6980.

Balaji Lakshminarayanan, Alexander Pritzel, and Charles Blundell. 2017. Simple and scalable predictive uncertainty estimation using deep ensembles. In *Advances in Neural Information Processing Systems*, pages 6402–6413.

Wuwei Lan and Wei Xu. 2018. Neural network models for paraphrase identification, semantic textual similarity, natural language inference, and question answering. *arXiv preprint arXiv:1806.04330*.

Jinhyuk Lee, Wonjin Yoon, Sungdong Kim, Donghyeon Kim, Sunkyu Kim, Chan Ho So, and Jaewoo Kang. 2019. Biobert: a pre-trained biomedical language representation model for biomedical text mining. *CoRR*, abs/1901.08746.

Stefan Lee, Senthil Purushwalkam, Michael Cogswell, David Crandall, and Dhruv Batra. 2015. Why m heads are better than one: Training a diverse ensemble of deep networks. *arXiv preprint arXiv:1511.06314.*

VI Levenshtein. 1966. Binary Codes Capable of Correcting Deletions, Insertions and Reversals. *Soviet Physics Doklady*, 10:707.

Chin-Yew Lin. 2004. Rouge: A package for automatic evaluation of summaries. In *ACL 2004.*

Miaofeng Liu, Yan Song, Hongbin Zou, and Tong Zhang. 2019a. Reinforced training data selection for domain adaptation. In *Proceedings of ACL, 2019.*

Nelson F Liu, Matt Gardner, Yonatan Belinkov, Matthew Peters, and Noah A Smith. 2019b. Linguistic knowledge and transferability of contextual representations. *arXiv preprint arXiv:1903.08855.*

Xiaodong Liu, Jianfeng Gao, Xiaodong He, Li Deng, Kevin Duh, and Ye-Yi Wang. 2015. Representation learning using multi-task deep neural networks for semantic classification and information retrieval. In *HLT-NAACL.*

Xiaodong Liu, Pengcheng He, Weizhu Chen, and Jianfeng Gao. 2019c. Multi-task deep neural networks for natural language understanding. *CoRR*, abs/1901.11504.

Edward Loper and Steven Bird. 2002. Nltk: The natural language toolkit. *CoRR*, cs.CL/0205028.

Mingming Lu, Yu Fang, Fengqi Yan, and Maozhen Li. 2019. Incorporating domain knowledge into natural language inference on clinical texts. *IEEE Access.*

Bill MacCartney and Christopher D Manning. 2009. *Natural language inference.* Stanford University Stanford.

Richard Maclin and David W. Opitz. 1999. Popular ensemble methods: An empirical study. *J. Artif. Intell. Res.*, 11:169–198.

Marco Marelli, Stefano Menini, Marco Baroni, Luisa Bentivogli, Raffaella Bernardi, Roberto Zamparelli, et al. 2014. A sick cure for the evaluation of compositional distributional semantic models. In *LREC*, pages 216–223.

Mark Neumann, Daniel King, Iz Beltagy, and Waleed Ammar. 2019. Scispacy: Fast and robust models for biomedical natural language processing. *CoRR*, abs/1902.07669.

Kishore Papineni, Salim Roukos, Todd Ward, and Wei-Jing Zhu. 2001. Bleu: a method for automatic evaluation of machine translation. In *ACL.*

Ankur P Parikh, Oscar Täckström, Dipanjan Das, and Jakob Uszkoreit. 2016. A decomposable attention model for natural language inference. *arXiv preprint arXiv:1606.01933.*

Jeffrey Pennington, Richard Socher, and Christopher Manning. 2014. Glove: Global vectors for word representation. In *Proceedings of the 2014 conference on empirical methods in natural language processing (EMNLP)*, pages 1532–1543.

Matthew E. Peters, Mark Neumann, Mohit Iyyer, Matt Gardner, Christopher Clark, Kenton Lee, and Luke Zettlemoyer. 2018. Deep contextualized word representations. In *Proc. of NAACL.*

Alec Radford. 2018. Improving language understanding by generative pre-training.

Alec Radford, Jeff Wu, Rewon Child, David Luan, Dario Amodei, and Ilya Sutskever. 2019. Language models are unsupervised multitask learners.

Alexey Romanov and Chaitanya Shivade. 2018. Lessons from natural language inference in the clinical domain. In *EMNLP.*

Yan Song, Chia-Jung Lee, and Fei Xia. 2017. Learning word representations with regularization from prior knowledge. In *Proceedings of the 21st Conference on Computational Natural Language Learning (CoNLL 2017)*, pages 143–152, Vancouver, Canada.

Yan Song, Shuming Shi, Jing Li, and Haisong Zhang. 2018. Directional skip-gram: Explicitly distinguishing left and right context for word embeddings. In *Proceedings of the 2018 Conference of the North American Chapter of the Association for Computational Linguistics: Human Language Technologies, Volume 2*, pages 175–180, New Orleans, Louisiana.

Haitian Sun, Bhuwan Dhingra, Manzil Zaheer, Kathryn Mazaitis, Ruslan Salakhutdinov, and William W Cohen. 2018. Open domain question answering using early fusion of knowledge bases and text. *arXiv preprint arXiv:1809.00782.*

Kai Sheng Tai, Richard Socher, and Christopher D. Manning. 2015. Improved semantic representations from tree-structured long short-term memory networks. In *ACL.*

Yi Tay, Anh Tuan Luu, and Siu Cheung Hui. 2018. Compare, compress and propagate: Enhancing neural architectures with alignment factorization for natural language inference. In *Proceedings of the 2018 Conference on Empirical Methods in Natural Language Processing*, pages 1565–1575.

Alex Wang, Amanpreet Singh, Julian Michael, Felix Hill, Omer Levy, and Samuel R. Bowman. 2018. Glue: A multi-task benchmark and analysis platform for natural language understanding. In *BlackboxNLP@EMNLP.*

Adina Williams, Nikita Nangia, and Samuel Bowman. 2018. A broad-coverage challenge corpus for sentence understanding through inference. In *Proceedings of the 2018 Conference of the North American Chapter of the Association for Computational Linguistics: Human Language Technologies, Volume 1*

(Long Papers), pages 1112–1122. Association for Computational Linguistics.

Rowan Zellers, Yonatan Bisk, Roy Schwartz, and Yejin Choi. 2018. Swag: A large-scale adversarial dataset for grounded commonsense inference. In *Proceedings of the 2018 Conference on Empirical Methods in Natural Language Processing (EMNLP)*.

Hongming Zhang, Yan Song, and Yangqiu Song. 2019a. Incorporating Context and External Knowledge for Pronoun Coreference Resolution. In *Proceedings of NAACL-HLT, 2019*.

Hongming Zhang, Yan Song, Yangqiu Song, and Dong Yu. 2019b. Knowledge-aware Pronoun Coreference Resolution. In *Proceedings of ACL, 2019*.

KU_ai at MEDIQA 2019: Domain-specific Pre-training and Transfer Learning for Medical NLI

Cemil Cengiz **Ulaş Sert** **Deniz Yuret**

Koç University
Artificial Intelligence Laboratory
İstanbul, Turkey
`ccengiz17,usert17,dyuret@ku.edu.tr`

Abstract

In this paper, we describe our system and results submitted for the Natural Language Inference (NLI) track of the MEDIQA 2019 Shared Task (Ben Abacha et al., 2019). As KU_ai team, we used BERT (Devlin et al., 2018) as our baseline model and pre-processed the MedNLI dataset to mitigate the negative impact of de-identification artifacts. Moreover, we investigated different pre-training and transfer learning approaches to improve the performance. We show that pre-training the language model on rich biomedical corpora has a significant effect in teaching the model domain-specific language. In addition, training the model on large NLI datasets such as MultiNLI and SNLI helps in learning task-specific reasoning. Finally, we ensembled our highest-performing models, and achieved 84.7% accuracy on the unseen test dataset and ranked 10th out of 17 teams in the official results.

1 Introduction

Natural Language Inference (NLI) is one of the central problems in artificial intelligence. It requires understanding two input sentences and forming an inference relationship between them. Concretely, given a premise sentence p, and a hypothesis sentence h, NLI is the task of determining the inference relationship from p to h. In MedNLI, this relationship is one of the *neutral*, *entailment* and *contradiction* labels. Therefore, our task can be considered as a three-class sentence pair classification problem.

In previous research, sequence encoders connected with a classifier head have been commonly used as NLI systems (Conneau et al., 2017). Traditionally, the encoder layer has been an RNN-based model such as LSTM (Hochreiter and Schmidhuber, 1997). Vaswani et al. (2017) proposed the Transformer as an alternative model to the RNN. Since the Transformer is based on a self-attention mechanism rather than recurrent layers, it is much faster to train in parallel and can capture distant dependencies better. Therefore, the recent models originated from Transformer replaced the RNN-based encoders in many systems trained for natural language understanding tasks such as NLI, Question Answering, Common Sense Reasoning (Radford et al., 2018), and Neural Machine Translation (Lakew et al., 2018). As a Transformer based model, BERT uses self-attention to capture the relationships within the text during encoding, which can include one or more sentences (Devlin et al., 2018). Therefore, it can learn a joint representation for a premise-hypothesis pair, which can be fed to a classifier layer to predict the inference relation between them.

Recent studies have explored different ways of inference prediction instead of a straightforward classifier layer. Liu et al. (2018) proposed to use an answer module performing multi-step inference by iteratively refining its prediction. Likewise, models that aim to solve multiple problems simultaneously have gained attention due to their impressive performances (Liu et al., 2019). However, we concentrated on a single task, and wanted to keep the prediction layer simple. Therefore, we used neither of these approaches.

To succeed in NLI, a system must have strong reasoning skills and a good understanding of the language (MacCartney, 2009). If a large annotated dataset is available, the system can be trained from scratch for NLI, learning both the language and reasoning simultaneously. However, such data is often not available. Without seeing many syntactic variation and inference relation combinations, learning both is a hard task. Separating the two by training a language model first, and adjusting it for NLI later is a more effective approach. Due

427

Proceedings of the BioNLP 2019 workshop, pages 427–436
Florence, Italy, August 1, 2019. ©2019 Association for Computational Linguistics

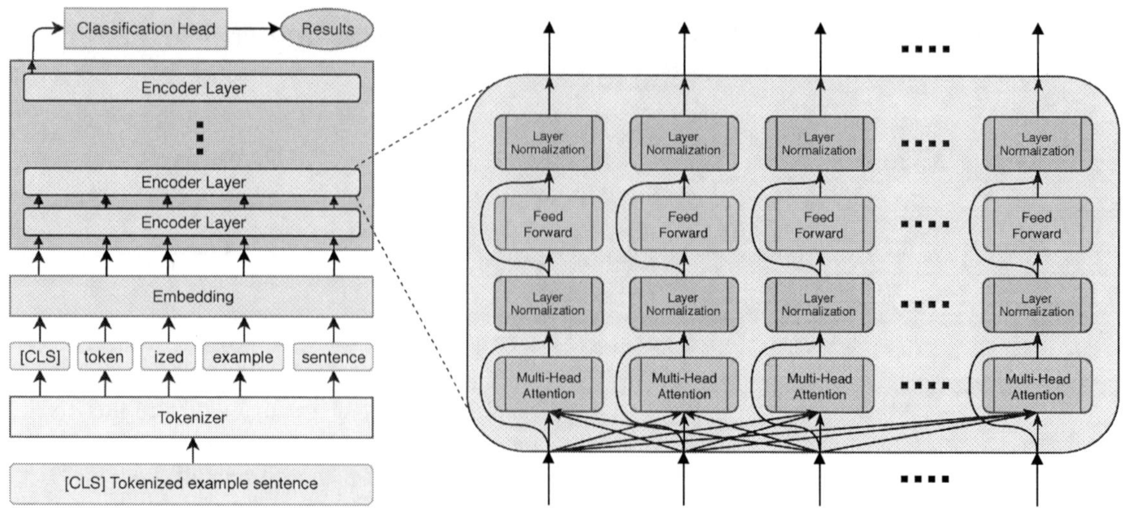

Figure 1: Baseline model architecture. On the left is the overview of the model, which includes the tokenizer, embedding layer, BERT encoder, and the classification head. The BERT encoder consists of twelve encoder layers. On the right are the details of what an encoder layer consists of. Each input and output of an encoder layer corresponds to a single token. Modules that are side-by-side share the same weights, only differing in inputs.

to the development of powerful language models that can be trained on unlabeled data in an unsupervised manner, many pre-trained context-aware encoders such as BERT (Devlin et al., 2018) and GPT (Radford et al., 2018) are publicly available. Most notably, Devlin et al. (2018) showed that BERT can be effectively used on many natural language understanding tasks, including NLI. Combining a pre-trained BERT encoder with a task-specific head, and then fine-tuning the entire model on the target task achieved state-of-the-art results on a number of tasks.

Directly applying BERT to NLI yields high accuracy if the dataset is large, and from a general domain (Phang et al., 2018). However, the dataset used in this shared task, MedNLI (Romanov and Shivade, 2018), is based on clinical notes (i.e. patient histories) and limited by size, thus it is a particularly challenging NLI task. To address this problem, we started with BERT, trained it on large NLI datasets, such as MultiNLI (Williams et al., 2018) and SNLI (Bowman et al., 2015), to support the inference reasoning, and then trained it further on our target task, MedNLI. This kind of intermediate training was shown to be effective when BERT is trained on a target with limited data (Phang et al., 2018). We call our approach *two-stage transfer learning*. In the first stage, we transfer the knowledge of the task, NLI, into our pre-trained model. During the second stage, we spe-

cialize our model on the MedNLI dataset. We hypothesized that this approach will produce higher accuracy on MedNLI compared to direct application of BERT.

By conducting extensive experiments, we explored the impact of different pre-trained model weights and transfer learning strategies. As a result, we signficantly improved the performance of BERT on MedNLI by initializing it from weights pre-trained on corpora close to clinical domain and applying two-stage transfer learning.

2 Model

Our baseline model is a BERT encoder (Devlin et al., 2018), combined with a classification head. The head outputs probabilities from a three-way softmax, corresponding to the three possible labels a sentence pair can have. The overview of this architecture can be seen from Figure 1.

2.1 BERT Encoder

We used BERT to encode our input tokens. Utilizing Transformer layers and self-attention (Vaswani et al., 2017), BERT looks at how the tokens are related, and outputs a hidden vector for each token inside of the input sequence. Therefore, BERT can process the sentence pair together as a whole sequence and output the encoded representation for all of it in one pass.

One of the reasons why we opted for a model

Weight	Model	Corpus	
Set	Size	Domain	Size
BERT$_{BASE}$	110M	Wikipedia	2.5B
		Books	0.8B
BERT$_{LARGE}$	340M	Wikipedia	2.5B
		Books	0.8B
BioBERT	110M	Biomedical	18.0B
		(+ BERT$_{BASE}$)	
SciBERT	110M	Biomedical	2.5B
		CompSci	0.6B

Table 1: Comparison of pre-trained BERT weight sets.

like the BERT encoder is that it completely avoids relying on recurrence and convolution operations. Replacing those with simple operations of self-attention (e.g. plain matrix multiplications) makes it more parallelizable and thus faster to train. Another reason is the strong starting point of BERT. It is a language modeling architecture, successfully trained on massive corpora. Lastly, there are a number of pre-trained weight sets for BERT from different domains. These weights are publicly available, which gave us the ability to test different starting points with minimal tinkering.

2.2 Pre-Trained Weights

We have considered four different pre-trained BERT weights as our starting point. From Devlin et al. (2018)'s work, we examined BERT$_{BASE}$ and BERT$_{LARGE}$. Both models were trained from scratch on English Wikipedia articles and books, and have the same vocabulary. However, the latter has more than triple the parameter size of the former.

Next we looked at BioBERT (Lee et al., 2019), which was trained on biomedical text. However, rather than randomly initialized weights, BioBERT utilizes BERT$_{BASE}$ as its starting point. Nevertheless, both versions use the same vocabulary for tokenization.

Lastly, we included SciBERT (Beltagy et al., 2019) in our experiments. It was trained on large-scale, annotated data from the scientific domain. It has the same size as BERT$_{BASE}$, and was trained from scratch, but uses a different vocabulary. Although the vocabulary size is the same as the original BERT, they have a total of 42% overlap between them. A direct comparison of the pre-trained BERT weights can be found in Table 1.

2.3 Input and Tokenization

During its pre-training process, one of the tasks BERT has to learn is the next sentence prediction (Devlin et al., 2018). It is a two-sentence classification task which requires predicting if the given sentences follow each other in the original text. Since our task is also a two-sentence classification task, we mimicked the inputs BERT receives during the pre-training. Thus, our input sentence pair is represented as "[CLS] *Premise* [SEP] *Hypothesis* [SEP]". The [CLS] and [SEP] are special tokens, denoting "Classification" and "Seperator" respectively. Since those are the tokens used during pre-training, they are kept in the same format to fully utilize BERT encoder's understanding of sentence pairs.

BERT uses WordPiece tokenizer (Wu et al., 2016), which divides the words in the input sequence into wordpieces (i.e. subwords). It maintains a good compromise between the character based representation's flexibility and the word based representation's efficiency. This balance improves the overall accuracy of the natural language system. Moreover, since it splits the infrequent words into wordpieces, it naturally handles the rare words problem (Wu et al., 2016).

2.4 Classification Head

To transform the token representations obtained from the encoder into label predictions, we used a small classification head. Following the convention of Devlin et al. (2018), we used the representation of the first token, [CLS], as the summary of the whole sequence. Then, this vector is linearly projected into a three-dimensional space such that each dimension represents the score for a label. Finally, a softmax operation is applied to convert the scores into class probabilities.

3 Datasets

The main dataset used in this shared task is MedNLI. Additionally, we used MultiNLI and SNLI to improve our accuracy via transfer learning.

3.1 MultiNLI and SNLI

MultiNLI and SNLI (Williams et al., 2018; Bowman et al., 2015) are general domain datasets, containing significantly more example pairs than MedNLI. The MultiNLI training dataset consists of five different genres of written and spoken

Dataset	Genre	Training Set Size
MedNLI	Patient Records	11, 232
MutliNLI	Fiction	77, 348
	Government	77, 350
	Slate	77, 306
	Telephone	83, 348
	Travel	77, 350
	Total	392, 702
SNLI	Image Captions	550, 152

Table 2: Comparison of NLI datasets by their genres and sizes.

1	Her a[$**$ Location $**$]e and PO intake have been normal.
2	on [$** 1 - 31 **$] Dr. [$** $Name (NI) $**$] documented that there was 1 positive ascitic fluid culture

Table 3: Two partial examples from MedNLI sentence pairs. Note the de-identification artifacts.

English, such as telephone conversations, travel guides and press releases from government websites. All examples in the SNLI training dataset are created from image captions, hence SNLI is regarded as a single genre. We have only used the training sets of MultiNLI and SNLI in our experiments to teach our model general domain NLI. A detailed comparison of the NLI datasets can be found in Table 2.

3.2 MedNLI

Our target dataset is MedNLI (Romanov and Shivade, 2018). We have used the provided splits without change. We trained our models on the training set, evaluated them on the development set. However, we did not use the testing set during training or hyperparameter selection.

MedNLI is created from text in the clinical domain, particularly patient records. To keep the confidentiality of various parties, names (of patients and places) and dates in the source texts have been de-identified. Therefore, the dataset contains artifacts, some examples of which are shown in Table 3.

In the example sequences, "a[$**$ Location $**$]e", "[$** 1 - 31 **$]" and "[$** $Name (NI) $**$]" are de-identification artifacts. This hurts the performance of our model since when WordPiece to-

1	her, a, [, $*$, $*$, location, $*$, $*$,], e, and, po, intake, have, been, normal, .
2	on, [, $*$, $*$, 1, $-$, 31, $*$, $*$,], dr, .,, [, $*$, $*$, name, (, ni,), $*$, $*$,], documented, that, there, was, 1, positive, as, ##cit, ##ic, fluid, culture

Table 4: The results of tokenizing the examples provided in the Table 3. The commas are used to separate different tokens.

1	Her ae and PO intake have been normal
	her, ae, and, po, intake, have, been, normal, .
2	on Dr. documented that there was 1 positive ascitic fluid culture
	on, dr, ., documented, that, there, was, 1, positive, as, ##cit, ##ic, fluid, culture

Table 5: The results of pre-processing the examples provided in Table 3, and their respective tokenizations.

kenizer segments the de-identified words into subwords, an excessive number of tokens are generated. The tokenization outputs of the examples are shown at Table 4.

Since BERT tokenizer treats the special characters such as "[" and "$*$" as wordpieces, the resulting tokenization contains an inflated number of unnecessary tokens. These tokens include the special characters and de-identified place-holders such as "Name", "Location". We suspected that WordPiece tokenizer makes the de-identification artifacts harmful to the performance since their tokenizations introduce too many erroneous tokens. To validate this observation, we performed some experiments where we removed the de-identified tokens from the MedNLI dataset while pre-processing. Resulting text and tokenizations of the sample sequences after the removal can be found in Table 5. As a result, the accuracies improved as shown in Section 5. Hence, we conducted the remaining experiments by performing this pre-processing step.

4 Training Strategy

4.1 Sequential Transfer Learning

Following the advice of Romanov and Shivade (2018), we experimented with transfer learning techniques to boost the accuracy of our model. In

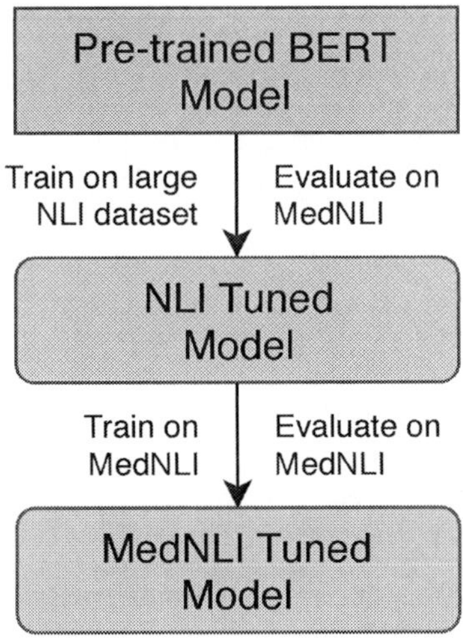

Figure 2: Two-stage sequential transfer learning strategy.

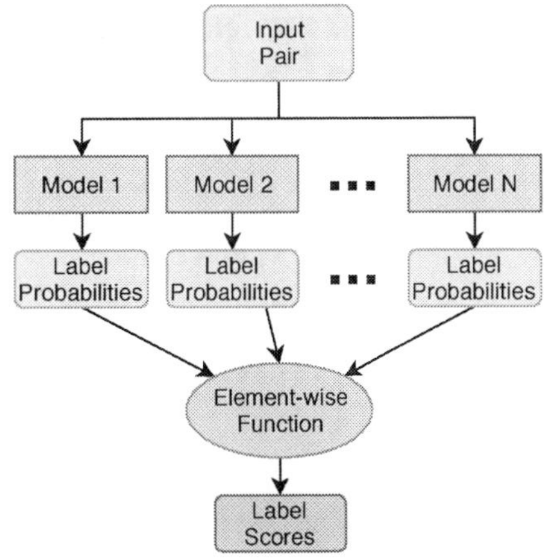

Figure 3: Ensembling procedure of N different models for an input premise-hypothesis pair.

fact, our baseline approach is itself a type of sequential transfer learning. We take a pre-trained BERT, append a classifier head on top of it, and fine-tune the whole model on the MedNLI training dataset. Since BERT is a powerful sequence encoder, this approach alone yielded better results on the MedNLI development dataset compared to the published baselines (Romanov and Shivade, 2018). However, because the dataset is limited in size, performances of these approaches are restricted. Therefore, instead of training our model directly on the target task, we added an intermediate training step. During this step, the model is trained on a large NLI dataset from a general domain such as MultiNLI and SNLI. Our aim is to help the model learn task specific reasoning skills using the large number of training examples. Then, the model is further fine-tuned on the MedNLI training dataset so that it can adapt its parameters to the clinical domain and MedNLI style. We call this strategy *two-stage sequential transfer learning*, which is shown in Figure 2. In both stages of the learning, the model is trained until its accuracy on the MedNLI development set is maximized.

Since there are multiple large and general domain NLI datasets available, we wanted to leverage them. Therefore, we also experimented with three-stage transfer learning. In the first two stages, we trained our model on MultiNLI and SNLI successively. In the third stage, we finally trained it on MedNLI. However, this configuration yielded slightly worse results compared to two-stage transfer, as discussed in Section 5. Therefore, we changed our method and utilized an ensembling approach to combine the benefits of the available datasets.

4.2 Ensembling

We perform ensembling on independently trained models to get the benefit of multiple datasets. First, these models are fully trained by two-stage transfer learning with different general domain datasets. After that, we compute a set of label probabilities for the sentence pairs using these models. Then, we combine the probabilities with an element-wise operation such as *sum*, *product*, or *max* to obtain a single score for each label. Figure 3 summarizes this process for a single input. Finally, we report the label corresponding to the highest score as our prediction for each example. This approach might be regarded as a soft version of *majority voting*, a commonly used ensembling method. As Section 5 shows, this approach yielded the best results obtained by our models.

5 Results and Discussion

5.1 Implementation Details

We used PyTorch (Paszke et al., 2017) as our deep learning framework and the BERT implementation provided by Hugging Face[1]. Our models are based on the *BertForSequenceClassification* class. We used the *BertAdam* optimizer from the same

[1]`https://github.com/huggingface/pytorch-pretrained-BERT`

Training Strategy	Training Set	Sequence Length	Batch Size	Optimizer	Initial Learning Rate
Direct Training					
	MedNLI	256	16	*BertAdam*	2×10^{-5}
Task pre-training					
	MultiNLI	256	32	*BertAdam*	2×10^{-5}
	SNLI	256	32	*BertAdam*	2×10^{-5}
Domain fine-tuning					
	MedNLI	256	16	*BertAdam*	2×10^{-5}

Table 6: The hyperparameters for different training settings.

repository, which imitates the Adam implementation of the original BERT.

We always evaluated our models according to the accuracy on the MedNLI development set. For all experiments, we trained a model until its accuracy in the last four epochs did not improve over its best accuracy. If a model kept improving, we stopped the training after 80 epochs. All model weights are updated during the training phases of the experiments.

The training procedure was stable, there were no serious failure cases with unexpectedly low accuracies. Nevertheless, to mitigate the possible negative effects of the randomness on the optimization, we repeated each experiment with three different seeds and selected the run with the best accuracy.

The primary hyperparameters of our model are the sequence length of the encoder layers, the batch size, and the initial learning rate of the optimizer, *BertAdam*. We kept the remaining hyperparameters such as dropout rate the same as the original implementation. To determine the sequence length, we counted the number of resulting tokens from MultiNLI, SNLI, and MedNLI sentence pairs after tokenization. Since the maximum token count is 256, we naturally set the sequence length to 256. When we trained the model on MedNLI, we used batches with size of 16. In contrast, when a large dataset (e.g. MultiNLI and SNLI) is used, we used batch size of 32 to speed up the training process. In the initial experiments, we tried $2 * 10^{-5}$, $3 * 10^{-5}$, and $5 * 10^{-5}$ as *BertAdam*'s initial learning rate which resulted in very similar accuracies. Nevertheless, since $2 * 10^{-5}$ generally yielded slightly better results, we conducted the remaining experiments with this initial learning rate. Table 6 summarizes the hyperparameters

Weight Set	Dev Accuracy
BERT$_{BASE}$	82.3%
BERT$_{LARGE}$	82.9%
BioBERT	83.4%
SciBERT	**83.7%**

Table 7: Results of directly training on MedNLI data, starting from various pre-trained weight sets. The highest accuracy is indicated with bold.

used in the training phases.

5.2 Experiments

We have conducted a number of experiments to test our model and the effectiveness of different training strategies. We report the resulting percent accuracies on the MedNLI development dataset.

To start with, we trained our model on MedNLI directly. We initialized it with various pre-trained weights to compare their performances. The results of the experiment can be seen from Table 7. It shows that the weights pre-trained on domains related to the task, such as biomedical or scientific data, have a noticeable advantage over the weights obtained from general text corpora, such as Wikipedia. Moreover, the effect of pre-training is more significant compared to the model size. BioBERT and SciBERT surpassed BERT$_{LARGE}$ although they are three times smaller.

Next, we tested the two-stage sequential transfer learning method using all combinations of pre-trained weights and rich NLI datasets mentioned before. Table 8 shows the results of this experiment. As expected, all models benefit from two-stage transfer learning. Moreover, the trend of specialized pre-trained weights having an advantage continues on this experiment as well. However, BioBERT outperforms SciBERT, unlike the pre-

Weight Set	MultiNLI	SNLI
BERT$_{BASE}$	82.9%	82.4%
BERT$_{LARGE}$	83.7%	84.4%
BioBERT	85.6%	**85.8%**
SciBERT	85.4%	85.6%

Table 8: Development set accuracies achieved by performing two-stage sequential transfer learning, utilizing different intermediary datasets, and starting from various pre-trained weights. The highest accuracy is indicated with bold.

	MultiNLI	SNLI
BERT$_{BASE}$	+0.7%	±0.0%
BERT$_{LARGE}$	+0.6%	+2.2%
BioBERT	+1.2%	+0.6%
SciBERT	+0.4%	+1.3%

Table 9: Accuracies gained by pre-processing the MedNLI dataset, after performing two-stage sequential transfer learning with different starting points and datasets.

vious results. We suspect that since BioBERT is pre-trained starting from BERT$_{BASE}$, it benefits more from general domain task training.

In order to test the effect of pre-processing described in Section 3.2, we repeated the two-stage training experiment without removing the de-identification artifacts. We compared the results of this experiment with the previous results to see how the accuracy is affected. The improvements obtained from the pre-processing can be found in Table 9. It increases the accuracy in all cases, except for BERT$_{BASE}$ - SNLI, where the accuracy is unchanged. Note that all other experiments are conducted with the pre-processing is enabled.

We have also tested the performance of three-stage sequential transfer learning on BioBERT. Training on SNLI and MultiNLI sequentially before MedNLI produced lower accuracies compared to the two-stage transfer learning experiment. Moreover, switching the training order of SNLI and MultiNLI did not change the resulting accuracy. We suspect that further training on a second intermediate dataset brings the model closer to a worse local optimum. A comparison between different training strategies on BioBERT can be found in Table 10.

Lastly, we experimented with ensembling. We tested four ensemble models, one for each pre-

Training Strategy	Dev Accuracy
Direct Training on MedNLI	83.4%
Two-Stage Transfer:	
MultiNLI → MedNLI	85.6%
SNLI → MedNLI	**85.8%**
Three-Stage Transfer:	
MultiNLI → SNLI → MedNLI	84.9%
SNLI → MultiNLI → MedNLI	84.9%

Table 10: Single model development set accuracies of different training strategies, starting from BioBERT. The highest accuracy is indicated with bold.

Weight Set	Sum	Product	Max
BERT$_{BASE}$	83.3%	83.4%	83.3%
BERT$_{LARGE}$	85.1%	85.0%	85.2%
BioBERT	**86.1%**	86.0%	**86.1%**
SciBERT	85.9%	85.8%	85.8%

Table 11: Development set accuracies resulting from ensembling two models starting from the same pre-trained weights. The highest accuracies are indicated with bold.

trained BERT variant. For each of these starting points, we combined the two models obtained from the two-stage transfer learning experiment. One of these models is trained with MultiNLI, the other with SNLI. We tried three different element-wise operations for ensembling, and compared their effects. Table 11 shows that this approach yielded better results compared to the two-stage transfer methods. Therefore, we effectively combined the benefits of training on different, rich datasets.

Among all the experiments, the best result we obtained is 86.1% accuracy by ensembling BioBERT models trained with two-stage transfer learning. That accuracy was obtained by combining the output probabilities with element-wise sum operation.

Therefore, we participated in the shared task with that ensembled model. Consequently, our model achieved 84.7% accuracy on the unseen test dataset reserved for the shared task.

5.3 Error Analysis

Following Romanov and Shivade (2018), we conducted a similar error analysis on MedNLI development set to understand how the different methods improve the baseline. We chose BioBERT for

Ground Truth Label	CON		ENT		NTR	
Predicted Label	ENT	NTR	CON	NTR	CON	ENT
Direct Training on MedNLI	33	22	19	84	20	53
Two-Stage Transfer:						
MultiNLI → MedNLI	16	14	24	77	22	48
SNLI → MedNLI	23	14	15	65	18	63
Three-Stage Transfer:						
MultiNLI → SNLI → MedNLI	19	17	18	83	22	51
SNLI → MultiNLI → MedNLI	21	20	13	73	24	59
Ensemble Model:						
MultiNLI + SNLI with Sum	20	15	16	67	20	56

Table 12: Label breakdown of errors made in MedNLI development set by different models. All models were trained with different strategies starting from BioBERT. "CON", "ENT" and "NTR" label abbreviations mean "Contradiction", "Entailment", and "Neutral" respectively.

Category	ABB	MED	NUM	WOR
Direct Training on MedNLI	32	126	30	43
Two-Stage Transfer:				
MultiNLI → MedNLI	24	113	29	35
SNLI → MedNLI	22	111	28	37
Three-Stage Transfer:				
MultiNLI → SNLI → MedNLI	23	116	30	41
SNLI → MultiNLI → MedNLI	25	116	30	39
Ensemble Model:				
MultiNLI + SNLI with Sum	23	106	30	35

Table 13: Category breakdown of errors made in MedNLI development set by different models. All models were trained with different strategies starting from BioBERT. The category abbreviations mean "Abbreviation", "Medical Knowledge", "Numerical Reasoning", and "World Knowledge" respectively.

the analysis since it is the starting point of our highest-scoring model. We compared the errors made by the models resulting from direct training, two-stage transfer learning, three-stage transfer learning, and ensembling.

In the first study, we concentrated on analyzing the distribution of the misclassified examples over labels, whose results are shown at Table 12. First thing to notice is that the errors mostly originated from confusion between the entailment and the neutral labels. For all models, this confusion causes approximately 60% of the errors. Two-stage transfer results show that intermediate training on a large dataset helps in identifying the contradiction relation. The error counts of the ensemble model are lower than or around the averages of the models that are ensembled.

The second analysis is separating the errors into four broad categories. These involve "Abbreviation", "Medical Knowledge", "Numerical Rea-soning", and "World Knowledge". "Abbrevation" represents the existence of medical abbrevations critical to decode the inference relation. "Medical Knowledge" refers to the requirement of reasoning with medical knowledge. "Numerical Reasoning" denotes that the inference type depends on the value of the number(s) present in the sentence pair. "Word Knowledge" indicates the need for common sense or general domain knowledge to understand the inference. We manually categorized each misclassified sentence pair.

Table 13 shows the results of the error categorization. "Numerical Reasoning" errors are almost the same across all models. We believe that this is because the scale of the numerical values highly depends on the context. Adding general domain knowledge does not seem to help in learning numerical scales on the medical domain. As the two-stage transfer results show, the intermediate training on a general domain NLI dataset decreases the error rates on the remaining three categories. However, training with three-stage transfer learning does not improve the "Medical Knowledge" and "World Knowledge" categories as much as the two-stage transfer. Although getting poorer results after training with more data seems counter-intuitive, Romanov and Shivade (2018) observed a similar trend on transfer learning using SNLI, and the genres of MultiNLI. In both findings, the model performance does not directly correlate with the size of the intermediate training data.

Lastly, most of the improvements introduced by ensembling fall under "Medical Knowledge". Since the component models are trained on different datasets on intermediate training step, the errors they made on the MedNLI development set differ. We suspect that the models are not very confident in some of their "Medical Knowledge" errors, hence the other model may correct these mistakes to an extent.

6 Conclusion

In this paper, we showed that a pre-trained BERT encoder, combined with a classifier head forms a strong baseline for MedNLI task. More importantly, we also demonstrated that the model's accuracy can be remarkably improved by utilizing a two-stage transfer learning strategy. The success of the final model depends on the initial BERT weights, as well as the particular transfer learning method. Finally, we showed that ensembling two separate models trained on different NLI datasets is more effective than using these datasets to train a single model. Our best model, BioBERT ensembled, achieved 86.1% accuracy on the MedNLI development set, and 84.7% accuracy on the unseen test dataset reserved for the MEDIQA 2019's NLI Shared Task.

While empirical results show that contextualized sequence encoders enhanced with transfer learning are strong, their performances might be further improved with external knowledge integration. As future work, we would like to extend BERT's contextualized word vectors using semantic relationships.

Acknowledgments

Cemil Cengiz is supported by Huawei Turkey R&D Center through the Huawei Graduate Research Support Scholarship.

References

Iz Beltagy, Arman Cohan, and Kyle Lo. 2019. Scibert: Pretrained contextualized embeddings for scientific text. *Computing Research Repository*, arXiv:1903.10676.

Asma Ben Abacha, Chaitanya Shivade, and Dina Demner-Fushman. 2019. Overview of the mediqa 2019 shared task on textual inference, question entailment and question answering. In *Proceedings of the BioNLP 2019 workshop, Florence, Italy, August 1, 2019*. Association for Computational Linguistics.

Samuel R. Bowman, Gabor Angeli, Christopher Potts, and Christopher D. Manning. 2015. A large annotated corpus for learning natural language inference. In *Proceedings of the 2015 Conference on Empirical Methods in Natural Language Processing*. Association for Computational Linguistics.

Alexis Conneau, Douwe Kiela, Holger Schwenk, Loc Barrault, and Antoine Bordes. 2017. Supervised learning of universal sentence representations from natural language inference data. In *Proceedings of the 2017 Conference on Empirical Methods in Natural Language Processing*. Association for Computational Linguistics.

Jacob Devlin, Ming-Wei Chang, Kenton Lee, and Kristina Toutanova. 2018. BERT: Pre-training of Deep Bidirectional Transformers for Language Understanding. *Computing Research Repository*, arXiv:1810.04805.

Sepp Hochreiter and Jrgen Schmidhuber. 1997. Long short-term memory. *Neural Computation*, 9(8):1735–1780.

Surafel Melaku Lakew, Mauro Cettolo, and Marcello Federico. 2018. A comparison of transformer and recurrent neural networks on multilingual neural machine translation. In *Proceedings of the 27th International Conference on Computational Linguistics*, pages 641–652, Santa Fe, New Mexico, USA. Association for Computational Linguistics.

Jinhyuk Lee, Wonjin Yoon, Sungdong Kim, Donghyeon Kim, Sunkyu Kim, Chan Ho So, and Jaewoo Kang. 2019. Biobert: a pre-trained biomedical language representation model for biomedical text mining. *Computing Research Repository*, arXiv:1901.08746.

Xiaodong Liu, Kevin Duh, and Jianfeng Gao. 2018. Stochastic answer networks for natural language inference. *Computing Research Repository*, arXiv:1804.07888.

Xiaodong Liu, Pengcheng He, Weizhu Chen, and Jianfeng Gao. 2019. Multi-task deep neural networks for natural language understanding. *Computing Research Repository*, arXiv:1901.11504.

Bill MacCartney. 2009. *Natural Language Inference*. Ph.D. thesis, Stanford University, Stanford, CA, USA. AAI3364139.

Adam Paszke, Sam Gross, Soumith Chintala, Gregory Chanan, Edward Yang, Zachary DeVito, Zeming Lin, Alban Desmaison, Luca Antiga, and Adam Lerer. 2017. Automatic differentiation in pytorch. In *NIPS-W*.

Jason Phang, Thibault Févry, and Samuel R. Bowman. 2018. Sentence encoders on stilts: Supplementary training on intermediate labeled-data tasks. *Computing Research Repository*, arXiv:1811.01088.

Alec Radford, Karthik Narasimhan, Tim Salimans, and Ilya Sutskever. 2018. Improving language understanding by generative pre-training.

Alexey Romanov and Chaitanya Shivade. 2018. Lessons from natural language inference in the clinical domain. In *Proceedings of the 2018 Conference on Empirical Methods in Natural Language Processing*, pages 1586–1596, Brussels, Belgium. Association for Computational Linguistics.

Ashish Vaswani, Noam Shazeer, Niki Parmar, Jakob Uszkoreit, Llion Jones, Aidan N Gomez, Łukasz Kaiser, and Illia Polosukhin. 2017. Attention is all you need. In I. Guyon, U. V. Luxburg, S. Bengio, H. Wallach, R. Fergus, S. Vishwanathan, and R. Garnett, editors, *Advances in Neural Information Processing Systems 30*, pages 5998–6008. Curran Associates, Inc.

Adina Williams, Nikita Nangia, and Samuel Bowman. 2018. A broad-coverage challenge corpus for sentence understanding through inference. In *Proceedings of the 2018 Conference of the North American Chapter of the Association for Computational Linguistics: Human Language Technologies, Volume 1 (Long Papers)*. Association for Computational Linguistics.

Yonghui Wu, Mike Schuster, Zhifeng Chen, Quoc V. Le, Mohammad Norouzi, Wolfgang Macherey, Maxim Krikun, Yuan Cao, Qin Gao, Klaus Macherey, Jeff Klingner, Apurva Shah, Melvin Johnson, Xiaobing Liu, Lukasz Kaiser, Stephan Gouws, Yoshikiyo Kato, Taku Kudo, Hideto Kazawa, Keith Stevens, George Kurian, Nishant Patil, Wei Wang, Cliff Young, Jason Smith, Jason Riesa, Alex Rudnick, Oriol Vinyals, Greg Corrado, Macduff Hughes, and Jeffrey Dean. 2016. Google's neural machine translation system: Bridging the gap between human and machine translation. *Computing Research Repository*, arXiv:1609.08144.

DUT-NLP at MEDIQA 2019: An Adversarial Multi-Task Network to Jointly Model Recognizing Question Entailment and Question Answering

Huiwei Zhou, Xuefei Li, Weihong Yao, Chengkun Lang, Shixian Ning
School of Computer Science and Technology
Dalian University of Technology
116024 Dalian, China
{zhouhuiwei, weihongy}@dlut.edu.cn
{lixuefei, kunkun, ningshixian}@mail.dlut.edu.cn

Abstract

In this paper, we propose a novel model called Adversarial Multi-Task Network (AMTN) for jointly modeling Recognizing Question Entailment (RQE) and medical Question Answering (QA) tasks. AMTN utilizes a pre-trained BioBERT model and an Interactive Transformer to learn the shared semantic representations across different task through parameter sharing mechanism. Meanwhile, an adversarial training strategy is introduced to separate the private features of each task from the shared representations. Experiments on BioNLP 2019 RQE and QA shared task datasets show that our model benefits from the shared representations of both tasks provided by multi-task learning and adversarial training, and obtains significant improvements upon the single-task models.

1 Introduction

With the rapid development of Internet and medical care, online health queries are increasing at a high rate. In 2012, 59% of U.S. adults looked for health information online[1]. However, it is always difficult for search engines to return relevant and trustworthy health information every time if the symptoms are not accurately described (Pletneva et al., 2012; Scantlebury et al., 2017). Therefore, many websites provide online doctor consultation services, which can answer questions or give advice from doctors or experts to the customers. Unfortunately, manually answering some simple queries or answering similar questions multiple times is quite time-consuming and wasteful. A Question Answering (QA) system that can automatically understand and answer the health care questions asked by customers is urgently needed (Wren, 2012).

To this end, BioNLP 2019 (Abacha et al., 2019) provides a series of challenging shared tasks, including: (1) Natural Language Inference (NLI) in the clinical domain; (2) Recognizing Question Entailment (RQE); (3) medical Question Answering (QA). This paper mainly focuses on RQE and QA task.

RQE task aims at identifying entailment relation between two questions in the context of QA (Abacha and Fushman, 2016), which can be represented as "a question Q1 entails a question Q2 if every answer to Q2 is also a complete or partial answer to Q1".

QA task aims at automatically filtering and improving the ranking of automatically retrieved answers (Abacha and Fushman, 2019). There are two targets for QA: (1) determining whether the given sentence could answer the given question; (2) ranking all the right answers according to their relevance to the question.

Neural networks and deep learning (DL) currently provide the best solutions for RQE and QA tasks. Among various neural networks, such as traditional Convolutional Neural Networks (CNN) (LeCun et al., 1998) and Long Short-Term Memory (LSTM) networks (Hochreiter and Schmidhuber, 1997), Transformer (Vaswani et al., 2017) has demonstrated superiority in multiple

---

[1] http://www.pewinternet.org/2013/01/15/health-online-2013/

Proceedings of the BioNLP 2019 workshop, pages 437–445
Florence, Italy, August 1, 2019. ©2019 Association for Computational Linguistics

natural language processing tasks (Verga et al., 2017; Shen et al., 2017; Yu et al., 2018). Transformer (Vaswani et al., 2017) is based solely on attention mechanisms and it can effectively capture the long-range dependencies between words.

More recently, the pre-trained language models, such as ELMo (Matthew et al., 2018), OpenAI GPT[2], and BERT (Devlin et al., 2018), have shown their effectiveness to capture the deep semantic and syntactic information of words. BioBERT (Lee et al., 2019) is one of the BERT-based pre-trained language model for biomedical domain, and it achieves great improvement in many biomedical tasks. For this reason, we believe that the pre-trained language models, especially the BioBERT, should be valid for RQE and QA under reasonable use.

Most previous researches train the model of RQE task or QA task separately based on a single training set. However, such single-task method cannot provide essential mutual supports between the two tasks. The inherent interactions between the two tasks might help us do even better on the RQE and QA tasks. RQE task can find Frequently Asked Questions (FAQs) similar to a consumer health question, providing consumers with appropriate FAQs and enabling QA systems to identify the right answers with greater precision and higher speed (Harabagiu and Hickl, 2006).

Multi-Task Learning (MTL) is a learning paradigm in machine learning and its aim is to leverage shared representations contained in multiple related tasks to help improve the generalization performance of all the tasks. MTL is usually done with parameter sharing of hidden layers. Hard parameter sharing is the most commonly used approach to MTL in neural networks. It is generally applied by sharing the hidden layers between all tasks, while keeping several task-specific output layers. However, it is difficult for MTL to distinguish the commonalities and differences between different tasks.

A common way to improve the robustness of the system is to train the system using different datasets through adversarial training (Goodfellow et al., 2014). Chen et al. (2017) propose a shared-private model, which extracts shared features and private features from multiple corpus, and introduces adversarial training for shared representation learning. Drawing on the practices of previous studies, we plan to use an adversarial multi-task framework to extract the noise-robust shared representation directly.

Considering the similarity between RQE and QA tasks, this paper proposes a novel Adversarial Multi-Task Network (AMTN) to jointly model these two tasks. Specifically, AMTN first utilizes BioBERT as an embedding layer to generate context-dependent word representations. Then, a common Interactive Transformer layer is introduced for sentence representation learning and inter-sentence relationship modeling, which allows knowledge transfer from other tasks. Finally, two specific classifiers are used for RQE and QA tasks respectively. Here, we only consider the target (1) of QA task for the multi-task learning to ensure the consistency between RQE and QA tasks. Furthermore, to prevent the shared and private feature spaces from interfering with each other, an adversarial training strategy is introduced to make the shared feature representations to be more compatible and task-invariant among different tasks. Experimental results show that our AMTN model is effective to improve the performance for both RQE and QA tasks upon the single-task models, which demonstrates the superiority of the adversarial multi-task strategy.

Our contributions can be summarized into two folds.

- A well-designed Interactive Transformer layer is introduced for sentence representation learning and inter-sentence relationship modeling.

- A novel adversarial multi-task strategy is introduced to jointly model RQE and QA tasks, in which multi-task learning is proposed for shared representation learning and adversarial training is used to force the shared representation purer and task-invariant.

2 Method

This section gives a detailed description of the proposed AMTN, which is shown in Figure 1. AMTN mainly consists of three parts: a shared

---

[2] https://s3-us-west-2.amazonaws.com/openai-assets/research-covers/language-unsupervised/language_understanding_paper.pdf

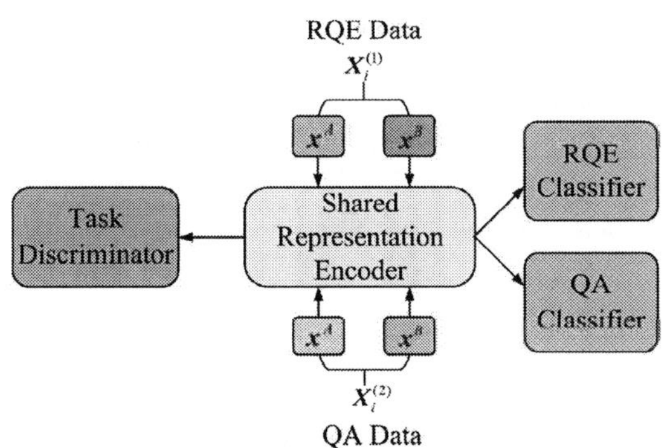

Figure 1: The framework of AMTN.

encoder, a task discriminator and two classifiers for the RQE task and the QA task, respectively.

Shared encoder is used to learn the shared semantic representations across different tasks through parameter sharing mechanism. Task discriminator is used to form an adversarial training with the shared encoder to separate the private features of each task from the shared representations. Two specific classifiers are applied to judge whether a sentence pair is an entailment relationship (RQE task) or a question-and-answer relationship (QA task).

Next, we will use four subsections to introduce our AMTN model in detail: Data Preprocessing, Shared Representation Learning, Task Specific Classifier and Adversarial Training.

2.1 Data Preprocessing

Define a data set $\{X_i^{(k)}, y_i^{(k)}\}_{i=1}^{N_k}$, where $X_i^{(k)}$ is the ith input for the kth task, $y_i^{(k)}$ is the corresponding labels of $X_i^{(k)}$, N_k is the number of training data in the kth task. In this paper, $k=1$ refers to RQE task, and $k=2$ refers to QA task. Each $X_i^{(k)}$ is composed of the concatenation of a unique [CLS] flag with a sentence pair $x^A = \{x_1^A, x_2^A, ..., x_n^A\}$ and $x^B = \{x_1^B, x_2^B, ..., x_m^B\}$, where n, m are the sequence lengths. Specially, since the answers of the QA task are too long, we intercept the first sentence of them as x^B .

2.2 Shared Representation Learning

We use the shared encoder to learn the shared representations as the input for the classifiers and the task discriminator. Figure 2 illustrates the architecture of shared encoder, which contains BioBERT Embedding Layer, Interactive Transformer Layer and Combination Layer.

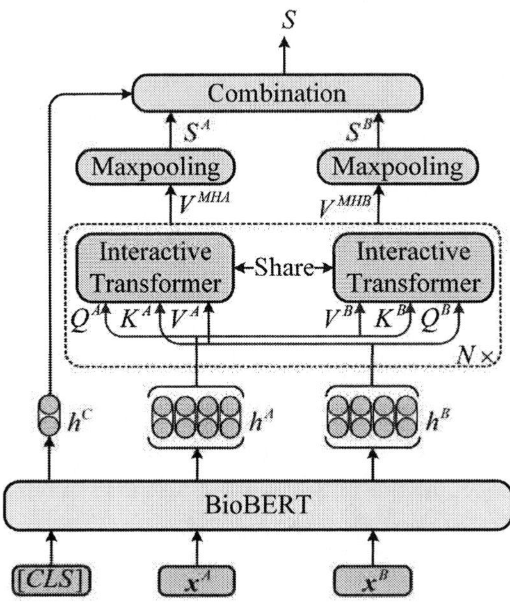

Figure 2: The architecture of the shared encoder.

BioBERT Embedding Layer: BioBERT is a domain specific language representation model pre-trained on large-scale biomedical corpora (Lee et al., 2019). It could effectively enhance the learning ability of encoding biomedical information.

We use BioBERT as an embedding layer, whose final hidden representation of each word is treated as word embedding. Given the sequence input X , the corresponding hidden representation sequence $H = \{h^C, h^A, h^B\}$ can be obtained through the BioBERT layer, where $h^A \in \mathbb{R}^{d_1 \times n}$, $h^B \in \mathbb{R}^{d_1 \times m}$ and $h^C \in \mathbb{R}^{d_1 \times 1}$ correspond to the sentence x^A, the sentence x^B and the unique [CLS] flag respectively, and d_1 is the output dimension of BioBERT.

Interactive Transformer Layer: To effectively capture the long dependency information and establish an interaction between the two sentences, the hidden representation sequences h^A and h^B are fed to an Interactive Transformer Layer. The Interactive Transformer consists of N blocks, each of which contains a multi-head attention with interactive process. Multi-head attention performs the scaled dot product attention multiple times on linearly projected query (Q), Key (K) and Value (V), which is shown in the following formula:

$$\text{Attention}(Q, K, V) = \text{softmax}(\frac{Q^T K}{\sqrt{d_K}})V \qquad (1)$$

where d_K is the dimension of K. Vaswani et al. (2017) point out that the input of softmax grows large in magnitude, pushing the softmax function into regions where it has extremely small gradients. Therefore, the dot productions are scaled by $1/\sqrt{d_K}$ to counteract this effect.

For the first sentence x^A, we take its hidden representation h^A as Q and h^B as K, V. In this way, the information flow inside features of the sentence x^B are dynamically conditioned on the features of the sentence x^A. The inputs for the first sentence x^A can be represented as:

$$
\begin{aligned}
Q^A &= \left\{ h_1^A, h_2^A, ..., h_n^A \right\}, \\
K^A &= \left\{ h_1^B, h_2^B, ..., h_m^B \right\}, \\
V^A &= \left\{ h_1^B, h_2^B, ..., h_m^B \right\}.
\end{aligned}
\qquad (2)
$$

For the second sentence x^B, we take h^B as Q and take h^A as K, V, which can be represented as:

$$
\begin{aligned}
Q^B &= \left\{ h_1^B, h_2^B, ..., h_m^B \right\}, \\
K^B &= \left\{ h_1^A, h_2^A, ..., h_n^A \right\}, \\
V^B &= \left\{ h_1^A, h_2^A, ..., h_n^A \right\}.
\end{aligned}
\qquad (3)
$$

Therefore, the multi-head attentions with interactive process for the given pair of sentences can be formulated as:

$$head_i^A = \text{Attention}(\mathbf{W}_i^Q Q^A, \mathbf{W}_i^K K^A, \mathbf{W}_i^V V^A) \qquad (4)$$

$$head_i^B = \text{Attention}(\mathbf{W}_i^Q Q^B, \mathbf{W}_i^K K^B, \mathbf{W}_i^V V^B) \qquad (5)$$

$$V^{MHA} = \mathbf{W}^H [head_1^A; head_2^A; ...; head_L^A] \qquad (6)$$

$$V^{MHB} = \mathbf{W}^H [head_1^B; head_2^B; ...; head_L^B] \qquad (7)$$

where $\mathbf{W}_i^Q \in \mathbb{R}^{d_{head} \times d_1}$, $\mathbf{W}_i^K \in \mathbb{R}^{d_{head} \times d_1}$, $\mathbf{W}_i^V \in \mathbb{R}^{d_{head} \times d_1}$ and $\mathbf{W}^H \in \mathbb{R}^{d_1 \times L d_{head}}$ are trainable shared parameters. $[head_1; head_2; ...; head_L]$ is a concatenation of outputs of L heads.

Different from the original Transformer, in which the input of Q, K and V are all the same, Interactive Transformer takes different sentences as the inputs of Q and K, V. In this way, we expect to effectively compute dependencies between any two words of the sentence pairs and encode the abundant semantic information for each sequence word.

Combination Layer: After modeling the association between the two sentences, we utilize the max pooling operation to obtain the final shared semantic representations of x^A and x^B respectively:

$$S^A = Maxpooling(V^{MHA}) \qquad (8)$$

$$S^B = Maxpooling(V^{MHB}) \qquad (9)$$

Then, we perform vector combination on S^A, S^B, and flag representation h^C through a dense layer to generate the sentence pair representation S for classification, which is calculated as follows:

$$
\begin{aligned}
S = \text{ReLU}(\mathbf{W}_0 [S^A; S^B; S^A - S^B; \\
S^A \odot S^B; h^C] + b_0)
\end{aligned}
\qquad (10)
$$

where $\mathbf{W}_0 \in \mathbb{R}^{d_0 \times 5 d_1}$ and $b_0 \in \mathbb{R}^{d_0 \times 1}$ are trainable parameters.

2.3 Task Specific Classifier

For each task, a specific classifier is employed to judge whether a sentence pair is an entailment relationship (RQE) or a question-and-answer relationship (QA). Each classifier is composed of a two-layer fully-connected neural network, which uses a ReLU nonlinearity after the first fully connected layer and a softmax nonlinearity after the second fully connected layer. It can be written as follows:

$$\hat{\mathbf{y}}_i^{(m)} = softmax\left(\mathbf{W}_2 \text{ReLU}\left(\mathbf{W}_1 S + b_1\right) + b_2\right) \qquad (11)$$

The classifier takes the sentence pair representation S as input and outputs a probability distribution to predict whether the current sentence pair is entailment relation or question-and-answer relation.

Both classifiers are trained by optimizing the cross-entropy loss as follows:

$$J_{classifier} = -\sum_{i=1}^{N_k} \sum_{j=1}^{C} \left(\mathbf{y}_{i,j}^{(k)} \log\left(\hat{\mathbf{y}}_{i,j}^{(k)}\right)\right) \qquad (12)$$

where C is the number of categories of classification label, $\hat{\mathbf{y}}_{i,j}$ is the predicted probability of the jth category of the ith sentence pair.

2.4 Adversarial Training

In order to make shared representations contain more common information and reduce the mixing of task-specific information, adversarial training is introduced into the above multi-task framework.

The goal of the proposed adversarial training strategy is to form an adversary with shared representation learning by introducing a task discriminator. In this paper, we take the shared encoder as generative network G and the task discriminator as discriminative model D , in which G needs to learn as much semantic information as possible from the shared data distribution between the two tasks and D aims to determine which task (RQE or QA) the input sentence belongs to by using the shared representations.

Specifically, we first use the shared encoder $G(X, \theta_s)$, which is mentioned in section 2.2, to get the sentence pair representation S. θ_s is the shared parameter need to be trained. Then, the shared representations will be fed to the task discriminator D to determine the task to which the current input belongs. D can be expressed by the following formula:

$$D(S, \theta_d) = softmax\left(\mathbf{W}_4 \, ReLU\left(\mathbf{W}_3 S + b_3\right) + b_4\right) \quad (13)$$

Besides the task loss for RQE and QA, we additionally introduce an adversarial loss J_{adv} to prevent task-specific feature from creeping into shared space and thus get a purer shared representation. The adversarial loss J_{adv} is trained in alternating fashion as shown below:

$$J_{adv} = \min_{\theta_s} \left(\max_{\theta_d} \left(\sum_{k=1}^{2} \sum_{i=1}^{N_k} \mathbf{t}_i^{(k)} \log\left[D\left(G\left(X, \theta_s\right), \theta_d\right)\right] \right) \right)$$

$$(14)$$

where $\mathbf{t}_i^{(k)}$ is the correct task label (RQE task or QA task) of the given sentence pair X. Here the basic idea is that, the shared representations learned by the shared encoder need to mislead the task discriminator. At the same time, the task discriminator needs to predict the task (RQE or QA) to which the data belongs as accurately as possible. The two are adversarial to each other and alternately optimized to separate the private features from the shared representations.

Finally, the shared encoder and the task discriminator reach a balance point and achieve mutual promotion.

3 Experiments

3.1 Dataset

Our experiments are conducted on the BioNLP RQE and QA shared tasks. The QA dataset contains a total of 3042 question-answer pairs: 1701 for training, 234 for validation, and 1107 for test. The RQE dataset contains a total of 9120 question pairs: 8588 for training, 302 for validation, and 230 for test. The statistic of the two datasets are shown in Table 1.

Task	Train	Validation	Test
RQE	8588	302	230
QA	1701	234	1107

Table 1: Statistic of sentence pairs in RQE and QA datasets.

3.2 Experimental settings and metric

In the shared encoder module, we use the pre-trained uncased BioBERT$_{base}$[3] for computational complexity considerations. The number of its Transformer blocks and multiple heads are both 12. For the Interactive Transformer, we use 3 blocks with 16 heads. The hidden layer dimension of BioBERT$_{base}$ and Interactive Transformer are both set to 768. We use a mini batch size of 8 and epoch of 30. Adam optimizer (Kingma and Ba, 2014) is used for both shared encoder and task discriminator to tune the parameters at the learning rates of $\lambda_1 = 1e-5$ and $\lambda_2 = 1e-4$, respectively. Specially, due to the small quantity of QA training data, we oversample it three times during training in order to balance the dataset of two tasks. The hyper-parameters settings used in this paper are shown in Table 2. The performance RQE and QA tasks are evaluated by the official evaluation scripts[4], which adopt accuracy as the evaluation metric.

<hr>

[3] https://github.com/naver/biobert-pretrained
[4] https://github.com/abachaa/MEDIQA2019/tree/master/Eval_Scripts

Hyper-parameters	Value
Pre-trained Model Heads	12
Pre-trained Model Blocks	12
Interactive Transformer Heads	16
Interactive Transformer Blocks	3
Hidden Layer Dimension	768
Epoch	30
Mini-batch	8
Learning Rate for Shared Encoder	1e-5
Learning Rate for Discriminator	1e-4

Table 2: Hyper-parameters settings.

3.3 Effects of the Adversarial Multi-Task Learning Strategy

This section first proposes two baseline strategies for comparison as described below:

- **Multi-Task**: Under this strategy, the architecture is constructed by removing the discriminator D from AMTN. We call it AMTN-Discriminator.

- **Single-Task**: Under this strategy, the architecture is constructed by removing the discriminator D from AMTN, and using the same classifier for the two tasks. We call it Single-Task Network (STN).

The results are shown in Table 3. From the table, we can see that single-task learning achieves the worst results, which is probably due to the simple model architecture. For the three methods using different dataset in Single-task learning, STN (QA+RQE) performs better than

STN (QA) and STN (RQE). It demonstrates that the two datasets have quite similar information distributions that could adequately complement each other and contribute to both RQE and QA tasks.

From the second block in Table 4, we can see that **Multi-Task** strategy performs clearly better than **Single-Task**. Note that, AMTN-Discriminator has an accuracy rate of 63.6% and 74.5% for RQE and QA tasks, which is the result of our submission in the task website. Multi-task learning jointly trains multiple sub-task models through a shared encoder. It can effectively capture the common features of the two task data, thereby promoting the generalization ability of RQE and QA tasks synchronously.

To explore the effects of the proposed adversarial multi-task strategy. Furthermore, we arm the above **Multi-Task** strategy with adversarial training, i.e. adding a discriminator to form the adversary with shared representation learning:

- **Adversarial Multi-Task**: Under this strategy, two architectures are constructed. One is our proposed AMTN. The other is a variant of AMTN, which adds a Private Encoder for each task to parallelly learn task-specific representations and shared representations.

Table 4 lists the comparison results. Compared with **Multi-Task**. AMTN achieves further improvement (0.7% and 1.3% accuracy for RQE and QA tasks respectively) with the help of additional task discriminator and the introduction

Strategy	Architecture	RQE	QA
Single-Task	STN (QA)	59.1	71.4
	STN (RQE)	50.0	61.0
	STN (QA+RQE)	61.7	71.4
Multi-Task	AMTN-Discriminator[†]	63.6	74.5

Table 3: Effects of multi-task learning strategy. All the results are reported by accuracy (%). STN (QA), STN (RQE) and STN (QA+RQE) represent STN trained on QA dataset, QRE dataset and both datasets, respectively. [†] indicates our submission results.

Strategy	Architecture	RQE	QA
Multi-Task	AMTN-Discriminator[†]	63.6	74.5
Adversarial Multi-Task	AMTN	**64.3**	**75.8**
	AMTN+Private Encoder	58.3	72.5

Table 4: Effects of adversarial multi-task learning strategy. All the results are reported by accuracy (%). [†] indicates our submission results. Bold font indicates the best performance.

of adversarial loss. We believe that the discriminator could strip private features from shared representations and make shared representations more general.

Finally, when we add a private encoder for each task, i.e. AMTN+Private Encoder, we can see that the performance is significantly reduced by 6.0% and 3.3% accuracy in RQE and QA tasks, respectively. Although private representation could provide task-specific information, it will introduce too many redundant parameters that could make the model prone to over-fitting, resulting in performance degradation.

3.4 Effects of the Shared Encoder

Our AMTN model uses **Interactive Transformer** as shared encoder to perform shared representation learning. To verify the effects of the shared encoder, we compare the **Interactive Transformer** with the following three baseline methods:

- **CNN** encoder: This method uses Convolutional Neural Network (CNN) to encode each sentence. 256 filters with window size of 3,4,5 are used in CNN, respectively.

- **Bi-LSTM** encoder: This method uses a single-layer bidirectional Long Short-Term Memory network (Bi-LSTM) to encode each sentence. The hidden layer dimension of each direction is set to 384.

- **Transformer** encoder: This method uses an original Transformer to encode each sentence. For sentence x^A, the three input (Q, K and V) of Transformer are all h^A. For sentence x^B, the three input (Q, K and V) of Transformer are all h^B. That is to say, there is no interaction between the two sentences in this encoder.

Note that, the final sentence representation is generated by max pooling on the output of the above shared encoder.

In addition, previous works in biomedical RQE and QA often use word embeddings trained on PubMed or PMC corpus. To verify the superiority of pre-trained language representation model, the above four shared encoders (including **Interactive Transformer**) are respectively equipped with the following three word representation methods:

- Word2Vec: Each word in a sentence is represented by word embeddings trained on PubMed abstracts and PubMed Central full-text articles (Wei et al., 2013) with Word2vec toolkit (Mikolov et al., 2013). The dimension of the pre-trained word embedding is 100. We use a transition matrix to convert its dimension to 768.

- BERT: The pre-trained BERT model is used to generate a hidden representation h of each word in the sentence as its word embedding. The purpose of this method is to increase the generalization ability of the Word2Vec and fully describe the character level, word level and sentence level information and even the relationship between sentences.

- BioBERT: Same as above, the pre-trained BioBERT model is used to generate a hidden representation h of each word in the sentence as its word embedding.

Encoder	Embedding	RQE	QA
CNN	Word2vec	57.4	56.5
	BERT	60.8	73.7
	BioBERT	63.0	74.9
Bi-LSTM	Word2vec	52.6	58.4
	BERT	62.2	74.1
	BioBERT	62.2	75.1
Transformer	Word2vec	54.4	60.8
	BERT	59.6	72.7
	BioBERT	59.6	74.0
Interactive Transformer	Word2vec	57.8	62.3
	BERT	61.7	73.5
	BioBERT	**64.3**	**75.8**

Table 5: Effects of shared encoder. All the results are reported by accuracy (%). Bold font indicates the best performance.

Table 5 lists all the results on the RQE and QA dataset. By analyzing Table 5, we obtain the following conclusions. On the one hand, we can find that BERT brings a qualitative leap to the performance of both RQE and QA tasks upon Wor2Vec. BioBERT enriches BERT with a large amount of biomedical information and achieves approximately 1% absolute accuracy improvement over the BERT on both the tasks. It

Error Type	Sentence Pair	Task	Gold/Prediction
Acronyms	Q1: … he went to hospital to have medical check-up with endoscopic ultrasonography, and found **GIST** with about 1cm in size … What are we supposed to do? … Q2: What are the treatments for **Gastrointestinal Stromal Tumor**?	RQE	Entailment/Contradiction
Ambiguous Samples	Q: Spina bifida; vertbral fusion; syrinx tethered cord. Can u help for treatment of these problem? A: Spina bifida (Complications): Spina bifida may cause minimal symptoms or only minor physical disabilities.	QA	True/False
Semantic Confusion	Q1: **What is** the **possibility** of **atypical pneumonia** occurring again less than a month after **treatment**? Q2: **What are** the possible treatments for atypical pneumonia?	RQE	Contradiction/Entailment

Table 6: Failure cases predicted by AMTN.

shows that pre-trained models could improve model robustness and uncertainty estimates.

On the other hand, among the four different encoders, **Interactive Transformer** shows the best results overall. **Interactive Transformer** could not only capture the long-range dependency information, but also establish an interaction between the given two sentences by the interactive process. The benefit of introducing the interactive process is that it can efficiently compute dependencies between any two words in a sentence pair and encode the rich semantic information for each sequence word.

3.5 Error Analysis

Although the proposed AMTN achieves great performance over strong baselines, some failure cases are also observed. We have carried out detailed statistics and analysis of these errors, and classified the possible causes into the following three categories.

The first error type is acronyms. Since most biomedical concepts have acronyms, e.g. "Gastrointestinal Stromal Tumor" vs. "GIST" in first sentence pair in Table 6, it is quite difficult for model to determine whether the two sentences focus on the same topic without any external knowledge, thus resulting in misclassification. This problem is also our concern for future work, e.g. how to integrate prior knowledge into the model.

The second error type is ambiguous samples, which means that the relationship between the sentences is fuzzy and difficult to judge, such as the QA sentence pair shown in the second block of Table 5. Its golden label is True, however, the answer sentence seems to be irrelevant to the question, thus leads to the wrong classification of our model.

The third error type is semantic confusion, which refers to the semantic misunderstanding caused by complex syntax or collocation of phrases. Take the sentence pair in third block of Table 6 as an example: Q1 contains almost all the words in Q2 ("possible", "atypical pneumonia", "treatments" and etc.), while the two sentences are of Contradiction relation. We believe that the sentence pair is quite confusing that AMTN does not really "understand" it.

4 Conclusion

In this paper, we propose an Adversarial Multi-Task Network to jointly model RQE and QA shared tasks. ATMN employs BioBERT and Interactive Transformer as the shared encoder to learn the shared representations across the two tasks. A discriminator is further introduced to form an adversarial training with the shared encoder for purer shared semantic representations. Experiments on BioNLP 2019 RQE and QA shared tasks show that our proposed AMTN model benefits from the shared representations of both tasks provided by multi-task learning and adversarial training, and gains a significant improvement upon the single-task models.

Acknowledgments

This work was supported by the National Natural Science Foundation of China (No. 61772109) and the Ministry of education of Humanities and Social Science research and planning Fund of China (No. 17YJA740076).

References

Natalia Pletneva, Alejandro Vargas, Konstantina Kalogianni, and Celia Boyer. 2012. Online health information search: what struggles and empowers the users? Results of an online survey. *Studies in health technology and informatics*, 180:843-847.

Arabella Scantlebury, Alison M. Booth, and Bec Hanley. 2017. Experiences, practices and barriers to accessing health information: a qualitative study. *International Journal of Medical Informatics*, 103:103-8.

Jonathan D. Wren. 2011. Question answering systems in biology and medicine--the time is now. *Bioinformatics*, 27(14):2025-2026.

Asma B. Abacha, Chaitanya Shivade, and Dina D. Fushman. 2019. Overview of the MEDIQA 2019 Shared Task on Textual Inference, Question Entailment and Question Answering. *In Proceedings of the MEDIQA 2019.* Asma B. Abacha and Dina D. Fushman. 2016. Recognizing question entailment for medical question answering. AMIA. In *Proceedings of the AMIA Symposium*, pages 310-318.

Asma B. Abacha and Dina D. 2019. A question-entailment approach to question answering. *arXiv preprint arXiv:1901.08079.*

Yann LeCun, Leon Bottou, Yoshua Bengio, and PatrickHaffner. 1998. Gradient-based learning applied to document recognition. In *Proceedings of the IEEE*, pages 2278-2324.

Sepp Hochreiter and Jürgen Schmidhuber. 1997. Long short-term memory. *Neural Computation*, 9(8):1735-1780.

Ashish Vaswani, Noam Shazeer, Niki Parmar, Jakob Uszkoreit, Llion Jones, Aidan N. Gomez, Łukasz Kaiser, and Illia Polosukhin. 2017. Attention is all you need. In *Advances in Neural Information Processing Systems*, pages 6000–6010.

Patrick Verga, Emma Strubell, Ofer Shai, and Andrew McCallum. 2017. Attending to all mention pairs for full abstract biological relation extraction. *arXiv preprint arXiv:1710.08312.*

Tao Shen, Tianyi Zhou, Guodong Long, Jing Jiang, Shirui Pan, and Chengqi Zhang. 2017. DiSAN: Directional self-attention network for rnn/cnn-free language understanding. *arXiv preprint arXiv:1709.04696.*

Adams Wei Yu, David Dohan, Minh-Thang Luong, Rui Zhao, Kai Chen, Mohammad Norouzi, and Quoc V. Le. 2018. QANet: Combining local convolution with global self-attention for reading comprehension. *arXiv preprint arXiv:1804.09541.*

Matthew E. Peters, Mark Neumann, Mohit Iyyer, Matt Gardner, Christopher Clark, Kenton Lee, and Luke Zettlemoyer. 2018. Deep contextualized word representations. *arXiv preprint arXiv:1802.05365.*

Jacob Devlin, Ming-Wei Chang, Kenton Lee, and Kristina Toutanova. 2018. Bert: Pre-training of deep bidirectional transformers for language understanding. *arXiv preprint arXiv:1810.048055.*

Jinhyuk Lee, Wonjin Yoon, Sungdong Kim, Donghyeon Kim, Sunkyu Kim, Chan Ho So, and Jaewoo Kang. 2019. Biobert: A pre-trained biomedical language representation model for biomedical text mining. *arXiv preprint arXiv:1901.08746.*

Sanda Harabagiu and Andrew Hickl. 2006. Methods for Using Textual Entailment in Open-Domain Question Answering. In *Proceedings of the 21st International Conference on Computational Linguistics and the 44th Annual Meeting of the Association for Computational Linguistics.* Association for Computational Linguistics, pages 905-912.

Rich Caruana. 1998. Multitask Learning. *Autonomous Agents and Multi-Agent Systems*, 27(1):95–133.

Ian Goodfellow, Jean P. Abadie, Mehdi Mirza, Bing Xu, David W. Farley, Sherjil Ozair, Aaron Courville, and Yoshua Bengio. 2014. Generative adversarial nets. In *Advances in Neural Information Processing Systems*, pages 2672–2680.

Xinchi Chen, Zhan Shi, Xipeng Qiu, and Xuanjing Huang. 2017. Adversarial Multi-Criteria Learning for Chinese Word Segmentation. In *Proceedings of the 55th Annual Meeting of the Association for Computational Linguistics (Volume 1: Long Papers)*, pages 1193-1203.

Diederik P. Kingma and Jimmy Ba. 2014. Adam: A method for stochastic optimization. *arXiv preprint arXiv:1412.6980.*

Chih-Hsuan Wei, Hung-Yu Kao, and Zhiyong Lu. 2013. PubTator: A web-based text mining tool for assisting biocuration. *Nucleic Acids Research*, 41(W1):W518-W522.

Tomas Mikolov, Ilya Sutskever, Kai Chen, Greg S Corrado, and Jeff Dean. 2013. Distributed representations of words and phrases and their compositionality. In *Advances in neural information processing systems*, pages 3111–3119.

DUT-BIM at MEDIQA 2019: Utilizing Transformer Network and Medical Domain-Specific Contextualized Representations for Question Answering

Huiwei Zhou, Bizun Lei, Zhe Liu, Zhuang Liu
School of Computer Science and Technology
Dalian University of Technology
116024 Dalian, China
zhouhuiwei@dlut.edu.cn
{leibzun, njjnlz, zhuangliu1992}@mail.dlut.edu.cn

Abstract

In medical domain, given a medical question, it is difficult to manually select the most relevant information from a large number of search results. BioNLP 2019 proposes Question Answering (QA) task, which encourages the use of text mining technology to automatically judge whether a search result is an answer to the medical question. The main challenge of QA task is how to mine the semantic relation between question and answer. We propose BioBERT Transformer model to tackle this challenge, which applies Transformers to extract semantic relation between different words in questions and answers. Furthermore, BioBERT is utilized to encode medical domain-specific contextualized word representations. Our method has reached the accuracy of 76.24% and spearman of 17.12% on the BioNLP 2019 QA task.

1 Introduction

In medical field, the professional vocabulary is large and the semantics are complex, which makes manually selecting answers to a medical question from search results time consuming. The question answering (QA) task proposed by BioNLP 2019 (BEN ABACHA et al., 2019) aims to automatically extract answers to a medical question by using text mining technology. This task consists of two objectives: one is to determine whether each candidate answer can be used as the correct answer to a question, and the other is to rank the retrieved answers according to the relevance to a question.

The nature of QA task is to match the meaning rather than only match words between question and answer sentences. Several QA approaches based on syntax information have been developed to match the meaning between question and answer. Wang et.al. (2007) propose a statistical syntax-based model that softly aligns a question sentence with a candidate answer sentence. Tymoshenko and Moschitti (2015) encode semantic knowledge directly into syntactic tree representations of a pair of questions and answers for answers ranking. However, all these models rely on dependency parsers, suffering from error propagation.

Neural network-based methods can automatically learn the inherent semantic features and have achieved good performance on QA task. Wang and Nyberg (2017) employ an attentional encoder-decoder model based on long short-term memory (LSTM) (Hochreiter and Schmidhuber, 1997) for answer ranking, and their model achieves the best performance of 63.7% average score on the TREC LiveQA 2017 challenge (Agichtein et al., 2017). Yang et al. (2017) use a convolutional neural network (CNN) model to classify a question into a restricted set of 10 question types and crawl relevant online web pages to find the answers. However, all the models described above neglect the long range dependency between words in question and answer, limiting their capacity when question and answer sequences are long.

Transformer (Vaswani et al., 2017) is a model based entirely on attention mechanisms and has achieved success on several natural language processing (NLP) tasks, such as machine translation (Vaswani et al., 2017) and language understanding (Devlin et al., 2018). Transformer uses multi-head attention mechanisms to effectively capture the long-range dependency

Proceedings of the BioNLP 2019 workshop, pages 446–452
Florence, Italy, August 1, 2019. ©2019 Association for Computational Linguistics

information in context sequences, which is vital for question answering task.

Recently, language models (LM) based on large-scale corpus pre-training have made great progress in several NLP tasks, such as machine translation and natural language inference (NLI). ELMo (Peters et al., 2018) learns two unidirectional LMs based on LSTM networks which is able to capture both sub-word information and contextual clues. OpenAI GPT (Radford et al., 2018) uses a left-to-right Transformer (Vaswani et al., 2017), which introduces minimal task-specific parameters and is trained on the downstream tasks by simply fine-tuning the pre-trained parameters. The major limitation of pre-trained model above is that they are unidirectional, which limits the choice of architectures that can be used during pre-training. BERT (Devlin et al., 2018) employs a bidirectional Transformer encoder to fuse both the left and the right context and can explicitly model the relationship of a pair of text. Thus, it can make progress in paired NLP tasks, such as NLI and QA. Based on the BERT architecture, BioBERT (Lee et al., 2019) is a domain-specific language representation model pre-trained on large-scale biomedical corpora and effectively transfers the knowledge from biomedical texts to biomedical text mining models.

Corpus of QA task proposed by BioNLP 2019 contains answers with long text, which requires models to capture the long range dependency information across words in both question and answer sentences. Thus, we propose BioBERT Transformer (BBERT-T) model based on Transformer to model the associations between question and answer. Specifically, question and answer sequences are first passed to BioBERT to generate medical domain-specific contextualized representations. Then, the question and answer representations are fed into two Transformers, respectively, to capture the long range dependency information and semantic relation between question and answer. Finally, a weighted cross entropy loss is applied to further improve the performance. Our method achieves accuracy of 76.24% and spearman of 17.12% on the BioNLP 2019 QA task.

2 System Description

In MEDIQA2019 medical Question Answering (QA) task, given a question q and n_a candidate answers $\{a^1, a^2, ..., a^{n_a}\}$, we need build model to rank all candidate answers and to recognize correct answers to the question. Let T be the set of all the question-answer pairs. For each question-answer pair (q, a), we use BioBERT to encode the contextual information, which improves the model generalization capability. Then we propose two Transformers to learn the long range dependency information between words in question and answer, respectively. In this section, we introduce our approach for QA task in two steps: (1) the preprocessing of the corpus; (2) the structure of the model.

2.1 Preprocessing

Firstly, we lowercase all the questions and answers. Then, following (Fajcik et al., 2019), for each text of questions and answers, we use the tokenizer, that comes from Hugging Face PyTorch re-implementation of BERT[1], to split input words into most frequent n-grams in the pre-training corpus, effectively representing text at the sub-word level. Next, following (Vaswani et al., 2017), (q, a) pair sequences are truncated to have at most 300 tokens. At last, we use the truncated pair sequence $[q_1, q_2, ..., q_{l_q}, a_1, a_2, ..., a_{l_a}]$ as input, where l_q is the length of question, l_a is the length of a candidate answer and $l_q + l_a \leq 300$.

2.2 BioBERT Transformer Model (BBERT-T)

Structure of the proposed model is shown in Figure 1, which is composed of three layers: (1) BioBERT layer; (2) Transformer layer; (3) classification and ranking layer. Take the question-answer pair sequence $[q_1, q_2, ..., q_{l_q}, a_1, a_2, ..., a_{l_a}]$ as input to the BioBERT layer, achieving the question representation and answer representation. Then the two representations are fed to Transformer layer to extract the long range dependency information between words in question and answer, respectively. Finally, the outputs of Transformer layer are passed to a max pooling layer to generate features used to perform classification and ranking.

---

[1] https://github.com/huggingface/pytorch-pretrained-BERT

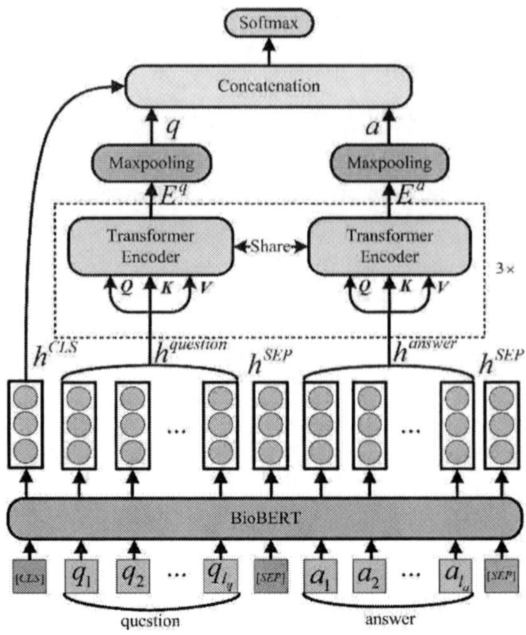

Figure 1: Architecture of BioBERT Transformer model.

The details of our model are described in the following subsections.

BioBERT layer: BioBERT (Lee et al., 2019) has achieved good performance after fine-tuning in several biomedicine NLP tasks. Therefore, we use the BioBERT to encode the question-answer pair sequence, which improves the model generalization capability. Following (Vaswani et al., 2017), given a question-answer pair sequence $[q_1,q_2,...,q_{l_q},a_1,a_2,...,a_{l_a}]$, we add [CLS] token as the first token to the sequence and separate question and answer sequence with [SEP] token to get the input sequence, i.e. $[CLS,q_1,q_2,...,q_{l_q},SEP,a_1,a_2,...,a_{l_a},SEP]$. BioBERT is used to encode the input sequence and the final layer output is used as the contextualized representation of the question-answer pair $H=[h^{CLS},h_1^q,...,h_{l_q}^q,h^{SEP},h_1^a,...,h_{l_a}^a,h^{SEP}]\in\mathbb{R}^{n\times d}$, where $n=l_q+l_{a^i}+3$ and d is the hidden layer dimension. Representations of questions and answers and the h^{CLS} will be used as inputs to the Transformer layer and the classification and ranking layer, respectively, which are described in following subsections. Note that all parameters of the BioBERT are fine-tuned during training.

Transformer layer: In this layer, two Transformers are applied to capture the long range

dependency information and semantic relation between question and answer. To help better understand this layer, we provide a brief overview of Transformer (Vaswani et al., 2017).

The key component of Transformer is the multi-head attention layer that allows the model to jointly attend to information from different representation sub-spaces at different positions. Formally, given the queries Q, keys K, values V, multi-head attention builds upon scaled dot product attention mapping a query and a set of key-value pairs to an output:

$$Att(Q,K,V) = \text{soft}\max(\frac{QK^T}{\sqrt{d}})V \quad (1)$$

where Q, K, V and output are all list of vectors with equal length, and d is the dimension size of K. Multi-head attention applies several scaled dot product attentions, which can be formulized as follows:

$$head_i = Att(W_i^Q Q, W_i^K K, W_i^V V) \quad (2)$$

$$MultiHead(Q,K,V)=W^H [head_1,head_2,\cdots,head_h] \quad (3)$$

where $W_i^Q \in \mathbb{R}^{d/h\times d}, W_i^K \in \mathbb{R}^{d/h\times d}$ and $W_i^V \in \mathbb{R}^{d/h\times d}$ are trainable parameter matrices and h represents the number of scaled dot product attention, or head. $[head_1,head_2,\cdots,head_h]$ is a concatenation of outputs of h heads. Note that, in this paper, we set $W^H \in \mathbb{R}^{d\times d}$ to a fixed identity matrix to reduce model complexity.

After multi-head attention layer, Transformer applies a two-layer full connection layer with ReLU activation:

$$H_q = \text{Re LU}(Q_{update}W_1 + b_1)W_2 + b_2 \quad (4)$$

where $W_1 \in \mathbb{R}^{d\times d}$, $W_2 \in \mathbb{R}^{d\times d}$, $b_1 \in \mathbb{R}^d$, $b_2 \in \mathbb{R}^d$ are trainable parameter matrices. $Q_{update} = MultiHead(Q,K,V)$ is the output of multi-head attention.

For both multi-head attention layer and the full connection layer, we use the residual concatenation (He et al., 2016) and Layer Normalization (Ba et al., 2016).

In our Transformer layer, we first split the contextualized representation $H=[h^{CLS},h_1^q,...,h_{l_q}^q,h^{SEP},h_1^a,...,h_{l_a}^a,h^{SEP}]$ from BioBERT into question representation

$$H^q = [h_1^q, h_2^q, ..., h_{l_q}^q] \in \mathbb{R}^{l_q \times d}$$ and answer representation $H^a = [h_1^a, h_2^a, ..., h_{l_a}^a] \in \mathbb{R}^{l_a \times d}$. Then question and answers representations are passed to two Transformers, respectively. Take the question for example, Question representation H^q forms the Q, K and V which are fed into a Transformer. The output of the Transformer for question is represented as E^q . In the same way, we can get E^a for answer. To establish the connection between question and answer, the two Transformers share the parameters. E^q and E^a will be used in the following classification and ranking layer.

2.3 Classification and Ranking Layer

In order to summarize the information of the questions and answers, we use the max pooling to generate question features $q \in \mathbb{R}^d$ and answer features $a \in \mathbb{R}^d$:

$$q = \max \text{pool}(E^q) \tag{5}$$

$$a = \max \text{pool}(E^a) \tag{6}$$

q and a are concatenated to form $[q, a]$ as the features for classification and ranking. To make full use of information about the relationship between question and answer, we further concatenate the classification embedding h^{CLS} from BioBERT layer to form the final features $[q, a, |q - a|, q \times a, h^{CLS}]$. Then, $[q, a, |q - a|, q \times a, h^{CLS}]$ is passed to a softmax layer to perform the classification. The softmax layer consists of a dense layer and a logistic regression classifier with a softmax function.

$$o = W_3[q, a, |q - a|, q \times a, h^{CLS}] + b_3 \tag{7}$$

$$p(y_t = j \mid T_t) = \text{soft}\max(W_o o + b_o) \tag{8}$$

where $W_3 \in \mathbb{R}^{5d \times d}$, $b_3 \in \mathbb{R}^d$, $W_o \in \mathbb{R}^{d \times 2}$ and $b_o \in \mathbb{R}^2$ trainable parameters, $j \in \{0,1\}$, and T_t represents tth training samples. We rank the answers according to the probability of being true answer (i.e. $p(y_t = 1 \mid T_t)$).

In order to make full use of the reference score in the training set and to carefully control the loss, this paper applies a weighted cross entropy loss.

For a candidate answer with the reference scores of 1, 2, 3, 4, the corresponding output labels should be 0, 0, 1, 1. We assign weights of 2, 1, 1, 2 to each label, respectively, when calculating the cross entropy loss:

$$\alpha = \begin{cases} 2, score = 1 \\ 1, score = 2 \\ 1, score = 3 \\ 2, score = 4 \end{cases} \tag{9}$$

$$loss = -\frac{1}{N} \sum_{t=1}^{N} \alpha \log p(y_t \mid T_t) \tag{10}$$

where N is the number of question-answer pairs.

3 Experiments

3.1 Dataset and Evaluation Metrics

Dataset: MEDIQA2019-Task3-QA task contains dataset of medical questions and the associated answers retrieved by CHiQA[2]. Table 1 describes the details of statistics of the dataset.

	Train	Dev	Test
Questions	208	25	150
Answers	1701	234	1107

Table 1: Statistics of dataset of QA task.

Evaluation Metrics: For BioNLP 2019 QA task, organizers employ two measurements: accuracy and spearman. The evaluation is reported by official evaluation toolkit[3], and accuracy is the main metric. For each experiment, we report the mean values with corresponding standard deviations over 3 repetitions.

3.2 Experimental Setup

The BioBERT we use includes 12 layers (i.e., Transformer blocks), and the dimension of hidden size is 768. The Transformer we use has 3 blocks, each of which contains 16 heads. For each head, the mapped Q, K, and V dimensions are 48. Thus, the input and output dimension of Transformer is 768. We use Adam (Kingma and Ba, 2015) with β_1 = 0.9, β_2 = 0.999 for optimization. The learning rate is 2e-5. The dropout rate is 0.5. The batch size is set to 4. The **BBERT-T**[4] is developed by

---

[2] https://chiqa.nlm.nih.gov/
[3] https://github.com/abachaa/MEDIQA2019/tree/master/Eval_Scripts

[4] https://github.com/ThreeTreeStar/Question-Answering

PyTorch[5]. We use the BioBERT module (Lee et al., 2019) without modifying. The Transformer is developed by ourselves. Computations are run on a single server computer equipped with a GPU.

3.3 Comparisons with baselines

To verify the effectiveness of our model, we compare **BBERT-T** with five baselines listed as follows.

w/o BioBERT: This variant does not use BioBERT. The processed sentence is embedded with pre-trained word embeddings released by (Moen et al., 2013). Then the embedded sequence input into the Transformer layer.

BBERT-LSTM: This variant replaces two Transformers with two BiLSTMs. Question representation H_q and answer representation H_a are passed to two BiLSTMs, respectively. Note that the two BiLSTMs share parameters.

BBERT-CNN: This variant replaces two Transformers with two CNNs with kernel size of $\{2, 3, 4\}$. Question representation H_q and answer representation H_a are passed to two CNNs, respectively. Note that the two CNNs share parameters.

BBERT-T (1 block): This variant uses Transformers with one block, rather than three blocks.

w/o CLS: In classification and ranking layer, we do not concatenate h^{CLS} with the output of max pooling q and a. We directly take $[q, a, |q - a|, q \times a]$ as input to softmax layer. The submitted results are from this model. After submitting the results, we found that **BBERT-T** concatenating h^{CLS} as features achieved higher results than **w/o CLS**.

From the results in Table 2, we can conclude followings. First, compared with **BBERT-T**, **BBERT-LSTM** replaces Transformer with BiLSTM which causes the accuracy to drop by 4.97% and the spearman to drop by 3.50%. **BBERT-CNN** replaces Transformer with CNN which causes the accuracy to drop by 4.45% and the spearman to drop by 3.59%. This indicates that long range dependency information extracted by our model is critical to QA task. After all, most of answers in the corpus of BioNLP 2019 QA task have long sequences and the semantic information may be distributed across long distance.

Model	Accuracy (%)	Spearman (%)	Time (sec)
w/o BioBERT	$51.39_{\pm0.56}$	$-18.48_{\pm3.39}$	132.83
BBERT-LSTM	$71.03_{\pm0.91}$	$6.86_{\pm4.29}$	399.73
BBERT-CNN	$71.55_{\pm1.73}$	$6.77_{\pm4.80}$	290.79
BBERT-T (1 block)	$73.83_{\pm0.25}$	$4.36_{\pm9.38}$	310.62
w/o CLS	$73.23_{\pm1.97}$	$7.80_{\pm5.50}$	369.88
BBERT-T	$76.0_{\pm1.30}$	$10.36_{\pm8.80}$	373.43
BBERT-T*	$76.24_{\pm1.31}$	$17.12_{\pm9.66}$	373.43

Table 2: Comparisons with baselines, * stands for using ensemble by averaging the last 4 epoch output probabilities. ± denotes standard deviation,

Second, compared with **BBERT-T**, **w/o BioBERT** decreases the accuracy by 24.61% and the spearman by 28.84%. This indicates that medical domain information is important for BioNLP 2019 QA task.

Third, comparing **BBERT-T** with **w/o CLS**, we can see that without the h^{CLS} feature decreases the accuracy and spearman. In BioBERT, h^{CLS} is originally used as features to classify whether given two sentences is adjacent. Similarly, in our model, h^{CLS} contains important information about the relationship between question and answer, which is critical features to QA task.

Finally, compared with **BBERT-T (1 block)**, **BBERT-T** has a higher complexity but achieves a better accuracy, which illustrates that the structure of three blocks is necessary.

3.4 Effects of architecture

To better understand the architecture of **BBERT-T**, we compare it with three variants:

w/o share: This variant uses two separate Transformers with different parameters .

BBERT-T (att): This variant replaces the max pooling with an attention mechanism. Take the question as example, we calculate the attention weight γ_i for the i th position in the output of Transformer E_q as follows:

$$\gamma_i = \text{soft max}(\tanh(W_\gamma E_i^q + b_\gamma)) \quad (11)$$

where $W_\gamma \in \mathbb{R}^d$ and b_γ are trainable parameters. Then the question features q is defined as follows:

---

[5] https://pytorch.org/

$$q = \sum_{i=1}^{i=l_q} \gamma_i h_i^q \qquad (12)$$

In the same way, the answer feature a is achieved.

BBERT-T (mean): This variant replaces the max pooling with a mean pooling.

From the results in Table 3, we can see that not

Model	Accuracy (%)	Spearman (%)
w/o share	$73.83_{\pm0.32}$	$12.90_{\pm3.29}$
BBERT-T (att)	$72.99_{\pm0.65}$	$16.55_{\pm5.08}$
BBERT-T (mean)	$73.62_{\pm0.56}$	$9.96_{\pm8.10}$
BBERT-T	$76.0_{\pm1.30}$	$10.36_{\pm8.80}$

Table 3: Effects of architecture.

sharing parameters between the two Transformers might lose connection between question and answer, leading to performance decrease. Using attention mechanism to generate question and answer features achieves a worse results than using max pooling. The reason might be that the self-attention structures of Transformer make each position of output equally important. Therefore, attention mechanism cannot learn the effective weight for each position. This can be further verified by similar accuracy of the mean pooling that gives equal weight of each position.

3.5 Effects of pre-training corpus knowledge

To explore the effects of large-scale pre-training corpus knowledge, we compare our BBERT-T with its two variants:

w/o Bio： This variant replaces BioBERT with BERT.

Model	Accuracy (%)	Spearman (%)
w/o BioBERT	$51.39_{\pm0.56}$	$-18.48_{\pm3.39}$
w/o Bio	$71.09_{\pm0.48}$	$18.89_{\pm8.10}$
BBERT-T	$76.0_{\pm1.30}$	$10.36_{\pm8.80}$

Table 4: Effects of large-scale pre-training corpus knowledge on performance on the QA dataset.

From Table 4, comparing **w/o Bio** with **w/o BioBERT**, we can conclude that the contextualized representations generating by BERT do provide the semantic information between question and answer. BioBERT, having same model structure as BERT, is pre-trained on large scale medical corpus, which could generate medical domain-specific representations. For BioNLP 2019 QA task in medical domain, applying medical domain-

specific representations is more effective than open domain representations.

3.6 Effect of reference loss

To investigate the effects of weighted cross entropy loss, we use the cross-entropy loss to train our model and the results are shown in Table 5.

Model	Accuracy (%)	Spearman (%)
Cross-entropy	$73.56_{\pm0.70}$	$7.76_{\pm4.86}$
BBERT-T	$76.0_{\pm1.30}$	$10.36_{\pm8.80}$

Table 5: Effects of large-scale pre-training corpus knowledge on performance on the QA dataset.

From Table 5, we can observe that cross-entropy gets worse results than weighted cross-entropy, which illustrates that weighted cross-entropy could take advantage of the reference score during training. The candidate answers with higher reference scores are more relevant. Weighted cross-entropy assigns a higher weight to the loss of the correct answer with higher reference score and loss of the incorrect answer with lower reference score, which makes the model more robust.

4 Conclusion

We propose BioBERT Transformer model which applies two Transformers to catch the association between question and answer. Experimental results show that our model benefits from the long range dependency information between words in question and answer and that medical domain-specific contextualized representations generated by BioBERT can effectively improve the performance of QA task. We evaluate on BioNLP 2019 QA test dataset with official evaluation toolkit. And our proposed method achieves the accuracy of 76.24% and spearman of 17.12%.

Acknowledgments

This work is supported by National Natural Science Foundation of China (No. 61772109) and the Humanities and Social Science Fund of Ministry of Education of China (No. 17YJA740076).

References

Asma Ben Abacha, Eugene Agichtein, Yuval Pinter and Dina Demner-Fushman1. 2017. Overview of the Medical Question Answering Task at TREC 2017 LiveQA. *text retrieval conference.*

Alec Radford, Karthik Narasimhan, Tim Salimans, and Ilya Sutskever. 2018. Improving language understanding by generative pre-training.

Ashish Vaswani, Noam Shazeer, Niki Parmar, Jakob Uszkoreit, Llion Jones, Aidan N Gomez, Lukasz Kaiser, and Illia Polosukhin. 2017. Attention is all you need. *In Advances in Neural Information Processing Systems,* pages 6000–6010.

Asma Ben Abacha, Chaitanya Shivade, and Dina Demner-Fushman. 2019. Overview of the MEDIQA 2019 Shared Task on Textual Inference, Question Entailment and Question Answering. ACL-BioNLP 2019.

Di Wang, and Eric Nyberg. 2017. Cmu oaqa at trec 2017 liveqa: A neural dual entailment approach for question paraphrase identi_cation. *In Proceedings of The Twenty-Sixth Text REtrieval Conference,* TREC 2017, Gaithersburg, Maryland, USA.

Jacob Devlin, Ming-Wei Chang, Kenton Lee, and Kristina Toutanova. 2018. Bert: Pre-training of deep bidirectional transformers for language understanding. *arXiv preprint arXiv:1810.04805.*

Jimmy Lei Ba, Jamie Ryan Kiros and Geoffrey E. Hinton. 2016. Layer normalization. *arXiv preprint arXiv:1607.06450.*

Jinhyuk Lee, Wonjin Yoon, Sungdong Kim, Donghyeon Kim, Sunkyu Kim, Chan Ho So, and Jaewoo Kang. 2019. Biobert: pre-trained biomedical language representation model for biomedicaltext mining. *arXiv preprint arXiv: 1901.08746.*

Kaiming He, Xiangyu Zhang, Shaoqing Ren and Jian Sun. 2016. Deep Residual Learning for Image Recognition. *computer vision and pattern recognition,*770-778.

Kateryna Tymoshenko and Alessandro Moschitti. 2015 Assessing the Impact of Syntactic and Semantic Structures for Answer Passages Reranking. *conference on information and knowledge management:* 1451-1460.

Kingma, Diederik P., and Jimmy Ba. 2014. Adam: A method for stochastic optimization. *arXiv preprint arXiv:1412.6980.*

Martin Fajcik, Lukáš, Burget, and Pavel Smrz. 2019. BUT-FIT at SemEval-2019 Task 7: Determining the Rumour Stance with Pre-Trained Deep Bidirectional Transformers. *arXiv preprint arXiv:1902.10126.*

Matthew E Peters, Waleed Ammar, Chandra Bhagavatula, and Russell Power. 2017. Semi-supervised sequence tagging with bidirectional language models. *arXiv preprint arXiv:1705.00108.*

Mengqiu Wang, Noah A. Smith and Teruko Mitamura. 2007. What is the Jeopardy Model? A Quasi-Synchronous Grammar for QA. *empirical methods in natural language processing,* 2007: 22-32.

Nitish Srivastava, Geoffrey Hinton, Alex Krizhevsky, Ilya Sutskever, Ruslan Salakhutdinov. 2014. Dropout: a simple way to prevent neural networks from overfitting. *Journal of Machine Learning Research,* 15(1): 1929-1958.

Peng Gao, Zhengkai Jiang, Haoxuan You, Pan Lu, Steven Hoi, Xiaogang Wang and Hongsheng Li. 2018. Dynamic Fusion with Intra- and Inter-Modality Attention Flow for Visual Question Answering. *arXiv: Computer Vision and Pattern Recognition.*

Peter Turney. 2000. Types of cost in inductive concept learning. *In Proceedings of the Cost-sensitive Learning Workshop at the 17th International Conference on Machine Learning,* Stanford, CA.

Rodrigo Nogueira and Kyunghyun Cho. 2019. Passage Re-ranking with BERT. *arXiv preprint arXiv:1901.04085 (2019).*

Sampo Pyysalo, Filip Ginter, Hans Moen, Tapio Salakoski, and Sophia Ananiadou. 2013. Distributional semantics resources for biomedical text processing.*Proceedings of LBM.*

Sepp Hochreiter and Jurgen Schmidhuber. 1997. Longshort-term memory. *Neural Computation,* 9:1735–1780.

Yuan Yang, Jingcheng Yu, Ye Hu, Xiaoyao Xu, and Eric Nyberg. 2017. Cmu livemedqa at trec 2017 liveqa: A consumer health question answering system. *In Proceedings of The Twenty-Sixth Text REtrieval Conference, TREC 2017,* Gaithersburg, Maryland, USA.

Dr.Quad at MEDIQA 2019: Towards Textual Inference and Question Entailment using contextualized representations

Vinayshekhar Bannihatti Kumar * **Ashwin Srinivasan*** **Aditi Chaudhary***
James Route **Teruko Mitamura** **Eric Nyberg**
{*vbkumar, ashwinsr, aschaudh, jroute, teruko, ehn*}@cs.cmu.edu
Language Technologies Institute
Carnegie Mellon University

Abstract

This paper presents the submissions by Team Dr.Quad to the ACL-BioNLP 2019 shared task on Textual Inference and Question Entailment in the Medical Domain. Our system is based on the prior work Liu et al. (2019) which uses a multi-task objective function for textual entailment. In this work, we explore different strategies for generalizing state-of-the-art language understanding models to the specialized medical domain. Our results on the shared task demonstrate that incorporating domain knowledge through data augmentation is a powerful strategy for addressing challenges posed by specialized domains such as medicine.

1 Introduction

The ACL-BioNLP 2019 (Ben Abacha et al., 2019) shared task focuses on improving the following three tasks for medical domain: 1) Natural Language Inference (NLI) 2) Recognizing Question Entailment (RQE) and 3) Question-Answering reranking system. Our team has made submissions to all the three tasks. We note that in this work we focus more on the task 1 and task 2 as improvements in these two tasks reflect directly on the task 3. However, as per the shared task guidelines, we do submit one model for the task 3 to complete our submission.

Our approach for both task 1 and task 2 is based on the state-of-the-art natural language understanding model MT-DNN (Liu et al., 2019), which combines the strength of multi-task learning (MTL) and language model pre-training. MTL in deep networks has shown performance gains when related tasks are trained together resulting in better generalization to new domains (Ruder, 2017). Recent works such as BERT (Devlin et al., 2018), ELMO (Peters et al., 2018) have shown

___
* equal contribution

the efficacy of learning universal language representations in providing a decent warm start to a task-specific model, by leveraging large amounts of unlabeled data. MT-DNN uses BERT as the encoder and uses MTL to fine-tune the multiple task-specific layers. This model has obtained state-of-the-art results on several natural language understanding tasks such as SNLI (Bowman et al., 2015), SciTail (Khot et al., 2018) and hence forms the basis of our approach. For the task 3, we use a simple model to combine the task 1 and task 2 models as shown in §2.5.

As discussed above, state-of-the-art models using deep neural networks have shown significant performance gains across various natural language processing (NLP) tasks. However, their generalization to specialized domains such as the medical domain still remains a challenge. Romanov and Shivade (2018) introduce a new dataset MedNLI, a natural language inference dataset for the medical domain and show the importance of incorporating domain-specific resources. Inspired by their observations, we explore several techniques of augmenting domain-specific features with the state-of-the-art methods. We hope that the deep neural networks will help the model learn about the task itself and the domain-specific features will assist the model in tacking the issues associated with such specialized domains. For instance, the medical domain has a distinct sublanguage (Friedman et al., 2002) and it presents challenges such as abbreviations, inconsistent spellings, relationship between drugs, diseases, symptoms.

Our resulting models perform fairly on the unseen test data of the ACL-MediQA shared task. On Task 1, our best model achieves +14.1 gain above the baseline. On Task 2, our five-model ensemble achieved +12.6 gain over the baseline and for Task 3 our model achieves a a +4.9 gain.

453

Proceedings of the BioNLP 2019 workshop, pages 453–461
Florence, Italy, August 1, 2019. ©2019 Association for Computational Linguistics

2   Approach

In this section, we first present our base model
MT-DNN (Liu et al., 2019) which we use for both
Task 1 and Task 2 followed by a discussion on
the different approaches taken for natural language
inference (NLI) (§2.3), recognizing question entailment (RQE) (§2.4) and question answer (QA)
(§2.5).

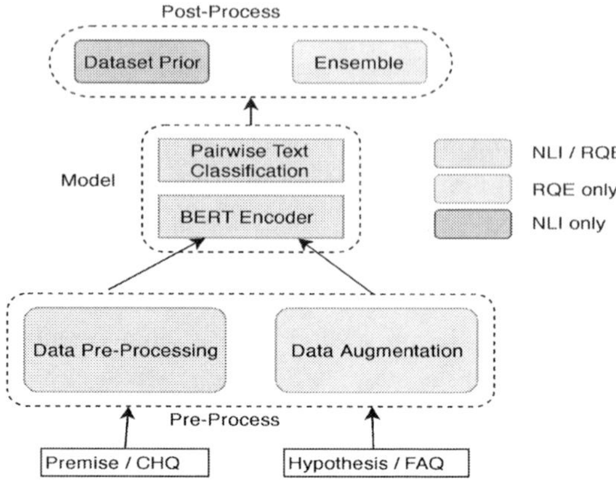

Figure 1: System overview for NLI and RQE task

2.1   Task 1 and Task 2 Formulation

Formally, we define the problem of textual
entailment as a multi-class classification task.
Given two sentences $\mathbf{a} = a_1, a_2..., a_n$ and $\mathbf{b} =
b_1, b_2, ..., b_m$, the task is to predict the correct la-
bel. For NLI, $\mathbf{a}$ refers to the *Premise* and $\mathbf{b}$ refers
to the *Hypothesis* and the label set comprises of
entailment, neutral, contradiction. For RQE, $\mathbf{a}$
refers to the *CHQ* and $\mathbf{b}$ refers to the *FAQ* and the
label set comprises of *True, False*.

2.2   Model Architecture

A brief depiction of our system is shown in Figure 1. We represent components which were used
for both NLI and RQE in Orange. An example of this is the Data Pre-processing component.
The RQE only components are shown in yellow
(eg. Data Augmentation). The components which
were used only for the NLI modules are shown in
Pink (eg. Dataset Prior). We base our model on
the state-of-the-art natural language understanding
model MT-DNN (Liu et al., 2019). MT-DNN is
a hierarchical neural network model which combines the advantages of both multi-task learning
and pre-trained language models. Below we describe the different components in detail.

	Train	Validation	Test
Entailment	3744	465	474
Contradiction	3744	465	474
Neutral	3744	465	474

Table 1: The number of train and test instances in each
of the categories of the NLI dataset.

Encoder: Following BERT (Devlin et al.,
2018), each sentence pair is separated by a [SEP]
token. It is then passed through a lexicon encoder
which represents each token as a continuous representation of the word, segment and positional
embeddings.   A multi-layer bi-directional transformer encoder (Vaswani et al., 2017) transforms
the input token representations into the contextual
embedding vectors. This encoder is then shared
across multiple tasks.

Decoder: We use the *Pairwise text classification output* layer (Liu et al., 2019) as our de-
coder.   Given a sentence pair $(\mathbf{a},\mathbf{b})$, the above
encoder first encodes them into $\mathbf{u}$ and $\mathbf{v}$ respectively.   Then a K-step reasoning is performed
on these representations to predict the final label.
The initial state is given by $\mathbf{s} = \sum_j \alpha_j \mathbf{u_j}$ where
$\alpha_j = \frac{\exp(\mathbf{w}^T \mathbf{u_j})}{\sum_i \exp(\mathbf{w_1}^T \mathbf{u_i})}$. On subsequent iterations $k \in
[1, K-1]$, the state is $\mathbf{s}^k = GRU(\mathbf{s}^{k-1}, \mathbf{x}^k)$ where
$\mathbf{x}_k = \sum_j \beta_j \mathbf{v_j}$ and $\beta_j = softmax(\mathbf{s}_{k-1} \mathbf{w_2}^T \mathbf{v})$.
Then a single-layer classifier predicts the label at
each iteration k:

$$P^k = softmax(\mathbf{w_3}^T[\mathbf{s}^k; \mathbf{x}^k; |\mathbf{s}^k - \mathbf{x}^k|; \mathbf{s}^k.\mathbf{x}^k])$$

Finally, all the scores across the K iterations are
averaged for the final prediction. We now describe
the modifications made to this model for each respective task.

2.3   Natural Language Inference

This task consists of identifying three inference relations between two sentences: Entailment, Neutral and Contradiction

Data: The data is based off the MedNLI dataset
introduced by Romanov and Shivade (2018). The
statistics of the dataset can be seen in Table 1.

Data Pre-Processing: On manual inspection of
the data, we observe the presence of abbreviations
in the premise and hypothesis. Since lexical overlap is a strong indicator of entailment by virtue of

pre-trained embeddings on large corpora, the presence of abbreviations makes it challenging. Therefore, we expand the abbreviations using the following two strategies:

1. *Local Context:* We observe that often an abbreviation is composed of the first letters of contiguous words. Therefore, we first construct potential abbreviations by concatenating first letter of all words in an sequence, after tokenization. For instance, for the premise shown below we get {CXR, CXRS, XRS, CXRSI, XRSI, RSI, etc}. This is done for both the premise and the hypothesis. We then check if this n-gram exists in the hypothesis (or the premise). If yes, then we replace that abbreviation with all the words that make up the n-gram. Now the model has more scope of matching two strings lexically. We demonstrate an example below:

   Premise: Her **CXR** was clear and it did not appear she had an infection.
   Hypothesis: **Chest X-Ray** showed infiltrates.

   Premise Modified: Her **Chest X-Ray** was clear and it did not appear she had an infection.

2. *Gazetteer:* If either the premise/hypothesis does not contain the abbreviation expansion or contains only partial expansion, the *Local Context* technique will fail to expand those abbreviations. Hence, we use an external gazetteer extracted from CAMC[1] to expand commonly occurring medical terms. There were 1373 entries in the gazetteer, covering common medical and clinical expansions. For instance,
   Premise: On arrival to the **MICU** , patient is hemodynamically stable .
   Premise Modified: On arrival to the **Medical Intensive Care Unit** , patient is hemodynamically stable .

We first performed the local context replacement as they are more specific to a given premise-hypothesis pair. If there was no local context match, then we did a gazetteer lookup. It is to be noted that one abbreviation can have multiple expansions in the gazatteer and thus we hypothesized that local context should get preference while expanding the abbreviation.

Training Procedure: For training the MT-DNN model, we use the same hyper-parameters provided by the authors (Liu et al., 2019). We train model for 4 epochs and early stop when the model reaches the highest validation accuracy.

Baselines: We use the following baselines similar to Romanov and Shivade (2018).

- *CBOW:* We use a Continuous-Bag-Of-Words (CBOW) model as our first baseline. We take both the premise and the hypothesis and sum the word embeddings of the respective statements to form the input layer to our CBOW model. We used 2 hidden layers and used softmax as the decision layer.

- *Infersent:* Inferesent is a sentence encoding model which encodes a sentence by doing a max-pool on all the hidden states of the LSTM across time steps. We follow the authors of Romanov and Shivade (2018) by using shared weights LSTM cell to get the sentence representation of the premise(U) and the hypothesis(V). We feed these representations U and V to an MLP to perform a 3 way prediction. For our experiments, we use the pre-trained embeddings trained on the MIMIC dataset by Romanov and Shivade (2018). We used the same hyperparameters.

- *BERT:* Since MT-DNN is based off of the BERT (Devlin et al., 2018) model as the encoder, we also compare results using just the pre-trained BERT. We used *bert-base-uncased* model which was trained for 3 epochs with a learning rate of 2e-5 and a batch size of 16 with a maximum sequence length of 128. WE used the last 12 pre-trained layers of the model.

2.3.1 Results and Discussion

In this section we discuss the results of all of our experiments on the NLI task.

Ablation Study: First, we conduct an ablation study to study the effect of abbreviation expansion. Table 2 shows the results of the two abbreviation expansion techniques for the Infersent model. We observe the best performance with the *Gazetteer* strategy. This is because most sentences in the dataset did not have the abbreviation

---

[1] https://www.camc.org/

Model Ablation	Accuracy
Infersent	78.8 +/- 0.06
Infersent + Local-Context	78.8 +/- 0.02
Infersent + Local-Context + Gazetteer	78.5 +/- 0.36
Infersent + Gazetteer	**79.1 +/ 0.14**

Table 2: The results reported in the table is mean and variance of the models averaged on 3 runs using different random seeds.

Model Ablation	Accuracy
CBOW	74.7
Infersent	79.1
BERT	80.4
Ours	**82.1**
(Ben Abacha et al., 2019) (Unseen Test)	71.4
Ours (Unseen Test)	79.6
Ours (Unseen Test) + Prior	**85.5**

Table 4: NLI results on the validation set.

matched through the local context match. Since expanding abbreviations helped increase lexical overlap, going forward we use the expanded abbreviation data for all our experiments henceforth. Table 3 shows the confusion matrix for the Infersent model. The rows represent the ground truth and the columns represent the predictions made by us. We can see that the model is most confused about the *entailment* and *neutral* classes. 82 times the model predicts *neutral* for *entailment* and 85 times vice versa. In order to address this issue, we add a prior on the dataset as a post processing step.

	Contradiction	Entailment	Neutral
Contradiction	396	43	26
Entailment	30	353	**82**
Neutral	23	**85**	357

Table 3: Confusion matrix for NLI classes for Infersent model. Rows denote the true labels and columns denote the model predictions.

Prior on the dataset: Our dataset analysis on the validation set revealed that there were three hypothesis for a given premise with mutually exclusive labels. Since we know that for a given premise there can only be one entailment because of the nature of the dataset, we post-process the model predictions to add this constraint. For each premise we collect the prediction probability for each of the hypothesis and pick the hypothesis having the highest probability for entailment. We perform the same selectional preference procedure on the remaining two classes. Such a post-processing ensures that each premise always has three hypotheses with mutually exclusive labels.

Table 4 documents the results of the different models on the validation set. We observe that our method gives the best performance among the three baselines. Based on these results, our final submission on the unseen data can be seen in the last row.

2.3.2 Error Analysis

We perform qualitative analysis of our model and bucket the errors into the following categories.

1. **Lexical Overlap:** From Table 6, we see that there is a high lexical overlap between the premise and hypothesis, prompting our model to falsely predict *entailment*.

2. **Disease-Symptom relation:** In the second example, we can see that our model lacks sufficient domain knowledge to relate *hyperglycemia* (a symptom) to *diabetes* (a disease). The model interprets these to be two unrelated entities and labels as *neutral*.

3. **Drug-disease relation:** In the final example we see that our model doesn't detect that the drug names in the premise actually entail the condition in hypothesis.

These examples show that NLI in the medical domain is very challenging and requires integration of domain knowledge with respect to understanding complex drug-disease or symptom-disease relations.

2.4 Recognizing Question Entailment

This task focuses on identifying entailment between two questions and is referred as recognizing question entailment (RQE). The task is defined as : "a question A entails a question B if every answer to B is also a complete or partial answer to A". One of the questions is called CHQ and the other FAQ.

Data: The data is based on the RQE dataset collected by Abacha and Dina (2016). The dataset statistics can be seen in Table 7.

Pre-Processing: Similar to the NLI task, we pre-process the data to expand any abbreviations in the CHQ and FAQ.

Type	CHQ	FAQ	Label
Train	What is the treatment for tri-iodothyronine thyrotoxicosis?	What is the treatment for T3 (triiodothyronine) thyrotoxicosis?	True
	Do Coumadin and Augmentin interact?	How do you inject the bicipital tendon?	False
Validation	sepsis. Can sepsis be prevented. Can someone get this from a hospital?	Who gets sepsis?	True
	medicine and allied. I LIKE TO KNOW RECENT THERAPY ON ARRHYTHMIA OF HEART	What is an Arrhythmia?	False

Table 5: Examples of question entailment from train and validation set.

Lexical Overlap	**Premise**	She is on a low fat diet
	Hypothesis	She said they also have her on a low salt diet.
	Ground truth	Neutral
	Prediction	Entailment
Disease-Symptom relation	**Premise**	Patient has diabetes
	Hypothesis	The patient presented with a change in mental status and hyperglycemia.
	Ground truth	Entailment
	Prediction	Neutral
Drug-Disease relation	**Premise**	She was treated with Magnesium Sulfate, Labetalol, Hydralazine and bedrest as well as betamethasone.
	Hypothesis	The patient is pregnant
	Ground truth	Entailment
	Prediction	Neutral

Table 6: Qualitative analysis of the outputs produced by our model. We categorize the errors into different buckets and provide cherry-picked examples to demonstrate each category.

Label	Train Set	Validation set
True	4655	129
False	3933	173

Table 7: The number of train and validation instances in each of the categories of the RQE dataset.

Training Procedure: The multi-task MT-DNN model gave the best performance for the NLI task, which motivated us to use it for the RQE task as well. We use the same hyperparamters as Liu et al. (2019) and train the model for 3 epochs.

Baselines: We compare our model with the following baselines:

- *SVM:* Similar to Abacha and Dina (2016), we use a feature based model SVM and Logistic Regression for the task of question entailment. We extract the features presented in Abacha and Dina (2016) to the best of our abilities. Their model uses lexical features such as word overlap, bigram proportion, Named Entity Recognition (NER) fea-

tures and features from the Unified Medical Concepts (UMLS) repository. Due to access issues, we only use the i2b2 [2] corpus for extracting the NER features.

- *BERT:* Like before, we compare our model with the pre-trained BERT model. For this task, we used the *bert-base-uncased* model and fine-tuned the last 12 layers for 4 epochs with learning rate 2e-5. A batch size of 16 was used.

2.4.1 Distribution Mismatch Challenges

The RQE dataset posed many unique challenges, the main challenge being that of distribution mismatch between the train and validation distribution. Table 5 shows some examples from the training and validation set which illustrate these challenges. We observe that in the training set, entailing examples always have high lexical overlap. There were about 1543 datapoints in the training set where the CHQ and FAQ were exact duplicates. The non-entailing examples in the training

---

[2]https://www.i2b2.org/NLP/DataSets/

set are completely un-related and hence the negative examples are not strong examples. Whereas in the validation set the negative examples also have lexical overlap. Furthermore, the nature of text in the validation set is more informal with inconsistent casing, punctuation and spellings whereas the training set is more structured. Furthermore, the length of the CHQ in the validation set is much longer than those observed in the training set. Therefore, we design our experimental settings based on these observations.

2.4.2 Data Augmentation

In order to address these challenges, we attempt to create synthetic data which is similar to our validation set. Another motivation for data augmentation was to increase the training size because neural networks are data hungry. Since most deep neural models rely on lexical overlap as strong indicator of entailment, we therefore use the UMLS features to augment our training set, but such that they help disambiguate the false positives. We use the following procedure for data augmentation:

1. We retrieve UMLS features for each question in the training, validation and test datasets, using the MetaMap [3] classifier.

2. We use the retrieved concept types and canonical names to create a new question-pair with the same label as shown in Figure 2, where the phrase *primary ciliary dyskinesia* has been replaced by its canonical name *kartaganer syndrome* and concept type *Disease or Syndrome*. Since BERT and MT-DNN have been trained on vast amount of English data including Wikipedia, the models are sensitive to language structure. Therefore, while augmenting data with UMLS features, we attempt to maintain the language structure, as demonstrated in Figure 2. Since UMLS provides the canonical features for each phrase in the sentence, we replace the found phrase with the following template < *UMLS Canonical name* >, *a* <*UMLS Concept Type*>.

Along with the synthetic data, we also experiment with another question entailment dataset Quora-Question Pairs (QQP). We describe the different training data used in our experiments:

1. *Orig:* Using only the provided training data.

2. *DataAug:* Using the validation set augmented with the UMLS features as discussed above. The provided training data was not used in this setting because of distribution mismatch. Despite the validation set being low-resources (300 sentences), MT-DNN has shown the capability of domain adaptation even in low-resource settings.

3. *QQP:* Quora Question pair [4](QQP) is a dataset which was released to identify duplicate questions on Quora. Questions are considered duplicates if the answer to one question can be be used as the answer to another question. We hypothesized that jointly training the model with the Quora-Question Pairs dataset should help as it is closest to our RQE dataset in terms of online forum data. We choose a subset of approx. 9k data points from QQP as this dataset has 400k training data points, in order to match the data points from the RQE training data. Along with this we use the validation set to train our model.

4. *Paraphrase:* Generated paraphrases of the *DataAug* using an off-the-shelf tool [5]. This was inspired by the observation that validation set was in-domain but since it was low-resourced, this tool provides a cheap way of creating additional artificial dataset.

2.4.3 Results and Discussion

The results over the validation set are in Table 9. We see that the MT-DNN model performs the best amongst all the other models. Addition of the *QQP* datasets did not add extra value. We hypothesize that this is due to lack of in-domain medical data in the QQP dataset.

The results of the MT-DNN model with the different training settings can be seen in Table 10. The test set comprises of 230 question pairs. We observe that the *DataAug* setting where the MT-DNN model is trained on in-domain validation set augmented with UMLS features, performs the best amongst all the strategies. Similar to the validation set, in this setting we also modify the test set with the UMLS features by augmenting it using the procedure of data augmentation described

---

[3]https://metamap.nlm.nih.gov

[4]https://data.quora.com/First-Quora-Dataset-Release-Question-Pairs

[5]https://paraphrasing-tool.com

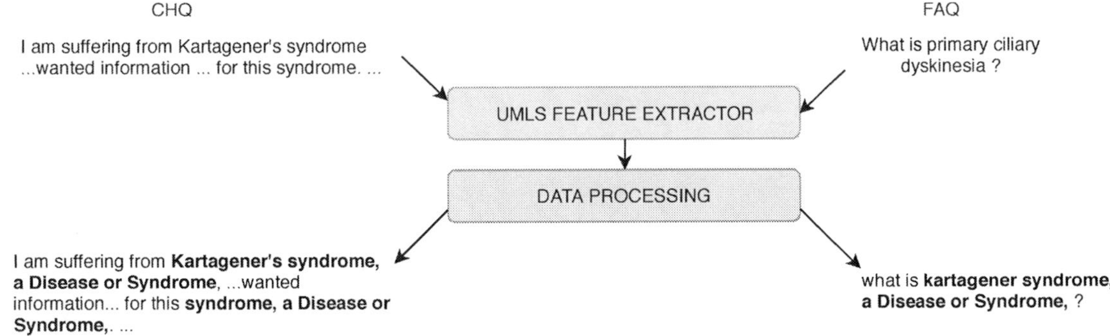

Figure 2: Data augmentation using domain knowledge for RQE.

Lexical Overlap	**CHQ**	Please i want to know the cure to Adenomyosis... I want to see a specialist doctor to help me out.
	FAQ	Do I need to see a doctor for Adenomyosis ?
	Ground truth	False
	Prediction	True
Multiple Questions	**CHQ**	Bipolar and Generalized Anxiety Disorder I read about Transcranial magnetic stimulation Therapy. Do you know anything about it? Has it had success? Also wondering about ECT? ... Is that true for mixed bipolar and generalized anxiety disorder along with meds? Have you ever heard of this?
	FAQ	How effective is Transcranial magnetic stimulation for GAD?
	Ground truth	True
	Prediction	False
Co-reference	**CHQ**	spina bifida; vertbral fusion;syrinx tethered cord. can u help for treatment of these problem.
	FAQ	Does Spina Bifida cause vertebral fusion?
	Ground truth	True
	Prediction	True

Table 8: Qualitative analysis of the outputs produced by our RQE model. We categorize the errors into different buckets and provide cherry-picked examples to prove our claim.

above. Therefore, the test set now comprises of 460 question pairs. We refer to the provided test set of 230 pairs as *original* and the augmented test set as *UMLS*. We submitted the outputs on both the original test set and the UMLS augmented test set and observe that the latter gives **+4.3** F1 gain over the original test set. We hypothesize that the addition of the UMLS augmented data in the training process helped the model to disambiguate false negatives.

Despite training data being about medical questions, it has a different data distribution and language structure. Adding it actually harms the model, as seen by the *+ Orig + DataAug + QQP* model. For our final submission, we took an ensemble of all submissions using a majority vote strategy. The ensemble model gave us the best performance.

Model	Accuracy	F1
Abacha and Dina (2016)	-	75.0
SVM	71.9	70.0
BERT	76.2	76.2
MT-DNN + Orig	78.1	**77.4**
MT-DNN + QQP	**80.8**	77.2

Table 9: Results on the RQE validation set.

	Model	F1
	Ben Abacha et al. (2019)	54.1
MT-DNN	+ Orig	58.9
	+ Orig + DataAug + QQP	60.6
	+ DataAug (UMLS)	**64.9**
	+ DataAug (original)	61.5
	+ DataAug + QQP (UMLS)	**64.9**
	Ensemble	**65.8**

Table 10: Results on the RQE test set.

	Questions	Avg answer count	Avg answer length
Train set 1	104	8	434.8
Train set 2	104	8	432.5
Validation set	25	9	420.4
Test set	150	7	418.0

Table 11: Dataset statistics for re-ranking task.

2.4.4 Error Analysis

Since we used the validation set for training the model, we cannot directly perform a standard error analysis. However, we manually analyze 100 question pairs from the test set and look at the different model predictions. We categorize errors into the following categories, as shown in Table 8.

1. **Lexical Overlap:** Most of the models we used above rely strongly on lexical overlap of tokens. Therefore, question-pairs with high orthography overlap have a strong prior for the *True* label denoting entailment.

2. **Multiple-Questions:** Often CHQ questions contained multiple sub-questions. We hypothesize that multiple questions tend to confuse the model. Furthermore, as seen in Table 8, the FAQ entails from two sub-questions in the CHQ. This shows that the model lacks the ability to perform multi-hop reasoning.

3. **Co-reference:** The model is required to perform entity co-reference as part of the entailment. In the example shown in Table 8, majority of our models marked this as entailment purely because of lexical overlap. However, there was a need for the model to identify co-reference between *these problem* and the problems mentioned in the previous sentence.

2.5 Question-Answering

In this section, we focus on building a re-ranker for question-answering systems. In particular, we attempt to use the NLI and RQE models for this task. In the ACL MediQA challenge, the question-answering system CHiQA [6] provides a possible set of answers and the task is to rank them in the order of relevance.

Data: The task-3 dataset comprises of 2 training sets and a validation set. The distribution of the data across train, validation and test was consistent

in terms of average number of answer candidates and average answer length per questio can be seen in Table 11.

2.5.1 Our Method

We implement the following re-ranking methods.

BM25: This is a ranking algorithm used for relevance based ranking given query. The formulation is given below:

$$\text{score}(D, Q) = \sum_{i=1}^{n} \text{IDF}(q_i) \cdot \frac{f(q_i, D) \cdot (k_1 + 1)}{f(q_i, D) + k_1 \cdot \left(1 - b + b \cdot \frac{|D|}{\text{avgd}}\right)} \tag{1}$$

$$\text{IDF}(q_i) = \log \frac{N - n(q_i) + 0.5}{n(q_i) + 0.5} \tag{2}$$

Here D is the answer. Q is a list of all words in the question. q_i refers to a single word. $f(q_i, D)$ is the term frequency of q_i in document D. avgd is the average answer length. The hyper-parameters used for this experiment were b = 0.75 and k1= 1.2. As shown in table 12 this gave an accuracy of 66.6 on the validation set.

NLI-RQE based model: In our second approach we leverage the pre-built NLI and RQE models from Task 1 and 2 by including the NLI and RQE scores for each question-answer pair as a feature. For instance, given a question, for each answer snippet we compute NLI scores for each sentence in the answer with the question. Since the answer snippet also contains sub-questions, we use the RQE scores to compute entailment with the question. This is illustrated below:

Question: "about uveitis. IS THE UVEITIS, AN AUTOIMMUNE DISEASE"

For the NLI scoring we would consider statements from the answer which might predict entail, contradict or neutral for the pair. Such as *Uveitis is caused by inflammatory responses inside the eye.*

Similarly we use the question phrases from the answer to give the particular answer a RQE score based on the number of entailments *Facts About Uveitis (What Causes Uveitis?)*

⁶https://chiqa.nlm.nih.gov/

Finally, we use the BM25 score for the given answer and concatenate with the above features and use SVM as the classifier.

Model	Accuracy %
BM-25	66.6
RQE+NLI+Source	**67.5**
Ben Abacha et al. (2019) (Unseen Test)	51.7
Ours	56.5

Table 12: Accuracy for task 3 on both validation set (top) and test set (bottom).

2.5.2 Results

Table 12 documents the results of our experiments. We observe that adding NLI and RQE as features show some improvement over the BM25 model.

3 Conclusion and Future Work

In this work, we present a multi-task learning approach for textual inference and question entailment tailored for the medical domain. We observe that incorporating domain knowledge for specialized domains such as the medical domain is necessary. This is because models such as BERT and MT-DNN have been pre-trained on large amounts of generic domains, leading to possible domain mismatch. In order to achieve domain adaptation, we explore techniques such as data augmentation using UMLS features, abbreviation expansion and observe a gain of +10.8 F1 for RQE. There are still many standing challenges such as incorporating common-sense knowledge apart from domain knowledge and multi-hop reasoning which pose an interesting future direction.

In the future, we also plan to explore other ranking methods based on relevancy feedback or priority ranking for task 3. We believe using MedQuad (Ben Abacha and Demner-Fushman, 2019) as training set could further help improve the performance.

Acknowledgement

We are thankful to the anonymous reviewers for their valuable suggestions.

References

Asma Ben Abacha and Demner-Fushman Dina. 2016. Recognizing question entailment for medical question answering. In *AMIA Annual Symposium Proceedings*, volume 2016, page 310. American Medical Informatics Association.

Asma Ben Abacha and Dina Demner-Fushman. 2019. A question-entailment approach to question answering. *arXiv e-prints*.

Asma Ben Abacha, Chaitanya Shivade, and Dina Demner-Fushman. 2019. Overview of the mediqa 2019 shared task on textual inference, question entailment and question answering. In *Proceedings of the BioNLP 2019 workshop, Florence, Italy, August 1, 2019*. Association for Computational Linguistics.

Samuel R Bowman, Gabor Angeli, Christopher Potts, and Christopher D Manning. 2015. A large annotated corpus for learning natural language inference. *arXiv preprint arXiv:1508.05326*.

Jacob Devlin, Ming-Wei Chang, Kenton Lee, and Kristina Toutanova. 2018. Bert: Pre-training of deep bidirectional transformers for language understanding. *arXiv preprint arXiv:1810.04805*.

Carol Friedman, Pauline Kra, and Andrey Rzhetsky. 2002. Two biomedical sublanguages: a description based on the theories of zellig harris. *Journal of biomedical informatics*, 35(4):222–235.

Tushar Khot, Ashish Sabharwal, and Peter Clark. 2018. Scitail: A textual entailment dataset from science question answering. In *Thirty-Second AAAI Conference on Artificial Intelligence*.

Xiaodong Liu, Pengcheng He, Weizhu Chen, and Jianfeng Gao. 2019. Multi-task deep neural networks for natural language understanding. *arXiv preprint arXiv:1901.11504*.

Matthew E Peters, Mark Neumann, Mohit Iyyer, Matt Gardner, Christopher Clark, Kenton Lee, and Luke Zettlemoyer. 2018. Deep contextualized word representations. *arXiv preprint arXiv:1802.05365*.

Alexey Romanov and Chaitanya Shivade. 2018. Lessons from natural language inference in the clinical domain. *arXiv preprint arXiv:1808.06752*.

Sebastian Ruder. 2017. An overview of multi-task learning in deep neural networks. *arXiv preprint arXiv:1706.05098*.

Ashish Vaswani, Noam Shazeer, Niki Parmar, Jakob Uszkoreit, Llion Jones, Aidan N Gomez, Łukasz Kaiser, and Illia Polosukhin. 2017. Attention is all you need. In *Advances in neural information processing systems*, pages 5998–6008.

Sieg at MEDIQA 2019: Multi-task Neural Ensemble for Biomedical Inference and Entailment

Sai Abishek Bhaskar*, Rashi Rungta*, James Route, Eric Nyberg, Teruko Mitamura
Language Technologies Institute, Carnegie Mellon University
{sabhaska, rashir, jroute, ehn, teruko}@cs.cmu.edu

Abstract

This paper presents a multi-task learning approach to natural language inference (NLI) and question entailment (RQE) in the biomedical domain. Recognizing textual inference relations and question similarity can address the issue of answering new consumer health questions by mapping them to Frequently Asked Questions on reputed websites like the NIH[1]. We show that leveraging information from parallel tasks across domains along with medical knowledge integration allows our model to learn better biomedical feature representations. Our final models for the NLI and RQE tasks achieve the 4th and 2nd rank on the shared-task leaderboard respectively.

1 Introduction

The MEDIQA challenge (Abacha et al., 2019) aims to improve textual inference and entailment in the medical domain to build better domain-specific Information Retrieval and Question Answering systems. There are three subtasks (NLI, RQE, QA), out of which we focus on -

1. **Natural Language Inference (NLI):** Identifying the three types of inference relations (Entailment, Neutral and Contradiction) between two sentences.

2. **Recognizing Question Entailment (RQE):** Predicting entailment between two questions (if every answer for question 1 is at least a partial answer for question 2) in the context of QA.

The task is motivated by the need to explore and develop better question answering systems in the medical domain. Identifying the type of

---

**denotes equal contribution*
[1]`https://www.nih.gov/`

correlation between questions as well as medical sentences will help the biomedical community cope with the increasing number of consumer health questions posted on community question answering websites, many of which have already been asked before and can easily be answered by linking them with a previously answered question by an expert.

In this paper, we start with a discussion of the previous work done on multi-task learning and textual inference and entailment in the biomedical domain in Section 2, followed by the dataset description in Section 3. The baselines and our proposed approach are detailed in Section 4 and 5 respectively. We conclude with the discussion of our results in Section 6 and a detailed error analysis in Section 7.

2 Related Work

2.1 Multi-Task Learning

Multi-task Learning (MTL) is inspired by the idea that it is useful to jointly learn multiple related tasks so that the knowledge gained in one task can benefit other tasks. Recently, there is growing interest in using deep neural networks (DNNs) to apply MTL to representation learning (Collobert et al. 2011, Liu et al. 2017). MTL provides an effective way to use supervised data from a number of related tasks and also provides for a regularization effect by not overfitting to a specific task, thus making the learned representations more robust.

2.2 Biomedical Textual Inference

The initial approaches for predicting inference relations between two sentences in the medical domain involved several neural architectures. (Ro-

Proceedings of the BioNLP 2019 workshop, pages 462–470
Florence, Italy, August 1, 2019. ©2019 Association for Computational Linguistics

manov and Shivade, 2018) details the curation of the MedNLI dataset, and describes multiple baseline approaches. A Feature-based, Bag-of-Words (BOW), the ESIM model (Chen et al., 2016) and the InferSent model (Conneau et al., 2017) being among them.

2.3 Biomedical Question Entailment

The initial work (Ben Abacha and Demner-Fushman, 2017), in addition to creating the working dataset for RQE, uses handcrafted lexical and semantic features as an input to traditional machine learning models like SVM, Logistic Regression, and Naive Bayes for question entailment in the clinical domain. The lexical features include word overlap, bigram similarity and best similarity from a set of 5 similarity measures (Levenshtein, Jaccard, Cosine, Bigram, Word Overlap) while semantic features include the number of overlapping medical entities and problems based on a CRF classifier trained across different corpora. Ben Abacha and Demner-Fushman 2019 use question analysis based features such as question type and focus recognition which helps identify the different focus points of consumer health questions such as information, symptoms, or treatments based on specific trigger words.

3 Datasets

3.0.1 NLI

MedNLI (Romanov and Shivade) is a dataset annotated by doctors for NLI in the clinical domain. It is available through the MIMIC-III derived data repository.

- Train: 11232 sentence pairs

- Validation: 1395 sentence pairs

- Test: 1422 sentence pairs

- Test (Leaderboard): 230 sentence pairs

Labels: {contradiction, entailment, neutral}
Evaluation Metric: Accuracy

Since the train, validation and test sets are from the same distribution, we combined them and took a subset of 90% to be the new training set and the rest 10% to be the held-out validation set.

3.0.2 RQE

The RQE dataset comprises of consumer health questions (CHQs) received by the National Library of Medicine and frequently asked questions (FAQs) collected from the National Institutes of Health (NIH) websites (Ben Abacha and Demner-Fushman, 2017).

- Training Set: 8,588 medical question pairs

- Test: 302 medical question pairs

- Test Set (Leaderboard): 230 medical question pairs

Labels: {true, false}
Evaluation Metric: Accuracy

On further analysis of the RQE train and test data, we found that the two datasets come from different distributions. The CHQs in the training set follow a more formal third person based language structure while CHQs in the test set are verbose with more colloquial language phrases. For example, a CHQ from the training set is - "How should I treat polymenorrhea in a 14 year-old girl?" while a CHQ from the test set is - "lupus. Hi, I want to know about Lupus and its treatment. Best, Mehrnaz".
In light of this, we modify our training set to contain 302 examples from the original training set, all the 302 examples in the test set and 930 questions from *icliniq* as explained in section 5.1.3. As with NLI, we took a subset of 90% to be the new training set and the rest 10% to be the held-out validation set.

4 Baselines

4.1 NLI

InferSent (Romanov and Shivade, 2018) is a sentence encoder model that has given near state-of-the-art results across the NLP (including NLI) and computer vision domains. For the MedNLI dataset, the model uses a Bi-directional LSTM with domain knowledge incorporated through retrofitting and attention. We use this InferSent model as our baseline for the NLI task. A re-implementation using data preprocessed with UMLS (5.2.3) and abbreviation expansion (5.2.5), along with different word embeddings (5.2.2) gives a slight bump in the accuracy value.

InferSent	Accuracy	Embeddings
Reported	78.3	MIMIC FastText
Re-implementation	79.3	PubMed MIMIC FastText

Table 1: Baseline accuracy values for NLI dev set

4.2 RQE

The SVM model described in Ben Abacha and Demner-Fushman 2017 is our RQE baseline. The input features are detailed in 2.3 and the corresponding metrics are shown in Table 2.

	P	R	F
SVM	75.0	75.2	75.0

Table 2: Baseline precision, recall and F1 values for RQE

5 Proposed Approach

5.1 Additional Datasets

Our hypothesis is that these parallel datasets will help our multi-task neural model capture salient biomedical features to help our main NLI and RQE tasks.

5.1.1 PubMed RCT

The Pubmed RCT dataset contains 2.3m sentences from 200k PubMed abstracts of randomized controlled trial (RCT) articles. We use the smaller subset of the sentences from 20k abstracts. The sentences are labeled based on their role in the abstract which belongs to one of the following five classes: background, objective, method, result, or conclusion. This single sentence classification is a parallel dataset for the NLI task.

5.1.2 MultiNLI

The MultiNLI dataset (Williams et al., 2017) contains 433k sentences which have been annotated with textual entailment information. This textual inference classification corpus forms one of the parallel datasets for the NLI task.

5.1.3 *icliniq.com* Questions

Given the limited size of the RQE dataset, we looked for ways to augment our data with additional examples from the same distribution.
We use data scraped from *icliniq.com*, which is an online doctor consultation platform. The website has a format where each question has a summary question, followed by the entire text entered by the user. We take the summary question to be the FAQ and the question text as the CHQ corresponding to the RQE task. 465 question pairs were scraped (Regin, 2017) and an equal number of negative examples is generated through negative sampling. This gives us a total of 930 additional question pairs. An example from icliniq is:

Q1 (CHQ): Hello doctor, I do not have a white half moon on my nails. Is there any thyroid issue? If yes, please suggest some treatment."
Q2 (FAQ): Does the absence of the white half moon on nails indicate a thyroid problem?
Gold Label: True

5.1.4 GARD Question Type

The dataset released by the Genetic and Rare diseases information center (Roberts et al., 2014) allows our model to learn question type information necessary for the RQE task. It contains 3137 questions each of which has one of 13 unique labels. Since the question type is an important hand-crafted feature while considering traditional ML approaches for the RQE task, we use this dataset so that our multi-task model can leverage this information. The merit of this approach is shown in Table 3.

5.1.5 Quora Question Pairs

The Quora Question Pairs dataset (Quora, 2017) contains more than 400k duplicate question pairs released by Quora, a popular community QA website. We hypothesize that using this as a parallel dataset for the RQE task will help us generalize better since Quora users adopt an informal and colloquial form of language which is similar to the language of CHQs.

5.2 Domain Knowledge Integration and Preprocessing

5.2.1 ScispaCy

We use ScispaCy (Neumann et al., 2019), a tool for practical biomedical/scientific text processing, based on the spaCy library to preprocess and incorporate domain knowledge in the NLI and RQE datasets. Its use is detailed in the subsequent sections.

5.2.2 Biomedical Word Vectors

We use the biomedical word vectors released by the NCBI BioNLP Research Group (Chen et al., 2018) as the word embeddings for the InferSent model for the NLI task. Fasttext (Bojanowski et al., 2017) was used to train 200-dimensional word vectors on PubMed abstracts and MIMIC III clinical notes.

5.2.3 UMLS Metamap

We use a python wrapper for UMLS Metamap (Aronson and Lang, 2010), called pyMetamap[2] to extract preferred names and CUIs (Concept Unique Identifiers) for medical entities from the UMLS Metathesaurus (Bodenreider, 2004). As a pre-processing step, we identify medical terms in the data using ScispaCy, and replace them with their *preferred_name* occurring with the highest score in UMLS.

Using ScispaCy helps us by acting as a filter against common terms like *patient* and *lab*, which would otherwise get identified to be medical entities.

In cases where the preferred name for a medical entity was exactly the term itself, we used the additional dataset MRCON (Rogers et al., 2012) to extract all entity names with the same CUI as the one for the entity identified initially. We created a set of these synonymous entities and picked the one which had the highest semantic similarity to the medical entities identified in the parallel sentence/question. We then append this identified synonymous entity's name to where the originally identified entity was found in the first sentence/question.

5.2.4 DrugBank

DrugBank (Wishart et al., 2017) is a bioinformatics and cheminformatics dataset containing detailed drug data for more than 12k drugs along with their synonyms, parent medical categories (i.e. what kind of drug it is) and pharmacological information.

Our use of DrugBank to augment the RQE and NLI datasets with domain knowledge is as follows:

- We load SciSpacy with two pretrained Spacy models. The first is a NER model trained on the BC5CDR corpus to identify drug names and the second is a general pipeline for biomedical data.

- From the first sentence, we extract drug names using the first SciSpacy model.

- From the second sentence in the particular sentence-pair, we extract biomedical terms and search for a string overlap with the relevant drug information from the Drugbank dataset.

- If a particular phrase exists in the drug information, we append this phrase after the drug name in the first sentence.

5.2.5 Abbreviation expansion

We use the Recognizing Abbreviation Definitions dataset (S Schwartz and Hearst, 2003) to construct an initial dictionary. To further augment it, we use the CAMC (Charleston Area Medical Center) medical word list[3]. In order to get an extended dictionary which took into account the several newly created acronyms, or those which are more colloquial than formal, we scraped the medical abbreviation Wikipedia pages and appended this to our dictionary. If more than one medical phrase was found for an abbreviation, we gave preference to the first one. On manual combing of the thus created dictionary, we edited/deleted entries which felt incorrect. For example, *FS* which was being mapped to *Flow Sheet* was changed to *Fingerstick*. As one of the preprocessing steps, ScispaCy is used to identify abbreviations in the text which are then appended with their corresponding expanded medical term.

---

[2] https://github.com/AnthonyMRios/pymetamap/

[3] https://www.camc.org/documents/patientlink/Abbreviations-List.pdf

5.2.6 Bio-BERT

BioBERT (Lee et al., 2019) uses the pretrained BERT base model and finetunes it for the biomedical domain by further training on PubMed abstracts and PMC full-text articles. We converted the Tensorflow version of the saved model weights to PyTorch using the PyTorch pretrained BERT library. The three variants of the BioBERT model based on the data used to finetune it are-

- PubMed abstracts (4.5B words)

- PMC full-text articles (13.5B words)

- Both PubMed abstracts and PMC full-text articles

The latter variant outperforms single dataset trained BioBERT with respect to most of the biomedical named entity recognition datasets but has mixed results for the relation extraction and question answering datasets as mentioned in (Lee et al., 2019).
We use the PubMed+PMC BioBERT v1.0 model (cased vocabulary) to initialize our MT-DNN architecture.

5.2.7 SciBERT

SciBERT (Beltagy et al., 2019), is another BERT based model for the scientific and biomedical domain which outperforms BioBERT by an average of 0.51 F1 score at biomedical named entity recognition, text classification and relation classification. It was trained on 1.14M papers from Semantic Scholar (Ammar et al., 2018) of 18% is from the computer science domain and 82% is from the biomedical domain. The full text of the papers are used, not just the abstracts.

There are four variants of SciBERT -

- Cased or Uncased

- BERT-Base vocab or scivocab (30k words, having a 42% overlap with BERT-Base vocab)

We use the recommended uncased scivocab SciB-ERT model to initialize our MT-DNN architecture. Our final model ensemble consists of SciB-ERT in addition to BioBERT as the two models were trained on different datasets and hence they will be able to capture different salient features of biomedical knowledge.

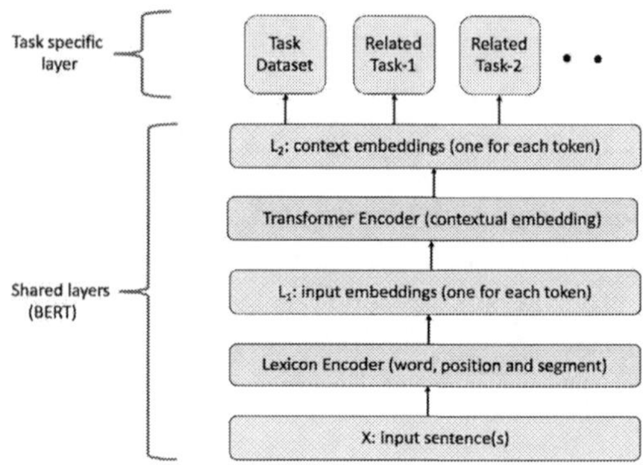

Figure 1: Architecture of the multi-task MT-DNN model

Datasets	Test Accuracy
RQE	58.2
RQE + GARD Question Type (GARD)	62.6
RQE + Quora Question Pairs (QQP)	66.0
RQE + QQP + GARD	66.0

Table 3: Parallel dataset results (values obtained post the shared task completion) for the RQE task using the MT-DNN base model.

5.3 Model

We are interested in leveraging multi-task learning across different datasets to improve the learning of the biomedical text representations. For the current work, we use the Multi-Task Deep Neural Networks for Natural Language Understanding (MT-DNN) introduced in Liu et al. 2019, which demonstrates the effectiveness of multi-task learning by beating the state-of-art on eight out of nine GLUE benchmark tasks (Wang et al., 2019). The architecture of our MT-DNN model is shown in Figure 1. Both the NLI and RQE tasks share the lower layers, while the top layers represent task-specific outputs. The input X, which is a word sequence (biomedical question for RQE and sentence text for NLI) is first represented as a sequence of embedding vectors, one for each word, in L_1. Then the transformer encoder captures the contextual information for each word via self-attention and generates a sequence of contextual embeddings in L_2. This is the shared semantic representation that is trained by the multiple task objectives. The lexicon encoder (L_1) and transformer encoder (L_2) pre-training involves the approach introduced in the BERT model (Devlin et al., 2018).

Model	Datasets	Domain Knowledge	Test Accuracy
InferSent Baseline (Romanov and Shivade 2018)	NLI (train set only)	UMLS	71.4
MT-DNN + MT-DNN(BioBERT)	NLI	UMLS	83.5
MT-DNN + MT-DNN(BioBERT) + MT-DNN(SciBERT)	NLI + MultiNLI + PubMed 20k RCT	UMLS	87.2
MT-DNN + MT-DNN(BioBERT) + MT-DNN(SciBERT) + InferSent	NLI + MultiNLI + PubMed 20k RCT	UMLS + DrugBank + Abbreviation Expansion	**91.1**

Table 4: Results for the NLI Task

Model	Datasets	Domain Knowledge	Test Accuracy
SVM Baseline (Ben Abacha and Demner-Fushman 2017)	RQE (train set only)	biomedical NER	54.1
MT-DNN + MT-DNN(SciBERT)	RQE	UMLS + DrugBank + Abbreviation Expansion	65.8
MT-DNN + MT-DNN(SciBERT)	RQE + GARD Question Type	UMLS + DrugBank + Abbreviation Expansion	66.7
MT-DNN + MT-DNN(BioBERT) + MT-DNN(SciBERT)	RQE + Quora Question Pairs + GARD Question Type	UMLS + DrugBank + Abbreviation Expansion	**70.6**

Table 5: Results for the RQE Task

5.3.1 Implementation details

The BERTAdam optimizer with a learning rate of 5e-5, batch size of 32, linear learning rate decay schedule with warm-up over 0.1 and gradient clipping is used. These hyperparameters are in accordance with those proposed in the MT-DNN work (Liu et al. 2019). In each epoch, a mini-batch from all the parallel datasets is taken and the model is updated.

The training procedure of the model consists of two stages: pretrained BERT model loading and multi-task fine-tuning. We use BioBERT (5.2.6), SciBERT (5.2.7) and the MT-DNN base model (pretrained on the GLUE benchmark tasks) to initialize our MT-DNN model variants.

6 Experiments and Results

The accuracy values obtained on the shared task's leaderboard are listed in Table 4 and Table 5 for the NLI and the RQE task respectively.

For the NLI task, Table 4, we see that an ensemble of the MT-DNN base model along with MT-DNN initialized with SciBERT and BioBERT keeping PubMed RCT and MultiNLI as the parallel datasets achieved a better accuracy than using only the NLI dataset with an MT-DNN base model, BioBERT ensemble.

To account for missing drug information and the lack of biomedical context around abbreviations in the input data, we preprocess our dataset by expanding medical abbreviations (5.2.5) and including DrugBank (5.2.4) information.

We see that taking a four-way ensemble of the MT-DNN base model, MT-DNN initialized with BioBERT, SciBERT and InferSent along with a three-pronged domain knowledge inclusion with MultiNLI and PubMed RCT as the parallel datasets gave us the best result of **91.1%** on the leaderboard. Our hypothesis behind this model ensemble was that since BioBERT and SciBERT are trained on different datasets, they will capture different features and hence taking an ensemble of these two models along with InferSent based on majority confidence scores will help us achieve a better accuracy than a single model. Our Infersent re-implementation results are shown in Table 1.

To demonstrate the usefulness of parallel datasets for the RQE task and for easy comparison with the results on the leaderboard (Table 5), we measure the test accuracy for different dataset combinations using the test dataset labels released by the task organizers post completion of the shared task. These results are shown in Table

467

Category	Premise	Hypothesis	Predicted	Gold Label
Numeric Reasoning	On transfer, patient VS were 102, 87/33, 100% on 60% 450 x 18 PEEP 5.	The patient's vitals were normal on transfer	neutral	contradiction
	Was given a 500cc bolus and responded to 89/50.	The patient was hypotensive.	neutral	entailment
	His initial BP at OSH 130/75, down to 93/63 after nitro.	The patient was initially normotensive.	contradiction	entailment
Inconclusive cases	The pt was discharged home [**2188-5-3**].	the patient was discharged with home medications	entailment	neutral
	On the floor, he is doing relatively well.	The patient is stable.	entailment	neutral
	His symptoms occur about every day to every other day and have been stable over the past year.	His symptoms are severe.	contradiction	neutral

Table 6: Error types observed during the qualitative analysis for the NLI Task

Category	Q1 (CHQ)	Q2 (FAQ)	Predicted	Gold Label
Understanding	milroy disease hello , my daughter has lymph edema her both legs and left hand is swelling , this problem now started when she was of 3 months now she is 16 months , her swelling is growing day by day , im clue less what to do and what kind of treatment i should do plz help and suggest us	Is walking good for lymphedema?	true	false
	If oleandor was ingested by touching the plant stems inner part and then directly eating without washing hands, how long would u exspect symptoms would start? And how severe would you say symptoms may get.	What are the symptoms of Oleander poisoning?	false	true
Multiple Questions	more information in relation to Ellis van creveld syndrome Specifically in later life can they have children has it ever been reported any researchcarried out and just as much information as possible to help my understanding of what I have Many thanks	What is Ellis-van Creveld syndrome?	true	false
	Achondroplasia research. Hello. We are students from [LOCATION] and we are doing a biology project of genetic diseases. We chose Achondroplasia as our disease to research. We have a few question and we are hoping you could answer them. Our questions are, can you tell if your child will have Achondroplasia when you are pregnant? When do people usually come in when they think something isn't right with their child? what are the worse cases of Achondroplasia you've ever seen? Thank you in advance. sincerely, [NAME]	How to diagnose Achondroplasia?	false	true

Table 7: Error types observed during the qualitative analysis for the RQE Task

3. We see that using only the RQE dataset got us an accuracy of 58.2% while using the GARD question type decomposition and Quora Question Pairs increased our accuracy by 4.3% and 7.8% respectively.

Building on the observation of variation in performance of the different parallel datasets, we see that having GARD question types as the parallel dataset gives us a slight boost in accuracy from 65.8% to 66.7% as shown in Table 5. Our best result of **70.6%** is obtained when we take an ensemble of the MT-DNN base model along with MT-DNN initialized with BioBERT and SciBERT, keeping Quora Question Pairs and GARD Question Type as the parallel datasets.

7 Error Analysis

7.1 NLI

Equivalent to the error analysis in Romanov and Shivade 2018, we present some of the representative examples from the Test set (using the gold labels released by the task organizers) in Table 6.

We broadly classify them into categories we felt they were closest to.

7.1.1 Numeric values

Example pairs where the premise is solely based on *numeric values* describing the patient's vitals are often classified incorrectly due to the several variations in the values used across examples. This can be seen in Example 1, 2 and 3 from the table. Most of such examples are often incorrectly predicted to be neutral by our model.

7.1.2 Inconclusive cases

We also come across examples where the sentences are not entirely conclusive, but the model assumes them to be, hence making an incorrect prediction. These examples are clubbed under the *Inconclusive cases* category.

Consider the case of Example 5 from Table 6, the hypothesis claims the patient to be stable, while the premise does not state this explicitly, thus leaving a margin for a less definite hypothesis. Our model predicts *entailment* for this pair, when the expected label is *neutral*.

7.2 RQE

Table 7 shows a few examples representative of the two broad categories of errors observed in the Test set (using the gold labels released by the task organizers) for the RQE task.

7.2.1 Understanding

The *CHQ* from Example 1 in the table is asking for treatment suggestions for the condition *lymphedema*, and the *FAQ* is a question verifying if walking is good for *lymphedema*. The expected label is *false*, while the model predicts it to be *true*. The two questions are semantically different because of which one does not entail the other, but the model might be confusing a suggestive question (*FAQ* in this example) to be a part of the broader question (*CHQ*) thus failing to understand the subtle difference between the two.

In Example 2, the *CHQ* asks about two questions related to the symptoms - how long they will take to occur, and how severe they would get. The *FAQ* inquires about what the symptoms are. These questions have the same focus, but could be understood as being different when compared semantically. However, since the answer to the *FAQ* might partially answer the *CHQ*, the expected label is true, while our model predicted this as false.

7.2.2 Multiple Questions

The other kind of errors we observed were when the *CHQs* had multiple questions within them. For instance in Example 3, in the *CHQ* the user seems to have decent knowledge about the said syndrome and wants more in-depth knowledge on the subject. The repeated questions about more *information* might have misled the model into predicting this as true, when the expected label was false.

In Example 4, we see that the several questions contained in the *CHQ* confuse our classifier to predict false when the *FAQ* is actually entailed.

8 Future Work

Going forward, this work could be improved by more intensive domain knowledge incorporation. To start with, using medical side effects relations from SIDER (Kuhn et al., 2015) and leveraging the ontology relations in UMLS (Bodenreider, 2004) would be appropriate steps to strengthen the proposed system. We would like to thank our anonymous reviewers for these inputs.

A large part of the success of this work can be attributed to preprocessing the input data to incorporate biomedical knowledge which, at the same time makes it harder to generalize this pipeline to other domains. Therefore, investigating the performance of our proposed approach in non-biomedical domains by training with different parallel datasets to enforce generalization is an interesting avenue for future research.

9 Conclusion

In this paper, we investigate various preprocessing pipelines along with parallel dataset combinations in a multi-task learning setup for efficient language processing in the biomedical domain. We demonstrate the effectiveness of using transformer based neural models for predicting natural language inference and recognizing question entailment in the medical domain which beat the baselines (as shown in Table 4 and Table 5) by a margin of 19.7% and 16.5% for the NLI and RQE tasks respectively.

References

Asma Ben Abacha, Chaitanya Shivade, and Dina Demner-Fushman. 2019. Overview of the mediqa 2019 shared task on textual inference, question entailment and question answering. *ACL-BioNLP 2019*.

Waleed Ammar, Dirk Groeneveld, Chandra Bhagavatula, Iz Beltagy, Miles Crawford, Doug Downey, Jason Dunkelberger, Ahmed Elgohary, Sergey Feldman, Vu Ha, Rodney Kinney, Sebastian Kohlmeier, Kyle Lo, Tyler Murray, Hsu-Han Ooi, Matthew E. Peters, Joanna Power, Sam Skjonsberg, Lucy Lu Wang, Chris Wilhelm, Zheng Yuan, Madeleine van Zuylen, and Oren Etzioni. 2018. Construction of the literature graph in semantic scholar. *CoRR*, abs/1805.02262.

Alan Aronson and Franois-Michel Lang. 2010. An overview of metamap: Historical perspective and recent advances. *Journal of the American Medical Informatics Association : JAMIA*, 17:229–36.

Iz Beltagy, Arman Cohan, and Kyle Lo. 2019. Scibert: Pretrained contextualized embeddings for scientific text.

Asma Ben Abacha and Dina Demner-Fushman. 2017. Recognizing question entailment for medical question answering. *American Medical Informatics Association*.

Asma Ben Abacha and Dina Demner-Fushman. 2019. A question-entailment approach to question answering. *CoRR*, abs/1901.08079.

Olivier Bodenreider. 2004. The unified medical language system (umls): Integrating biomedical terminology. *Nucleic acids research*, 32.

Piotr Bojanowski, Edouard Grave, Armand Joulin, and Tomas Mikolov. 2017. Enriching word vectors with subword information. *Transactions of the Association for Computational Linguistics*, 5:135–146.

Qian Chen, Xiaodan Zhu, Zhen-Hua Ling, Si Wei, and Hui Jiang. 2016. Enhancing and combining sequential and tree LSTM for natural language inference. *CoRR*, abs/1609.06038.

Qingyu Chen, Yifan Peng, and Zhiyong Lu. 2018. Biosentvec: creating sentence embeddings for biomedical texts. *CoRR*, abs/1810.09302.

Ronan Collobert, Jason Weston, Léon Bottou, Michael Karlen, Koray Kavukcuoglu, and Pavel P. Kuksa. 2011. Natural language processing (almost) from scratch. *CoRR*, abs/1103.0398.

Alexis Conneau, Douwe Kiela, Holger Schwenk, Loïc Barrault, and Antoine Bordes. 2017. Supervised learning of universal sentence representations from natural language inference data. *CoRR*, abs/1705.02364.

Jacob Devlin, Ming-Wei Chang, Kenton Lee, and Kristina Toutanova. 2018. BERT: pre-training of deep bidirectional transformers for language understanding. *CoRR*, abs/1810.04805.

Michael Kuhn, Ivica Letunic, Lars Juhl Jensen, and Peer Bork. 2015. The SIDER database of drugs and side effects. *Nucleic Acids Research*, 44(D1):D1075–D1079.

Jinhyuk Lee, Wonjin Yoon, Sungdong Kim, Donghyeon Kim, Sunkyu Kim, Chan Ho So, and Jaewoo Kang. 2019. Biobert: a pre-trained biomedical language representation model for biomedical text mining. *CoRR*, abs/1901.08746.

Xiaodong Liu, Pengcheng He, Weizhu Chen, and Jianfeng Gao. 2019. Multi-task deep neural networks for natural language understanding. *CoRR*, abs/1901.11504.

Xiaodong Liu, Yelong Shen, Kevin Duh, and Jianfeng Gao. 2017. Stochastic answer networks for machine reading comprehension. *CoRR*, abs/1712.03556.

Mark Neumann, Daniel King, Iz Beltagy, and Waleed Ammar. 2019. Scispacy: Fast and robust models for biomedical natural language processing.

Quora. 2017. Quora question pairs - kaggle. *Kaggle.com*.

Lasse Regin. 2017. Medical question answer data. *https://github.com/LasseRegin/medical-question-answer-data*.

Kirk Roberts, Halil Kilicoglu, Marcelo Fiszman, and Dina Demner-Fushman. 2014. Decomposing consumer health questions. *Proceedings of BioNLP 2014*, pages 29–37.

Willie Rogers, Francois-Michel Lang, and Cliff Gay. 2012. Metamap data file builder. *metamap.nlm.nih.gov*, page 6.

Alexey Romanov and Chaitanya Shivade. Lessons from natural language inference in the clinical domain.

Alexey Romanov and Chaitanya Shivade. 2018. Lessons from natural language inference in the clinical domain. *CoRR*, abs/1808.06752.

Ariel S Schwartz and Marti Hearst. 2003. A simple algorithm for identifying abbreviation definitions in biomedical text. *Pacific Symposium on Biocomputing. Pacific Symposium on Biocomputing*, 4:451–62.

Alex Wang, Amanpreet Singh, Julian Michael, Felix Hill, Omer Levy, and Samuel R. Bowman. 2019. GLUE: A multi-task benchmark and analysis platform for natural language understanding. In the Proceedings of ICLR.

Adina Williams, Nikita Nangia, and Samuel R. Bowman. 2017. A broad-coverage challenge corpus for sentence understanding through inference. *CoRR*, abs/1704.05426.

David S Wishart, Yannick D Feunang, An C Guo, Elvis J Lo, Ana Marcu, Jason R Grant, Tanvir Sajed, Daniel Johnson, Carin Li, Zinat Sayeeda, Nazanin Assempour, Ithayavani Iynkkaran, Yifeng Liu, Adam Maciejewski, Nicola Gale, Alex Wilson, Lucy Chin, Ryan Cummings, Diana Le, Allison Pon, Craig Knox, and Michael Wilson. 2017. DrugBank 5.0: a major update to the DrugBank database for 2018. *Nucleic Acids Research*, 46(D1):D1074–D1082.

IIT-KGP at MEDIQA 2019: Recognizing Question Entailment using Sci-BERT stacked with a Gradient Boosting Classifier

Prakhar Sharma, Sumegh Roychowdhury
Indian Institute of Technology Kharagpur,
India
{prakharsharma, sumegh01}@iitkgp.ac.in

Abstract

The number of people turning to the Internet to search for a diverse range of health-related subjects continues to grow and with this multitude of information available, duplicate questions become more frequent and finding the most appropriate answers becomes problematic. This issue is important for question-answering platforms as it complicates the retrieval of all information relevant to the same topic, particularly when questions similar in essence are expressed differently, and answering a given medical question by retrieving similar questions that are already answered by human experts seems to be a promising solution. In this paper we present our novel approach to detect question entailment by determining the type of question asked rather than focusing on the type of the ailment given. This unique methodology makes the approach robust towards examples which have different ailment names but are synonyms of each other. Also it enables us to check entailment at a much more fine-grained level.

1   Introduction

Seeking health-related information is one of the top activities of todays online users via both personal computers and mobile devices. In all, 80 percent of Internet users, or about 93 million Americans, have searched for a health-related topic online, according to a study released on 16th July 2018 by the Pew Internet & American Life Project (Weaver, 2016) .Thats up from 62 percent of Internet users who said they went online to research health topics in 2001, the Washington research firm found. China (Guo et al., 2018) also has 194.76 million Internet health users in 2016, increased 28.0% compared with that in 2015. Despite the widespread need, the search engines often fail in returning relevant and trustworthy health information (Natalia et al., 2012)(Arabella

et al., 2017). In this paper we try to bridge this gap by predicting entailment between questions. We particularly tackle this problem by checking entailment of a given consumer health question (CHQ) with most similar Frequently Asked Question (FAQ). Given two general English sentences this Question Entailment system can conclude whether answer of one question implies the other question's answer.

Q1: "Can you mail me patient information about Glaucoma, I was recently diagnosed and want to learn all I can about the disease."

Q2: "What is glaucoma?"

In the above two questions the answer of Q1 implies the answer of Q2. (Entailment)

Detecting Question Entailment is a challenging task as it involves an amalgamation of tasks like Question Answering and Textual Entailment (Abacha and Demner-Fushman, 2019). Question answering is used to generate answers for both the questions and then checking textual entailment between the answers to give predictions possibly integrating Named Entity Recognition(NER) to our advantage. In this paper, we experiment on the MEDIQA 2019 task (Ben Abacha et al., 2019) by presenting an all-together different approach **QSpider** which overcomes these challenges by detecting question types instead of treating it like a pure Textual entailment or Question answering task.

Attempts have been made to tackle this problem, the most notable one being (Abacha and Demner-Fushman, 2016) which is the baseline for this task. The Baseline method uses supervised methods like SVM, Logistic Regression, Naive Bayes and used manual feature engineering but it fails to explore over the semantic space of the sentence. In this paper, we propose our model **QSpider** to tackle this problem.

QSpider is a staged system consisting of state-

Proceedings of the BioNLP 2019 workshop, pages 471–477
Florence, Italy, August 1, 2019. ©2019 Association for Computational Linguistics

of-the-art model **Sci-BERT** used as a multi-class classifier aimed at capturing both question types and semantic relations stacked with a **Gradient Boosting Classifier** which checks for entailment. QSpider achieves an accuracy score of **68.4%** which outperforms the baseline model (54.1%) by an accuracy score of 14.3%.

2   Related Work

2.1   Quora Question Pairs

Quora Question Pairs[1] is a binary classification task where the goal is to determine if two questions asked on Quora are semantically equivalent (Chen et al., 2018). Several works are done on this task with best performing ones being (MT-DNN (Liu et al., 2019), DIIN (Gong et al., 2018)). With MT-DNN's model incorporating a pre-trained bidirectional transformer language model similar to BERT (Devlin et al., 2018) while the fine-tuning part is leveraging multi-task learning. The DIIN model uses encoders to encode both the sentences and uses an interaction layer on top of it which is fed into a feature extraction layer. Finally the output layer decodes the acquired features to give predictions.

2.2   Recognizing Question Entailment

While textual entailment in open-domain has been extensively addressed in the literature, RQE has been less addressed for more restricted and specialized fields such as the medical domain. In *Recognizing Question Entailment for Medical Question Answering* (Abacha and Demner-Fushman, 2016) lexical features like Word Overlap and Bigram Similarity measures are used. It also tried to account for semantic features by using Negation Scope for Q1 and Q2, recognizing medical entities of 3 type: Problem, Treatment and Test. A different approach of using entailment in the QA problem is done in both the Pascal-RTE Challenge (Dagan et al., 2007), and in the CLEFAVE task (Kouylekov et al., 2006), by considering a question Q turned into an affirmative sentence as the hypothesis, and a text passage containing a candidate answer A as the text (i.e.systems have to decide whether A supports, or entails, Q).

3   Task Description & Dataset

The objective of this task is to identify entailment between two questions in the context of Question Answering. We use the following definition of question entailment: *Question A entails a Question B if every answer to B is also a complete or partial answer to A.* So, basically we need to predict, given two questions, if they entail each other or not.

The training corpus of MEDIQA 2019 RQE Shared Task (Ben Abacha et al., 2019) consists of 8,588 training pairs, containing 54.2% positive pairs. The remaining pairs (3,933) are negative examples collected by associating a random short form of NLM dataset question (JW et al., 2000) having at least one common keyword and at least one different keyword for each original question. The validation test corpus contains 302 pairs of questions consisting of 173 negative pairs and 129 positive pairs. Also the hidden test set had in total 230 pairs of questions of which 115 (50%) were true pairs and rest (115) false pairs. The question pairs in validation and hidden test set had its first question a Consumer asked Health Question (CHQ) and second question a Frequently Asked Question (FAQ). Upon doing an elementary analysis of the task dataset, we observe there are examples in validation and test set where medical entities are not in same form (either synonyms or abbreviation) in both questions but they still entail each other and vice versa.

Validation Set	**Positive**	**Negative**
Same Medical Entity	112	54
Different Medical Entity	17	119

Test Set	**Positive**	**Negative**
Same Medical Entity	87	101
Different Medical Entity	28	14

Table 1: Dataset Statistics : Positive means Entailment & Negative means Not Entailment.

We **additionally** used an annotated corpus of consumer health questions (Roberts et al., 2014) to build our question type prediction classifier. The corpus consists of 1,467 consumer-generated requests for disease information, containing a total of 2,937 questions. The dataset has these requests classified into 13 question types or classes namely: *Anatomy, Cause, Complication, Diagnosis, Information, Management, Manifestation,*

Other effects, PersonOrg, Prognosis, Susceptibility, Other, Not Disease.

4 Models

In this section we will discuss about the various approaches we have used for building our Question Entailment detection model.

- **Dependency Tree-LSTM** (Tai et al., 2015): A generalization of LSTM (Long Short-Term Memory) to tree-structured network topologies. The model was aimed to capture the syntactic relations between two questions.

- **BERT$_{\text{Large, uncased}}$** (Devlin et al., 2018): BERT which stands for Bidirectional Encoder Representations from Transformers is designed to train deep bidirectional representations by jointly conditioning on both left and right context in all layers. Language models have demonstrated that rich, unsupervised pre-training is an integral part of many language understanding systems. Hence, we try fine-tuning BERT to obtain better results on this task.

- **Bio-BERT** (Lee et al., 2019): Domain specific language representation model based on BERT and pre-trained on large-scale biomedical corpora.

- **Sci-BERT + Hinge loss** (Beltagy et al., 2019) : A pre-trained contextualized embedding model based on BERT to address the lack of high-quality, large-scale labeled scientific data, fine-tuned with a Hinge loss function. This outperformed all other systems during the validation phase.

Now we describe all our approaches in-detail.

4.1 Dependency Tree-LSTM

We refer to a Child-Sum Tree-LSTM(Tai et al., 2015) applied to a dependency tree as a Dependency Tree-LSTM. We produced dependency parses[2] of the questions in the dataset for our Dependency Tree-LSTM model. Each Tree-LSTM unit (indexed by j) contains input and output gates i_j and o_j, a memory cell c_j and hidden state h_j. The difference between the standard LSTM unit and Tree-LSTM units is that gating

---

[2]Dependency parses produced by the Stanford Neural Network Dependency Parser (Chen and Manning, 2014)

vectors and memory cell updates are dependent on the states of possibly many child units. Additionally, instead of a single forget gate, the Tree-LSTM unit contains one forget gate f_{jk} for each child k. This allows the Tree-LSTM unit to selectively incorporate information from each child. Each Tree-LSTM unit takes an input vector x_j. We took, each x_j as a vector representation of a word in a sentence. The input word at each node depends on the tree structure used for the network.

We first produce sentence representations h_L and h_R for **question1** and **question2** respectively in the pair using a Tree-LSTM model over question's parse tree. Given these sentence representations, we calculate the entailment probability $\hat{p}_\theta$ using a neural network that considers both the distance and angle between the pair (h_L, h_R):

$$\text{h}_\times = h_L \odot h_R,$$

$$\text{h}_+ = |h_L - h_R|,$$

$$\text{h}_s = \sigma\left(W^{(\times)}h_\times + W^{(+)}h_+ + b^{(h)}\right),$$

$$\hat{p}_\theta = \text{softmax}\left(W^{(p)}h_s + b^{(p)}\right),$$

We want $\hat{p}_\theta$ given model parameters θ to be close to the p. Here y denotes whether it is an entailment. Hence we decide the cost function as the regularized KL-divergence between p and $\hat{p}_\theta$:

$$\text{p}_i = ||\, i - y\,| - 1\,| \qquad i = \{0, 1\}$$

$$\text{J}(\theta) = \frac{1}{m}\sum_{k=1}^{m}\text{KL}\left(p^{(k)}\,\middle\|\,\hat{p}_\theta^{(k)}\right) + \frac{\lambda}{2}\|\theta\|_2^2,$$

where m is the number of training pairs and the superscript k indicates the k-th sentence pair.

4.2 BERT

We chose BERT$_{\text{Large, uncased}}$ as our underlying BERT model. It consists of 24-layers, 1024-hidden, 16-heads, and 340M parameters. It was trained on the BookCorpus (800M words) and the English Wikipedia (2,500M words). The two input sentences in form of **question1** and **question2** were first tokenized with the BERT basic tokenizer to perform punctuation splitting, lower casing and invalid characters removal. The maximum sequence length was defined as 128, with shorter sequences padded and longer sequences truncated to this length.

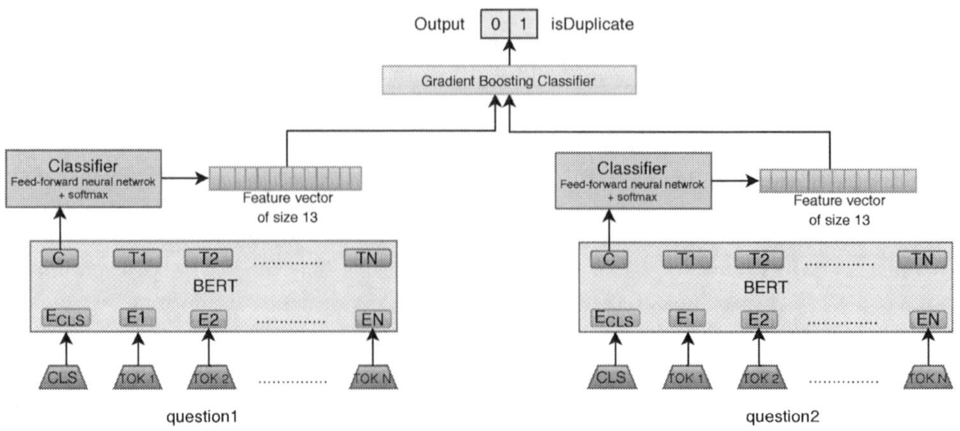

Figure 1: Model - QSpider

We used the PyTorch implementation from *pytorch-pretrained-bert*[3] which had the BERT tokenizer, positional embeddings, and pre-trained BERT model. Following the recommendation for fine-tuning in the original BERT approach (Devlin et al., 2018), we trained our classifier with a batch size of 32 for 5 epochs. The dropout probability was set to 0.1 for all layers, and Adam optimizer was used with a learning rate of 2e-5 with Binary Cross Entropy Loss as the loss function defined below:

$$\text{L}_{entropy}(x, class) = -\log\left(\frac{\exp(x[class])}{\sum_j \exp(x[j])}\right)$$

4.3 Sci-BERT + Hinge loss

We then tried using domain specific variants of BERT such as Bio-BERT (Lee et al., 2019) and Sci-BERT (Beltagy et al., 2019). Bio-BERT was pre-trained on biomedical domain corpora (e.g., PubMed abstracts, PMC full-text articles), whereas Sci-BERT consists of a custom-made vocabulary (Sci-Vocab) which consists of frequently observed words and subwords in scientific text which may differ from those occurring in general domain text. Sci-BERT outperformed Bio-BERT in this task. Since for binary classification tasks, both Hinge Loss and Cross-Entropy Loss are widely used, we tried incorporating both of these losses in our model. In this task, Hinge Loss did give a better accuracy as reported below. Hence, we focused on finetuning Sci-BERT by changing the loss function from Binary Cross Entropy

to Hinge Loss(used in SVMs) which resulted in an increase of accuracy approximately by 2% on the validation set.

$$\text{L}_{hinge}(f) = \sum_{j \neq y_i} max(0, s_j - s_{y_i} + 1)$$

We discuss further about this in later sections.

5 QSpider - System Description

QSpider is a staged system consisting of Sci-BERT (Beltagy et al., 2019) stacked with a Gradient Boosting Classifier which performed the best in the hidden test set among all the models described above. This model aims at capturing question types and use them as features to detect question entailment. We trained a multi-label classifier (as a question can fall in more than one class) on a annotated corpus (13 question types as mentioned above in *Section 3*) of consumer health questions (Roberts et al., 2014). For example:

Q: "Can you mail me patient information about Glaucoma, I was recently diagnosed and want to learn all I can about the disease." .

Qtype: "Information" .

Since the available annotated dataset was not sufficient to build our question type classifier model hence we used pre-trained language model to efficiently learn from this small dataset. We used Sci-BERT as language model here as it can easily detect the semantic feature of question. After training on this dataset we predicted on our original Train, Validation and Test dataset.

We used Scibert-scivocab-uncased as the vo-

---

[3]https://github.com/huggingface/pytorch-pretrained-BERT

cabulary for our model. A vector of 1's and 0's of length 13 (number of question types) was obtained for each question in our Train, Validation and Test set. We horizontally stacked these vectors for **question1** and **question2** and used them as feature vector of shape 26 for our next model. Next we use these feature to train our **Gradient Boosting Classifier**[4], which predicts whether the two questions are an entailment or not. We further fine-tuned our Gradient Boosting Classifier by keeping the number of estimators as 5000, to obtain the optimal performance on our hidden Test set without overfitting.

6 Results

This section discusses regarding the results of various approaches we applied in this task. Since the training data-set (Ben Abacha et al., 2019) had less training examples, the systems were made to learn from the training data and tested on the validation data for validation results while for test results the systems learned from the training + validation data and tested on training data. *Table 2* represents the accuracy of the systems described on the validation and test data.

Taking $BERT_{large}$ as our baseline, it gives an accuracy of 76.2% outperforming Tree-LSTM (64%) and QSpider (62.0%) on validation set. The more domain-specific models like Bio-BERT (77.6%) and *Sci-BERT + Hinge Loss (80.5%)* gave a significant boost. Also, Sci-BERT + Hinge Loss was the **best performing system** among all participants during Validation phase. For the test set $BERT_{large}$ gives an accuracy of *48.1%* and similar models like Bio-BERT and Sci-BERT + Hinge Loss gives an accuracy of 49.6% and 51.3% respectively. Here the more syntatic models like Tree-LSTM (60.2%) perform much better. Our model *Qspider (68.4%)* performs the best here and **3rd** overall among all participating systems.

7 Error Analysis

The training examples were much easy to check for entailment, with most of the positive pairs having common sub-strings or having similar syntactic structure. As discussed earlier in *Section 3*, in the validation set, out of 302 examples, 112 examples had same same medical entities which also

Model	Valid	Test
Tree-LSTM	64.0	60.2
$BERT_{large, uncased}$	76.2	48.1
Bio-BERT	77.6	49.6
Sci-BERT + Hinge Loss	**80.5**	51.3
QSpider	62.0	**68.4**

Table 2: Accuracy results for various models.

entail each other and 119 examples with different medical entities which do not entail each other (refer *Table 2*) because of which attention models like BERT gained a huge success by focusing more on entity name. It is also evident (refer *Table 3*) that these models fail in those cases where there is same medical entity on both sides but the pair is not an entailment.

Validation Set	Correct	Wrong
Same Medical Entity (Positive)	112	0
Same Medical Entity (Negative)	1	53
Different Medical Entity (Positive)	13	4
Different Medical Entity (Negative)	117	2
Test Set	**Correct**	**Wrong**
Same Medical Entity (Positive)	87	0
Same Medical Entity (Negative)	1	100
Different Medical Entity (Positive)	24	4
Different Medical Entity (Negative)	6	8

Table 3: Number of Correct and Wrong predictions made by Sci-BERT on the task dataset. Positive means Entailment & Negative means Not Entailment.

The Hidden Test set had more than 80 % pairs (refer *Table 2*) where there are same medical entity in both questions but still more than 50% pairs among these does not entail each other. Remaining examples are even more complicated like pairs having medical entity names as synonyms/abbreviated forms of each other. **This caused a huge drop in accuracy of attention based models like BERT**. Here is where QSpider comes to the rescue, by not only focusing on syntactic but also on semantic to capture the type of question asked and also not giving high attention to entity name.

QSpider on the other side didn't perform equally well in the validation set since there are considerable number of examples having different medical entities in **question1** and **question2**. We didn't give any attention to entity name while designing QSpider keeping in mind the Test set. This is the reason QSpider doesn't perform well on the Validation set but gives good results on the Test set.

8 Conclusion and Future Work

In this paper we discussed regarding various deep learning approaches and our final model **QSpider**. It is evident from the results that even with very small sized data type we were able to generate satisfactory predictions for question type. There is a scope of improvement with the increase in the question type data. We can see that question type plays an important role in capturing question entailment but if the questions has same type but different medical entity name then our system might mis-classify. Since, our Test dataset didn't have such examples with different medical entities, hence we didn't integrate this with **QSpider** then.

We plan to integrate our model with detection of medical entity names of the questions and append them to our existing feature vector to capture difficult examples. Currently, we are using question types as discrete and independent classes which we pass onto the Gradient Boosting Classifier. But in reality, any question asked cannot be always classified into a particular question type. It always consists of a blend of various types of question. So we plan upon using the GloVe embeddings of the question classes (as mentioned in above sections) as extended features to be passed onto our classifier.

9 Acknowledgements

We would like to thank everyone in the organizing committee and the reviewers for reviewing our paper and providing their valuable feedback. We also acknowledge support from the **Computer Science and Engineering Department** of the **Indian Institute of Technology Kharagpur, India** for providing us with the computational resources required for running our models for this task. Lastly, we would also like to thank **T.Y.S.S. Santosh**[5] for proof-reading the paper and providing other valuable suggestions towards improving the paper.

References

Asma Ben Abacha and Dina Demner-Fushman. 2016. In recognizing ques- tion entailment for medical question answer- ing.

Asma Ben Abacha and Dina Demner-Fushman. 2019. A question-entailment approach to question answering.

Scantlebury Arabella, Booth A, and Hanley B. 2017. Experiences, practices and barriers to accessing health information: A qualitative study.

Iz Beltagy, Arman Cohan, and Kyle Lo. 2019. Scibert: Pretrained contextualized embeddings for scientific text.

Asma Ben Abacha, Chaitanya Shivade, and Dima Demner-Fushman. 2019. Overview of the mediqa 2019 shared task on textual inference, question entailment and question answering, acl-bionlp 2019. In *Proceedings of the BioNLP 2019 workshop, Florence, Italy, August 1, 2019*. Association for Computational Linguistics.

Danqi Chen and Christopher D. Manning. 2014. A fast and accurate dependency parser using neural networks.

Z. Chen, H. Zhang, X. Zhang, and L. Zhao. 2018. Quora question pairs.

Ido Dagan, Bill Dolan, Danilo Giampiccolo, and Bernardo Magnini. 2007. The third pascal recognising textual entailment challenge.

Jacob Devlin, Ming-Wei Chang, Kenton Lee, and Kristina Toutanova. 2018. Bert: Pre-training of deep bidirectional transformers for language understanding.

Yichen Gong, Heng Luo, and Jian Zhang. 2018. Natural languagev inference over interaction space.

Haihong Guo, Xu Na, and Jiao Li. 2018. Qcorp: an annotated classification corpus of chinese health questions.

Ely JW, Osheroff JA, Gorman PN, Ebell MH, Chambliss ML, and Pifer EA. 2000. A taxonomy of generic clinical questions: classification study. *British Medical Journal*, 321:429–432.

Milen Kouylekov, Matteo Negri, Bernardo Magnini, and Bonaventura Coppola. 2006. Towards entailment-based question answering: Itc-irst at clef 2006.

[5]https://scholar.google.ca/citations?user=aYytWsAAAAAJ&hl=en

Jinhyuk Lee, Wonjin Yoon, Sungdong Kim, Donghyeon Kim, Sunkyu Kim, Chan Ho So, and Jaewoo Kang. 2019. Biobert: a pre-trained biomedical languagerepresentation model for biomedical text mining.

Xiadong Liu, Pengcheng He, Weizhu Chen, and Jianfeng Gao. 2019. Multi-task deep neural networks for natural language understanding.

Pletneva Natalia, Vargas A, Kalogianni K, and Boyer C. 2012. Online health information search: what struggles and empowers the users?

Kirk Roberts, Kate Masterton, Marcelo Fiszman, Halil Kilicoglu, and Dina Demner-Fushman. 2014. Annotating question types for consumer health questions.

Kai Sheng Tai, Richard Socher, and Christopher D. Manning. 2015. Improved semantic representations from tree-structured long short-term memory networks.

Jane Weaver. 2016. More people search for health online.

A   Supplemental Material

This is the link to our classifier code for **QSpider** which can be used to reproduce the results claimed in the Results section above for **QSpider** - `https://github.com/Team-IIT-KGP/Qspider`. The README section is updated for instructions to run the code.

ANU-CSIRO at MEDIQA 2019:
Question Answering Using Deep Contextual Knowledge

Vincent Nguyen
Australian National University
CSIRO Data61
vincent.nguyen@anu.edu.au

Sarvnaz Karimi
CSIRO Data61
Sydney, Australia
sarvnaz.karimi@csiro.au

Zhenchang Xing
Australian National University
Canberra, Australia
zhenchang.xing@anu.edu.au

Abstract

We report on our system for textual inference and question entailment in the medical domain for the ACL BioNLP 2019 Shared Task, MEDIQA. Textual inference is the task of finding the semantic relationships between pairs of text. Question entailment involves identifying pairs of questions which have similar semantic content. To improve upon medical natural language inference and question entailment approaches to further medical question answering, we propose a system that incorporates open-domain and biomedical domain approaches to improve semantic understanding and ambiguity resolution. Our models achieve 80% accuracy on medical natural language inference (6.5% absolute improvement over the original baseline), 48.9% accuracy on recognising medical question entailment, 0.248 Spearman's rho for question answering ranking and 68.6% accuracy for question answering classification.

1 Introduction

Medical health search is the second most searched thematic query, representing 5% of all queries on Google (Cocco et al., 2018). However, many queries are semantically identical and are potentially already answered by experts (Abacha and Demner-Fushman, 2016). However, these questions may not be directly retrievable due to semantic ambiguity involving abbreviations (Wu et al., 2017), patient colloquialism (Graham and Brookey, 2008) or esoteric terminology (Lee et al., 2019). Furthermore, in regards to disease, temporality is a key factor in determining the relevance of retrieved answers (Lee et al., 2019). For example, it is more appropriate to retrieve answers relating to the *summer cold* in the summer.

As a means to retrieve these questions that are already answered by experts, question entailment has been proposed to discern relationships between pairs of questions. Recognising

Question Entailment (RQE) is the task of determining the relationship between a question pair, $RQE(Q_1, Q_2)$, as either entailment or not entailment, where Abacha and Demner-Fushman (2016) define question entailment as the situation where "a question, Q_1, entails another question, Q_2, if every answer to Q_2 is also a complete or partial answer to Q_1."

Natural Language Inference (NLI) is determining the relationship between pairs of sentences, not just questions. NLI is the task of determining whether a *hypothesis*, H, is inferred (entailment), not inferred (contradiction) or neither (neutral), given a *premise*. In the context of question answering (QA), it can be used to validate if the answer can be inferred from the question.

Though RQE and NLI have thrived in the open-domain setting (Bowman et al., 2015; Rajpurkar et al., 2016), there are unique challenges in applying these tasks directly to the biomedical question answering field. Previous models in the medical domain that used NLI and RQE relied on models which were shallowly bidirectional (Romanov and Shivade, 2018) or rule-based approaches with shallow keyword matching techniques (Abacha and Demner-Fushman, 2016) which would not generalise well.

The MEDIQA (Ben Abacha et al., 2019) challenge, as part of the ACL BioNLP workshop, aims to further research efforts in NLI and RQE by introducing their applications to Biomedical QA.

In this paper, we detail our approach in MEDIQA which addresses some of the problems with biomedical text such as utilising deep contextual relationships between words within a sentence for semantic understanding and ambiguity associated with esoteric terminology, abbreviations, and patient colloquialism. We combine biomedical and open-domain approaches as a means to improve generalisation and bridge the gap between patient colloquialism and biomedical terminology.

Proceedings of the BioNLP 2019 workshop, pages 478–487
Florence, Italy, August 1, 2019. ©2019 Association for Computational Linguistics

2 Datasets

MEDIQA 2019 (Ben Abacha et al., 2019) provides datasets to be used for three different tasks.

Task 1: Natural Language Inference The MEDNLI dataset is used for this task (Romanov and Shivade, 2018). A collection of 11232 medical premise-hypothesis pairs are used for training, 2817 pairs for validation and 405 for testing. We preprocessed the text to remove punctuation, that were designed to ensure patient anonymity as a means to reduce noise while ensuring that sentence integrity was not broken.

For example, *cerebrovascular accident in [**2948**]* → *cerebrovascular accident in 2948.* Furthermore, we expand all medical abbreviations using the ADAM database (Wu et al., 2017). For example, *On arrival to the ED T97 BP 184/94 HR 92* → *On arrival to the emergency department Temperature 97 Blood Pressure 184/94 Heart rate 92.*

Task 2: Recognizing Question Entailment For RQE, a collection of 8588 medical question pairs for training, 302 pairs for validation (Abacha and Demner-Fushman, 2016) and 230 pairs for testing is released. The RQE collection aims to match consumer health questions from the National Library of Medicine with Frequently Asked Questions (FAQs) from NIH websites.

Task 3: Question Answering Two separate training datasets were provided from the MEDIQA challenge (Ben Abacha et al., 2019):

LiveQAMed: 104 consumer health questions covering different types of questions about diseases and drugs alongside their associated answers.

Alexa: 104 simple questions about the most frequent diseases and associated answers.

No external data was used for any of the tasks as a conscious decision in order to assess the fine-tuning performance of our models. However, external data has shown to be useful in knowledge-based approaches (Romanov and Shivade, 2018) and we leave this as future work.

3 Our System

Due to the similarity of our approaches in the three tasks, we first describe a shared model that was utilised by all the tasks. Our approach extends upon the current state-of-the-art models (Lee et al.,

---

Algorithm 1: Ensemble Approach for NLI, RQE and QA

Input: Training Data, $x \in X$, Test Data, $z \in Z$, Hyperparameters Θ, Pre-trained Models M_{Brt} and M_{Bio}
Output: Label Predictions, $y \in Y$
$X \leftarrow PreprocessText(X);$
$Z \leftarrow PreprocessText(Z);$
while *numEpochs < totalEpochs* **do**
  for $b_x \in X$ **do**
    $//b_x$ is a minibatch of X
    $M_{BioFT} \leftarrow Train(M_{Bio}, b_x, \Theta);$
    $M_{BrtFT} \leftarrow Train(M_{Brt}, b_x, \Theta);$
    $//M_{FT}$ denotes the fine-tuned model
  end
  $numEpochs$++;
end
for $x \in X$ **do**
  $Pred^x_{Bio} \leftarrow Predict(M_{BioFT}, x);$
  $Pred^x_{Brt} \leftarrow Predict(M_{BrtFT}, x);$
  $//Pred$ is the softmax score outputs from each model
  $SVM \leftarrow Train(Pred^x_{Bio} \oplus Pred^x_{Brt})$
end
$Pred^Z_{Bio} \leftarrow Predict(M_{BioFT}, Z);$
$Pred^Z_{Brt} \leftarrow Predict(M_{BrtFT}, Z);$
$Y = Predict(SVM, Pred^Z_{Bio} \oplus Pred^Z_{Brt});$
return Y

2019; Devlin et al., 2019) in the open-domain and apply them to the MEDIQA biomedical tasks. As the state-of-the-art models currently employ transfer learning, we modelled an ensemble transfer learning approach used in the medical computer vision domain (Menegola et al., 2017; Kumar et al., 2017).

BERT As part of our strategy to combine open-domain approaches to a biomedical focused one, we elected to use a current state-of-the-art open-domain approach, *BERT* (Devlin et al., 2019), that is based on deeply bidirectional, unsupervised language representation that has been trained on Wikipedia.

BioBERT From the biomedical focused approach, we used *BioBERT* (Lee et al., 2019), a version of *BERT* that has been pre-trained using additional biomedical datasets, including *PubMED* and *PMC*.

Table 1: Hyperparamters used for each run for Tasks 1 & 2.

Run	Model	Task 1			Task 2		
		Learning Rate	Batch Size	Epochs	Learning Rate	Batch Size	Epochs
1	BioBERT	2e-5	64	1	2e-5	64	1
	BERT	8e-6	32	1	8e-6	32	1
2	BioBERT	2e-5	64	40	2e-5	64	40
	BERT	8e-6	32	40	8e-6	32	40
3	BioBERT x3	2e-5	64	40	2e-5	64	40
	BERT	8e-6	32	40	8e-6	32	40
4	BioBERT x3	-	-	-	2e-5	64	-
	BERT	-	-	-	2e-5	32	-
5	BioBERT x3	1e-6	32	100	1e-6	32	100
	BERT	1e-6	32	100	1e-6	32	100

Table 2: Tokenisation statistics for all Tasks.

Task	Statistic	Training	Validation	Testing
1	Average Sequence Length	386	190	64
2	Average Sequence Length	176	276	230
3	Average Sequence Length	605	632	582
	Portion of Docs >512 Sequence Length	0.32	0.37	0.32

Support Vector Machine We combined our predictions from our open-domain and biomedical domain approaches using a support vector machine (Cortes and Vapnik, 1995), which here, is akin to using a data-driven weighting function.

Learning-to-Rank We also used learning-to-rank models such as LambdaRank (Burges et al., 2007) and RankNet (Burges et al., 2005), which were implemented in Tensorflow Ranking[1] for the ranking portion of the challenge.

Sentence Embeddings When encoding our features into sentence embeddings, we used bert-as-service[2] in conjunction with BioBERT to create context-rich embeddings of text. In one of our post-challenge runs, we used a biomedical word2vec word embedding model (Chiu et al., 2016).

Hyperparameters For all three tasks, we experimented with batch sizes (2^N, $n \in \{3, 4, 5, 6, 7\}$) and learning rates ($A \times 10^B$, $A \in \{1, 2, 3...10\}$, $B \in \{2, 3, 4, 5, 6\}$) and selected the parameters that maximised performance on the validation set. We used the default sequence length of 64 for training, validation and testing of all three tasks.

Algorithm For the classification tasks in the challenge, we used an ensemble approach (see Algorithm 1). First, the text training data, X, and testing data, Z, is preprocessed. This preprocessing is done differently depending on the submission and task. Preprocessing includes punctuation removal and abbreviation expansion. This training data is used to train the BERT and BioBERT models using hyperparameters, Θ. The softmax scores for each training example, X, predicted by the final fine-tuned models are concatenated (denoted by $\oplus$ and used to train an SVM). The final predictions for the testing set, Z, are collected by first using the fine-tuned models to predict the softmax scores. These softmax scores are concatenated and fed as input into the SVM which outputs predictions, Y, for the test set.

Task 1: Natural Language Inference

The models were trained as follows: For the first and second run, BERT[3] is trained for a single epoch with a learning rate of 8e-6 with a batch size of 32, while the BioBERT[4] models were trained with a learning rate of 2e-5 with a batch size of 64. The models had their predictions combined

---

[1] Tensorflow Ranking
[2] Bert-As-Service Sentence Embeddings
[3] BERT Base Model
[4] BioBERT Pretrained Models

via an SVM (sklearn-pandas, version 1.8.0) with a penalty of 1.0, RBF kernel and gamma with the 'auto' parameter, which was then used as a data-driven weighting function. The code used for this portion was based on the following code from the BERT repository.[5]

For run 1, we established a baseline approach with no preprocessing and the models were trained only for one epoch. From run 2 onwards, preprocessing was done to the text to remove punctuation used for patient anonymity and expand medical abbreviations as mentioned previously. For runs 3 and 5, instead of using a single BioBERT model, the three variants of BioBERT were trained individually using the same parameters as in run 2. However, in the fourth run, early stop validation was used to select the best models that maximised validation accuracy. However, we excluded this run because it had the same predictions as run 3. In the final run, the learning rate was lowered and trained over a larger number of epochs.

Task 2: Recognizing Question Entailment

We use the same runs as Task 1. However, we did not do any preprocessing for any runs as it did not have any benefit on the validation set.

Task 3: Question Answering Task 3 was a 2-part challenge where answer snippets needed to be ranked and classified as relevant or irrelevant.

Algorithm 2: Ensemble Approach for Ranking QA

Input: Alexa Training Data, T_A, LiveQA Training Data, T_L, Test Data, Z
Output: Ranked List, RL
while *numEpochs < totalEpochs* **do**
    for $b_a, b_l \in (T_A, T_L)$ **do**
        $//b_a$ is a minibatch of T_A
        $M_{Alexa} \leftarrow Train(FE(b_a), \Theta)$;
        $M_{LiveQA} \leftarrow Train(FE(b_l), \Theta)$;
        //FE is a feature extractor that
          vectorizes input
    end
    $numEpochs$++;
end
$RL_{Alexa} \leftarrow Predict(M_{Alexa}, Z)$;
$RL_{LiveQA} \leftarrow Predict(M_{LiveQA}, Z)$;
$RL \leftarrow RankScore(RL_{Alexa}, RL_{LiveQA})$
return RL

---

[5] Sentence Classification Bert Code

In this task, for the ranking task, we mainly used an ensemble of two separate learning-to-rank models that were trained on LiveQA and Alexa (see Algorithm 2). We used the following features as input to the model:

1. BioBERT sentence embedding of Question
2. BioBERT sentence embedding of Answer
3. BioBERT sentence embedding of Entailed Answer from MedQUAD
4. NLI predictions over all candidates summed
5. NLI predictions over all candidates averaged

The first two features were embeddings that were encoded using BioBERT, as mentioned previously. The third feature was found through the following steps:

1. Use BM25 (Stephen Robertson, 1994) to find the question candidates in MedQUAD, M, which are most related to a Question, Q.

2. Set a cut-off value, ρ to minimise the number of candidates for RQE/NLI. For the challenge, we set $rho = 4$.

3. Predict the question entailment between all questions, Q and candidates M using the RQE model, $pred_{rqe}(Q, m) = RQE(Q, m \in M)$.

4. Retain all candidate answers, R, that had questions predicted to be entailed to the Question.

5. Perform NLI on the answers in the original ranked list, L, and all candidate answers extracted from MedQUAD, $pred_{nli}(l, r) = NLI(l \in L, r \in R)$.

6. Use the answer with the highest BM25 score for the third feature.

The fourth and fifth features were performed by summing NLI predictions, $\sum pred_{nli}(l \in L, r \in R)$, and averaging, $\frac{1}{|R|} \sum pred_{nli}(l \in L, r \in R)$.

The features were fed into Tensorflow learning-to-rank models (RankNet for run 1 and LambdaRank for runs 3 and 4) with 2307 features using the Adam optimizer (Kingma and Ba, 2015), a group size of 2 and a learning rate of 0.001.

We ensembled predictions from the two models in two different ways. We used simple averaging for Run 1. However, for subsequent runs, we used *RankScore* (Li et al., 2013), which we define as:

Table 3: Results for all 3 tasks in the MEDIQA shared task, additional post challenge runs are included. **Note:** With the exception of Task 1, all post challenge runs were evaluated using the official evaluation script.

	Task 1	Task 2	Task 3		
Run	**Accuracy**	**Accuracy**	**Accuracy**	**Spearman's Rho**	**Precision@1**
1	0.751	0.481	0.581	0.093	0.580
2	0.800	0.485	0.584	0.122	0.640
3	0.796	0.481	0.584	-0.007	0.520
4	-	0.489	0.584	-0.043	0.533
5	0.768	0.485	0.577	0.162	0.593
Post Challenge Runs					
Task	**Description**		**Accuracy**	**Spearman's Rho**	**Precision**
1	Run 5 + Maximum Sequence Length (Validation Set)		0.827 (+0.016)	-	-
2	Run 5 + Maximum Seq. Length		0.489 (+0.004)	-	-
3	Run 5 (Corrected Submission)		0.686 (+0.109)	0.0513 (-0.111)	0.771 (+0.178)
3	Run 5 (Corrected Submission) + Max Seq. Length		0.663 (-0.023)	0.0971 (+0.046)	0.749 (-0.022)
3	Run 1 with word2vec embedding		-	0.284 (+0.189)	-
3	Run 5 (Corrected Submission) with UMLS concept expansion		0.659 (-0.027)	0.0200 (-0.0313)	0.749 (-0.022)

$R_s(d \in D) = 1/d_r$. We use RankScore to score each item in the ranked lists of Alexa and LiveQA models. We then combine the items by summing the documents RankScore from each model and sorting.

For classification, the same architecture from Tasks 1 and 2 for Runs 3 - 5 was used (4-ensemble with SVM layer). For runs 2 and 5, we use softmax scores output from the classification to rank documents.

4 Results and Discussion

Ensembles have been successfully utilised in other biomedical domains (Kumar et al., 2017; Brijesh and Zahid, 2011), with the main idea behind using these being to incorporate complementary strengths of the members of the ensemble. Thus, BERT is used in conjunction with BioBERT in order to correct the mistakes that the model makes by injecting non-domain specific knowledge. This idea was supported in our baseline experiments on task 1 where BioBERT scored 0.7913 on validation, while BERT scored 0.7715 on validation, but ensembling resulted in a higher final score of 0.7950.

NLI Baseline System Problems Our baseline system made characteristic mistakes on the validation set, which is shown in Table 4 for Task 1. We found that our system had trouble with *numerical interpretation* and, for instance, was not able to determine the difference between type 1 and type 2 diabetes. Furthermore, this problem is exacerbated when *abbreviations and numerical interpretation* are required in phrases such as *T97 BP 184/94*. Thus, to aid the system in disambiguating abbreviations, we expanded all abbreviations using the ADAM database of common clinical abbreviations and resulted in an 0.049 increase in accuracy. Furthermore, the system would struggle with medical forms of *negation*. However, due to the use of BERT/BioBERT, conventional techniques such as NegEx or removal would break sentence integrity and reduce comprehension, thereby affecting word context, and thus were not viable. Furthermore, punctuation, in terms of patient anonymisation, is also a problem as the punctuation does not carry meaningful semantic content and will confuse the classifiers.

RQE Baseline System Problems In task 2, we found that our baseline system made similar mistakes for different reasons (see Table 5). We found examples of what we consider *near miss* where the definition of partial entailment depends on interpretation. For example, in this question, the user

Table 4: Common mistakes made by the baseline system in Task 1.

Type	Premise	Hypothesis
Numerical Interpretation	PAST MEDICAL HISTORY: Type 2 diabetes mellitus.	the patient has type 1 diabetes
Abbreviation and Numerical Interpretation	On arrival to the ED T97 BP 184/94 HR 92 RR 24 88% on RA ->98% on NRB.	The patient was hypertensive in the ED
Negation	He denied headache or nausea or vomiting.	He has no head pain
Semantic Gap	HISTORY OF PRESENT ILLNESS:,The patient is a 54 year old male with endstage renal disease secondary to type 1 diabetes who presents for kidney transplant from wife.	patient is on insulin

wants information on hypertension (high blood pressure). However, according to the gold standard, this is not a form of entailment, partial or otherwise. We hypothesise that this lies on the borderline of the entailment definition or may be due to bias. Furthermore, our system struggles with *abbreviations*. However, the examples in the second task dataset are more related to problems with co-reference resolution where abbreviations appear in the original question but not in the FAQ question.

Furthermore, phrases like "come out of" should be aligned to terms such as "discharge", which is an example of a *semantic gap* and require common sense comprehension. This is problematic as BERT is known to struggle with this sort of reasoning (Talmor et al., 2018). Also, we did not adjust the *sequence length* parameter (set to 64), which may have been a source of error. However, a later investigation through a post-challenge run that shows that only Task 1 benefits from an increase in sequence length (see Table 3). Finally, *patient colloquialism* presents a unique challenge where "hole in lung" is to be interpreted as "pleurisy" (lung inflammation). Although we did not address this complex problem, it could be potentially solved through crowd-sourcing of medical forum data. This may be suitable as an area to investigate for future work.

We found that in all our submissions on the test set of the challenge, although our system was able to achieve high results on the validation set of 79%, the models were not well suited for the test set. Our model predicted entailment 92% of the time on the test set, suggesting that the model is overfitting, even though our baseline was trained for only one epoch. We found that the cases where

the models make errors are cases where the question contains words such as *diagnosis* and the disease is mentioned, but the semantic content of the question might be about *treatment* rather than the diagnosis. This is very different from the training and validation datasets that were provided, which were much more straightforward and did not require as much comprehension. An example illustrating this difficulty is *Question A: Glaucoma: Can you mail me patient information about Glaucoma, I was recently diagnosed and want to learn all I can about the disease."* and *Question B: How is glaucoma diagnosed?*

Question Answering Submission Problems
For the third Task, we incorrectly trained our models to recognise documents with a relevance score of one as irrelevant. In contrast, the task is defined to classify documents of relevance score one and two as irrelevant. By fixing this error, we found that we had over a 10% increase in accuracy (Table 3). However, interestingly, we found that the ranking quality (shown through Spearman's Rho) decreased. Upon investigation, we found two reasons why this problem occurred: (1) our system was able to differentiate the relevance of one from the other three labels much better than differentiating between labels of one/two against three/four. This was reflected in the validation accuracy of our initial incorrect model, which achieved an accuracy of 95% whereas the corrected model scores only 70% on the validation set, (2) we found that the longer the models were trained, the worse the ranking quality became. We hypothesise that the problem is due to how cross entropy loss and softmax functions work. Since the models are minimising KL-Divergence, the softmax scores become more extreme, falling close to 1 or very

Table 5: Common mistakes made by the baseline system in Task 2.

Type	Question A	Question B
Near Miss	I want more information on Hypertension and fibromyalgia, I seem to be getting only topics on diabetes and I do not have this. I enjoy reading the current info.	What is high blood pressure?
Abbreviation	Hi I have retinitis pigmentosa for 3years, Im suffering from this disease. Please intoduce me any way to treat mg eyes such as stem cell ... Thank you	Are there treatments for RP?
Semantic Gap	Which drug we I take to stop water come out of my nipple	How to Treat Nipple Discharge
Sequence Length	... The problem is my binocular vision is not good enough ... is there any operation that can fix this?	What is Vision Therapy When and why is it needed [for binocular vision]?
Patient Colloquialism	Cure for hole in lung. I certainly would like to request for medical for hole in the lung	How Are Pleurisy and Other Pleural Disorders Treated?

close to 0. This results in the differences between scores of the documents to be very low (forming dense clusters) which reduces ranking quality as the ranking becomes more sensitive to noise and uncertainty (Siddhant and Lipton, 2018).

Question Answering Baseline System Problems
Due to the error of our submissions for Task 3, we will not discuss the mistakes that occurred within the challenge for the pointwise ranking runs. Instead, we will look at the mistakes that the post-challenge run encountered for those. However, for the pairwise runs within the challenge, we found that it performed much worse than expected. We attribute this ranking deficit to two important factors.

The first is that BERT sentence embeddings are not useful to represent sentences because the vector space is too condensed (vector representations are very close together). The second is that our vector representations were too large, with BERT sentence embeddings producing embeddings up to 800 dimensions. Using 3 of these embeddings results in a very large input which would take too long to train or hinder convergence. This effect was observed in a post-challenge run where we used Chiu et al. (2016)'s biomedical word2vec embeddings and achieved a much higher Spearman's Rho. The second factor was that the LambdaLoss (Burges et al., 2007) function was not a suitable objective function as the RankNet model performed better.

From Table 2, we find that Task 3 is more ver-

bose than the other two tasks and presents unique challenges as almost a third of the documents will have information loss due to the limitation of maximum sequence length by BERT being 512 due to quadratic memory explosion (Liu et al., 2018). However, we did a post-challenge run where we increased the sequence length with no noticeable difference. This is because the majority of information in these long sequence can be safely discarded. Furthermore, the BERT truncation strategy is to truncate from the end of the sentence, implying that the important information is typically at the start of the answer.

We also find that there are unique challenges in Task 3 due to the use of real patient questions shown in Table 6. We found that problems such as *typos, grammar and spellings* mistakes were not directly fixed by the BERT/BioBERT ensemble as the collections were pretrained on academic or formal language (Pubmed, PMC and Wikipedia). However, problems such as *synonyms* (for example, abetalipoproteinemia and Bassen-Kornzweig syndrome) which should be addressed by the model were also not addressable due to a limitation in the vocabulary of the models, which is discussed below. Furthermore, we found cases of *near miss*, for example, the model identifies anemia and treatment options, but it is not the target disease of the question. To address these problems, we use a heuristic to expand UMLS terms in the question and answer, and add these to the start of the sentence to combat the mentioned problems.

Table 6: Common mistakes made by baseline system in Task 3

Type	Question	Answer
Typo	abetalipoproteimemia hi, I would like to know if there is any support for those suffering with abetalipoproteinemia ... keen to learn how to get it diagnosed...	abetalipoproteinemia: Abetalipoproteinemia is an inherited disorder that affects the absorption of dietary fats, cholesterol, and fat-soluble vitamins...
Synonyms	abetalipoproteimemia hi, I would like to know if there is any support for those suffering with abetalipoproteinemia...	Bassen-Kornzweig syndrome (Exams and Tests): There may be damage to the retina of the eye (retinitis pigmentosa). Tests that may be done to help diagnose this condition include...
Near miss	about thalassemia treatment sir,my friend is suffering from thalassemia ,in that majorly red blood anemia,white blood anemia and the blood is comming out from mouth when she got cough .her condition is very severe...	Anemia (Treatment): Anemia treatment depends on the cause. - Iron deficiency anemia. Treatment for this form of anemia...
Grammar and spelling mistakes	Absence seizures Does any damage occurre from these spells. Mental or physical	Seizures: A seizure is a sudden, uncontrolled electrical disturbance in the brain. It can cause changes in your behavior, movements or feelings, and in levels of consciousness. If you have two...
Semantic Gap	Bad Breath I have very bad breath and at times it can make myself and others sick. I need some advice as to what I need to do.	Breath odor (Home Care): Use proper dental hygiene, especially flossing. Remember that mouthwashes are not effective in treating the underlying problem...

We found that the model performs better on the validation set than any of the post-challenge runs (79% accuracy, a 5% absolute increase over the other runs), but did not perform substantially better on the test set (see Table 3).

Problems with Underlying Models One problem in using models such as *BERT* and *BioBERT* is the limitation in the maximum sequence length. This is demonstrated in the test portion of the challenge, where test set answers were much longer than those seen in the training and validation collection. These sequences were longer than the 512 sequence length limit allowed by the *BERT* architecture, which is constrained due to a problem known as the quadratic memory explosion (Liu et al., 2018) leading to exponentially longer training times and memory usage.

Though there are ways to overcome these restrictions such as striding the sentences pairs and labels, this results in contextual information being lost and label imbalance. This restriction also hinders the encoding of long-range dependencies between sequences as only contexts within a fixed length can be considered (Dai et al., 2019).

In addition, we use *BioBERT* as a means of contributing deep clinical contextual understanding of sentences. However, we find that during WordPiece Tokenisation (Devlin et al., 2019), medical terms are *always* split into their sub-word representations as they are out-of-vocabulary, e.g., *arthralgias* $\rightarrow$ *art hra al gia s*. Wordpiece tokenisation relies on the idea that morphemes carry meaning. However, due to the use of this non-medical vocabulary, specific medical related mor-

phemes are not being learned. For instance, arthr-
(where - denotes prefix), means joints and -algias
means pain, so the correct tokenisation should be
arthralgias → *arthr algia s* so that the model can
currently learn the semantic meaning behind the
morpheme. We find that these limitations hindered
the use of these models and their application to the
MEDIQA tasks.

We emphasise that there is a real-world application with the models and methods in this challenge. However, if we were to scale our approach to real-world application, we would require
external data. Therefore for future work, given
more time, we would like to use external datasets
such as emrQA (Pampari et al., 2018) and explore
multi-task learning due to the similarity of the
three tasks and aim to incorporate other medical
tasks for a better generalisation of the biomedical
question answering. We would also want to train
the BERT models on biomedical-focused vocabulary and additional data in the future as a baseline
to compare against multi-task learning.

5   Conclusions

In this shared task, we use and improve upon NLI
and RQE techniques for medical question answering. Our approach involves utilising deep contextual relationships between words emphasising
semantic understanding and resolving ambiguity.
We combine biomedical and open-domain strategies to improve generalisation and bridge the gap
between the open-domain and biomedical domain
question answering.

Acknowledgements

This research is supported by the Australian Research Training Program and the CSIRO Postgrad-
uate Scholarship.

References

Ben Abacha and Demner-Fushman. 2016. Recognizing Question Entailment for Medical Question Answering. *American Medical Informatics Association
Annual Symposium Proceedings*, 2016:310–318.

Asma Ben Abacha, Chaitanya Shivade, and Dina
Demner-Fushman. 2019. Overview of the MEDIQA
2019 shared task on textual inference, question entailment and question answering. In *Proceedings of
the BioNLP 2019 workshop*, Florence, Italy. Association for Computational Linguistics.

Samuel Bowman, Gabor Angeli, Christopher Potts, and
Christopher Manning. 2015. A large annotated corpus for learning natural language inference. In *Proceedings of the 2015 Conference on Empirical Methods in Natural Language Processing (EMNLP)*, Lisbon, Portugal. Association for Computational Linguistics.

Verma Brijesh and Hassan Syed Zahid. 2011. Hybrid
ensemble approach for classification. *Applied Intelligence*, 34(2):258–278.

Chris Burges, Tal Shaked, Erin Renshaw, Ari Lazier,
Matt Deeds, Nicole Hamilton, and Greg Hullender.
2005. Learning to rank using gradient descent. In
*Proceedings of the 22Nd International Conference
on Machine Learning*, ICML '05, pages 89–96, New
York, NY. ACM.

Christopher Burges, Robert Ragno, and Quoc Le. 2007.
Learning to rank with nonsmooth cost functions. In
*Advances in Neural Information Processing Systems
19*, pages 193–200.

Billy Chiu, Gamal Crichton, Anna Korhonen, and
Sampo Pyysalo. 2016. How to train good word embeddings for biomedical NLP. In *Proceedings of
the 15th Workshop on Biomedical Natural Language
Processing*, pages 166–174, Berlin, Germany.

Anthony Cocco, Rachel Zordan, David Taylor, Tracey
Weiland, Stuart Dilley, Joyce Kant, Mahesha Dom-
bagolla, Andreas Hendarto, Fiona Lai, and Jennie
Hutton. 2018. Dr Google in the ED: searching for
online health information by adult emergency department patients. *The Medical Journal of Australia*, 209:342–347.

Corinna Cortes and Vladimir Vapnik. 1995. Support-
vector networks. *Machine Learning*, 20(3):273–
297.

Zihang Dai, Zhilin Yang, Yiming Yang, Jaime Carbonell, Quoc Le, and Ruslan Salakhutdinov. 2019.
Transformer-xl: Attentive language models beyond
a fixed-length context. *Computing Research Repository*, abs/1901.02860.

Jacob Devlin, Ming-Wei Chang, Kenton Lee, and
Kristina Toutanova. 2019. BERT: Pre-training of
deep bidirectional transformers for language understanding. In *Proceedings of the Conference of
the North American Chapter of the Association for
Computational Linguistics: Human Language Technologies*, Minneapolis, MN.

Suzanne Graham and John Brookey. 2008. Do patients
understand? *The Permanente journal*, 12(3):67–69.

Diederik Kingma and Jimmy Ba. 2015. Adam: A
method for stochastic optimization. In *3rd International Conference on Learning Representations*, San
Diego, CA.

Ashnil Kumar, Jinman Kim, David Lyndon, Michael Fulham, and Dagan Feng. 2017. An ensemble of fine-tuned convolutional neural networks for medical image classification. *IEEE Journal of Biomedical and Health Informatics*, 21(1):31–40.

Jinhyuk Lee, Wonjin Yoon, Sungdong Kim, Donghyeon Kim, Sunkyu Kim, Chan Ho So, and Jaewoo Kang. 2019. BioBERT: A pre-trained biomedical language representation model for biomedical text mining. *arXiv e-prints*, page arXiv:1901.08746.

Vincent Li, Paul Thomas, and David Hawking. 2013. Merging algorithms for enterprise search. *ACM International Conference Proceeding Series*, pages 42–49.

Peter Liu, Mohammad Saleh, Etienne Pot, Ben Goodrich, Ryan Sepassi, Lukasz Kaiser, and Noam Shazeer. 2018. Generating wikipedia by summarizing long sequences. *Computing Research Repository*, abs/1801.10198.

Afonso Menegola, Julia Tavares, Michel Fornaciali, Lin Li, Sandra Fontes de Avila, and Eduardo Valle. 2017. Recod titans at isic challenge 2017. *Computing Research Repository*, abs/1703.04819.

Anusri Pampari, Preethi Raghavan, Jennifer Liang, and Jian Peng. 2018. emrqa: A large corpus for question answering on electronic medical records. *Computing Research Repository*, abs/1809.00732.

Pranav Rajpurkar, Jian Zhang, Konstantin Lopyrev, and Percy Liang. 2016. Squad: 100, 000+ questions for machine comprehension of text. *Computing Research Repository*, abs/1606.05250.

Alexey Romanov and Chaitanya Shivade. 2018. Lessons from natural language inference in the clinical domain. *Computing Research Repository*, abs/1808.06752.

Aditya Siddhant and Zachary Lipton. 2018. Deep bayesian active learning for natural language processing: Results of a large-scale empirical study. In *Proceedings of the 2018 Conference on Empirical Methods in Natural Language Processing*, pages 2904–2909, Brussels, Belgium.

Susan Jones Micheline Hancock-Beaulieu Mike Gatford Stephen Robertson, Steve Walker. 1994. Okapi at trec-3. In *Proceedings of the Third Text REtrieval Conference (TREC 1994)*, Gaithersburg, MD.

Alon Talmor, Jonathan Herzig, Nicholas Lourie, and Jonathan Berant. 2018. Commonsenseqa: A question answering challenge targeting commonsense knowledge. *Computing Research Repository*, abs/1811.00937.

Yonghui Wu, Joshua Denny, Rosenbloom Trent, Randolph Miller, Dario Giuse, Lulu Wang, Carmelo Blanquicett, Ergin Soysal, Jun Xu, and Hua Xu. 2017. A long journey to short abbreviations: developing an open-source framework for clinical abbreviation recognition and disambiguation (CARD). *Journal of the American Medical Informatics Association*, 24(e1):e79–e86.

MSIT_SRIB at MEDIQA 2019: Knowledge Directed Multi-task Framework for Natural Language Inference in Clinical Domain

Sahil Chopra[1], Ankita Gupta[2], and Anupama Kaushik[1]

[1]Maharaja Surajmal Institute of Technology, Delhi .
[2] Samsung Research Institute, Bangalore.
{sahilchopra, anupama}@msit.in
gupta.ankita@samsung.com

Abstract

In this paper, we present Biomedical Multi-Task Deep Neural Network (Bio-MTDNN) on the NLI task of MediQA 2019 challenge (Ben Abacha et al., 2019). Bio-MTDNN utilizes "transfer learning" based paradigm where not only the source and target domains are different but also the source and target tasks are varied, although related. Further, Bio-MTDNN integrates knowledge from external sources such as clinical databases (UMLS) enhancing its performance on the clinical domain. Our proposed method outperformed the official baseline and other prior models (such as ESIM and Infersent on dev set) by a considerable margin as evident from our experimental results.

1 Introduction

The task of natural language inference (NLI) intends to determine whether a given hypothesis can be inferred from a given premise. This task also referred to as recognizing textual entailment (RTE), is one of the most prevalent tasks among NLP researchers . It has been one of the significant components for several other language applications such as Information Extraction (IE), Question Answering (QA) or Document Summarization. For example, Harabagiu and Hickl (2006) argue that RTE can enable QA systems to identify correct answers by allowing filtering and re-ranking them w.r.t a given question. Another approach is proposed by Ben Abacha and Demner-Fushman (2016), whereby the authors employ RTE in IE/QA domain to answer a given question (queried by a consumer) by retrieving similar questions that are already well responded by professionals.

In order to address this simple yet challenging task of NLI, several open domain datasets have been proposed, with Stanford Natural Language Inference (SNLI) (Bowman et al., 2015) and MultiNLI (Williams et al., 2018) being the most popular ones. They serve as a standard to assess recent NLI systems. However, there have been only a few resources available in specialized domains such as biomedical or medicine. Language inference in the medical domain is extremely complex and remains less explored by the ML community. This scantiness of adequate resources (in terms of datasets) can be attributed to the fact that patient's data is sensitive, is accessible to authorized medical professionals only, and requires domain experts to annotate it, unlike generic domains where one can rely on crowd-sourcing based techniques to acquire annotations.

To this end, Ben Abacha et al. (2019) released a new dataset made available through MIMIC-III derived data repository, named MedNLI, for NLI in the clinical domain which has been annotated by experts. Along these lines, the MediQA 2019 challenge aims to foster the development of appropriate methods, techniques and standards for inference/entailment in the medical domain, specifically on MedNLI dataset through a shared task. The task intends to recognize three inference relations between two sentences: Entailment, Neutral and Contradiction.

Previous research associated with the present task, such as work by Romanov and Shivade (2018) analyzed several state-of-the-art open domain models for NLI on the MedNLI dataset. The same has been utilized as a baseline for comparison in the above mentioned shared task. Prior to this, efforts have been made towards the automatic construction of RTE datasets (Ben Abacha and Demner-Fushman, 2016; Abacha et al., 2015), application of active learning on small RTE data (Shivade et al., 2015).

Our approach to solving the NLI task on the

Proceedings of the BioNLP 2019 workshop, pages 488–492
Florence, Italy, August 1, 2019. ©2019 Association for Computational Linguistics

MedNLI data is based on leveraging transfer learning paradigm integrated with direct incorporation of domain-specific knowledge from medical knowledge bases (KB). Unlike Romanov and Shivade (2018) which utilizes transfer learning to utilize standard NLI models (such as InferSent and ESIM trained specifically on NLI task only) in the clinical domain, we employ Mutli-task learning (MTL) framework with domain adaptation to learn representations across multiple natural language understanding (NLU) tasks. This approach not only leverages vast amounts of cross-task data but also benefits from a regularization effect that leads to better generalization and facilitates adaptation to new tasks and domains. Besides domain adaptation, we also directly infuse domain specific knowledge from database of medical terminologies so as to enable the system to perform well in the clinical domain.

The rest of the paper is organized as follows: Section 2 describe the details of our approach. Section 3 demonstrates the experimental results. We conclude in Section 4.

2 Approach

This section elaborates on the various methods we experimented with for the NLI task. In order to establish a simple baseline first, we utilize a feature-based system. The extracted features include word containment (Lyon et al., 2001) and Jaccard similarity (unigram, bigram, and trigram) based features. We also use similarity measure of distributed sentence representations obtained using universal sentence encoder (Cer et al., 2018). We consider Levenshtein, and Euclidean distance, negations and cosine function as similarity measures. In order to find the n-grams, we utilize NLTK and scispaCy tokenizer (Neumann et al., 2019). We train a 3-class logistic regression classifier with above-mentioned features to output the inference relations. Apart from this baseline, We now elaborate on the transfer learning and external knowledge integration based method in the following subsections.

2.1 Transfer Learning

Given the vast amounts of data available in the open-domain NLU tasks, we leverage them to attack the NLI task on MedNLI. Given a source domain D_S, a corresponding source task T_S, as well as a target domain D_T and a target task T_T, the objective of transfer learning is to learn the target conditional probability distribution $P(Y_T|X_T)$ in D_T with the information gained from D_S and T_S where $D_S \neq D_T$ and/or $T_S \neq T_T$. X and Y are feature and label space respectively.

We consider the scenario when $D_S \neq D_T$ (D_S being open-domain and D_T being clinical domain) and $T_S \neq T_T$, with two possibilities for target task T_T. In the first scenario, we consider a single related T_T and in the second scenario we leverage multi-task framework where we augment the T_T with multiple but related NLU tasks. For both the scenarios, we utilize the method of sequential transfer where a model is pre-trained on the large source domain data and fine-tuned on limited target domain data (clinical here). Next, we describe the neural network based models that we utilize.

2.1.1 Bi-CNN-MI

We leverage Bi-CNN-MI model (Yin and Schütze, 2015) to realize the single transfer task scenario. This DNN model is trained on a similar NLU task of paraphrase identification (PI) which is formalized as a binary classification task: for given two sentences, determine whether they both convey roughly the same meaning.

Bi-CNN-MI compares two sentences on multiple levels of granularity (word, short n-gram, long n-gram and sentence) and learns corresponding sentence representations using a convolutional neural network (CNN) based Siamese network. It also captures the sentence interactions between two sentences by computing an interaction matrix at each level of granularity. This model has been reported to outperform various earlier approaches on PI (Yin and Schütze, 2015).

We leverage this model for sequential transfer by learning the model parameters on the PI task and fine-tuning them on MedNLI dataset. Note that the classification task in MedNLI can also benefit by capturing interactions at various levels of granularity making it related to PI task but at the same time different from PI as the objective of MedNLI is not only to determine if a pair of sentences convey the same meaning but also segregate if they oppose each other or are unrelated.

2.1.2 MT-DNN

In the second scenario of transfer learning, we augment the target task T_T by various related NLU tasks and train the model to perform on all of them. This approach not only leverages exten-

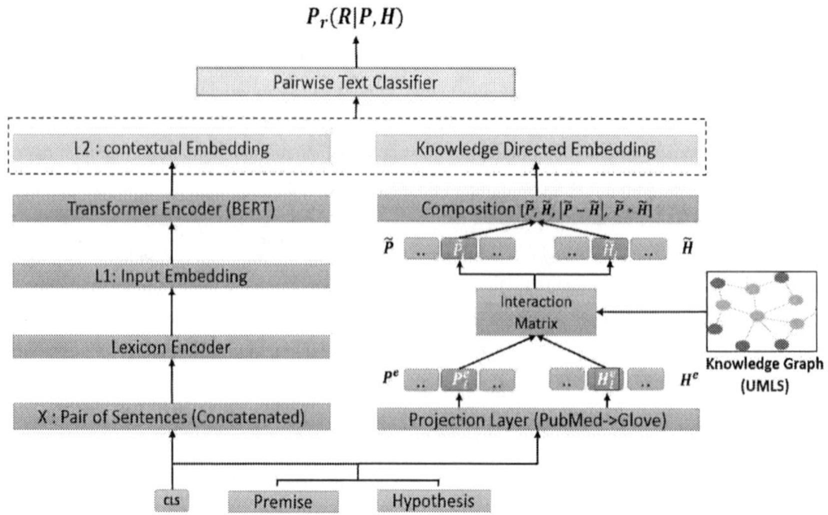

Figure 1: Architecture Diagram for Bio-MTDNN. $P_r(R|P,H)$ denotes the probability of inference relation (R) between Premise (P) and Hypothesis (H).

sive amounts of data on multiple tasks but also enables the regularization effect leading to better generalization ability. Essentially, we want to use the knowledge acquired by learning from related tasks to do well on a target task. For this approach, we utilize MT-DNN (Liu et al., 2019) which combines MTL with pre-trained language model (BERT) to improve the text representations.

The MT-DNN model combines four types of NLU tasks: single-sentence classification (sentiment classification, grammatical acceptability), pairwise text classification (NLI on several corpus and PI), text similarity scoring (STS-B), and relevance ranking (QNLI). Note, the pairwise text classification task is the NLI task that we originally intended to address in MedNLI.

The model architecture of MT-DNN involves lower layers that are shared across all tasks, while the top layers represent task-specific outputs. The input X, comprising of premise P and hypothesis H is concatenated and represented as a sequence of embedding vectors (Layer L1). The transformer encoder (BERT) then captures contextual information in the second layer (L2). This is the shared semantic representation that is trained by the multi-task objectives.

MT-DNN trained on all of the above-mentioned tasks on open-domain datasets is then fine-tuned by MedNLI dataset. In this fine-tuning step, we update the shared weights and weights associated with only the pairwise text classification task. Essentially, we first try to capture the knowledge

from several related tasks in NLU followed by adapting the model to the clinical domain.

2.2 Knowledge from External Sources

Medical texts often hold relations between entities which require domain-specific knowledge for the analysis. For example, the knowledge that pneumonia is a lung disease may not be evident from the clinical text directly. In such a scenarios, incorporation of external knowledge which conveys such relationships can help. We utilize UMLS database (restricted to the SNOMED-CT terminology) represented as a graph where clinical concepts are nodes, connected by edges representing relations, such as synonymy, parent-child, etc. Next we discuss the details of the mechanism to incorporate this external knowledge, thus elaborating our Bio-MTDNN model architecture.

2.2.1 Bio-MTDNN

We propose Bio-MTDNN model which integrates domain knowledge on top of the MT-DNN model in a way similar to how interactions are captured in Bi-CNN-MI model. Specifically, we calculate the interaction matrix $I \epsilon R^{N \times M}$ between all pairs of tokens P_i and H_j in the input premise (length N) and hypothesis (length M) respectively. The value in each cell is the length of the shortest path l_{ij} between the corresponding concepts of the premise and the hypothesis in SNOMED-CT. This matrix is then utilized to generate knowledge attended representations, $\tilde{P}$ and $\tilde{H}$. Each token $\tilde{P}_i$ of the premise is a weighted sum of the embedding H_j^e of the relevant tokens H_j of the hypothesis, weights derived from the interaction matrix. Finally, the two knowledge directed representations (averaged over the token representations) of the premise $\tilde{P}$ and hypothesis $\tilde{H}$ are composed together using elementary operations (concatenation, multiplication and subtraction) and fed to a single feed forward layer. This composed representation is then concatenated with the L2 layer of MT-DNN before passing it to the task-specific layers.

In the above process, the creation of knowledge directed representations relies upon the input token embeddings of premise (P_j^e) and hypothesis (H_j^e). One of the simplest options for token embeddings is to use GloVe embeddings (Pennington et al., 2014). However, these embeddings are not specific to the clinical domain and may result in many tokens being mapped to the embedding of the unknown (UNK) token. To alleviate this issue, we learned a non-linear transformation (Sharma et al., 2018) that maps words from PubMed (Pyysalo et al., 2013) to GloVe subspace. We train the DNN using the common words in both the embeddings. We obtain the transformed embeddings for all the words in the PubMed that are not present in the GloVe by using inference step of the learned DNN.

Note that, here we cannot utilize the embeddings learned in the first layer (L1) of MT-DNN as they incorporate segment embeddings of the premise and hypothesis concatenated together. Thus, the L1 layer of MT-DNN learns the interactions between premise and hypothesis in an end-to-end manner. However, what we are trying is to learn these interactions which are directed by the knowledge obtained from UMLS enabling Bio-MTDNN to incorporate external information.

3 Experiments and Results

3.1 Setup and Implementation Details

For the feature-based system we used Logistic Regression classifier from the scikit-learn library (Pedregosa et al., 2011). We use publicly available implementations for Bi-CNN-Mi[1] and MT-DNN[2]. For external knowledge integration, the required medical concepts in SNOMED-CT were identified in the premise and hypothesis sentences using MetaMap by Aronson and Lang

---

[1]https://github.com/chantera/bicnn-mi

[2]https://github.com/namisan/mt-dnn

Model	Dataset	
	Dev	**Test**
MT-DNN + External Knowledge	81.2	81.3
MT-DNN	80.1	80.5
Infersent	73.5	-
ESIM	73.1	-
Official Baseline		71.4
Features Baseline	51.9	49.4
Bi-CNN-MI	54.1	53.6

Table 1: Experimental Results

(2010). We used glove and PubMed word embeddings and used DNN (Sharma et al., 2018) for non-linear projection. In all experiments we report the average result (on the dev set) of 5 different runs, with the same hyperparameters and different random seeds. For the best performing systems, we also report the results on the test set.

3.2 Results and Discussions

Table 1 mentions the experimental results for all the systems. Bio-MTDNN performs best among all the systems with 81.2% accuracy on the dev set. Integration of external knowledge in Bio-MTDNN helped the system to outperform the MT-DNN performance (with 80.1% accuracy). The multi-task learning framework boosted the performance of both the systems. We submitted results from Bio-MTDNN for the challenge which obtained 81.3% accuracy on the test set.

In order to compare against other transfer learning based approaches (Romanov and Shivade, 2018), we also mention the results of Infersent and ESIM (note that for both these models, $D_S \neq D_T$ and $T_S = T_T$, unlike the scenarios we considered). It can be observed that Bio-MTDNN outperforms both ESIM and Infersent with significant margins. This can be attributed to the external knowledge incorporation and ability of MTL framework which empowers the model to learn better shared representations. However, contrary to the expectations, Bi-CNN-MI model performs very poorly on the dev dataset with only 54.1% accuracy, only slightly better than feature based baseline which achieves 51.9 % accuracy. This may be attributed to the possibility that the knowledge gained by Bi-CNN-MI when trained on PI task (although a related task to NLI) is not sufficient for the model to be able to segregate contradicting premise and hypothesis.

4 Conclusion

In this paper, we introduce Bio-MTDNN , which is a knowledge directed, multi-task learning based language inference model for biomedical text mining. While MT-DNN was built for general purpose language understanding, Bio-MTDNN effectively leverages domain specific knowledge from UMLS as demonstrated by our experimental study. We presented our results on the MedNLI dataset under MediQA challenge. Incorporation of knowledge from external sources such as UMLS gives performance advantage to Bio-MTDNN. Our proposed system outperformed the official baseline and other prior models (ESIM and Infersent on dev set) by a great margin.

References

Asma Ben Abacha, Duy Dinh, and Yassine Mrabet. 2015. Semantic analysis and automatic corpus construction for entailment recognition in medical texts. In *Conference on Artificial Intelligence in Medicine in Europe*, pages 238–242. Springer.

Alan R Aronson and François-Michel Lang. 2010. An overview of metamap: historical perspective and recent advances. *Journal of the American Medical Informatics Association*, 17(3):229–236.

Asma Ben Abacha and Dina Demner-Fushman. 2016. Recognizing question entailment for medical question answering. In *AMIA Annual Symposium Proceedings*, volume 2016, page 310. American Medical Informatics Association.

Asma Ben Abacha, Chaitanya Shivade, and Dina Demner-Fushman. 2019. Overview of the mediqa 2019 shared task on textual inference, question entailment and question answering. In *Proceedings of the BioNLP 2019 workshop, Florence, Italy, August 1, 2019*. Association for Computational Linguistics.

Samuel R Bowman, Gabor Angeli, Christopher Potts, and Christopher D Manning. 2015. A large annotated corpus for learning natural language inference. *arXiv preprint arXiv:1508.05326*.

Daniel Cer, Yinfei Yang, Sheng-yi Kong, Nan Hua, Nicole Limtiaco, Rhomni St John, Noah Constant, Mario Guajardo-Cespedes, Steve Yuan, Chris Tar, et al. 2018. Universal sentence encoder. *arXiv preprint arXiv:1803.11175*.

Sanda Harabagiu and Andrew Hickl. 2006. Methods for using textual entailment in open-domain question answering. In *Proceedings of the 21st International Conference on Computational Linguistics and the 44th annual meeting of the Association for Computational Linguistics*, pages 905–912. Association for Computational Linguistics.

Xiaodong Liu, Pengcheng He, Weizhu Chen, and Jianfeng Gao. 2019. Multi-task deep neural networks for natural language understanding. *CoRR*, abs/1901.11504.

Caroline Lyon, James Malcolm, and Bob Dickerson. 2001. Detecting short passages of similar text in large document collections. In *Proceedings of the 2011 conference on empirical methods in natural language processing (EMNLP)*, pages 118–125.

Mark Neumann, Daniel King, Iz Beltagy, and Waleed Ammar. 2019. Scispacy: Fast and robust models for biomedical natural language processing.

Fabian Pedregosa, Gaël Varoquaux, Alexandre Gramfort, Vincent Michel, Bertrand Thirion, Olivier Grisel, Mathieu Blondel, Peter Prettenhofer, Ron Weiss, Vincent Dubourg, et al. 2011. Scikit-learn: Machine learning in python. *Journal of machine learning research*, 12(Oct):2825–2830.

Jeffrey Pennington, Richard Socher, and Christopher Manning. 2014. Glove: Global vectors for word representation. In *Proceedings of the 2014 conference on empirical methods in natural language processing (EMNLP)*, pages 1532–1543.

Sampo Pyysalo, Filip Ginter, Hans Moen, Tapio Salakoski, and Sophia Ananiadou. 2013. Distributional semantics resources for biomedical text processing. In *Proceedings of the 5th International Symposium on Languages in Biology and Medicine*.

Alexey Romanov and Chaitanya Shivade. 2018. Lessons from natural language inference in the clinical domain. *arXiv preprint arXiv:1808.06752*.

Vasu Sharma, Nitish Kulkarni, Srividya Pranavi, Gabriel Bayomi, Eric Nyberg, and Teruko Mitamura. 2018. Bioama: Towards an end to end biomedical question answering system. In *Proceedings of the BioNLP 2018 workshop*, pages 109–117.

Chaitanya Shivade, Courtney Hebert, Marcelo Lopetegui, Marie-Catherine De Marneffe, Eric Fosler-Lussier, and Albert M Lai. 2015. Textual inference for eligibility criteria resolution in clinical trials. *Journal of biomedical informatics*, 58:S211–S218.

Adina Williams, Nikita Nangia, and Samuel Bowman. 2018. A broad-coverage challenge corpus for sentence understanding through inference. In *Proceedings of the 2018 Conference of the North American Chapter of the Association for Computational Linguistics: Human Language Technologies, Volume 1 (Long Papers)*, pages 1112–1122.

Wenpeng Yin and Hinrich Schütze. 2015. Convolutional neural network for paraphrase identification. In *Proceedings of the 2015 Conference of the North American Chapter of the Association for Computational Linguistics: Human Language Technologies*, pages 901–911.

UU_TAILS at MEDIQA 2019: Learning Textual Entailment in the Medical Domain

Noha S. Tawfik
Arab Academy for Science & Technology
Alexandria 1029, Egypt
noha.abdelsalam@aast.edu
Utrecht University
3584CC Utrecht, The Netherlands
n.s.tawfik@uu.nl

Marco R. Spruit
Utrecht University
3584 CC Utrecht, The Netherlands
m.r.spruit@uu.nl

Abstract

This article describes the participation of the *UU_TAILS* team in the 2019 MEDIQA challenge intended to improve domain-specific models in medical and clinical NLP. The challenge consists of 3 tasks: medical language inference (NLI), recognizing textual entailment (RQE) and question answering (QA). Our team participated in tasks 1 and 2 and our best runs achieved a performance accuracy of 0.852 and 0.584 respectively for the test sets. The models proposed for task 1 relied on BERT embeddings and different ensemble techniques. For the RQE task, we trained a traditional multilayer perceptron network based on embeddings generated by the universal sentence encoder.

1 Introduction

Detecting semantic relations between sentence pairs is a long-standing challenge for computational semantics. Given two snippets of text: Premise P and Hypothesis H, textual entailment recognition determines if the meaning of H can be inferred from that of P (Dagan et al., 2013). The significance of modeling text inference is evident since it evaluates the capability of Natural language Processing (NLP) to grasp meaning and interprets the linguistic variability of the language. Natural language inference (NLI) tasks, also known as Recognizing Textual Entailment (RTE) require a deep understanding of the semantic similarity between the hypothesis and the premise. Moreover, they overlap with other linguistic problems such as question answering and semantic text similarity. The recent years witnessed regular organization of shared tasks targeting the RTE/NLI task, which consequently led to advances in the field. More complex models were developed that rely on deep neural networks, this was feasible with the availability of large amounts of annotated datasets such as SNLI and MultiNLI (Bowman et al., 2015; **?**). However, most models fail to generalize across different NLI benchmarks (Talman and Chatzikyriakidis, 2018). Additionally, they do not perform accurately on domain-specific datasets. This is specifically true in the medical and clinical domain. Compared to open domain data, the language used to describe biomedical events is usually complex, rich in clinical semantics and contains conceptual overlap. And hence, it is difficult to adapt any of the former models directly.

The MEDIQA challenge (Ben Abacha et al., 2019) addresses the above limitations through its three proposed tasks. The first task aims at identifying inference relations between clinical sentence pairs and introduces the medical natural language inference benchmark dataset *MedNLI* (Romanov and Shivade, 2018). Its creation process is similar to the creation of the gold-standard SNLI dataset with adaptation to the clinical domain. Expert annotators were presented 4,638 premises extracted from the MIMIC-III database (Johnson et al., 2016) and were asked to write three hypotheses with a true, false and neutral description of the premise. The final dataset comprises 14,049 sentence pairs divided into 11,232, 1,395 and 1,422 for training, development and testing respectively. An additional test batch was provided by the challenge organizers with 405 unlabelled instances. to the biomedical domain.

Similarly, the second task, Recognizing Question Entailment (RQE), tackles the problem of finding duplicate questions by labeling questions based on their similarity (Ben Abacha and Demner-Fushman, 2016). Extending the earlier NLI definition, the authors define question entailment as "Question A entails Question B if every answer to B is also a correct answer to A exactly or partially". The dataset is specifically designed to

493

Proceedings of the BioNLP 2019 workshop, pages 493–499
Florence, Italy, August 1, 2019. ©2019 Association for Computational Linguistics

find the most similar frequently asked question (FAQ) to a given question. The training set was constructed from the questions provided by family doctors on the National Library of Medicine (NLM) platform resulting in 8,588 question pairs where 54.2% are positive pairs. For validation, two sources of questions were used: validated questions from the NLM collections and FAQs retrieved from the National Institutes of Health (NIH) website. The validation set has 302 pairs of questions with 42.7% pairs positively labelled. The test set for the challenge was balanced and comprised of 230 question pairs.

The rest of the paper is organized as follows: Section 2 briefly discusses related work. We limit our summary to textual inference research in the biomedical domain only. In Section 3, we describe our proposed model and the implementation details for both tasks. In Section 4, we show the experiment results of our proposed models. Finally, we conclude our analysis of the challenge, as well as some additional discussions of the future directions in Section 5.

2   Related Work

In (Ben Abacha and Demner-Fushman, 2016), the authors introduce a baseline model for the RQE dataset. The feature-based model relies on negation, medical concepts overlap and lexical similarity measures to detect entailment among medical question pairs. Romanov and Shivade conducted multiple experiments on the MedNLI dataset to evaluate the transferability of existing methods in adapting to clinical RTE tasks (Romanov and Shivade, 2018). The best performing was the bidirectional LSTM encoder of the inferSent. Their findings also showed that transfer learning over the larger SNLI set did not improve the results. In a previous work, we tried to model textual entailment found in biomedical literature by restructuring an existing YES/NO question-answering dataset extracted from PubMed(2019). The newly formed dataset aligned with standard NLI datasets format. Further on, we combined hand-crafted features with the inferSent model to detect inference.

To the best of our knowledge, other than the work previously mentioned, there has been minimal research conducted directly on the textual entailment task in the biomedical domain. Below, we summarize scattered attempts to extract contradictions and conflicting statements found in medical documents. Sarafraz et al. (2012), extracted negated molecular events from biomedical literature using a hybrid of machine learning features and semantic rules. Similiarly, De Silve et al. (2017), extracted inconsistencies found in miRNA research articles. The system extracts relevant triples and scores them according to an appositeness metric suggested by the authors. Alamri et al. (2016), introduced a dataset of 259 contradictory claims that answer 10 medical questions related to cardiovascular diseases. Their proposed model relied on n-grams, negation, sentiment and directionality features while in (Tawfik and Spruit, 2018), the authors exploited semantic features and biomedical word embeddings to detect contradictions using the same dataset. Zadrozny et al. (2018) suggested a conceptual framework based on the mathematical sheaf model to highlight conflicting and contradictory criteria in guidelines published by accredited medical institutes. It transforms natural language sentences to formulas with parameters, creates partial order based on common predicates and builds sheaves on these partial orders.

3   Exploratory Embedding Analysis

With the fast developmental pace of text embedding methods, there is a lack of unified methodology to assess these different techniques in the biomedical domain. We attempted to conduct a comprehensive evaluation of different text representations for both tasks, prior to submission of round 2 of the challenge. We use the *MedSentEval*[1] toolkit, a python-based toolkit that supports different embedding techniques including tradiional word embeddings like GloVe and FastText, contextualized embeddings like Embeddings from Language Models (ELMO) and Bidirectional Encoder Representations from Transformers (BERT) and dedicated sentence encoders such as inferSent and Universal Sentence Encoder (USE). To evaluate the sentence representations fairly, we adopt a straightforward method that extracts embeddings from different techniques and feeds them to a logistic regression classifier. Our analysis showed that for the NLI task, embeddings from the inferSent model achieved the best performance. This is not surprising, and aligns

[1]https://github.com/nstawfik/
MedSentEval

with the results reported by the benchmark creator (Romanov and Shivade, 2018). Moreover, we notice that embeddings acquired from language models such as ELMO and BERT, were the second best performing with minimal accuracy difference. For the *RQE* task, the transformer encoder of the USE model outperformed all other methods by a clear margin followed by inferSent trained with GloVe embeddings. This might be contributed to the multi-type training data employed by USE with questions and entailment sentence pairs among others. As observed in the General Language Understanding Evaluation (GLUE) benchmark dataset, BERT-based models are currently the state-of-the art models for the NLI task. Accordingly, we have tried to further investigate the performance of BERT in the biomedical NLI domain. We also employed USE and inferSent sentence embeddings for task 2.

Bidirectional Encoder Representations from Transformers BERT is a neural model developed by Google, that makes heavy use of language representation models designed to pre-train deep bidirectional representations (Devlin et al., 2018). It is trained in an unsupervised manner over an enormous amount of publicly available plain text data. Language Modeling (LM) serves as an unsupervised pre-training stage that can generate the next word in a sentence with knowledge of previous words in a sentence. BERT is different from other LM-based models because it targets a different training objective, it uses masked language modeling instead of traditional LM. It replaces words in a sentence randomly and inserts a "masked" token. The transformer generates predictions for the masked words by jointly conditioning on both left and right context in all layers.

Universal Sentence Encoder USE is referred to as "universal" since, in theory, it is supposed to encode general properties of sentences given the large size of datasets it is trained on (Cer et al., 2018). The multi-task learning encoder uses several annotated and unannotated datasets for training. Training data consisted of supervised and unsupervised sources such as Wikipedia articles, news, discussion forums, dialogues and question/answers pairs. It has two variants of the encoding architectures; The transformer model is designed for higher accuracy, but the encoding requires more memory and computational time. The

Deep Averaging Network (DAN) model on the other hand is designed for speed and efficiency, and some accuracy is compromised. When integrated in any downstream task, USE should be able to represent sentences efficiently without the need for any domain specific knowledge. This is a great advantage when limited training resources are available for specific tasks.

4 Methods

4.1 Task 1: Natural Language Inference (NLI)

Experimental Settings We take advantage of two newly released BERT models trained on different biomedical data. The following models were initialized from the original bert-base-uncased setting pre-trained with 12 transformer layers, hidden unit size of d=768, 12 attention heads and 110M parameters.

- SciBERT[2] trained on a random sample of 1.14M scientific articles available in the semantic scholar repository. The training data consists of full-text papers from the biomedical and computer sciences domain with a 2.5B and 0.6B word count, respectively (Beltagy et al., 2019).

- ClinicalBERT[3] trained on approximately 2M clinical records. The training data consists of intensive care notes distributed among 15 types available in the MIMIC database (Alsentzer et al., 2019).

We combined both training and evaluation records to form a new training set of 12627 sentence pairs. The original test set was used for evaluation and development. We experimented with all models in pytorch, using the HuggingFace[4] re-implementation of the original BERT python package. We convert the SciBERT models to make it compatible with PyTorch. We use the fine-tuning script to train the model on the MEDNLI dataset in an end-to-end fashion. We trained a total of 30 models with variations of the model configuration. All models with accuracy less than 0.786

---

[2]The pre-trained weights for for the SciBERT model are available at `https://github.com/allenai/scibert`

[3]The pre-trained weights for the ClinicalBERT model are available at `https://github.com/EmilyAlsentzer/clinicalBERT`

[4]`https://github.com/huggingface/pytorch-pretrained-BERT`

Hyperparameter	Value
Learning rate	3e-5, 2e-5, 5e-5
Sequence length	64, 128
Number of Epochs	3
Batch Size	8, 16

Table 1: Hyperparameters values for training BERT models

on development data were discarded. The threshold value was set to the best accuracy achieved for the MedNLI dataset as reported in the paper. Table 1 list the hyperparameters for this set of experiments, the values for other parameters were kept the same as the original BERT model.

4.1.1 BERT Ensemble Model

Rather than using only a single model for predictions, ensemble techniques can be considered as a useful method to boost the overall performance. A key factor in ensembling is how to blend the results. We experimented with different systems in terms of size and fusion technique in order to increase performance accuracy:

- Drop-out Averaging: All BERT models are added into the candidate ensemble set. Iteratively, we randomly drop one model at a time. With each dropout, we test the ability of the new ensemble set to improve the overall performance by calculating the ensemble's accuracy for the development set by averaging the output probabilities for each class. The process has been repeated until no improvements were observed and the best performing set is chosen as the final ensemble set.

- STACKING BERT 1: A meta learner trained on the predictions generated from all base models and optimally combine them to form the final decision. We train three classifiers, by using five-fold cross validation, including a K-Nearest Neighbor (KNN), a linear Support Vector Machine (SVM) and Naive Bayesian (NB). The classifiers were implemented through the scikit-learn library [5] and we also apply the grid search method for parameter tuning (Pedregosa et al., 2011).

- STACKING BERT 2: We create a second level ensemble stacking. In this level, we train a logistic regression classifier on top of

the combined predictions generated from the first level stacking stacking BERT phase.

4.2 Task 2: Recognizing Question Entailment (RQE)

Experimental Settings We use the transformer-based architecture of the USE encoder as it was proven to yield better results. USE was implemented through its TF hub module [6]. For all pairs, each input question was embedded separately and then their combined embedding vector is formed as $(u, v, | u - v |, u * v)$, which is a concatenation of the premise and hypothesis vectors and their respective absolute difference and hadamard product. We experiment with both logistic regression and multilayer perceptron on top of the generated input representations. The MLP consists of a single hidden layer of 50 neurons using the adam optimizer and a batch size of 64.

5 Results & Discussion

5.1 Task 1: Natural Language Inference(NLI)

The best performing single BERT model achieved 0.828 for the evaluation set. Table 2 shows results of each model ensemble used for the NLI task. For the first run, we only averaged predictions generated by the ClinicalBERT model. The drop-out ensembling resulted in 12 models in total. For the second run, we used KNN classification over predictions from all trained BERT models. The remaining 3 runs use a second level logistic regression classifier while varying the first level classification model. We can observe consistent improvement from successive ensembling from one to two stacking levels. Our five runs showed substantial improvement in the performance over the original baseline with accuracy gain ranging from 10.6% to 13.8%. By the end of the challenge, 42 teams submitted a total of 143 runs to the NLI task. our top performing submission ranked the 12th over all teams [7]. Its corresponding model could be viewed as a three-stage architecture with 2 level stacking ensemble as illustrated in figure 1.

All runs submitted relied solely on BERT text rep-

---

[5] https://scikit-learn.org/

[6] The TF version of the USE model is available at https://tfhub.dev/google/universal-sentence-encoder-large/3

[7] Leaderboard for the NLI task: https://www.aicrowd.com/challenges/mediqa-2019-natural-language-inference-nli/leaderboards (accessed 1st of June 2019)

Submission	Model	Accuracy	
		Dev	Test
1	Drop-out BERT AVG: 12 models with averaging ensemble	0.836	0.820
2	Stacking BERT 1: KNN	0.846	0.840
3	Stacking BERT 2: KNN followed by LR	0.847	0.847
4	Stacking BERT 2: (KNN/SVM/NB) followed by LR	0.849	0.852
5	Stacking BERT 2: Linear SVM followed by LR	0.846	0.823

Table 2: Results of our team runs on the MEDIQA challenge for the NLI task.

Submission	Model	Accuracy	
		Dev	Test
1	USE embeddings with LR Classifier	0.770	0.584
2	USE embeddings with MLP Classifier (1 hidden layer with 50)	0.778	0.580

Table 3: Results of our team runs on the MEDIQA challenge for the RQE task.

resentations without any external features. Initially, we assumed that training our models with more than just embedding features should help classification and improve overall performance. We used the predictions generated by the dropout averaging ensemble as extra features to further fine-tune a second-level BERT model. The model hyperparameters settings were the same as the best performing single base model. We did not find this experiment to yield any gains in the evaluation phase, compared to ensemble models, with only 0.815 accuracy for the development set. This was also affirmed post submission, with the release of the gold-labels. The accuracy for the test set was only 0.812.

5.2 Task 2: Recognizing Question Entailment (RQE)

Table 3 shows our two submitted runs for task 2. Even though our approach for this task was much simpler than task 1, we still managed to achieve a considerably good accuracy outperforming the baseline by 4.3%. The final results show that our team ranked the 23rd among all 54 participants[8]. Due to time constraints we were unable to fully investigate all models described in section 3, nor conduct a suitably thorough hyperparameter search for the MLP. However, we were able to conduct more evaluations post submission. We trained the inferSent Bi-LSTM encoder on the

MedNLI data using GloVe embeddings. We then used the trained model to generate embeddings for the RQE data, and used the same MLP architecture to generate predictions. Despite the similarity of both tasks and the potential benefit from transfer learning, the model achieved an accuracy of 0.623 and 0.532 for dev and test set respectively.

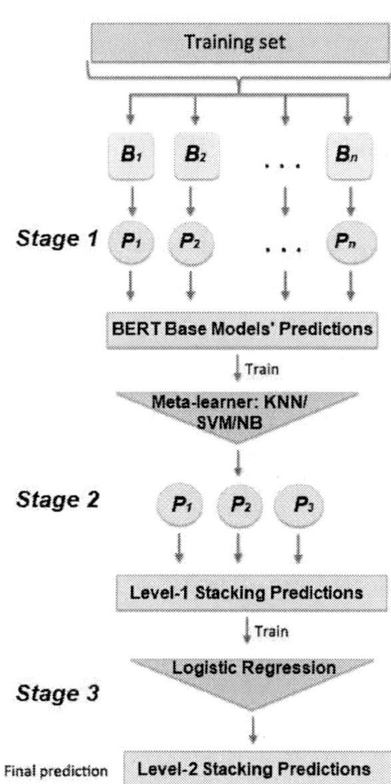

Figure 1: Overview of the ensemble architecture of the best run for the NLI task.

497

6 Conclusion

In this paper, we presented our solution for textual entailment detection in the clinical domain. Our proposed approach for the NLI task relies on BERT contextual embeddings features and machine learning algorithms such as KNN, SVM and LR for ensembling. We use two different pretrained BERT weights to train the base models and generate corresponding probabilities for the test set. Then, we adopt a 5-fold stacking strategy to learn and combine predictions. In the third and final level of the ensemble, we use a logistic regression over the outputs from level-1 stacking, to predict the final class labels. A future extension of our model is to use BERT in feature extraction mode instead of fine-tuning the end-to-end model on the MedNLI dataset. This would allow the selection of layers from which to extract embeddings and/or the combination of multiple layers. In the former scenario, different neural networks could be used to generate the base model predictions before applying ensemble techniques.

For the RQE task, we train an MLP classifier on top of USE embeddings. The results obtained were promising, given the simplicity of the model. More complex and deeper networks could be employed with the combination of USE embeddings. We also experimented with transfer learning by training the inferSent model on MedNLI before fine-tuning on the RQE corpus. While this approach did not improve the results, we aim at further investigating other inferSent architectures and training on clinical word embedding.

References

Abdulaziz Alamri. 2016. *The Detection of Contradictory Claims in Biomedical Abstracts*. Ph.D. thesis, University of Sheffield.

Emily Alsentzer, John R. Murphy, Willie Boag, Wei-Hung Weng, Di Jin, Tristan Naumann, and Matthew B. A. McDermott. 2019. Publicly Available Clinical BERT Embeddings. *arXiv e-prints*, page arXiv:1904.03323.

Iz Beltagy, Arman Cohan, and Kyle Lo. 2019. SCIBERT: Pretrained Contextualized Embeddings for Scientific Text. Technical report.

Asma Ben Abacha and Dina Demner-Fushman. 2016. Recognizing Question Entailment for Medical Question Answering. *AMIA ... Annual Symposium proceedings. AMIA Symposium*, 2016:310–318.

Asma Ben Abacha, Chaitanya Shivade, and Dina Demner-Fushman. 2019. Overview of the MEDIQA 2019 Shared Task on Textual Inference, Question Entailment and Question Answering. In *ACL-BioNLP*, Florence.

Samuel R. Bowman, Gabor Angeli, Christopher Potts, and Christopher D. Manning. 2015. A large annotated corpus for learning natural language inference. In *Proceedings of the 2015 Conference on Empirical Methods in Natural Language Processing (EMNLP).*, Lisbon. Association for Computational Linguistics.

Daniel Cer, Yinfei Yang, Sheng-yi Kong, Nan Hua, Nicole Limtiaco, Rhomni St John, Noah Constant, Mario Guajardo-Céspedes, Steve Yuan, Chris Tar, Yun-Hsuan Sung, Brian Strope, and Ray Kurzweil Google Research Mountain View. 2018. Universal Sentence Encoder. In *Proceedings of the 2018 Conference on Empirical Methods in Natural Language Processing: System Demonstrations*, pages 169–174, Brussels. Association for Computational Linguistics.

Ido Dagan, Dan Roth, Mark Sammons, and Fabio Massimo Zanzotto. 2013. Recognizing Textual Entailment: Models and Applications. *Synthesis Lectures on Human Language Technologies*, 6(4):1–220.

Jacob Devlin, Ming-Wei Chang, Kenton Lee, Kristina Toutanova Google, and A I Language. 2018. BERT: Pre-training of Deep Bidirectional Transformers for Language Understanding. *arXiv e-prints*.

Alistair E.W. Johnson, Tom J. Pollard, Lu Shen, Liwei H. Lehman, Mengling Feng, Mohammad Ghassemi, Benjamin Moody, Peter Szolovits, Leo Anthony Celi, and Roger G. Mark. 2016. MIMIC-III, a freely accessible critical care database. *Scientific Data*, 3:160035.

F Pedregosa, G Varoquaux, A Gramfort, V Michel, B Thirion, O Grisel, M Blondel, P Prettenhofer, R Weiss, V Dubourg, J Vanderplas, A Passos, D Cournapeau, M Brucher, M Perrot, and E Duchesnay. 2011. Scikit-learn: Machine Learning in Python. *Journal of Machine Learning Research*, 12:2825–2830.

Alexey Romanov and Chaitanya Shivade. 2018. Lessons from Natural Language Inference in the Clinical Domain. *Proceedings of the 2018 Conference on Empirical Methods in Natural Language Processing*, pages 1586–1596.

Noha S. Tawfik and Marco R. Spruit. 2019. Towards Recognition of Textual Entailment in the Biomedical Domain. In *International Conference on Applications of Natural Language to Information Systems*, Manchester. Springer.

Farzaneh Sarafraz. 2012. *Finding conflicting statements in the biomedical literature*. Ph.D. thesis, University of Manchester.

Nisansa de Silva, Dejing Dou, and Jingshan Huang. 2017. Discovering inconsistencies in pubmed abstracts through ontology-based information extraction. In *ACM Conference on Bioinformatics, Computational Biology, and Health Informatics (ACM BCB)*.

Aarne Talman and Stergios Chatzikyriakidis. 2018. Testing the Generalization Power of Neural Network Models Across NLI Benchmarks. Technical report.

Noha S. Tawfik and Marco R. Spruit. 2018. Automated Contradiction Detection in Biomedical Literature. In *Proceedings of international Conference on Machine Learning and Data Mining in Pattern Recognition*, pages 138–148. Springer, Cham.

Wlodek Zadrozny and Luciana Garbayo. 2018. A Sheaf Model of Contradictions and Disagreements. Preliminary Report and Discussion. In *International Symposium on Artificial Intelligence and Mathematics,*, Florida.

UW-BHI at MEDIQA 2019: An Analysis of Representation Methods for Medical Natural Language Inference

William R. Kearns[†], **Wilson Lau**[†], and **Jason A. Thomas** [†]

[†]Department of Biomedical Informatics and Medical Education, University of Washington
850 Republican Street
Seattle, WA
{kearnsw, wlau, thomasjt}@uw.edu

Abstract

Recent advances in distributed language modeling have led to large performance increases on a variety of natural language processing (NLP) tasks. However, it is not well understood how these methods may be augmented by knowledge-based approaches. This paper compares the performance and internal representation of an Enhanced Sequential Inference Model (ESIM) between three experimental conditions based on the representation method: Bidirectional Encoder Representations from Transformers (BERT), Embeddings of Semantic Predications (ESP), or Cui2Vec. The methods were evaluated on the Medical Natural Language Inference (MedNLI) subtask of the MEDIQA 2019 shared task. This task relied heavily on semantic understanding and thus served as a suitable evaluation set for the comparison of these representation methods.

1 Introduction

This paper describes our approach to the Natural Language Inference (NLI) subtask of the MEDIQA 2019 shared task (Ben Abacha et al., 2019). As it is not yet clear the extent to which knowledge-based embeddings may provide task-specific improvement over recent advances in contextual embeddings, we provide an analysis of the differences in performance between these two methods. Additionally, it is not yet clear from the literature the extent to which information stored in contextual embeddings overlaps with that in knowledge-based embeddings for which we provide a preliminary analysis of the attention weights of models that use these two representation methods as input. We compare BERT fine-tuned to MIMIC-III (Johnson et al., 2016) and PubMed to Embeddings of Semantic Predications (ESP) trained on SemMedDB and a baseline that uses Cui2Vec embeddings trained on clinical and biomedical text.

Two recent advances in the unsupervised modeling of natural language, Embeddings of Language Models (ELMo) (Peters et al., 2018) and Bidirectional Encoder Representations from Transformers (BERT) (Devlin et al., 2018), have led to drastic improvements across a variety of shared tasks. Both of these methods use transfer learning, a method whereby a multi-layered language model is first trained on a large unlabeled corpus. The weights of the model are then frozen and used as input to a task specific model (Peters et al., 2018; Devlin et al., 2018; Liu et al., 2019). This method is particularly well-suited for work in the medical domain where datasets tend to be relatively small due to the high cost of expert annotation.

However, whereas clinical free-text is difficult to access and share in bulk due to privacy concerns, the biomedical domain is characterized by a significant amount of manually-curated structured knowledge bases. The BioPortal repository currently hosts 773 different biomedical ontologies comprised of over 9.4 million classes. SemMedDB is a triple store that consists of over 94 million predications extracted from PubMed by SemRep, a semantic parser for biomedical text (Rindflesch and Fiszman, 2003; Kilicoglu et al., 2012). These available resources make a strong case for the evaluation of knowledge-based methods for the Medical Natural Language Inference (MedNLI) task (Romanov and Shivade, 2018).

2 Related Work

In this section, we provide a brief overview of methods for distributional and frame-based semantic representation of natural language. For a more detailed synthesis, we refer the reader to the

500

Proceedings of the BioNLP 2019 workshop, pages 500–509
Florence, Italy, August 1, 2019. ©2019 Association for Computational Linguistics

review of Vector Space Models (VSMs) by Turney and Pantel (2010).

2.1 Distributional Semantics

The distributed representation of words has a long history in computational linguistics, beginning with latent semantic indexing (LSI) (Deerwester et al., 1990; Hofmann, 1999; Kanerva et al., 2000), maximum entropy methods (Berger et al., 1996), and latent Dirichlet allocation (LDA) (Blei et al., 2003). More recently, neural network methods have been applied to model natural language (Bengio et al., 2003; Weston et al., 2008; Turian et al., 2010). These methods have been broadly applied as a method of improving supervised model performance by learning word-level features from large unlabeled datasets with more recent work using either Word2Vec (Mikolov et al., 2013; Pavlopoulos et al., 2014) or GloVe (Pennington et al., 2014) embeddings. Recent work has learned a continuous representation of Unified Medical Language System (UMLS) (Aronson, 2006) concepts by applying the Word2Vec method to a large corpus of insurance claims, clinical notes, and biomedical text where UMLS concepts were replaced with their Concept Unique Identifiers (CUIs) (Beam et al., 2018).

Models that incorporate sub-word information are particularly useful in the medical domain for representing medical terminology and out-of-vocabulary terms common in clinical notes and consumer health questions (Romanov and Shivade, 2018). Most approaches use a temporal convolution over a sliding window of characters and have been shown to improve performance on a variety of tasks (Kim et al., 2015; Zhang et al., 2015; Seo et al., 2016; Bojanowski et al., 2017).

Embeddings from Language Models (ELMo) computes word representations using a bidirectional language model that consist of a character-level embedding layer followed by a deep bidirectional long short-term memory (LSTM) network (Peters et al., 2018). Bidirectional Encoder Representations from Transformers (BERT) replaces the each forward and backward LSTMs with a single Transformer that simultaneously computes attention in both the forward and backward directions and is regarded as the current state-of-the-art method for language representation (Vaswani et al., 2017; Devlin et al., 2018). This method additionally substitutes two new unsupervised training objectives in place of the classical language models, i.e., masked language modeling (MLM) and next sentence prediction (NSP). In the case of MLM, a percentage of the words in the corpus are replaced by a [MASK] token. The task is then for the system to predict the masked token. For NSP, the task is given two sentences, $s1$ and $s2$, from a document to determine whether $s2$ is the next sentence following $s1$.

While ELMo has been shown to outperform GloVe and Word2Vec on consumer health question answering (Kearns and Thomas, 2018), BERT has outperformed ELMo on various clinical tasks (Si et al., 2019) and has been fine-tuned and applied to the biomedical literature and clinical notes (Alsentzer et al., 2019; Huang et al., 2019; Si et al., 2019; Lee et al., 2019). BERT supports the transfer of a pretrained general purpose language model to a task-specific application through fine-tuning. The next sentence prediction objective in the pretraining process suggests this method would be inherently suitable for NLI. In addition, BERT utilizes character-based and WordPiece tokenization (Wu et al., 2016) to learn the morphological patterns among inflections. The subword segmentation such as *##nea* in the word *dyspnea* makes it capable to understand the context of an out-of-vocabulary word making it a particularly suitable representation for clinical text.

2.2 Frame-based Semantics

FrameNet is a database of sentence-level frame-based semantics that proposes human understanding of natural language is the result of frames in which certain roles are expected to be filled (Baker et al., 1998). For example, the predicate *"replace"* has at least two such roles, the thing being replaced and the new object. A sentence such as *"The table was replaced."* raises the question *"With what was the table replaced?"*. Frame-based semantics is a popular approach for semantic role labeling (SRL) (Swayamdipta et al., 2018), question answering (QA) (Shen and Lapata, 2007; Roberts and Demner-fushman, 2016; He, 2015; Michael et al., 2018), and dialog systems (Larsson and Traum, 2000; Gupta et al., 2018).

Vector symbolic architectures (VSA) are an approach that seeks to represent semantic predications by applying binding operators that define a directional transformation between entities (Levy and Gayler, 2008). Early approaches included binary spatter code (BSC) for encoding structured knowledge (Kanerva, 1996, 1997) and Holographic Embeddings that used circular convolution as a binding operator to improve the scalability of this approach to large knowledge graphs (Plate, 1995). The resurgence of neural network methods has focused attention on extending these methods as there is a growing interest in leveraging continuous representations of structured knowledge to improve performance on downstream applications.

Knowledge graph embeddings (KGE) are one approach that represents entities and their relationships as continuous vectors that are learned using TransE/R (Bordes and Weston, 2009), RESCAL (Nickel et al., 2011), or Holographic Embeddings (Plate, 1995; Nickel et al., 2015). Stanovsky et. al (2017) showed that RESCAL embeddings pretrained on DbPedia improved performance on the task of adverse drug reaction labeling over a clinical Word2Vec model. RESCAL uses tensor products whose application to representation learning dates back to Smolensky (1986; 1990) that used the inner product and has recently been applied to the bAbI dataset (Smolensky et al., 2016; Weston et al., 2016). Embeddings of Semantic Predications (ESP) are a neural-probabilistic representational approach that uses VSA binding operations to encode structured relationships (Cohen and Widdows, 2017). The Embeddings Augmented by Random Permutations (EARP) used in this paper are a modified ESP approach that applies random permutations to the entity vectors during training and were shown to improve performance on the Bigger Analogy Test Set by up to 8% against a fastText baseline (Cohen and Widdows, 2018).

3   Methods

In this section, we provide details on the three representation methods used in this study, i.e. BERT, Cui2Vec, and ESP. We continue with a description of the inference model used in each experiment to predict the label for a given hypothesis/premise pair.

3.1   Representation Layer

There are many publicly available biomedical BERT embeddings which were initialized from the original BERT Base models. BioBERT was trained on PubMed Abstracts and PubMed Central Full-text articles (Lee et al., 2019). In this study, we applied ClinicalBERT that was initialized from BioBERT and subsequently trained on all MIMIC-III notes (Alsentzer et al., 2019).

For Cui2Vec, we used the publicly available implementation from Beam et al. (2018) that was trained on a corpus consisting of 20 million clinical notes from a research hospital, 1.7 million full-text articles from PubMed, and an insurance claims database with 60 million members.

For ESP, we used a 500-dimensional model trained over SemMedDB using the recent Embeddings Augmented by Random Permutations (EARP) approach with a 10^{-7} sampling threshold for predications and a 10^{-5} sampling threshold for concepts excluding concepts that had a frequency greater than 10^6 (Cohen and Widdows, 2018).

To apply Cui2Vec and ESP, we first processed the MedNLI dataset (Romanov and Shivade, 2018) with MetaMap to normalize entities to their concept unique identifier (CUI) in the UMLS (Aronson, 2006). MetaMap takes text as input and applies biomedical and clinical entity recognition (ER), followed by word sense disambiguation (WSD) that links entities to their normalized concept unique identifiers (CUIs). Entities that mapped to a UMLS CUI were assigned a representation in Cui2Vec and ESP. Other tokens were assigned vector representations using fastText embeddings trained on MIMIC-III data (Bojanowski et al., 2017; Romanov and Shivade, 2018).

3.2   Inference Model

For all experiments, we used the AllenNLP implementation (Gardner et al., 2018) of the Enhanced Sequential Inference Model (ESIM) architecture (Chen et al., 2017). This model encodes the premise and hypothesis using a Bidirectional LSTM (BiLSTM) where at each time step the hidden state of the LSTMs are concatenated to represent its context. Local inference between the two sentences is then achieved by aligning the relevant information between words in the premise and hypothesis. This alignment based on soft attention is implemented by the inner product between the encoded premise and encoded hypothesis to produce an attention matrix (Figure 1 and 2). These attention values are used to create a weighted representation of both sentences. An enhanced representation of the premise is created by concatenating the encoded premise, the weighted hypoth-

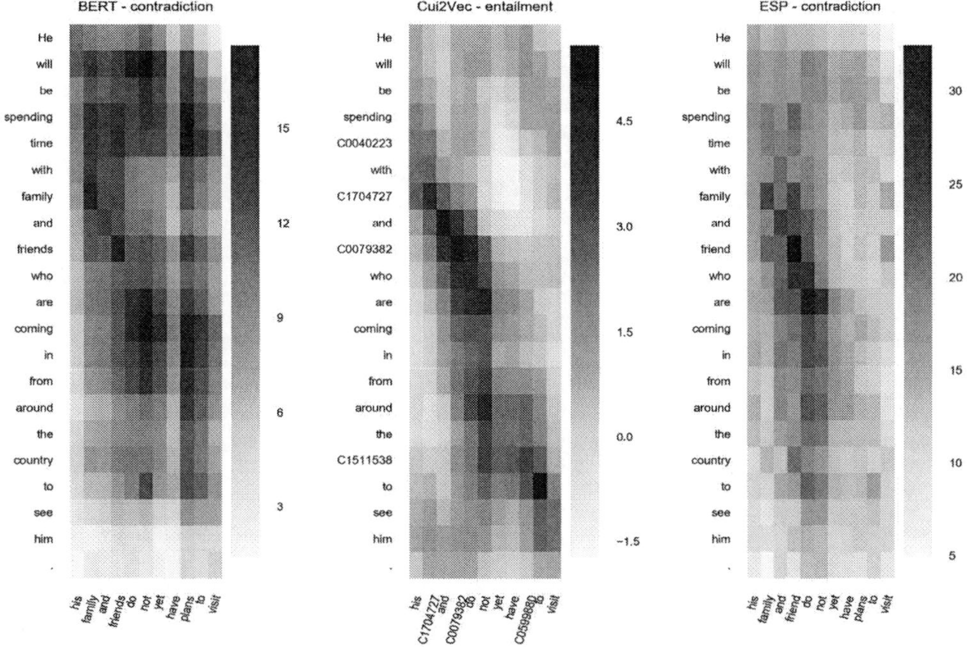

Figure 1: An example of a correct BERT prediction demonstrating its general domain coverage and contextual embedding. Premise: *"He will be spending time with family and friends who are coming in from around the country to see him."* Hypothesis: *"his family and friends do not yet have plans to visit."*

esis, the encoded premise minus the weighted hypothesis, and the element-wise multiplication of the encoded premise and the weighted hypothesis. The enhanced representation of the hypothesis is created similarly. This operation is expected to enhance the local inference information between elements in each sentence. This representation is then projected into the original dimension and fed into a second BiLSTM inference layer in order to capture inference composition sequentially. The resulting vector is then summarized by max and average pooling. These two pooled representations are concatenated and passed through a multi-layered perceptron followed by a sigmoid function to predict probabilities for each of the sentence labels, i.e. *entailment*, *contradiction*, and *neutral*.

4   Results

The ESIM model achieved an accuracy of 81.2%, 65.2%, and 77.8% for the MedNLI task using BERT, Cui2Vec, and ESP, respectively. Table 1 shows the number of correct predictions by each embedding type. The BERT model has the highest

accuracy on predicting *entailment* and *contradiction* labels, while the ESP model has the highest accuracy on predicting *neutral* labels. However, the difference is only significant in the case of *entailment*.

To evaluate the ability to set a predictive threshold for use in clinical applications, we sought to measure the certainty with which the model made its predictions. To achieve this goal, we used the predicted probabilities of each embedding type on their respective subset of correct predictions such that. We found the predicted probability of ESP to be much higher than the others as depicted in Figure 3. ESP's minimum predicted probability as well as the variance of its distribution is the lowest among all embedding types.

4.1   Error Analysis

To examine the relationship between embedding prediction performance and hypothesis focus, we first annotated the test set for:

- hypothesis focus (e.g. *medications, procedures, symptoms*, etc.)

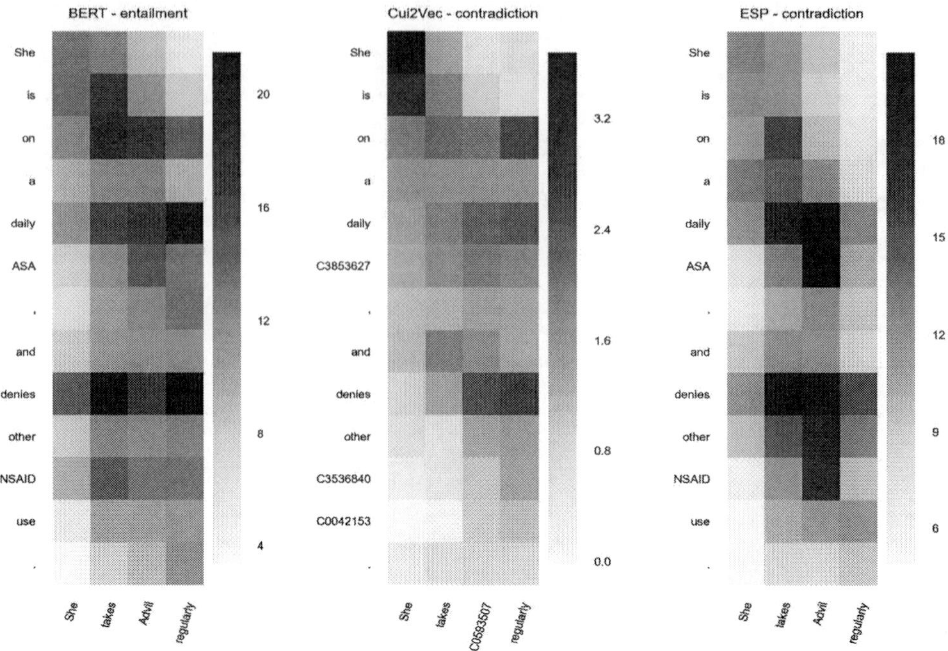

Figure 2: An example of a correct ESP prediction demonstrating its ability to associate Advil as a subclass of NSAIDs. Premise: *"She is on a daily ASA, and denies other NSAID use."* Hypothesis: *"She takes Advil regularly."*

| Label | **Embedding Type** | | |
	BERT	**Cui2Vec**	**ESP**
Entailment	**82.22% (n=111)**	60.00% (n=81)	71.85% (n=97)
Contraction	**88.15% (n=119)**	74.81% (n=101)	87.41% (n=118)
Neutral	73.33% (n=99)	60.74% (n=82)	**74.07% (n=100)**

Table 1: Model accuracy for each label by embedding type.

- hypothesis tense (e.g. *past, current, future*)

4.1.1 Focus

A total of eleven, non-mutually exclusive hypothesis focus classes were arrived at by consensus of the three authors after an initial blinded round of annotation by two annotators. The remaining data was annotated by one of these annotators. We provide definitions of the classes and their overall counts in Table 2. The classes are: *State, Anatomy, Disease, Process, Temporal, Medication, Clinical Finding, Location, Lab/Imaging, Procedure,* and *Examination.*

We then performed Pearson's chi-squared test with Yates' continuity correction on 2x2 contingency tables for each embedding sentence pair prediction (correct or incorrect) with each hypothesis focus (presence or absence) using the *chisq.test* function in R software and results reported in Table 3.

The only significant relationships between hypothesis focus and embedding accuracy were found between BERT and *Disease* (p-value = 0.01) and Cui2Vec and *Disease* (p-value = 0.01) through Pearson's Chi-squared test with Yates' continuity correction. Both embeddings achieved higher accuracy on sentence pairs with a hypothesis focus labeled *Disease* (BERT=90.4%; Cui2Vec=76.6%) than without (BERT=78.5%; Cui2Vec=61.7%).

4.1.2 Tense

Each hypothesis was annotated for tense into one of three mutually exclusive classes: *Past, Current,*

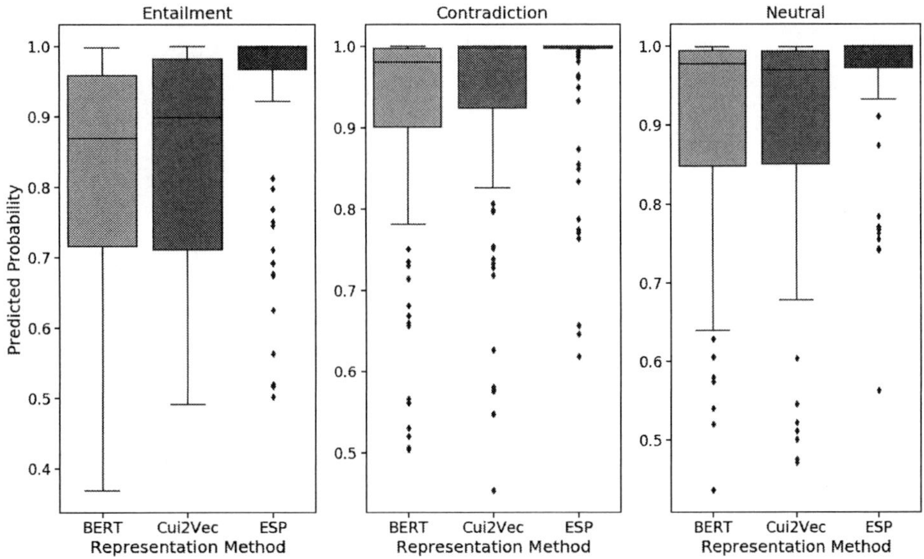

Figure 3: Distribution of predicted probability of the gold label from the subset of correct predictions for each representation method.

Hypothesis Focus	Definition	Count(%)
State	Patient state or symptoms (e.g. *"...has high blood pressure..."*)	251 (62.0)
Anatomy	Specific body part referenced (e.g. *"... has back pain"*)	115 (28.4)
Disease	Similar to state, but a defined disease (e.g. *"...has Diabetes"*)	95 (23.5)
Process	Events like transfers, family visiting, scheduling, or vague references to interventions (e.g. *"...received medical attention"*)	52 (12.8)
Temporal	Reference to time (e.g. *"...initial blood pressure was low"*) besides tense or history	51 (12.6)
Medication	Any reference to medication (e.g. *"antibiotics"*, *"fluids"*, *"oxygen"*, *"IV"*) including administration and patient habits	32 (7.9)
Clinical Finding	Results of an exam, lab/image, procedure, or a diagnosis	28 (6.9)
Location	Specific physical location specified (e.g. *"...discharged home"*)	28 (6.9)
Lab/Imaging	Laboratory tests or imaging (e.g. *histology, CBC, CT scan*)	24 (5.9)
Procedure	Physical procedure besides Lab/Image or exam (e.g. *"intubation"*, *"surgery"*, *"biopsies"*)	14 (3.5)
Examination	Physical examination or explicit use of the word exam(ination)	3 (0.7)

Table 2: Hypothesis foci definitions, examples, and count for all 405 hypotheses in the test set.

and *Future*. Test set hypotheses were predominantly *Current* (n=273; 67.4%) or *Past* (n=131; 32.3%) tense. Only one hypothesis (0.2%) was *Future* tense. A subset (n=22; 7.9%) of the *Current* tense hypotheses explicitly described patient history (e.g. *"The patient has a history of PE"*).

5 Discussion

Our preliminary analysis, identified several patterns from the attention heatmaps that differentiated the three representation methods. We describe two here and provide the entire set of attention matrices along with supplemental analysis on Github [1].

[1]https://kearnsw.github.io/
MEDIQA-2019/

| | Embedding Type | | | | | | | | |
| Focus | BERT | | | Cui2Vec | | | ESP | | |
	(+)	(-)	p-value	(+)	(-)	p-value	(+)	(-)	p-Value
Anatomy	93	22	1	73	42	0.74	90	25	0.99
Clinical Finding	24	4	0.71	16	12	0.47	24	4	0.42
Disease	85	9	**0.01**	72	22	**0.01**	78	16	0.21
Examination	3	0	0.93	2	1	0.58	3	0	0.82
Lab/Imaging	30	7	1	22	15	0.55	31	6	0.48
Location	21	7	0.53	14	14	0.12	19	9	0.28
Medication	27	5	0.81	24	8	0.30	28	4	0.25
Procedure	12	2	0.93	7	7	0.35	11	3	1
Process	41	11	0.78	35	17	0.85	40	12	1
State	198	53	0.16	158	93	0.27	191	60	0.36
Temporal	38	12	0.41	37	13	0.22	41	9	0.56

Table 3: Results from chi-squared (with Yates' continuity correction) test of correct(+) and incorrect(-) predictions by embedding and hypothesis focus type.

The coverage of entities and their associations was characteristic of BERT predictions (Figure 1). BERT associated *"spending time"* with *"plans"* in addition to the lexical overlap of the word *"family"* which is attended by each experimental condition in this example. All three embeddings identified the contradictory significance of the word *"not"* in the hypothesis. However, BERT associated it with both spans *"will be"* and *"are coming"* in the premise, which led to the correct prediction. Cui2Vec over-attended the lexical match of the words *"and"*, *"to"* and *"C0079382"*, which led to the wrong prediction.

The ESP model recognized hierarchical relationships between entities, e.g. *"Advil"* and *"NSAIDs"* (Figure 2). In this example, the ESP approach attends to the daily use of *"ASA"* (acetyl-salicylic acid), i.e. aspirin, and the patient denying the use of *"other NSAIDs"*. This pattern was recognized multiple times in our analysis and provides a strong example of how continuous representations of biomedical ontologies may be used to augment contextual representations.

6 Limitations

The results presented in this paper compare a single model for each representation method fine-tuned to the development set. However, it is well known that the weights of the same model may vary slightly between training runs. Therefore, a more comprehensive approach would be to present the average attention weights across multiple training runs and to examine the weights at each attention layer of the models which we leave for future work.

7 Conclusion

We have presented our analysis of representation methods on the MedNLI task as evaluated during the MEDIQA 2019 shared task. We found that BERT embeddings fine-tuned using PubMed and MIMIC-III outperformed both Cui2Vec and ESP methods. However, we found that ESP had the lowest variance and highest predictive certainty, which may be useful in determining a minimum threshold for clinical decision support systems. *Disease* was the only hypothesis focus to show a significant positive relationship with embedding prediction accuracy. This association was present for BERT and Cui2Vec embeddings - but not ESP. Overall, contradiction was the easiest label to predict for all three embeddings, which may be the result of an annotation artifact where contradiction pairs had higher lexical overlap often differentiated by explicit negation. However, overfitting on the negation can lead to lower accuracy on other entailment labels. Further, our preliminary results indicate that recognition of hierarchical relationships is characteristic of ESP suggesting that they can be used to augment contextual embeddings which, in turn, would contribute lexical coverage including sub-word information. We propose combining these methods in future work.

Acknowledgments

We would like to acknowledge Trevor Cohen for sharing the Embeddings of Semantic Predications used in this study. Author Jason A. Thomas' work was supported, in part, by the National Library of Medicine (NLM) Training Grant T15LM007442. This work was facilitated, in part, through the use of the advanced computational, storage, and networking infrastructure managed by the Research Computing Club at the University of Washington and funded by an STF award.

References

Emily Alsentzer, John R. Murphy, Willie Boag, Wei-Hung Weng, Di Jin, Tristan Naumann, and Matthew B. A. McDermott. 2019. Publicly available clinical BERT embeddings. *CoRR*, abs/1904.03323.

Alan R Aronson. 2006. Metamap: Mapping text to the umls metathesaurus. *Bethesda MD NLM NIH DHHS*, pages 1–26.

Collin F. Baker, Charles J. Fillmore, and John B. Lowe. 1998. The berkeley framenet project. In *Proceedings of the 36th Annual Meeting of the Association for Computational Linguistics and 17th International Conference on Computational Linguistics - Volume 1*, ACL '98/COLING '98, pages 86–90, Stroudsburg, PA, USA. Association for Computational Linguistics. https://doi.org/10.3115/980845.980860.

Andrew L. Beam, Benjamin Kompa, Inbar Fried, Nathan P. Palmer, Xu Shi, Tianxi Cai, and Isaac S. Kohane. 2018. Clinical concept embeddings learned from massive sources of medical data. *CoRR*, abs/1804.01486.

Asma Ben Abacha, Chaitanya Shivade, and Dina Demner-Fushman. 2019. Overview of the mediqa 2019 shared task on textual inference, question entailment and question answering. In *Proceedings of the BioNLP 2019 workshop, Florence, Italy, August 1, 2019*. Association for Computational Linguistics.

Yoshua Bengio, Réjean Ducharme, Pascal Vincent, and Christian Janvin. 2003. A neural probabilistic language model. *J. Mach. Learn. Res.*, 3:1137–1155.

Adam L. Berger, Vincent J. Della Pietra, and Stephen A. Della Pietra. 1996. A maximum entropy approach to natural language processing. *Comput. Linguist.*, 22(1):39–71.

David M. Blei, Andrew Y. Ng, and Michael I. Jordan. 2003. Latent dirichlet allocation. *J. Mach. Learn. Res.*, 3:993–1022.

Piotr Bojanowski, Edouard Grave, Armand Joulin, and Tomas Mikolov. 2017. Enriching word vectors with subword information. *Transactions of the Association for Computational Linguistics*, 5:135–146. https://doi.org/10.1162/tacl_a_00051.

Antoine Bordes and Jason Weston. 2009. Learning Structured Embeddings of Knowledge Bases. *Artificial Intelligence*, (Bengio):301–306. https://doi.org/10.1016/j.procs.2017.05.045.

Qian Chen, Xiaodan Zhu, Zhen-Hua Ling, Si Wei, Hui Jiang, and Diana Inkpen. 2017. Enhanced LSTM for natural language inference. In *Proceedings of the 55th Annual Meeting of the Association for Computational Linguistics (Volume 1: Long Papers)*, pages 1657–1668, Vancouver, Canada. Association for Computational Linguistics. https://doi.org/10.18653/v1/P17-1152.

Trevor Cohen and Dominic Widdows. 2017. Embedding of semantic predications. *Journal of Biomedical Informatics*, 68:150–166. https://doi.org/10.1016/j.jbi.2017.03.003.

Trevor Cohen and Dominic Widdows. 2018. Bringing order to neural word embeddings with embeddings augmented by random permutations (EARP). In *Proceedings of the 22nd Conference on Computational Natural Language Learning*, pages 465–475, Brussels, Belgium. Association for Computational Linguistics.

Scott C. Deerwester, Susan T. Dumais, Thomas K. Landauer, George W. Furnas, and Richard A. Harshman. 1990. Indexing by latent semantic analysis. *JASIS*, 41:391–407.

Jacob Devlin, Ming-Wei Chang, Kenton Lee, and Kristina Toutanova. 2018. BERT: pre-training of deep bidirectional transformers for language understanding. *CoRR*, abs/1810.04805.

Matt Gardner, Joel Grus, Mark Neumann, Oyvind Tafjord, Pradeep Dasigi, Nelson F. Liu, Matthew Peters, Michael Schmitz, and Luke Zettlemoyer. 2018. AllenNLP: A deep semantic natural language processing platform. In *Proceedings of Workshop for NLP Open Source Software (NLP-OSS)*, pages 1–6, Melbourne, Australia. Association for Computational Linguistics.

Sonal Gupta, Rushin Shah, Mrinal Mohit, Anuj Kumar, and Mike Lewis. 2018. Semantic parsing for task oriented dialog using hierarchical representations. *CoRR*, abs/1810.07942.

Luheng He. 2015. Question-Answer Driven Semantic Role Labeling : Using Natural Language to Annotate Natural Language. *Emnlp2015*, (September):643–653. https://doi.org/10.18653/v1/D15-1076.

Thomas Hofmann. 1999. Probabilistic latent semantic indexing. In *Proceedings of the 22Nd Annual International ACM SIGIR Conference on Research and Development in Information Retrieval*, SIGIR '99, pages 50–57, New York, NY, USA. ACM. https://doi.org/10.1145/312624.312649.

Kexin Huang, Jaan Altosaar, and Rajesh Ranganath. 2019. ClinicalBERT: Modeling Clinical Notes and Predicting Hospital Readmission. *arXiv:1904.05342 [cs]*. ArXiv: 1904.05342.

Alistair E W Johnson, Tom J Pollard, Lu Shen, Liwei H Lehman, Mengling Feng, Mohammad Ghassemi, Benjamin Moody, Peter Szolovits, Leo Anthony Celi, and Roger G Mark. 2016. MIMIC-III, a freely accessible critical care database. pages 1–9.

Pentti Kanerva. 1996. Binary spatter-coding of ordered k-tuples. In *Proceedings of the 1996 International Conference on Artificial Neural Networks*, ICANN 96, pages 869–873, London, UK, UK. Springer-Verlag.

Pentti Kanerva. 1997. Fully distributed representation. In *In Proceedings Real World Computing Symposium (Report TR-96001)*, pages 358–365.

Pentti Kanerva, Jan Kristoferson, and Anders Holst. 2000. Random indexing of text samples for latent semantic analysis. In *In Proceedings of the 22nd Annual Conference of the Cognitive Science Society*, pages 103–6. Erlbaum.

William R Kearns and Jason A Thomas. 2018. Resource and response type classification for consumer health question answering. *AMIA Annual Symposium proceedings. AMIA Symposium*, 2018:634–643.

Halil Kilicoglu, Dongwook Shin, Marcelo Fiszman, Graciela Rosemblat, and Thomas C. Rindflesch. 2012. SemMedDB: A PubMed-scale repository of biomedical semantic predications. *Bioinformatics*, 28(23):3158–3160. https://doi.org/10.1093/bioinformatics/bts591.

Yoon Kim, Yacine Jernite, David Sontag, and Alexander M. Rush. 2015. Character-aware neural language models. *CoRR*, abs/1508.06615.

Staffan Larsson and David R. Traum. 2000. Information state and dialogue management in the trindi dialogue move engine toolkit. *Nat. Lang. Eng.*, 6(3-4):323–340. https://doi.org/10.1017/S1351324900002539.

Jinhyuk Lee, Wonjin Yoon, Sungdong Kim, Donghyeon Kim, Sunkyu Kim, Chan Ho So, and Jaewoo Kang. 2019. BioBERT: a pre-trained biomedical language representation model for biomedical text mining. *arXiv:1901.08746 [cs]*.

Simon D. Levy and Ross Gayler. 2008. Vector symbolic architectures: A new building material for artificial general intelligence. In *Proceedings of the 2008 Conference on Artificial General Intelligence 2008: Proceedings of the First AGI Conference*, pages 414–418, Amsterdam, The Netherlands, The Netherlands. IOS Press.

Xiaodong Liu, Pengcheng He, Weizhu Chen, and Jianfeng Gao. 2019. Multi-task deep neural networks for natural language understanding. *CoRR*, abs/1901.11504.

Julian Michael, Gabriel Stanovsky, Luheng He, Ido Dagan, and Luke Zettlemoyer. 2018. Crowdsourcing question-answer meaning representations. In *Proceedings of the 2018 Conference of the North American Chapter of the Association for Computational Linguistics: Human Language Technologies, Volume 2 (Short Papers)*, pages 560–568, New Orleans, Louisiana. Association for Computational Linguistics. https://doi.org/10.18653/v1/N18-2089.

Tomas Mikolov, Ilya Sutskever, Kai Chen, Greg Corrado, and Jeffrey Dean. 2013. Distributed representations of words and phrases and their compositionality. In *Proceedings of the 26th International Conference on Neural Information Processing Systems - Volume 2*, NIPS'13, pages 3111–3119, USA. Curran Associates Inc.

Maximilian Nickel, Lorenzo Rosasco, and Tomaso A. Poggio. 2015. Holographic embeddings of knowledge graphs. *CoRR*, abs/1510.04935.

Maximilian Nickel, Volker Tresp, and Hans-Peter Kriegel. 2011. A three-way model for collective learning on multi-relational data. In *Proceedings of the 28th International Conference on International Conference on Machine Learning*, ICML'11, pages 809–816, USA. Omnipress.

Ioannis Pavlopoulos, Aris Kosmopoulos, and Ion Androutsopoulos. 2014. Continuous Space Word Vectors Obtained by Applying Word2Vec to Abstracts of Biomedical Articles. http://bioasq.lip6.fr/info/BioASQword2vec/.

Jeffrey Pennington, Richard Socher, and Christopher Manning. 2014. Glove: Global vectors for word representation. In *Proceedings of the 2014 Conference on Empirical Methods in Natural Language Processing (EMNLP)*, pages 1532–1543, Doha, Qatar. Association for Computational Linguistics. https://doi.org/10.3115/v1/D14-1162.

Matthew E. Peters, Mark Neumann, Mohit Iyyer, Matt Gardner, Christopher Clark, Kenton Lee, and Luke Zettlemoyer. 2018. Deep contextualized word representations. *CoRR*, abs/1802.05365.

T. A. Plate. 1995. Holographic reduced representations. *IEEE Transactions on Neural Networks*, 6(3):623–641. https://doi.org/10.1109/72.377968.

Thomas C Rindflesch and Marcelo Fiszman. 2003. The interaction of domain knowledge and linguistic structure in natural language processing: interpreting hypernymic propositions in biomedical text. *Journal of Biomedical Informatics*, 36(6):462–477. https://doi.org/10.1016/j.jbi.2003.11.003.

Kirk Roberts and Dina Demner-fushman. 2016. Annotating Logical Forms for EHR Questions. In *Proceedings of the 10th International Conference on Language Resources and Evaluation*, Section 3, pages 3772–3778.

Alexey Romanov and Chaitanya Shivade. 2018. Lessons from natural language inference in the clinical domain. *CoRR*, abs/1808.06752.

Min Joon Seo, Aniruddha Kembhavi, Ali Farhadi, and Hannaneh Hajishirzi. 2016. Bidirectional attention flow for machine comprehension. *CoRR*, abs/1611.01603.

Dan Shen and Mirella Lapata. 2007. Using semantic roles to improve question answering. In *Proceedings of the 2007 Joint Conference on Empirical Methods in Natural Language Processing and Computational Natural Language Learning (EMNLP-CoNLL)*, page 12–21.

Yuqi Si, Jingqi Wang, Hua Xu, and Kirk Roberts. 2019. Enhancing Clinical Concept Extraction with Contextual Embedding. *arXiv:1902.08691 [cs]*. ArXiv: 1902.08691.

P. Smolensky. 1986. Parallel distributed processing: Explorations in the microstructure of cognition, vol. 1. chapter Information Processing in Dynamical Systems: Foundations of Harmony Theory, pages 194–281. MIT Press, Cambridge, MA, USA.

P. Smolensky. 1990. Tensor product variable binding and the representation of symbolic structures in connectionist systems. *Artif. Intell.*, 46(1-2):159–216. https://doi.org/10.1016/0004-3702(90)90007-M.

Paul Smolensky, Moontae Lee, Xiaodong He, Wen tau Yih, Jianfeng Gao, and Li Deng. 2016. Basic reasoning with tensor product representations. *CoRR*, abs/1601.02745.

Gabriel Stanovsky, Daniel Gruhl, and Pablo Mendes. 2017. Recognizing mentions of adverse drug reaction in social media using knowledge-infused recurrent models. In *Proceedings of the 15th Conference of the European Chapter of the Association for Computational Linguistics: Volume 1, Long Papers*, pages 142–151, Valencia, Spain. Association for Computational Linguistics.

Swabha Swayamdipta, Sam Thomson, Kenton Lee, Luke Zettlemoyer, Chris Dyer, and Noah A. Smith. 2018. Syntactic scaffolds for semantic structures. *CoRR*, abs/1808.10485.

Joseph Turian, Lev-Arie Ratinov, and Yoshua Bengio. 2010. Word representations: A simple and general method for semi-supervised learning. In *Proceedings of the 48th Annual Meeting of the Association for Computational Linguistics*, pages 384–394, Uppsala, Sweden. Association for Computational Linguistics.

Peter D. Turney and Patrick Pantel. 2010. From frequency to meaning: Vector space models of semantics. *Journal of Artificial Intelligence Research*, 37:141–188. https://doi.org/10.1613/jair.2934.

Ashish Vaswani, Noam Shazeer, Niki Parmar, Jakob Uszkoreit, Llion Jones, Aidan N Gomez, Ł ukasz Kaiser, and Illia Polosukhin. 2017. Attention is all you need. In I. Guyon, U. V. Luxburg, S. Bengio, H. Wallach, R. Fergus, S. Vishwanathan, and R. Garnett, editors, *Advances in Neural Information Processing Systems 30*, pages 5998–6008. Curran Associates, Inc.

Jason Weston, Antoine Bordes, Sumit Chopra, and Tomas Mikolov. 2016. Towards ai-complete question answering: A set of prerequisite toy tasks. *CoRR*, abs/1502.05698.

Jason Weston, Frédéric Ratle, and Ronan Collobert. 2008. Deep learning via semi-supervised embedding. In *Proceedings of the 25th International Conference on Machine Learning*, ICML '08, pages 1168–1175, New York, NY, USA. ACM. https://doi.org/10.1145/1390156.1390303.

Yonghui Wu, Mike Schuster, Zhifeng Chen, Quoc V. Le, Mohammad Norouzi, Wolfgang Macherey, Maxim Krikun, Yuan Cao, Qin Gao, Klaus Macherey, Jeff Klingner, Apurva Shah, Melvin Johnson, Xiaobing Liu, ukasz Kaiser, Stephan Gouws, Yoshikiyo Kato, Taku Kudo, Hideto Kazawa, Keith Stevens, George Kurian, Nishant Patil, Wei Wang, Cliff Young, Jason Smith, Jason Riesa, Alex Rudnick, Oriol Vinyals, Greg Corrado, Macduff Hughes, and Jeffrey Dean. 2016. Google's Neural Machine Translation System: Bridging the Gap between Human and Machine Translation. *arXiv:1609.08144 [cs]*. ArXiv: 1609.08144.

Xiang Zhang, Junbo Jake Zhao, and Yann LeCun. 2015. Character-level convolutional networks for text classification. *CoRR*, abs/1509.01626.

Saama Research at MEDIQA 2019: Pre-trained BioBERT with Attention Visualisation for Medical Natural Language Inference

Kamal Raj Kanakarajan , Suriyadeepan Ramamoorthy, Vaidheeswaran Archana,
Soham Chatterjee, Malaikannan Sankarasubbu[*]
SAAMA AI Research Lab, Chennai, India
{kamal.raj, suriyadeepan.ramamoorthy, archana.iyer, soham.chatterjee,
malaikannan.sankarasubbu}@saama.com

Abstract

Natural Language inference is the task of identifying relation between two sentences as entailment, contradiction or neutrality. MedNLI is a biomedical flavour of NLI for clinical domain. This paper explores the use of Bidirectional Encoder Representation from Transformer (BERT) for solving MedNLI. The proposed model, BERT pre-trained on PMC, PubMed and fine-tuned on MIMIC-III v1.4, achieves state of the art results on MedNLI (83.45%) and an accuracy of 78.5% in MEDIQA challenge. The authors present an analysis of the attention patterns that emerged as a result of training BERT on MedNLI using a visualization tool, bertviz.

1 Introduction

Natural Language Inference (NLI) is a fundamental task in Natural Language Processing in which the objective is to determine if the hypothesis is true (entailment), false (contradiction) or undetermined (neutral), given a premise. Entailment, Contradiction and Neutral (semantic independence) are semantic concepts that represent the relationship between sentences. The ability to infer these relations between sentences or pieces of text, is crucial in tasks like Information Retrieval, Semantic Parsing, Commonsense Reasoning, etc. NLI, like most NLP tasks, is challenging due to the ambiguous nature of natural language. A particular meaning can be expressed in multiple linguistic forms. This calls for methods that can capture meaningful semantic concepts from text.

Stanford Natural Language Inference (SNLI) corpus (Bowman et al., 2015) is a collection of sentence pairs labeled for entailment, contradiction, and semantic independence. It contains approximately 550,000 labeled hypothesis/premise pairs. Multi-Genre Natural Language Inference (Multi-NLI) corpus (Williams et al., 2017) contains 433,000 samples, covering a wide range of (10) genres of written and spoken English. Multi-NLI, in its complexity, is closer to Natural Language than SNLI.

MedNLI (Romanov and Shivade) is a dataset for natural language inference in clinical domain, analogous to SNLI. Romanov et al in (Romanov and Shivade), used InferSent (Conneau et al., 2017), a bidirectional LSTM based model for achieving an accuracy of 73.5% in MedNLI. In (Jin et al., 2019), Jin et al make use of BioBERT (Lee et al., 2019), a biomedical version of BERT along with pre-trained LMs(Language Models) as feature extractors, to achieve an accuracy of 81.7% on MedNLI.

This work uses BERT pre-trained on PMC and PubMed corpus, and fine-tuned on MIMIC-III v1.4 data (BioBERT) to solve MedNLI. This approach achieves new state of the art results when evaluated on MedNLI test set (83.4%). Evaluation on MEDIQA (Ben Abacha et al., 2019) test set (Shivade, 2019) results in an accuracy of 78.5 %.

2 Data

2.1 MedNLI

MedNLI or Medical Natural Language Inference is a publicly annotated dataset in the clinical domain. MedNLI was created as a NLI dataset comparable to SNLI, adjusted for the clinical domain (Table 1).

---

[*]Equal Contribution: Kamal had sole access to MIMIC and MEDIQA data, focussed on the algorithm development and implementation. Suriyadeepan and Archana focussed on the attention visualisation and writing. Soham and Malaikannan focussed on reviewing

Proceedings of the BioNLP 2019 workshop, pages 510–516
Florence, Italy, August 1, 2019. ©2019 Association for Computational Linguistics

#	Premise	Hypothesis	Label
1	ALT, AST and lactate were elevated as noted above	patient has abnormal fits	entailment
2	Chest x-ray showed mild congestive heart failure	The patient has complaints of cough	neutral
3	During Hospitalisation, patient became progressively more dyspnic requiring BiPaP and then a NRB	The patient is on room air	contradiction

Table 1: Examples from Development set of MedNLI

Dataset Size	
Training Pairs	11232
Development Pairs	1395
Test Pairs	1422
MEDIQA	405
Average Sentence Length in Token	
Premise	20.0
Hypothesis	5.8
Maximum Sentence Length in Tokens	
Premise	202
Hypothesis	20

Table 2: Data Statistics

2.2 Deriving from MIMIC-III v.1.4

While adapting the structure of SNLI, MedNLI derives its data from the MIMIC III v.1.4 dataset (Johnson et al., 2016). The MIMIC-III v.1.4 dataset consists of around 2,078,705 clinical notes written by healthcare professionals. These notes contain the de-identified records of 38,597 patients.

Annotations were done by two board-certified radiologists and two additional clinicians pursuing their residency programs.

2.3 Dataset Statistics

The MedNLI dataset used over 4 clinicians working on a total of 4,683 premises over a period of 6 weeks with 14,049 unique sentence pairs. The dataset was then split into training, development, and test sets. The class distribution is even across all classes, throughout training, development and test sets (Table 2).

2.4 MEDIQA Shared Task

MEDIQA(Ben Abacha et al., 2019) is a shared task which is part of BIONLP 2019. It was cre-ated by using an annotation technique similar to MedNLI. It serves as an additional test for the MedNLI data. It contains 405 premise-hypothesis pairs. These pairs were randomly sampled from records, segmented from *Past Medical History* section with a simple rule-based method.MedNLI train set is used to train the model and hyper parameter are tuned based MedNLI development and test set accuracy. MedNLI and MEDIQA test set follows the same label mapping.

3 BERT

3.1 Description

Bidirectional encoder representation from transformer (Vaswani et al., 2017) is a language representation model which performs on a wide range of NLP tasks such as question answering and language inference. The architecture of the BERT leverages the use of pre-trained deep bidirectional representations. Existing pre-trained language representations include feature-based (ELMO) (Peters et al., 2018) and fine-tuning approach (OpenAI GPT) (Radford et al., 2018) . However, these models are severely restricted due to their unidirectional nature. BERT uses masked language models to enable pre-trained deep bidirectional representations.

The BERT model the authors experimented with, is $BERT_{BASE}$. The model is composed of 12 transformer blocks with a hidden size of 768 and 12 attention heads. The feed-forward/filter size is 4 times the hidden size. For fine tuning on MedNLI, a classification layer is added and all the parameters of the final model are fine-tuned jointly as per the original paper (Devlin et al., 2018).

3.2 BERT on MedNLI

BERT displays a clear supremacy over contemporary architectures (Radford et al., 2018) (Peters et al., 2018) on several NLP tasks. BERT's use of bidirectional encoders is a characteristic feature that separates it from other architectures.

Natural language inference requires learning the relationship between two sentences which is not supported by naive language models. Thus, BERT which is pre-trained on binarized next sentence prediction is vital for NLI.

MedNLI is built based on GLUE (General Language Understanding Evaluation) dataset (Wang et al., 2018). The goal as of before, with inference is to predict how the first sentence is related to the next in terms of entailment, contradiction or neutral. MedNLI is a sequence level task. The model needs to learn a minimum number of parameters and is used with an additional output layer with BERT.

4 Experiments

All the experiments in this paper are done with BERT pre-trained on unlabelled biomedical data-BioBERT (Lee et al., 2019). Three pretrained models are available: One model is trained only on PubMed articles, one is trained on PMC articles and one trained on both PubMed and PMC articles.

BioBERT trained on PubMed articles was also finetuned with MIMIC III v1.4 III v1.4-III (Johnson et al., 2016) notes. MIMIC III v1.4-III is a de-identified biomedical corpus compared to PubMed articles. All the 18 HIPAA (Atchinson and Fox, 1997) identifiers are removed and masked with unique PHI (Protected Health Information) tags in MIMIC III v1.4. The reason for fine-tuning BERT on MIMIC III v1.4 is because the MedNLI is a small subset derived from MIMIC III v1.4 database. No special preprocessing for PHI elements present in MIMIC III v1.4 data was done. Furthermore, MedNLI training data also contain sentences with PHI mask similar to MIMIC III v1.4. Finally the fine-tuned model is trained on MedNLI dataset. Evaluation is performed on MedNLI test and dev sets. The trained model has also been evaluated on MEDIQA test set. The results are presented in table 4.

Fine tuning on BioBERT was done using TensorFlow with three GeForce GTX 1080Ti GPUs for 2 weeks. The model on MIMIC III v1.4 is trained with maximum sequence length 128 with batch size 32 and learning rate 2e-5 for 200,000 steps. The sequence length is limited such that it can fit into GPU memory. The pretraining data from MIMIC III v1.4 is prepared using scripts from the original BERT github repository (Devlin et al., 2018) with the default parameters. Further fine tuning on MedNLI task is done with one GeForce GTX 1080Ti GPU with 11 GB of RAM. One epoch on MedNLI takes around 3 minutes on a single GPU[1].

4.1 Hyperparameter Search

Learning Rate	2e-5,3e-5,5e-5 .
Max Sequence Length	128
Batch Size	16, 32
Warmup Proportion	0.1-0.3
Number of Epochs	3, 4, 5

Table 3: Hyperparameters

All of hyperparameter search is done with a fixed random seed of 42. Each iteration took an average of 3-4 minutes. A variant of Adam optimizer which selectively avoids applying weight decay to normalization layers, proposed in BERT(Devlin et al., 2018) paper is used. Only learning rate is tuned while all the other hyperparameters like β_1, β_2, L_2 weight decay are fixed at 0.9, 0.999 and 0.01 respectively.

4.2 Results

The results of the experiments with BERT pre-trained on PubMed, PMC and fine-tuned on MIMIC III v1.4, are tabulated in Table 4. Pre-training on PubMed and PMC gives similar results. Pretraining on both PubMed and PMC gives a slight increase in accuracy in both dev and test sets. Finally, BioBERT-MIMIC III v1.4, BERT pre-trained on PubMed, fine-tuned on MIMIC III v1.4 outperforms other models by roughly a 2% margin, and marks a new state-of-the-art for MedNLI. The same model when evaluating on MEDIQA (Ben Abacha et al., 2019) test set, gives an accuracy of 78.5%.

5 Visualizations

Vig et al (Vig, 2019) uses a visualization tool, bertviz (Vig, 2019), presents 6 patterns of attention observed in BERT. Each attention pattern as explained in (Vig, 2019), provides with intuition regarding the underlying mechanics of the model. In deep learning models which are notoriously

---

[1]The code is available at https://github.com/kamalkraj/biobert

Model	Dev(%)	Test(%)	MEDIQA(%)
BioBERT-PubMed	83.42	80.74	78.3
BioBERT-PMC	83.05	81.07	77.8
BioBERT-PubMed + PMC	83.22	81.92	78.1
BioBERT-MIMIC III v1.4	**85.16** (SOTA)	**83.45** (SOTA)	78.5

Table 4: Comparision of Results

known for their opaque nature, these intuitions offer a peek behind the curtains. bertviz, was subsequently used to visualize BioBERT-MIMIC III v1.4 before and after training on MedNLI task. In this section, some of the interesting patterns are presented which were observed by comparing and contrasting attention patterns before and after finetuning on MedNLI task. The distinct patterns that emerge from fine-tuning are heavily dependent on the nature of the task (NLI).

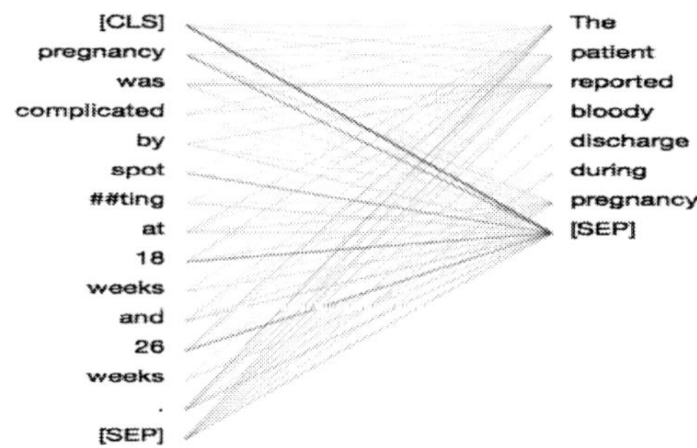

Figure 1: Distribution of Attention-Before

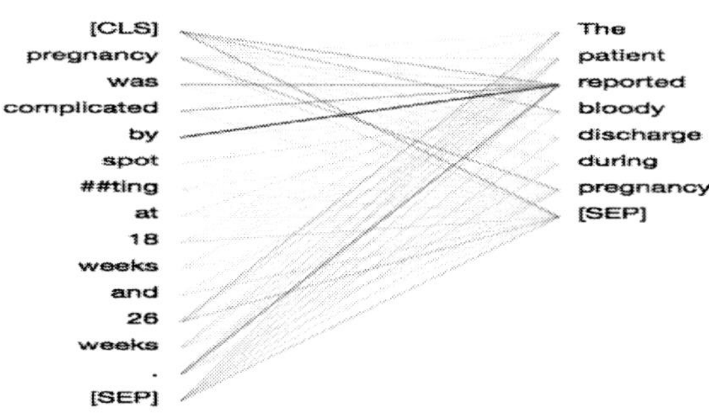

Figure 2: Distribution of Attention-After

1. **Distribution of Attention** Before training, majority of the attention is focused on the delimiter token of the second sentence [SEP], as seen in figure 1. After fine-tuning on MedNLI, the attention is distributed all over the second sentence as observed in figure 2. The dense connections seen in the figure, could be perceived as a natural consequence of fine-tuning the network on a NLI

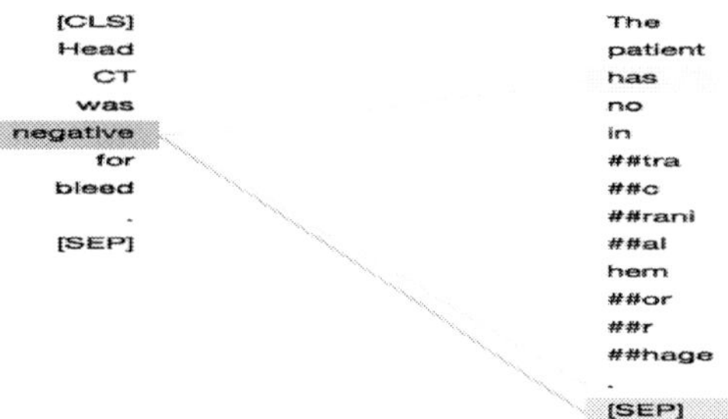

Figure 3: Word Similarity-Before

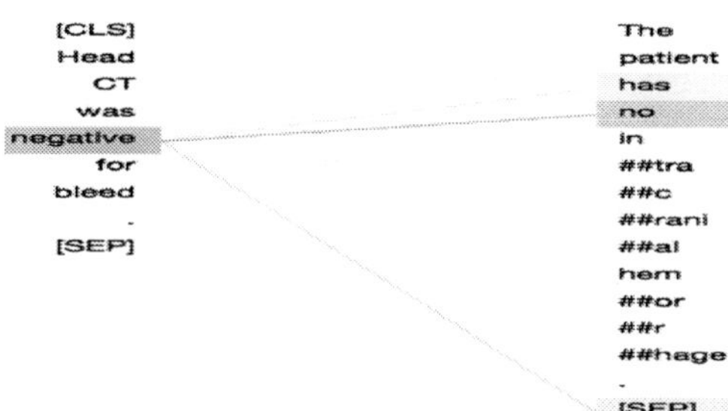

Figure 4: Word Similarity-After

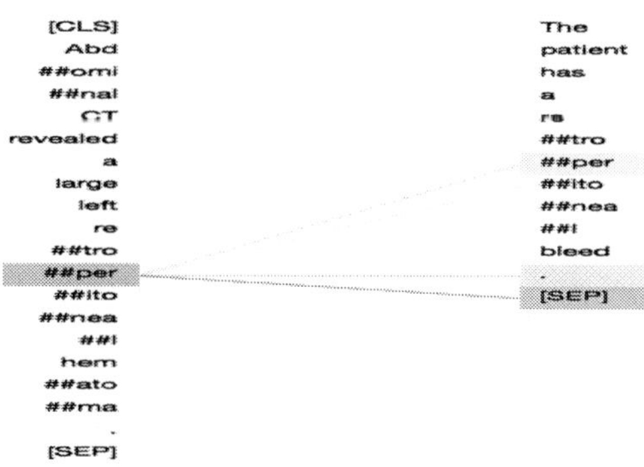

Figure 5: Tokenized Words-Before

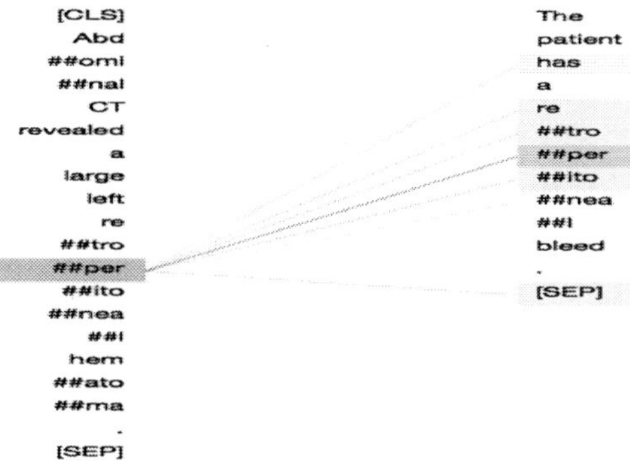

Figure 6: Tokenised Words-After

task, where establishing connections between two sentences is crucial.

2. **Word Similarity** It can be observed that

words similar to source word gets more attention. Notice the words *negative* and *no*, expressing similar sentiments (negative), connected via attention flow in figure 4. Word-level similarity, although not always, is a good indicator of entailment. Upon encountering sentences with similar words, it is reasonable for a network to be biased towards entailment.

3. **Tokenized Words** In BERT, OOV (Out of Vocabulary) words are identified and split into segments. This way, the morphological information is maintained, which comes in handy in tasks such as textual entailment where word-level similarity is an important aspect to notice. Before fine-tuning, the OOV (Out of Vocabulary) words split into multiple tokens receive weak attention from source tokens, as observed in figure 5. After fine-tuning on MedNLI, a strong attention flow between the tokenized words across two sentences can be seen. As mentioned above, these connections as seen in figure 6, help in identifying word-level similarity between sentences.

The authors have presented a error analysis study based on attention patterns in Appendix A. Based on the intuitions gained from error analysis, the authors propose a list of changes that could improve the performance of the model. A limitations of the proposed approach and a list of possible improvements are presented in Appendix B.

6   Conclusion

In this paper, a variant of BERT, fine-tuned on MIMIC III v1.4, is proposed to solve the task of MedNLI, a Natural Language Inference task designed for clinical domain. The experiments include evaluation of BERT pre-trained on PMC, PubMed and MIMIC III v1.4, on MedNLI test and dev sets, and MEDIQA test set for MedNLI. State-of-the-Art results (83.45%) in MedNLI is achieved by pre-training BERT on PubMed followed by fine-tuning on MIMIC III v1.4. The same model achieves an accuracy of 78.5% on MEDIQA test set. The authors have identified distinct attention patterns in BERT trained on MedNLI and have explored the origin and significance of those patterns in the context of NLI.

Acknowledgments

The authors would like to thank Bhuvana Kundumani for reviewing the manuscript and for providing her technical inputs. The authors would also like to extend their gratitude to Saama Technologies Inc. for providing the perfect research and innovation environment.

References

Brian K Atchinson and Daniel M Fox. 1997. From the field: The politics of the health insurance portability and accountability act. *Health Affairs*, 16(3):146–150.

Asma Ben Abacha, Chaitanya Shivade, and Dina Demner-Fushman. 2019. Overview of the mediqa 2019 shared task on textual inference, question entailment and question answering. In *Proceedings of the BioNLP 2019 workshop, Florence, Italy, August 1, 2019*. Association for Computational Linguistics.

Samuel R Bowman, Gabor Angeli, Christopher Potts, and Christopher D Manning. 2015. A large annotated corpus for learning natural language inference. *arXiv preprint arXiv:1508.05326*.

Alexis Conneau, Douwe Kiela, Holger Schwenk, Loic Barrault, and Antoine Bordes. 2017. Supervised learning of universal sentence representations from natural language inference data. *arXiv preprint arXiv:1705.02364*.

Jacob Devlin, Ming-Wei Chang, Kenton Lee, and Kristina Toutanova. 2018. Bert: Pre-training of deep bidirectional transformers for language understanding. *arXiv preprint arXiv:1810.04805*.

Qiao Jin, Bhuwan Dhingra, William W Cohen, and Xinghua Lu. 2019. Probing biomedical embeddings from language models. *arXiv preprint arXiv:1904.02181*.

Alistair E. W. Johnson, Tom J. Pollard, Lu Shen, Liwei H. Lehman, Mengling Feng, Mohammad Ghassemi, Benjamin Moody, Peter Szolovits, Leo Anthony Celi, and Roger G. Mark. 2016. Mimic-iii, a freely accessible critical care database. *Scientific Data*, 3:160035 EP –. Data Descriptor.

Jinhyuk Lee, Wonjin Yoon, Sungdong Kim, Donghyeon Kim, Sunkyu Kim, Chan Ho So, and Jaewoo Kang. 2019. Biobert: a pre-trained biomedical language representation model for biomedical text mining. *arXiv preprint arXiv:1901.08746*.

Matthew E. Peters, Mark Neumann, Mohit Iyyer, Matt Gardner, Christopher Clark, Kenton Lee, and Luke Zettlemoyer. 2018. Deep contextualized word representations. In *Proc. of NAACL*.

Alec Radford, Karthik Narasimhan, Time Salimans, and Ilya Sutskever. 2018. Improving language understanding with unsupervised learning. Technical report, Technical report, OpenAI.

Alexey Romanov and Chaitanya Shivade. Lessons from natural language inference in the clinical domain.

Chaitanya Shivade. 2019. Mednli for shared task at acl bionlp 2019,physionet.

Ashish Vaswani, Noam Shazeer, Niki Parmar, Jakob Uszkoreit, Llion Jones, Aidan N Gomez, Łukasz Kaiser, and Illia Polosukhin. 2017. Attention is all you need. In *Advances in neural information processing systems*, pages 5998–6008.

Jesse Vig. 2019. Visualizing attention in transformer-based language representation models. *arXiv preprint arXiv:1904.02679*.

Alex Wang, Amapreet Singh, Julian Michael, Felix Hill, Omer Levy, and Samuel R Bowman. 2018. Glue: A multi-task benchmark and analysis platform for natural language understanding. *arXiv preprint arXiv:1804.07461*.

Adina Williams, Nikita Nangia, and Samuel R Bowman. 2017. A broad coverage challenge corpus for sentence understanding through inference. *arXiv preprint arXiv:1704.05426*.

A   Error Analysis

The authors have studied the misclassified examples in MedNLI (test set) and MedQA (task set). 70% of the misclassified examples are falsely labelled as Contradiction. The confusion matrix consisting of the count of misclassified examples for both the sets are presented in figures 9 and 10. The common pattern that exists in misclassified examples, is the model's lack of understanding of certain tokens that are crucial for relating the premise to the hypothesis. Consider the example presented below.

Premise : "Reports lack of appetite but no n/v." Hypothesis : "the patient denies nausea and vomiting"

The abbreviation *n/v* in the premise expands to nausea and vomiting. The hypothesis contains the expanded form *nausea and vomiting*. It is clear from observing the attention pattern (figure 7) that the model doesn't identify *n/v* and *nausea and vomiting* as same concepts. When the abbreviation in the premise was expanded to nausea and vomiting, the model identified them as same concepts which is clearly evident from figure 8. Based

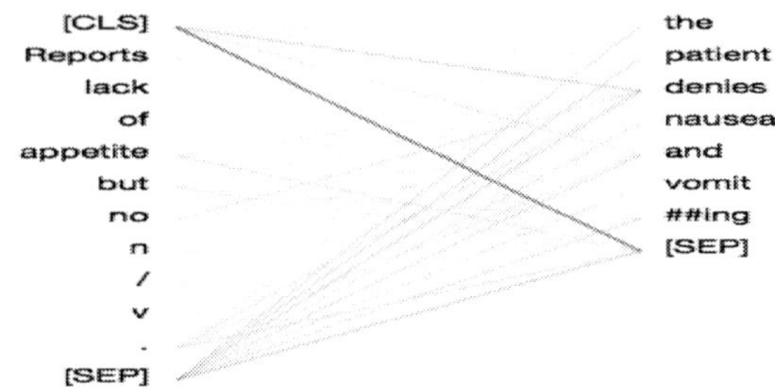

Figure 7: Attention distribution pattern for the example presented in section A without preprocessing

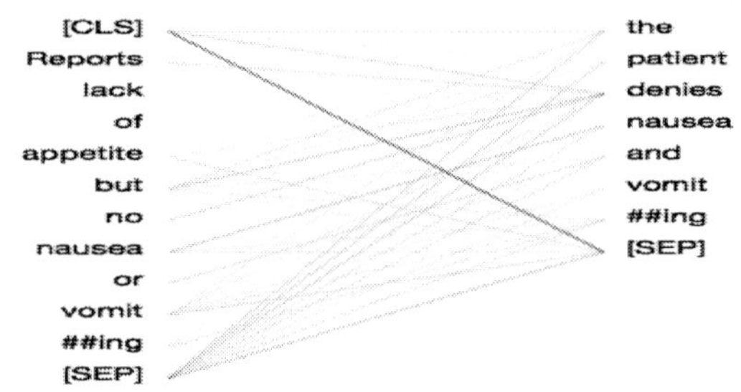

Figure 8: Attention distribution pattern for the example presented in section A after preprocessing

on this intuition, the authors suggest a preprocessing step, that expands and normalizes abbreviated terms.

B   Limitations and Future Work

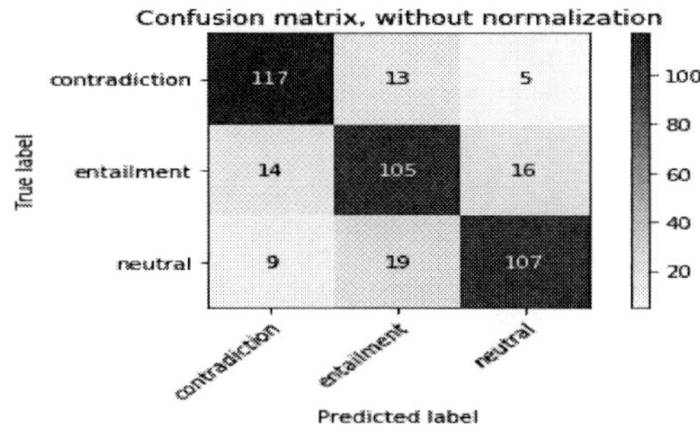

Figure 9: Confusion matrix of misclassified examples from MEDIQA test set

One of the limitations of this work is the lack of text preprocessing. The only preprocessing step followed by the authors is tokenization. In domain-specific tasks like Medical NLI, it would be beneficial to identify and normalize medical concepts which could be represented in more than one form. The other significant limitation is that the sentences are tokenized based on a 30,000 size vocabulary derived from Wikipedia corpus. Al-

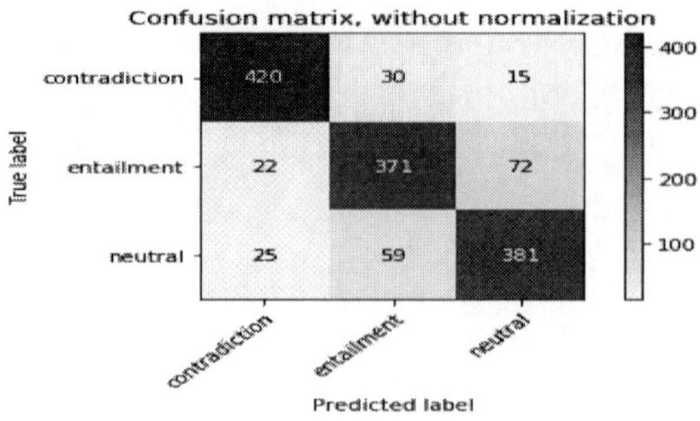

Figure 10: Confusion matrix of misclassified examples from MedNLI task set

though the fine-tuning is done on Pubmed, the commonly occurring medical terms are identified as unknown words and split into tokens.

The authors suggest a preprocessing step that identifies and normalizes medical concepts. The vocabulary could be built based on PubMed corpus which ensures that most common medical terms are part of the vocabulary. Along those lines, the authors suggest the use of entity embeddings to learn medical concepts and make use of the information contained in them.

516

IITP at MEDIQA 2019: Systems Report for Natural Language Inference, Question Entailment and Question Answering

Dibyanayan Bandyopadhyay[1]   **Baban Gain**[1]   **Tanik Saikh**[2]   **Asif Ekbal**[2]

Government College Of Engineering And Textile Technology, Berhampore[1]

Indian Institute of Technology Patna[2]

{dibyanayan, gainbaban}@gmail.com[1]

{1821cs08, asif}@iitp.ac.in[2]

Abstract

This paper presents the experiments accomplished as a part of our participation in the MEDIQA challenge, an (Abacha et al., 2019) shared task. We participated in all the three tasks defined in this particular shared task. The tasks are *viz. i. Natural Language Inference (NLI) ii. Recognizing Question Entailment(RQE)* and their application in medical *Question Answering (QA)*. We submitted runs using multiple deep learning based systems (runs) for each of these three tasks. We submitted five system results in each of the NLI and RQE tasks, and four system results for the QA task. The systems yield encouraging results in all the three tasks. The highest performance obtained in NLI, RQE and QA tasks are 81.8%, 53.2%, and 71.7%, respectively.

1 Introduction

Natural Language Processing (NLP) in biomedical domain is an essential and challenging task. With the availability of the data in electronic form it is possible to apply Artificial intelligence (AI), machine learning and deep learning technologies to build data driven automated tools. These automated tools will be helpful in the field of medical science. An ACL-BioNLP 2019 shared task, namely the MEDIQA challenge aims to attract further research efforts in NLI, RQE and their application in QA in medical domain. The motivation of this shared task is in a need to develop relevant methods, techniques, and gold standard data for inference and recognizing question entailment in medical domain and their application to improve domain specific Information Retrieval (IR) and Question Answering (QA) systems. The MEDIQA has defined several tasks related to *Natural Language Inference, Question Entailment and Question Answering* in medical do-

main. We participated in all the three tasks defined in this shared task. We offer multiple systems for each the tasks. The workshop comprises of three tasks namely *viz. i. Natural Language Inference (NLI): This task involves in identifying three inference relations between two sentences: i.e. Entailment, Neutral and Contradiction (Romanov and Shivade, 2018) ii. Recognizing Question Entailment (RQE): This task focuses on identifying entailment relation between two questions in the context of QA. The definition of question entailment is as follows: "a question A entails a question B if every answer to B is also a complete and/or partial answer to A" (Abacha and Demner-Fushman, 2019) and iii. Question Answering (QA): The goal of this task is to filter and improve the ranking of automatically retrieved answers. The existing medical QA system namely CHiQA is applied to generate the input ranks. (Harabagiu and Hickl, 2006; Abacha et al., 2017; Abacha and Demner-Fushman, 2019).* We participated in all the three tasks defined above and submitted the results. Our proposed systems produce encouraging results.

2 Proposed Method

We propose multiple runs for each of the three tasks. The following subsections will discuss the methods applied to tackle each of these tasks.

2.1 Natural Language Inference

In the task 1 the system has to decide the entailment relationship between a pair of texts i.e. either they are *Entailment, Contradiction or Neutral*. The input to this task are sentence pairs and as output we wish to get the entailment relation between those two pieces of texts. We propose five runs for this task. The following set of hyper parameters are applied for the following runs. **Batch**

Proceedings of the BioNLP 2019 workshop, pages 517–522

Florence, Italy, August 1, 2019. ©2019 Association for Computational Linguistics

Size = 32, **Learning Rate** = 2e-5, **Maximum Sequence Length** = 128, **number of epochs** = 10. The following points will discuss the approaches (i.e. runs).

Run 1: Our first proposed method is based on a BioBERT (Lee et al., 2019) model, i.e. a Bidirectional Encoder Representation from Transformer model pre-trained on biological data (both on PubMed abstracts and PMC full-text articles). After getting the vector corresponding to the special classification token ([CLS]) from final hidden layer of this mode, we use it for classification. We use 2 dense (feed forward) layers and a softmax activation function at the end. Only the feed forward part is trained end to end for 10 epochs after getting output vector from BioBERT. In this method no fine tuning is used. This method yields an accuracy of 60.8%.

Run 2: The second approach is based on the Bidirectional Encoder Representation from Transformer (BERT) model (bert-base-uncased) (Devlin et al., 2018). We make use of this to get the embedding of the inputs to this system. Instead of using 2 feed forward layers at the end, we choose to use only 1 feed forward layer. The full model is then trained in an end to end manner. All of the parameters of BERT and the last feed forward layer are fine-tuned jointly for 10 epochs to maximize the log-probability of the correct label (Entailment, Neutral or Contradiction). This model produces an accuracy of 71.7%.

Run 3: We use a BioBERT model for this system. This BioBERT model is pre-trained on PubMed abstracts only. We apply only one feed forward layer at the end. The full model is fine tuned as described for the system in run 2. This method gives output with an accuracy of 77.1%.

Run 4: The system proposed in this run is same as the model of Run 3. The differences between them are as follows:

- The BioBERT model we used is a pre-trained model on both the PubMed abstracts and PMC full-text articles instead of only PubMed abstracts as in case of in run 3.

- Here, we combine the full dataset of MedNLI (14049 sentence pairs) for training. Whereas in the previous run we made use only 11232 sentence pairs for training.

Following these changes in run 4, the accuracy increases from 77.1% in run 3 to 80.3% in run 4.

Run 5: This model is the combination of three BioBERT models. Two of them are pre-trained on both the PubMed abstracts and PMC full text articles and the third one is pre-trained only on PubMed abstracts. We fine tune each of the models following the fine tuning process of run 4. We ensemble their predictions by voting each of them for a sentence pair. The label which gets the most vote is selected for final prediction. The accuracy increases to 81.8%.

2.2 Recognizing Question Entailment (RQE)

Recognizing Question Entailment is an important task. The objective of this task is to identify entailment between the two questions in the context of Question-Answering(QA). We use the following definition of question entailment: a question A entails a question B if every answer to B is also a complete or partial answer to A. We make use of the dataset provided by the task organizers'. We submit five runs which are broadly based on two approaches. The approaches are as follows:

- One is based on Siamese architecture (Mueller and Thyagarajan, 2016). This Siamese is based on the recurrent architectures for learning sentence similarity (Mueller et al., 2016). Here we feed the two questions (inputs) to two Bidirectional Long Short Term Memory (Bi-LSTM) (Hochreiter and Schmidhuber, 1997), respectively. Both of their weights are initialized to the same. After obtaining the last hidden representations from both of these Bi-LSTMs, we concatenate them. This vector represents our input sentence pair. We feed this vector to a feed forward neural network layer. At the end, there is a softmax layer to perform a 2-class classification (to Yes/No).

- In another one, we train (fine-tuned) a BioBERT model as described in the NLI task. We used the BioBERT model to perform a 3-way classification of a sentence pair into entailment, neutral or contradiction. The same approach is used here in RQE to classify a pair of questions into Yes or No. The hyper-parameters used in fine-tuning the BioBERT model are same as of task 1 (NLI) except here the training iteration is (i.e. epoch) 5. This is done so because the loss is decreasing rapidly between two training epochs, indicating over fitting on the train set.

Our proposed approaches are based on these methods with slight variations. The following points will show them.

Run 1: Each question pair is having two questions (namely chq and faq). We assume the first question as the premise and the second one as the hypothesis. We extract these two questions from the training set. We obtain the vector representation of each word using Gensim Word2Vec (Řehůřek and Sojka, 2010). The vector size is 50. Then vector representations of the words for both the question are fed to Siamese Network of Bidirectional LSTMs. We train the model with 50 epoch and achieve 53.2% accuracy in test set.

Run 2: In this method we make use of the BioBERT model. The model is pre-trained on PubMed abstracts and PMC full text articles. This task is essentially a sentence pair classification task. For each sentence pair we obtain a vector corresponding to ([CLS]) token at the last layer. The vector is subsequently fed into two dense layers followed by a final layer having softmax activation function layer. No fine-tuning is used here. We obtain an accuracy of 50.6%.

Run 3: Here also we use BioBERT, but it is pre-trained on PubMed abstracts only. We fine tune the model on the RQE training set consisting of 8588 pairs. A feed forward layer with final layer with softmax activation is used at the end for 2-way classification. We obtain an accuracy of 48.1%.

Run 4: Instead of training word vector representation from scratch using Gensim Word2Vec, as in run 1, we obtain the vector representations of words from a trained Google News corpus (3 billion running words) word vector model (3 million 300-dimension English word vectors). The architecture is same as what is there in the run 1. We obtain an accuracy of 50.2%.

Run 5: We use a BioBERT model pre-trained on both PubMed abstracts and PMC full text articles. Then we fine tune on the RQE train set. Everything else is same as in run 3. The accuracy decreases to 48.9%.

2.3 Question Answering (QA)

The objective of this Question Answering task is to filter and improve the ranking of automatically retrieved answers. The input ranks are generated by the existing medical QA system *CHiQA*. We use BERT to predict the reference score between pairs and BM25 (Robertson et al., 2009) to rank between them. First of all, the BERT is used as a sentence pair classifier model. The first token of every sequence is always the special classification token ([CLS]). The final hidden state (i.e., output of Transformer) corresponding to this token is used as the aggregate sequence representation for classification tasks. This final hidden state is a 768 dimensional vector (for bert-base) representing the input sentence pair. This vector is fed subsequently into one or more feed-forward layers with softmax activation function layer. We fine tune the whole system for 10 epochs to predict the reference score of test dataset. The predicted values of these reference score are subsequently used in the BM25 model. *All the hyper parameters setting are same as in Task-1 except here a batch size of 28 is used because the maximum length of sequence is increased from 128 to 256.* It is to be noted that, it is too much memory consuming to train a BERT model with a batch size of 32 and a maximum sequence length of 256. We propose four runs to combat this problem.

Pre-processing: The training dataset is divided into two files (QA-TrainingSet1-LiveQAMed and TrainingSet2-Alexa) both are containing 104 questions. They had 8.80 and 8.34 answers for every question on average, respectively. For a question with N answers are converted to N pairs with each pair containing the question, one of the answers, and their reference score.

Run 1: All reference scores from training dataset are replaced by 1 if it is 3 or 4, and with 0 otherwise. The range of reference score (as given) is between 1 to 4 in dataset. Here we fine tune a BERT model (bert-base-uncased) with a feed forward layer at the end for 10 epochs to classify a sentence pair into 2 labels (0 or 1). The trained model is then used to predict reference score of test set. From the predicted result obtained from BERT, BM25 score for every question and their corresponding answers are calculated. All answers for a question whose predicted labels are 1 ('YES') are sorted in decreasing order of their BM25 scores. After that all the 'YES' labels are retrieved, and the same procedure is applied for all answers for the same question whose predicted label is 0 (NO). The obtained accuracy is 57.3%.

Run 2: All reference scores are kept intact, i.e. between 1 to 4. Here we use the BERT model (bert-base-uncased) and fine tune it on train set with a feed forward net at the end. From the predicted result obtained from BERT, pairs whose reference score is 4 or 3 are marked as 'YES' and whose reference score are 2 or 1 are marked as 'NO'. BM25 score for every question and their corresponding answer is calculated. All answers for a question whose predicted label is 4 are sorted by decreasing order of their BM25 score. Same procedure is applied for all answers for the same question whose predicted label is 3,2 and 1, respectively. We obtain an accuracy of 65.1% in this run.

Run 3: Here the validation dataset is also included to the training set. We merged them. Instead of using a BERT model here we use BioBERT model which is pre-trained on PubMed abstracts and PMC full text articles. We fine tune this model as explained in run 1. The rest of the procedures are same as in run 2. The accuracy increases to 67.8%.

Run 4: This method is an ensemble of 5 BioBERT (PubMed-PMC) models and fine tuned on the train dataset. Each of the models is then evaluated on the validation set (which is included in training set of Run 3). It is seen that one of those models performs well than the ensemble of 5 models. The model is then used to predict reference score of the test set. The rest of the procedures is same as what is there in the Run 3. The accuracy is 71.7% for this run.

3 Experiments, Results and Discussions

We submitted system results (runs) for all the three tasks. In all these tasks, we make use of the dataset released as a part of this shared task. In the following we discuss the dataset, evaluation results and the necessary analysis of the results obtained.

Data: In the NLI task, the training and test instances are having 14049 and 405 number of sentence pairs, respectively. In task 2 (i.e. RQE), the training set is having 8588 number of pairs, out of which 4655 and 3933 pairs are having *True* and *False* class, respectively. The validation and test set are having 302 (true: 129 and false: 173) and 230 (true: 115 and false: 115) number of instances. In the QA task, training sets are provided

Runs	Result(Accuracy(%))
1	60.8
2	71.7
3	77.1
4	80.3
5	**81.8**

Table 1: Submission results of all the five runs for the NLI task (Task-1)

from two domains *viz. LiveQAMed and ii. Alexa,* each having 104 number of questions and at an average of 8.80 and 8.34 number of answers per question. There are 25 number of questions and at an average of 10.44 answers per question are there in the validation set. The test set for this task is having 150 question pairs and on an average 8.5 answer per question.

Task 1(NLI): In the first task, we propose five runs. In all the tasks, we make use of either BERT or BioBERT models. We merge the input sentences pairs into a single sequence having maximum length of 128. They are separated by a special token ([SEP]). The first token of every sequence is always a special classification token ([CLS]). The final hidden state (i.e., output of Transformer) corresponding to this token is used as the aggregate sequence representation for the classification tasks. This final hidden state is a 768 dimensional vector (for bert-base) representing the input sentence pair. This vector is fed subsequently into one or more feed-forward layers with soft-max activation at the end for 3-way classification (Entailment, Neutral or Contradiction). The results for this task are shown in the Table 1. We have discussed the way we can use a BERT model to perform sentence classification in medical domain. It is observed that an absolute improvement of 5.4% in accuracy has been achieved by using a BioBERT (pre-trained on PubMed abstracts) model in run 3 instead of using the original BERT-base-uncased model (Pre-trained on Wikipedia and Book Corpus (as used in run 2)). The increase in result may be the effect of BioBERT, because the other experimental set up remain same. The reason for using 1 feed forward layer at the end of BERT models in all the runs except the run 1 (no fine tuning), using only one feed forward layer was putting the model into an under fitting state. While in case of fine tuning a large model, one feed forward is enough

Runs	Result(Accuracy(%))
1	**53.2**
2	50.6
3	48.1
4	50.2
5	48.9

Table 2: Submission results in all the five runs for the RQE Task (Task-2)

Runs	Results			
	Accuracy(%)	Spearmans Rho	MRR	Precision
1	57.3	0.053	0.8241	0.5610
2	65.1	0.042	0.7811	0.7235
3	67.8	0.034	0.8366	0.7421
4	**71.7**	0.024	0.8611	0.7936

Table 3: Results obtained in all the four runs for the QA Task (Task - 3), where, MRR: Mean Reciprocal Rank

as suggested by (Devlin et al., 2018). Up to run 3, we make use of 11232 sentence pairs for the training. Those sentence pairs are same as the one used to train several models used in (Romanov and Shivade, 2018). We use the remaining 2817 sentence pairs for validation. The validation set accuracy is always around 3-4% higher than the test case accuracy for all the runs up to run 3. For getting the higher accuracy we combine all the 14049 pairs in the subsequent run. We get the accuracy of 81.8 % which is the highest among all the proposed methods. As per our knowledge, in the official results of NLI task we stand at 12th position among the 17 official teams which participated for the NLI task.

Task 2 (RQE): In the second task i.e. task of Recognising Question Entailment, we propose five runs. The results are shown in the Table 2. It is interesting to note the variation in accuracy for the different runs. Siamese architecture performs much better here. Another peculiarity is that fine tuning BERT hurts the performance while using pre-trained BERT embedding without fine tuning seems to be more useful. This is concluded by observing the results of run 2 and run 3. In run 2, we used only pre-trained BERT embedding for ([CLS]) token for classification, whereas in run 3, we fine tuned the BERT model. The highest accuracy is achieved by a Siamese Model consisting of 2 Bi-LSTMs with shared weights and a dense layers. In this task, 12 teams submitted their systems, and we stood the 10th position.

Task 3 (QA): In this task, we offer 4 runs to tackle the problem. The results for this are shown in the Table 3. As we can see from the above discussions, the systems we build for this task comprises of two components, they are BERT and BM25. The BERT is used to predict the reference score of the test dataset. We rank the predicted scores using BM25. The BM25 part of the system is same for all the runs. In this task, participants are encouraged to compute the Mean Reciprocal Rank (MRR), Precision, and Spearman's Rank Correlation Coefficient as the evaluation measures in addition to Accuracy. We actually used BioBERT instead of original BERT from the run 3, which increases the accuracy with an absolute margin of 2.7% (65.1 to 67.8%). Using BioBERT we observe an improvement in MRR by 5.5%. Our best run with an accuracy of 71.7% attains the position of 6th among 10 teams in the official result.

4 Conclusion and Future Work

In this paper, we present our system details and the results of various runs that reported as a part of our participation in the MEDIQA challenge. In this shared task three tasks, namely *viz. i. Natural Language Inference ii. Question Entailment and iii. Question Answering* were introduced in the medical domain. We offer multiple systems (runs) for each of these tasks. Most of the proposed models are based on BERT/Bio-BERT embedding and BM25. These models yields encouraging performance in all the tasks.In future we would like to extend our work as follows:

- Detailed analysis of the top-scoring models to understand their techniques and findings.

- We can do the task of NLI by fostering an *Embedding from Language model (EMLo)* based model and do a comparative analysis with BERT based model.

References

Asma Ben Abacha, Eugene Agichtein, Yuval Pinter, and Dina Demner-Fushman. 2017. Overview of the medical question answering task at TREC 2017 LiveQA. In *Proceedings of The Twenty-Sixth Text REtrieval Conference, TREC*, pages 15–17.

Asma Ben Abacha and Dina Demner-Fushman. 2019. A question-entailment approach to question answering. *arXiv preprint arXiv:1901.08079.*

Asma Ben Abacha, Chaitanya Shivade, and Dina Demner-Fushman. 2019. Overview of the MediQA 2019 shared task on textual inference, question entailment and question answering. In *Proceedings of the BioNLP 2019 workshop, Florence, Italy, August 1, 2019.* Association for Computational Linguistics.

Jacob Devlin, Ming-Wei Chang, Kenton Lee, and Kristina Toutanova. 2018. BERT: pre-training of deep bidirectional transformers for language understanding. *CoRR*, abs/1810.04805.

Sanda Harabagiu and Andrew Hickl. 2006. Methods for using textual entailment in open-domain question answering. In *Proceedings of the 21st International Conference on Computational Linguistics and 44th Annual Meeting of the Association for Computational Linguistics*, pages 905–912, Sydney, Australia. Association for Computational Linguistics.

Sepp Hochreiter and Jürgen Schmidhuber. 1997. Long short-term memory. *Neural computation*, 9(8):1735–1780.

Jinhyuk Lee, Wonjin Yoon, Sungdong Kim, Donghyeon Kim, Sunkyu Kim, Chan Ho So, and Jaewoo Kang. 2019. BioBERT: pre-trained biomedical language representation model for biomedical text mining. *arXiv preprint arXiv:1901.08746.*

Jonas Mueller and Aditya Thyagarajan. 2016. Siamese recurrent architectures for learning sentence similarity. In *Thirtieth AAAI Conference on Artificial Intelligence.*

Radim Řehůřek and Petr Sojka. 2010. Software Framework for Topic Modelling with Large Corpora. In *Proceedings of the LREC 2010 Workshop on New Challenges for NLP Frameworks*, pages 45–50, Valletta, Malta. ELRA.

Stephen Robertson, Hugo Zaragoza, et al. 2009. The probabilistic relevance framework: BM25 and beyond. *Foundations and Trends® in Information Retrieval*, 3(4):333–389.

Alexey Romanov and Chaitanya Shivade. 2018. Lessons from natural language inference in the clinical domain. In *Proceedings of the 2018 Conference on Empirical Methods in Natural Language Processing*, pages 1586–1596, Brussels, Belgium. Association for Computational Linguistics.

LasigeBioTM at MEDIQA 2019: Biomedical Question Answering using Bidirectional Transformers and Named Entity Recognition

Andre Lamurias* and Francisco M. Couto

LASIGE, Faculdade de Ciências, Universidade de Lisboa, Portugal

Abstract

Biomedical Question Answering (QA) aims at providing automated answers to user questions, regarding a variety of biomedical topics. For example, these questions may ask for related to diseases, drugs, symptoms, or medical procedures. Automated biomedical QA systems could improve the retrieval of information necessary to answer these questions. The MEDIQA challenge consisted of three tasks concerning various aspects of biomedical QA. This challenge aimed at advancing approaches to Natural Language Inference (NLI) and Recognizing Question Entailment (RQE), which would then result in enhanced approaches to biomedical QA.

Our approach explored a common Transformer-based architecture that could be applied to each task. This approach shared the same pre-trained weights, but which were then fine-tuned for each task using the provided training data. Furthermore, we augmented the training data with external datasets and enriched the question and answer texts using MER, a named entity recognition tool. Our approach obtained high levels of accuracy, in particular on the NLI task, which classified pairs of text according to their relation. For the QA task, we obtained higher Spearman's rank correlation values using the entities recognized by MER.

1 Introduction

Question Answering (QA) is a text mining task for which several systems have been proposed (Hirschman and Gaizauskas, 2001). This task is particularly challenging in the biomedical domain since this is a complex subject as answers may not be as straightforward compared to other domains. However, clinical and health care information systems could benefit greatly from automated

biomedical QA systems, which could improve the retrieval of information necessary to answer these questions.

To help progress on this topic, the MEDIQA challenge proposed three tasks in the biomedical domain(Ben Abacha et al., 2019):

1. Natural Language Inference (NLI) - classify the relation between two sentences as either entailment, neutral or contradiction;

2. Recognizing Question Entailment (RQE) - classify if two questions are entailed with each other or not;

3. Question Answering (QA) - classify which answers are correct for a given answer and rank them.

We applied the same approach to all three tasks since they all could be modelled as text classification tasks. The objectives of the tasks were to classify pairs of text: sentence-sentence (NLI), question-question (RQE), and question-answer (QA). For the NLI task, we had three possible labels for each pair (entailment, neutral or contradiction), while the RQE task was a binary classification. For the QA task, each pair should be given a reference score representing how well the question is answered, which ranged between 1 and 4.

QA is a complex task that involves various components, and can be approached in several ways. While real-world scenarios require the retrieval of correct answers from larger databases, the QA task of this challenge simplified this problem by providing up to 10 answers retrieved by the medical QA system CHiQA. This system also provided a ranking to each answer, however, we observed that this ranking did not follow the manual ranking in most cases. We also observed that the retrieved

---

*alamurias@lasige.di.fc.ul.pt

Proceedings of the BioNLP 2019 workshop, pages 523–527
Florence, Italy, August 1, 2019. ©2019 Association for Computational Linguistics

answers could consist of one or more sentences. While in some QA scenarios, systems are required to select the text span that contains the answer (Rajpurkar et al., 2016), in this case it was only requested to re-rank the retrieved answers and classify which ones were correct. Although specific ranking algorithms exist (Radev et al., 2000), due to the nature of the task and the fact that the other two tasks involved comparison of text, we decided to train a classifier that compared each question with a potential answer, i.e., we predicted how good a text is at answering a given question.

Our approach uses pre-trained weights as a starting point, to fine-tune deep learning models based on the Transformer architecture for each of the challenge tasks (Vaswani et al., 2017). We used the BioBERT weights, trained on PubMed abstracts and PMC full articles, as the type of text should be more similar to the challenge data than the standard BERT models, which were trained on Wikipedia and BookCorpus. Furthermore, we incorporated other datasets into the RQE and QA tasks, and enrich the training data with semantic information obtained using MER (Minimal Named-Entity Recognizer) (Couto and Lamurias, 2018), a high computing performance named entity recognition tool.

2   Related Work

Deep learning approaches have lead to state-of-the-art results in various text mining tasks. These approaches make use of intermediary representations of the data to then fine-tune the weights to different tasks. Various models have been proposed, and, recently, the most successful ones have been based around the Transformer architecture (Vaswani et al., 2017). An advantage of this type of models is that we can use pre-trained weights such as those provided by BERT (Devlin et al., 2018) as a starting point to train a model for a specific task. These weights are tuned on large corpora using the Transformer architecture and have been shown to be effective language models. Different models were made available by the authors, with two variations of the architecture, and whether the true case and accent markers of the tokens are taken into account.

Due to the effectiveness of the BERT architecture, it has been already adapted for other domains. Lee et al. (2019) presented a model specific to biomedical language, which was trained on a large-scale biomedical corpora: 200k PubMed abstracts, 270k PMC full texts, and a combination of these two. Although the BioBERT models use the same vocabulary as the BERT models, the same WordPiece tokenization is performed. This way, even if biomedical documents contain words that were not in the original vocabulary, the tokenizer will separate these words into frequent subwords, minimizing out-of-vocabulary issues and keeping compatibility with the original models. The authors tested these models on several biomedical text mining tasks, obtaining competitive performance when compared with other state-of-the-art models.

One of the most common text mining tasks is entity recognition. This task is important because it is often the first step to other tasks, such as entity linking and relation extraction. MER is a simple but efficient approach to entity recognition, which uses vocabularies that can be extracted from ontologies to identify and link entities. MER focuses on simplicity and flexibility to reduce the processing time and the time necessary to adapt to other domains and entity types.

3   Methodology

3.1   Data Preparation

We participated in the three tasks using the same approach by modeling each one as a text classification problem. We used the training data of each tasks as document pairs, where a document could be a sentence, paragraph, question or answer. The NLI and RQE data had obvious labels, while for the QA data we used the reference scores. However, to distinguish between correct answers with more detail, we also incorporated the manually assigned ranks to the answers with reference scores 3 and 4:

$$\text{FinalScore} = \text{ReferenceScore} + \frac{11 - \text{Rank}}{10}$$

As there are up to 10 possible answers to each question, the final score will range between 1 and 5.

We removed instances where each element of the pair contained the same text, which happened sometimes in the RQE training set. Furthermore, we performed named entity recognition using MER to identify several types of entities mention in both questions and answers. We used MER since it can provide reliable entity mention annotations at a reasonable speed. We appended the textual labels of the terms recognized to the end of the document, as a list separated by whitespaces. Since MER matches ontology concepts, if the synonym of a concept was recognized, it was converted to its main label.

We recognized terms from the: Human Phenotype Ontology, Disease Ontology, Chemical Entities of Biological Interest (ChEBI) ontology and Gene Ontology. Our objective was to add to each text a list of the entities that could summarize that text. We chose those ontologies because the questions were about biomedical subjects, and therefore the ontologies chosen should reflect the main domains of the data. The ontologies that we used comprise a total of 350,233 terms.

We also explored additional sources of data to train the classifiers, for the RQE and QA tasks. Regarding the RQE task, we employed the NLI dataset since it also contained entailment relations. Even though these datasets were generated from different corpora and the NLI dataset and for different purposes, we considered that additional data could still improve the results. To this end, we transformed the NLI dataset so that all entailment relations were labeled as positive, and the neutral and contradiction as negative.

For the QA task, we added one of the suggested MedQuAD datasets, namely the Cancer-Gov dataset. Although all these additional datasets had a similar structure, we did not have time to train and test which ones would be more helpful for this task. These datasets contained only examples of correct answers, which we assigned the reference score 4, since it could skew the trained classifier towards higher scores. To balance this, we generated incorrect answers from the other QA of the same document. We assumed that if an answer was correct for one question, it would be incorrect for the other questions about the same topic. To make sure this was true, we took into account the "qtype" parameter of each question, since it is unlikely that questions of different types would have the same answers. This parameter indicated the nature of the question in the context of the main topic of the document. For example, a document about a specific cancer type could have the following "qtypes": information, symptoms, exams and tests, outlook, and treatment.

Run	Training data	Dev	Test
1	NLI training set	0.836	0.724

Table 1: Accuracy obtained on the NLI task.

3.2 System architecture

We adapted the pytorch implementation of BERT[1]. As such, we used the WordPiece tokenization and Adam optimizer that are implemented by default. We used the BioBERT PubMed+PMC pretrained weights, which are based on the bert-base-cased model. The authors chose this model as many biomedical entities are case sensitive. We initially tested with the standard BERT weights, and observed an improvement when using the BioBERT weights instead. A model fine-tuned to the clinical domain, which is the domain of the documents of this challenge, would be more appropriate, but not such pre-trained model was available at the time.

Using the data previously described, we trained variations of the same model, focusing mostly on the RQE and QA tasks. These variations consisted of the additional datasets previously described, but also different training parameters, such as initial training rate, number of epochs, batch size and maximum sequence length. We started with the default values and made incremental changes to understand if we could improve the results on the validation set, while training just with the provided training set. After setting the best parameters, we then trained the classifiers on the additional datasets.

For the NLI, we tested only the baseline approach, which consisted in using the BioBERT weights fine-tuned for the task.

4 Results and discussion

We submitted one run to the NLI task, three runs to the RQE task and four runs to the QA task. We focused mainly on studying the effect of different training data on the performance of the classifiers.

We evaluated on the development sets that were provided for each task, and then submitted our predictions for the test sets. The scores obtained for the development and test sets of each task are shown in tables 1, 2 and 3, as well as the differences between each run.

---

[1] `https://github.com/huggingface/pytorch-pretrained-BERT`

Run	Training data	Dev	Test
1	RQE training set	0.732	0.481
2	RQE training set + NER	**0.752**	0.481
3	RQE and NLI training set	0.749	**0.485**

Table 2: Accuracy obtained on the RQE task.

We can see that the accuracy obtained during the development phase was considerably higher than on the test set. This could have been due to the test set containing other type of questions from the development set, or due to over-fitting of the hyper-parameters on the development set, which limited the performance of the model. Both on the test and development set, we obtained high accuracy on the NLI task, for which we submitted only one run. The NLI data was generated by asking experts to give one example of each class (neutral, entailment and contradiction) to a series of statements. As such, this dataset is highly regular and the model was able to learn from it.

On the development set, we can see that adding the named entities recognized by MER (Run 2 of RQE and QA) improved the accuracy. However, this effect did not occur on the test set; for the RQE task, it did not change the accuracy and it decreased the accuracy of the QA task. On the other hand, adding external training data (Run 3) had a positive effect on the test set results of both tasks, improving the accuracy of the QA task.

For the QA, we also trained a classifier using both the training and development datasets (Run 4). We could not evaluate this classifier on the development set since it had already seen those examples and the results would have been biased. However, this classifier achieved the best test set accuracy and Mean Reciprocal Rank (MRR) of the four runs submitted to this task.

The best results obtained with our approach were on the NLI task. However, we considered the QA task to be the main task of the challenge and put most effort into it in terms of exploration hyper-parameter tuning. Since the organizers considered the accuracy to be the main metrics, we optimized our system to that metric. While the MRR was high on all three runs, the Spearman's coefficient was generally much lower. This means that although our system was able to detect correct answers to a certain degree, their ranking matched poorly with the gold standard.

5   Conclusions and Future Work

For the MEDIQA challenge, we developed a system that could be used for the 3 proposed tasks with minimal changes. This was possible due to the recently introduced Transformer architecture, along with pre-trained weights that severely reduce the training time necessary to generate a language representation model. The training data provided for each task was used to train classification models for each task. We also explored external datasets to improve the models of the RQE and QA tasks. We observed that adding more data to train the model leads to better results on the test set, as expected.

In the future we will improve the capacity of the models to classify new data by adding more external training data. We observed that Runs 3 and 4 of the QA task achieved higher scores, which could have been due to the larger training set employed to train the models. While for Run 3 we used only one additional set, there were 9 more available of the same type, which were not used due to time constraints. A similar strategy could be used to find more pairs of questions with an entailment relation.

Another way to enrich the training set would be to automatically retrieve the descriptions of the entities identified in the text, or their ancestors, as they also provide useful information about entities. A similar approach was shown to improve the results of a relation extraction task using deep learning (Lamurias et al., 2019).

References

Asma Ben Abacha, Chaitanya Shivade, and Dina Demner-Fushman. 2019. Overview of the mediqa 2019 shared task on textual inference, question entailment and question answering. In *Proceedings of the BioNLP 2019 workshop, Florence, Italy, August 1, 2019*. Association for Computational Linguistics.

F. Couto and A. Lamurias. 2018. MER: a shell script and annotation server for minimal named entity recognition and linking. *Journal of Cheminformatics*, 10(58).

Jacob Devlin, Ming-Wei Chang, Kenton Lee, and Kristina Toutanova. 2018. BERT: Pre-training of deep bidirectional transformers for language understanding. *arXiv preprint arXiv:1810.04805*.

Lynette Hirschman and Robert Gaizauskas. 2001. Natural language question answering: the view from here. *natural language engineering*, 7(4):275–300.

Run	Training data	Dev			Test		
		Accuracy	Spearman	MRR	Accuracy	Spearman	MRR
1	QA training set	0.782	0.067	0.760	0.585	**0.220**	0.843
2	QA training sets + NER	**0.791**	**0.198**	0.840	0.551	0.026	0.733
3	QA training sets + CancerGov	0.756	0.183	**0.920**	0.600	0.201	0.870
4	QA training and dev sets	-	-	-	**0.637**	0.211	**0.910**

Table 3: Results obtained on the QA task. Spearman: Spearman's Rank Correlation Coefficient; MRR: Mean Reciprocal Rank.

A. Lamurias, Diana Sousa, L. Clarke, and F. Couto. 2019. BO-LSTM: classifying relations via long short-term memory networks along biomedical ontologies. *BMC Bioinformatics*, 20(10).

Jinhyuk Lee, Wonjin Yoon, Sungdong Kim, Donghyeon Kim, Sunkyu Kim, Chan Ho So, and Jaewoo Kang. 2019. Biobert: a pre-trained biomedical language representation model for biomedical text mining. *arXiv preprint arXiv:1901.08746*.

Dragomir R Radev, John Prager, and Valerie Samn. 2000. Ranking suspected answers to natural language questions using predictive annotation. In *Proceedings of the sixth conference on Applied natural language processing*, pages 150–157. Association for Computational Linguistics.

Pranav Rajpurkar, Jian Zhang, Konstantin Lopyrev, and Percy Liang. 2016. Squad: 100,000+ questions for machine comprehension of text. *arXiv preprint arXiv:1606.05250*.

Ashish Vaswani, Noam Shazeer, Niki Parmar, Jakob Uszkoreit, Llion Jones, Aidan N Gomez, Łukasz Kaiser, and Illia Polosukhin. 2017. Attention is all you need. In *Advances in neural information processing systems*, pages 5998–6008.

NCUEE at MEDIQA 2019:
Medical Text Inference Using Ensemble BERT-BiLSTM-Attention Model

Lung-Hao Lee*, Yi Lu, Po-Han Chen, Po-Lei Lee and Kuo-Kai Shyu
Department of Electrical Engineering, National Central University, Taiwan
Pervasive Artificial Intelligence Research (PAIR) Labs, Taiwan
*lhlee@ee.ncu.edu.tw

Abstract

This study describes the model design of the NCUEE system for the MEDIQA challenge at the ACL-BioNLP 2019 workshop. We use the BERT (Bidirectional Encoder Representations from Transformers) as the word embedding method to integrate the BiLSTM (Bidirectional Long Short-Term Memory) network with an attention mechanism for medical text inferences. A total of 42 teams participated in natural language inference task at MEDIQA 2019. Our best accuracy score of 0.84 ranked the top-third among all submissions in the leaderboard.

1   Introduction

Natural Language Inference (NLI) is the task of determining whether a given hypothesis is true (entailment), false (contradiction) or undetermined (neutral) by inferring a given premise. The Stanford Natural Language Inference (SNLI) corpus is a well-known dataset and serves as a benchmark for NLI system evaluations (Bowman et al., 2015). However, it is restricted to a single text genre. Therefore, the MedNLI dataset, which is annotated by doctors and grounded in patients' medical histories, was built to perform NLI tasks in the clinical domain (Romanov and Shivade, 2018). In addition to feature-based methods and bag-of-words (BOW) models, other experiments have tested several modern neural networks-based models for the specialized and knowledge intensive field of medicine, including InferSent (Conneau et al., 2017) and ESIM (Chen et al., 2017)

The MEDIQA challenge focuses on attracting research efforts in Natural Language Inference (NLI), Recognizing Question Entailment (RQE) and their applications in medical Question Answering (QA). The MEDIQA challenge includes three tasks: 1) NLI: identifying three inference relations between two medical sentences, that is, entailment, neutral and contradiction. 2) RQE: identifying entailment between two questions in the context of QA. 3) QA: filtering and improving the input ranks of retrieved answers, generated by the medical QA system CHiQA. The reuse of NLI and/or RQE systems for this task is highly recommended.

Under the policies of the MEDIQA challenge, we only participated in the first NLI task. Recently, a new method of pre-training language representations named BERT (Bidirectional Encoder Representations from Transformers) has obtained groundbreaking results on a wide array of natural language processing tasks (Devlin et al., 2018). This achievement motivates us to explore using a BERT based model to tackle the textual inference problem in the medical domain.

This paper describes the NCUEE (National Central University, Dept. of Electrical Engineering) system for the NLI task of the MEDIQA challenge at the ACL-BioNLP 2019 workshop. Our solution explores a BERT-based model, in which the BiLSTM network with attention mechanism is integrated for textual inference. The input sentence-pair is represented as a sequence of words. Each word refers to distributed vectors from a pre-trained BERT to form as an embedding matrix. The datasets provided by the task organizers are used to train the BiLSTM network with attention model for the prediction task. The output is a value from 0 to 1 representing the estimated class probability. The class with the highest probability (that is, one of entailment, neutral and contradiction) will be regarded as the inference result. Our best accuracy score of 0.84 ranked in the top-third of all 42 submissions in the leaderboard.

Proceedings of the BioNLP 2019 workshop, pages 528–532
Florence, Italy, August 1, 2019. ©2019 Association for Computational Linguistics

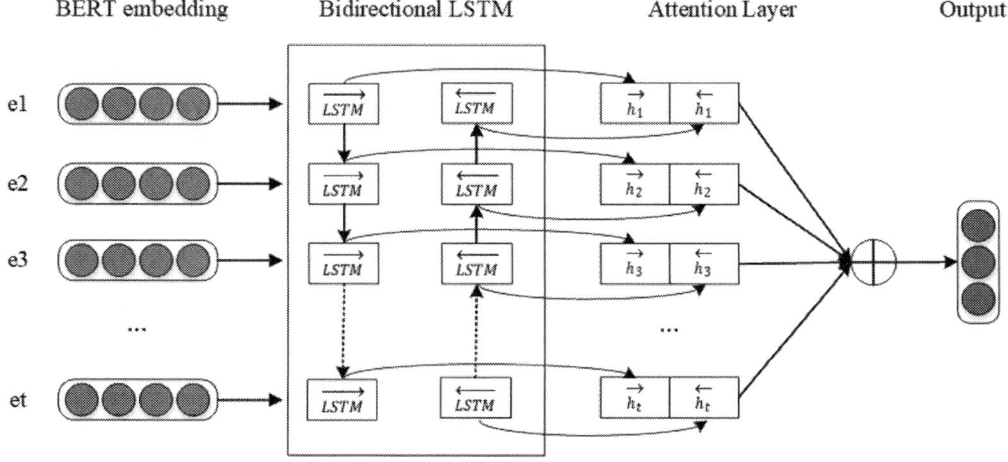

Figure 1: Our BERT-BiLSTM-Attention architecture for the NLI task.

The rest of this paper is organized as follows. Section 2 describes the NCUEE system for the NLI task. Section 3 presents the evaluation results and performance comparisons. Conclusions are finally drawn in Section 4.

2 The NCUEE System

Figure 1 shows our BERT-BiLSTM-Attention architecture for the NLI task. Our model consists of a BERT embedding layer, a BiLSTM layer and an attention layer. An input sentence-pair is represented as a sequence of words. Each word refers to a row looked up in a word embedding matrix from the last layer of the original BERT (Devlin et al., 2018). In this NLI task, we model the sentence-pairs using Bidirectional LSTM (Graves et al., 2013), an extension of the traditional LSTM, to train two LSTMs on the input pairs. The second LSTM is a reversed copy of the first one, so that we can take full advantage of both past and future input features for a specific time step. Consequently, we leverage the word attention mechanism to capture the distinguishing influence of the words and then form a dense vector (Yang et al., 2017). We then use final softmax activation function to classify the sentence-pairs to obtain the probability that belongs to each of the three classes.

During the training phase, if a sentence-pair (premise vs. hypothesis) is true (entailment), the class is assigned as 1, and 0 otherwise (contradiction). If both sentences are neutral without specific relationships, the class is assigned as 2. All training sentence pairs and their accompanying classes are used for training our BERT-BiLSTM-Attention model.

To classify a sentence pair during the test phase, we use the output probability as an indicator for classification. The class with the highest probability will be regarded as the inference result. In addition, ensemble strategies have been widely used in various research fields because of their good performance. This work uses a simple but efficient ensemble strategy called majority voting that involves selecting the class which has a majority, that is, more than half the votes from various trained models.

3 Evaluation

3.1 Data

The datasets were mainly provided by task organizers (Ben Abacha et al., 2019). The sentence-pairs for the NLI task were collected from the MedNLI dataset (Romanov and Shivade, 2018). The training, validation and test datasets were comprised of data from an independent set of sentence-pairs. During the system development phase, the training and validation sets respectively consisted of 12,627 and 1,422 sentence-pairs, for designing and implementing the system. In total, only 405 sentence-pairs in the test dataset were used for final performance evaluation.

The pre-trained word vectors are publicly available for download at the official BERT website [1]. We also used pre-trained weights of BioBERT (Lee et al., 2019), a language representation model for the biomedical domain and publicly available at the GitHub site[2].

3.2 Results

During system development phase, in addition to pre-trained word vectors, we only use the training set to train the system parameters and evaluate the result on the validation set.

In the first set of experiments, the following fine-tuning BERT models were compared to demonstrate their performance for classification.

- BERT-Base, Uncased: 12-layer, 768-hidden, 12-heads, 110M parameters.

- BERT-Base, Cased: 12-layer, 768-hidden, 12-heads, 110M parameters.

- BERT-Large, Uncased: 24-layer, 1024-hidden, 16-heads, 340M parameters.

- BERT-Large, Cased: 24-layer, 1024-hidden, 16-heads, 340M parameters.

Google Research has released the BERT-Base and BERT-Large models (12-layer/24-layer Transformer). Uncased means that the text has been lowercased before WordPiece tokenization and any accept markers have also been removed. Cased means that the true case and accent markers are preserved. The same setups are used for comparisons. The maximum sequence is 128. The training batch size is 32. The learning rate is 2e-5. The number of training epochs is 10.

Table 1 shows the results. The BERT-Large models achieved relatively better accuracy than the BERT-Base models, regardless of case-sensitivity.

The best accuracy was obtained by the Uncased BERT-Large model.

In the second set of experiments, the objective is to compare the performance of both BERT and BioBERT models. The BioBERT is based on the same vocabulary as the Cased BERT-Base model. Hence, we selected the pre-trained weights of BioBERT with PubMed 200K and PMC 270K comparing with the Cased BERT-Base model. Note that both models have been fine-tuned with optimal parameter settings.

Table 2 shows the performance comparisons, where the BioBERT outperforms the BERT model, suggesting that the BioBERT model is more suitable for biomedical text mining tasks through incorporating biomedical corpora such as PubMed and PMC.

In the third set of experiments, we evaluated our proposed model based on the previous results. Since the BERT-Large and BioBERT-Base achieved better accuracy, we fine-tuned these two models and sought to identify seek the optimal system parameters. Moreover, we adopted the last layers of these two models as the word embedding to integrate the BiLSTM network with attention mechanism. The setups of the BiLSTM is follows. The hidden size is 256. The dropout rate is 0.5.

Table 3 compares the results. The BioBERT models outperformed the BERT models, regardless of whether BioBERT was used as the word embedding or fine-tuning its original model. In addition, our integrated architecture with the BiLSTM-Attention was found to produce a slight

Models	Accuracy
Fine-tuning BERT-Base, Cased	0.792
Fine-tuning BioBERT-Base, Cased	0.822

Table 2: Results of BERT vs. BioBERT

Models	Accuracy
Fine-tuning BERT-Base, Uncased	0.759
Fine-tuning BERT-Base, Cased	0.751
Fine-tuning BERT-Large, Uncased	0.796
Fine-tuning BERT-Large, Cased	0.793

Table 1: Results of fine-tuning BERT models

Models	Accuracy
Fine-tuning BioBERT-Base	0.822
BioBERT-Base + BiLSTM-Attention	0.824
Fine-tuning BERT-Large	0.809
BERT-Large + BiLSTM-Attention	0.809

Table 3: Results of BERT-BiLSTM with attention

---

[1] https://github.com/google-research/bert

[2] https://github.com/naver/biobert-pretrained

performance enhancement. The best accuracy was obtained by the BioBERT-Base + BiLSTM-Attention model.

3.3 Comparisons

During final testing phase of the NLI task, we used the training set to train the models and the validation set for parameter optimization. Each participating team was allowed to submit a maximum of 5 runs for each task. We submitted the four abovementioned models accompanying with the ensemble model. For our ensemble strategy, we have trained the models 5 times using the BioBERT-Base + BiLSTM-Attention. The final inference result is the majority voting of the class with the highest probability.

Table 4 shows the results of our testing models. In addition to the BioBERT-Base model, the other models achieved promising accuracy. As expected, our ensemble strategy has the better performance. Our ensemble BioBERT+BiLSTM-Attention model achieved a high accuracy score of 0.84, ranking it in the top-third of all 42 participating teams participated the NLI task in the leaderboard. After excluding invalid submissions, including those did not report their team information (name, affiliation, and so on) and/or submit their working notes papers, our best accuracy score of 0.84 ranked the 11th among all 17 valid submissions.

Models	Accuracy
BioBERT-Base	0.786
BioBERT-Base + BiLSTM-Attention	0.805
BERT-Large	0.805
BERT-Large + BiLSTM-Attention	0.808
Ensemble BioBERT+BiLSTM-Attetion	0.840

Table 4: Results of our testing modes.

The test set (405 instances) is extremely small, but the testing period (15 days) is relatively long. Human intervention can be used to manipulate the results. In addition, the test set is arranged with an obvious pattern. One premise always accompanies with three hypotheses respectively denoting each class (entailment, contradiction and neutral). Base on this observation, it's easy to upgrade the final ranking of this task through a post-editing rule. For example, if two testing instances in the three-class group are predicted as the same class, the former is changed as the class entailment. Consequently, in the same condition, if the class entailment has been determined, then the former is changed as the class contradiction. With this post-editing rule, our best model can be enhanced to achieve a high accuracy score of 0.975.

4 Conclusions

This study describes the NCUEE system in the ACL-BioNLP'19 shared task, including system design, implementation and evaluation. We present our first exploration of this research topic in medical text inference. Future work will exploit other textual features to improve performance.

Acknowledgments

This study is partially supported by the Ministry of Science and Technology, under the grant MOST 108-2218-E-008-017-MY3 and MOST 108-2634-F-008-003- through Pervasive Artificial Intelligence Research (PAIR) Labs, Taiwan.

References

Alex Graves, Abdel-rahman Mohamed, and Geoffrey Hinton. 2013. Speech recognition with deep recurrent neural networks. In *Proceedings of the 2013 IEEE International Conference on Acoustics, Speech and Signal Processing*. Institute of Electrical and Electronics Engineers, pages 6645-6649. https://doi.org/10.1109/ICASSP.2013.6638947

Alexey Romanov, and Chaitanya Shivade. 2018. Lessons from natural language inference in the clinical domain. In *Proceedings of the 2018 Conference on Empirical Methods in Natural Language Processing*. Association for Computational Linguistics, pages 1586-1596. https://www.aclweb.org/anthology/D18-1187

Alexis Conneau, Douwe Kiela, Holger Schwenk, Loic Barrault, and Antoine Bordes. 2017. Supervised learning of universal sentence representations from natural language inference data. In *Proceedings of the 2017 Conference on Empirical Methods in Natural Language Processing*. Association for Computational Linguistics, pages 670-680. https://www.aclweb.org/anthology/papers/D/D17/D17-1070/

Asma Ben Abacha, Chaitanya Shivade, and Dina Demner-Fushman. 2019. Overview of the MEDIQA 2019 shared task on textual inference, question entailment and question answering. In *Proceedings of the 2019 Workshop on Biomedical Natural Language Processing*. Association for Computational Linguistics.

Jacob Devlin, Ming-Wei Chang, Kenton Lee, and Kristina Toutanova. 2018. BERT: Pre-training of deep bidirectional transformers for language understanding. *arXiv Preprint*. Cornell University, https://arxiv.org/abs/1810.04805v1

Jinhyuk Lee, Wonjin Yoon, Sungdong Kim, Donghyeon Kim, Sunkyu Kim, and Chan So. BioBERT: a pre-trained biomedical language representation model for biomedical text domain. *arXiv Preprint*. Cornell University, https://arxiv.org/abs/1901.08746

Qian Chen, Xiaodan Zhu, Zhen-Hua Ling, Si Wei, Hui Jiang, and Diana Inkpen. 2017. Enhanced LSTM for natural language inference. In *Proceedings of the 55th Annual Meeting of the Association for Computational Linguistics (Volume 1: Long Papers)*. Association for Computational Linguistics, pages 1657-1668. https://www.aclweb.org/anthology/papers/P/P17/P 17-1152/

Samuel R. Bowman, Gabor Angeli, Christopher Potts, and Christopher D. Manning. 2015. A large annotated corpus for learning natural language inference. In *Proceedings of the 2015 Conference on Empirical Methods in Natural Language Processing*. Association for Computational Linguistics, pages 632-642. https://aclweb.org/anthology/papers/D/D15/D15-1075/

Zichao Yang, Diyi Yang, Chris Dyer, Xiaodong He, Alex Smola, and Eduard Hovy. 2017. Hierarchical attention networks for document classification. In *Proceedings of the 2017 Conference of the North American Chapter of the Association for Computational Linguistics: Human Language Technologies*. Association for Computational Linguistics, pages 1480-1489. https://www.aclweb.org/anthology/N16-1174

ARS_NITK at MEDIQA 2019: Analysing Various Methods for Natural Language Inference, Recognising Question Entailment and Medical Question Answering System

Anumeha Agrawal, Rosa Anil George, Selvan Sunitha Ravi, Sowmya Kamath S and Anand Kumar M

Department of Information Technology
National Institute of Technology Karnataka, Surathkal, India 575025
{anumehaagrawal29, rosageorge97@gmail.com, sunitha98selvan}@gmail.com,
{sowmyakamath, m_anandkumar}@nitk.edu.in

Abstract

This paper includes approaches we have taken for Natural Language Inference, Question Entailment Recognition and Question-Answering tasks to improve domain-specific Information Retrieval. Natural Language Inference (NLI) is a task that aims to determine if a given *hypothesis* is an entailment, contradiction or is neutral to the given *premise*. Recognizing Question Entailment (RQE) focuses on identifying entailment between two questions while the objective of Question-Answering (QA) is to filter and improve the ranking of automatically retrieved answers. For addressing the NLI task, the UMLS Metathesaurus was used to find the synonyms of medical terms in given sentences, on which the InferSent model was trained to predict if the given sentence is an entailment, contradictory or neutral. We also introduce a new Extreme gradient boosting model built on PubMed embeddings to perform RQE. Further, a closed-domain Question Answering technique that uses Bi-directional LSTMs trained on the SquAD dataset to determine relevant ranks of answers for a given question is also discussed. Experimental validation showed that the proposed models achieved promising results.

1 Introduction

Recent studies have shown that patient-specific data can be utilized for the development of intelligent Healthcare Information Management Systems (HIMS), that support a wide range of supporting applications that enhance healthcare delivery platforms. The application of natural language processing, sophisticated data modeling, and predictive algorithms make it a highly interesting area of research. Patient data is continuously generated in large volume and variety, given the multiple modalities, it is available in (e.g., discharge summaries, physician's notes, clinical reports, lab reports etc). With an abundance of such diverse information sources available in the medical domain, sophisticated solutions that can adapt to the heterogeneity and specific manifold nature of health-related information are a critical requirement for HIMS development.

In clinical text, a commonly occurring problem would be to understand the correlation and association between various factors like disease, symptoms, diagnoses and treatment. Clinical text is inherently unstructured and written in natural language, and hence is prone to significant issues in effective interpretability and utilization. Challenges like paraphrase detection, anaphora resolution, natural language inference etc must be effectively dealt with in order to extract useful knowledge that can be used to build intelligent decision support applications. Such support systems require extensive evidence-based analysis, and context-sensitive processing, in order to enable higher-level functionalities like clinical question-answering. Thus, dealing with such issues is paramount importance.

Natural Language Inference is used to determine whether a given *hypothesis* can be inferred from a given *premise* (Ben Abacha et al. (2019)). The three inference relations to be identified between the statements are Entailment, Neutrality and Contradiction. If a statement is a true description of the other then it is labelled *Entailment*. If it is a false description then it is labelled *Contradiction*, otherwise, it is considered to be *Neutral*. The goal of Recognizing Question Entailment(RQE) is to retrieve answers to a premise question by retrieving inferred or entailed questions, called hypothesis questions that already have associated answers. Therefore, we define the entailment relation between two questions as: a question A en-

533

Proceedings of the BioNLP 2019 workshop, pages 533–540
Florence, Italy, August 1, 2019. ©2019 Association for Computational Linguistics

tails a question B if every answer to B also correctly answers A (Abacha and Demner-Fushman, 2016). RQE is particularly relevant due to the increasing numbers of similar questions posted online (Luo et al., 2015). For Question Answering, the input ranks are generated by the medical QA system CHiQA. Extracting certain elements of a question like the question type and focus is the main approach in question answering. If the question happens to contain multiple subquestions then an answer will be considered complete only if all sub-questions are answered. The rest of this paper is organized as follows: Section 2 presents a summarization on relevant existing research done in the area of interest. We discuss the Proposed Architecture for NLI, RQE and QA in Section 3. Section 4 presents the results and performance of the various models for each task, followed by error analysis, conclusion and references.

2   Related Work

There has been considerable research in the field of Medical Question Answering Systems. Incorporating QA systems with NLI and RQE give a machine the ability to better understand a query and fetch precise answers.

Modeling natural language inference is a complicated task but with the introduction of MedNLI (Romanov and Shivade, 2018; Goldberger et al., 2000), a new publicly available expert annotated dataset for NLI it has become possible to train models in order to achieve state-of-the-art performance. Chen et al. (2017) experimented with the SNLI corpus (Bowman et al., 2015) and MultiNLI corpus (Williams et al., 2017) to train complex models, to increase the performance of the neural network based NLI models with external knowledge. Most previous works on NLI worked on relatively small datasets, Chen et al. (2016) designed an approach to merge the modeling ability of neural networks with extra external inference knowledge. The advantage of using external knowledge is more significant when the training data is of limited size and is beneficial as more information is obtained. They obtained good results with this approach.

Romanov and Shivade (2018) presented a systematic comparison of various open domain models for NLI on MedNLI and studied the applicability of transfer learning techniques from the open domain to the clinical domain. They discussed their experimentation with a feature-based system in order to establish a baseline performance on MedNLI. Models that were explored include the Bag of Words model, InferSent (Conneau et al., 2017), ESIM (Enhanced Sequential Inference Model)(Chen et al., 2016). Other techniques included those that employed transfer learning, use of word embeddings and knowledge integration. In our work, we built on the work of these authors, by adapting their models and benchmarking them on different features, various word embeddings and clinical domain-oriented knowledge base to predict the relationship between the hypothesis and premise. Abacha and Demner-Fushman (2016) developed a method where RQE is applied to find a frequently asked question similar to consumer health questions, in order to answer consumer health questions with the answers given to similar FAQs. Groenendijk and Stokhof (1984) define an entailment relation between two questions Q_1, Q_2 if every proposition giving an answer to Q_1 is also giving an answer to Q_2. In our case, we used a supervised machine learning approach to determine whether or not a question Q_2 can be inferred from a question Q_1 by modeling the medical context's syntactic and semantic features, including complex relationships like negation, medical entities like disease, symptom, diagnoses and treatment etc. Abacha and Demner-Fushman (2016) used the NLM (National Library of Medicine) collection of 4,655 clinical questions asked by family doctors to construct the training corpus for RQE. For test pairs, two types of test data were collected - pairs of manually validated questions from the NLM collections and pairs of questions including FAQs retrieved online with a manual search of NIH websites. Four different statistical learning algorithms, SVM, Logistic Regression, Naive Bayes and J48, were used for RQE on the feature vector created. They reported the best results using the SVM classifier in the form of 75% F-measure values.

Abacha and Demner-Fushman (2019) studied question entailment in the medical domain and the effectiveness of the end-to-end RQE-based QA approach is calculated by evaluating the relevance of the retrieved answers. They benchmarked machine learning and deep learning approaches to RQE using different kinds of datasets, including textual inference, question similarity and entailment in both the open and clinical domains. The RQE methods (i.e. deep learning model and logistic regression classifier) are evaluated using two datasets of sentence pairs (SNLI and multiNLI), and three datasets of question pairs (Quora, Clinical-QE, and SemEval-cQA). They analyzed two methods for RQE: a deep learning model and Logistic Regression Classifier. Deep learning models achieve good results on open-domain and clinical datasets but delivered a lower performance on consumer health questions. When trained and tested on the same corpus, the Deep learning model with GloVe embeddings (Pennington et al. (2014)) gave the best results. Logistic Regression gave the best Accuracy on the Clinical-RQE dataset. When tested on our test set (850 medical CHQs-FAQs pairs), Logistic Regression trained on Clinical-QE gave the best performance.

Question answering (QA) is a crucial task that requires both natural language processing and domain related knowledge. Many Question Answering systems have been developed around the Question Answering dataset from Stanford (SQuAD) (Rajpurkar et al., 2016). The public leaderboard on the SQuAD website displays many deep learning models built for the task. Since the seminal work by Rajpurkar et al. (2016), many researchers have proposed different architectures for the task. The main feature of the dataset is that the answers are present as a span in the reverence document. The present state-of-the-art model is an AoA neural network by Cui et al. (2016), with an F1 score of 89.281, EM score of 82.482 and also outperforms the performance of humans.

3 Proposed Approaches

In this section, a detailed discussion on the various models designed for addressing the NLI, RQE and QA tasks, are presented.

3.1 Natural Language Inference

The first model proposed for NLI, a recurrent neural network (RNN) method is designed. We use the content of the two sentences to determine the two new rows $sentence_{id_1}$ and $sentence_{id_2}$ respectively which are formed using 300 dimensional glove embeddings. This feature vector created has been passed through a RNN with 300 nodes. Once this model was trained, we were able to get an accuracy of 67.1% with the test dataset.

The second model is the InferSent Model,

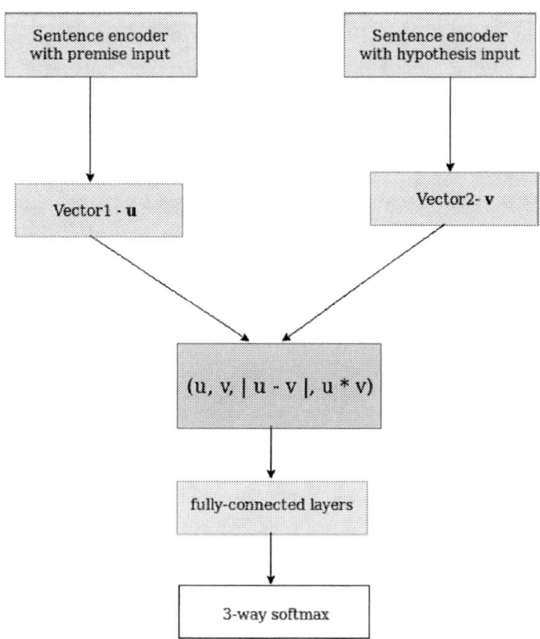

Figure 1: The Infersent Architecture adopted for NLI task

which is a sentence embedding method that provides semantic representations for English sentences. As shown in Figure 1, the architecture centralizes on the idea that two sentences (premise input and hypothesis input) will be transformed by sentence encoder (same weights). After that, it leverages three matching methods to recognize relations between premise input and hypothesis input. The three matching methods are: concatenation of two vectors, product two vectors element-wise and absolute difference of two vectors. Conneau et al. (2017) proposed the model which is trained using GloVe word embeddings(Pennington et al., 2014). In our work, we used the MedNLI (Romanov and Shivade, 2018; Goldberger et al., 2000) along with different word embeddings such as 300D GloVe embeddings, MIMIC clinic data embeddings (Johnson et al., 2016), Wikipedia (english) embeddings, the combination of Wikipedia english and MIMIC clinical data embeddings and even with the combination of 300D GloVe with BioASQ (Tsatsaronis et al., 2015) and MIMIC embeddings. All the techniques were set with number of training epochs as 100 and were trained on GPUs.

We also designed a novel technique to extract the semantic aspect of the clinical terms, for which we used the UMLS Metathesaurus (Aronson, 2001). The Metathesaurus is the largest component of UMLS that is organized by concept, or meaning, and links similar names for the term from over two hundred different vocabularies. The Metathesaurus is able to identify useful and relevant relationships between the various medical and non-medical concepts while preserving basic meaning and relationships from each vocabulary. We made use of the MetaMap tool for recognizing UMLS concepts in text. It can map medical texts to the UMLS Metathesaurus, using which we generated the synonyms for the terms that are not stop words and all synonyms have been generated with the use of UMLS Metathesaurus and NLTK corpus wordnet synsets. With this technique we were able to generate the highest accuracy yet of 87.7% on the test dataset given for the MediQA shared task.

3.2 Question Answering Task

The objective is to filter and improve the ranking of automatically retrieved answers, and the workflow employed is shown in Figure 2. Each question consists of several possible answers - relevant or irrelevant and are ranked based on the medical QA system CHiQA. We propose a system based on Question Answering model called Deeppavlov.ai (Burtsev et al., 2018). The context based question answering model uses SQuAD dataset to predict the answer. For every possible question, there are multiple answers and answer URLs associated with it. We scrape the content from the URL links and use that as context to the Deeppavlov model. The model takes in a question and a context to predict the answer. The answer provided in the AnswerText is a subset of the URL context. In case an answer does not have a URL associated with it, we use the AnswerText as the context.

Next, the model provides a score for every answer. This helps us determine how relevant the answer is for a question. We pass all the answers pertaining to a question to the model and obtain a score. We rank the answers based on the score obtained. We also need to determine if an answer is relevant or irrelevant. Based on the training dataset we set the threshold for relevance. If the answer has a score above the threshold then it is relevant, otherwise, it is irrelevant. This threshold is taken as the average of the scores of all answers belonging to a question. The threshold can be fur-

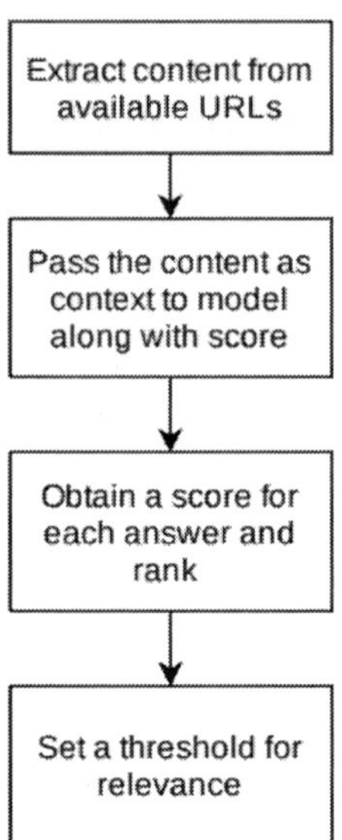

Figure 2: The proposed Question Answering workflow

ther improved based on the mode and median of the scores.

3.3 Recognizing Question Entailment

For the next task, the stop words were removed and word stemming using the Porter algorithm (Porter, 1980) was performed for all (Q_1, Q_2) training pairs, to extract relevant features. We create the feature vector using lexical features and semantic features. The semantic features used are Negativity and Positivity, and Named entities count. The lexical features used are Jaccard similarity, Mover's distance and Bigram overlap (Abacha and Demner-Fushman, 2016). We used scikit-learn library (Pedregosa et al., 2011) for all the machine learning models. We also used NLTK (Natural Language Toolkit) (Loper and Bird, 2002) to find ngrams, Wordnet (pri, 2010) and StanfordNERTagger. Wordnet is a lexical database which can be used to find synonyms. StanfordNERTagger is used to find named entities in the text.

Four different models were benchmarked for the RQE task on the test dataset - Support Vector Machine (SVM), Logistic Regression Classifier (LRC), AdaBoost Classifier and XGBoost with PubMed Embeddings.

Support Vector Machine for RQE: We use word overlap, common bigrams, Jaccard similarity, cosine similarity and Levenshtein distance as the features(Abacha and Demner-Fushman

(2016)). We also calculate the Word Mover's distance and this is included in the feature vector. We pass this feature vector through a SVM model. In order to enhance the performance and alter it for the medical domain, we use PubMed (Pyysalo et al. (2013)) 200D embeddings to find the word vectors.

Logistic Regression Classifier for RQE: - The same feature vector that was used for the Support Vector Machine task is being used here. In addition to that, the feature list also includes the maximum and average values obtained with these measures and the question length ratio (length(P Q)/length(HQ)).The morphosyntactic feature indicating the number of common nouns and verbs between P Q and HQ is also used (Abacha and Demner-Fushman (2019)).

K-Nearest Neighbors Classifier for RQE: Using the same feature vectors that was used in the Logistic Regression Classifier, we have used the K-nearest neighbors classifier for RQE. In order to obtain the value of K resulting with the highest accuracy, we have ran the algorithm for K ranging from 5 to 70. With the varying values of K, the accuracy is measured and the highest accuracy measure was with K=47.

Ada Boost Classifier for RQE: We use a new approach which uses the Ada Boost Classifier. Adaptive Boosting uses results from weak learner algorithms and combines it into a weighted sum which represents the final output of the boosted classifier. Using the same feature set as above we pass the feature vector through the ensemble based model. AdaBoost produces better results as it is adaptive. This algorithm works better than the single classifiers as it pools the prediction of multiple classifiers and reduces model bias and variance.

XGBoost with PubMed embeddings for RQE: We present a new approach for RQE in the medical domain using XGBoost with PubMed embeddings. XGBoost (Chen and Guestrin (2016)) is an optimized distributed gradient boosting library designed to be highly efficient, flexible and portable. It implements machine learning algorithms under the Gradient Boosting framework and is an ensemble model. Ensemble learning offers a systematic solution to combine the predictive power of multiple learners. The result is a single model which gives the aggregated output from several models.

In XGboost, the ensemble trees are constructed much faster than any other ensemble classifier as it makes use of distributed computing. The feature vector created above is passed through the Extreme Gradient boosting algorithm and a new feature based on the PubMed 200D embedding is added to the feature vector for calculating the similarity between the two medical questions. This is one unique feature which can capture relations between medical terms and thus gives high accuracy.

4 Experimental Results and Discussion

We performed several experiments to benchmark the relative performance of the various proposed models for the three different tasks - NLI, RQE and QA. We analyzed the accuracy obtained using the standard metrics defined for the three tasks. The datasets provided from the ACL-BioNLP'19 Shared Task (Ben Abacha et al., 2019) were used for the experimental studies.

For the NLI task, we used the MedNLI dataset (Romanov and Shivade, 2018; Goldberger et al., 2000) built on different word embeddings such as 300D GloVe embeddings (Pennington et al. (2014)) , MIMIC clinic data embeddings (Johnson et al. (2016)) , Wikipedia English embeddings, combination of Wikipedia English and MIMIC clinical data embeddings and even with the combination of 300D GloVe with BioASQ (Tsatsaronis et al. (2015)) and MIMIC embeddings. All the techniques were set with number of training epochs as 100 and were trained on GPUs. The observed performance for the MediQA released test dataset is tabulated in Table 1.

Table 1: Performance of the InferSent model for NLI task when different embeddings are used

Embeddings used	Accuracy
Wiki English	71.4%
MIMIC	71.7%
GloVe with BioASQ and MIMIC	72.4%
Wiki English with MIMIC	74.4%

The accuracy obtained with different methods is listed in Table 2. The RNN based model for NLI was trained for 30 epochs and a validation accuracy of 67.1% was achieved. Further accuracy can be improved by using a Bidirectional LSTM or Bidirectional GRU. The InferSent Model for NLI with MIMIC (Johnson et al., 2016) embeddings that were trained for 100 epochs gave an accu-

Table 2: Comparative performance of the proposed approaches for the NLI, RQE and QA tasks

Methology Used	Task	Accuracy
RNN	NLI	67.1%
Infersent+MIMIC	NLI	71.7%
Infersent+MIMIC+Wiki	NLI	74.4%
UMLS Metathesaurus	NLI	87.7%
SVM	RQE	62%
Logistic Regression	RQE	64.5%
KNN	RQE	62.4%
Naive Bayes	RQE	65%
Ada Boosting	RQE	66%
XgBoost	RQE	66.7%
Closed domain QA	QA	53.6%

racy of 71.7% This is because the medical context of the sentences was taken care of by the MIMIC word embeddings. The InferSent Model for NLI with MIMIC and Wikipedia english words embeddings gave an accuracy of 74.4% when trained for 100 epochs. This is because the medical and grammatical concepts were given special emphasis during the modeling phase. The model built on UMLS Metathesaurus and NLTK wordnet synsets model achieved an accuracy of 87.7% on the test data and 93.2% on validation data.

In the case of the RQE task, the SVM model was trained using a few features like semantic features, bigram overlap, word movers distance and cosine similarity. An accuracy of 62% achieved. The Logistic Regression model was trained using several handcrafted features and an accuracy of 64.5% was achieved. The KNN algorithm was also used for this classification task and an accuracy of 62.4% was obtained with K=47. The Naive Bayes model was fine-tuned and trained using the constructed feature vector. This gave an accuracy of 64%. The Naive Bayes model feature vector was modified again to include a feature which will consider the content of both the questions, which improved the accuracy by 1%. The AdaBoost classifier was used and this ensemble based method performed better than the naive methods and gave an accuracy of 66%. The XG-Boost method performed the best and gave an accuracy of 66.7% on the test set.

As can be seen from Table 2, the closed domain question answering model gave an accuracy of 53.6% which is much above the baseline fixed at 51%. This accuracy was achieved because this method focuses on finding the specific answer in the given context which is more relevant to a given question. Based on the scores obtained from the closed domain model, the answers have been ranked accordingly. The model achieved an accuracy of 53.6%, precision of 55.9% and Mean Reciprocal Rank of 62.93%.

4.1 Discussion

Based on the experimental results, we hereby present several observations and insights into the proposed models. In the case of the NLI task, we performed error analysis for the InferSent model by varying the embeddings and incorporating UMLS Metathesaurus and found that the error ratio also varies. The ratio of the error rate between neutral, entailment and contradiction was observed to be 5 : 4 : 3, when the MIMIC word embeddings (Johnson et al. (2016)) were used. However, it changed to 2 : 1 : 1 when the UMLS Metathesaurus with WordNet (pri, 2010) synsets are used. Thus, it can be concluded that the neutral label was the hardest to predict and differentiating between entailment and neutrality is also challenging. In our current implementation, if similar terms are present in the hypothesis and premise, then the label of entailment is still predicted, whereas the statements could actually be neutral. We also noticed that, by using clinical domain-specific embeddings, the predictions become more accurate.

Table 4 shows the Precision and Recall values for RQE using XgBoost. XgBoost provides a parallel tree boosting which improves the accuracy. Also, it uses continued training so it can further boost an already fitted model on new data, thus a significant improvement in accuracy is observed.

Table 3: Confusion Matrix for NLI

Label	True	False
Entailment	121	14
Neutral	114	21
Contradiction	128	15

Table 4: Confusion Matrix for RQE

Parameters	True	False
Precision	0.68	0.66
Recall	0.65	0.69

5   Concluding Remarks

In this paper, several techniques for the NLI, RQE and QA tasks were discussed. For addressing the NLI task, the UMLS Metathesaurus was used to find the synonyms of medical terms in given sentences, on which the InferSent model was trained to predict if the given sentence is an entailment, contradictory and neutral. We also designed a new Extreme gradient boosting model built on PubMed embeddings to perform RQE. Further, a closed-domain Question Answering technique that uses Bi-directional LSTMs trained on the SquAD dataset to determine relevant ranks of answers for a given question was also presented. Among the proposed models, the UMLS Metathesaurus and NLTK wordnet synsets model achieved the highest accuracy of 87.7% on the test dataset provided by the MediQA Challenge (Ben Abacha et al., 2019). For RQE, the highest accuracy of 66.7% was achieved using the XGBoost method. For the QA task, we achieved an accuracy of 53.6%, precision of 55.9% and Mean Reciprocal Rank of 62.93%.

As future work, we intend to extend the textual inference model for the clinical domain to develop decision support applications so that treatment methods can be simplified by grouping similar diseases and problems together. This can be achieved by using RQE which can aid in analyzing if two different health conditions are similar enough to have the same treatment. The model can also be trained on MedQuAD dataset (Abacha and Demner-Fushman (2019)) to improve the accuracy so that the model can perform more accurately in real-world hospital scenarios.

References

Princeton university "about wordnet." wordnet. princeton university. [online]. 2010.

Asma Ben Abacha and Dina Demner-Fushman. 2016. Recognizing question entailment for medical question answering. In *AMIA Annual Symposium Proceedings*, volume 2016, page 310. American Medical Informatics Association.

Asma Ben Abacha and Dina Demner-Fushman. 2019. A question-entailment approach to question answering. *arXiv preprint arXiv:1901.08079*.

Alan R Aronson. 2001. Effective mapping of biomedical text to the umls metathesaurus: the metamap program. In *Proceedings of the AMIA Symposium*, page 17. American Medical Informatics Association.

Asma Ben Abacha, Chaitanya Shivade, and Dina Demner-Fushman. 2019. Overview of the mediqa 2019 shared task on textual inference, question entailment and question answering. In *Proceedings of the BioNLP 2019 workshop, Florence, Italy, August 1, 2019*. Association for Computational Linguistics.

Samuel R Bowman, Gabor Angeli, Christopher Potts, and Christopher D Manning. 2015. A large annotated corpus for learning natural language inference. *arXiv preprint arXiv:1508.05326*.

Mikhail Burtsev, Alexander Seliverstov, Rafael Airapetyan, Mikhail Arkhipov, Dilyara Baymurzina, Nickolay Bushkov, Olga Gureenkova, Taras Khakhulin, Yuri Kuratov, Denis Kuznetsov, et al. 2018. Deeppavlov: Open-source library for dialogue systems. In *Proceedings of ACL 2018, System Demonstrations*, pages 122–127.

Qian Chen, Xiaodan Zhu, Zhen-Hua Ling, Diana Inkpen, and Si Wei. 2017. Neural natural language inference models enhanced with external knowledge. *arXiv preprint arXiv:1711.04289*.

Qian Chen, Xiaodan Zhu, Zhenhua Ling, Si Wei, Hui Jiang, and Diana Inkpen. 2016. Enhanced lstm for natural language inference. *arXiv preprint arXiv:1609.06038*.

Tianqi Chen and Carlos Guestrin. 2016. Xgboost: A scalable tree boosting system. *CoRR*, abs/1603.02754.

Alexis Conneau, Douwe Kiela, Holger Schwenk, Loic Barrault, and Antoine Bordes. 2017. Supervised learning of universal sentence representations from natural language inference data. *arXiv preprint arXiv:1705.02364*.

Yiming Cui, Zhipeng Chen, Si Wei, Shijin Wang, Ting Liu, and Guoping Hu. 2016. Attention-over-attention neural networks for reading comprehension. *arXiv preprint arXiv:1607.04423*.

Ary L Goldberger, Luis AN Amaral, Leon Glass, Jeffrey M Hausdorff, Plamen Ch Ivanov, Roger G Mark, Joseph E Mietus, George B Moody, Chung-Kang Peng, and H Eugene Stanley. 2000. Physiobank, physiotoolkit, and physionet: components of a new research resource for complex physiologic signals. *Circulation*, 101(23):e215–e220.

Jeroen Antonius Gerardus Groenendijk and Martin Johan Bastiaan Stokhof. 1984. *Studies on the Semantics of Questions and the Pragmatics of Answers*. Ph.D. thesis, Univ. Amsterdam.

Alistair EW Johnson, Tom J Pollard, Lu Shen, H Lehman Li-wei, Mengling Feng, Mohammad Ghassemi, Benjamin Moody, Peter Szolovits, Leo Anthony Celi, and Roger G Mark. 2016. Mimic-iii, a freely accessible critical care database. *Scientific data*, 3:160035.

Edward Loper and Steven Bird. 2002. Nltk: The natural language toolkit. In *In Proceedings of the ACL Workshop on Effective Tools and Methodologies for Teaching Natural Language Processing and Computational Linguistics. Philadelphia: Association for Computational Linguistics.*

Jake Luo, Guo-Qiang Zhang, Susan Wentz, Licong Cui, and Rong Xu. 2015. Simq: Real-time retrieval of similar consumer health questions. *J Med Internet Res.*

Fabian Pedregosa, Gaël Varoquaux, Alexandre Gramfort, Vincent Michel, Bertrand Thirion, Olivier Grisel, Mathieu Blondel, Peter Prettenhofer, Ron Weiss, Vincent Dubourg, et al. 2011. Scikit-learn: Machine learning in python. *Journal of machine learning research*, 12(Oct):2825–2830.

Jeffrey Pennington, Richard Socher, and Christopher D. Manning. 2014. Glove: Global vectors for word representation. In *Empirical Methods in Natural Language Processing (EMNLP)*, pages 1532–1543.

M.F. Porter. 1980. An algorithm for suffix stripping. *Program*, 14(3):130–137.

Sampo Pyysalo, Filip Ginter, Hans Moen, Tapio Salakoski, and Sophia Ananiadou. 2013. Distributional semantics resources for biomedical text processing.

Pranav Rajpurkar, Jian Zhang, Konstantin Lopyrev, and Percy Liang. 2016. Squad: 100,000+ questions for machine comprehension of text. *arXiv preprint arXiv:1606.05250.*

Alexey Romanov and Chaitanya Shivade. 2018. Lessons from natural language inference in the clinical domain. *arXiv preprint arXiv:1808.06752.*

George Tsatsaronis, Georgios Balikas, Prodromos Malakasiotis, Ioannis Partalas, Matthias Zschunke, Michael R. Alvers, Dirk Weissenborn, Anastasia Krithara, Sergios Petridis, Dimitris Polychronopoulos, Yannis Almirantis, John Pavlopoulos, Nicolas Baskiotis, Patrick Gallinari, Thierry Artiéres, Axel-Cyrille Ngonga Ngomo, Norman Heino, Eric Gaussier, Liliana Barrio-Alvers, Michael Schroeder, Ion Androutsopoulos, and Georgios Paliouras. 2015. An overview of the bioasq large-scale biomedical semantic indexing and question answering competition. *BMC Bioinformatics.*

Adina Williams, Nikita Nangia, and Samuel R Bowman. 2017. A broad-coverage challenge corpus for sentence understanding through inference. *arXiv preprint arXiv:1704.05426.*

Author Index

Association for Computational Linguistics
209 N. Eighth Street
Stroudsburg, Pennsylvania 18360

ISBN 978-1-5108-9103-6